Origin, Evolution, and Modern Aspects of Biomineralization in Plants and Animals

Origin, Evolution, and Modern Aspects of Biomineralization in Plants and Animals

Edited by
Rex E. Crick
The University of Texas at Arlington
Arlington, Texas

SPRINGER SCIENCE+BUSINESS MEDIA, LLC

Library of Congress Cataloging-in-Publication Data

International Symposium on Biomineralization (5th : 1986 : Arlington, Tex.)
Origin, evolution, and modern aspects of biomineralization in plants and animals / edited by Rex E. Crick.
p. cm.
"Based on the proceedings of the Fifth International Symposium on Biomineralization, held May 18-23, 1986, at the University of Texas at Arlington, Arlington, Texas"--T.p. verso.
Includes bibliographical references.
ISBN 978-1-4757-6116-0 ISBN 978-1-4757-6114-6 (eBook)
DOI 10.1007/978-1-4757-6114-6
1. Biomineralization--Congresses. 2. Mineral cycle (Biogeochemistry)--Congresses. I. Crick, Rex E. II. Title.
QH512.I566 1986
574.19'214--dc20 90-6763
CIP

Based on the proceedings of the Fifth International Symposium on Biomineralization, held May 18-23, 1986, at The University of Texas at Arlington, Arlington, Texas

DEDICATION

This volume is dedicated to the memory of Dr. Robert M. Garrels, whose visionary research in the fields of geochemistry and biochemistry significantly advanced the integration of these disciplines and opened avenues for others to follow. Dr. Garrels opened the symposium with the keynote lecture *Some factors influencing biomineralization in Earth history,* which appears here as the opening article.

PREFACE

The Fifth International Biomineralization Symposium was held in May 1986 at The University of Texas at Arlington, Arlington, Texas. The chosen theme was the origin, evolution and modern aspects of biomineralization in plants and animals. Thus, the symposium was designed to bring together experts in ocean and atmospheric chemistry, geochemistry, paleontology, biology, medicine and related fields to share accumulated knowledge and to broaden research horizons. The contents of this volume reflect the diversified interests and views of contributors from these fields. Topics range from contrasting views of the origin of ocean chemistry, the cause or causes for the biomineralization among plants and animals, the evolution of style and structure of biomineralization, and the role of inorganic and organic compounds in biomineralization.

It was clear from those gathered in Arlington that the efforts of all researchers in any aspect of biomineralization can be strengthened and extended by greater exposure to the work of others in allied fields. At the time of this printing, several collaborative efforts have grown from interest and contacts developed during the symposium.

Rex E. Crick

ACKNOWLEDGEMENTS

The symposium would not have occurred with the financial support of The Organized Research Fund of The University of Texas at Arlington and The Sea Grant Program administered by Texas A & M University. The staff of the Department of Geology of The University of Texas at Arlington were largely responsible for providing a pleasant atmosphere for learning.

This volume would not have been possible without the dedication of Pat Cowen during countless hours of typing and retyping manuscripts, the resources of the Department of Geology, contributing authors, and reviewers.

CONTENTS

OCEAN CHEMISTRY: ITS HISTORY AND RELATIONSHIP TO BIOMINERALIZATION SYSTEMS

BIOMINERALIZATION WITHIN THE INVERTEBRATES

THE STYLE AND STRUCTURE OF BIOMINERALIZATION: ANCIENT & MODERN

ROLE OF PHOSPHATE IN BIOMINERALIZATION

ORGANIC CHEMISTRY AND CALCIFICATION

INORGANIC AND ISOTOPE CHEMISTRY OF BIOMINERALS

BIOMINERALIZATION IN THE FUNGI, PLANTAE, MONERA AND PROTISTA

SOME FACTORS INFLUENCING BIOMINERALIZATION IN EARTH HISTORY

Robert M. Garrels

Department of Marine Science, University of South Florida, St. Petersburg, FL 33701

ABSTRACT. The history of the earth from first life is reviewed in terms of various factors that could influence biomineralization. The first record of life, prokaryotic organisms, seems to coincide with the first occurrences of sedimentary rocks. Carbon isotopes and the organic content of sediments 3.6 billions of years old are like those of today, attesting to oxygen production because of carbon dioxide reduction and organic Carbon deposition. However, oxygen sinks were so abundant that atmospheric oxygen levels remained low (10^{-5} Today?) until perhaps 2.0 billions of years ago. Thus there was a very long period (until 2.5 – 3.8 billion years ago) during which oxidation of iron, sulfur and primordial gases was taking place, increasing the reservoirs of oxygen – sulfates and ferric iron compounds and decreasing oxygen sinks – chiefly ferrous silicates, ferrous carbonates and ferrous sulfides.

Until about 2.0 billion years ago ferrous iron cycled in the hydrologic cycle with calcium, magnesium, and manganese. Sulfides were oxidized irreversibly to sulfates in sedimentary systems; the resultant sulfates may have been reduced by circulation of sea water through basaltic ocean floor, but bacterial reduction of sulfates apparently was not important until 2.7 billion years ago, and perhaps not until about 2.3 billion years ago.

The abundance of siderite in sedimentary deposits until about 2.0 billion years can be interpreted as a result of irreversible sulfur oxidation and perhaps, therefore, generation of a high methane, carbon dioxide atmosphere, which would give an important greenhouse effect, and high global temperatures (40 – 60° C?).

Many lines of evidence point to 2.0 billion years as the time of the transition from low oxygen levels in the atmosphere (but perhaps coincident with continued high $O2$ production), and important bacterial reduction of sulfate in the oceans, converting an irreversible oxygen sink to a system in which sulfur oxidation and reduction was reciprocal to carbon oxidation and reduction during sedimentary cycling.

Little is known about the important transitional period from about 2.0 billion years until about 0.7 billion years ago when the first metazoa appeared and the redox system of the world became dominated by sulfur and carbon, with iron taking a subsidiary role.

Since the beginning of the Phanerozoic (here assumed at 0.7 billion years), the earth surface system seems to have lost any truly continuous long term secular trends, but has oscillated over very long time spans between a system with a net increase in sulfate and a decrease in organic carbon and a system reversing that trend. The evolution of vascular plants makes its mark in the sedimentary record with an increase in the ratio of reduced carbon to oxidized carbon in sediments and perhaps an increase in atmospheric O_2.

We are on the verge of understanding some emerging Phanerozoic correlations. Reversals in the trend of $\delta^{34}S$ with time (over very long intervals) seem to be related to evaporites and phosphorites. Increases in the ratio of terrestrial to marine sediments seem to correlate with changes in $\delta^{13}C$ of marine organic carbon, and perhaps with increases in atmospheric oxygen levels.

Origin, Evolution, and Modern Aspects of Biomineralization in Plants and Animals
Edited by Rex E. Crick
Plenum Press, New York

Finally, it must be recognized that diagenetic changes in sediments tend to obscure primary secular changes. Perhaps research will permit separation of the effects of these processes that affect the overall mineralogy, chemistry, and biotic content of ancient sediments.

SOME ASPECTS OF EARTH HISTORY WITH CONSEQUENCES FOR BIOMINERALIZATION

Introduction

Knowledge of the interactions between organisms and their environments to produce minerals is disseminated through a vast literature. Here some of the pertinent geochemical history of the earth is presented.

Interpretation of the older Precambrian rests upon a few remnants of the rock record. The rock record available for sampling decreases exponentially from the present to the past. Archean rocks (>2.6 billion years) of sedimentary origin probably constitute far less than 1% of those originally deposited. In most instances their mineralogy is now different from the original, but careful sampling can commonly provide good data on their initial elemental composition.

Despite the problems engendered by selective loss of sediments through time by erosion and metamorphism, discovery of "protected" Precambrian sediments here and there provides a few samples of the past in which the rocks discovered are hardly altered from their initial state, and have sufficient bulk to permit comparison of their facies with those of modern rocks.

The oldest rocks known today were deposited 3.8 billion years ago at Isua, Greenland. They are highly metamorphosed, but contain carbon with an isotopic signature markedly different from primordial mantle carbon (Holland, 1984:360), a circumstance quite consistent with the influence of life presumably prokaryotic organisms (bacteria). The Isua rocks were certainly deposited in liquid water, assuring a world at 3.8 billion years ago that was neither frozen nor boiling.

The ancient areas of the earth where rocks of the Archean Eon (3.8 billion to about 2.6 billion years) are exposed have provided many surprises. Chemical analyses of 400 Archean shales or slates, from a dozen localities in Canada, have bulk chemical compositions close to those of the last few hundred million years. Analyses of 400 Early Proterozoic shales and slates are strikingly similar to the Archean ones. Table 1 presents the chemical analyses of these rocks. The average chemical composition of shaly rocks of the Phanerozoic Eon (the last 600 million years) are shown for comparison. There seem to be no significant differences in elementary chemical composition. If anything, Phanerozoic shales are more like Archean shales than Early Proterozoic shales.

To a first approximation, then, the elemental composition of sedimentary rocks has not changed with time, although the distribution of the elements in the various mineral reservoirs has fluctuated markedly. At some very early stage of the earth, organic carbon was absent, as were significant quantities of hematite and gypsum. With the advent of life and photosynthesis, carbon was reduced from a formal valence of +4 (carbon dioxide and carbonates) to a formal valence of zero in stored organic matter (CH_2O). Reduction of carbon took place at the expense of oxidation of iron and sulfur in iron silicates, carbonates, and sulfides. Figures 1 and 2 show that the percentage of organic carbon in sediments has been of the same order of magnitude (0.5 – 1.5 weight percent) since at least 3.5 billion years ago; that the $\delta^{13}C$ composition of carbonates; shows little trend; and that the difference between the $\delta^{13}C$ of carbonates and that of contemporary organic carbon, has averaged about 25 per mil. Although there is tremendous scatter in the $\delta^{13}C$ values of the organics, there seems to be little trend, except perhaps for a lessening of the scatter toward the present. The greater range of $\delta^{13}C$ in the early days of the earth may be because of the bacterial nature of the biomass. According to Funnell (1983), photosynthetic bacteria have $\delta^{13}C$ values between -28 and -39 per mil, cyanobacteria range from -8 to -22 per mil, and methanogenesis results in a range from +6 to -31 per mil. After the development of land plants, only 400 million years ago, the stored organics became chiefly a mixture of marine flora and fauna with $\delta^{13}C$ values from about -18 to -23 per mil, plus a terrestrial component of vascular tissue with values around -26 to -29 per mil.

TABLE 1. Chemical compositions of some average shales (Cameron & Garrels, 1980).

MAJOR ELEMENTS IN SHALES

	Archean	Early Proterozoic	CO_2 - free Phanerozoic
SiO_2	60.52	65.00	63.11
Al_2O_3	17.63	16.22	18.80
Fe_2O_3	7.91	6.37	7.79
MgO	3.47	2.51	3.11
CaO	2.16	0.52	0.07
Na_2O	2.62	1.44	1.27
K_2O	2.84	4.83	3.69
C_{org}	0.77	1.68	0.87
S	0.69	0.65	0.38
TiO_2	0.76	0.76	0.89
TOTAL	99.37	99.98	99.98

Measurements of the organic carbon content of Precambrian sediments are not adequate to use to calculate the organic carbon reservoir changes from 3.5 billion to 2.0 billion years. Presumably there was a zero organic carbon at some early time. Here it is interpreted that the organic carbon reservoir grew from early days until about 2.0 billion years ago.

The rationale for this interpretation is as follows. The sulfur isotopes of Archean and Early Proterozoic sediments show no significant fractionation relative to primordial mantle sulfur (Cameron, 1982; Skyring & Donnelly, 1982). One explanation, accepted here, is that there was no significant bacterial reduction of sulfate until about 2.3 billion years. On the other hand, siderite and iron silicates were oxidized to hematite or magnetite, which were back-reduced in a more or less closed cycle. In the absence of sulfate reduction, methane and carbon dioxide were produced by fermentation of organic matter.

While organic matter accumulated as a residue from photosynthesis leaving oxygen behind in the atmosphere, the oxygen oxidized ferrous iron in carbonates and silicates, as well as sulfides, chiefly pyrite. The siderite and ferrous silicate oxidized to magnetite and hematite during weathering and transportation. A nearly steady-state redox system probably developed. In the absence of sulfate reduction, fermentative decay of organic material produces methane and carbon dioxide, so there probably was a significant methane component in the atmosphere. This, plus relatively high carbon dioxide,

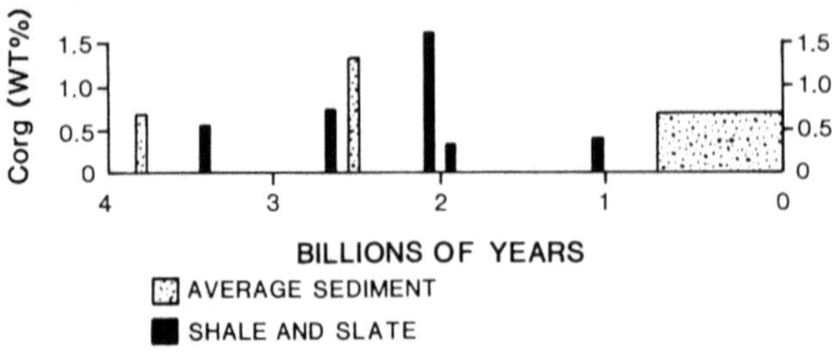

FIGURE 1. Organic carbon content of shales and slates as a function of age (adapted from Schidlowski, 1982:107).

required for observed equilibrium between siderite and ferrous silicate (greenalite), provided the necessary "greenhouse" to offset the chill of a less luminous early sun.

Oxygen levels would have been held very low by the ferrous and the irreversible sulfide sinks. Various estimates (*cf.* Holland, 1984:411) place the level in the atmosphere at 10^{-4} to 10^{-5} bars. The flux, however, may have been a significant fraction of that of today.

There were undoubtedly many depositions of minerals by organisms in the early days, but it is almost impossible to document them, because the only organisms were single-celled. As fossils they are almost impossible to find. The generally accepted documentation of bacteria in the Early Archean is a tribute to the careful researches and extensive sampling of the small residual rock record of that ancient Eon.

However, there is certainly the possibility that chemical autotrophic bacteria could have been responsible for much of the Archean pyrite in sediments. Pyrite is widespread in Archean sediments; in fact the reduced sulfur content of Archean shales is considerably higher than that of younger ones. A simple explanation is the addition of volcanic H_2S to oxygen deficient sedimentary basins with high concentrations of dissolved ferrous iron. Chemical autotrophs today, at the effluence of cycled seawater from hot basalts at ridge crests and in similar environments, oxidize H_2S to sulfur and make themselves from CO_2:

$$2H_2S + CO_2 = CH_2O + 2S + H_2O.$$

The availability of sulfur and H_2S and ferrous iron makes formation of pyrite highly energetic:

$$H_2S + S + Fe^{2+} = FeS_2 + 2H^+.$$

This pyrite, if formed as supposed, can be counted a biomineralization - secretion of sulfur by the chemautotrophs followed by reaction to pyrite. A little free oxygen is required by today's chemautotrophs to make pyrite. There probably was enough oxygen available during the Archean to fulfill this function.

The early procaryotes may have deposited various minerals within their cells; some of the modern ones deposit magnetite (Petersen *et al.*, 1986; Stoltz *et al.*, 1986), a very common mineral of the Early Archean.

In summary of the Archean and Early Proterozoic, the organic carbon reservoir increased as did the ferric oxide and sulfate reservoirs. Oxygen in the atmosphere was low, methane and carbon dioxide probably many times today's levels. Temperatures

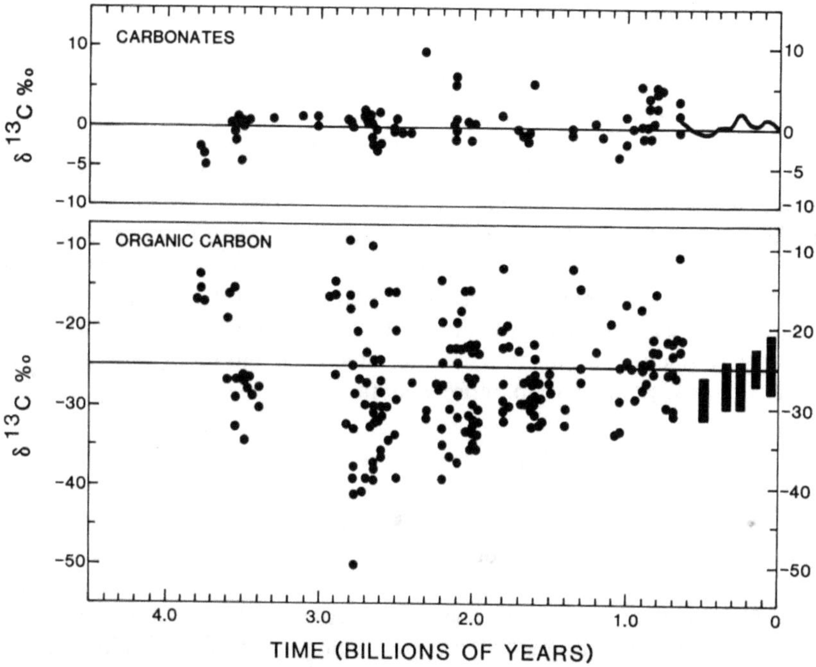

FIGURE 2. Carbon isotope composition of organic matter in sedimentary rocks as a function of age (adapted from Schidlowski *et al.*, 1983:158).

may have been higher than today. Ferrous iron carbonates and silicates were important cycling components of the sedimentary mass. They, plus the accumulating ferric oxides, are preserved in part as the Banded Iron Formations of Archean and Early Proterozoic age. Biomineralization may have been restricted to deposition of compounds within cells perhaps chiefly pyriteand magnetitemagnetite.

"End of the Early Proterozoic: The beginning of the modern world"

The beginning of significant bacterial sulfate reduction changed the iron-sulfur system from an irreversible oxygen sink to a reversible system. Methane and siderite and sedimentary ferrous silicate went out the window as the ferrous iron cycling through these minerals was captured by pyrite. Loss of methane and the carbon dioxide from cycling siderite lowered temperatures because of diminished "greenhousing". Evidence of glaciations appeared in the rocks. The reduction of sulfate was a new oxygen source. Oxygen and carbon dioxide reversed their roles; oxygen rose from a trace gas to a significant atmospheric component; carbon dioxide did the reverse. The "new world" eliminated the old iron oxide-carbonate-silicate steady state system. During Phanerozoic time the reversible global redox system is described by the reaction

$$15H_2O + 4FeS_2 + 8CaCO_3 + 7MgCO_3 + 7SiO_2 \leftrightarrow 15CH_2O + 8CaSO_4 + 7MgSiO_3 + 2Fe_2O_3$$

No gases are involved in this equation, which means that if it were strictly obeyed oxygen and carbon dioxide levels in the ocean-atmosphere system would be fixed.

The time required to switch from the ferrous silicate-ferrous carbonate system to the ferrous sulfide system, with concomitant rise of oxygen to an approximation of its present level, does not yet have good restraints.

The invention by bacteria of ways to reduce sulfate at low temperatures apparently was necessary for the rise of oxygen, the elimination of banded iron formations, the evolution of eukaryotes and then metazoa. The invention can be compared to the discovery of fire, the use of gun powder, and the nuclear bomb. Ironically, the bacteria that made the discovery in a sense relegated bacteria in general to secondary roles in global affairs, but the operation of the present system depends upon them as much as any ancient system did. Imagine a world with no bacterial decay! The revolution at about 2.0 billion years also changed the earth from a system with continuous secular redox change to one that seems to have fluctuated back and forth over the last five or six hundred million years with no entirely continuous secular changes in the exogenic cycle.

Little is currently known of the evolutionary consequences of the "turn-around" at about 2.0 billion years. The first eukaryotic organisms appeared somewhere between 1.4 and 2.0 billion years (Cloud, 1983:29). Eukaryotes, of course, are required precursors of metazoa. The oldest fossils of complex organisms, with specialized tissues, date back to 800 million years. The first metazoa with shells appeared in a number of phyla at about 570 million years, as if they had all learned together. "Shell school" remains a puzzle of dimensions comparable to the origin of photosynthesis. It is almost as if some one thing lacking, or some one thing too much, had been added to or subtracted from the ocean.

It is known that free oxygen in the atmosphere-ocean system is required for shell formation, but preliminary modeling of Precambrian oxygen levels gives no indication that there was some sudden increase in oxygen at the beginning of the Cambrian. Shell formation is a useful method of using calcium that has to be eliminated from organisms during growth. Speculation were Precambrian oceanic calcium levels too high for shell formation?

From the Cambrian to the Present, or perhaps even from the Vendian (700–800 million years), the physical history of the earth seems to be expressible as one great cycles (see Bambach *et. al.*, 1985). In Vendian time the continents were high and together; there was glaciation and low sea level. Then the continents spread apart, and by Cambrian time there were a number of small dispersed continents. Sea levels in the early Paleozoic were generally high. By Devonian time, some 200 million years after the beginning of the Cambrian, the continents came together again to form the super continent of Pangea. From early Paleozoic time until Permian time, when the continents broke up again, there was a general trend to larger and higher land masses, lowering of sea level, abundant evaporite deposits ($CaSO_4$ and NaCl), and a trend toward more intense, sporadic glaciations.

From Cambrian to Permian there was a general lowering of the $\delta^{34}S$ of the oceans, suggesting an increase in the storage of oxygen in evaporites and perhaps an increase in atmospheric oxygen levels. A great deal of carbonate (in shells) was stored in the shallow seas that rhythmically invaded and withdrew from the continents.

The ending of the Permian was a time of great and rapid changes. Pangea began to rift and break up, the biota was convulsed with the greatest extinctions of all geologic history. Temperatures rose, sea levels rose.

Aspects of biomineralization during the Phanerozoic

First of all, the Phanerozoic Eon (here stretching from 700 million years ago to Today) is pretty clearly representative of a basically steady state world. Most of the chemical species we can measure that we think have not changed since their time of deposition show fluctuations on very long time scales from values that were like today's values 700 million years ago. In other words, we see a global system in which major trends have occurred over intervals of several hundreds of millions of years, but these trends seem to be countered before extremes are reached, and a reverse trend of comparable magnitude sets in. Current guesses are that global temperatures ranged between $+5^{\circ}$ C and 20° C, with about 15° C as a mean (today's global average is 12° C).

From Vendian to Cambrian the $\delta^{34}S$ of the oceans increased from about +17 ppt to about +33 ppt. Such a change is consistent with the accumulation of salinity in the oceans (probably chiefly sulfate and chloride), perhaps because of low sea levels and the absence of suitable extensive peripheral shallow water areas for deposition of evaporites.

INFLUENCE OF SALINITY ON BIOMINERALIZATION

Fluctuations in oceanic salinity no doubt have played an important role in controlling the composition of the marine biota, and hence their deposition of inorganic compounds. Our knowledge of Precambrian ocean compositions is highly speculative. Many reconstructions of ancient oceans assume oceans basically like those of today. This assumption is probably a serious mistake. Veizer (1983a) has perhaps best described the situation. The strontium isotopes of Archean sediments indicate dominance of oceanic compositions by seawater reactions with rocks of mantle composition, with relatively small contributions from the continents. For instance, it is not beyond the realm of permissible speculation to believe that Archean and Early Proterozoic oceans were fairly acid with a pH of about 5, and that as a result, their trace metal contents could have been at levels high enough to be lethal to eukaryotes today. Procaryotes, by the way, are known for their tolerances to elements like copper, lead, cobalt, and mercury. So high acidity and salinity of Precambrian oceans may have been barriers to eucaryote evolution.

Salinities must have varied considerably during Phanerozoic time, as indicated by the large fluctuations of sulfur isotopes in sulfate evaporite deposits which mirror the seawater from which they come. Modeling of the fluctuations of the reservoirs of sedimentary sulfate suggests the possibility of considerable variation in oceanic sulfate concentrations (Garrels & Lerman, 1984). Holland (1984:444) has determined the limits of the possible sulfate range, but that range is still severalfold. Variation in oceanic sulfate is also required by the fact that evaporite deposits containing gypsum or anhydrite occur sporadically in time. The conditions required for such deposits are not continuously present. In general large areas of shallow sea peripheral to the oceans proper must exist, and these areas must have a net of evaporation over precipitation. Therefore during many millions of years sulfate is added to the ocean by streams and not removed

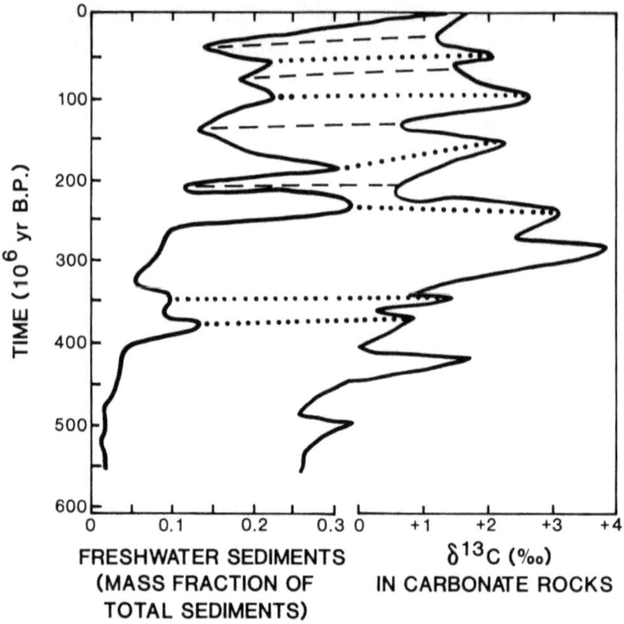

FIGURE 3. Plots of the mass ratio of freshwater sediments and of $\delta^{13}C$ for marine carbonates from data of Ronov (1976) and Veizer (1983b). The dashed lines indicate approximate correlation between the minima of the two curves, and the dotted lines connect their maxima.

as evaporites, so it is unlikely that all this added sulfate is removed by reduction to sulfide and fixation by iron. Increase of oceanic sulfate would increase oceanic calcium, which in turn would influence the carbonate depositional system and the pH. For example, evaporite deposits of gypsum or anhydrite are relatively unimportant from Cretaceous to the Miocene. During the Miocene, in a few million years, a volume of sulfate evaporites was deposited that is estimated to be several times that of the preceding 40 or 50 million years (Ronov, 1976).

In summary, salinity changes of the oceans, not currently well documented, but being actively researched today, may have been responsible for drastic changes in the biota, and presumably in their mineral products.

Changes in the structure of carbonate shells from calcite to aragonite have been attributed by many workers to changes in ocean pH. Changes in salinity would drive changes in pH, as well as changes in concentrations of other elements that might influence the polymorph of $CaCO_3$ deposited by a given organism.

Nutrient phosphate and ratio of terrestrially derived organic matter to marine organics

The availability of nutrients is one of the most important factors in biomineralization. For siliceous organisms, such as diatoms, both silica and phosphorus supply are overriding considerations. Diatoms can lower silica concentrations in ocean water far below the level in equilibrium with the tests of the diatoms. For a given level of the nutrient phosphorus, diatoms can utilize silica for their shells at levels far below silica saturation. Carbonate secreters, on the other hand, at the same phosphorus level, cannot lower the dissolved $CaCO_3$ even to saturation with their shells. This means that the biota has utilized the available dissolved phosphorus chiefly for soft tissues, and the shell makers have not been sufficiently successful to remove the calcium being added by streams, so calcium becomes supersaturated with respect to inorganic $CaCO_3$.

Changes in the makeup of the biota, with increase in the proportion of shell-forming species, may well have caused changes in the saturation state of the ocean, and influenced the crystalline structure and composition of the carbonate minerals secreted as shells.

Current studies of fluid inclusions in evaporites of different ages may give direct evidence of the compositions of ancient seas, and will provide "ground truth" for modeling of the geologic past (H.D. Holland, personal communication, 1986).

On the time scale of geologic Periods, oceanic phosphate can be considered to be in an essentially steady state condition. The residence time of phosphorus is only about 100,000 years, so in general the flux of phosphorus into the ocean is equal to that out of the ocean. Undoubtedly the flux of phosphorus is a function of land area and of temperature, among other variables, so the burial rate of organics would fluctuate even at a fixed burial ratio of organic carbon to phosphorus. However, most factors that increase the flux of phosphorus in and out of the ocean also increase the rate at which oxygen is consumed. At the moment variations in the carbon/phosphorus ratio, reflecting changes in the ratio of the terrestrial component of buried sediments to the total sediment mass, seems to be a major factor in the production rate of oxygen. Figure 3 shows the ratio of terrestrial sediments to total sediments compared with $\delta^{13}C$ of carbonates as a function of sediment age.

We are still a long way from integrating individual nutrient cycles into cycling of the total sedimentary system. Perhaps the brief discussion here will show how important such an effort will be in understanding the interplay of organisms and environment.

ACKNOWLEDGEMENTS

Part of the work presented here was supported by the National Science Foundation, EAR 8515872. The manuscript was read critically by Terri L. Woods and was prepared by Valerie Joy Kimball. I am indebted to many people for discussions of this and related materials through the years. Support for the symposium was provided through grants from The Organized Research Fund of The University of Texas at Arlington and The Sea Grant Program administered by Texas A&M University.

REFERENCES

BAMBACH, R.K., FISCHER, A.G. & GARRELS, R.M., 1985. Physical-chemical environments and their geologic history. *In* The Evolution of Complex and Higher Organisms, (eds. D. Milne, D. Raup, J. Billingham, K. Niklaus, & K. Padian), Chapter 4, pp. 67-91. NASA Spec. Publ. 478. Washington, DC: NASA Scient. Tech. Infor. Bran..

CAMERON, E.M., 1982. Sulphate and sulphate reduction in early Precambrian oceans. Nature, 296:145-148.

CAMERON, E.M. & GARRELS, R.M., 1980. Geochemical compositions of some Precambrian shales from the Canadian Shield. Chem. Geol., 28:181-197.

CLOUD, P., 1983. Early biogeologic history: the emergence of a paradigm. *In* Earth's Earliest Biosphere Its Origin and Evolution, (ed. J. W. Schopf), Chapter 1, pp. 14-31. Princeton, New Jersey: Princeton Univ. Press.

FUNNELL, B.M., 1983. Palaeophysiology: the geological context. *In* Dimensions of Palaeophysiology, (ed.. B.M. Funnell). J. Geol. Soc. (Lond.), 140:1-4.

GARRELS, R.M. & LERMAN, A., 1984. Coupling of the sedimentary sulfur and carbon cycles--an improved model. Amer. J. Sci., 284:989-1007.

HOLLAND, H.D., 1984. The Chemical Evolution of the Atmosphere and Oceans. 582 pp. Princeton, New Jersey: Princeton Univ. Press.

PETERSEN, N., VON DOBENECK, T. & VALI, H., 1986. Fossil bacterial magnetite in deep-sea sediments from the South Atlantic Ocean. Nature, 320, 611-615.

RONOV, A.B., 1976. Global carbon geochemistry, volcanism, carbonate accumulation, and life. Translated from Geokhimiya, No. 8:1252-1277, USSR.

SCHIDLOWSKI, M., 1982. Content and isotopic composition of reduced carbon in sediments. *In* Mineral Deposits and the Evolution of the Biosphere, (eds. H.D. Holland & M. Schidlowski), pp. 103-122. Dahlem Konferenzen, Berlin, Heidelberg, New York: Springer-Verlag.

SCHIDLOWSKI, M., HAYES, J.M. & KAPLAN, I.R., 1983. Isotopic inferences of ancient biochemistries: carbon, sulfur, hydrogen, and nitrogen. *In* Earth's Earliest Biosphere Its Origin and Evolution, (ed. J. W. Schopf), Chapter 7, pp. 149-186. Princeton, New Jersey: Princeton Univ. Press.

SKYRING, G.W. & DONNELLY, T.H., 1982. Precambrian sulfur isotopes and a possible role for sulfite in the evolution of biological sulfate reduction. Precambr. Res., 17:41-61.

STOLTZ, J.F., CHANG, S.B.R. & KIRSCHVINK, J.L., 1986. Magnetotactic bacteria and single-domain magnetite in hemipelagic sediments. Nature, 321, 849-851.

VEIZER, J., 1983a. Geologic evolution of the Archean-Early Proterozoic Earth. *In* Earth's Earliest Biosphere Its Origin and Evolution, (ed. J.W. Schopf), Chapter 10, pp. 240-259. Princeton, New Jersey: Princeton Univ. Press.

VEIZER, J., 1983b. Trace elements and isotopes in sedimentary carbonates. *In* Carbonates: Mineralogy and Chemistry, (ed. R.J. Reeder), Chapter 8, pp. 265-299, vol. 11, Reviews in Mineralogy. Washington, D.C.: Mineral. Soc. Amer.

BIOMINERALIZATION AND TENTATIVE LINKS TO PLATE TECTONICS

Fred T. Mackenzie and Catherine R. Agegian

Department of Oceanography and Hawaii Institute of Geophysics, University of Hawaii, Honolulu, Hawaii 96822

ABSTRACT. Tectonic processes appear to be a principal forcing function controlling some aspects of the sedimentary cycling of materials during the geologic past. Chemical trends observed in the sedimentary rock record for the past 80 Ma substantiate the hypothesis that plate tectonics play a major role in influencing, on a long-time scale, several variables of ocean-atmosphere-sediment chemistry. The apparent long-term decrease in sea level since the Late Cretaceous (80 Ma) was probably caused by a global decrease in sea-floor spreading rate and ridge volume. This waning of tectonic activity may have resulted in reduced volcanic and diagenetic-metamorphic CO_2 release to the earth's surface environment, giving rise to lower atmospheric CO_2 levels and temperatures. The carbonate saturation state and the Mg/Ca, Sr/Ca, Na/Ca, $^{87}Sr/^{86}Sr$, and $^{143}Nd/^{144}Nd$ ratios of sea water may have progressively increased during the sea level fall. It is postulated that the global environmental response to increased tectonic activity probably is opposite to that proposed for the past 80 Ma.

Physical and chemical changes in earth's surface environment induced by plate tectonic processes may have been the cause of some observed changes in the composition of skeletal and non-skeletal carbonates during the Phanerozoic. The inferred mineralogy of carbonate ooids and cements appears to vary in an oscillatory manner between aragonite and calcite, perhaps reflecting cyclic variations in sea water carbonate saturation state and Mg/Ca ratio. Inferred variations in skeletal carbonate mineralogy suggest a progressive increase with decreasing geologic age in the deposition of skeletal high-magnesian calcite and aragonite, marked by the appearance of two principal groups of modern reef-building organisms, hermatypic corals (aragonite) and coralline algae (high-magnesian calcite), in the mid to late Mesozoic (200 Ma). This evolutionary mineralogical change may reflect a long-term trend of decreasing atmospheric CO_2 and increasing sea water carbonate saturation state and Mg/Ca ratio.

Experimental evidence indicates that the solubility of both skeletal and non-skeletal Mg-calcites is a function of the Mg content of the calcite. The Mg content of the coralline alga, *Porolithon gardineri*, grown in controlled microcosm environments, increased with increasing temperature and calcite saturation state of sea water. Experimental data confirm the hypothesis that skeletal and non-skeletal carbonate compositions can be a function of sea water carbonate saturation state, Mg/Ca ratio, and temperature.

It appears that the fossil record of biomineralization may be linked to internal earth processes through the influence of plate tectonic events on the physico-chemistry of the surface environment. Gaian and extraterrestrial processes as forcing functions for biological change need to be considered in light of earth's plate tectonic history.

FIGURE 1. Three general causal mechanisms proposed to govern short- and long-term environmental changes at earth's surface. Figure modified from Saint-Exupery (1943).

INTRODUCTION

There are numerous explanations for short- and long-term changes in earth's surface environment. These explanations can be grouped into three general causes, as depicted in Figure 1. Extraterrestrial causes depend in part on agents arriving from outside the planetary atmosphere e.g., cometary or asteroid impact. Biologically-mediated processes, in their ultimate form, the Gaia Hypothesis (Lovelock & Margulis, 1974), are thought to regulate the geochemistry of the exogenic system so as to favor the survival of life. These processes are to a significant degree, but not exclusively, driven by changes in the intensity of the sun's energy. Physical, chemical and biological changes in the earth's surface environment may also be influenced by the dynamics of internally-driven processes, such as variations in internal heat production and subsequent effects on plate tectonics.

In this paper we will develop the relationships between plate tectonics and sea level and resultant changes in atmosphere-ocean chemistry and climate, and subsequent effects on biomineralization. The paper is partially a compendium and synthesis of the work of many authors, including ourselves (see references), but our emphasis is on the tentative links between plate tectonics and biomineralization.

PLATE TECTONICS AND GLOBAL SEA LEVEL

Although there are several ways to produce changes in global sea level, major changes over a long-time period, on the order of hundreds of millions of years, can be interpreted in terms of modern plate tectonic concepts. The generalized sea level curve for the Phanerozoic, as derived from stratigraphic data by Vail *et al.* (1977), is shown in Figure 2. The curve is more complex in detail, but of particular interest is the apparent n x 10^8 years cyclic pattern of sea level change. This generalized sea level curve correlates well with the percentage continental flooding curves for North America and the Soviet Union (Hallam, 1977, 1981; Turcotte & Burke, 1978), indicating that there has been considerable variation in sea level during the Phanerozoic following the trends of Figure 2. These long-term variations probably reflect changes in the volume of mid-ocean ridges (*e.g.*, Hays & Pitman, 1973; Southam & Hay, 1977; Steckler, 1984) which are related to sea floor spreading rates. A rapid sea floor spreading rate gives rise to large ridge volume and thus high global sea level, a low spreading rate leads to low ridge volume and low global sea level.

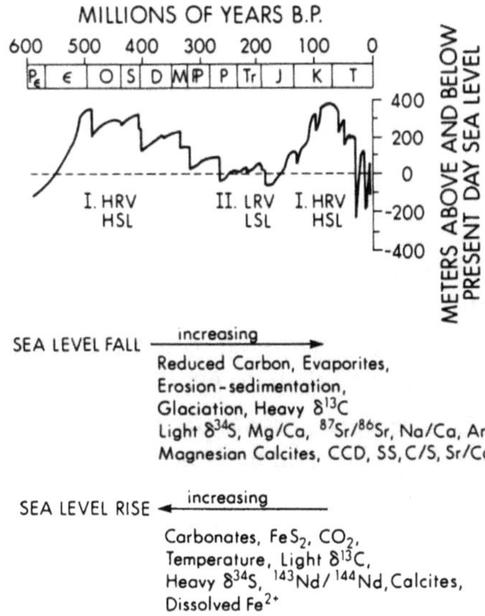

FIGURE 2. Generalized sea level curve for the Phanerozoic (modified after Vail *et al.*, 1977) identifying submergent [high sea level (HSL), high ridge volume (HRV)] and emergent [low sea level (LSL), low ridge volume (LRV)] episodes. Predicted changes in the isotopic and elemental composition and the deposition of sedimentary materials are summarized for periods of falling and rising sea level.

SEA LEVEL VARIATION & OCEAN-ATMOSPHERE-SEDIMENT CHANGES

The long-term rise and fall of sea level, as depicted in Figure 2, are accompanied by changes in the exogenic cycle of the ocean-atmosphere-sediment-biosphere system. The anticipated changes and their relationship to sea level rise and fall on a time period of $n \times 10^8$ years are summarized in Figure 2. The trends shown are those predicted for a model in which plate interactions and sea-floor spreading are the major causal factors influencing long-term changes in sea level. A variety of other factors, however, may be important, particularly on time scales shorter than 10^8 years. These factors which are not mutually independent of themselves or plate tectonic factors are, for example, variations in climate and paleooceanographic conditions, continental topographic relief, volcanic activity, rock type weathered and so forth.

It is not the intent of this paper to discuss each of the changes shown in Figure 2 in detail; the reader should consult the reference list for more explicit interpretations of the data. For example, Holser (1984) discusses in detail variations in mineral-inventories, isotope ratios and trace metals vs. sediment geologic age, and problems associated with interpretation of these data. He points out, in contrast to the model of Figure 2, that the maximum observed in the $^{87}Sr/^{86}Sr$ ratio of sedimentary materials in the early Paleozoic (a time of high sea level) cannot be strictly interpreted on the basis of plate tectonics and sea floor spreading considerations. We have portrayed in Figure 2 our interpretation of changes anticipated if plate tectonics and sea floor spreading were the main causal factors producing the changes.

Sea level rise

During a long period of sea level rise brought about by increasing sea floor spreading rate and ridge volume, the following scenario would result, climaxing in the submergent (Mackenzie & Pigott, 1981) or greenhouse (Fischer, 1982) modes of high sea levels of the middle Paleozoic and late Cretaceous.

1. The $\delta^{13}C$ values of carbonate rocks would decrease and the $\delta^{34}S$ values of evaporitic sulfate minerals would increase. These changes reflect transfers of sedimentary carbon from the reduced organic carbon reservoir to the oxidized inorganic carbon reservoir in limestones. Sea level rise results in less restricted environments of carbonate deposition, and an increase in the ratio of fluxes of inorganic carbon to organic carbon to the sea floor.

2. With sea level rise, sulfur is transferred from the oxidized sulfur reservoir in evaporitic sulfate rocks to the reduced reservoir of pyrite in sediments. The ratio of the fluxes of inorganic sulfate to pyritic sulfur to the sea floor decreases.

3. CO_2 levels progressively increase resulting from an increased rate of CO_2 production from diagenetic and metamorphic reactions at subduction zones and input of volcanic CO_2 at ridges. Increased atmospheric CO_2 gives rise to an increased "greenhouse" effect, and the average temperature of the earth increases during progressive flooding of the continents.

4. The $^{143}Nd/^{144}Nd$ ratio would increase as the input of hydrothermal fluids at ridges increased during sea level rise. ^{143}Nd is enriched in these fluids whereas the other major source of chemical constituents for the oceans, the rivers, is depleted in this isotope relative to ^{144}Nd.

5. Low Mg-calcite (< 5 mol% $MgCO_3$) skeletal materials and marine cements would be favored over aragonites and high Mg-calcites (> ~10 mol% $MgCO_3$) as sea level rose, because the progressive increase in the CO_2 of the ocean-atmosphere system would lead to a decrease in the saturation state of sea water. Thus, more environments of the oceans would have saturation states lower than those necessary to precipitate high Mg-calcites, and perhaps in some areas aragonite.

6. Dissolved iron concentrations in sea water would increase because of a lower redox state of the global ocean resulting from development of an extensive oxygen minimum layer. A thick and extensive O_2 layer would come about from decreased oceanic mixing during times of high temperatures when ocean vertical temperature gradients would be more uniform.

Sea level fall

During a long period of sea level fall as a result of decreased sea floor spreading rate and ridge volume, a scenario somewhat opposite to that of sea level rise would develop. The climax of the sea level fall scenario is that of the oscillatory (Mackenzie & Pigott, 1981) or ice-house (Fischer, 1982) modes of low sea levels of the late Precambrian-early Cambrian, Permo-Triassic, and late Cenozoic.

1. Carbon-sulfur isotope exchanges are opposite to those of rising sea level: reduced carbon and evaporitic sulfur fluxes increase relative to their counterparts of limestone carbon and pyritic sulfur.

2. Erosion-sedimentation fluxes increase because of increasing land area. More nutrients enter the ocean owing to erosion of previously submerged continental areas.

The increased nutrient load of the ocean gives rise to increased organic production and accumulation rates of sedimentary organic carbon.

3. Decreasing temperatures brought about by lower levels of atmospheric CO_2 increase the probability of continental glaciations whose direct cause may be changes in orbital parameters.

4. Slower sea floor spreading rates give rise to increasing $^{87}Sr/^{86}Sr$, Mg/Ca, Na/Ca and Sr/Ca ratios in sea water. These changes arise from variations in fluxes of chemical constituents involved with river water input and basalt-sea water reactions.

$^{87}Sr/^{86}Sr$ ratios. As sea level falls, chemical exchange between the oceanic basalt crust and sea water decreases, whereas the riverine input to the ocean increases. River waters weathering old sialic rocks from continental interiors have high $^{87}Sr/^{86}Sr$ ratios relative to waters derived from volcanic and intrusive rocks of plate margins and ocean basins. Thus, as sea floor spreading rates decrease, input of low $^{87}Sr/^{86}Sr$ fluids to the ocean from hydrothermal activity at ridges and from submarine weathering decrease, whereas the river input of high $^{87}Sr/^{86}Sr$ river waters increases, giving rise to sea water with a progressively higher $^{87}Sr/^{86}Sr$ ratio. Sedimentary materials formed from these waters are progressively enriched in $^{87}Sr/^{86}Sr$ with falling sea level.

Mg/Ca, Sr/Ca, and Na/Ca ratios. -- These ratios progressively increase as sea level falls. Hydrothermal reactions between sea water and basalt at ridges result in chemical exchanges such that Mg^{2+} and Na^{+} are removed from the sea water and Ca^{2+} added. If decreasing sea level implies decreasing intensity of hydrothermal activity at submarine ridges, then progressively less Mg^{2+} and Na^{+} will be removed from the ocean and Ca^{2+} added as sea level falls. Thus, a time of low sea level will correspond to a time of relatively high Mg/Ca and Na/Ca ratios in sea water, and sedimentary materials derived from the water. Similarly the Sr/Ca ratio of sea water will rise as sea level falls, because of decreasing Ca^{2+} input to the ocean from ridges.

5. The carbon/sulfur ratio of sediments will increase as sea level falls and tectonic activity subsides. Lowering of sea level will favor continental sites of organic matter accumulation. These sedimentary environments lack dissolved sulfate leading to little pyrite formation and higher C/S ratios in sediments.

6. Aragonitic and high Mg-calcite minerals will become progressively more important in skeletons of marine organisms and as marine cements as sea level falls. This change is due to lowering of atmospheric CO_2 levels and increasing global ocean carbonate saturation levels leading to more environments sufficiently saturated to precipitate these minerals. Also, progressive increase in the Mg/Ca ratio of sea water as sea level falls will favor formation of high Mg-calcites (and aragonite) over low Mg-calcite.

7. The global carbonate compensation depth will become progressively deeper as sea level falls, tectonic activity subsides, CO_2 levels fall and temperatures drop. The higher sea water saturation state, a result of lower atmospheric CO_2, promotes deepening of the CCD.

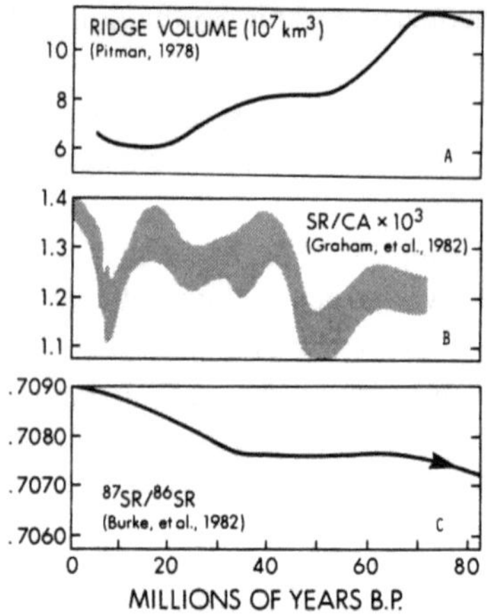

FIGURE 3. Trends in ridge volume (▪A) and Sr/Ca (▪B) and $^{87}Sr/^{86}Sr$ (▪C) ratios in sedimentary materials during the Cenozoic. Trends in ▪A and▪B are possibly related to a progressive decrease in the hydrothermal input of Ca- and ^{86}Sr-rich fluids from mid-ocean ridges.

EVIDENCE FROM THE LAST 80 MA OF EARTH HISTORY

What is the evidence for tectonically-controlled variations in atmosphere-ocean chemistry on the time scale of hundreds of millions of years? During the last decade a wealth of data on the chemical properties of Cenozoic sedimentary rocks has been acquired. Some of these data exhibit trends with geologic age that can be correlated with the sea level curve for the past 80 Ma. Since the last great sea level transgression of the Late Cretaceous, most evidence points to a pattern of long-term sea level fall (Figs. 2,4). This pattern of sea level fall presumably was caused by a long-term progressive decrease in ocean ridge volume for the past 80 Ma of earth history (Fig. 3a). Thus, the overall pattern of sea level fall apparently has as its causal mechanism, plate tectonic processes. It appears that some long-term trends in sedimentary rock properties can be correlated with the sea level fall curve. Consequently, these trends can be reasonably argued to be a result of chemical changes in the ocean-atmosphere-sediment system induced by plate tectonic mechanisms. These trends and their tie to sea level and ridge volume provide some justification for the relationships shown in Figure 2. What are some of these trends and their link to place tectonics?

Sr content, Sr/Ca and $^{87}Sr/^{86}Sr$ ratios of sedimentary materials

The Sr content, Sr/Ca and, $^{87}Sr/^{86}Sr$ ratios of sedimentary materials progressively increase with decreasing geologic age during the past 80 Ma of earth history (Figs. 3b,c & 5). Although there are other causes for these trends, it is likely that the overall trends represent decreasing sea floor spreading rates during the Cenozoic. The progressive decrease in plate tectonic activity manifests itself in progressive decrease in hydrothermal inputs of Ca- and ^{86}Sr-rich fluids from mid-ocean ridges. This change leads to sea water and sea water precipitates progressively enriched in Sr and ^{87}Sr during the Cenozoic.

$\delta^{18}O$ of benthic foraminifera

Benthic foraminifera become progressively enriched in $\delta^{18}O$ with decreasing geologic age during the Cenozoic (Fig. 4). This trend points to global ocean bottom water temperatures of about 13°C (Fig. 4b) at the equator 80 Ma ago, whereas today's temperatures are near 1°C. A potential cause of the $\delta^{18}O$ trend, and hence temperature, is a decrease in plate tectonic activity. The interpretation is that, as ocean ridge volume, and hence sea level (Fig. 4a) decreased, the rate of addition of volcanic and metamorphic CO_2 to the ocean-atmosphere system decreased, as did the concentration of atmospheric CO_2. These changes led to progressively lower global mean temperatures with decreasing geologic age during the Cenozoic. If the model of Figure 2 is correct, the decline in atmospheric CO_2 during the Cenozoic implies a progressive increase in the saturation state of the global ocean, and hence a deepening of the CCD (Fig. 4c).

Mg content of pelagic carbonates

The original Mg content of pelagic carbonates has varied during the last 140 Ma (Fig. 5). The main cause of this variation appears to be due to variations in mid-ocean ridge hydrothermal activity. During times of intense activity, Mg^{2+} is abstracted from sea water and Ca^{2+} is added. Because the oceanic residence times for Mg^{2+} and Ca^{2+} are more than an order of magnitude different (40 Ma *vs* 2 Ma), the Mg/Ca ratio decreases slowly. Conversely, during times of decreased plate tectonic activity the Mg content and Mg/Ca ratio of sea water slowly increase.

Neodinium isotopes

Although data are still somewhat sparse, the $^{143}Nd/^{144}Nd$ ratio of sedimentary materials appears to decrease slightly with decreasing rock age for the past 50 Ma (Fig. 6a). Nd isotopes can be used to characterize the relative importance of submarine hydrothermal inputs and continental inputs to the ocean. Hydrothermal fluids have a relatively high $^{143}Nd/^{144}Nd$ ratio compared to continental river inputs. Thus the slight enrichment of ^{144}Nd in sedimentary materials observed for the past 50 Ma suggests an increased flux of continental waters to the ocean relative to hydrothermal fluids over this time period.

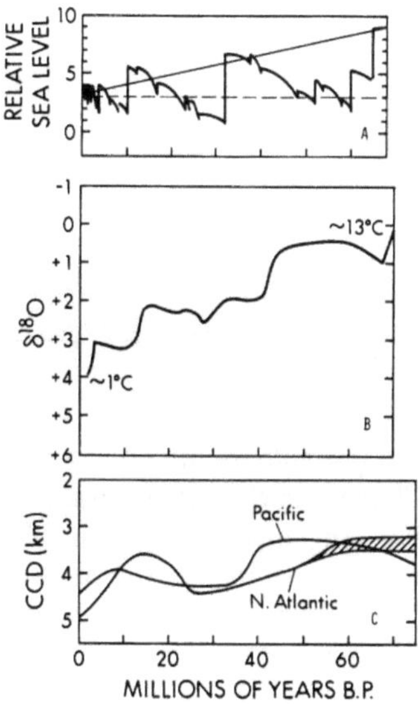

FIGURE 4. Comparison of trends through the Cenozoic. ▪A. Relative sea level, after Vail *et al.* (1977). ▪B. The $\delta^{18}O$ content of benthic foraminifera (Savin *et al.*, 1975). ▪C. Progressive change of the North Atlantic and Pacific carbonate compensation depth (CCD) (van Andel, 1975). As ocean ridge volume and sea level decreased, the rate of addition of volcanic and metamorphic CO_2 to the ocean-atmosphere system decreased, as did the concentration of atmospheric CO_2. The decline in atmospheric CO_2 during the Cenozoic may have resulted in a progressive increase in the saturation state of the global ocean and hence a deepening of the carbonate compensation depth.

Na/Ca ratio of foraminifera

An increase in the Na/Ca ratio of foraminifera with decreasing sediment age is observed during the Cenozoic (Fig. 6b). Although part of this trend may be diagenetic, decreased Na^+-Ca^{2+} exchange between basalts and sea water at elevated temperature along ridge systems may have resulted as tectonic activity slowed during the Cenozoic. In particular, reduced inputs of Ca^{2+} to the ocean resulting from hydrothermal processes would tend to lead to an increase in the Na/Ca ratio of sea water, and hence in foraminifera, during the last 80 Ma of earth history.

In conclusion, there are chemical trends in the sedimentary rock record for the past 80 Ma that substantiate the hypothesis that, on a long-time scale, plate tectonics plays

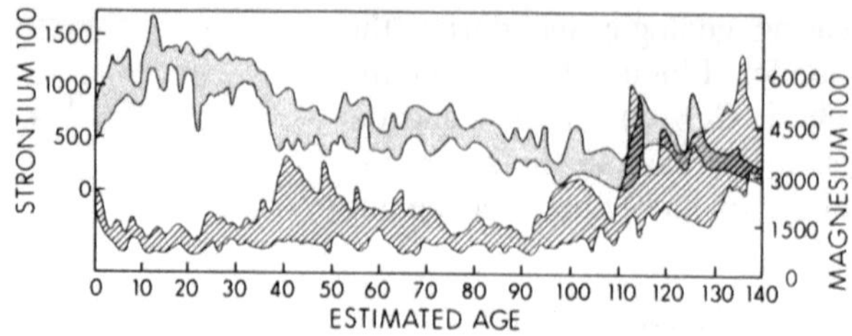

FIGURE 5. Variations in the Mg and Sr content of pelagic carbonates during the past 140 Ma (after Renard, 1986).

a major role in influencing several variables of ocean-atmosphere-sediment chemistry. Changes in chemistry induced by tectonic processes can affect biomineralization processes. In particular, tectonically-controlled variations in atmospheric CO_2 levels, and hence earth surface temperature, and sea water Mg/Ca ratio, carbonate chemistry and pH, and hence carbonate saturation state, may influence the type of carbonate minerals precipitated.

PHANEROZOIC COMPOSITIONAL HISTORY OF SKELETAL AND NON-SKELETAL CARBONATE MINERALS

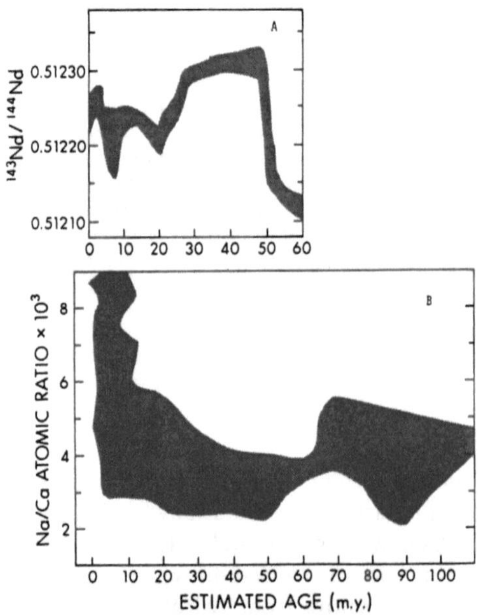

FIGURE 6. Generalized trends during the Cenozoic. ▪A. The $^{143}Nd/^{144}Nd$ ratio in sedimentary materials (after Palmer & Elderfield, 1986). ▪B. Na/Ca ratio of foraminifera (after Graham *et al.*, 1982).

The original mineralogy and chemistry of skeletal and non-skeletal carbonate minerals over geologic time have been inferred for ooids, cements, and fossils. A number of factors are used to assess the original composition of the carbonate phase, including degree of recrystallization, texture, and fabric of the carbonate host. Although the procedures used to infer composition *vs* geologic age for these carbonate components are still somewhat controversial and the data sparse, some general patterns have emerged, and have been used to interpret the history of ocean-atmosphere chemistry.

Non-skeletal components

It appears that the carbonate mineralogy of ooids has changed during the Phanerozoic. Because carbonate ooids reflect to a significant extent the physical chemistry of the water from which they were deposited, these carbonate grains may be good indicators of ocean-atmosphere environmental changes. A tentative synthesis of the data for the inferred mineralogy of carbonate ooids, from the works of Sandberg (1975, 1983), Wilkinson *et al.* (1985), and Mackenzie & Pigott (1981) is shown in Figure 7. An oscillating trend in the relative percentage of calcite *vs* aragonite ooid mineralogy may be superimposed on a long-term, evolutionary increase in ooids with an inferred original aragonitic mineralogy. The cyclic trend in ooid mineralogy can be roughly correlated to the sea level curve. Increased continental flooding corresponds to times of increased propensity for calcitic ooid deposition, whereas sea level fall and continental emergence correspond to an increase in aragonitic ooid deposition. The suggested evolutionary trend is also roughly correlated to the Phanerozoic continental flooding-curve. An overall decrease in the extent of continental flooding corresponds to an overall increased probability of aragonitic ooid deposition.

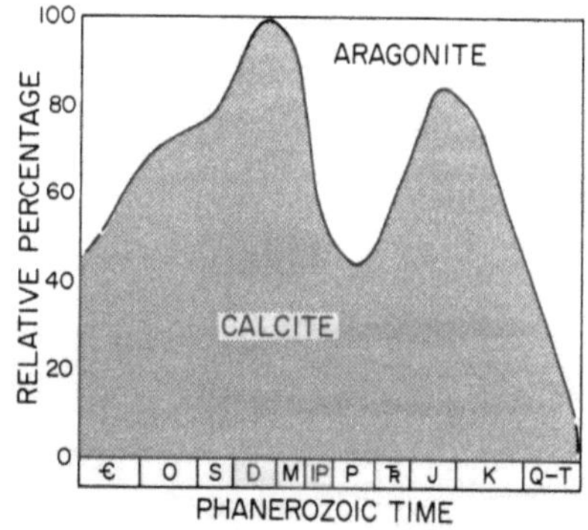

FIGURE 7. Relative percentage of inferred mineralogical composition of ooids during Phanerozoic time. This summary represents our interpretation of the data of Sandberg (1975, 1983), Wilkinson *et al.* (1985), and Mackenzie & Pigott (1981).

If global eustasy and ooid mineralogy are correlated, then further inferences can be made concerning ocean-atmosphere chemistry. As mentioned previously, rising sea level implies increased global tectonic activity giving rise to high atmospheric CO_2 levels and earth surface temperatures. Rising atmospheric CO_2 levels would lead to lowered carbonate saturation states and Mg/Ca ratios of sea water favoring the precipitation of calcite low in Mg over the more soluble high Mg-calcite and aragonite. The reverse situation would be true for global eustatic conditions of falling sea levels. The evolutionary trend in ooid mineralogy may reflect a long-term Phanerozoic trend of decreasing continental flooding, lower atmospheric CO_2 level, higher Mg/Ca sea water ratio, and increasing saturation state of ocean waters.

It should be further pointed out that the long-term cyclic trend in ooid mineralogy also correlates with cyclic trends observed for the inferred mineralogy of carbonate cements (Sandberg, 1983), and for the abundance of oolitic ironstones in Phanerozoic sedimentary sequences (Wilkinson *et al.*, 1985). The distribution of oolitic ironstones shows that these sediments are abundant during times of continental submergence. Wilkinson *et al.* (1985) contend that their abundance reflects high atmospheric CO_2 levels and temperature, decreased oceanic mixing, and a laterally more extensive oxygen minimum zone. The lower redox state of the global oceans would lead to increased ferrous iron concentrations in sea water, and thus, a higher probability for formation of oolitic ironstones.

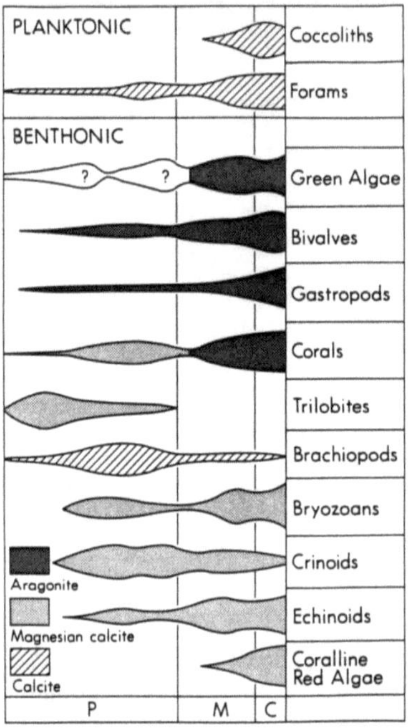

FIGURE 8. Mineralogical evolution of benthic and planktonic organisms during the Phanerozoic based on summaries of Milliken & Pigott (1977) and Wilkinson (1979).

The trends in the inferred mineralogy of non-skeletal carbonate components may reflect ocean-atmosphere environmental changes. If so, are these possible changes reflected in the biomineralization record of fossil organisms?

Skeletal components

Mineralogy *vs* geologic age trends for fossil calcareous organisms are present in the Phanerozoic rock record, but interpretation of these trends in terms of ocean-atmosphere chemistry is not straight-forward. The inferred mineralogy of major organism groups for the Phanerozoic is shown in Figure 8. If biomineralization processes in marine organisms were governed solely by inorganic environmental factors, we might anticipate that organisms with a low-Mg calcitic skeleton would be predominant during active sea floor spreading events, *i.e.*, periods of high atmospheric PCO_2, high temperature, low sea water carbonate saturation state, low sea water Mg/Ca ratio. During periods of low global tectonic activity, corresponding to periods of low atmospheric CO_2, low temperature, high sea water carbonate saturation state, and high sea water Mg/Ca ratio, aragonitic and high-Mg calcitic skeletons in organisms would be favored. The oscillatory trend seen in non-skeletal carbonate components, however, is not clearly apparent in the mineralogy of fossil organisms. Overall, there is a general increase in the number of major groups of calcareous organisms, *i.e.* coccolithophorids, pteropods, hermatypic corals and coralline algae. Organisms with all mineralogies are well represented over the past 100 Ma. It is noteworthy, however, that the major groups of pelagic and benthic organisms contributing to carbonate sediments in today's ocean first appeared in the fossil record during the middle Mesozoic and became progressively more abundant. There is a possibility that the relative abundance of specific organism groups may have followed cyclic trends. What is most evident from Figure 8 is a long-term increase in the production of biogenic carbonates dominated by calcite, Mg-calcite, and aragonite mineralogies. This trend could possibly represent a long-term, evolutionary trend in ocean-atmosphere chemistry, favoring development of today's metastable carbonate mineral skeletal assemblage; that is, overall decrease in atmospheric CO_2 and earth surface temperature, and increase in the carbonate saturation state and Mg/Ca ratio of ocean water.

EXPERIMENTAL BIOMINERALIZATION AS A CLUE TO OCEAN-ATMOSPHERE CHEMISTRY

Experimental studies on inorganic and biogenic carbonate precipitates can be used in conjunction with trends in the composition of skeletal and non-skeletal carbonates in the rock record to decipher earth's surface chemical history. In this section, we will concentrate on the effects of changes in the physico-chemical properties of sea water on the precipitation and composition of Mg-calcites. Figure 9 is a compilation of data showing the relationship between the Mg/Ca ratio of an aqueous solution and the Mg content of inorganic carbonate precipitates. There is a moderately good positive correlation between the Mg/Ca ratio and the Mg content of the precipitate. Thus we would anticipate that the Mg content of a Mg-calcite precipitate formed at or near saturation with sea water would directly reflect the Mg content of sea water.

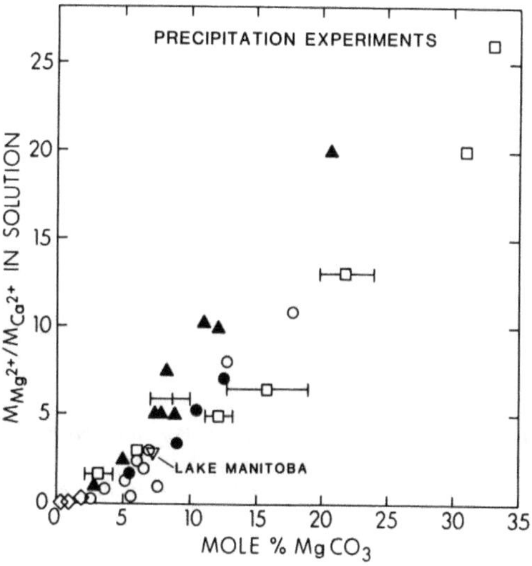

FIGURE 9. $MgCO_3$ content of Mg-calcites precipitated from solutions with different molar ratios of Mg and Ca. See Mackenzie *et al.* (1983) for references to data.

It is now well established that the Mg content of skeletal Mg-calcites decreases with increasing latitude (see Chave, 1954; Mackenzie *et al.*, 1983, for review). Furthermore, the Mg content of marine Mg-calcite cements also decreases with latitude, as well as depth, in the ocean (*e.g.*, Mackenzie *et al.*, 1983; Wilkinson & Given, 1986; Burton & Walter, 1987). Because the Mg/Ca ratio of today's ocean water is constant, it is likely that the change in Mg content of Mg-calcite cements with latitude and depth is a function of the temperature and saturation state of the ambient environment, with temperature perhaps being the most important variable (Burton & Walter, 1987). For skeletal Mg-calcite precipitates, growth rate as well as temperature and saturation state, are variables cited as influencing Mg content.

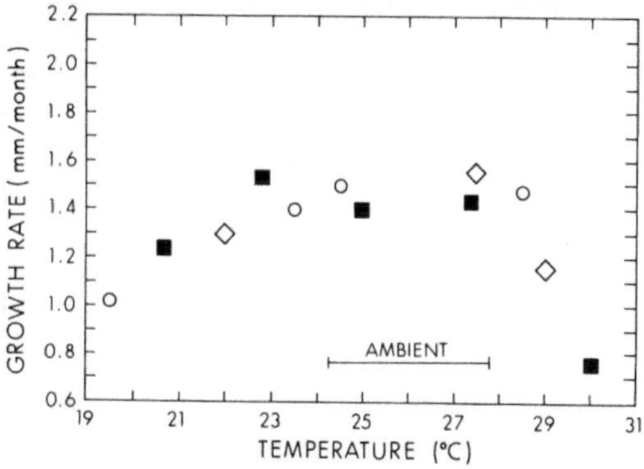

FIGURE 10. Average growth rate of *Porolithon gardineri* at different temperatures. Bar indicates average seasonal variation in temperature of "ambient" growth conditions in the natural environment.

To determine what variables are important in controlling the composition of skeletal Mg-calcites, preliminary microcosm experiments were performed using the coralline alga, *Porolithon gardineri*, grown in

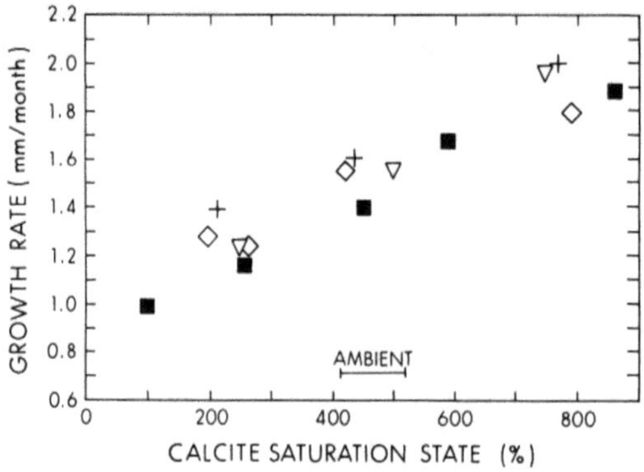

FIGURE 11. Average growth rate of *Porolithon gardineri* at different calcite saturation states. Bar indicates average measured range in calcite saturation state for "ambient" growth conditions in the natural environment.

sea water at different sea water temperatures and saturation states. Details of the experimental procedures and results are given in Agegian (1985). Results important to our discussion are portrayed in Figures 10, 11, 12, and 13. The important conclusions are:

1. The growth of *Porolithon gardineri* is strongly influenced by temperature and carbonate saturation state. Growth rate above about 28°C was dramatically reduced.

2. The Mg content of the calcite skeleton of the alga increased with increasing temperature.

3. The Mg content increased in a curvilinear manner in the alga calcite skeleton with increasing saturation state.

It appears from observational and experimental evidence that increasing sea water temperature and carbonate saturation state favor increasing Mg contents of both inorganic and biogenic calcites. Furthermore, unless temperature increases become lethal, increased calcite saturation state favors increased growth rates of skeletal Mg-calcite, and hence increased Mg content. We would predict from these results that Mg-calcite biominerals and cements would exhibit higher Mg contents at higher temperature, saturation states, and sea water Mg/Ca ratios than at lower values of these variables. The global tectonic model presented here for carbonate precipitate mineralogical changes predicts that a global CO_2 increase driven by increased tectonic activity leads to an increase in temperature, but a decrease in saturation state. The two variables are not directly correlated, as their influence on the Mg content of Mg-calcite precipitates predicts. Thus if the global tectonic model is correct and carbonate precipitate composition over geologic time is responding to environmental changes induced by tectonics, the Mg/Ca ratio and the carbonate saturation state of sea water, which are directly correlable and influence the Mg content of precipitates in the same direction, are the most important variables governing inorganic and biogenic precipitate composition over geologic time. For aragonite *vs* calcite precipitates, the tectonic model is in accord with the direction of

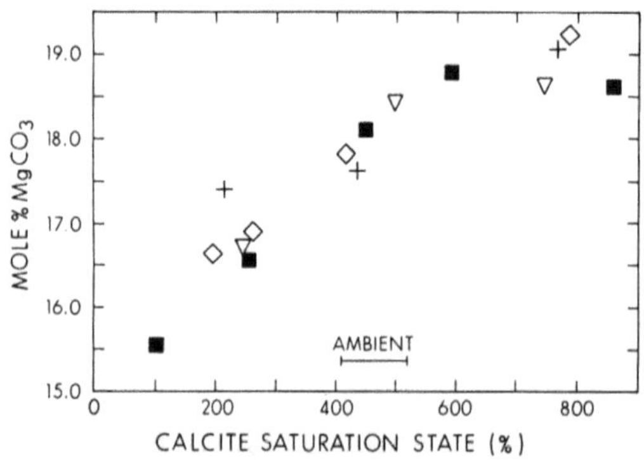

FIGURE 12. Average Mg content of *Porolithon gardineri* grown at different temperatures. The ambient temperature range of the natural environment in which this alga lives is also shown.

chemical changes that favor one phase over the other; *e.g.*, high tectonic activity leads to a high atmospheric CO_2 level and high temperature, decreased sea water saturation state and Mg/Ca ratio, which in turn favor calcite biominerals and inorganic precipitates over aragonitic phases (*e.g.*, Chave, 1954; Mackenzie *et al.*, 1983; Burton & Walter, 1987).

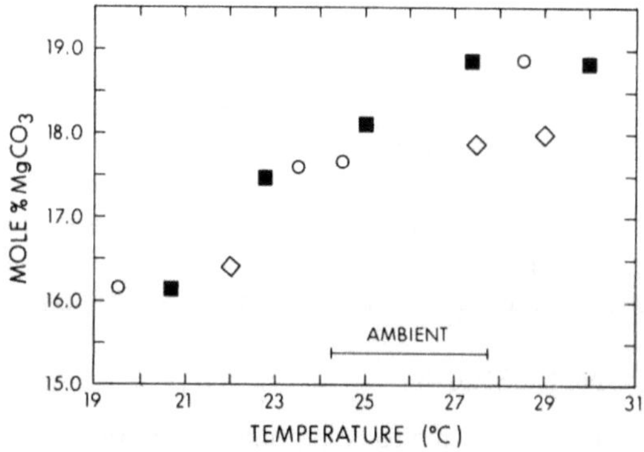

FIGURE 13. Average Mg content of *Porolithon gardineri* grown at different calcite saturation states. The ambient temperature range of this alga's natural environment is also shown.

One further and intriguing aspect of the experimental microcosm studies on *Porolithon gardineri* is worth mentioning. During the course of these studies, the carbonate saturation state of the sea water in which the alga grew was determined. Details of the calculation are given in Agegian (1985). Figure 14 shows the negative logarithm of the ion activity product of the sea waters in which *P. gardineri* was grown as a function of the Mg content of the skeleton precipitated by the alga. The -log IAP is defined as

$$a^{(1-x)}Ca^{2+}\ a^{(x)}Mg^{2+}\ a CO_3^{2-},$$

where (x) is the mole fraction of Mg in the alga skeleton, and (a) is the activity of the various subscripted dissolved species. Three stability curves for the Mg-calcite solid solution are also plotted in Figure 14, the lower the -IAP value, the less stable, or more soluble, the phase.

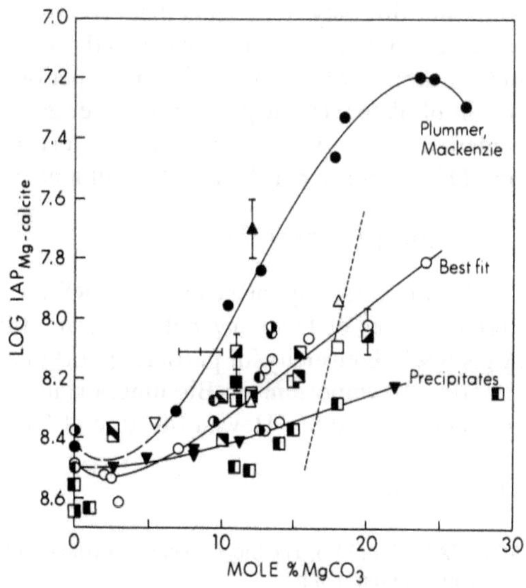

FIGURE 14. Mg-calcite solubilities expressed as -log IAP Mg-calcite from various experiments (see Mackenzie *et al.*, 1983, for references). Three curves are drawn through (1) dissolution experiments of Plummer & Mackenzie (1974), (2) the general trend of the dissolution experiment results, and (3) the precipitation experiment results of Mucci & Morse (1984). Straight dashed line shows the -log IAP of sea water in which the coralline alga *Porolithon gardineri* was grown as a function of the Mg content of the skeleton precipitated by the alga (Agegian, 1985).

It can be seen from Figure 14 that the -log IAP of the sea waters from which a range of compositions of *Porolithon gardineri* skeletal calcite was precipitated defines a trend with composition that is very different than the trends obtained from dissolution experiments of biogenic or synthetic inorganic phases, and experiments involving precipitation of inorganic Mg-calcites. Thus, although the composition of the precipitated skeleton of *P. gardineri* depends on the physico-chemical factors of temperature and carbonate saturation state, vital or kinetic effects apparently play a strong role in governing the composition of the skeletal precipitate. This result leads us to conclude that the skeletal composition of *P. gardineri* cannot be used directly to decipher ambient environmental conditions of growth without experiments on growth conditions. This statement is certainly true for this alga, and perhaps coralline algae and other invertebrates in general. The lack of experimentation certainly limits present interpretations of environmental factors based on skeletal composition alone.

CONCLUSIONS

We would like to re-emphasize the importance of global tectonic processes in driving atmospheric CO_2 changes. The resulting effects on temperature (climate), and sea water saturation state, and the coincident changes in sea water Mg/Ca ratio may have had a profound influence on the biomineralization of carbonate minerals. We suggest that the abundance and composition of skeletal and non-skeletal carbonate minerals may be controlled on a long-term, evolutionary time scale, as well as on shorter cyclic time scales, by changes in ocean-atmosphere chemistry induced by variations in global tectonic activity. Rich *et al.* (1985) have pointed out a significant correlation between sea-floor spreading rate fluctuation, as a measure of global tectonic activity, and marine taxonomic diversity. It appears that evolutionary events are influenced by internal earth processes, and there is a link between these processes and biological phenomena. Gaia and extraterrestrial events as forcing functions for biological change need to be viewed in light of plate tectonic processes. Over geologic time, the earth's internal energy, the fundamental component driving global tectonic activity, may be progressively decreasing. This decrease may be a factor influencing marine organism skeletal evolution.

ACKNOWLEDGEMENTS

This work was supported by ACS-PRF Grant 17371-AC2. We thank The Organized Research Fund of The University of Texas at Arlington and The Sea Grant Program of Texas A&M University for partial support of our travel and housing during the Fifth International Symposium on Biomineralization held at The University of Texas at Arlington in May, 1986. Hawaii Institute of Geophysics Contribution No. 1902.

REFERENCES

AGEGIAN, C.R., 1985. The Biogeochemical Ecology of *Porolithon gardineri* (Foslie). Unpubl. Ph.D. Thesis, Univ. Hawaii, 178 p.

BURKE, W.H., DENISON, R.E., HETHERINGTON, E.A., KOEPNICK, R.B., NELSON, H.F., & OTTO, J.B., 1982. Variation of sea water $^{87}Sr/^{86}Sr$ throughout Phanerozoic time. Geology, 10:516-519.

BURTON, E.A., & WALTER, L.M., 1987. Relative precipitation rates of aragonite and Mg-calcite from seawater: Temperature or carbonate ion control? Geology, 15:111-114.

CHAVE, K.E., 1954. Aspects of the biogeochemistry of magnesium. 1. Calcareous marine organisms. J. Geol., 62:266-283.

FISHER, A.G., 1982. Long-term climatic oscillations recorded in stratigraphy. *In* Climate in Earth History (J.C. Crowell, ed.), Nat. Acad. Sci., Washington, D.C., 97-104.

GRAHAM, D.W., BENDER, M.L., WILLIAMS, D.F., & KEIGWIN, JR., L.D., 1982. Strontium-calcium ratios in Cenozoic planktonic foraminifera. Geochim. Cosmochim. Acta., 46:1281-1292

HALLAM, A., 1977. Secular changes in marine inundation of USSR and North America through the Phanerozoic. Nature, London., 269:769-772.

HALLAM, A., 1981. Facies Interpretation and the Stratigraphic Record. W.H. Freeman and Company, San Francisco., 291 p.

HAYS, J.D. & PITMAN, W.C., 1973. Lithospheric plate motion, sea-level changes & climatic and ecological consequences. Nature, London., 246:18-22.

HOLSER, W.T., 1984. Gradual and abrupt shifts in ocean chemistry during Phanerozoic time. *In* Patterns of Change in Earth Evolution (H.D. Holland & A.F. Trendall, eds.), Springer-Verlag, New York, 123-144.

LOVELOCK, J.E., & MARGULIS, L., 1974. Atmospheric homeostasis by and for the biosphere. Tellus., 26:1-10.

MACKENZIE, F.T. & PIGOTT, J.D., 1981. Tectonic controls of Phanerozoic sedimentary rock cycling. J. Geol. Soc. Lond., 138:183-196.

MACKENZIE, F.T., BISCHOFF, W.D., BISHOP, F.C., LOJENS, F.C., SCHOONMAKER, J. & WOLLAST, R., 1983. Magnesian calcites: Low-temperature occurrence, solubility, and solid-solution behavior. *In* Carbonates: Mineralogy and Chemistry (R.J. Reeder, ed.), Rev. Miner., v. II, Mineralogical Society of America., 97-144.

MUCCI, A., & MORSE, J.W., 1984. The solubility of calcite in sea water solutions at various magnesium concentration, I_t = 0.697 m at 25°C and one atmosphere total pressure. Geochim. Cosmochim. Acta, 48:815-822.

PALMER, M.R., & ELDERFIELD, H., 1986. Rare earth elements and neodinium isotopes in ferromanganese oxide coating of Cenozoic foraminifera from the Atlantic Ocean. Geochim. Cosmochim. Acta., 50:409-417.

PITTMAN, W.C., 1978. Relationship between eustasy and stratigraphic sequences of passive margins. Geol. Soc. Amer. Bull., 89:1389-1403.

PLUMMER, L.N., & MACKENZIE, F.T., 1974. Predicting mineral solubility from rate data: Application to the dissolution of magnesian calcites. Amer. J. Sci., 24:61-83.

RENARD, M., 1986. Pelagic carbonate chemostratigraphy (Sr, Mg, ^{18}O, ^{13}C). Mar. Micropaleont., 10:117-164.

RICH, J.E., JOHNSON, G.L., JONES, J.E., & CAMPSIE, J., 1985. A significant correlation between fluctuations in sea-floor spreading rates and evolutionary pulsations. Nordita, Nordisk Institut Teoretisk Atomfysik, Copenhagen, Denmark, 17 p.

SAINT-EXUPERY, ANTOINE DE, 1943. The Little Prince. Harcourt, Brace, New York., 91 p.

SANDBERG, P.A., 1975. New interpretations of Great Salt Lake ooids & of ancient non-skeletal carbonate mineralogy. Sedimentology., 22:497-537.

SANDBERG, P.A., 1983. An oscillating trend in Phanerozoic non-skeletal carbonate mineralogy. Nature, 305:19-22.

SAVIN, S.M., DOUGLAS, R.G., & STEHLI, F.G., 1975. Tertiary marine paleotemperatures. Geol. Soc. Amer. Bull., 86:1499-1510.

SOUTHAM, J.R., & HAY, W.W., 1977. Time scales and dynamic models of deep-sea sedimentation, J. Geophy. Res., 82:3825-3842.

STECKLER, M., 1984. Changes in sea level. *In* Patterns of Change in Earth Evolution (H.D. Holland & A.F. Trendall, eds.), Springer-Verlag, New York., 103-122.

TURCOTTE, D.L., & BURKE, K., 1978. Global sea-level changes and the thermal structure of the earth. Ear. Planet. Sci. Lett., 41:341-346.

VAIL, P.R., MITCHUM, JR., R.M., & THOMPSON. S., 1977. Seismic stratigraphy and global changes of sea level. 4. Global cycles of relative changes of sea level. Amer. Assoc. Pet. Geol., Mem. 26:83-97.

VAN ANDEL, TJ. H., 1975. Mesozoic/Cenozoic calcite compensation depth and the global distribution of calcareous sediments. Ear. Planet. Sci. Lett., 26:187-194.

WILKINSON, B.H., 1979. Biomineralization, paleoceanography, and the evolution of calcareous marine organisms. Geology, 7:524-527.

WILKINSON, B.H., & GIVEN, R.K., 1986. Secular variation in abiotic marine carbonates: constraints on Phanerozoic atmospheric carbon dioxide contents and oceanic Mg/Ca ratios. J. Geol., 94:321-334.

WILKINSON, B.H., OWEN, R.M., & CARROLL, A.R., 1985. Submarine hydrothermal weathering, global eustasy, and carbonate polymorphism in Phanerozoic marine oolites. J. Sed. Petrol., 55:171-183.

ADDITIONAL BIBLIOGRAPHY

ARTHUR, M.A., & SCHLANGER, S.O., 1979. Cretaceous oceanic anoxic events as causal factors in development of reef-reservoired giant oil fields. Amer. Assoc. Pet. Geol. Bull., 63:870-885.

BERNER, R.A., & RAISWELL, R., 1983. Burial of organic carbon and pyrite sulfur in sediments over Phanerozoic time: A new theory. Geochim. Cosmochim. Acta, 47:855-862.

BERNER, R.A., LASAGA, A.C., & GARRELS, R.M., 1983. The carbonate-silicate geochemical cycle and its effect on atmospheric carbon dioxide over the past 100 million years. Amer. J. Sci., 283:641-683.

BERRY, W.B.N., & WILDE, P., 1978. Progressive ventilation of the oceans: An explanation of the distribution of the lower Paleozoic black shales. Amer. J. Sci., 278:257-275.

BOYLE, E.A., & KEIGWIN, L.D., 1985. Comparison of Atlantic and Pacific paleochemical records for the last 215,000 years: Changes in deep ocean circulation and chemical inventories. Ear. Planet. Sci. Lett., 76:135-150.

BROECKER, W.S., 1982. Ocean chemistry during glacial time. Geochim. Cosmochim. Act, 46:1689-1705.

CHAMBERLAIN, T.C. 1898. The influence of great epochs of limestone formation upon the constitution of the atmosphere. J. Geol., 6:609-621.

CLAYPOOL, G.E., HOLSER, W.T., KAPLAN, I.R., SAKAI, H., & ZAK, I., 1972. Sulfur and oxygen isotope geochemistry of evaporite sulfates. Abstract, Geol. Soc. Amer. , 4:473.

CLAYPOOL, G.E., HOLSER, W.T., KAPLAN, I.R., SAKAI, H., & ZAK, I., 1980. The age curves of sulfur and oxygen isotopes in marine sulfate and their mutual interpretation. Chem. Geol., 28:199-259.

CRONBLAD, H.G., & MALMGREN, B.A., 1981. Climatically controlled variation of Sr and Mg in Quaternary planktonic foraminifera. Nature., 291:61-64.

FISCHER, A.G., & ARTHUR, M.A., 1977. Secular variations in the pelagic realm. Soc. Econ. Paleont. Miner., Spec. Publ. 25:19-50.

GARRELS, R.M., & MACKENZIE, F.T., 1971. Evolution of Sedimentary Rocks, W.W. Norton and Co. Inc., New York. 397 p.

GARRELS, R.M., & PERRY, E.A., 1974. Cycling of carbon, sulfur, and oxygen through geologic time. *In* The Sea, 5 (Goldberg, E.D., ed.), Wiley-Interscience, New York, 303-336.

GARRELS, R.M., LERMAN, A., & MACKENZIE, F.T., 1976. Controls of atmospheric O_2 and CO_2: Past, present, and future. Amer. Scientist, 64:306-315.

HESS, J., BENDER, M.L., & SCHILLING, J.-G., 1986. Evolution of the ratio of strontium-87 to strontium-86 in sea water from Cretaceous to present. Science, 231:979-984.

HOLLAND, H.D., 1973. Systematics of the isotopic composition of sulfur in the oceans during the Phanerozoic and its implications for atmospheric oxygen. Geochim. Cosmochim. Acta, 37:2605-2616.

HOLLAND, H.D., & TRENDALL, A.F., 1984. Patterns of Change in Earth Evolution. Springer-Verlag, New York, 431 p.

HOLSER, W.T., & KAPLAN, I.R., 1966. Isotope geochemistry of sedimentary sulfates. Chem. Geol., 1:93-135.

KEITH, M.L. 1982. Violent volcanism, stagnant oceans and some inferences regarding petroleum, strata-bound ores and mass extinctions. Geochim. Cosmochim. Acta, 46:2621-2637.

KOEPNICK, R.B., BURKE, W.H., DENISON, R.E., HETHERINGTON, E.A., NELSON, H.F., OTTO, J.B., & WAITE, L.E., 1985. Construction of the sea water $^{87}Sr/^{86}Sr$ curve for the Cenozoic and Cretaceous: Supporting Data. Chem. Geol., 58:55-81.

LI, Y.-H., 1972. Geochemical mass balance among lithosphere, hydrosphere, and atmosphere. Amer. J. Sci., 272:119-137.

RONOV, A.B., 1964. Common tendencies in the chemical evolution of the earth's crust, ocean, and atmosphere. Geochemistry, 8:715-743.

SCHLANGER, S.O., & JENKYNS, H.C., 1976. Cretaceous oceanic anoxic events--causes and consequences. Geol. Mijnb., 55:179-184.

SHAW, H.F., & WASSERBURG, G.J., 1985. Sm-Nd in marine carbonates and phosphates: Implications for Nd isotopes in sea water and crustal ages. Geochim. Cosmochim. Acta, 49:503-518.

VEIZER, J., FRITZ, P., & JONES, B., 1986. Geochemistry of brachiopods: Oxygen and carbon isotopic records of Paleozoic oceans. Geochim. Cosmochim. Acta, 50:1679-1696.

VEIZER, J., HOLSER, W.T., AND WILGUS, C.K., 1980. Correlation of $^{13}C/^{12}C$ and $^{34}S/^{32}S$ secular variations. Geochim. Cosmochim. Acta, 44:579-587.

VEIZER, J., & JANSEN, S.F., 1979. Basement and sedimentary recycling and continental evolution. J. Geol., 87:341-370.

ZIEGLER, A.M., SCOTESE, C.R., MCKERROW, W.S., JOHNSON, M.E., & BAMBACH, R.K., 1979. Paleozoic paleogeography. Ann Rev. Ear. Planet. Sci., 7:473-502.

THE SODA OCEAN CONCEPT AND ITS BEARING ON BIOTIC EVOLUTION

Stephan Kempe[1], Józef Kazmierczak[2] and Egon T. Degens[1]

[1]*SCOPE/UNEP International Carbon Unit, Geological-Paleontological Institute, University of Hamburg, Bundesstrasse 55D-2000 Hamburg 13, Federal Republic of Germany;* [2]*Institute of Paleobiology, Polish Academy of Science, Zwirki i Wigury 93, PL-02089 Warszawa, Poland*

ABSTRACT. The soda ocean hypothesis (Kempe & Degens, 1985) suggests an early alkaline ocean of high pH and low calcium and magnesium concentrations. The dissolved carbonates were gradually lost during the Precambrian, leaving the present sodium chloride ocean. The Precambrian paleontological record and the calcium physiology of living cells implicate that the stepwisebuildup of calcium in the ancient ocean was of primary importance for the generation of multicellular life and the onset of biocalcification.

INTRODUCTION

Proton and electron donor-acceptor reactions govern the aqueous environment on earth. If one speculates on the composition of the early ocean, then these thermodynamic processes set boundaries within which any geologically and biochemically plausible scenario has to be discussed (Degens, 1973; Walker, 1982; Holland, 1984).

It seems to be the general consent that the early ocean was of a reducing state, *i.e.* it contained no SO_4^{2-}, and no Fe^{3+} but rather Fe^{2+} and other reducing species. Concerning the bulk composition of the proto-ocean most authors conclude, that it was not much different from that of today, *i.e.*, Na^+ and Cl^- were the major ions plus K^+, Mg^{2+}, Ca^{2+}, and HCO_3^- constituting a neutral or slightly acid state (Conway, 1943; Wedepohl, 1963; Garrels & Mackenzie, 1971; Maisonneuve, 1982; Holland, 1984). Ronov (1967) suggested a composition dominated by Mg^{2+} and Ca^{2+}.

Some atmospheric chemists postulated that the early atmosphere must have had a high CO_2 partial pressure (PCO_2) to yield a strong greenhouse effect in order to compensate for the faint early sun in their radiation models (Hart, 1978; Kuhn & Kasting, 1983; Kasting, 1984). This, in consequence, would call for an acid ocean in equilibrium with such an atmosphere. Henderson-Sellers & Cogley (1982) noted that a PCO_2 of orders of magnitudes higher than the present atmospheric level (PAL = 300 ppmv) would not be stable in the presence of water but would react with silicates (*i.e.*, Urey reactions; Urey, 1951) during planetesimal impacts to form carbonates. The other extreme, a very low PCO_2, was tested by Henderson-Sellers & Cogley (1982) with a

Origin, Evolution, and Modern Aspects of Biomineralization in Plants and Animals
Edited by Rex E. Crick
Plenum Press, New York

radiation model including cloud feedback for a solar flux of 0.8 its present value and a PCO_2 of PAL. Due to a lower albedo caused by the absence of land masses, temperatures still stayed above freezing in these models and contradicted the hypothesis of a frozen early earth (Siever, 1977). Henderson-Sellers & Cogley (1982) further note that neither NH_3 nor CH_4 could have been major components in the atmosphere due to their photochemical instability.

THE SODA OCEAN HYPOTHESIS

In contrast to the above picture of a neutral or acid ocean, we suggested a highly alkaline early ocean containing mainly dissolved sodium carbonates ("soda") (Kempe & Degens, 1985). This "soda ocean" hypothesis (SOH) rests on the occurrence of modern soda lakes, on thermodynamic, kinetic and mass balance arguments and offers intriguing aspects to account for the Precambrian geochemical and biological evolution. We think that the emergence and eventual disappearance of a soda ocean is responsible for substantial variation in the calcium concentration of the sea which in our opinion is the single most important geochemical factor for the development of multicellular life and the onset of biomineralization (Kazmierczak *et al.*, 1985; Degens *et al.*, 1986a, b).

Soda lakes: modern analogs of the ancient ocean?

Figure 1 depicts the distribution of modern soda lakes and their association to volcanic areas aligned to active plate boundaries. We consider a "soda lake" to be high in dissolved carbonates (in equivalents) relative to alkaline earth elements, that is, part of the charge balance of the carbonates is supplied by alkaline elements:

$$HCO_3^- + CO_3^{2-} > Ca^{2+} + Mg^{2+} \text{ and } Na^+ + K^+ > Cl^- + SO_4^{2-} \quad (1)$$

Lakes of this type which have an outlet,(Lake Taupo, New Zealand) or which are geologically young (*e.g.*, lakes in recent calderas) show very low alkalinities and consequently do not display high pH values. Figure 2 illustrates the relationship of the total alkalinity with pH for a variety of lakes. Apparently, concentrations above 10^{-2} eq/l result in increased pH values. It is important to note that due to the high solubility of sodium carbonates, soda concentrations can range over three orders of magnitude. The function of pH versus total alkalinity can not be precisely defined due to varying proportions of bicarbonate and carbonate ions.

The largest soda lake is Lake Van in Eastern Anatolia (volume = 607 km^3). It ranks fourth following the Caspian Sea, Lake Aral and Lake Issykul (USSR) among closed lakes on earth (Kempe, 1977). Its pH is 9.55 with an alkalinity of 147 meq/l. It is less saline than seawater but shows a marked relative enrichment in alkaline elements versus alkaline earth elements relative to seawater (Table 1). Due to a low calcium concentration, phosphate is high, at least for unpolluted lake water (0.5 mg/l). Lake Van is surrounded by the andesitic Anatolian volcanos. Table 1 also lists the chemistry of one of the most extreme alkaline environments known today, Lake Magadi, situated in the volcanic area of the East African Rift (Jones *et al.*, 1977). Total alkalinity can be as high as 3,890 meq/l and its pH is 10.5. Sodium carbonates and silicates precipitate from this lake (Eugster, 1967, 1969). The dilute end of the "soda lake population" is marked by Lake Taupo, which collects water from dacitic and rhyolitic volcanic terrains in Northern New Zealand and feeds the Waikato River (Bryers, 1985). Its waters are very dilute and

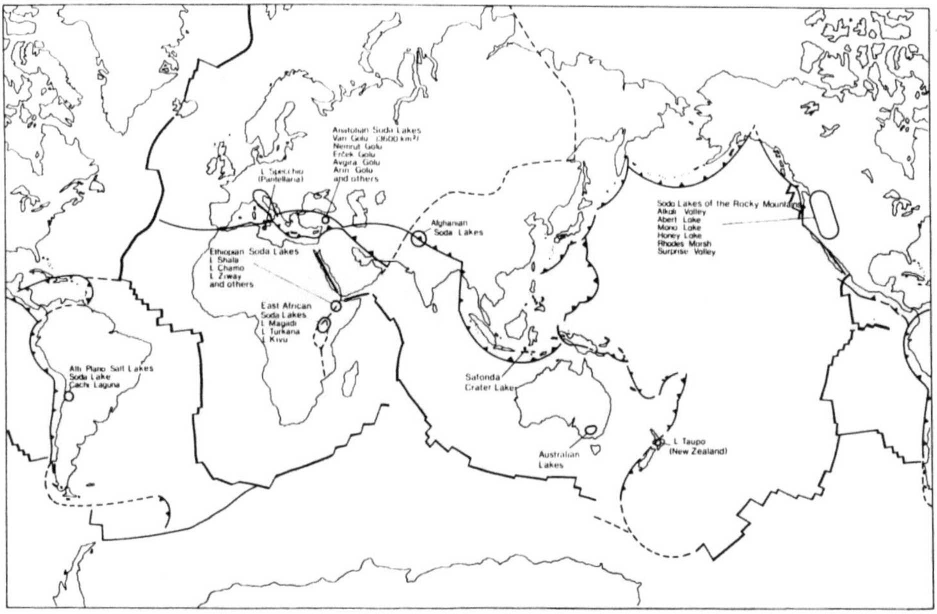

FIGURE 1. Distribution of modern soda lakes. Note their close association to volcanic active regions of the world.

TABLE 1. Chemical composition of three soda lakes, an alkaline seawater lake and seawater

	LAKE and LOCATION				
	Van[1] Turkey	Magadi[2] E-Africa	Taupo[3] New Zeal.	Satonda[4] Indonesia	Seawater average[5]
pH	9.55	10.5	7.9	8.55	ca.8.2(surface)
g salt/l	21.7	355.6	0.095	31.1	33.5
Na meq/l	337.	6,250.	0.62	440.	470.
K meq/l	13.	55.	0.054	50	54.
Li meq/l	0.22	–	–	–	0.014
Mg meq/l	7.8	0.	0.21	93.	107.
Ca meq/l	0.4	0.	0.29	9.6	20.5
Sr meq/l	0.02	–	–	–	0.3
Mg/Ca	19.5	–	0.72	9.75	5.22
CO_3meq/l	111.	3,780.	0.	Alk.:3.7	2.35
HCO_3meq/l	35.9	111.	0.75	–	–
Cl meq/l	154.	2,470.	0.26	507.	548.3
SO_4meq/l	48.8	40.	0.12	n.d.	56.5
PO_4 mg/l	0.5	84.	–	–	0.19 (average)
SiO_2 mg/l	70.	1,400.	–	–	12. (average)

[1]Kempe, 1977; [2]Jones *et al.*, 1977; [3]Bryers, 1985; [4]Kempe, unpubl; [5]Sverdrup, *et al.*, 1970

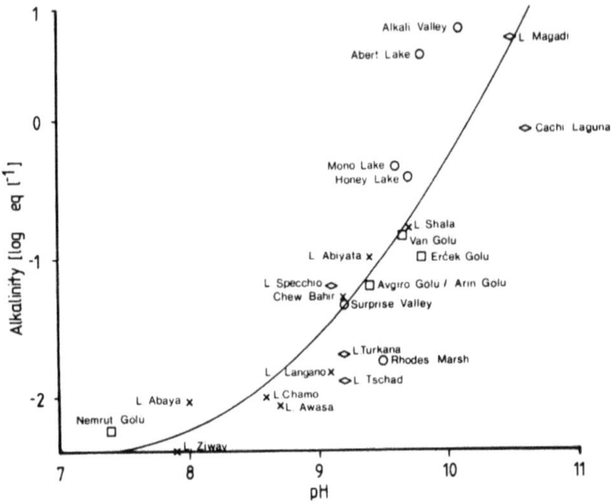

FIGURE 2. Scatter plot of pH and total alkalinity for various soda lakes.

do not show a raised pH but are nevertheless clearly of a soda chemistry [our precondition (1) holds (compare Table 1)]. Next in Table 1 is the chemistry of the crater lake of Satonda Island (Sumbawa, Indonesia). The lake is filled with seawater but has no direct communication with the open sea other than by percolation through its volcanic tuff. Here alkalinity and pH are much higher than in seawater while the calcium concentration is markedly reduced.

These chemical examples illustrate that wherever fresh igneous silicates are exposed to intensive weathering, possibly enhanced by hydrothermal CO_2 release, solutions emerge in which part of the dissolved carbonates are charge balanced by alkaline elements. Average continental surface runoff does not exhibit this characteristic, instead solutions show a $Ca^{2+} + Mg^{2+} > HCO_3^- + CO_3^{2-}$ relationship (Garrels & Mackenzie, 1971) because the composition of average sediments is characterized by an excess of calcium and magnesium versus sodium and potassium. Volcanic areas certainly represent better models of the environment of early earth than continents covered by sediments, considering especially the relative late appearance of granodioritic continents only 3 Ga ago (Veizer *et al.*, 1982; Veizer, 1983).

Thermodynamic, kinetic and mass-balance considerations of the SOH

As shown above, alkaline environments are common features in volcanic regions. In fact, alkaline terminal lakes develop by evaporation from diluted waters obeying our condition (1). Eugster & Hardie (Hardie & Eugster 1970; Eugster & Hardie, 1978) have documented this for many of these lakes and Garrels & Mackenzie (1967) and Hardie & Eugster (1970) have used thermodynamic models to theoretically follow this concentration by evaporation. As a first step, the water is concentrated until the ion-activity-product of calcite is exceeded. Consequently calcite, then aragonite, protodolomite and magnesite could precipitate. The Mg/Ca ratio increases (compare also

Table 1) and magnesium silicates would precipitate. Eventually a brine rich in Na^+, Cl^-, and CO_3^{2-} species is left whereas SO_4^{2-} would either be lost by reduction or remain in solution. The main event, the surpassing of the calcite ion-activity-product, essentially depends on the CO_2 pressure of the solution. Lakes in general have a PCO_2 close to atmospheric levels (Kempe, 1982). Thus, at a possibly higher primordial PCO_2 calcite would not precipitate as rapidly and the brine would not become as alkaline as under PAL of CO_2.

The question is, how fast does the silicate weathering proceed? The weathering of an average plagioclase is formulated as

$$Na_{0.62}\,Ca_{0.38}\,Al_{1.38}\,Si_{2.62}\,O_8 + CO_2 + H_2O = HCO_3^- + SiO_2 + 0.6\,Na^+ + 0.4\,Ca^{2+} + \text{kaolinite} \quad (2).$$

The mineral balance of annual average river transport gives one estimate of the rate of silicate weathering (Kempe, 1979). Annually 0.1×10^{15}g C are consumed on the continents for silicate weathering (0.7×10^6g C/km^{2}a). With an atmospheric content of 700×10^{15}g CO_2-C, present continental weathering would consume an amount of PAL CO_2 within 7,000 years only. This weathering, however, does proceed in the soil, albeit under a PCO_2 much higher (between 0.03 and 0.003 atm) than PAL. Nevertheless, and without the aid of plants, silicate weathering on continents would, within a geologically short time, sequester all available CO_2 from the atmosphere leaving a PCO_2 which is determined by the rate of input from volcanoes versus the rate of silicate weathering on abiotic rock surfaces.

The outflow of Lake Taupo gives another figure of the rate of silicate weathering within its volcanic sediments: 30.2×10^6g C/km^{2}a. From the salt-age of Lake Van, a rate of 1×10^6 g C/km^{2}a can be estimated (Kempe & Degens, 1985). Assuming primordial conditions similar to Lake Taupo for the whole earth, (*i.e*,. an ocean floor covered with fresh volcanic tuff and in contact by CO_2) bicarbonate would rise in the sea at a global rate of 15×10^{15}g C/a and would reach the alkaline threshold (1 gC/kg H_2O, see Fig. 2) within 0.1×10^6a. Further weathering would subsequently cause calcium and magnesium carbonates to precipitate and the water to become alkaline with ever increasing amounts of dissolved alkaline carbonates.

This model rate of CO_2 consumption has to be checked against the average release rate of CO_2 from the mantle in the course of earth history. A lower boundary is obtained if all crustal, biotic and oceanic carbon is taken together and divided by the age of the earth: 65.5×10^{21}g C/4.5 Ga = 0.015×10^{15}g C/a. In short, the release rate is 1000 times lower relative to that sequestered by a "Lake Taupo" earth.

Release of CO_2 from the mantle was not constant through time. An early and rapid outgassing of volatiles now present in crust, hydrosphere and atmosphere is commonly assumed (Fanale, 1971; Thomsen, 1980; Allègre, 1985). Volatiles of cometary origin were added. Degassed carbon and chloride amounts to 65.5×10^{21} and 52×10^{21}g, respectively (Kempe & Degens, 1985). Thus the mol-ratio of these two major acids is 3.7 in favor of carbonic acid. The composition of the early komateitic crust has been estimated by Rogers (1978): 5.1% MgO, 4.6% CaO, 4.9% Na_2O and 1.1% K_2O (in mol%). Assuming that these elements were titrated by a mixture of carbonic and hydrochloric acid, then any H_2CO_2/HCl mol-ratio above 1.1 would yield carbonates in excess of alkaline earth concentration. In time, alkaline conditions would be established according to equation 1. The argument receives support from the fact that alkaline earth

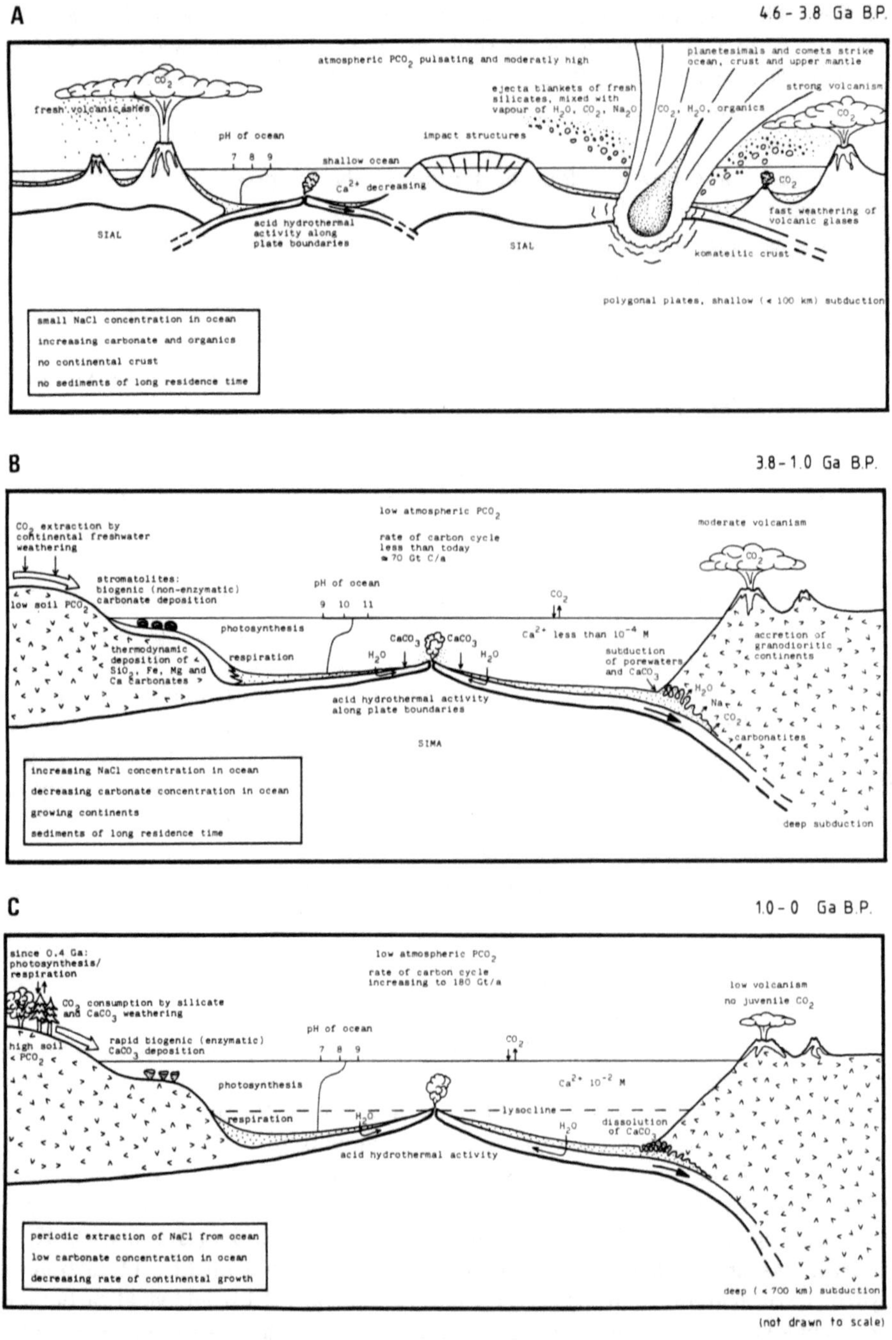
A
4.6 - 3.8 Ga B.P.
atmospheric PCO2 pulsating and moderatly high
planetesimals and comets strike ocean, crust and upper mantle
CO2
fresh volcanic ashes
ejecta blankets of fresh silicates, mixed with vapour of H2O, CO2, Na2O
CO2, H2O, organics
strong volcanism
pH of ocean
7 8 9
impact structures
shallow ocean
Ca2+ decreasing
fast weathering of volcanic glases
acid hydrothermal activity along plate boundaries
SIAL
SIAL
komateitic crust
polygonal plates, shallow (< 100 km) subduction
small NaCl concentration in ocean
increasing carbonate and organics
no continental crust
no sediments of long residence time
B
3.8 - 1.0 Ga B.P.
low atmospheric PCO2
rate of carbon cycle less than today ≈ 70 Gt C/a
CO2 extraction by continental freshwater weathering
moderate volcanism
stromatolites: biogenic (non-enzymatic) carbonate deposition
low soil PCO2
pH of ocean
9 10 11
CO2
photosynthesis
thermodynamic deposition of SiO2, Fe, Mg and Ca carbonates
respiration
CaCO3
CaCO3
H2O
H2O
Ca2+ less than 10-4 M
subduction of porewaters and CaCO3
accretion of granodioritic continents
Na
carbonatites
acid hydrothermal activity along plate boundaries
SIMA
increasing NaCl concentration in ocean
decreasing carbonate concentration in ocean
growing continents
sediments of long residence time
deep subduction
C
1.0 - 0 Ga B.P.
since 0.4 Ga: photosynthesis/ respiration
low atmospheric PCO2
rate of carbon cycle increasing to 180 Gt/a
CO2 consumption by silicate and CaCO3 weathering
low volcanism
no juvenile CO2
high soil PCO2
rapid biogenic (enzymatic) CaCO3 deposition
pH of ocean
7 8 9
CO2
photosynthesis
Ca2+ 10-2 M
lysocline
respiration
H2O
H2O
dissolution of CaCO3
acid hydrothermal activity
periodic extraction of NaCl from ocean
low carbonate concentration in ocean
decreasing rate of continental growth
deep (< 700 km) subduction
(not drawn to scale)

elements tend to form secondary silicate minerals such as epidote, clinozoisite, ferroactinolite and tremolite (Reed, 1982) whereas alkaline ions are preferentially mobilized. In a crust of the Rogers' composition sufficient sodium is present in a layer 2–4 km thick to bind all CO_2 as sodium carbonate. A peridotitic crust contains the same amount of sodium in a layer 18–36 km thick. Consequently, cometary impacts would expose enough fresh silicates, and rapidly neutralize any amount of degassed acid volatiles.

Thus, thermodynamic, kinetic, and mass-balance considerations favor a fast build-up of alkaline carbonates in the early ocean. The first 700 Ma of planet earth would suffice to equilibrate silicates, water and acid volatiles (Fig. 3a). At times PCO_2 may have been high due to a sudden release of CO_2 from impacting bolides and evaporation of large volumes of the early carbonate ocean.

However, once a soda ocean became established, subduction of seawater as porefluid and its incorporation into continental crust, and the formation of organic carbon by photosynthesis could remove dissolved carbonates from the ocean. Both processes can only lead to a substantial decrease in the total alkalinity of seawater providing a long-term storage is available, *i.e.*, after the onset of cratonization. Numerical modelling shows that various combinations of the two processes can cause a decrease of the carbonate concentration in seawater within the time-scale of a billion years (Kempe & Degens, 1985). These models, however, leave wide margins within which parameters such as degassing constants, subduction rates, residence times of organics etc. can change. They do not allow a detailed reconstruction of the evolution of ocean's chemistry. They only provide the conclusion that the soda ocean could have been lost in time to meet the onset of biomineralization. Figure 3b shows pertinent features of this phase in the evolution of the ocean.

One of the thermodynamic consequences of an alkaline environment is the extremely low concentration of divalent cations; calcium concentration measurements from soda lakes indicate values of around 10^{-4} mol/l or less. As most of these analyses have been made after long storage of samples, values could, however, be even lower. Measurements available so far do not account for organic complexes and thermodynamic calculations for such high pH ranges and high alkalinities have not as yet been made. Certainly the concentration of the free calcium ion is much lower than quoted.

Mg/Ca ratios are higher than in sea water and many alkaline lakes reportedly deposit dolomite. Also, due to the lack of calcium, the phosphate concentrations are much higher in soda lake environments than in fresh or sea water. Inasmuch as SiO_2 is highly soluble at pH-values exceeding 9, the silica concentrations may reach high values. All factors combined could have favored the formation of Precambrian primary dolomite, dolostone, phosphorite, siderite, jaspilite and their composites, such as the Banded Iron Formations, for which Eugster (1969) postulated an alkaline milieu. Also the absence of massive gypsum deposits with a true marine origin until about 800 Ma could have been due not only to the missing sulfate in an anaerobic environment, but also due to the missing calcium.

FIGURE 3. Geologic scenarios for the three main periods in world ocean history. ▪A. The rise of the soda ocean (4.6–3.8 Ga). ▪B. The soda ocean age (3.8–0.8 Ga). ▪C. The halite ocean age (0.8–0 Ga). Individual features are not drawn to scale.

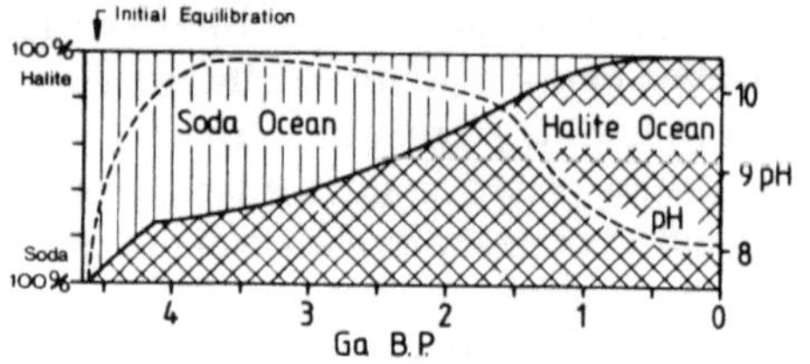

FIGURE 4. Evolution of ocean chemistry during geologic time according to the soda ocean hypothesis. pH-curve deduced from total alkalinity and pH-relation of present-day soda lakes (from Kempe & Degens, 1985).

Even though the removal of the dissolved carbonate from the ocean may have been very gradual, the pH value would have altered dramatically towards the end of this removal period when the concentration sank below 10^{-2} M, and would have caused most of the carbonate ions to change to bicarbonate ions. Similarly at this point the calcium would have increased more rapidly than during the 3 Ga of gradual carbonate removal. Thus, thermodynamically, one has to postulate a quite sharp "titration" endpoint of the soda ocean (Fig. 4).

Before that endpoint, calcium carbonate would have been extracted solely by non-enzymatic reactions, either inorganically as a chemical precipitate or physiochemically by algal mats (stromatolites). At acid hydrothermal vents, massive deposits of calcium carbonate could have been formed in the deep-sea. Upon subduction these deposits could have become carbonatite magmas (Fig. 3b). After the titration endpoint only enzymatic reactions control calcium carbonate deposition in various skeletal structures (Fig. 3c).

BIOTIC IMPLICATIONS OF THE SODA OCEAN HYPOTHESIS

In cellular systems, calcium occupies also a crucial position. The ubiquity of calcium as a structural and functional element in cell physiology is well documented (*e.g.* Degens, 1976; Erulkar, 1981; Lehninger, 1982; Kretsinger, 1983; Campbell, 1983; Marmé, 1985; Carafoli & Penniston, 1985). The central role of calcium is principally attributable to its affinity for oxygen ligands in complex biochemical molecules such as proteins. The notion even exists that calcium is the second messenger in the metabolic and genetic apparatus. In short, an ion of such importance in cellular regulation requires strict control devices permitting calcium to be present at the right spot, in the right concentration, and at the right time.

All cells maintain a cytosolic calcium content close to 10^{-7} M (Kretsinger, 1983). This level appears to be a universal prerequisite in all three superkingdoms of life, archaebacteria, eubacteria, and eukaryotes. In all three major lines of descent the same basic processes, the same genetic code, and similar biochemical pathways are utilized. Yet, important details of just every process appear to differ in one way or another in each of the superkingdoms (Woese & Fox, 1977; Woese *et al.*, 1978). These findings suggest that the three lines of descent diverged before the level of complexity usually associated with the prokaryotic cell was reached, a level of organization called the "progenote". Thus, those features of the cell that have to do with refining molecular functions, coping with a large genome, etc., evolved independently in the three primary lines of descent as each reached a more complex level of organization (Woese & Fox, 1977). Even the most ancestral species of today must be considered a highly derived form in relation to their ancestral progenotic phenotypes. Consequently, statements on the chemistry of the primordial atmosphere and hydrosphere or the energy pathways which are based on extant life studies are meaningless.

Our approach is just the opposite. We are using chemical, physical and geological data to outline a model on the state of air and water *in statu nascendi* and their subsequent development across 4 Ga. Origin and evolution of life are examined in the light of this development. Once life enters the terrestrial scene, our soda ocean model has to be constantly adjusted to account for the multiple feedbacks within the air-water-earth-life supersystem. In one word, the evolution of the four listed spheres is tightly bound and consequently we are dealing with an irreversible thermodynamic system. Thus, the evolution of life should be examined in relation to that of air, water and earth.

In the following, we will use temporal changes of calcium levels in the Precambrian sea as deduced from the SOH and examine the response of organisms to that dramatic shift from an early Archaean level being possibly as low as cytosol level (10^{-7} M), due to the abundance of calcium complexing biotic and abiotic polymers in the seawater, to the present-day seawater calcium concentration of 10^{-2} M reached during the late Precambrian.

The early Precambrian microbiota contain forms assigned to bacteria, cyanobacteria (*e.g.*, Awramik *et al.*, 1983; Knoll & Barghoorn, 1977), yeast-like organisms (Pflug, 1978), and problematic unicells called acritarchs (*e.g.*, Hofmann, 1976, 1985; Pflug & Reitz, 1985), part of which may be related to green algae. Cyanobacteriac are exclusively alkalophilic microorganisms (Brock, 1973) with extremely low requirements for Ca^{2+} (Gerloff & Fishbeck, 1969) but with essential need for Na^{+} (Kratz & Myers, 1955). The same is true for many unicellular green algae (O'Kelley, 1968; Gerloff & Fishbeck, 1969). Yeasts are particularly good candidates for early eukaryotes which may have originated in a calcium poor environment. Unlike other eukaryotes, yeasts are equipped with highly efficient calcium sequestering mechanisms which permit them to live in calcium limited conditions with calcium concentrations as incredibly low as 2×10^{-9} M and still trap the ion in a chelated form (Kovác, 1985).

During the transition from the Archaean to the Proterozoic an increase in calcium levels to 10^{-5} or 10^{-4} M, due to alkalinity decrease and consumption of dissolved Ca-complexing polymers, may be envisaged. This increase was associated with the rapidly progressing cratonization of the lithosphere during the early Proterozoic (Veizer & Compston, 1976; Veizer, 1985). Increased continentality supplied the widespreading epicontinental seas with high calcium amounts by riverine runoff. The calcium shifts in the environment promoted, in our opinion, an advancement of calcium regulation and calcium extrusion systems in eukaryotic cells. This was manifested first of all by the evolution of a whole family of cytosolic calcium binding proteins of which troponin C, parvalbumin and calmodulin are the most common species (Kretsinger, 1976). A major function of calmodulin is to mobilize the calcium extrusion system when intracellular calcium concentration is well above the normal level. A high-affinity calcium binding protein serves also as a carrier for shipping the calcium surplus outside the cell via the endoplasmic reticulum and Golgi apparatus.

The dramatic increase of calcium concentrations in the marine environment during the Proterozoic was probably of great importance for enhancing cell aggregation, cell-to-cell adhesion and other cell contact phenomena leading to multicellular organization. A calcium level not lower than 10^{-5} M is, for instance, required to keep slime molds cell aggregates together (Bonner, 1971). Similarly, in sponges *Microciona* and *Haliclona* the cells dissociate when calcium in the water falls below 10^{-5} M and they reaggregate after calcium level is raised above 10^{-5} M (Kretsinger, 1977). Extracellular calcium concentration is the primary factor controlling the colony formation in the green alga *Coelastrum*. Media with calcium levels about 2×10^{-4} M support only formation of unicells. The percentage of colonies increase when the calcium level is raised to 2×10^{-3} M; 10^{-2} M calcium has an inhibitory effect on *Coelastrum* growth (Chan, 1976). Even in highly developed metazoan tissues, cells fail to adhere when the extracellular calcium falls below 10^{-4} M (*e.g.*, Gilula & Epstein, 1976; Franchi & Camatini, 1985). The presence of well-defined coenobial algae in the *ca.* 1.9 Ga old Gunflint Iron Formation and Michigamme Formation (Kazmierczak, 1976; LaBerge *et al.*, 1984) may be taken as indication that the above-mentioned critical calcium level needed for cell adhesion was achieved by that time. The abundance and variety of colonial algae increased significantly in the late Proterozoic deposits (Timofeev, 1956; Tynni & Uutela, 1984; Lindgren, 1982).

An increase in extracellular calcium and other divalent cations concentrations promotes in some cell aggregates the cell fusion phenomena (Ahkong *et al.*, 1975; Knutton & Pasternak, 1979). At this point one can speculate that the calcium level increase in the ancient ocean resulted in the case of some heterotrophic cell aggregates, in the appearance of polyenergid organization (*e.g.*, ciliates) and, in the case of some algal cell aggregates, in coenocytic organization (*e.g.*, siphonaleans). Orthodox views of biospheric evolution require that a very long entirely prokaryotic period preceded the appearance of eukaryotic life (Cloud, 1976; Schopf, 1978; Knoll, 1983). However, recent data of a paleontologic nature suggest that not only have algal eukaryotes been present in the earth's biosphere since early Proterozoic time (Darby, 1974; Kazmierczak, 1976; LaBerge *et al.*, 1984; Hofmann & Grotzinger, 1985; Pflug & Reitz, 1985), but also probable evidence for small metazoans preserved in the sediment either as burrows (Kauffman & Steidtmann, 1981) or as fecal pellets (Robbins *et al.*, 1985).

Interesting conclusions concerning the influence of increasing calcium levels in the ancient ocean on the biotic development can be drawn from the Precambrian record of acritarchs. These unicellular eukaryotes of unknown systematic affinities (? green or yellow-green algae) appear first as relatively small (10 – 15 μm) spherical bodies in sediments *ca.* 2.4 Ga old (Hofmann, 1976). In rocks about 1.9 Ga old, acritarchs are represented together with sphaeromorphs by slightly spiny forms (Hofmann & Grotzinger, 1985; Pflug & Reitz, 1985). In the late Proterozoic (from *ca.* 1 Ga) acritarchs increased considerably in number and diversity, and close to the Cambrian boundary reached remarkably large sizes (> 500 μm) and a variety of shapes (for review see: Vidal & Knoll, 1982; Vidal, 1984; Wang, 1985). This trend may reflect the response of acritarch algae to increased calcium levels in their environments. It has been known since Stegmann's work (Stegmann, 1940) that high calcium deficit will strongly influence the size of *Chlorella* cells. Cells growing in calcium deficient water are significantly smaller. Kylin & Das (1967) and Trainor (1969) have shown that extracellular calcium levels have a profound effect on the shape of cells in the green alga *Scenedesmus*. Above 10^{-3} M

Ca^{2+}, spiny, spindle-like cells develop, whereas below 10^{-3} M Ca^{2+}, spherical and spinless cells dominate controlled cultures.

Excess calcium in the environment forces many unicellular algae to excrete large amounts of Ca^{2+}-affine organic matter dominated by aspartic acid (Kempe, 1976; Ittekkot, 1982; Lee & Bada, 1977; Degens & Ittekkot, 1986). A similar strategy to bind toxic excess of transitional metal-ions outside the cell with specific metal chelators is known in certain species of algae (see Huntsman & Sunda, 1980, for review). The intensive production of metal-complexing substances in metal overloaded environments is often associated with a considerable increase in algal cell sizes or changes in cell morphologies (*e.g.* Ericson, 1972; Canterford, 1980; Thomas *et al.*, 1980). In the light of these observations the giant acritarchs like *Chuaria* and related forms thriving in the late Proterozoic seas (Ford & Breed, 1973; Hofmann, 1977, 1985) can be interpreted as a direct response of algal unicells to dramatically increasing calcium and other metal ions concentrations in the environment. At the end of the Precambrian, the calcium stress exercised on the biota was additionally aggravated by high amounts of phosphate in the seawater as documented by large Vendian phosphorite deposits (Sheldon, 1981; Cook & Shergold, 1984). Phosphate stimulates calcium migration into cells (Cotmore *et al.*, 1971) and simultaneous excess of Ca^{2+} and PO_4^{3-} in the cell surrounding may have toxic effects (Hellman & Anderson, 1978). Unicellular algae respond in such a situation by copious release of extracellular substances (Chrost, 1978) and/or increase of cell size (Friebele *et al.*, 1978). Many metazoans reacted against the almost synchronous Ca^{2+} and PO_4^{3-} excess in the marine environment at the end of the Vendian by depositing first Ca-phosphatic skeletons and shortly hereafter, as phosphate levels lowered, calcareous skeletons (Kazmierczak *et al.*, 1985). In such an interpretation the sudden onset of biocalcification close to the Precambrian/Cambrian boundary loses its mysterious character and is merely an extension over 3 Ga of the adjustment of the biosphere to the environmental build-up of the very essential but at the same time dangerous calcium cation. Biocalcification is only one of many strategies invented by the biosphere to cope with metal stress. This environmental strategy is and was undoubtedly effective in terms of the habitat detoxification but may hamper further evolution by not always acting as advantageous functional and energetic constraints.

ACKNOWLEDGEMENTS

Authors acknowledge continued support for various carbon-cycle related projects through the German Ministry for Research and Technology, the Scientific Committee on Problems of the Environment (SCOPE), the United Nations Environmental Programme (UNEP) and the Polish Academy of Sciences. Participation in The Fifth International Symposium on Biomineralization was made possible by funds provided through The Organized Research Fund of the The University of Texas at Arlington and The Sea Grant Program administered by Texas A & M University.

REFERENCES

AHKONG, Q.F., TAMPION, W. & LUCY, J.A., 1975. Promotion of cell fusion by divalent cation ionophores. Nature, 256:208-209.

ALLÉGRE, C., 1985. The evolving Earth system. Terra Cognita, 5:5-14.

AWRAMIK, S.M., SCHOPF, J.W. & WALTER, M.R., 1983. Filamentous fossil bacteria from Archean of Western Australia. Precambr. Res., 20:357-374.

BONNER, J. T., 1971. Aggregation and differentiation in the cellular slime molds. Ann. Rev. Microbiol., 25:78-92.

BROCK, T.D., 1973. Lower pH limit for the existence of blue-green algae: evolutionary and ecological implications. Science, 179:480-483.

BRYERS, G.G., 1985. Water quality of Lake Taupo and the Waikato River - a general overview. *In* Transport of Carbon and Minerals in Major World Rivers Vol. III (eds. E.T. Degens, S. Kempe & R. Herrera). Mitt. Geol.-Palaeontol. Inst. Univ. Hamburg, 58:525-537.

CAMPBELL, A.K., 1983. Intracellular Calcium. Its Universal Role as Regulator. 556 pp. Chichester: J. Wiley & Sons Ltd.

CANTERFORD, G.S., 1980. Formation and regeneration of abnormal cells of the marine diatom *Ditylum brightwellii* (West) Grunow. J. Mar. Biol. Assoc. U.K., 60:243-253.

CARAFOLI, E. & PENNISTON, J.T., 1985. The calcium signal. Scientific American, 253:50-58.

CHAN, K.Y., 1976. Control of colony formation in *Coelastrum microporum*) (Chlorococcales, Chlorophyta). Phycologia, 15:149-154.

CHROST, R.J., 1978. Extracellular release in *Chlorella vulgaris* culture and role of bacteria accompanying algae in this process. Acta Microbiol. Polon., 27:55-62.

CLOUD, P., 1976. Beginnings of biospheric evolution and their biogeochemical consequences. Paleobiology, 2:351-387.

CONWAY, E.J., 1943. The chemical evolution of the ocean. Proc. Roy. Irish Acad.. Sect. B, XLVIII, 161-212.

COOK, P. & SHERGOLD, J.H., 1984. Phosphorus, phosphorites and skeletal evolution at the Precambrian/Cambrian boundary. Nature, 308:231-236.

COTMORE, J.M., NICHOLS, JR., J. & WUTHIER, R.E., 1971. Phospholipid calcium phosphate complex: enhanced calcium migration in the presence of phosphate. Nature, 172:1339-1341.

DARBY, D.G., 1974. Reproductive modes of *Huroniospora microreticulata* from cherts of the Precambrian Gunflint Iron Formation. Geol. Soc. Amer. Bull., 85:1595-1596.

DEGENS, E.T., 1973. Accounting for the salts in the sea. Nature, 243:504-507.

DEGENS, E.T., 1976. Molecular mechanisms of carbonate, phosphate and silica deposition in the living cell. Top. Curr. Chem., 64:1-112.

DEGENS, E.T. & ITTEKKOT, V., 1986. Ca^{2+}-stress, biological response and particle aggregation in the aquatic habitat. Netherl. J. Sea Res., 20:109-116.

DEGENS, E.T., KAZMIERCZAK, J. & ITTEKKOT, V., 1986a. Cellular response to Ca^{2+} stress and its geological implications. Acta Palaeontol. Polon., 30 (for 1985), 115-135.

DEGENS, E.T., KAZMIERCZAK, J. & ITTEKKOT, V., 1986b. Biomineralization and the carbon isotope record. Tschermaks Mineral. Petrogr. Mitt., 35:117-126.

ERICSON, S.J., 1972. Toxicity of copper to *Thalassiosira pseudonana* in enriched inshore seawater. J. Phycol., 8:318-323.

ERULKAR, S.D., 1981. The versatile role of calcium in biological systems. Interdisc. Sci. Rev., 6:323-332.

EUGSTER, H.P., 1967. Hydrous sodium silicates from Lake Magadi, Kenya: Precursors of bedded chert. Science, 157:1177-1180.

EUGSTER, H.P., 1969. Inorganic bedded cherts from the Magadi Area, Kenya. Contr. Mineral. Petrol., 22:1-31.

EUGSTER, H.P. & HARDIE, L.A., 1978. Saline lakes. *In* Lakes- Chemistry, Geology, Physics (ed. A. Lerman). pp. 237-293. New York: Springer Verlag.

FANALE, F.P., 1971. A case for catastrophic early degassing of the Earth. Chem. Geol., 8:79-105.

FORD, T. D. & BREED, W. J., 1973. The problematic fossil *Chuaria*. Palaeontology, 16:535-550.

FRANCHI, E. & CAMATINI, M., 1985. Evidence that a Ca^{2+}-chelator and a calmodulin blocker interfere with the structure of inter-Sertoli junctions. Tiss. Cell, 17:13-25.

FRIEBELE, E.S., CORREL, D.L. & FAUST, M.A., 1978. Relationship between phytoplankton cell size and the rate of orthophosphate uptake: *in situ* observations of an estuarine population. Mar. Biol., 45:39-52.

GARRELS, R.M. & MACKENZIE, F.T., 1967. Origin of the chemical composition of some springs and lakes. *In* Equilibrium Concepts in Natural Water Systems. pp. 222-242. Amer. Chem. Soc., Adv. Chem., 67.

GARRELS, R.M. & MACKENZIE, F.T., 1971. Evolution of Sedimentary Rocks. 397 pp. New York: W.W. Norton & Comp.

GERLOFF, G. C. & FISHBECK, K. A., 1969. Quantitative cation requirements of several green and blue-green algae. J. Phycol., 5:109-114.

GILULA, N. B. & EPSTEIN, M. L., 1976. Cell-to-cell communication, gap junction and calcium. Sym. Soc. Exper. Biol., 30:257-272.

HARDIE, L.A. & EUGSTER, H.P., 1970. The evolution of closed-basin brines. Mineral. Soc. Amer., Spec. Publ., 3:273-290.

HART, M.H., 1978. The evolution of the atmosphere of the Earth. Icarus, 33:23-39.

HELLMAN, B. & ANDERSON, T., 1978. Calcium and pancreatic β-cell function. IV. Evidence that glucose and phosphate stimulate Ca^{45} incorporation into different intracellular pools. Biochim. Biophys. Acta, 541:483-491.

HENDERSON-SELLERS, A. & COGLEY, J.G., 1982. The Earth's early hydrosphere. Nature, 298:832-835.

HOFMANN, H. J., 1976. Precambrian microflora, Belcher Islands, Canada: significance and systematics. J. Paleont., 50:1040-1073.

HOFMANN, H.J., 1977. The problematic fossil *Chuaria* from the late Precambrian Uinta Mountain Group, Utah. Precambr. Res., 4:1-11.

HOFMANN, H.J., 1985. Precambrian carbonaceous megafossils. *In* Paleoalgology: Contemporary Research and Applications,(eds. D.F.Toomey & M.H.Nitecki), pp. 20-33. Berlin: Springer-Verlag.

HOFMANN, H.J. & GROTZINGER, J. P., 1985. Shelf-facies microbiota from the Odjick and Rocknest formations (Epworth Group; 1.89 Ga), northwestern Canada. Can. J. Ear. Sci, 22:1781-1792.

HOLLAND, H.D., 1984. The Chemical Evolution of the Atmosphere and Oceans. 582 pp. Princeton, N.J.: Princeton Univ. Press.

HUNTSMAN, S.A. & SUNDA, W.G., 1980. The role of trace metals in regulating the growth of phytoplankton. *In* The Physiological Ecology of Phytoplankton (ed. I. Morris), pp. 285-328. Boston: Blackwell Scientific.

ITTEKKOT, V., 1982. Variations of dissolved organic matter during a plankton bloom: qualitative aspects based on sugar and amino acid analyses. Mar. Chem., 11:143-158.

JONES, B. F., EUGSTER, H.P. & RETTIG, S.L., 1977. Hydrochemistry of the Lake Magadi basin, Kenya. Geochim. Cosmochim. Acta, 41:53-72.

KASTING, J.F., 1984. The Evolution of the prebiotic atmosphere. Origins of Life, 14:75-82.

KAUFFMAN, E.G. & STEIDTMANN, J.R., 1981. Are these the oldest metazoan trace fossils? J. Paleont., 55:923-947.

KAZMIERCZAK, J., 1976. Devonian and modern relatives of the Precambrian *Eosphaera*: possible significance for the early eukaryotes. Lethaia, 9:39-50.

KAZMIERCZAK, J., ITTEKKOT, V. & DEGENS, E.T., 1985. Biocalcification through time: environmental challenge and cellular response. Palaeontol. Zeit., 59:15-33.

KEMPE, S., 1976. Zur Geohydrochemie des Alsterbeckens. Mitt. Geol.-Palaeontol. Inst. Univ. Hamburg, 47:199-238.

KEMPE, S., 1977. Hydrographie, Warven-Chronologie und organische Geochemie des Van Sees, Ost-Tuerkei. Mitt. Geol.-Palaeontol. Inst. Univ. Hamburg, 47:125-228.

KEMPE, S., 1979. Carbon in the rock cycle. *In* The Global Carbon Cycle, SCOPE Rep. 13 (eds. B.Bolin, E.T. Degens, S. Kempe & P. Ketner). pp. 343-377. Chichester: J. Wiley & Sons.

KEMPE, S., 1982. Long-term records of CO_2 pressure fluctuations in fresh waters. *In* Transport of Carbon and Minerals in Major World Rivers, vol.I (ed. E.T. Degens). Mitt. Geol.-Palaeontol. Inst. Univ. Hamburg, 52:91-332.

KEMPE, S. & DEGENS, E.T., 1985. An early soda ocean? Chem. Geol., 53:95-108.

KNOLL, A. H., 1983. Biological interactions and Precambrian eukaryotes. *In* Biotic Interactions in Recent and Fossil Benthic Communities, (eds. M. J. S. Tevesz & P. L. McCall), pp. 251-283. New York: Plenum Publ. Corp.

KNOLL, A. H. & BARGHOORN, E. S., 1977. Archean microfossils showing cell division from the Swaziland System of South Africa. Science, 198:396-398.

KNUTTON, S. & PASTERNAK, C. A., 1979. The mechanism of cell-cell fusion. Trends Biochem. Sci., 4:220-223.

KOVÁC, L., 1985. Calcium and *Saccharomyces cerevisiae*. Biochimica et Biophysica Acta, 840:317-323.

KRATZ, W. & MYERS, J., 1955. Nutrition and growth of several blue-green algae. Amer. J. Botany, 42:282-287.

KRETSINGER, R.H., 1976. Calcium binding proteins. Ann. Biochem., 45:239-262.

KRETSINGER, R.H., 1977. Why does calcium play an informational role unique in biological systems? *In* Metal-Ligand Interactions in Organic Chemistry and Biochemistry (9th Jerusalem Symposium), pt. 2:(eds. B. Pullman & N. Goldbleum), pp. 257-263. Dordrecht: E. Lidel Publ. Co.

KRETSINGER, R.H., 1983; A comparison of the roles of calcium in biomineralization and in cytosolic signalling. *In* Biomineralization and Biological Metal Accumulation (eds. P. Westbroek & E. W. de Jong), pp. 123-131. Dordrecht: D. Reidel Publ. Co.

KUHN, W. R. & KASTING, J.F., 1983. Effects of increased CO_2 concentrations on surface temperature of the early Earth. Nature, 301:53-55.

KYLIN, A. & DAS, G., 1967. Calcium and strontium as micronutrients and morphogenetic factors for *Scenedesmus*. Phycologia, 6:201-210.

LABERGE, G.L., ROBBINS, E.I. & SCHMIDT, R.G., 1984. New microfossil evidence of eukaryotes in early Proterozoic rocks (1.9 B.Y.) from the Lake Superior Region, North America. 27th Intern. Geol. Congr., Moscow, August 4-14:1984, Abstracts, vol. 1:sect. 01 to 03. pp. 241. Moscow: Nauka.

LEE, C. & BADA, J.L., 1975. Amino acids in the equatorial Pacific, Sargasso Sea and Biscayne Bay. Limn. Oceanog., 22:502-510.

LEHNINGER, A.L., 1982. Principles of Biochemistry.1005 pp. New York: Worth Publ. Inc.

LINDGREN, S., 1982. Algal coenobia and leiospheres from the Upper Riphean of the Turukhansk region, eastern Siberia. Stockholm Contr. Geol., 38:1-20.

MAISONNEUVE, J., 1982. The composition of the Precambrian ocean waters. Sedimentary Geology, 31:1-11.

MARMÉ, D., 1985. Calcium and Cell Physiology. 390 pp. Berlin: Springer-Verlag.

O'KELLEY, J., 1968. Mineral nutrition of algae. Ann. Rev. Plant Physiol., 19:89-112.

PFLUG, H.D., 1978. Yeast-like microfossils detected in oldest sediments of the earth. Naturwissenschaften, 65:611-615.

PFLUG, H.D. & Reitz, E., 1985. Earliest phytoplankton of eukaryotic affinity. Naturwissenschaften, 72:656-657.

REED, M.H., 1982. Calculation of multicomponent chemical equilibria and reaction processes in systems involving minerals, gases and an aqueous phase. Geochim. Cosmochim. Acta, 46:513-528.

ROBBINS, E.I., PORTER, K.G. & HABERYAN, K.A., 1985. Pellet microfossils: possible evidence for metazoan life in early Proterozoic time. Proc. Nat. Acad. Sci. U.S.A., 82:5809-5813.

ROGERS, J.J.W., 1978. Inferred composition of early Archaean crust and variation in crustal composition through time. *In* Archean Geochemistry (eds. B.F. Windley & S.M. Naqvi.) pp. 25-39. Amsterdam: Elsevier.

RONOV, A.B., 1968. Probable changes in the composition of seawater during the course of geological time. Sedimentology, 10:25-43.

SCHOPF, J.W., 1978. The evolution of the earliest cells. Sci. Amer., 239:85-102.

SHELDON, R.P., 1981. Ancient marine phosphates. Ann Rev. Ear. Planet. Sci., 9:251-284.

SIEVER, R., 1977. Early Precambrian weathering and sedimentation: An impressionistic view. *In* Chemical Evolution of the Early Precambrian (ed. C. Ponnamperuma), pp. 13-23. New York: Academic Press.

STEGMANN, G., 1940. Die Bedeutung der Spurenelemente fuer *Chlorella*. Zeit. Bot., 35:385-422.

SVERDRUP, H.U., JOHNSON, M.W. & FLEMING, R.H., 1970. The Oceans, their Physics, Chemistry, and General Biology. 1087 pp. Engelwood Cliffs, N.J.: Prentice-Hall.

THOMAS, W. H., HALLIBAUGH, J. T. & SEIBERT, D. L. R., 1980. Effects of heavy metals on the morphology of some marine phytoplankton. Phycologia, 19:202-209.

THOMSEN, L., 1980. ^{129}Xe on the outgassing of the atmosphere. J. Geophy. Res., 85:4374-4378.

TIMOFEEV, B.V., 1966. Microphytological Investigations of Ancient Formations. Acad. Sci. USSR, Lab. Precambrian Geol. 145 pp. Leningrad: Nauka (in Russian).

TRAINOR, F.R., 1969. *Scenedesmus* morphogenesis. Trace elements and spine formation. J. Phycol., 5:185-190.

TYNNI, R. & UUTELA, A., 1984. Microfossils from the Precambrian Muhos Formation in Western Finland. Geol. Surv. Finland Bull., 330:5-38.

UREY, H.C., 1951. The origin and development of the Earth and other terrestrial planets. Geochim. Cosmochim. Acta, 1:209-277.

VEIZER, J., 1983. Geological evolution of the Archean-early Proterozoic Earth. *In* Earth's Earliest Biosphere (ed. J.W. Schopf), pp. 240-259. Princeton N.J.: Princeton Univ. Press.

VEIZER, J., 1985. Carbonates and ancient oceans: isotopic and chemical record on time scales of 10^7-10^9 years. Geophys. Monog. Amer. Geophys. Union, 32:595-601.

VEIZER, J. & COMPSTON, W., 1976. $^{87}Sr/^{86}Sr$ in Precambrian carbonates as an index of crustal evolution. Geochim. Cosmochim. Acta, 40:905-915.

VEIZER, J., COMPSTON, W., HOEFS, J., & NIELSEN, H., 1982. Mantle buffering of the early oceans. Naturwissenschaften, 69:173-180.

VIDAL, G.K., 1984. The oldest eukaryotic cells. Sci. Amer., 250:48-57.

VIDAL, G. & KNOLL, A.H., 1982. Radiations and extinctions of plankton in the late Proterozoic and early Cambrian. Nature, 297:57-60.

WALKER, J., 1983. Possible limits on the composition of the Archean ocean. Nature, 302:518-520.

WANG, F., 1985. Middle-Upper Proterozoic and lowest Phanerozoic microfossil assemblages from SW China and contiguous areas. Precambr. Res., 29:33-43.

WEDEPOHL, K.H., 1963. Einige Ueberlegungen zur Geschichte des Meerwassers. Fortsch. Geol. Rheinland Westfalen, 10:129-150.

WOESE, C.R. & FOX, G.E., 1977. Phylogenetic structure of the prokaryotic domain: the primary kingdoms. Proc. Nat. Acad. Sci. U.S.A., 74:5088-5091.

WOESE, C.R., MAGRUM, L.J. & FOX, G. E., 1978. Archaebacteria. J. Mol. Evol., 11:245-252.

ORGANISMIC AUTONOMY IN BIOMINERALIZATION PROCESSES

Klaus Vogel[1] and Wolfgang F. Gutmann[2]

[1] *J.W. Goethe University of Frankfurt, Geologisch-Paläontologisches Institut, D 6000 Frankfurt, Germany;* [2] *Senckenberg-Museum, D 6000 Frankfurt, Germany*

ABSTRACT. The energy budget of an organism is subject to selection. Based on this fact we present the following points:

1. Uptake of ions (mineral substances), their involvement in metabolism and their release are strictly controlled by the organism. The control requires energy.

2. Building-up mineral deposits which are larger than necessary for times of deficiency, charges the energy budget. For utilization as skeletal parts they must be secreted at or transported to special sites. The organism must often synthesize organic membranes for their protection or as templates. These and other related activities cost energy.

3. It follows that the incipient stages of biomineralization and the deposition of well-organized skeletal structures were already backed by selective advantages.

4. As primary selective advantages of biomineralization of unicellular organisms, control of cell shape, scaffolding for activities of cytoplasmatic fibres and rhizopodia and stabilisation of nucleus and organelles are discussed.

5. It results that the origin and evolution of biomineralization is controlled by the pre-existing biochemical organization and biomechanical construction of the organism. Biomineralization reflects the autonomy of the organism and is more than a mirror of evolution of ocean chemistry and more than detoxification.

INTRODUCTION

Evolution and the process of biomineralization have been discussed primarily from the view-points of biochemistry and environment. Although skeletal structures apparently have a considerable mechanical impact on the organism, biomechanical preconditions for skeletal evolution as well as biomechanical results, consequences and selective advantages of skeletal formation have been almost entirely neglected.

The approach presented here is based on two principles: (1) *like engines, organisms consume matter and energy* and (2) *organisms are mechanical constructions.*

Consumption of matter and energy

Development and maintenance of the organismic machinery as well as their feasibility in relation to the environment depend on energetic costs. Matter and energy are acquired by the activity of the organismic machine, the motor activity being provided by conversion of chemical energy into mechanical energy. As the result of over-production of progeny, organisms compete for matter and energy. Selective pressure forces the evolutionary sequence of organisms to economize and to optimize the energy gathering and energy converting machinery. Consequently, energy consumption and the building-up of energy-requiring-structures can be increased only if rewarded by some selective advantage (Gutmann & Peters, 1973). Such a view of evolution is prescribed by the law of energy conservation. This states that energy can neither be created nor destroyed. If this is true, the concept that any organismic structure or activity could be selectively neutral must be refuted.

Organisms are mechanical constructions

Their shape as well as every movement or deformation of their body obeys the laws of mechanics. It follows that introduction of a mechanical part, such as a skeletal element, must have mechanical consequences. New skeletal elements which interfere with essential functions of the organisms cannot be evolutionarily successful. Only new skeletal elements that do not hinder vital organismic functions but support or even improve them will prevail and become established. If organisms are conceived as soft hydraulic constructions and as mechanical entities, it follows that all attempts to explain the incorporation of mineral structures or other skeletal elements must refer to mechanical mechanisms. However, this concept has escaped reductionistic approaches based on principles of biochemistry and molecular biology.

Relations between the origin of biomineralization and biomechanics have already been discussed for some important metazoan groups (Gutmann, 1974; Vogel & Gutmann, 1981). One example is the shell of the earliest molluscs (Fig. 1): Such shells could not evolve before the dorsal muscle wall of their annelid-like predecessor had been liberated from (strong) muscular activity. Otherwise, the shell would have interfered with the activity of the muscle wall. The formation of the skeletal structures could have happened concurrently with the reinforcement of the ventral part of the muscle tube and its gradual transformation into the molluscan foot. The primary advantage of the skeletal elements was that they served as rigid counterparts to the dorsoventral muscles and that their metabolic inactivity was more economical than that of muscular and connective tissue structures. Thus the costly hydrostatic skeleton was partly replaced by a more efficient muscle skeleton system. We argue that no physiological process nor any environmental change would have produced (shell-bearing) molluscs (and accordingly - brachiopods, trilobites etc.) at the Precambrian-Cambrian boundary if there had not been appropriate forerunner constructions "at disposal".

ORIGIN OF SKELETONS AS BY-PRODUCTS OF DETOXIFICATION?

To our knowledge, the necessity of biomechanical prerequisites for the origin of skeletal structures in the Protista has not been stressed. Quite recently, attempts were made to explain biocalcification in both Protozoa and Metazoa as a form of mineral detoxification (Kazmierczak *et al.*, 1985). We believe that this hypothesis is incomplete and that the origin of biomineralization and biomineralization processes cannot be un-

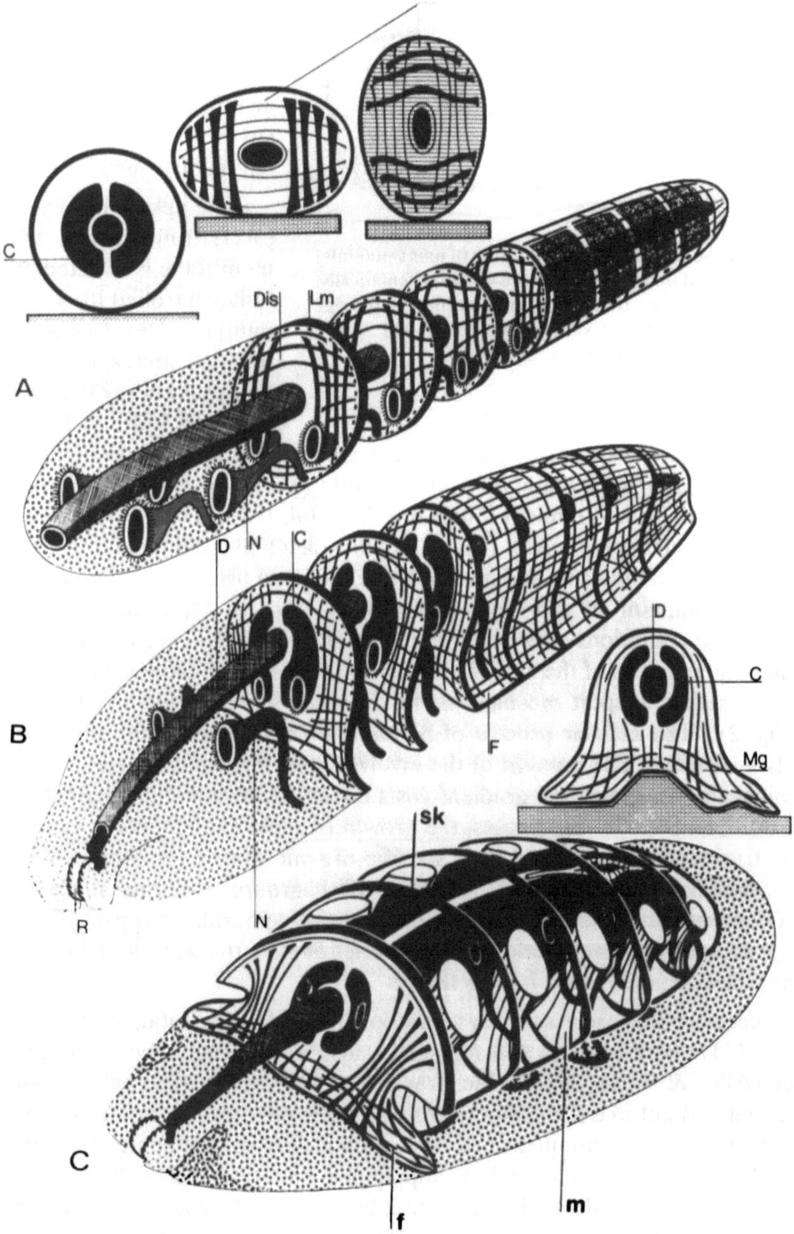

FIGURE 1. Shell formation in molluscs. ▪ A. Like most lower metazoans the ancestors of molluscs were certainly highly deformable worms that were unable to develop skeletons without forfeiting their mobility. ▪ B. Improvement of creeping abilities on firm substrate resulted in the formation of a flat muscular foot and reduced deformation of the dorsal side. ▪ C. Here skeletal parts could evolve for "low cost" stabilization of shape and as antagonists for dorsoventral muscles. **C**: coelom, **D**: intestine, **Dis**: dissepiments, **F**, **f**: foot, **m**: dorsoventral muscles, **R**: radula, **sk**: skeletal bars. After Gutmann, 1974 and Vogel & Gutmann, 1981.

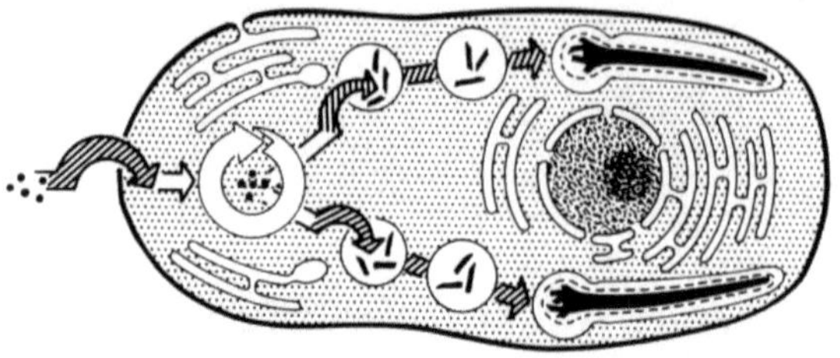

FIGURE 2. Uptake, migration and utilization of ions (minerals) that eventually build up skeletal parts are shown schematically. Hatched arrows: energy consuming mechanisms as passage through cell membrane and vesicle membrane, and vesicle transport to specified sites (see text for details).

derstood solely as simple reactions to changes in ocean chemistry. Acceptance of such an interpretation would not concur with the aforementioned energetic principles.

The uptake of ions from the environment by the cell membrane is effected and carefully controlled by so-called ion pumps. Development, maintenance and activity of these pumps is energy-dependent (for Ca^{2+} see Carafoli & Penniston, 1986:80), especially when working against concentration gradients. In diatoms for instance, these pumps must maintain a 30 to 60 fold Si-ion concentration within the cell as compared with its environment (Sullivan & Volcani, 1981). However, even when ions follow the concentration gradient (Ca^{2+}), we must keep in mind that it is the activity of the organism which builds up and maintains an adequate osmotic condition. Thus even "passive" mechanism of ion-migration require energy. Material, structures and physiochemical conditions on which these mechanisms depend are generated by energy consuming activities of the cell. This also holds true for the following diffusion barriers and vesicle transport mechanisms for instance by microtubuli (Allen & Weiss, 1987)(Fig. 2). The cellular process of biomineralization begins with the secretion of minerals in vesicles. The passage of the involved ions through the vesicle membranes and against the concentration gradient costs energy (Carafoli *et al.*, 1984; Carafoli & Penniston, 1986:80). In many cases, the growth of skeletal elements in vesicles is apparently triggered or organized by a specific organic matrix (diatoms, amoeba with siliceous skeleton: Volcani, 1981; Harrison, 1981; groups of Foraminifera - *Heterostegina, Globorotalia*: Hemleben *et al.*, 1986; Coccolithophorida: Tappan, 1980). Skeletal development from a multitude of small vesicles or from the deposition in one or more single vesicles certainly is not free of charge.

It is necessary that the vesicles are transported not to accidental, but to appropriate positions. No matter how this well arranged transport is brought about, energy must be supplied (Allen & Weiss, 1987). The structures which actuate the control of the building material and enforce the order of skeletal elements require matter and energy for their formation. One consequence of this action is that several energy barriers are intercalated between environmental ion supply and excretion of a skeletal part (Fig. 2). Hence, the evolution of skeletal parts cannot be accounted for as simple by-products of excretion or detoxification processes. This explanation is insufficient and not conclusive.

If our first principle that selection forces the evolutionary sequence of organisms to economize and to optimize the biological system is correct, then it must also be true with regard to the development of skeletal parts in protists. Skeletal parts can evolve only if their incipient stages are backed already by a selective advantage. If mineral materials utilized in skeletal elements were only a burden, selective mechanisms would favor strict closure of the cell membrane to prevent passage of such material.

If one considers the metabolic requirements of mobile cells, further arguments against any type of uncontrolled disposal of mineral material can be established. The ion pumps which control the passage of at least calcium ions through the cell membrane must be very ancient evolutionary "inventions". They must have functioned most precisely long before skeletal structures appeared. Since Ca^{2+} regulation and precise control of its concentration in the cell are essential for actomyosin contraction (and many other physiological processes), the corresponding cell mechanisms must have been in existence for as long as organisms have been capable of changing their form and of moving actively, [*i.e.*, from the incipient stages of motoric mechanisms on the basis of actomyosin fibrils (Bereiter-Hahn, in print)]. An excess or deficiency of Ca^{2+} will momentarily interfere with motoric activity. Less information is available on the function of Si in cell metabolism. However, since Si ions also seem to play an important role in diverse physiological processes (Sullivan & Volcani, 1981), strict control mechanisms for this ion may also have existed for a long time.

If the above conceived physiological interpretation is true, then it is most economical for the cell to pump only that quantity of ions through the cell membrane which is actually required for metabolism. This does not exclude the storage of small mineral deposits which can be tapped during a mineral deficiency or used as a neutralization mechanism during ion surplus in the cell (*e.g.*, after damage of a cell membrane). These deposits do not depend on a preformed organic matrix and the transport of their vesicles must not follow a well organized pattern that leads to an orderly skeleton.

Biomineralization of protists brings about further disadvantages: it renders more difficult copulation and proliferation and hinders plankton by increase in weight.

Once again, we reach the conclusion that the origin of skeletons solely as an accidental by-product of detoxification or of an environmental pressure is extremely improbable.

BIOMECHANICAL PREREQUISITES AND EFFECTS OF SKELETAL STRUCTURES

We must now raise the question as to the selective advantages which were brought about by the formation of skeletal parts early in their evolution. Or, based on our first principle, we might ask whether the introduction of rigid elements into the mechanical fabric of the cell could result in an economization or optimization. Within the frame of our second principle which states that organisms can be explained only if we take into account their constructional system, we will seek possible selective advantages of skeletal parts for the mechanical construction of the cell.

Skeletal structures of protist cells may fulfill a variety of functions. Within the wide spectrum of these, three aspects will be discussed as follows: (1)Stabilization of organelles and the nucleus in the cell; (2) Control of cell shape by low cost inorganic elements; and (3) Division of functions by a rigid mechanical frame.

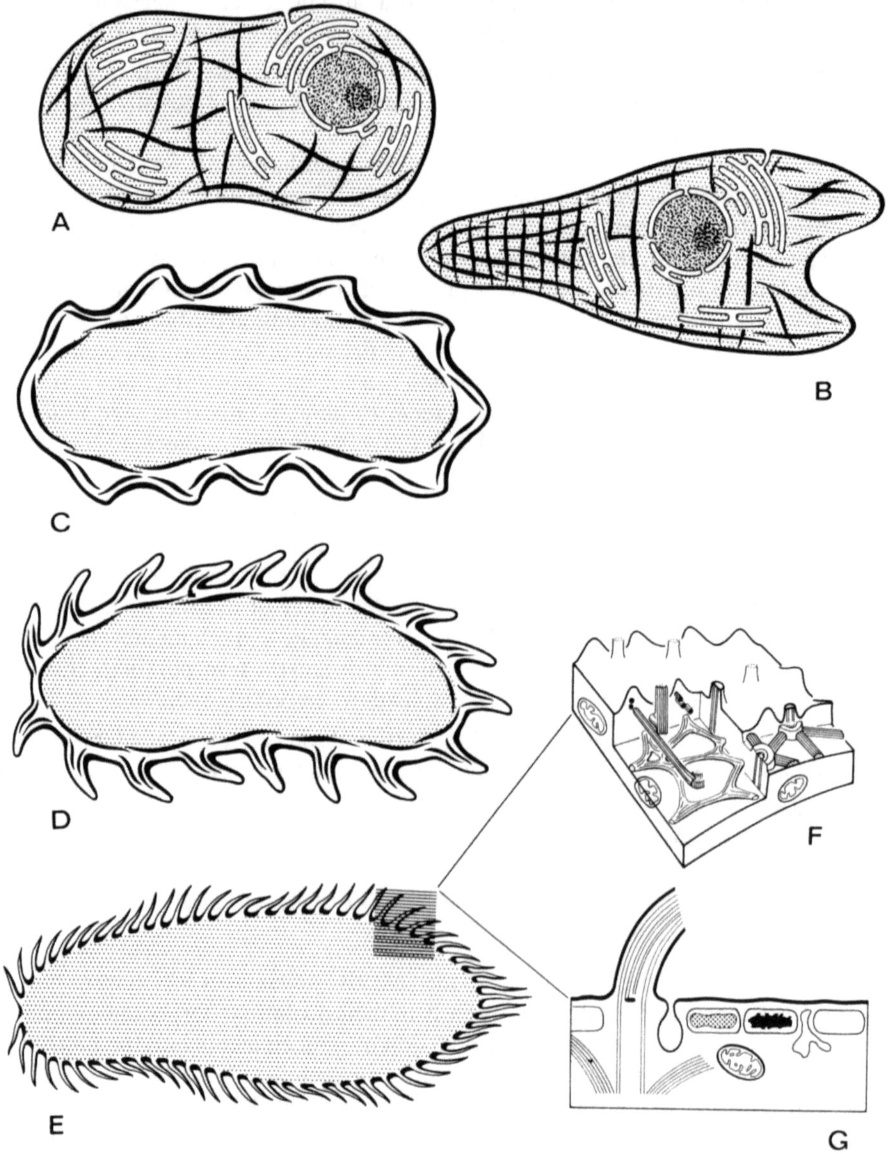

FIGURE 3. Shape and deformation of cells, separation of motoric cytoplasmatic movement from interior metabolic plasmatic currents. ▪ A. Typical cell enclosing fluid, endoplasmatic reticulum, and form-enforcing fibrous structures. ▪ B. Motoric activity is generated hydraulically by deforming and shifting the content. ▪ C. A fibrillar and tubular network in the cortex, which is illustrated in (▪F) permits relatively independent motoric activity of the cell's superficial parts. This stage may lead to a typical ciliate: ▪ D, & ▪ E, ▪ F & ▪ G. A vesicular skeleton (▪G) may substitute a fibrillar and tubular network (▪F) for separation of interior and exterior cytoplasmic movements as well as for support of the cilia. The vesicles may contain organic or inorganic deposits (pointed: organic, black: inorganic (pointed resp. black in ▪G). (▪F) &, (▪G) after Hausmann (1985, Fig. 82 & 83).

Stabilization of organelles and the nucleus in the cell.

In cells the cytoplasm is continuously circulating. Controlled intracellular movement enables these eucaryotic organisms to ingest particulate food by means of phagocytosis, to cycle nutrients within their cells, and to expel waste products efficiently. All these movements are energy consuming because every type of movement performed depends on the effective suppression of unwanted movements by active mechanical bracing structures.

To understand motoric activity as determined by correlated stabilization and deformation, it is necessary to conceive the cell body as a highly deformable hydraulic system which is composed of fluid surrounded by flexible membranes. Any determinate shape of the cell is actively generated by a system of fibrous structures operating on the fluid filling; every movement is a result of deformation processes and concomitant active stabilization mechanisms (Fig. 3). Thus, the motoric activity has two components, a dynamic part consisting of the deforming activity, and a static part which is responsible for shape control and the suppression of all other deformations possible in a hydraulic system while the one required movement is being performed.

It is evident that the formation of stiff or unyielding structures within the cell may effect a selective gain. It may be advantageous for the functions of the nucleus and perhaps some organelles to maintain an approximate central position in the cell and to be exempt from drift into the intensely moving and reorganizing outer areas of the cell (Fig. 4). In these areas, bulky structures such as organelles or a nucleus would further reduce the capability of shape control and would not allow the formation of delicate pseudopodial nets. Anchorage of nucleus and organelles in skeleton free cells is accomplished by energy consuming activities of the "cytoskeleton", *e.g.*, microtubuli and microfilaments (Dyson, 1978:457; filaments as karyophores in certain Ciliates: Fig. 4E).

The evidence presented here is sufficient to show that non-skeletal stabilizing structures can be formed in the cytosol. However, an appropriate inert skeleton could bring about the same result more economically. The central capsule and the siliceous spherical skeletons of radiolarians (Fig. 4), the tests of foraminifers (Fig. 5), and even a vesicular skeleton as found in ciliates (Fig. 3), all can fulfill this function; see below.

Control of cell shape by low cost rigid elements.

A globular cell shape is automatically attained in a sufficiently filled cell in which the membrane is held tight over the fluid filling by underlying contractile fibrils (Bonik *et al.*, 1978; Gutmann & Bonik, 1981; Bereiter-Hahn, 1985). Any alteration of shape or motoric deformation of the basic cell form must be generated by the internal bracing structures of fibrils which form a mechanical coherent whole with the membrane (Fig. 3). These fibrous structures were misleadingly called cytoskeleton.

Effective movements are based on the cooperation of two mechanisms, (1) deformations as required for the motoric effect, and (2) an effective bracing of all body parts such that deformations other than those demanded are hindered (hydraulic system). These activities have to be energized.

Energy can be economized if undesired deformations are prevented by inert rigid elements instead of costly cytoskeleton activities. In Radiolaria and Foraminifera for instance a mineralized skeleton keeps the shape of the central part of the plasm body

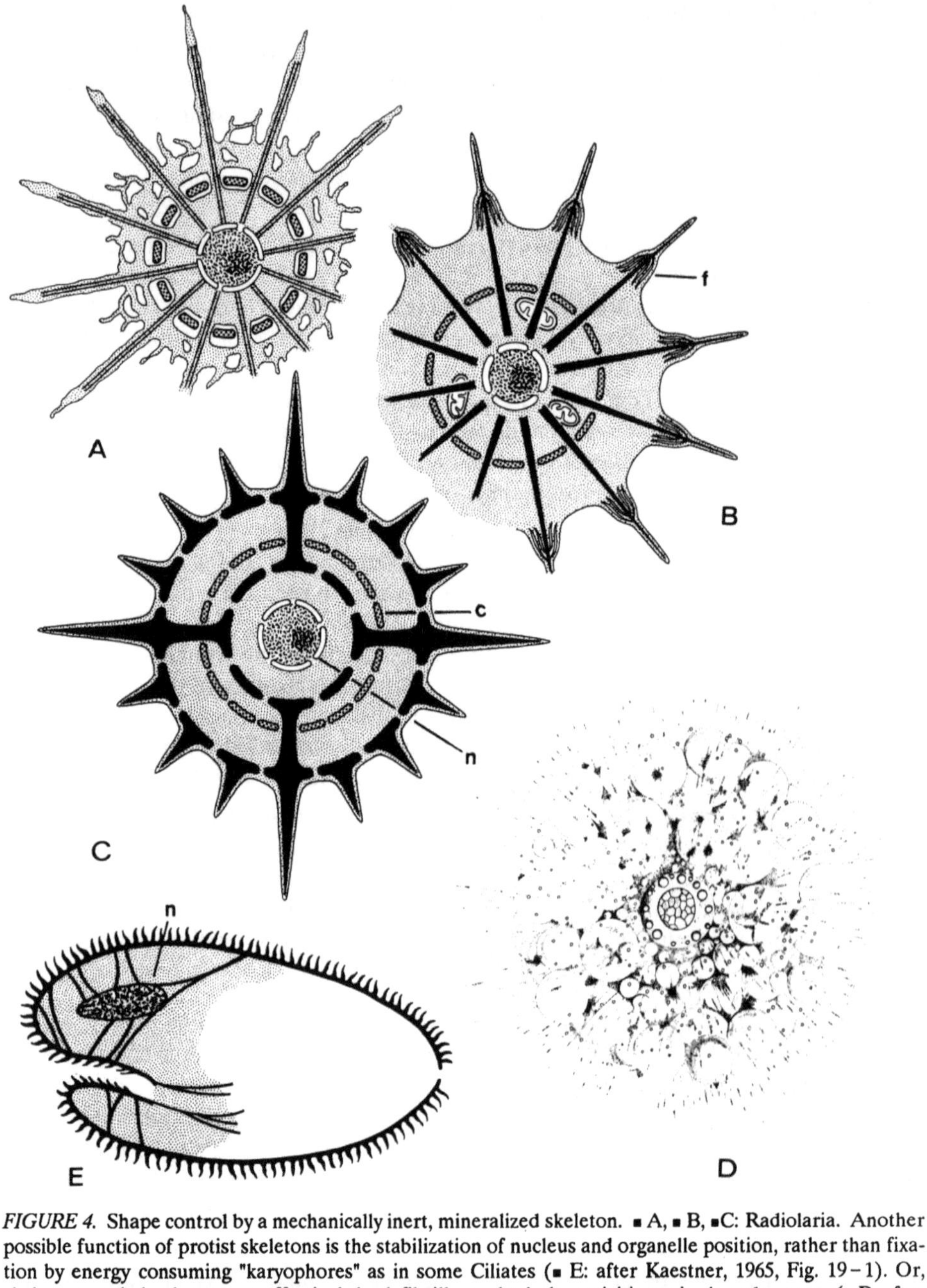

FIGURE 4. Shape control by a mechanically inert, mineralized skeleton. ■ A, ■ B, ■C: Radiolaria. Another possible function of protist skeletons is the stabilization of nucleus and organelle position, rather than fixation by energy consuming "karyophores" as in some Ciliates (■ E: after Kaestner, 1965, Fig. 19–1). Or, skeletons or skeletal parts can effectively back fibrillar and tubular activities as in *Acanthometron* (■ B: after Grell, 1968, Fig. 369). Shape control and fixation of nucleus and organelle positions can also be achieved by integration of turgescent vacuoles as in *Thalassophysa* (■ D: after Trégouboff, 1953, Fig. 222). **c**: capsular wall, **f**: filaments, **n**: nucleus.

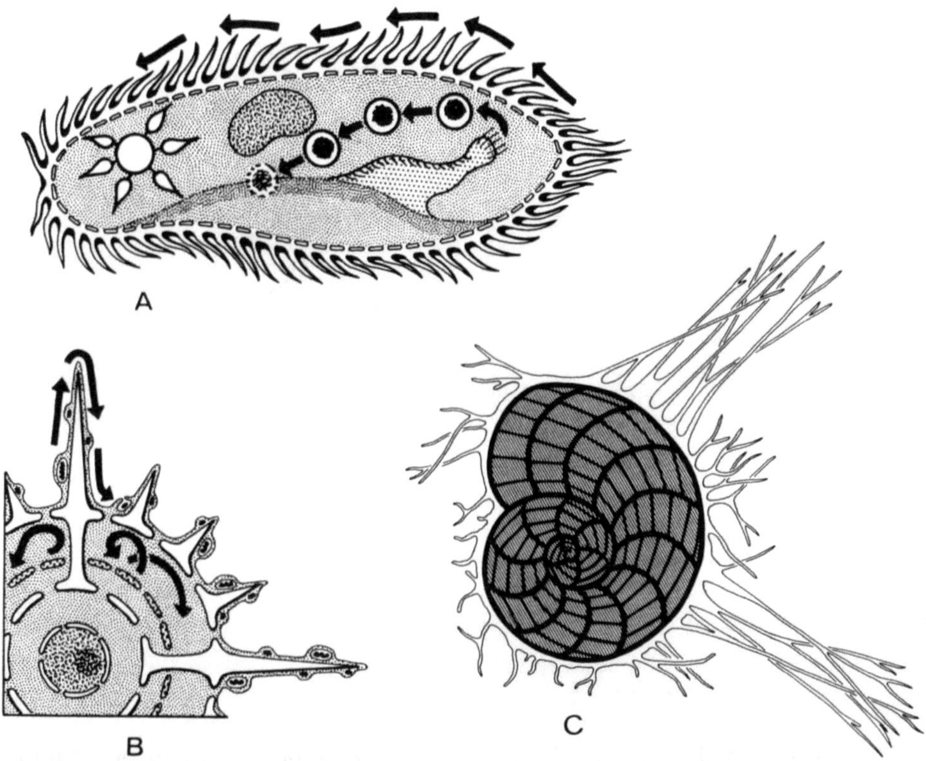

FIGURE 5. Form control and enhancement of division of labour and functions by skeletal scaffolds. ▪ A. Ciliate with vesicular skeleton (black arrows = inner and outer cytoplasmic currents). ▪ B. Radiolaria. ▪ C. Foraminifera (*Heterostegina*, after Röttger, 1973, Fig. 1 & 2).

undeformed, whereas the ektoplasm can build up, remove and rebuild a complicated network of pseudopodia relative independently (Fig. 4,5,6).

Division of functions by a rigid mechanical frame.

Shape control by skeletal structures apparently offers two additional and interrelated advantages. While these structures largely relieve the ectoplasmatic layer of its function to ensure form control over the entire body, the internal movements become independent of the locomotory and predatory movements of the whole cell. This facilitates the formation of protoplasmatic differentiations for propulsion, predation or anchoring on the outer parts of the cell body. For instance, cilia can function on the external side of a stiff grid that keeps the whole body under control. Such a condition can be observed in Ciliata where an energy consuming grid of microtubuli and fibrous structures and deposition of organic or inorganic material in vesicles below the cilia provides a stabilized, but not totally rigid framework (Fig. 3).

Skeletal stabilization of the superficial layers of the cell body is not intimately coupled with a cilial covering, but may have advantageous effects in connection with other modes of propulsion (Fig. 6). The body frame of dinoflagellata which possess only one or two long flagellae is made rigid by cellulose plates which lie close to the cell surface. In coc-

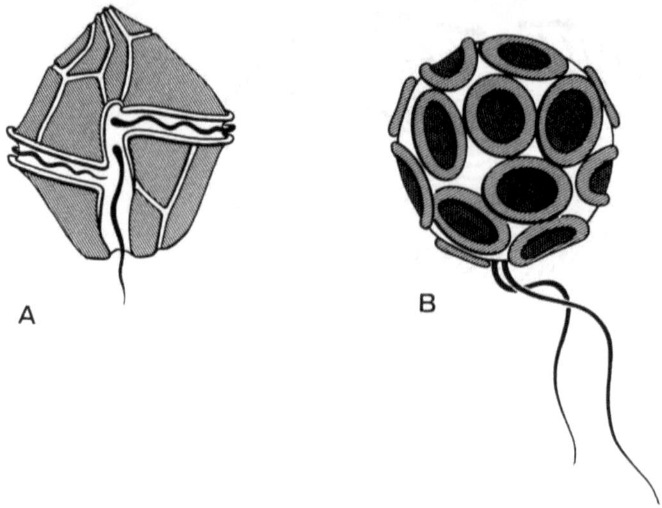

FIGURE 6. Stabilization of the soft and deformable cell body by skeletal plates increases propulsion efficiency of the flagella. ▪ A. dinoflagellates, ▪ B. coccolithophorids.

colithophorids, a high degree of stiffness is attained by a stabilizing armour of calcareous plates. When the cell body is held stiff by skeletal structures, propulsion by flagellae becomes more effective since irregular deformations are rigorously suppressed. Placed in an evolutionary perspective, the Ciliata, Dinoflagellata, and coccolithophorids may be acknowledged as stages in the gradual formation of skeletal structures. The incipient stage would be a vesicle skeleton (Fig. 3G). By increasing deposition of organic and inorganic skeletal parts, complete skeletal armors could be attained finally.

To prevent misunderstanding of the skeletal structures, it should be emphasized that the cytoplasm is not necessarily inserted at the skeletal elements. Direct insertion of protoplasmatic structures such as fibrils and microtubles is only rarely observed.

Even if no direct insertion of intracellular tensile elements is present, the skeletal structures can serve as supporting frames for the movements of the cells and for the deformation of delicate shape differentiations in the ectoplasmatic layer. The skeletal structures allow the movements and deformations of the ectoplasmatic portions of the cells to become independent of internal motions (Fig. 5).

In all cases, the skeletal structures are formed within the soft body system of the cell, and the cell is arranged in and around the skeletal frame, such that both are inseparately interconnected. In this way, the rigidness of the skeletal structures provide a supporting effect for the cell. As a result of skeletal frame formation, size limitations for single protozoan cells can be overcome. A protoplasmatic system which has been made rigid by a skeletal apparatus can attain considerable size, as demonstrated by many Foraminifera.

CONCLUSIONS

The above mentioned constructional aspects of skeleton formation concern the machinery of organisms. These have seldom been discussed as evolutionarily relevant. On the one hand, physiological studies were mainly concerned with the physiological mechanisms, and on the other hand, morphological interpretations were regarded with respect to adaptation to external environmental requirements. These include protection, increase of friction surface, increase of food and energy gathering surface, and improvement of capabilities to anchor on a substrate. By pointing out the biomechanical prerequisites and effects of skeletal structures, the predominance of the organismic construction in its interaction with the environment during the evolutionary history has been stressed. In this way, post-Darwinian aspects of evolutionary theory can be exploited (Gutmann & Bonik, 1981). The constructional preconditions and limitations of the soft-body constructions determine where and in which way skeletal structures can develop. Thus attained advantages elucidate prerequisites for the exploitation of potential environmental conditions. Biomechanical prerequisites are as important as those of a biochemical nature in sustaining life and reproduction during the course of evolutionary transformation. Appropriate evolutionary reconstructions should contain both principles.

ACKNOWLEDGEMENTS

For helpful advice in biochemical and cytologial questions we express our thanks to Dr. J. Bereiter-Hahn, Frankfurt a. M. We gratefully acknowledge the technical assistance of A. Siebel for the drawings and the tailoring of our English text by P. Rismiller. Participation in the Symposium was made possible by funds from The Organized Research Fund of The University of Texas at Arlington, The Sea Grant Program administered by Texas A & M University and the German Research Foundation.

REFERENCES

ALLEN, A.D. & WEISS, D.G., 1987. Microtubuli als interzelluläres Transport system. Spektrum der Wissenschaft, 4/1987:76-85.

ANDERSON, O.R., 1983. Radiolaria. 355 pp. New York: Springer.

BEREITER-HAHN, J., 1985. Architecture of tissue cells. The structural basis which determines shape and locomotion of cells. Acta biotheor., 34:139-148.

BEREITER-HAHN, J., (in press). Compartmention of calcium and energy metabolic pathways: Implications for eukaryote evolution and control of cell proliferation. New York Academy of Science.

BONIK, K., GRASSHOFF, M., GUTMANN, W.F., & MAIER, W., 1978. Hydraulik als Grundlage der Morphologie aller tierischen Lebewesen. Natur und Museum, 108:162-174.

CARAFOLI, E., INESI, G. & ROSEN, B.P., 1984. Calcium Transport across biological membranes. *In* Sigel, H. Metal ions in biological systems, 17, 532 pp. New York, Basel: M. Dekker Inc.

CARAFOLI, E. & PENNISTON, J.T., 1986. Das Calcium-Signal, Spektrum der Wissenschaft, 1986, No. 1:76-85.

DYSON, R.D., 1978. Cell biology: A molecular approach. 616 pp. Boston, London, Sydney, Toronto: Allyn and Bacon.

GRELL, K.G., 1968. Protozoologie, 2. Aufl., 511 pp. Berlin, Heidelberg, New York: Springer.

GUTMANN, W.F., 1972. Die Hydroskelett-Theorie. Aufsätze und Reden der Senckenbergischen naturforschenden Gesellschaft, 21, 91 pp. Frankfurt/Main: W. Kramer.

GUTMANN, W.F, 1974. Die Evolation der Mollusken-Konstruktion: ein phylogenetisches Modell. Aufsatze u. Reden der Senckenberg. naturforschenden Gesellschaft, 25:11-84.

GUTMANN, W.F. & PETERS, D.S., 1973. Konstruktion und Selektion: Argumente gegen einen morphologisch verkürzten Selektionismus. Acta biotheor., 22:151-180.

GUTMANN, W.F. & BONIK, K., 1981. Kritische Evolutionstheorie. 227 pp. Hildesheim: Gerstenberg.

HARRISON, F.W., DUNKELBERGER, D., WATABE, N. & STUMP, A.B., 1981. Ultrastructure and Deposition of silica in Rhizopod Amþbae. *In* Simpson, T.L., & Volcani, B.E. (Eds.). Silicon and Siliceous Structures in Biological Systems, 281-294. New York, Heidelberg, Berlin: Springer.

HAUSMANN, K., 1985. Protozoologie, 351 pp. Stuttgart, New York: G. Thieme.

HEMLEBEN, CH., ANDERSON, O.R., BERTHOLD, N. & SPINDLER, M., 1986. Calcification and chamber formation in Foraminifera - a brief overview. *In* Leadbeater, B.S.C. & Riding, R. (eds.). Biomineralisation in lower plants and animals. Sys. Assoc., spec. vol. 30:237-249.

KAESTNER, A., 1965. Lehrbuch der Speziellen Zoologie. I.: Wirbellose, 1. Teil, 845 pp. 2. Aufl. Stuttgart: Fischer.

KAZMIERCZAK, J., ITTEKKOT, V. & DEGENS, E.T., 1985. Biocalcification through time: environmental challenge and cellular response, Paläont. Zeitschr., 59:15-33.

NIGGLI, V., SIGEL, E. & CARAFOLI, E., 1982. The purified Ca2+ pump of human erythrocyte membranes catalyzes an electroneutral Ca^{2+}-H^{+} exchange in reconstituted liposomal systems. J. Biol. Chem., 257:2350-2356.

RÖTTGER, R., 1973. Die Ektoplasmahülle von *Heterostegina depressa* (Foraminifera: Nummulitidae). Mar. Biol., 21:127-138.

SIMPSON, T.L. & VOLCANI, B.E. (Eds.), 1981. Silicon and Siliceous Structures in Biological Systems, 587 pp. New York, Heidelberg, Berlin: Springer.

SULLIVAN, C.W. & VOLCANI, B.E., 1981. Silicon in the cellular metabolism of diatoms. *In* Simpson, T.L. & Volcanic, B.E. Silicon and siliceous structures in Biological Systems, 587 pp. New York, Heidelberg, Berlin: Springer.

TAPPAN, H., 1980. The paleobiology of plant protists. 128 pp. San Francisco: Freeman.

TRÉGOUBOFF, G., 1953. Classe des Radiolaires (Radiolaria J. Müller 1858, emend.). *In* Grassé, P.-P. (Ed.). Traité de Zoologie. 1:321-436. Paris: Masson et Cie.

VOGEL, K. & GUTMANN, W.F., 1981: Zur Entstehung von Metazoen-Skeletten an der Wende vom Präkambrium zum Kambrium. Festschrift der wissenschaftlichen Gesellschaft der J.W. Goethe-Universität Frankfurt am Main. 517-537. Wiesbaden: W. Steiner.

VOLCANI, B.E., 1981. Cell wall formation in diatoms: morphogenesis and biochemistry. *In* Simpson, T.L., & Volcani, B.E., (Eds.). Silicon and Siliceous Structures in Biological Systems. 156-200. New York, Heidelberg, Berlin: Springer.

WEBER K. & OSBORN, M., 1985. Die Moleküle des Zellskeletts. Spektrum der Wissenschaft, 1985, No. 12:102-114.

BIOMINERALIZATION MECHANISMS

Ted McConnaughey[1]

Oceanography WB-10, University of Washington, Seattle, WA 98195 USA; [1]present address: Department of Botany, University of California, Davis, CA 95616]

ABSTRACT. Biological calcification, silicification, and the precipitation of transition metal oxides appear to have much in common. Three features are discussed: the role of the boundary membrane, ATP driven electroneutral transport of cations to the mineralization site, and passive electroneutral concentration of anions. Ion concentrations at the site of $CaCO_3$ and $Ca_5OH(PO_4)_3$ precipitation are modeled, and some relationships between mineralization and cytosolic chemistry and evolution are explored.

INTRODUCTION

Many organisms precipitate more than one mineral, sometimes from the same tissues. Calcite and aragonite polymorphs of $CaCO_3$ co-occur in many phyla. Mammalian bone and tooth formation apparently begins with a carbonate rich phase and involves the enzyme carbonic anhydrase, which catalyzes HCO_3^- formation from CO_2 (Nakahara & Kakei, 1983; Kakei & Nakahara, 1984, 1985). Bone formation also passes through stages rich in silicon (Carlisle, 1981, 1986). Limpets and chitons posess $CaCO_3$ shells and SiO_2 or Fe_3O_4 teeth, which pass through developmental stages rich in Ca^{2+} and PO_4^{3-} (Towe and Lowenstam, 1967; Runham *et al.*, 1969; Mann, this volume). Sclerosponges sequentially precipitate $CaCO_3$ and SiO_2. Mitochondria sometimes precipitate Ca^{2+} phosphates and carbonates (Lehninger, 1983), and also accumulate silicon (Carlisle, 1986) and Fe^{2+} (Flatmark and Romslo, 1975). Each mineralogical variation or elemental accumulation may involve its own specific biochemistry, but they appear likely to have much in common.

Three main premises underly the present analysis: 1) semipermeable membranes separate the precipitating mineral from the cell cytosol; 2) divalent cations are usually pumped (directly or indirectly) into the mineralizing medium, at the expense of ATP, and in electroneutral exchange for protons; and 3) anion transport is passive and electroneutral, and ultimately driven by pH gradients created during cation transport. Figure 1 summarizes the roles of these processes during the precipitation of $CaCO_3$, $Ca_5OH(PO_4)_3$, SiO_2, and Fe_3O_4.

Origin, Evolution, and Modern Aspects of Biomineralization in Plants and Animals
Edited by Rex E. Crick
Plenum Press, New York

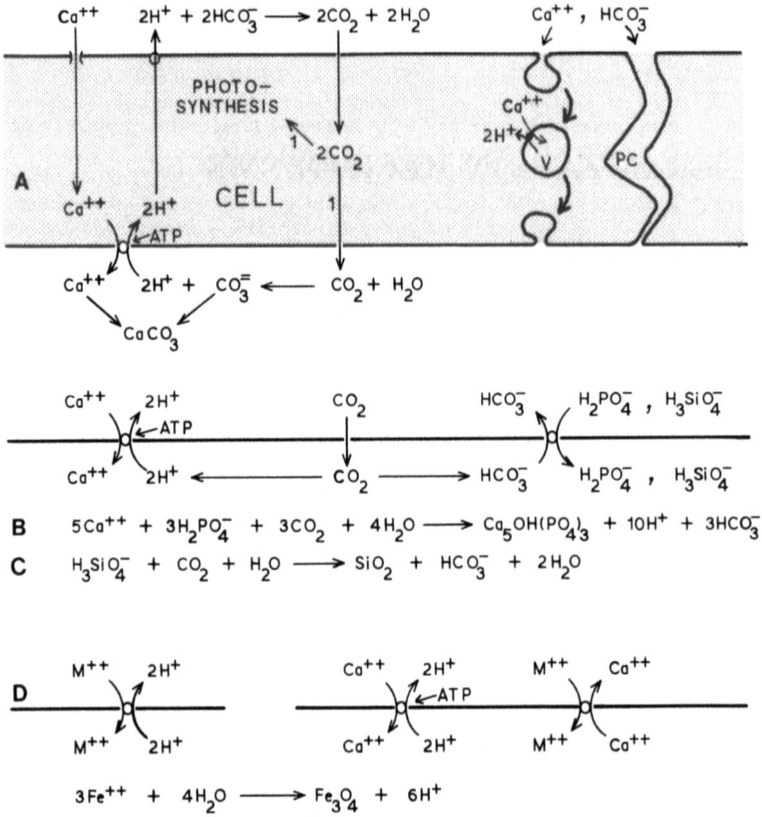

FIGURE 1. Postulated transport mechanisms associated with the biological precipitation of $CaCO_3$, $Ca_5OH(PO_4)_3$, SiO_2, & Fe_3O_4. ▪A. Calcium carbonate. **V,** vacuolar and **PC,** pericellular routes. ▪B. Calcium hydroxyapatite. ▪C. Silica. ▪D. Magnetite.

Semipermeable membranes

Biological mineralization usually occurs near the cell membrane or intracellularly, within membrane enclosed vesicles. Membrane properties are therefore important to biomineralization.

Phospholipid membranes are permeable to small, uncharged molecules such as CO_2 (Gutknecht *et al.*, 1977). Ions generally require enzymatic facilitation to cross membranes. Ion impermeability allows membranes to separate fluids containing different concentrations of a particular ion. ATP driven ion pumps are usually responsible (directly or indirectly) for generating the concentration differences. Ion transport often exhibits "ping-pong" kinetics, in which the enzyme first transports one ion across the membrane, then transports another ion (usually having the same electrical charge) in the opposite direction (Gutknecht & Walter, 1982; Haynes & Mandveno, 1987).

Some materials may reach the site of mineralization by fluid routes such as vacuoles and pericellular channels (Fig. 1). Such morphological features are sometimes evident in microscopic studies (*e.g.*, Johnston, 1980). Chemical analysis of the precipitates may also suggest involvement of relatively non-specific fluid transport. For example, coral skeletons incorporate particles of silt, stains, and most elements which occur in ambient waters (Barnard *et al.*, 1974; Sherer & Seitz, 1980; Buddemeier *et al.*, 1981; Ohde & Kitano, 1981). Fluid transport does not however increase mineral saturation states.

Membranes may direct the growth of precipitates by orienting the precipitating molecules, although contact inhibition of crystal growth often appears more likely (see for example, Wheeler & Sikes, 1984).

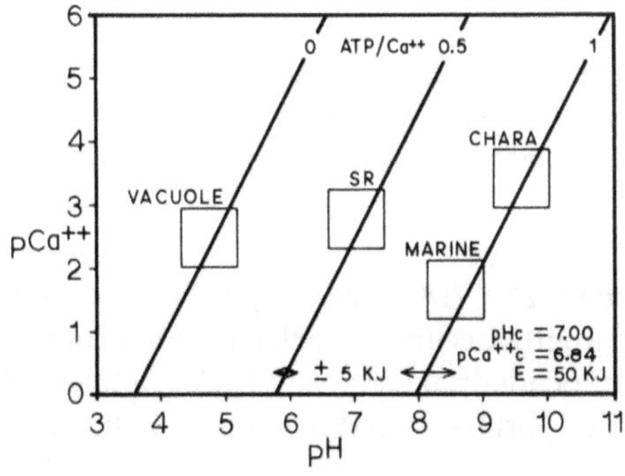

FIGURE 2. Thermodynamic limits to extracytosolic Ca^{2+} and H^+ concentrations, at equilibrium with the cytoplasm *via* ATP driven $2H^+/Ca^{2+}$ exchange. Diagonal lines are n=0, 0.5, and 1 ATP/Ca^{2+} stoichiometries. Calculations assume E=50 KJ/mol and cytoplasmic pH=7, pCa=6.84.

Cation transport

Ca^{2+} is the principle cation involved in biomineralization. The enzyme Ca^{2+}-ATPase has been implicated in Ca^{2+} transport to the site of mineral deposition in coccolithophorids (Klaveness, 1976; Okazaki *et al.*, 1984), calcareous macroalgae (Okazaki, 1977), corals (Isa *et al.*, 1980; Kingsley & Watabe, 1985), bird eggs (Coty & McConkey, 1982), and bones (Shen *et al.*, 1983).

When reconstituted into artificial membranes, Ca^{2+}-ATPase purified from human erythrocytes exchanges Ca^{2+} for $2H^+$ with $2H^+/Ca^{2+}$/ATP stoichiometry (Niggli *et al.*, 1982). $4H^+/2Ca^{2+}$/ATP stoichiometry occurs in Ca^{2+}-ATPase from mammalian sarcoplasmic reticulum (see Haynes & Mandveno, 1987). mH^+/Ca^{2+} exchange (where "m" is an unknown stoichiometric factor) has been suggested for radish Ca^{2+}-ATPase (Rasi-Coldogno *et al.*, 1987).

$2H^+/Ca^{2+}$ exchange pumping should raise both the pH and Ca^{2+} concentration in the extracytosolic calcifying fluid. The thermodynamic limit to $2H^+/Ca^{2+}$ pumping is given by the Nernst equation, which can be simplified in the case of electroneutral $2H^+/Ca^{2+}$ exchange to the following:

$$([Ca^{2+}{}_o]/[Ca^{2+}{}_i])/([H^+{}_o]/[H^+{}_i])^2 = \exp(-nE/RT) \qquad (1)$$

where subscripts "o" and "i" refer to concentrations outside and inside the cytosol, E is the free energy of ATP hydrolysis under physiological conditions, R is the gas constant, T is Kelvin temperature, and "n" is the stoichiometry of ATP/Ca^{2+}. For simplicity, ion activity coefficients are assumed to be equal on both sides of the membrane, and in much of this paper, concentration brackets would more correctly be represented by activities. Figure 2 shows the result of the above calculation for n=0, 0.5, and 1, assuming E = 50 KJ/mol and cytosolic free Ca^{2+} concentrations of 145 nM (based on the alga *Nitellopsis*; Miller & Sanders, 1987) and cytosolic pH of 7.

Superimposed on this calculation are approximate fields of pH and pCa^{2+} observed in the central vacuoles of charophytes (*e.g.* Moriyasu *et al.*, 1984), the mammalian sarcoplasmic reticulum, the alkaline bands of the fresh water alga *Chara* (Lucas *et al.*, 1983), and carapace fluids of marine crabs (Wood & Cameron, 1985). The stoichiometry for n=1 ATP/Ca^{2+} appears appropriate for marine crabs (and other marine organisms) and for the alkaline bands of *Chara* while the more economical n=0.5 may apply in mammalian bones and teeth. Passive (n=0) $2H^+/Ca^{2+}$ exchange may occur across the tonoplast membranes of plants (Schumaker & Sze, 1985).

Other divalent cations (M^{2+}) may also be exchanged for protons, either through ATP dependent $2H^+/M^{2+}$ exchange, or by coupling passive M_1^{2+}/M_2^{2+} exchange with active $2H^+/M_2^{2+}$ exchange. ATP dependent Fe^{2+} uptake by rat liver mitochondria apparently occurs in exchange for protons, by a mechanism similar to those used in mitochondrial Ca^{2+} and Mn^{2+} uptake (Flatmark & Romslo, 1975). Mg^{2+} and Mn^{2+} transport by rat erythrocytes is ATP coupled and electroneutral (Feray & Garay, 1987).

Anion transport

Anions, such as PO_4^{3-}, lack neutral species at physiological pH (pK_{1P}=2.2) and require enzymatic assistance to cross phospholipid membranes at high rates. Anions whose neutral acids are not particularly permeable in the lipid phase of a membrane, perhaps including $Si(OH)_4$, permeability estimated at 10^{-10} m/s, (Raven, 1983) may also cross membranes largely by facilitated anion transport.

Anion transport apparently occurs as a passive exchange of electroneutral combinations (nH^+, A^{-n}) across the membrane in contexts such as the human erythrocyte (Fortes, 1977; Jennings, 1978; Wieth, 1979; Gutknecht & Walter, 1982), mammalian kidney (Ullrich *et al.*, 1984; Low *et al.*, 1984), and mitochondria (Papa *et al.*, 1969; Coty & Pedersen, 1975). Anion exchange exhibits trans-stimulation and cis-inhibition by other anions, *i.e.* anion B^{-n} stimulates transport of anion A^{-n} if they occur on opposite sides of the membrane, and inhibits transport of anion A^{-n} if they occur on the same side of the membrane. This suggests obligatory anion exchange.

The thermodynamic limit to passive electroneutral anion exchange is described by the Nernst equation, analogous to equation 1 where E=0, *e.g.* $[A^-_o]/[A^-_i]=[B^-_o]/[B^-_i]$. A high concentration of B^- on one side of the membrane can therefore concentrate A^- through passive exchange diffusion.

CO_2 is always present in a physiological context, and since CO_2 diffuses rapidly through membranes, its concentration is generally similar on both sides of a membrane. The equilibrium concentration of HCO_3^- is given by $[HCO_3^-]=K_1[CO_2]/[H^+]$, so HCO_3^- concentrates on the alkaline side of the membrane. Substituting this for $[B^-]$, the result is the Gibbs-Donnan equilibrium distribution of anions:

$$[A^{-n}_o]/[A^{-n}_i]=([H^+_i]/[H^+]_o)^n=10^{n(pH_o-pH_i)} \quad (2)$$

Accordingly, passive electroneutral anion exchange concentrates anions on the alkaline side of a membrane. Polyvalent anions are most strongly concentrated. A pH gradient of 1 unit can concentrate HCO32- ×10, CO_3^{2-} ×100, and PO_4^{3-} ×1000.

CALCIUM CARBONATE PRECIPITATION

Membranes and ion transport properties lend themselves to a simple description of $CaCO_3$ precipitation (Fig. 1a). Ca^{2+}-ATPase exchanges $2H^+$ for Ca^{2+}, raising both the Ca^{2+} concentration and pH of the extracytosolic calcifying fluid. The alkaline solution absorbs CO_2, releasing protons which are exchanged for more Ca^{2+}. The net reaction is $Ca^{2+}+CO_2+H_2O=CaCO_3+2H^+$. Evidence for Ca^{2+}-ATPase involvement and

$2H^+/Ca^{2+}$ exchange by Ca^{2+}-ATPase has already been cited. Biochemical and isotopic studies support CO_2 utilization.

The enzyme carbonic anhydrase (CA) catalyzes CO_2 hydration, so the association of CA with calcification (reviewed by Simkiss, 1976; Borowitzka, 1977, Pentecost, 1980) suggests that calcification uses CO_2. CA need not occur at the site of calcification to have this effect, since CA also accelerates CO_2 entry into and through cells (see discussions of facilitated diffusion in Bauer *et al.*, 1980).

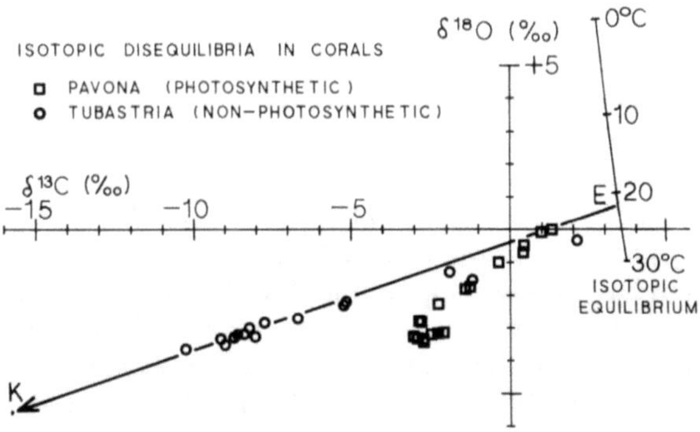

FIGURE 3. Oxygen and carbon isotopic disequilibria in two corals. Coral data are compared to aragonite precipitated at isotopic equlibrium with ambient water (E), and with $CaCO_3$ precipitated under conditions where kinetic isotope effects during CO_2 hydration control the isotopic composition of the precipitate (K).

Oxygen and carbon isotopic disequilibria in biological $CaCO_3$ provide more direct evidence for CO_2 utilization. When $CaCO_3$ precipitates slowly from HCO_3^-, O and C isotopes partition themselves between $CaCO_3$ and solution in a manner which minimizes the free energy of the system (O'Neil *et al.*, 1969). Many biological carbonates contain lower $^{18}O/^{16}O$ and $^{13}C/^{12}C$ ratios than would occur at equilibrium (Fig. 3). This heavy isotope depletion apparently results from kinetic descrimination against the heavy isotopes ^{18}O and ^{13}C during the hydration and hydroxylation of CO_2 (McConnaughey, 1986).

A laboratory simulation of CO_2 based calcification produced results somewhat more extreme than those observed in most biological carbonates ("K" in Fig. 3). The lesser disequilibria in nature, and variation in the degree of disequilibrium within individual skeletons, reflects partial isotopic equilibration. Equilibration apparently occurs through fluid transport of HCO_3^-, and by back-reaction of HCO_3^- to produce CO_2 within the calcifying solution, followed by CO_2 exchange across the calcifying membrane. Both mechanisms produce linear correlations between oxygen and carbon isotope ratios, extending between point "K" and equilibrium. Carbonates from photosynthetic organisms generally have higher $^{13}C/^{12}C$ ratios than those of non-photosynthetic organisms because photosynthesis preferentially removes ^{12}C from the system (Vogel, 1980).

A seawater model

A model depicting ion concentrations at the site of $CaCO_3$ precipitation can be constructed from equations describing the speciation of dissolved CO_2 and alkalinity. CO_2 dissolution does not change alkalinity, while $2H^+/Ca^{2+}$ exchange and $CaCO_3$ precipitation add or subtract 2 units of alkalinity respectively. Alkalinity is therefore approximated by:

$$[Alk]=[HCO_3^-]+2[CO_3^{2-}]+[OH^-]-[H^+]+n[X^{-n}]=[Alk]^i+2([Ca^{2+}]-[Ca^{2+}]^i) \quad (3)$$

where "i" denotes initial conditions and X^{-n} represents any other base which can be protonated in the pH range of interest. Initial conditions might be set to those observed in the intercellular fluids, since there is often some fluid transport to the site of calcification. Seawater conditions are used here, as might apply in a coral. $[HCO_3^-]$ and $[CO_3^{2-}]$ are estimated as $[CO_2]K_1/[H^+]$ and $[CO_2]K_1K_2/[H^+]^2$, where K_1 and K_2 are the first and second ionization constants of CO_2. Equilibrium constants are given values appropriate for seawater of 25° C, 35°/oo salinity (Millero, 1979). The CO_2 concentration in seawater equilibrated with the atmosphere is 10^{-5} M; intracellular CO_2 concentrations may be higher or lower depending on the balance of CO_2 uptake, regeneration, and exchange with the environment. Seawater pH is taken as 8.23, and [Alk]i and $[Ca^{2+}]$i are set to 2.2 mEq/*l* and 10 mM. The only "other base" considered here is $B(OH)_4^-$.

The concentrations of Ca^{2+} and CO_3^{2-} in this system are shown in the top portion of Figure 4, as a function of pH. Calculations are shown for 3 levels of CO_2: normal ambient (10^{-5} M), 0.1 times ambient, and 10x ambient. $[Ca^{2+}]$ is relatively constant at pH values below those of ambient seawater (pH=8.23), but increases as pH rises. $[CO_3^{2-}]$ increases with pH at all pH values. Although Ca^{2+} transport energetically drives the system, increasing $[CO_3^{2-}]$ is mainly responsible for increasing the saturation state of $CaCO_3$ ($S=[Ca^{2+}][CO_3^{2-}]/Ksp$, where Ksp is the solubility product for $CaCO_3$) (Fig. 4, middle). This is possible because the membrane is impermeable to leakage by ions.

Fluid transport may supplement trans-membrane fluxes of Ca^{2+} and CO_2 toward the skeleton. The minimum contribution from trans-membrane fluxes can be estimated by subtracting the initial concentrations (in seawater) from the totals calculated by equation 3. Results are shown in the third portion of Figure 4. Trans-membrane CO_2 fluxes are more important than trans-membrane Ca^{2+} fluxes because seawater contains less C than Ca^{2+}. From the coral isotopic data in Figure 3, it would appear that trans-membrane CO_2 fluxes account for as much as 70% of the skeletal carbon. This corresponds to a maximum pH some 0.5 pH units higher than ambient seawater.

At about this pH, the energy required for $2H^+/Ca^{2+}$ exchange across the membrane also equals the yield from ATP hydrolysis (Fig. 4, bottom). These calculations assume intracellular $pH_i=7$, $[Ca^{2+}{}_i]$ = 145 nM. Ca^{2+} and H+ gradients across the membrane may then approach the thermodynamic limits for Ca^{2+} ATPase. pH elevations of about 0.4 units above haemolymph values (or 1 unit above skeletal muscle values) have been estimated for the carapace fluids of crabs (Wood and Cameron, 1985).

One hundred fold $CaCO_3$ supersaturations are also reached at about 0.5 pH unit elevations. This should cause rapid $CaCO_3$ precipitation, so larger pH elevations units would not normally occur.

Different polymorphs of $CaCO_3$ may precipitate in different situations largely for kinetic reasons. In general, aragonite seems to be kinetically favored over the thermodynamically stable calcite when reactants are supplied rapidly, at higher Mg^{2+} concentrations, and at higher temperatures (Reddy & Wang, 1980; Given & Wilkinson, 1985; Burton & Walter, 1987). Seawater contains about 5 times as much Mg^{2+} as Ca^{2+}, and Mg^{2+} may reach the site of skeletal precipitation by fluid transport.

Regulation of calcification

Ca^{2+} influx to cells occurs through voltage sensitive Ca^{2+} "channels", which close when cytosolic free Ca^{2+} levels exceed threshold values less than 1 uM (Eckert & Chad, 1984). Ca^{2+} efflux *via* Ca^{2+}-ATPase is regulated by various allosteric modulators such as calmodulin (Rega & Garrahan, 1986). Calcification may then be regulated by turning on or off Ca^{2+}-ATPase, with Ca^{2+} influx being self-regulated.

As support for this scenario, Miller & Sanders (1987) observed that cytosolic free Ca^{2+} levels in *Nitellopsis* decrease when the alga is illuminated, and return to higher levels in the dark. Control of cytosolic free Ca^{2+} levels through Ca^{2+} influx appears unlikely since changes in membrane potential exhibited different kinetics than changes in cytosolic free Ca^{2+} concentrations, and because the photosynthetic inhibitor DCMU prevented the light induced drop in cytosolic free Ca^{2+} concentrations, but did not prevent light induced hyperpolarization of the membrane.

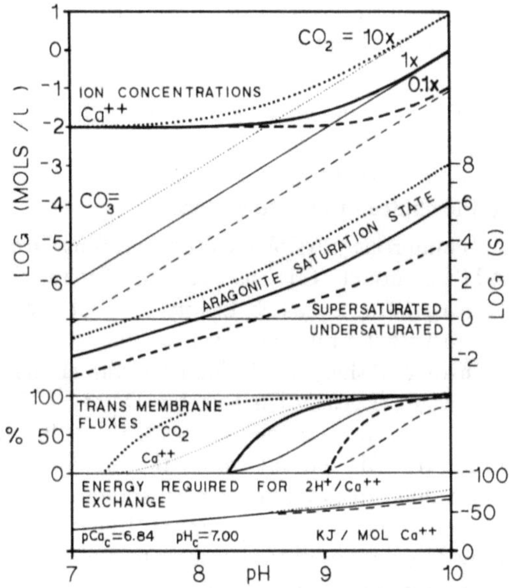

FIGURE 4. Seawater based model depicting the development of CaCO3 supersaturation, as a function of the pH of the mineralizing fluid. **Top**, Ca^{2+} and CO_3^{2-} concentrations. **Second**, Aragonite saturation state (scale on right). **Third**, Fractional contribution of trans-membrane Ca^{2+} and CO_2 fluxes to skeletal $CaCO_3$. **Bottom**, Energy required for $2H^+/Ca^{2+}$ exchange across the mineralizing membrane.

Calcification in charophytes

Calcification in characean algae such as *Chara* and *Nitellopsis* is generally understood in terms of localized OH^- efflux from the plant (or H^+ influx), which converts extracellular HCO_3^- to CO_3^{2-} and causes $CaCO_3$ supersaturation (reviewed by Borowitzka, 1977). The model preferred here is analogous to that shown in Figure 1: $2H^+/Ca^{2+}$ exchange raises both the pH and Ca^{2+} concentration in the pericellular fluid, and CO_2 diffuses from the cytoplasm (pH about 7.5) to the alkaline pericellular fluid (pH about 10), to yield CO_3^{2+} + $2H^+$. Membrane physiology in plants is largely understood through experiments on characeans (Tazawa *et al.*, 1987), so it is useful to examine this conflict.

Stable isotope evidence supports the use of CO_2 in calcification. Characean $CaCO_3$ seems to be depleted in heavy isotopes compared to equilibrium $CaCO_3$ (Forbes, 1987). Since calcification occurs in alkaline environments where HCO_3^- is far more abundant than CO_2, the use of CO_2 implies that HCO_3^- cannot readily reach the site of calcification. It appears that once a $CaCO_3$ sheath has built up around a portion of the plant, HCO_3- diffusion to the calcified plant surface is blocked. Ca^{2+} diffusion toward the calcified plant surface should be similarly impeded, hence Ca^{2+}, like CO_2, must also come from inside the plant. The present model therefore appears reasonable.

Calcification in characeans is an inducible phenomenon which occurs in hard water, alkaline environments. Most physiological experiments have been done on non-calcified plants, often cultured in relatively soft, acidic water. Some of the apparent problems with the current hypothesis (mainly concerning measured Ca^{2+} fluxes) may be related to the choice of sample materials.

Numerous physiological observations are nevertheless consistent with the Ca^{2+}-ATPase model. Ca^{2+} is necessary for the formation of alkaline pericellular regions (Lucas, 1976; Lucas & Dainty, 1977; Lucas, 1979; Bisson, 1984), and pH banding, photosynthetic rate, and cyclosis (which is Ca^{2+} mediated; Hayama *et al.*, 1979; Williamson & Ashley, 1982) decrease with similar kinetics when plants are placed in Ca^{2+} free water (Lucas, 1976; Lucas & Dainty, 1977). Large internal Ca^{2+} reserves (Tazawa *et al.*, 1976) presumably support Ca^{2+} efflux under these conditions. Electrogenic H^+ influx (or OH^- efflux) within the alkaline band appears unlikely since variations in pericellular pH sometimes correspond rather poorly to electrical fluxes and variations in membrane electrical potential (Lucas & Nuccitelli, 1980; Coster *et al.*, 1985). When they do correspond (in non-calcified plants) this is probably because the membrane is more permeable to electrogenic Ca^{2+} influx within the alkaline zone. Kishimoto *et al.* (1984) used electrical evidence to suggest the possibility of electroneutral proton influx having $2H^+/ATP$ stoichiometry, and Luhring & Tazawa (1985) raised the possibility of nH^+/Ca^{2+} exchange. Finally, as pointed out earlier, cytoplasmic free Ca^{2+} concentrations drop when plants are exposed to light (Miller & Sanders, 1987), which is when alkalinization of the pericellular environment occurs.

Calcification promotes aquatic photosynthesis

The ratio of calcification to photosynthesis often approaches 1:1 in coccolithophorids (Paasche, 1964; Sikes *et al.*, 1980), seaweeds (Borowitzka & Larkum, 1976; Pentecost, 1980), corals (Barnes & Taylor, 1973; Davies, 1984), and foraminifera (Duguay & Taylor, 1978). Thus calcification appears to be loosely coupled to photosynthesis. This relationship is summarized by the following equations representing calcification, HCO_3^- utilization, photosynthesis, and the net reaction:

In photosynthetic organisms, calcification may be viewed mainly as a source of protons. These protons convert extracellular HCO_3^- to CO_2, which is taken up for photosynthesis (Fig. 1a). $CaCO_3$ production may be incidental.

CALCIUM PHOSPHATE PRECIPITATION

Calcium phosphates may precipitate much like $CaCO_3$, but with the addition of electroneutral $H_2PO_4^-/HCO_3^-$ exchange across the calcifying membrane (Fig. 1b). Hydroxyapatite (HAP) precipitation is represented as:

$$5Ca^{2+} + 3H_2PO_4^- + 3CO_2 + 4H_2O = Ca_5OH(PO_4)_3 + 3HCO_3^- + 10H^+$$

Ca^{2+}-ATPase exchanges 5 Ca^{2+} for 10 H^+, and an anion transport protein exchanges 3 $H_2PO_4^-$ for 3 HCO_3^-. CO_2 ionization provides HCO_3^-.

OH^- (from H_2O dissociation) could fulfill the role postulated for HCO_3^-, but HCO_3^- is about 10^4 times more abundant under physiological conditions. Carbonic anhydrase, which catalyzes CO_2 hydration, also occurs in close association with precipitating bones

$$CO_2 + Ca^{2+} + H_2O = CaCO_3 + 2H^+ \qquad (4)$$

$$2HCO_3^- + 2H^+ = 2CO_2 + 2H_2O \qquad (5)$$

$$CO_2 + H_2O = CH_2O + O_2 \qquad (6)$$

$$2HCO_3^- + Ca^{2+} = CaCO_3 + CH_2O + O_2 \qquad (7)$$

and teeth (Wuthier, 1984; Kakei & Nakahara, 1985). Such observations, in addition to the cis-inhibition and trans-stimulation of anion transport mentioned earlier, suggest $HCO_3^-/H_2PO_4^-$ exchange. The accumulation of HCO_3^- in the relatively alkaline extracytoplasmic fluid drives the passive accumulation of phosphate (eq. 2).

The ionic composition of a calcium phosphate precipitating system can be modeled from equation 3, where "other bases" include the phosphate anions $H_2PO_4^-$, HPO_4^{2-}, and PO_4^{3-}. The initial composition of the extracellular fluid is taken as that of human blood plasma, having a pH of 7.4, $[Ca^{2+}]$ = 1.35 mM, carbonate alkalinity of 25.8 mEq/*l*, and phosphate alkalinity of 2.2 mEq/*l*. In this model (Fig. 5), the concentration of Ca^{2+} increases much more rapidly than that of PO_4^{3-} as pH increases above 7.4. This is because the system is well buffered by carbonate and phosphate species, and considerable $2H^+/Ca^{2+}$ exchange must occur to increase pH very much. Both HAP and calcite are supersaturated in blood plasma, and saturation state is highly sensitive to pH (Fig. 5, middle). Small elevations of pH can therefore be expected (and are found; Wuthier, 1984) at the site of bone mineralization.

The energy required for $2H^+/Ca^{2+}$ exchange is shown in the bottom of Figure 5, calculated as before. Since this energy is around 25 KJ/mol Ca^{2+}, the overall stoichiometry of Ca^{2+} pumping in this situation may be $4H^+/Ca^{2+}$/ATP, as in the sarcoplasmic reticulum.

Since $CaCO_3$ and $Ca_5OH(PO_4)_3$ reach saturation at about the same pH, it may be reasonable to suppose that the concentrations of inorganic C, P, and Ca are all regulated by mineral solubilities.

Ca^{2+} salts of other sparingly soluble ions such as oxalate may prcipitate in much the same manner as HAP. The selection of anion may depend on substrate specificities of membrane anion transporters, although these enzymes sometimes exhibit low substrate speficity (*e.g.* Low *et al.*, 1984). It is also possible that cellular chemistry, mineral solubilities, and kinetics determine the mineral phase.

SILICA

As pointed out at the beginning of this article, silica (SiO_2) sometimes precipitates from biological systems which concentrate Ca^{2+} and PO_4^{3-}. Silicon also accumulates

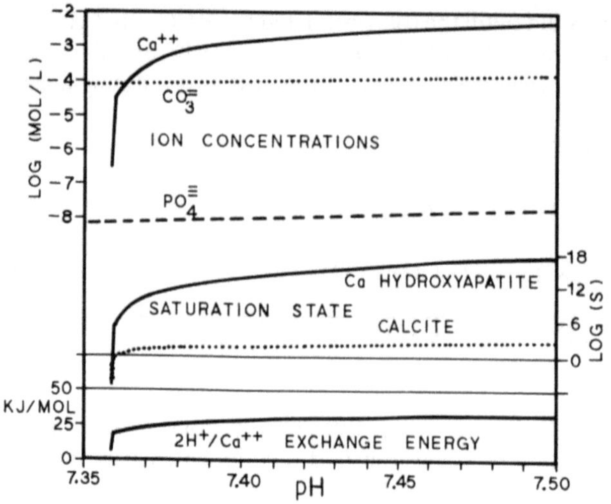

FIGURE 5. Model depicting the development of $Ca_5OH(PO_4)_3$ supersaturation, based on human blood plasma. **Top**, Ca^{2+}, CO_3^{2-}, and PO_4^{3-} concentrations. **Middle**, Saturation state with respect to $CaCO_3$ and $Ca_5OH(PO_4)_3$. **Bottom**, Energy required for $2H^+/Ca^{2+}$ exchange.

in relatively alkaline, Ca^{2+} and PO_4^{3-} rich cellular organelles such as the nucleus and mitochondria (Volcani, 1983; Carlisle, 1981, 1986). Riedel & Nelson (1985) observed that the half saturation constants for silicon uptake by diatoms are pH independent when expressed in terms of the $Si(OH)_3O^-$ anion. Such observations suggest that silicate anions concentrate much like phosphate.

$Si(OH)_4$ readily polymerizes in solution (Alvarez & Sparks, 1985), yielding complexes of multiple negative charge and lower apparent pK values. The pK of silica gel, for example, is about 6.8, compared to about 9.5 for $Si(OH)_4$. Silicate polymers should concentrate rather strongly on the alkaline side of a membrane, and probably cannot move through anion exchange proteins because of their size. These factors may allow fairly strong concentration of silicon on the alkaline side of a semipermeable membrane.

The precipitation of SiO_2 may be represented by:

$$Si(OH)_3O^- + CO_2 = SiO_2 + H_2O + HCO_3^- \qquad (8)$$

where $Si(OH)_3O^-$ is exchanged for HCO_3^-, which is derived from CO_2 hydration. Ca^{2+} may make silicate more susceptible to nucleophilic attack and thereby assist polymerization.

OXIDES OF TRANSITION METALS

As discussed earlier, ATP dependent transport of divalent transition metals in direct or indirect exchange for protons appears likely in several systems. High extracytosolic pH, which would result from $2H^+/M^{2+}$ exchange, promotes spontaneous oxidation of the metal. Precipitation releases protons in 2:1 ratio to the metal, as required for $2H^+/M2+$ exchange:

$$3Fe^{2+} + 4H_2O \rightarrow Fe_3O_4 + H_2 + 6H^+ \qquad (9)$$

$$Mn^{2+} + 1/2O_2 + H_2O \rightarrow MnO_2 + 2H^+ \qquad (10)$$

Once oxidized, the transition metal is no longer transportable by a divalent ion carrier, and is trapped at the site of mineralization.

SUBCELLULAR ORGANELLES

Mineral solubility constraints apply with respect to sub-cellular compartments as well as the extracellular medium. Alkaline, Ca^{2+} rich subcellular compartments include mitochondria (reviewed by Fiskum, 1984), the nucleus, and endoplasmic reticulum (*e.g.* Williams *et al.*, 1985). Evidence has long existed for connections between biomineralization and such organelles. Klaveness (1976), for example, pointed out that coccolithophorids and other flagellated algae apparently form their mineralizing vesicles in close association with the membranes of these organelles.

Mitochondria

Mammalian mitochondria precipitate $CaCO_3$ and Ca^{2+} phosphates and organophosphates (such as ATP) under pathological conditions, and apparently do this under some normal conditions as well (Lehninger *et al.*, 1978; Lehninger, 1983). Mitochondrial mineralization represents a possible means of sequestering Ca^{2+}, storing PO_4^{3-} and perhaps adenine nucleotides, and perhaps nucleating calcium phosphates.

Mitochondria accumulate Ca^{2+} by two pathways (Lehninger *et al.*, 1978), both of which will support mineralization. Ca^{2+} uptake normally occurs by electrodiffusion into the matrix, which is negatively charged through respiration. Rapid ATP driven Ca^{2+} uptake also occurs. Ca^{2+} efflux from rat liver mitochondria occurs in electroneutral exchange for $2H^+$ (Fiskum & Lehninger, 1979). Ca^{2+} efflux in exchange for Na^+ may also occur (Fiskum, 1984), but since there is also rapid Na^+/H^+ exchange, the net effect is the same.

Ca^{2+} efflux is thought to waste the energy spent in Ca^{2+} uptake, and this waste is presumably tollerated for its contribution to cell Ca^{2+} regulation (Fiskum, 1984). It appears possible however that mitochondrial Ca^{2+} efflux conserves respiratory energy, with a stoichiometry of $4H^+/2Ca^{2+}/ATP$. If this interpretation is valid, then mitochondrial Ca^{2+} cycling may supplement or substitute for H^+ motive oxidative phosphorylation. Since the solubilities of Ca^{2+} salts constrain pH and Ca^{2+} concentration in the mitochondrial matrix, large membrane concentration gradients are only possible because cytosolic free Ca^{2+} concentrations are quite low.

The cell nucleus

The extent of Ca^{2+} and pH elevation in the nucleus is unclear. Williams *et al.*, (1985) suggested that free Ca^{2+} concentrations in the nuclei of smooth muscle cells are about twice as high as in the cytoplasm. These nuclei also appear to be relatively alkaline, as evidenced in that study by a tendency for the anionic Ca^{2+} binding indicator dye Fura-2 to concentrate there. The nuclei of muscle cells are fairly quiescent metabolically, hence likely to show relatively modest elevations of Ca^{2+} and pH. Total Ca^{2+} is 16–32 times more concentrated in the dinocaryotic nuclei of dinoflagellates than in the rest of the cell (Sigee, 1983).

Insofar as biomineralization is concerned, the combination of divalent cations with DNA may produce the ultimate biomineral. DNA is an enormous polyanion, having a 1^- charge per nucleotide. The DNA of bacteria (Sigee *et al.*, 1985) and dinoflagellates (Kearns & Sigee, 1980) evidently binds large amounts of Ca^{2+}, Fe^{2+}, Ni^{2+}, Cu^{2+}, Zn^{2+}, and Mn^{2+}. In dinoflagellates, divalent metals approximately titrate the negative charges on the DNA. Ca^{2+} makes up some 34–84% of the metal load, but the lower abundance of the transition metals in solution implies considerably higher tendencies toward concentration in the chromatin. This may be due to stronger binding to DNA.

EVOLUTION

Nucleotide triphosphates bear 3^- to 4^- charges under physiological conditions, so their polymerization with highly anionic nucleic acids requires stabilization from such sources as bound metals. Metal binding proteins may have originated as catalysts of nucleic acid synthesis. Modern proteins which bind DNA often contain divalent metals, evolutionarily conserved amino acid sequences which apparently bind divalent metals, and ATP binding sites (Berg, 1986; Doolittle *et al.*, 1986).

One may image an evolutionary sequence involving the following steps:

1.Metals complex nucleic acids. Transition metals bind more tightly than calcium.

2.Proteins which bind metals and nucleotide triphosphates catalyze nucleic acid synthesis. Metals compete with H^+ for binding sites, possibly giving rise to electroneutral substitution of M^{2+} for $2H^+$.

3.Conformational changes associated with ATP hydrolysis allow proteins to pump metals across membranes.

4.Fluctuating nuclear Ca^{2+} concentrations modulate genomic activity by titrating transition metals from the DNA or from metal binding nucleoproteins.

5.Cationic histone proteins replace transition metals for DNA binding. The phosphorylation of histones (promoted by Ca^{2+}) reduces DNA-Histone binding, and activates the genome.

Through such a scenario, Ca^{2+} becomes a "switch" for genomic activity. High nuclear Ca^{2+} levels promote DNA replication and transcription; low Ca^{2+} levels inactivate the genome. In primitive eukaryotes the chromatin is condensed with transition metals, while histones play this role in higher eukaryotes. The need for such a switch is obvious; a single, relatively non-specific switch can control thousands of genes, located on several chromosomes, more efficiently than can thousands of more specialized switches.

One indication that Ca^{2+} plays such a role is the observation that the entire genome is inactivated and condensed during cell division, when the nuclear membrane is absent in many eukaryotes, and the chromosomes are exposed to the low free Ca^{2+} concentrations of the cytosol. High free Ca^{2+} concentrations and pH (possibly associated with $2H^+/Ca^{2+}$ exchange) are also associated with genome activation in a variety of normal and cancerous cells (reviewed by Geering, 1985). Once a genomic switch based on fluctuating Ca^{2+} and pH levels developed, the switch was used to synchronize other cellular processes.

Fe^{2+} and Mn^{2+} were apparently more abundant in the primitive biosphere than they are now (Holland & Shidlowski, 1982). They presumably concentrated around the DNA of primitive organisms, as in modern bacteria and dinoflagellates. Internal precipitation of metal oxides might have occurred, particularly if $2H^+/Ca^{2+}$ exchange across the boundary membranes alkalinized the genome. Mechanisms for reducing transition metals might therefore have become useful. This may have motivated electron transport across the boundary membrane, while making possible the electrogenic transport of metals against membrane concentration gradients.

Metal coupled cycles of oxidative phosphorylation, as postulated above for mitochondria, are then possible. At any rate, H^+ motive ATP synthesis probably arose from precursors having different functions, and redox reactions involving transition metals are obvious candidates. A metaliferous origin for H^+ motive ATP synthesis is suggested by the nearly ubiquitous involvement of divalent metals in the cytochromes, iron-sulfur proteins, and the light absorbing complex of photosynthesis. Mitochrondria evidently accumulate Fe^{2+} (Flatmark & Romslo, 1975) and can accumulate Ca^{2+} (Lehniger et al., 1978) in exchange for $2H^+$, while chloroplasts show Ca^{2+} dependent ATP synthesis (Vambutus & Racker, 1965; Barr et al., 1982; Becker & Brand, 1985).

Although biomineralization might appear to be a sidelight in the evolution of metal transport systems, it is hardly insignificant: a large proportion of plants and animals contain mineral parts. Biogenic carbonates are among the most abundant crustal rocks (Milliman, 1974), and their precipitation and dissolution cycles largely control the pH of natural waters and CO_2 content of the atmosphere. This in turn affects photosynthesis and global temperatures. Metaliferous bacteria were apparently among the early inhabitants of our planet, and may be responsible for the immense Precambrian transition metal deposits mined commercially today (Holland & Shidlowski, 1982). Hydrogen gas produced during magnetite production would be largely lost to space from an anoxic earth, facilitating the deposition of oxidized metals on the planet's surface. This ultimately would have allowed photosynthetically generated O_2 to accumulate in the atmosphere, instead of being consumed in the oxidation of crustal iron.

ACKNOWLEDGEMENTS

Participation in The Fifth International Symposium on Biomineralization was possible by funds supplied through The Organized Research Fund of The University of Texas at Arlington and The Sea Grant Program administered by Texas A & M University.

REFERENCES

ALVAREZ, R. & SPARKS, D.L., 1985. Polymerization of silicate anions at low concentrations. Nature, 318:649-651.

BARNARD, L.A., MACINTYRE, I.G., & PIERCE, J. W., 1974. Possible environmental index in tropical reef corals. Nature, 252:219-220.

BARR, R., TROXEL, K.S., & CRANE, F. L., 1982. Calmodulin antagonists inhibit electron transport in photosystem II of spinach chloroplasts. Biochem. Biophys. Res. Comm., 104:1182-1188.

BAUER, C., GROS, G., & BARTELS, H., (eds.), 1980. Biophysics and Physiology of Carbon Dioxide. 458 pp. Berlin: Springer-Verlag.

BECKER, D. W.& BRAND, J.J., 1985. Anacystis nidulans demonstrates a photosystem II cation requirement satisfied only by Ca^{2+} or Na^+. Plant Physiology, 79:552-558. BERG, J.M., 1986. Potential metal-binding domains in nucleic acid binding proteins. Science, 232:485-487.

BOROWITZKA, M.A., 1977. Algal calcification. Oceanogr. Mar. Biol. Ann. Rev., 15:189-223.

BOROWITZKA, M.A. & LARKUM, A. W.D., 1976. Calcification in the green alga *Halimeda* II. J. Exp. Bot., 27:864-878.

BUDDEMEIER, R.W., SCHNEIDER, R.C., & SMITH, S.V., 1981. The alkaline earth chemistry of corals. Proc. 4th Inter. Coral Reef Sym., Manilla, 2:81-86.

BURTON, E.A. & WALTER, L.M., 1987. Relative precipitation rates of aragonite and Mg calcite from seawater: temperature and carbonate ion control. Geology, 15:111-114.

BYGRAVE, F.L., 1977. Mitochondrial calcium transport. Curr. Top. Bioenergetics, 6:259-318.

CAMERON, J.N. & WOOD, C.M., 1985. Apparent H^+ excretion and CO2 dynamics accompanying carapace mineralization in the blue crab (*Callinectes sapidus*) following moulting. J. Exp. Biol., 114:181-196.

CAMPBELL, A.K., 1983. Intracellular calcium: its universal role as regulator. 556 pp. New York: Wiley.

CARAFOLI, E., INESI, G., & ROSEN, B.P., 1984. Calcium transport across biological membranes. *In* Metal Ions in Biological Systems. (ed. H. Sigel), pp. 129-186. New York: Marcel Dekker.

CARLISLE, E.M., 1981. Silicon in bone formation. *In* Silicon and Siliceo E. Volcani), pp. 69-94. New York: Springer-Verlag.

CARLISLE, E.M., 1986. Silicon as an essential trace element in animal nutrition. *In* Silicon Biochemistry (ed. D. Evered & M. O'Connor), pp. 123-135. Chichester: John Wiley & Sons.

COSTER, H.G.L., CHILCOTT, T.C., & OGATA, K., 1985. Fluctuations in the electrical properties of *Chara* and the spatial structure of the electro-chemical characteristics. *In* Inorganic Carbon Uptake by Aquatic Photosynthetic Organsism (ed. W. J. Lucas and J. A. Berry), pp. 255-269. Rockville: Amer. Soc. Plant Physiol.

COTY, W.A. & MCCONKEY C., Jr., 1982. A high-affinity calcium stimulated ATPase activity in the hen oviduct shell gland. Arch. Biochim. Biophys., 219:444.

COTY, W.A. & PEDERSEN, P.L., 1975. Phosphate transport in rat liver mitochondria. Kinetics, inhibitor sensitivity, energy requirements, and labelled components. Molec. Cellu. Biochem., 9:109-124.

DOOLITTLE, R.F., JOHNSON, M.S., HUSAIN, I., VAN HOUTEN, B., THOMAS, D.C., & SANCAR, A., 1986. Domainal evolution of a prokaryotic DNA repair protein and its relationship to active-tansport proteins. Nature, 323:451-453.

DUGUAY, L.E. & TAYLOR, D.L., 1978. Primary production and calcification by the soritid foraminifer *Archais angulatus* (Finchtel & Moll). J. Protozool., 25:356-361.

ECKERT, R. & CHAD, J.E., 1984. Inactivation of Ca channels. Prog. Biophys. Molec. Biol., 44:215-267.

FERAY, J.-C. & GARAY, R., 1987. A one-to-one Mg^{2+}:Mn^{2+} exchange in rat erythrocytes. J. Biol. Chem., 262:5763-5768.

FISKUM, G., 1984. Physiological aspects of mitochondrial calcium regulation. *In* Metal Ions in Biological Systems v. 17: Calcium and its Role in Biology. (ed. H. Sigel), pp. 187-214. New York: Marcel Dekker.

FISKUM, G. & LEHNINGER, A.L., 1979. Regulated release of Ca^{2+} from respiring mitochondria by $Ca^{2+}/2H^+$ antiport. J. Biol. Chem., 254:6236-6239.

FISKUM, G., REYNAFARJE, B., & LEHNINGER, A. L., 1979. The electric charge stoichiometry of respiration-dependent Ca^{2+} uptake by mitochondria. J. Biol. Chem., 254:6288-6295.

FLATMARK, T. & ROMSLO, I., 1975. Energy dependent accumulation of iron by isolated rat liver mitochondria - requirement for a unidirectional flux of Fe(II) across inner membrane. J. Biol. Chem., 250:6432.

FORBES, J., 1987. Carbon and oxygen isotopic composition of holocene lake sediments from Okanogan county, Washington. Thesis, University of Washington, Seattle, 103 pp.

FORTES, P.A.G., 1977. Anion movements in red blood cells. *In* Membrane Transport in Red Cells, (ed. J. C. Ellory & V. L. Lew), pp. 175-195. London: Academic Press.

GEERING, K., 1986. Intracellular ionic changes in cell activation: regulation of DNA, RNA, and protein synthesis. Cur. Top. Membr. Trans., 27:221-259.

GIVEN, R.K. & WILKINSON, B.H., 1985. Kinetic control of morphology, composition, and mineralogy of abiotic sedimentary carbonates. J. Sed. Petr., 55:109-119.

GUTKNECHT, J., BISSON, M.A., & TOSTESON, F.C., 1977. Diffusion of carbon dioxide through lipid bilayer membranes. Effects of carbonic anhydrase, bicarbonate, and unstirred layers. J. Gen. Physiol., 69:779-794.

GUTKNECHT, J. & WALTER, A., 1982. Anion and proton transport through lipid bilayers and red cell membranes. *In* Chloride Transport in Biological Membranes (ed. J. A. Zadunaisky), pp. 91-109. New York: Academic Press.

HAYNES, D. H. & MANDVENO, A., 1987. Computer modeling of Ca^{2+} pump function of Ca^{2+}-Mg^{2+}-ATPase of Sarcoplasmic Reticulum. Physiol. Rev., 67:244-284.

HOLLAND, H.D. & SCHIDLOWSKI, M., eds., 1982. Mineral deposits and the evolution of the biosphere. 336 pp. Berlin: Springer-Verlag.

ISA, Y., NORIKATSU, I., & YAMAZATO, K., 1980. Evidence for the occurrence of Ca^{2+} dependent adenosine triphosphatase in the hermatypic coral *Acropora hebes*, Dana. Tech. Rep. Sesoko Mar. Sci. Lab. Ryukyu, 7:19-25.

JENNINGS, M.L., 1978. Characteristics of CO_2-independent pH equilibration in human red blood cells. J. Membr. Biol., 40:365-391.

JOHNSTON, I.S., 1980. The ultrastructure of skeletogenesis in hermatypic corals. Inter. Rev. Cytology, 67:171-214.

KAKEI, M. & NAKAHARA, H., 1984. Histochemical localization of carbonic anhydrase activity in epiphyseal growth cartilage and calvaria of the rat. Jap. J. Oral Biol., 26:554-558.

KAKEI, M. & NAKAHARA, H., 1985. Electroimmunoblotting study of carbonic anhydrase in developing enamel and dentin of the rat incisor. Jap. J. Oral Biol., 27:357-361.

KEARNS, L.P. & SIGEE, D.G., 1980. The occurrence of period IV elements in dinoflagellate chromatin: An X-ray microanalytical study. J. Cell Sci., 46:113-127.

KINGSLEY, R.J. & N. WATABE, 1985. Ca-ATPase localization and inhibition in the gorgonian *Leptogorgia virgulata* (Lamarck). J. Exp. Mar. Biol. Ecol., 93:157-167.

KISHIMOTO, U., KAMI-IKE, N., TAKEUCHI, Y, & OHKAWA, T., 1984. A kinetic analysis of the electrogenic pump of *Chara corallina:* I. Inhibition of the pump by DCCD. J. Membr. Biol., 80:175-183.

KLAVENESS, D., 1976. *Emiliania huxleyi* (Lohman) Hay and Mohler. 3. Mineral deposition and the origin of the matrix during coccolith formation. Protistologica, 12:217-224.

LEHNINGER, A.L., REYNAFJE, B., VERCESI, A. & TEW, W. P., 1978. Characteristics of energy-dependent calcium influx and efflux systems in mitochondria. *In* The Proton and Calcium Pumps. (Ed. G. F. Azzone), pp. 203-214. Amsterdam: Holland Biomedical Press.

LEHNINGER, A.L., 1983. The possible role of mitochondria and phosphocitrate in biological calcification. *In* Biomineralization and Biological Metal Accumulation (ed. P. Westbroek & E.W. de Jong) pp. 107-121. Dordrecht: D. Reidel.

LOW, I., FRIEDRICH, T., BURCKHARDT, G., 1984. Properties of an anion exchanger in rat renal basolateral membrane vesicles. Amer. J. Physiol., 246:F334-F342.

LUCAS, W.J., 1976. The influence of Ca^{2+} and K^+ on $H^{14}CO_3^-$ influx in internodal cells of *Chara corallina*. J. Exp. Bot., 27:32-42.

LUCAS, W.J., 1979. Alkaline band formation in *Chara corallina.* Plant Physiology, 63, 248-254.

LUCAS, W.J. & DAINTY, J., 1977. HCO_3^- influx across the plasmalemma of *Chara corallina*. Divalent cation requirement. Plant Physiology, 60:862-867.

LUCAS, W.J., KEIFER, D.W., & SANDERS, D., 1983. Bicarbonate transport in *Chara corallina*: Evidence for cotransport of HCO_3^- with H^+. J. Membr. Biol., 73:263-274.

LUCAS, W.J. AND NUCCITELLI, R., 1980. HCO_3^- and OH^- transport across the plasmalemma of *Chara. planta*, 150:120-131.

LUHRING, H. & TAZAWA, M., 1985. The effect of cytoplasmic Ca^{2+} on the membrane potential and membrane resistance of *Chara plasmalemma*. Plant Cell Physiol., 26:635-646.

MCCONNAUGHEY, T.A., 1986. Oxygen and carbon isotope disequililbria in Galapagos corals: Isotopic thermometry and calcification physiology. Ph.D. Dissertation, University of Washington, Seattle.

MILLER, A.J. & SANDERS, D., 1987. Depletion of cytosolic free calcium induced by photosynthesis. Nature, 326:397-400.

MILLERO, F.J., 1979. The thermodynamics of the carbonate system in seawater. Geochim. Cosmochim. Acta, 43:1651-1661.

MILLIMAN, J.D., 1974. Marine Carbonates. 375 pp. Berlin: Springer-Verlag.

MORIYASU, Y., SHIMMEN, T., & TAZAWA, M., 1984. Vacuolar pH regulation in *Chara australis*. Cell Struct. Func., 9:225-234.

NAKAHARA, H. & KAKEI, M., 1983. The central dark line in developing enamel crystallite: An electron microscopic study. Bull. Josai Dental Univ., 12:1-7.

NIGGLI, V., SIGEL, E., & CARAFOLI, E., 1982. The purified Ca^{2+} pump of human erythrocyte membranes catalyzes an electroneutral Ca^{2+}-H^+ exchange in reconstituted liposomal systems. J. Biol. Chem., 257:2350-2356.

OHDE, S. & KITANO, Y., 1981. Behavior of minor elements in the transformation of coral aragonite to calcite. Proc. 4th Inter. Coral Reef Sym., Manila, 2:91-94.

OKAZAKI, M., 1977. Some enzymatic properties of Ca^{2+}-dependent adeonosine triphosphatase from a calcareous marine alga, *Serraticardia maxima* and its distribuiton in marine algae. Botanica Marina 20:347-354.

OKAZAKI, M., FUJII, M., USUDA, Y., & FURUYA, K., 1984. Soluble Ca^{2+}-activated ATPase and its possible role in calcification of the coccolithophorid *Cricosphaera roscoffensis* var. *haptonemofera* (Haptophyta). Botanica Marina, 27:363-369.

O'NEIL, J.R., CLAYTON, R.N., & MAYEDA, T.K., 1969. Oxygen isotope fractionation in divalent metal carbonates. J. Chem. Phys., 51:5547-5558.

PAPA, S., LOFRUMENTO, N.E., LOGLISCI, M., & QUAGLIARIELLO, E., 1969. On the transport of inorganic phosphate and malate in rat-liver mitochondria. Biochim. Biophy. Acta, 189:311-314.

PENTECOST, A., 1978. Calcification and photosynthesis in *Corallina officinalis* L. using the $^{14}CO_2$ method. Brit. Phycol. J., 13:383-390.

PENTECOST, A., 1980. Calcification in plants. Inter. Rev. Cytology, 62:1-27.

RASI-COLDOGNO, F., PUGLIARELLO, M.C., & DE MICHELIS, M.I., 1987. The Ca^{2+}-transport ATPase of plant plasma membrane catalyzes a nH^+/Ca^{2+} exchange. Plant Physiology, 83:994-1000.

RAVEN, J.A., 1983. The transport and function of silicon in plants. Biol. Rev., 58:179-207.

REDDY, M. M. & WANG, K. K., 1980. Crystallization of calcium carbonate in the presence of metal ions.1. Inhibition by magnesium ion at pH 8.8 and 25° C. J. Crystal Growth, 50:470-480.

REGA, A.F. & GARRAHAN, P.J., 1986. The Ca^{2+} Pump of Plasma Membranes. 175 pp. Boca Raton: CRC Press.

RIEDEL, G.F. & NELSON, D.M., 1985. Silicon uptake by algae with no known silicon requirement. II. Strong pH dependence of uptake kinetic parameters in Phaeodactylum tricornutum (Bacillariophyceae). J. Phycol., 21:168-171.

RUNHAM, N.W., THORNTON, P.R., SHAW, D.W., & WAYTE, R.C., 1969. The mineralization and hardness of the radular teeth of the limpet *Patella vulgata* L. Zeit. Zellforsch., 99:608-626.

SCHERER, M. & SEITZ, H., 1980. Rare-earth element distribution in Holocene and Pleistocene corals and their redistribution during diagenesis. Chem. Geol., 28:279-289.

SCHUMAKER, K.S. & SZE, H., 1985. A Ca^{2+}/H^+ antiport system driven by the proton electrochemical gradient of a tonoplast H^+-ATPase from oat roots. Plant Physiology, 79:1111-1117.

SHEN, V., KOLHLER, G., & PECK, W.A., 1983. A high affinity, calmodulin-responsive (Ca^{2+}, Mg^{2+})-ATPase in isolated bone cells. Biochim. Biophys. Acta, 727:230.

SIGEE, D.C., 1983. Localized uptake and extraction of $calcium^{45}$ in dinoflagellate nuclei: An autoradiographic study. Protoplasma 117:185-195.

SIGEE, D.C., 1985. The dinoflagellate chromosome. Adv. Bot. Res., 12:205-264.

SIKES, C.S. & WHEELER, A.P., 1983. A systematic approach to some fundamental questions of carbonate precipitation. *In* Biomineralization and Biological Metal Accumulation (ed. P. Westbroek and E.W. de Jong), pp. 285-289. Dordrecht: D. Reidel.

SIMKISS, K., 1976. Cellular aspects of calcification. *In* The Mechanisms of Calcification in Invertebrates. (ed. N. Watabe and K. M. Wilbur). pp. 1-31. Columbia: University of South Carolina Press.

STROHL, W.R. & TOUVINEN, O.H., eds. 1984. Microbial Chemoautotrophy. 355 pp. Columbus: Ohio State University Press.

TAZAWA, M., KIKUYAMA, M., & SHIMMEN, T., 1976. Electric characteristics and cytoplasmic streaming of characeae cells lacking tonoplast. cEll Struct. Func., 1:165-176.

TAZAWA, M., SHIMMEN, T., & MIMURA, T., 1987. Membrane control in the characeae. Ann. Rev. Plant Physio., 38:95-117.

TOWE, K.M. & LOWENSTAM, H.A., 1967. Untrastructure and development of iron mineralization in the radular teeth of *Cryptochiton stelleri* (Mollusca). J. Ultrastruc. Res., 17:1-13.

ULLRICH, K.J., RUMRICH, G., & KLOSs, S., 1984. Contraluminal sulfate transport in the proximal tubule of the rat kidney. I. Kinetics, effects of K^+, Na^+, Ca^{2+}, H^+, and anions. Pflugers Archives 402:264-271.

VAMBUTUS, V.K. & RACKER, E., 1965. Partial resolution of enzymes catalyzing photophosphorylation. I. Stimulation of photophosphorylation by a preparation of a latent Ca^{2+}-dependent adenosine triphosphatase from chloroplasts. J. Biol. Chem., 240:2660.

VOGEL, J.C., 1980. Fractionation of the stable carbon isotopes during photosynthesis. pp. 111-135 in Sitzungsberichte der Heidelberger Akademie der Wissenschaften 1980, 3. Abhandlung. Berlin: Springer-Verlag.

VOLCANI, B.E., 1983. Aspects of silification in biological systems. *In* Biomineralization and Biological Metal Accumulation (ed. P. Westbroek and E. W. de Jong), pp. 389-406. Dordrecht: D. Reidel.

WHEELER, A.P. & SIKES, C.S., 1984. Regulation of carbonate calcification by organic matrix. Amer. Zool., 24:933-944.

WIETH, J.O., 1979. Bicarbonate exchange through the human red cell membrane determined with ^{14}C bicarbonate. J. Physiol., 294:521-539.

WILLIAMS, D.A., FOGARTY, K.E., TSIEN, R.Y., & FAY, F. S., 1985. Calcium gradients in single smooth muscle cells revealed by the digital imaging microscope using Fura-2. Nature, 318:558-561.

WILLIAMSON, R.E. & ASHLEY, C.C., 1982. Free Ca^{2+} and cytoplasmic streaming in the alga Chara. Nature, 296:647-651.

WOOD, C.M. & CAMERON, J.N., 1985. Temperature and the physiology of intracellular and extracellular acid-base regulation in the blue crab *Callinectes sapidus.* J. Exp. Biol., 114:151-180.

WUTHIER, R.E., 1984. Calcification of vertebrate hard tissues. Metal Ions in Biological Systems, 17:411-472.

THE EVOLUTION OF MINERAL SKELETONS

Bruce Runnegar[1]

Department of Geology and Geophysics, University of New England, Armidale, New South Wales 2351, Australia; [1] Present address: Institute of Geophysics and Planetary Physics & Department of Earth and Space Sciences, University of California, Los Angeles, CA 90024

ABSTRACT. All of the major skeletal minerals (calcite, magnesian calcite, aragonite, apatite and opal) and most of the major kinds of skeletal materials (fibrous apatite, spherulitic aragonite, nacre, crossed-lamellar aragonite, foliated calcite, echinoderm stereom etc.) appeared abruptly in many different kinds of organisms at the beginning of the Phanerozoic (570 million years to Recent). Almost all of the clades known to form mineral skeletons originated in the Cambrian Period of earth history (570 – 505 million years ago) and hardly any kinds of marine organisms developed new types of mineral skeletons after that time. Recent work on the earliest skeletal fossils provides little support for the view that phosphatic skeletons were more common than carbonate skeletons when hard parts first developed.

INTRODUCTION

Many different kinds of organisms deposit crystalline or amorphous minerals inside or outside their cells (Lowenstam, 1981). These compounds are the building-blocks of rigid skeletons but they also serve to rid cells of unwanted salts, to store useful ions, or to form parts of sensory organs that are used for sight, orientation and navigation. In sizable animals these functions are clearly discrete and they are often performed by different kinds of compounds. For example, some birds have phosphate endoskeletons, carbonate eggshells and gravity sensors, and a navigation system that in part depends upon magnetite.

The non-skeletal biominerals represent an insignificant volume of all but the smallest organisms. As a result, they are rarely fossilized and the overt history of biomineralization is largely provided by the fossil record of skeletal materials. This record begins at the end of the Precambrian, some 570 million years (Ma) ago, when many different kinds of animals and plants suddenly began to form mineral skeletons (Fig. 1).

The reason for the abruptness of this event remains obscure after more than a century of study. Some argue that it needs no explanation because it is the natural consequence of the explosive radiation of the multicelled animals (Metazoa) which had originated less than 100 Ma previously (S.J. Gould's "Sigmoid Fraud"; Stanley, 1976; Gould, 1977; Sepkoski, 1978). The other idea is that metazoan animals have a much longer Precambrian history (Durham, 1978; Glaessner, 1984) but were not preserved until they

Origin, Evolution, and Modern Aspects of Biomineralization in Plants and Animals
Edited by Rex E. Crick
Plenum Press, New York

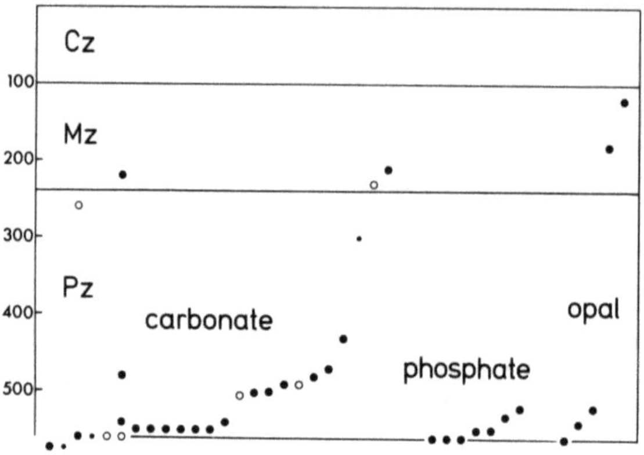

FIGURE 1. First appearance of mineral skeletons in different clades of organisms plotted against geological time in millions of years (Pz = Paleozoic, Mz = Mesozoic, Cz = Cenozoic). In the carbonate group, large filled circles represent calcitic skeletons, open circles aragonitic skeletons, and small filled circles calcareous skeletons of unknown original mineralogy. Changes in mineralogy within clades are shown above first appearances in some cases. Reading from left to right the clades are: Cyanobacteria, anabaritids, Archaeocyatha/Sphinctozoa, Coeloscleritophora, Hyolitha, Mollusca, Calcarea (Porifera), Brachiopoda, Echinodermata, Trilobita, Crustacea, Tentaculita, red algae, green algae, tabulate corals, receptaculitids, Bryozoa, sclerosponges, Foraminifera, rugose corals, octocorals, vertebrate eggshells, scleractinian corals, calcareous nannoplankton; Conodonta, Brachiopoda, tommotiids, etc., Ostracoda, *Hadimopanella*, Conulata, Vertebrata; Hexactinellida (Porifera), Chrysophyceae, Radiolaria, diatoms, silicoflagellates.

became large (Nicol, 1966), muscular (Towe, 1970) and/or mineralized (Rhoads & Morse, 1971; Lowenstam & Margulis, 1980). Thus the end of the Precambrian and the beginning of the Phanerozoic can be interpreted in two ways: either as an "explosion of animals" or what S. Bengtson and K.M. Towe independently have called an "explosion of fossils".

Nearly all of the mineral skeletons manufactured by organisms are made from one or two of five common inorganic compounds: calcite [the rhombohedral polymorph of calcium carbonate ($CaCO_3$)], magnesian calcite (a solid solution of $MgCO_3$ in $CaCO_3$), aragonite (orthorhombic $CaCO_3$, normally with a small amount of $SrCO_3$ in solid solution), apatite (a family of compounds of the general formula $Ca_5(PO_4)_3OH,F$) and opal (amorphous hydrated SiO_2). Each of these materials is readily available, sparingly soluble in water and non-toxic, but each is also soft and brittle, and with the exception of opal, is relatively dense (2.7 – 3.2). How then are these substances used by different kinds of organisms to build sizable, strong and light-weight structures, and when did these innovations in skeletal construction first take place?

STRUCTURE AND FUNCTION OF SKELETAL BIOMINERALS

Most skeletons can be regarded either as stiffened walls, scaffoldings or shells. Stiffened walls are mineralized after they have been shaped by membranes or layers of cells. Some examples are the calcified sheaths of filamentous cyanobacteria (Riding, 1977), the chicken eggshell and the septa of *Nautilus* (Ward and Chamberlain, 1983). They are analogous to architectural structures built by spraying concrete onto preformed molds. Scaffoldings are produced when a number of linear and often similar units are joined to produce static or dynamic structures such as the siliceous framework of hexactinellid sponges (Reif & Robinson, 1976) or the limbs of vertebrates.

Incremental growth results in straight, curved and often conical structures which may be hollow as in the case of snail shells or solid like the open-rooted incisor teeth of some herbivorous mammals. Most are scale-invariant objects that look the same at all magnifications. The best-known example is the logarithmic spiral of the *Nautilus* shell (Raup, 1966) but most other shells, scales, tests and teeth are built in a similar fashion.

In some cases two or more methods of construction may be used in a single skeleton. For example, the corona of echinoids is composed of columns of roughly hexagonal plates that not only increase in number during growth but also become larger by adding increments to the edges of each previously formed plate (Raup, 1968). However, the shape of the whole test is best modeled as a stiffened wall or "pneu" (Seilacher, 1979).

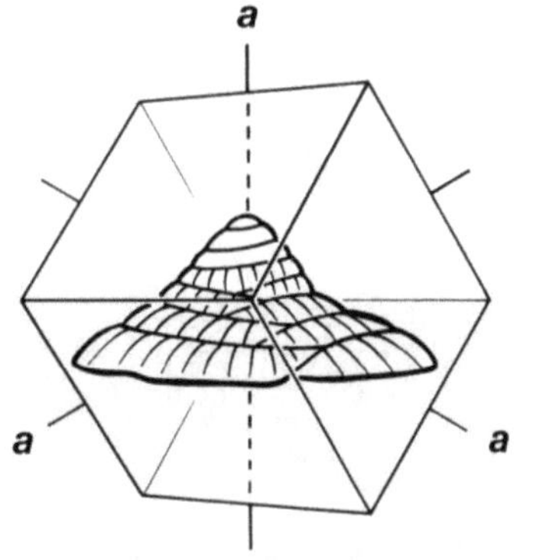

FIGURE 2. The skeleton of the foraminiferan *Patellina* is constructed from a single crystal of calcite. One *a*-axis of the crystal is colinear with the axis of coiling of the test.

Because of its non-crystalline, isotropic nature (Sanders, 1964) and intracellular method of formation (Simpson & Volcani, 1981), opaline silica has had limited potential as a skeletal material except in very small organisms (radiolarians, diatoms, silicoflagellates, etc.). In contrast, the salts of calcium form anisotropic crystals that may be shaped and assembled in a variety of ways by habit-modifying macromolecules. A few very small organisms are able to sculpt their skeletons from single crystals (Fig. 2) but this method of construction is not available to larger animals and plants because large crystals are difficult to grow, unsuitable in shape and mechanically weak. Most crystalline skeletons are therefore composites of organic polymers interspersed with a mineral phase. Two-component composite materials are able to resist both plastic deformation and brittle fracture and therefore have much greater strength than either component alone (Cottrell & Kelly, 1966).

Another way of increasing the strength of the mineral phase is to grow the crystals in the form of very thin ($10^2 - 10^3$ nm) and very long ($>10^5$ nm) fibers (Gordon, 1964). These are likely to be relatively free of dislocations and therefore much stronger than larger crystals. Fibers of these dimensions are found in some of the strongest and most successful skeletal materials: vertebrate enamel ($10^1 \times 10^5$ nm; Daculsi et al., 1984), brachiopod shells ($10^4 \times 10^6$ nm; Sass & Monroe, 1967), gastropod shells ($10^3 \times 10^6$ nm; MacClintock, 1967) and bivalve hinge ligaments ($10^2 \times -10^5$ nm; Marsh & Sass, 1980).

Perhaps for this reason, the crystals of calcium phosphate found in vertebrate bones and teeth and in most other phosphatic skeletons are extremely small and are therefore difficult to study. In contrast, aragonite and calcite form sizable crystals in many different kinds of skeletons and it is easier to understand the biological controls on the formation of these kinds of materials.

CARBONATE SKELETONS

Crystallography

Calcite and aragonite are the two common polymorphs of $CaCO_3$. Calcite forms crystals with rhombohedral symmetry and is stable at normal temperatures and pres-

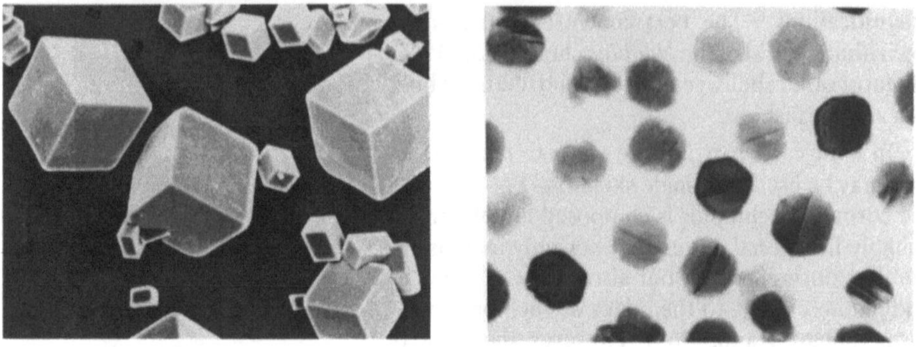

FIGURE 3. Synthetic calcite crystals (left, × 200) have an equidimensional shape; the form is the negative rhombohedron {10$\bar{1}$4}. In contrast, the carbonate crystals of a molluscan bivalve ligament (right, × 90000) illustrate the typical fibrous habit of aragonite. From Runnegar (1984) and Marsh & Sass (1980; copyright by AAAS), republished with permission.

sures. Aragonite is metastable at low pressures (<4 kb) and forms orthorhombic crystals that are frequently twinned and pseudohexagonal in cross-section (Fig. 3). Both minerals are composed of sheets of calcium atoms that alternate with sheets of CO_3 groups in planes lying perpendicular to the crystallographic *c*-axis (Reeder, 1983). Each oxygen atom is linked to two calcium atoms in calcite and to three calcium atoms in aragonite. The latter arrangement occupies less space and accounts for the greater specific gravity of aragonite (2.94 versus 2.71 for calcite) and its stability at high pressures.

Alkaline earth elements with an ionic radius greater than about 1Å form orthorhombic carbonates of the aragonite type and thus fit easily into the aragonite lattice. Those with ions smaller than Ca^{2+} (0.99 Å) form rhombohedral carbonates of the calcite type and also occur in solid solution in calcite. Thus aragonite commonly contains some Sr^{2+} (1.13 Å) but little Mg^{2+} (0.65 Å) whereas the converse is true for calcite. It is one of the coincidences of Nature that only $CaCO_3$ can form both orthorhombic and rhombohedral polymorphs at normal temperatures and pressures.

Although pure aragonite is thermodynamically unstable at normal pressures, about two-thirds of the $CaCO_3$ precipitated in modern oceans is aragonite; the rest is magnesian calcite, a solid solution of $MgCO_3$ in $CaCO_3$. Dry aragonite may survive almost indefinitely (Brown et al., 1962) and has done so in several bitumenous shales of Carboniferous age (ca. 300 Ma) where organic matter in the sediment seems to have protected the aragonite from the catalytic action of underground water (Kennedy & Hall, 1967). In porous rocks, aragonite is quickly converted to or replaced by calcite with a consequent destruction of the original microstructure. As a result, it has been difficult to determine the original nature of many carbonate skeletons of Paleozoic age (570–248 Ma).

$CaCO_3$ precipitated from pure solutions tends to form calcite at temperatures of less than 40°C and aragonite at higher temperatures (Wray & Daniels, 1957). The presence of foreign ions such as Mg^{2+} and those of other metals also promotes the formation of aragonite although only those ions larger than 1 Å are commonly incorporated in the aragonite lattice. Magnesium is particularly important in this regard as it is abundant in sea water. It seems to inhibit the nucleation of calcite by binding strongly to the crystal nuclei. Being small and strongly hydrated, Mg^{2+} will halt further growth of the calcite nuclei and aragonite will precipitate instead.

Synthetic calcite usually forms small rhombohedral crystals less than about 6 μm in size (Fig. 3). All of the crystal faces belong to a single closed form, the positive rhombohedron $\{10\bar{1}4\}$, and they are parallel to the three directions of perfect cleavage found in calcite crystals. Similar rhombohedra often line cavities in carbonate rocks but more slowly deposited crystals exhibit other forms. Two of the most common varieties are "dog-tooth" and "nail-head" spar; the former is a combination of prism and scalenohedron and the latter prism plus negative rhombohedron. Thus the normal habit of natural crystals of inorganic calcite is equidimensional or prismatic.

Aragonite crystals are almost invariably fibrous (Fig. 3). Each fiber is pseudo-hexagonal in cross-section, being bounded laterally by four faces of the prism $\{110\}$ and two faces of the pinacoid $\{010\}$. Growth is slow in these directions and almost all new material is added to the exposed tips of the fibers which pack together into radial aggregates. Thus in simple inorganic systems aragonite is characterized by its fibrous habit and spherulitic growth (Fig. 4). Magnesian calcite also commonly forms fibrous crystals, probably because the Mg^{2+} ion poisons the prism faces of the developing calcite crystals (Folk, 1974).

Geometrical considerations

Except where skeletons are carved from a single crystal or amorphous solid, small crystals must be packed together to make a larger structure. These building-blocks will generally be similar in size and shape and there are geometrical constraints on the ways they can be assembled.

Most skeletons are hollow structures in which the organisms reside. These hollow structures can be built in two main ways, either by assembling prefabricated modules in the shape required or by growing the mineral inwards from a preformed substrate that surrounds the body. Both of these methods of construction have severe geometrical constraints.

Skeletons formed from sculptured crystals

Because of its fibrous nature, skeletons of this kind are never formed of aragonite. The best examples of this method of construction are to be found in the foraminiferan *Patellina*, the calcareous sponges, and the echinoderms. All of these animals build their skeletons from calcite.

The test of *Patellina* (Fig. 2) resembles the shell of a minute snail except that it is divided internally by partitions called septa. The whole skeleton is composed of a single crystal of calcite which has one of its *a*-axes colinear with the axis of coiling of the test. This interpretation of the structure has been confirmed by polarized light microscopy, X-ray diffraction and secondary crystal overgrowths (Towe et al., 1974).

The unbranched or branched spicules of calcareous sponges and alcyonarian octocorals are also sculpted from crystals of calcite (Jones, 1955; Bengtson, 1981). Similar spicules are found in the larvae of sea urchins (Okazaki & Inoué, 1976; Inoué & Okazaki,

FIGURE 4. Two-dimensional computer-drawn simulation of the spherultic growth of aragonite fibers. The program allows equally-spaced radial lines to grow incrementally until they intersect. Program and simulation by F.A. Shaw.

1977). In fact, each of the plates and spines of an echinoderm skeleton can be regarded as a highly porous, somewhat disordered, single crystal of calcite (Raup, 1959; Towe, 1967; Donnay & Pawson, 1969; Blake et al., 1984).

One feature shared by *Patellina*, the calcareous sponges and the echinoderms is that their skeletons are composed of high-magnesian calcite. It has been shown that if echinoid larvae are deprived of magnesium, the spicules will begin to develop crystal faces and that these will disappear if the larvae are returned to normal sea water. It is also known that when echinoid spines are being regenerated after breakage, the amount of magnesium in the calcite increases systematically from the center of the spine to its edge (Davies et al., 1972). These observations are consistent with the idea the magnesium is used to shape the crystals by selectively poisoning appropriate parts of the lattice (O'Neill, 1981). Folk (1974) has pointed out that the influence of the smaller Mg^{2+} ion will be greatest at the edges of the sheets of calcium atoms and carbonate groups so that it will be the prism faces of calcite that are most easily slowed or stopped from growing. This may explain why the c axis of calcite is commonly colinear with the long axes of sponge spicules and echinoid spines but perpendicular to the coronal plates of many echinoids. In each of these cases it is the edges at right angles to the *c*-axis which need to be shaped most during growth.

This hypothesis has been rejected by Blake et al. (1984) on the grounds that the magnesium content of modern crinoid calcite is unexpectedly homogeneous over distances as small as 100 nm. However, if echinoderm calcite is constructed from precisely aligned fibrous crystallites about 100 nm in width (O'Neill, 1981), it may still be possible that the form and habit of the crystallites is controlled through the incorporation of magnesium into the calcite lattice. The shape of the skeleton may then be due to the mode of packing and differential growth of the fibrous crystallites.

In a sense, the echinoderm skeleton (Fig. 5) may also be viewed as a two-phase cubic "emulsion" of mineral (stereom) and living tissue (Donnay & Pawson, 1969). In such structures the interface between the two phases approximates a surface of zero mean curvature as might be formed in a well-shaken mixture of equal parts of oil and water (Schoen, 1970; Mackay, 1985).

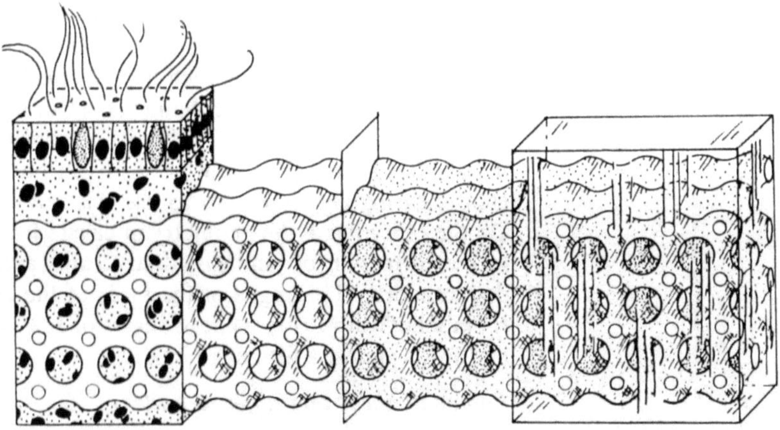

FIGURE 5. Schematic view of the nature and fossilization of echinoderm stereom. In the living animal (left), the calcite and living tissue form an interpenetrating structure that can be viewed as an approximation of a cubic "emulsion". After death and decay of the living tissue the stereom is annealed and often infilled with secondary calcite in crystallographic continuity with the original biomineral (right). From D. Nichols, 1979, Echinodermata, in Encyclopedia of Paleontology (eds., R.W. Fairbridge & D. Jablonski), Stroudsburg, Pa: Dowden, Hutchinson & Ross; republished with permission).

Modular constructions

The best example of modular construction comes from the group of unicellular golden-brown algae that manufacture calcitic plates called coccoliths within their cells. The coccoliths are then exported to the cell surface where they assemble into a calcareous exoskeleton, the coccosphere.

Braarudosphaera bigelowi is one of the most spectacular of all coccolith-bearing algae. It encysts within a regular dodecahedron formed of twelve equal-sized calcitic pentaliths (Fig. 6). Each pentalith is composed of five radial segments set at approximately 72° to each other. The surfaces of the segments correspond to the rhombohedral form $\{10\bar{1}4\}$ and the *c*-axes of the five crystals are consistently inclined in a counter-clockwise array (Runnegar, in prep.). The junctions between the segments do not coincide with low-index lattice planes.

Braarudosphaera bigelowi appears to be descended from the Early Cretaceous species *B. africana* which has star-shaped pentaliths formed of five $\{10\bar{1}4\}$ rhombohedra (Fig. 6). Truncation of the points of the star to produce the pentagonal pentalith of *B. bigelowi* was apparently achieved by making use of the $\{10\bar{1}0\}$ prism direction, which is almost coplanar with the $\{10\bar{1}4\}$ lattice plane of the adjacent segment (Fig. 7). The 58° bevel of the proximal edges of each pentalith which is required in the assembled coccosphere was apparently developed from initial crystallographic angles of 69° and 75°. By analogy with well-studied living relatives, it is probable that the architecture of the pentaliths of *B. bigelowi* is specified by an organic base-plate or "scale". This base-plate is likely to have some form of pentameral symmetry which does not exist in the calcite lattice.

There are only a few regular polyhedra that can be constructed from flat subunits of equal size and shape. The simplest of these are the five Platonic solids (tetrahedron,

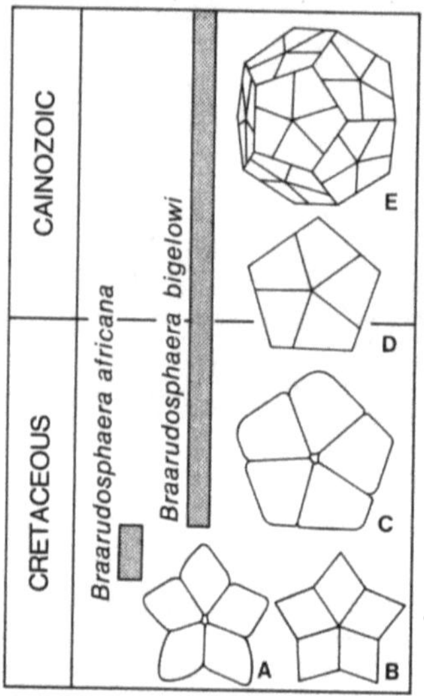

FIGURE 6. The evolution of *Braarudosphaera bigelowi* from *B. africana*. **A**, *B. africana*; **B**, model of a pentalith of *B. africana*; **C**, *B. bigelowi*; **D**, model of a pentalith of *B. bigelowi*; **E**, model of the coccosphere of *B. bigelowi*.

cube, octahedron, dodecahedron and icosahedron). For biological purposes the last two are the most useful because they begin to approach the spherical shape of many isolated cells. The protein coats of some viruses are formed of triangular subunits assembled into icosahedral arrays (Rayment et al., 1982) but no coccolith-bearing alga is known to use this arrangement. On the other hand, the geometry of calcite lends itself to the formation of pentagonal structures and hence dodecahedra. The edges of the faces of the cleavage rhombohedron $\{10\bar{1}4\}$ meet in angles of 102° and 78°, and 78° is almost a fifth of a circle.

Carbonate shells and stiffened walls

Most of the hollow mineralized structures formed by invertebrate animals may, at least in principle, be classified as shells. In these structures growth occurs at linear margins (Clark, 1976) or on previously formed surfaces. The ultimate strength of the construct depends primarily on its shape and the crack-stopping properties of its component materials.

In the simplest and most primitive kinds of shells the carbonate is nucleated on the inner surface of an enclosing unmineralized membrane (epitheca, periostracum etc.) and it grows inwards towards the living animal in an essentially inorganic fashion. If the mineral is aragonite or magnesian calcite the crystallites will normally be fibrous and the gross structure of the skeleton will consist of a series of spherulitic pseudoprisms (Fig. 4). Such pseudoprisms are molded by surface forces rather than chemical bonds and they are found in both inorganic and unstructured biologic deposits (Bryan & Hill, 1941; Taylor, 1973). They begin as hemispherical aggregates which may be nucleated at random on the substrate, and they grow competitively to achieve a uniform size and polygonal cross-section by obeying the rules of soap-bubble geometry. Spherulitic microstructures of this kind are found, for example, in the walls and septa of coral skeletons (Bryan & Hill, 1941), in the outer layers of mollusc shells (Towe & Thompson, 1972) and in vertebrate eggshells (Packard et al., 1984).

These simple fibrous microstructures are strong in compression but are weak in tension because cracks penetrate easily through the layers parallel to the fibers. This defect may be overcome by inclining the fibers to the surface of the shell or by constructing the shell from inclined fibers of more than one preferred orientation (Currey, 1980). Molluscan crossed-lamellar shell structure is the best example of this kind of skeletal material.

Another way of making shells stronger is to deposit the mineral in thin layers parallel to the shell surface. This requires a modification of the habit and/or form of the carbonate crystals to produce shapes that are rarely, if ever, found in non-biological systems. Molluscan nacre and its calcitic equivalents are good examples of such layered

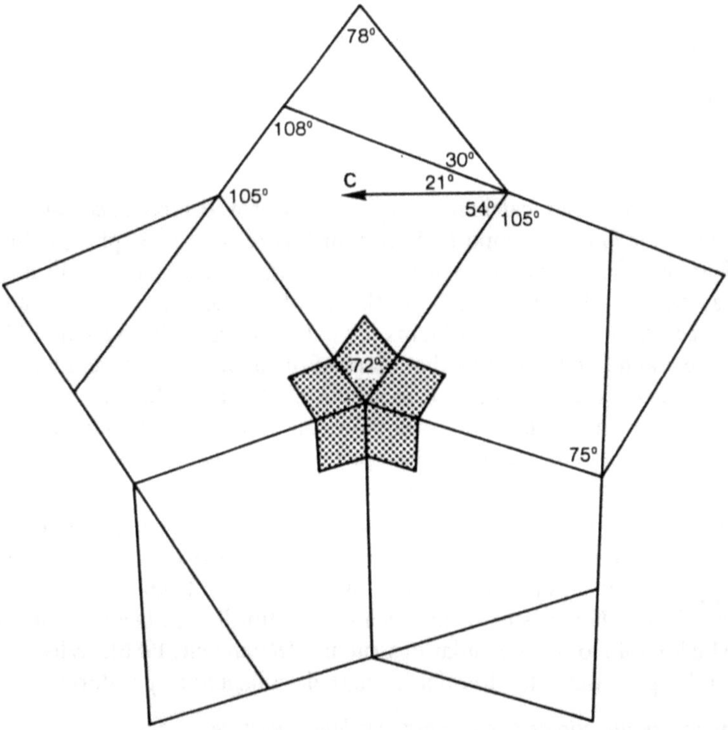

FIGURE 7. Crystallographic model of a pentalith of *Braarudosphaera africana* showing how the edges of the faces of adjacent segments may be projected to produce the pentalith of *B. bigelowi*. The arrow represents the projection of the *c*-axis of calcite on to one segment. The surfaces of the segments are the form $\{10\bar{1}4\}$.

skeletal carbonates (Fig. 8). The great strength of nacre (Currey, 1980) allows *Nautilus* to inhabit depths of 500–600 m where the water pressure is about 60 atmospheres (Ward et al., 1984). The origin of this and other layered structures may be explained as follows.

The spherulitic aragonitic prisms of the outer layer of the *Nautilus* shell present the {001} faces of innumerable fibers to the secreting surface of the mantle (Fig. 9). The fundamental difference between the unmodified aragonite fibers (Fig. 3) and the flat nacre tablets of the inner shell layers (Fig. 10) lies in the difference in habit of the crystals not in the difference in form. Both kinds of crystals are frequently bounded by {010} + {110} and terminated by {001} (Wada, 1972; Marsh & Sass, 1980) but in nacre growth on {001} is very slow whereas in the fibers it is very fast.

It seems likely that proteins with the repetitive amino acid sequences $(Asp-Gly)_n$ or $(Asp-Ser)_n$ may be involved in limiting growth on the {001} direction of aragonite (Weiner & Traub, 1980, 1984). Proteins of this kind have been isolated from the organic matrices of a variety of skeletal carbonates. They are believed to occur as β-pleated sheets, to bind calcium, and to have inter-residue dimensions that would fit well with the arrays of calcium atoms in the surfaces of nacre tablets and calcite folia (Weiner & Traub, 1980, 1984; Runnegar, 1984). It is therefore not difficult to envisage how layered structures such as nacre and foliated calcite may have developed rapidly and convergently in different lineages.

PHOSPHATE SKELETONS

As a skeletal material, calcium phosphate has proved most successful as either a dermal armour or internal support. Partly for this reason, most phosphatic skeletons consist of many different parts (bones, teeth, sclerites etc.) that may disassociate and disperse at death. Reconstruction of the whole skeleton, or more generally the scleritome, presents particular problems for paleontologists (Bengtson, 1985). These problems are magnified by the fact that many of the animal groups that had phosphatic skeletons are now extinct or reduced in number. It is even difficult in many cases to determine whether the skeleton was originally composed of calcium phosphate because secondary apatite may coat, fill or replace primary carbonate in subtle ways (Runnegar & Bentley, 1983; Prévô & Lucas, 1986).

Little is known about the mechanism of formation of phosphatic skeletons in animals other than vertebrates. However, the phosphatic deposits of invertebrates do not, in general, appear to be intimately associated with fibrillar collagen as in vertebrate bone (Glimcher, 1984). This is surprising because the fibrillar collagens of all metazoans appear to be homologous and similar in structure (Runnegar, 1986). Why, for example, did the benthic graptolites fail to mineralize their collagenous periderm?

Secondary phosphate copies of carbonate skeletal structures

Although secondary phosphatization may obscure an original carbonate mineralogy, it may also preserve anatomical and topographical features that would otherwise be obliterated by decay and diagenesis. Such features include the bodies and appendages of ostracods and other Cambrian (ca. 520 Ma) arthropods (Müller, 1979, 1982), myosin fibers in the muscles of Eocene (ca. 50 Ma) vertebrates, and the aragonitic microstructures of the shells of Cambrian molluscs (Runnegar, 1985a). In many cases it is clear that the phosphate coats and fillings must have been formed at or immediately after death (e.g., Müller, 1979) perhaps in a manner analogous to dental calculus (Lustmann et al., 1976). We need to know much more about the role of bacteria in the deposition of phosphate from sea water (O'Brien et al., 1981) and in the decay process (Gristina et al., 1985).

The grain size of the secondary phosphate fillings of fossil shells is often beyond the resolution of a scanning electron microscope, so the surfaces of phosphatic internal molds may have a topographic fidelity comparable to that of a carbon replica prepared for transmission electron microscopy. As a result, it has been possible to identify a number of different kinds of aragonitic and calcitic microstructures in Cambrian molluscs (Figs. 10–11; Runnegar, 1985a, b), either through the use of phosphatic internal molds or the phosphatic fillings of shell borings made by endolithic algae. Endolithic algae frequently expose the edges of individual crystallites during the construction of their tunnels and so the nature of the original microstructure may be revealed in secondary casts of the borings (Poulicek et al., 1981).

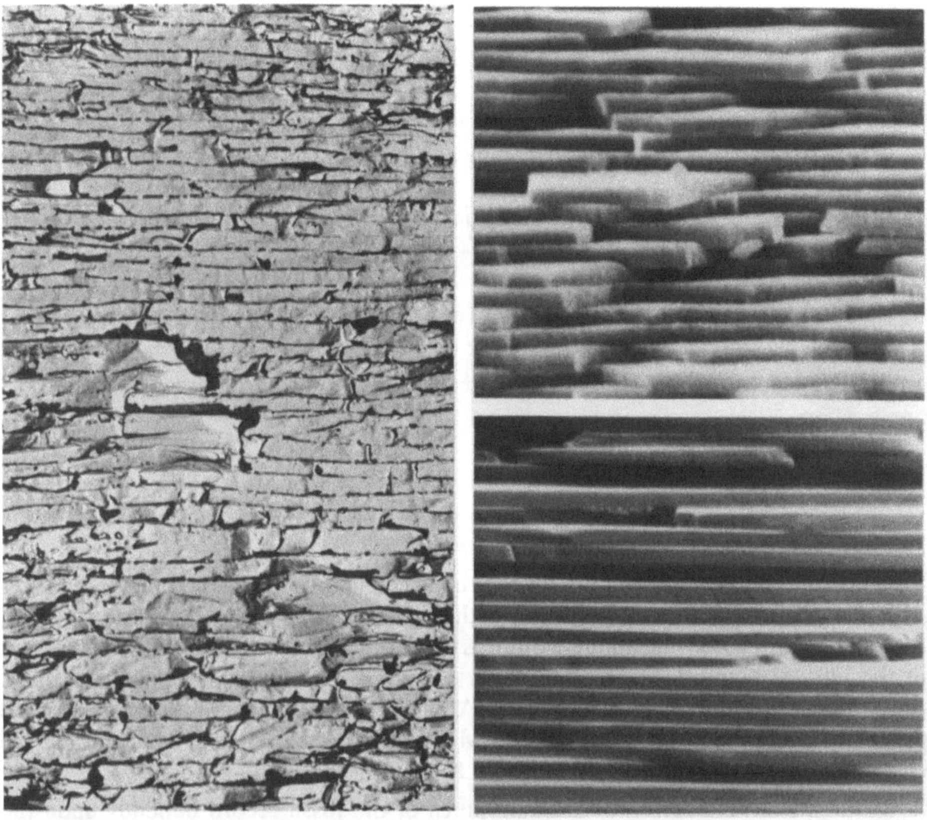

FIGURE 8. Layered calcitic and aragonitic skeletal microstructures. Left, cross-bladed calcite from the Permian productoid brachiopod *Wyndhamia clarkei*, TEM replica × 2000; print by courtesy of Dr. J.D. Armstrong. Upper right, nacreous aragonite from the living bivalve mollusc *Neotrigonia margaritacea*, × 8000. Lower right, foliated calcite from the living bivalve mollusc *Anomia descripta*, × 8000.

It has recently been realized that carbonate internal molds may also replicate the topographic features of original aragonitic microstructures, and that these replicas may survive subsequent diagenesis of the fossil and the matrix. It is therefore likely that the study of phosphatic and other natural replicas of original calcareous microstructures will prove to be a powerful tool for the analysis of the skeletal biominerals of other extinct groups of organisms. This kind of preservation may be widespread in Cambrian and post-Cambrian rocks.

HISTORY OF SKELETAL BIOMINERALIZATION

A current and widely held view is that many different kinds of animals and plants suddenly developed mineral skeletons at the beginning of the Paleozoic, but it was those that used phosphate rather than carbonate or silica that diversified first (Rhodes & Bloxam, 1971; Lowenstam & Margulis, 1980; Lowenstam, 1981). This "phosphate to carbonate transition" is said to have occurred within clades such as the Ostracoda,

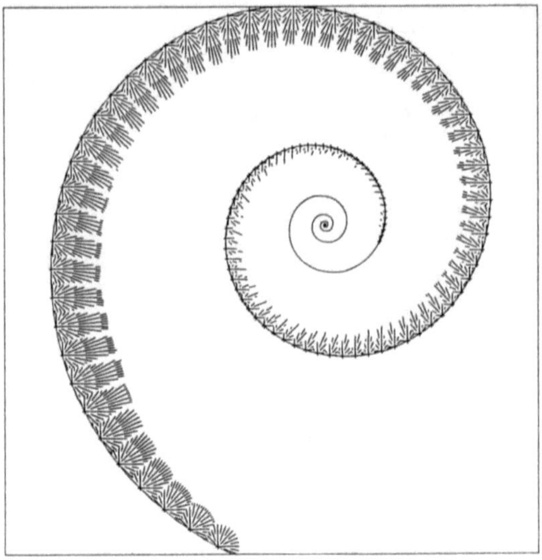

FIGURE 9. Computer-drawn simulation of the growth of the outer shell layer of *Nautilus*. Aragonite fibers represented by straight lines grow inwards from the periostracum (solid curved line) towards the mantle surface. Program and simulation by F.A. Shaw.

Brachiopoda and Cnidaria but also by the replacement through extinction of organisms with phosphatic skeletons by organisms with calcareous hard parts. Lowenstam (1981), for example, estimated that about two-thirds of the organisms with hard parts that existed in the earliest Cambrian had phosphatic skeletons; twenty million years later this proportion was reduced to about a half. After that time carbonate skeletons became dominant and aragonitic materials slowly replaced calcitic ones throughout the remainder of the Phanerozoic (Wilkinson, 1979).

Is this picture correct? Because so much of the fossil record depends upon the existence of mineral skeletons, it is important to distinguish between the evolution of mineralization within clades and the first appearance of higher taxa when attempting to identify historical changes of the type described above. For example, the aragonitic skeletons of the scleractinian corals may either be regarded as an independent development from an unmineralized ancestor (Fig. 1; Oliver, 1980) or a within-lineage conversion of a calcitic skeleton of Paleozoic type. Given the anatomical and developmental differences between the Scleractinia and the Rugosa as well as the stratigraphic gap in the Early Triassic (ca. 245 Ma), an independent origin of the scleractinian corals is more probable (Oliver, 1980). Consequently, there is no clear case for a within-lineage transformation of a calcitic skeleton to an aragonitic one.

Similarly, a suggested evolutionary succession from phosphatic to calcareous hard parts within the coelenterates (Lowenstam, 1981) depends upon the dubious taxonomic decision to place the extinct Conulata within the Cnidaria (Oliver, 1984). We are therefore left with only two possible cases of the phyletic replacement of phosphatic skeletons by calcareous ones. These are the secondary origin of calcitic articulate brachiopods from phosphatic inarticulates and the derivation of calcareous ostracods from pre-existing phosphatic forms.

Neither of these cases is clear cut. In a rigorous analysis of the early history of the Brachiopoda, Rowell (1982) provided good evidence that the phylum is a monophyletic

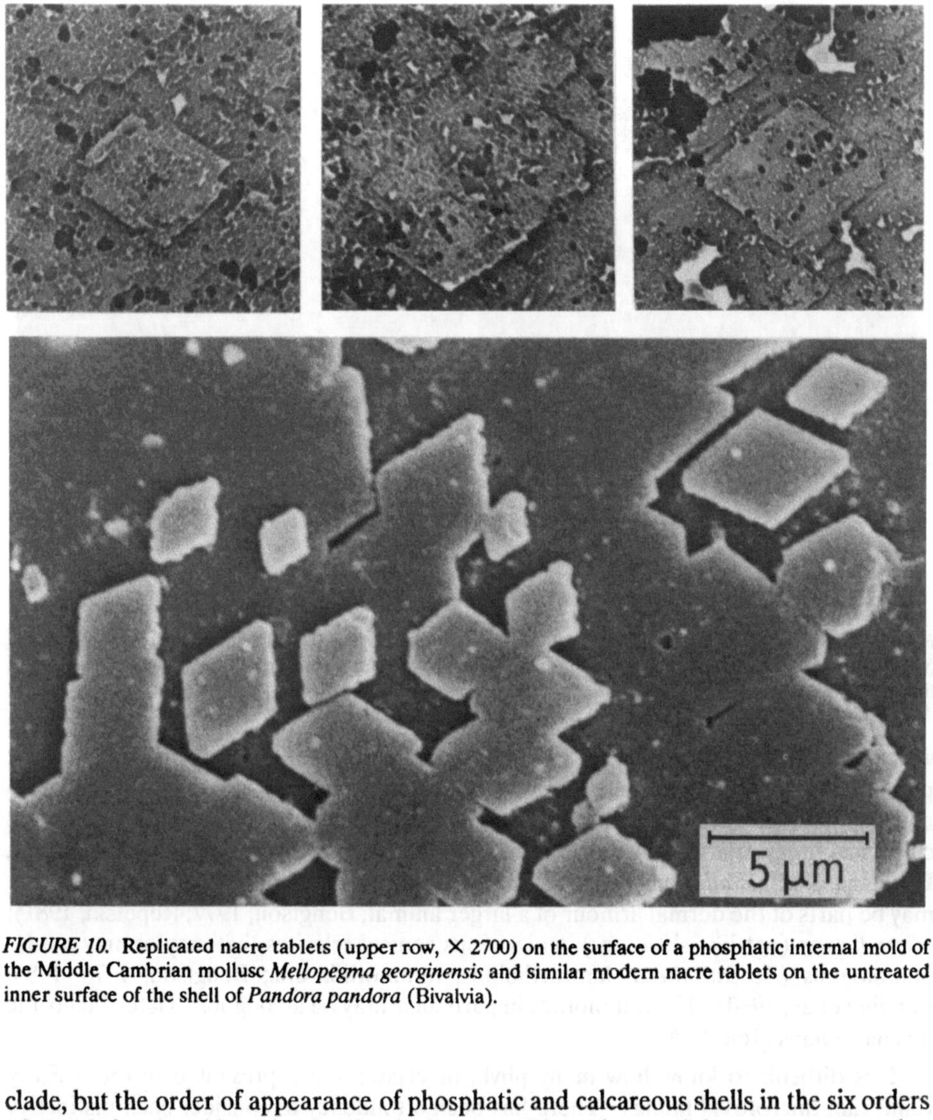

FIGURE 10. Replicated nacre tablets (upper row, × 2700) on the surface of a phosphatic internal mold of the Middle Cambrian mollusc *Mellopegma georginensis* and similar modern nacre tablets on the untreated inner surface of the shell of *Pandora pandora* (Bivalvia).

clade, but the order of appearance of phosphatic and calcareous shells in the six orders of Cambrian brachiopods remained obscure. There is certainly no consensus that brachiopods with calcitic microstructures evolved from those with phosphatic shells (e.g., Goryanskiy & Popov, 1985).

A comparable uncertainty surrounds the early history of the bivalved Crustacea (Jones & McKenzie, 1980; Briggs, 1983). Both phosphatic and calcareous forms appear to have co-existed during the Cambrian and there is, as yet, no clear evidence for the derivation of one group from the other.

On the other hand, it seems clear that more different kinds of animals formed phosphatic skeletons in the Early Cambrian than at any later time (Fig. 1). Groups

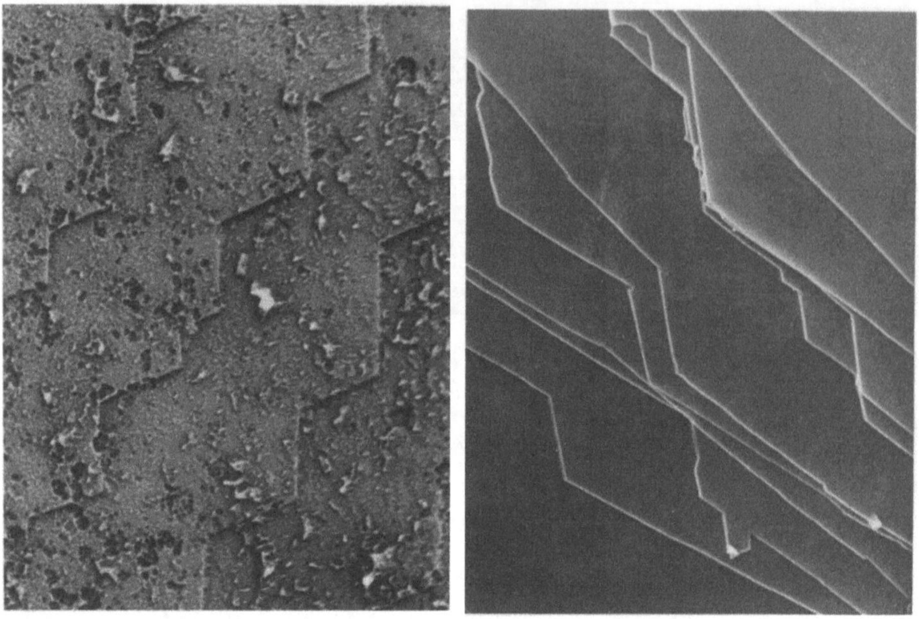

FIGURE 11. Replicated foliated calcite (left, × 4000) on the surface of a phosphatic internal mold of the Middle Cambrian mollusc *Pseudomyona queenslandica* and a fractured surface of the foliated calcite of the living bivalve mollusc *Placuna placenta* (right, × 2000) for comparison.

which subsequently disappeared from the fossil record either through extinction or perhaps demineralization include: protoconodonts (?grasping structures of chaetognath-like animals and probable early conodonts; Bengtson, 1983); hyolithellids (small cylindrical tubes that may or may not be allied to the disk-like operculum *Mobergella*; Bengtson, 1968); *Hadimopanella* (=*Lenargyrion*, 50–500 nm button-shaped objects that may be parts of the dermal armour of a larger animal; Bengtson, 1977; Repetski, 1981); *Microdictyon* (net-like skeletons); and various conical sclerites that may or may not be homologous (*Eccentrotheca*, lapworthellids, tommotiids etc.; Bengtson, 1970, 1983; Landing et al., 1980). The tommotiids in particular may be analogous to later Paleozoic machaeridians (Jell, 1979).

It is difficult to know how many phyla or classes are represented in these Early Cambrian microfossil faunas. There has been a tendency to consider each distinctive morph a separate higher taxon and a representative of an independently evolved mineralizing clade. Counted in this way in any compilation, obscure and short-lived taxa such as Fomitchella (Bengtson, 1983) may be given equal rank to diverse and long-lived groups such as the Vertebrata. It is therefore necessary to attempt to equate taxa of similar rank and perhaps comparable diversity, biomass and longevity when assessing the significance of this Early Cambrian phosphogenic event (Cook & Shergold, 1984).

At the present state of knowledge it is not possible to determine the biological affinities of many of the enigmatic and disarticulated calcareous and phosphatic microfossils found in Early Cambrian rocks. It is therefore difficult to know how many different clades developed the ability to begin forming mineral skeletons at this time. It is, however, possible to obtain some idea of the relative abundance of phosphatic and calcareous clades by dealing only with the groups that are complex, diverse or well understood. Judged in this way, clades with phosphatic skeletons have always been less numerous than clades with calcareous skeletons (Fig. 1). The principal difference between the situation in the Early Cambrian and later times is that a number of early clades of both types have become extinct. These extinct groups include the phosphatic taxa mentioned above plus the Conodonta and the Conulata. The principal groups with calcareous skeletons that disappeared during the Cambrian were the Archaeocyatha, the Coeloscleritophora (calcareous sclerites called sachitids and chancelloriids; Bengtson & Conway Morris, 1984) and the anabaritids (conical tubes with three-fold symmetry). The calcareous clades that survived beyond the Cambrian include the calcareous sponges, hyoliths, molluscs, brachiopods, echinoderms, trilobites and other arthropods, tentaculites and coralline algae.

A second major radiation of organisms with calcareous skeletons (corals, receptaculitids, bryozoans, sclerosponges and foraminiferans) began in the Ordovician (ca. 505–438 Ma) at a time when the calcareous brachiopods and the molluscs were also radiating in an unprecedented way. Given the diversity and complexity of the Late Cambrian and Early Ordovician ecosystems it is unlikely that this transformation of the skeletonized biota can be attributed to a single cause. On the other hand, global changes in atmospheric composition or sea water chemistry may have made it easier for marine organisms to form calcareous skeletons (Sandberg, 1983) just as tropical waters do today. Relatively few new kinds of mineral skeletons appeared after the early Paleozoic (Fig. 1), although some innovations such as the evolution of the calcareous plankton in the Mesozoic must have had a significant effect on the method and rate of removal of $CaCO_3$ from ocean waters.

The number of kinds of organisms that have evolved the ability to form siliceous skeletons is limited (Fig. 1). Siliceous (hexactinellid) sponges appear in latest Precambrian or earliest Cambrian strata and are abundant thereafter in certain environments. The first Radiolaria and siliceous algae are also Cambrian in age (White, 1986; Allison, 1981). So far as is known, there are no extinct higher taxa of silica-secreting organisms.

The other notable thing about the history of biomineralization is that almost all of the kinds of skeletons discussed in the first part of this article appeared within a few tens of millions of years of the beginning of the Cambrian (Figs. 10–12; James & Klapper, 1983). Innovations such as molluscan nacre (Fig. 10; Runnegar, 1985a), echinoderm stereom (Fig. 12; Berg-Madsen, 1986) and the characteristic fabric of the secondary layer of articulate brachiopod shells (Williams, 1968) were being manufactured by at least the middle of the Early Cambrian if not earlier. Hardly any new kinds of skeletal materials were evolved after the Cambrian and the only kinds of mineral skeletons that appeared during the invasion of the land were the carbonate eggshells of snails and vertebrates.

It is the abrupt and synchronous development of a great variety of mineral skeletons in distantly related groups of organisms (cyanobacteria, protistans, metazoans and plants) which points to an extrinsic cause for the Cambrian explosion. One possible

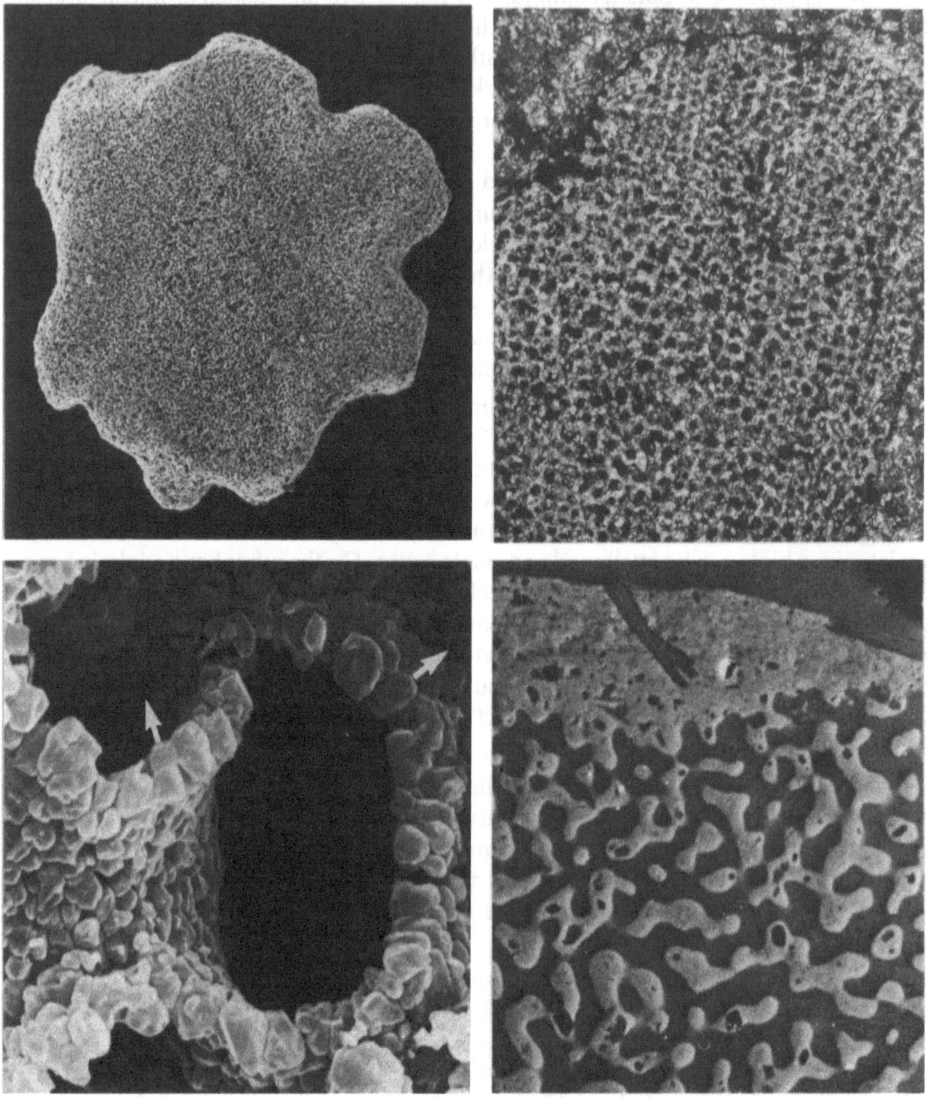

FIGURE 12. Echinoderm stereom from the Early Cambrian Parara Limestone, South Australia. The pores in the calcite stereom are partly filled with phosphate which appears dark in cross-polarized light (upper right) and light in an SEM view of a polished surface (lower right). The phosphate is actually a thin coat that is rigid enough to remain intact when both the stereom and the enclosing carbonate matrix are dissolved in dilute acetic acid. An enlargement (lower left, × 5000) of the phosphate copy of one echinoderm plate (upper left) shows that the secondary apatite crystals did not fill the space occupied by the living tissue (arrows).

scenario for this event, an increase in the oxygen content of the atmosphere, is reviewed elsewhere (Runnegar, 1982).

ACKNOWLEDGEMENTS

I thank M.T.C. Runnegar for help with the literature, F.A. Shaw, B.A. Stait and K.A. Stait for technical assistance, and the Australian Research Grants Scheme for financial support. Participation in the Symposium was made possible by grants from The Organized Research Fund of The University of Texas at Arlington and The Sea Grant Program administered by Texas A & M University.

REFERENCES

ALLISON, C.W., 1981. Siliceous microfossils from the Lower Cambrian of northwest Canada: possible source for biogenic chert. Science, 211, 53-55.

BENGTSON, S., 1968. The problematic genus *Mobergella* from the Lower Cambrian of the Baltic area. Lethaia, 1, 325-351.

BENGTSON, S., 1970. The Lower Cambrian fossil *Tommotia*. Lethaia, 3, 363-392.

BENGTSON, S., 1977. Early Cambrian button-shaped phosphatic microfossils from the Siberian Platform. Palaeontology, 20, 751-762.

BENGTSON, S., 1981. *Atractosella*, a Silurian alcyonarian octocoral. Journal of Paleontology, 55, 281-294.

BENGTSON, S., 1983. The early history of the Conodonta. Fossils and Strata, 15, 5-19.

BENGTSON, S., 1985. Taxonomy of disarticulated fossils. Journal of Paleontology, 59, 1350-1358.

BENGTSON, S. & CONWAY MORRIS, S., 1984. A comparative study of Lower Cambrian *Halkieria* and Middle Cambrian *Wiwaxia*. Lethaia, 17, 307-329.

BERG-MADSEN, V., 1986. Middle Cambrian cystoid (sensu lato) stem columnals from Bornholm, Denmark. Lethaia, 19, 67-80.

BLACK, M., 1972. Crystal development is Discoasteraceae and Braarudosphaeraceae (planktonic algae). Palaeontology, 15, 476-489.

BLAKE, D.F., PEACOR, D.R. & ALLARD, L.F., 1984. Ultrastructural and microanalytical results from echinoderm calcite: implications for biomineralization and diagenesis of skeletal material. Micron and Microscopica Acta, 15, 85-90.

BRIGGS, D.E.G., 1983. Affinities and early evolution of the Crustacea: the evidence of Cambrian fossils. *In* Crustacean Phylogeny (ed. F.R. Schram), pp. 1-22. Rotterdam: A.A. Balkema.

BROWN, W.H., FYFE, W.S. & TURNER, F.J., 1962. Aragonite in California glaucophane schists and the kinetics of the aragonite-calcite transformation. Journal of Petrology, 3, 566-582.

BRYAN, W.B. & HILL, D., 1941. Spherulitic crystallization as a mechanism of skeletal growth in hexacorals Proceedings of the Royal Society of Queensland, 52, 78-91.

COOK, P.J. & SHERGOLD, J.S., 1984. Phosphorus, phosphorites and skeletal evolution at the Precambrian-Cambrian boundary. Nature, 308, 231-236.

COTTRELL, A.H. & KELLY, A., 1966. The design of strong materials. Endeavour, 94, 27-32.

CLARK, G.R., 1976. Shell growth in the marine environment: approaches to the problem of marginal calcification. American Zoologist, 16, 617-626.

CRENSHAW, M.A., 1972. The inorganic composition of molluscan extrapallial fluid. Biol. Bull., Mar. Biol. Lab., Woods Hole, Massachusetts, 143:506-512.

CURREY, J.D., 1980. Mechanical properties of the mollusc shell. *In* The Mechanical Properties of Biological Materials (eds. J.F.V. Vincent & J.D. Currey), pp. 75-97. Cambridge: Cambridge University Press.

DACULSI, G., MENANTEAU, J., KEREBEL, L.M. & MITRE, D., 1984. Length and shape of enamel crystals. Calcified Tissue International, 36, 550-555.

DAVIES, T.T., CRENSHAW, M.A., & HEATFIELD, B.M. The effect of temperature on the chemistry and structure of echinoid spine regeneration. Journal of Paleontology, 46:874-883.

DONNAY, G. & PAWSON, D.L., 1969. X-ray diffraction studies of echinoderm plates. Science, 166, 1147-1150.

DURHAM, J.W., 1978. The probable metazoan biota of the Precambrian as indicated by the subsequent record. Annual Review of Earth and Planetary Sciences, 6, 21-42.

FOLK, R.L., 1978. The natural history of crystalline calcium carbonate: effect of magnesium content and salinity. Journal of Sedimentary Petrology, 44, 40-53.

GIVEN, R.K. & WILKINSON, B.H., 1984. Kinetic control of morphology, composition, and mineralogy of abiotic sedimentary carbonates. J. Sed. Petrol., 55:109-119.

GLAESSNER, M.F., 1984. The Dawn of Animal Life. 244 pp. Cambridge: Cambridge University Press.

GLIMCHER, M.J., 1984. Recent studies of the mineral phase in bone and its possible linkage to the organic matrix by protein-bound phosphate bonds. Philosophical Transactions of the Royal Society of London, B304, 479-508.

GORDON, J.E., 1964. Whiskers. Endeavour, 23, 8-12.

GORYANSKIY, V. YU & POPOV, L.E., 1985. Morfologiya, sistematicheskoe polozhenie i proiskhozhdenie bezzamkovykh braxipod s karbonatoy rakovinoy. Paleontologicheskiy Zhurnal, 1985/3, 3-15.

GOULD, S.J., 1977. Ever Since Darwin. Reflections in Natural History. New York: Norton.

GRISTINA, A.G., OGA, M., WEBB, L.X. & HOBGOOD, C.D., 1985. Adherent bacterial colonization in the pathogenesis of osteomylitis. Science, 228, 990-993.

INOUÉ, S., OKAZAKI, K., 1977. Biocrystals. Scientific American, 236/4, 82-92.

JAMES, N.P. & KLAPPA, C.F., 1983. Petrogenesis of Early Cambrian reef limestones, Labrador, Canada. Journal of Sedimentary Petrology, 53, 1051-1096.

JELL, J.S., 1974. The microstructure of some scleractinian corals. Proceedings of the Second International Coral Reef Symposium, vol. 2. Brisbane: Great Barrier Reef Committee.

JELL, P.A., 1979. *Plumlites* and the machaeridian problem. Alcheringa, 3, 253-259.

JONES, P.J. & MCKENZIE, K.G., 1980. Queensland Middle Cambrian Bradoriida (Crustacea): new taxa, palaeobiogeography and biological affinities. Alcheringa, 4, 203-225.

JONES, W.C., 1955. Crystalline properties of spicules of *Leucosolenia complicata.* Quarterly Journal of Microscopical Science, 96, 129-149.

KENNEDY, W.J. & HALL, A., 1967. The influence of organic matter on the preservations of aragonite in fossils. Proceedings of the Geological Society of London, 1643, 253-255.

LANDING, E., NOWLAN, G.S. & FLETCHER, T.P., 1980. A microfauna associated with Early Cambrian trilobites of the Callavia Zone, northern Antigonish Highlands, Nova Scotia. Canadian Journal of Earth Science, 17, 400-418.

LEHNINGER, A.L., 1983. Principles of Biochemistry. Worth Publishers, New York: 1011 pp.

LOWENSTAM, H.A., 1981. Minerals formed by organisms. Science, 211, 1126-1131.

LOWENSTAM, H.A. & MARGULIS, L., 1980. Evolutionary prerequisites for early Phanerozoic calcareous skeletons. BioSystems, 12, 27-41.

LUSTMANN, J., LEWIN-EPSTEIN, J. & SHEYER, A., 1976. Scanning electron microscopy of dental calculus. Calcified Tissue Research, 21, 47-55.

LUTZ, R.A. & RHOADS, D.C., 1977. Anaerobiosis and a theory of growth line formation. Science, 198:1222-1227.

MACCLINTOCK, C., 1967. Shell structure of patelloid and bellerophontid gastropods (Mollusca). Peabody Museum of Natural History, Yale University, Bulletin, 22, 1-140.

MACKAY, A.L., 1985. Periodic minimal surfaces. Nature, 314, 604-606.

MACKENZIE, F., BISCHOFF, W.D., BISHOP, F.C., LOIJENS, M., SCHOONMAKER, J., & WOLLAST, R., 1983. Magnesian calcites: low temperature occurrence, solubility and solid-solution behavior. *In* Reviews in Mineralogy, vol. 2 (ed. R.J. Reeder), Min. Soc. Amer.:97-144.

MARSH, M.E. AND SASS, R.L., 1980. Aragonite twinning in the molluscan bivalve hinge ligament. Science, 208, 1262-1263.

MÜLLER, K.J., 1979. Phosphatocopine ostracodes with preserved appendages from the Upper Cambrian of Sweden. Lethaia, 12, 1-27.

MÜLLER, K.J., 1982. Weichteile von Fossilien aus dem Erdaltertum Die Naturwissenschaften, 69, 245-249.

NICOL, D., 1966. Cope's rule and Precambrian and Cambrian invertebrates. Journal of Paleontology, 40, 1397-1399.

O'BRIEN, G.W., HARRIS, J.R., MILNES, A.R. & VEEH, H.H., 1981. Bacterial origin of east Australian continental margin phosphorites. Nature, 294, 442-444.

OKADA, M., 1943. Studies on the periodic pattern of hard tissues in animal body. Shanghai Evening Post, Medical Ed. September, 1943:26-31.

OKAZAKI, K. & INOUÉ, S., 1976. Crystal property of the larval sea urchin spicule. Development, Growth and Differentiation, 18, 413-434.

OLIVER, W.A., 1980. The relationship of the scleractinian corals to the rugose corals. Paleobiology, 6, 146-160.

OLIVER, W.A., 1984. *Conchopeltis*: its affinities and significance. Palaeontographica Americana.,54, 95-104.

O'NEILL, P.L., 1981. Polycrystalline echinoderm calcite and its fracture mechanics. Science, 213, 646-648.

PACKARD, M.J., HIRSCH, K.F. & IVERSON, J.B., 1984. Structure of shells from eggs of kinosternid turtles. Journal of Morphology, 181, 9-20.

POULICEK, M., JASPAR-VERSALI, M.F. & GOFFINET, G., 1981. Étude expérimentale de la dégradation des coquilles de mollusques au niveau des sédiments marins, Bulletin de la Société Royale des Sciences de Liége, 50/11-12, 513-518.

PRÉVÔT, L. & LUCAS, J., 1985. Microstructure of apatite-replacing carbonate in synthesized and natural samples. Journal of Sedimentary Petrology, 56, 153-159.

RAUP, D.M., 1959. Crystallography of echinoid calcite. Journal of Geology, 67, 661-674.

RAUP, D.M., 1966. Geometric analysis of shell coiling; general problems. Journal of Paleontology, 40, 1178-1190.

RAUP, D.M., 1968. Theoretical morphology of echinoid growth. Paleontological Society Memoir, 2, 50-63.

RAYMENT, I., BAKER, T.S., CASPAR, D.L.D. & MURAKAMI, W.T., 1982. Polyoma virus caspid structure at 22.5 Å resolution. Nature, 295, 110-115.

REEDER, R.J., 1983. Carbonates: mineralogy and chemistry. Reviews in Mineralogy, 11, 1-394.

REIF, W. & ROBINSON, J.A., 1976. On functional morphology of the skeleton in lychnisc sponges (Porifera, Hexactinellida). Paläontologische Zeitschrift. 50, 57-69.

REPETSKI, J.E., 1981. An Ordovician occurrence of *Utahphospha* MÜller & Miller. Journal of Paleontology, 55, 395-400.

RHOADS, D.C. & MORSE, J.W., 1971. Evolutionary and ecologic significance of oxygen dificient marine basins. Lethaia, 4, 413-428.

RHODES, F.T.H. & BLOXAM, T.W., 1971. Phosphatic organisms in the Paleozoic and their evolutionary significance. Proceedings of the North American Paleontological Convention, Part K, 1485-1513.

RIDING, R., 1977. Calcified *Plectonema* (blue-green algai), a Recent example of *Girvanella* from Aldabra atoll. Palaeontology. 20, 33-46.

ROSENBERG, G.D., 1980. An ontogenetic approach to the environmental significance of bivalve shell chemistry. *In* Skeletal Growth of Aquatic Organisms (ed. D.C. Rhoads & R.A. Lutz), Plenum, New York: 133-168.

ROSENBERG, G.D., ASHTON, M., HEWITT, R., & SIMMONS, D.J., 1980. Application of normalized power spectra to the analysis of chemical and structural growth patterns. *In* Skeletal Growth of Aquatic Organisms (ed. D.C. Rhoads & R.A. Lutz), Plenum, New York: 675-686.

ROSENBERG, G.D. & SIMMONS, D.J., 1980. Rhythmic dentinogenesis in the rabbit incisor: circadian, ultradian and infradian periods. Calc. Tiss. Intern., 32:29-44.

ROWELL, A.J., 1982. The monophyletic origin of the Brachiopoda. Lethaia, 15, 299-307.

RUNNEGAR, B., 1982. The Cambrian explosion: animals or fossils? Journal of the Geological Society of Australia, 29, 395-411.

RUNNEGAR, B., 1984. Crystallography of the foliated calcite shell layers of bivalve molluscs. Alcheringa, 8, 273-290

RUNNEGAR, B., 1985a. Shell microstructures of Cambrian molluscs replicated by phosphate. Alcheringa, 9, 245-257.

RUNNEGAR, B., 1985b. Early Cambrian endolithic algae. Alcheringa, 9, 179-182.

RUNNEGAR, B., 1986. Molecular palaeontology. Palaeontology 29, 1-24.

RUNNEGAR, B., in prep. Crystallography and biomineralization of the dodecahedral coccosphere of *Braarudosphaera bigelowi* (Gran and Braarud).

RUNNEGAR, B. & BENTLEY, C., 1983. Anatomy, ecology and affinities of the Australian Early Cambrian bivalve *Pojetaia runnegari* Jell. Journal of Paleontology, 57, 73-92

SANDBERG, P.A., 1983. An oscillating trend in Phanerozoic non-skeletal carbonate mineralogy. Nature, 305, 19-22.

SANDERS, J.V., 1964. Colours of precious opal. Nature, 204, 1151-1153.

SASS, D.B. & MONROE, E.A., 1967. Shell-growth in Recent terebratuloid Brachiopoda. Palaeontology, 10, 298-306.

SCHOEN, A.H., 1970. Infinite periodic minimal surfaces without self-intersections. NASA Technical Note, D-5541, 1-92.

SEILACHER, A., 1979. Constructional morphology of sand dollars. Paleobiology, 5, 191-221.

SEPKOSKI, J.J., 1978. A kinetic model of Phanerozoic taxonomic diversity. I. Analysis of marine orders. Paleobiology, 4, 223-251.

SIMPSON, T.L. & Volcani, B.E. (eds.), 1981. Silicon and Siliceous Structures in Biological Systems. New York: Springer-Verlag.

SMITH, A.B., 1980. Stereom microstructure of the echinoid test. Papers in Palaeontology, 25, 1-81.

STANLEY, S.M., 1976. Fossil data and the Precambrian-Cambrian evolutionary transition. American Journal of Science, 276, 56-76.

TAYLOR, J.D., 1973. The structural evolution of the bivalve shell. Palaeontology, 15, 73-87.

TOWE, K.M., 1967. Echinoderm calcite: single crystal or polycrystalline aggregate? Science, 157, 1048-1050.

TOWE, K.M., 1970. Oxygen-collagen priority and the early metazoan fossil record. Proceedings of the National Academy of Science, United States of America, 65, 781-788.

TOWE, K.M., BERTHOLD, W. & APPLEMAN, D.E., 1977. The crystallography of *Patellina corrugata* Williamson: a-axis preferred orientation. Journal of Foraminiferal Research, 7, 58-61.

TOWE, K.M. & THOMPSON, G.R., 1972. The structure of some bivalve shell carbonates prepared by ion-beam thinning. Calcified Tissue Research, 10, 38-48.

WARD, P., CARLSON, B., WEEKLY, M. & BRUMBAUGH, B., 1984. Remote telemetry of daily vertical and horizontal movement of *Nautilus* in Palau. Nature, 309, 248-250.

WARD, P.D. & CHAMBERLAIN, J., 1983. Radiographic observation of chamber formation in *Nautilus pompilius*. Nature, 304, 57-59.

WADA, K., 1972. Nucleation and growth of aragonite crystals in the nacre of some bivalve molluscs. Biomineralization, 6, 141-159.

WEINER, S. & TRAUB, W., 1980. X-ray diffraction study of the insoluble organic matrix of mollusk shells. Federation of European Biochemical Societies Letters, 111, 311-316.

WEINER, S. & TRAUB, W., 1984. Macromolecules in mollusc shells and their functions in biomineralization. Philosophical Transactions of the Royal Society of London, B304, 425-434.

WHITE, R.D., 1986. Cambrian Radiolaria from Utah. Journal of Paleontology, 60, 778-779.

WILBUR, K.M. & BERNHARDT, A.M., 1984. Effects of amino acids, magnesium, and molluscan extrapallial fluid on crystallization of calcium carbonate: *in vitro* experiments. Biol. Bull., Mar. Biol. Lab., Woods Hole, Massachusetts, 166:251-259.

WILBUR, K.M. & SALEUDDIN, A.S., 1983. Shell formation. *In* The Mollusca, vol. 4 (ed. A.S.M. Saleuddin & K.M. Wilbur), Academic Press, New York: 235-287.

WILKES, D.A. & CRENSHAW, M.A., 1979. Formation of a dissolution layer in molluscan shells. Scan. Elec. Micros., 2:469- 474.

WILKINSON, B.H., 1979. Biomineralization, paleoceanography and the evolution of calcareous marine organisms. Geology, 7, 524-527.

WILLIAMS, A., 1968. Shell structure of the billingsellacean brachiopods. Palaeontology, 11, 486-490.

WRAY, J.L. & DANIELS, F., 1957. Precipitation of calcite and aragonite. Journal of the American Chemical Society, 79, 2031-2034.

EVOLUTION OF BIOMINERALIZATION SYSTEMS WITHIN THE CEPHALOPODA

Rex E. Crick

Department of Geology, The University of Texas at Arlington, Arlington, TX 76019

ABSTRACT. Chemical data from unaltered nautiloid and ammonoid shells for a period of 300 million years serve to document several aspects of change in physiochemical systems of these groups: 1) ammonoids were more efficient than coeval coiled nautiloids in the precipitation of aragonite, 2) the Sr and Mg chemistry of early ammonoid skeletons was much difference than later descendents, and 3) the shell chemistry of early nautiloids is not markedly different from their modern descendent, *Nautilus*.

INTRODUCTION

The evolution of the physiochemical system of the Cephalopoda is considered by observing trends within and among various groups, some related and some not. The purpose of this study is to determine if the chemistry of unaltered skeletal carbonate of fossil cephalopods exhibits trends suggestive of changes toward more efficient system of carbonate production or toward a better form of structural material.

Within the Mollusca, and perhaps among all marine invertebrates, the extinct subclass Ammonoidea and extinct orders of Nautiloidea possessed the most sophisticated physiochemical systems with respect to carbonate production. There is little doubt that the earliest Cambrian representatives of the Nautiloidea used aragonite for shell material (Grandjean *et al.*, 1964) while coeval mollusks and other groups chose calcium phosphate, calcite, or silica. The greater ductile nature of layered and laminated aragonite was selected for early in cephalopod evolution in an effort to provide greater mobility with no loss in support or protection. The choice of aragonite was clearly advantageous since this shell material was the sole skeletal material for 500 million years of cephalopod evolution and continues today with *Nautilus*, the only surviving ectocochleate cephalopod.

A review of the evolutionary patterns within the Cephalopoda indicates that the need to find the most efficient solution to the problem of regulating the buoyancy of a chambered cone containing body tissue and gas filled chambers was the driving force in the macroevoluton of the group (Crick, 1988). Since all successful solutions to this problem required the efficient precipitation of calcium carbonate as a counterweight to buoyancy (most nautiloids) or as a coiled cone (some nautiloids and most ammonoids),

Origin, Evolution, and Modern Aspects of Biomineralization in Plants and Animals
Edited by Rex E. Crick
Plenum Press, New York

physiochemical systems utilized for carbonate production would have also been subjected to pressures of selection.

MATERIAL AND METHODS

The fossil and recent material which provided samples for this work came from personal field collections, collections of colleagues and museum collections. Specimens were used only when identification was possible to the generic level, and then only those that could provide a minimum of 5 samples for analysis. Because unaltered skeletal material is relatively rare there was no opportunity for unbiased geographic representation. For the same reason, the age representation is also uneven, with the Carboniferous providing more nautiloid genera and the Cretaceous providing more ammonoid genera. No unaltered specimens of Triassic age were available. With this exception, the full range of the ammonoids is represented from their first appearance in the Devonian through to their disappearance in the Late Cretaceous. Unaltered specimens for the early portion of nautiloid history, Cambrian, Ordovician, Silurian ages, were not available. Unaltered specimens were available for Devonian to Permian and Jurassic to Recent.

The procedure of analysis was the same for all samples and consisted of removing foreign material from shell material via air-abrasive, powdering the sample to less than 100 μm, and using atomic absorption spectrophotometry to determine the concentrations of Mg, Sr and Mn. The relative concentrations of Mg and Sr were used to access the changes over geologic time in physiochemical systems, while the concentration of Mn was used as an indication of the presence of diagenesis. The rationale for the use of Mn as an indicator of diagenesis is explained in Brand and Veizer (1980) and the technique modified for use with skeletal carbonate by Crick and Ottensman (1983). In general, a concentration greater than 100 μm of Mn is indicative of diagenesis sufficient to alter the concentrations of trace and minor elements such as Mg and Sr.

BIOMINERALIZATION IN CEPHALOPODS

The following is restricted to ectocochleate cephalopods and not meant for the Class Cephalopoda in general. Because we can only draw inference about the details of biomineralization in extinct groups, the details of biomineralization in the only living ectocochleate cephalopod, *Nautilus*, is used to provide information about these extinct groups. Use of these data is warranted because biomineralization of the type used to construct the skeleton is believed to be a very basic and conservative physiochemical function.

Like those of other mollusks, the physiochemical system of cephalopods governs the precipitation of calcium carbonate (Wilbur, 1976; Crenshaw, 1982). The physiochemical system of *Nautilus* and perhaps all ectocochleate cephalopods is or was possibly the most complex among marine invertebrates, because it involves the production of several minerals: 1) precipitation of aragonite from an extracellular fluid to form the shell; 2) precipitation of aragonite from another extracellular fluid to form the septum; 3) precipitation of high-Mg calcite by buccal papillae to form the beaks of the masticatory apparatus (Crick *et al.*, 1985); and 4) precipitation of statoconia, comprised of aragonite and amorphous calcium phosphate, in the statocyst. Although the shell and septum are aragonite, they are deposited by different portions of the mantle epithelium under different conditions and at different rates. The ultrastructure and elemental content of the shell and septum also differ.

To add to the complexity, the shell consists of three distinct layers of aragonite: the outer prismatic layer, the nacreous layer, and the inner prismatic layer. Of the two prismatic layers, the outer layer is the thinnest. The nacreous layer forms the largest volume of shell aragonite with the nacreous/prismatic layer thickness ratio being 3:1 along the ventral portion of the shell.

The balance in timing maintained among the various physiochemical activities is an impressive aspect of the cephalopod animal. Continuous precipitation of the shell occurs at a rate that may vary during ontogeny (Saunders, 1983). Septal aragonite is produced episodically in response to growth with the subsequent movement of the animal forward in the living chamber. Uroliths are produced in the renal appendages and their appearance may correspond to changes in the physiochemical system as it prepares to produce fluid form which septal aragonite will be precipitated (Crick *et al.*, 1985).

Additional details of cephalopod biomineralization can be found in Crick and Mann (1987).

TRACE ELEMENTS AND BIOMINERALIZATION

Factors that affect elemental concentrations in skeletal carbonate can be grouped into three general categories: 1) crystallographic controls, 2) environmental controls, and 3) physiological controls. Categories 1 and 2 are not viable concerns with cephalopods and the rationale for rejection of these categories is explained in Crick *et al.* (1987) and Crick and Mann (1987). The category of physiological controls is responsible for the minor and trace element makeup of cephalopod skeletons.

The physiological system of mollusks is controlled genetically (Wilbur, 1972); differences in the ionic composition of extracellular fluid and the elemental concentration of shell carbonate reflect genetic differences among physiochemical systems. Physiological aspects of cephalopods involve both the extrapallial fluid, in which shell aragonite was produced, and the transfer of materials from this fluid to the site of crystal formation (Wilbur, 1972, 1976). Ions that form the aragonite crystals of the shell and septum of cephalopods pass from the mantle cavity, containing the external medium and food, to the tissue and blood sinus, and finally into one or more extrapallial spaces, located between the growing surface of the shell and the mantle epithelium. Skeletal carbonate will contain concentrations of elements in proportion to their concentrations in the extracellular fluid. The compositional regulation of the extracellular fluids by the mantle cells is genetically controlled and is expected to vary with species (Wilbur & Owen, 1964). Such differences have been observed among extant bivalve genera (Wilbur, 1972) and among extinct congeneric species of nautiloid cephalopods (Crick & Ottensman, 1983). Thus long-term shifts in the way that physiochemical systems produced aragonite may be seen in the relative concentrations of the major trace elements of Sr and Mg.

ANALYTICAL RESULTS

The results of the analyses are listed in Table 1 along with the taxonomic assignment, geologic age and geography for each of the genera. Figure 1 compares the Mg and Sr contents of ammonoid shells with geologic age. Both Mg and Sr show real trends with time. Sr increased by more than an order of magnitude from the Late Devonian to the Late Cretaceous and Mg decreased by about 75% over the same time period, however much of this decrease occurred between the Late Devonian and Early Carboniferous. While seemingly anomalous, the Mg concentrations were consistent with small standard error over 9 specimens, 5 for *Sandbergeroceras* and 4 for *Manticoceras*.

Figure 2 illustrates the concentrations of Mg and Sr in skeletal carbonate of Devonian to Permian and Jurassic to Recent nautiloid cephalopods. Specimens were not available for Cambrian, Ordovician, Silurian and Triassic periods. Magnesium content decreased slightly with time while Sr increased by approximately 60%.

The two sets of data are compared in Figure 3. Neither set of data extends to the beginning or first appearance of the group. Although the ammonoids first appeared in the Devonian, the chemical data begins with Late Devonian specimens, a difference of some 40 million years. Approximately 140 million years are missing from the earliest portion of the nautiloid record. Figure 3 shows that the Sr content was higher in ammonoid skeletons than in nautiloid skeletons, and that during the Jurassic and Cretaceous the Sr content of ammonoid and nautiloid shells diverged significantly.

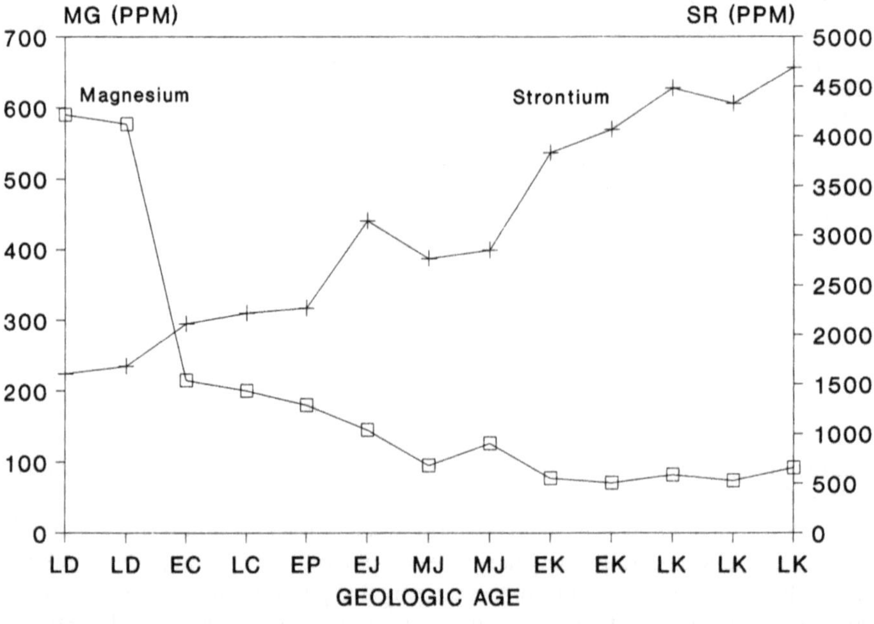

FIGURE 1. Sr and Mg contents of ammonoid cephalopod shells as a function of time. Age abbreviations are D, Devonian; C, Carboniferous; P, Permian; J, Jurassic; K, Cretaceous; L, Late; M, Middle; E, Early.

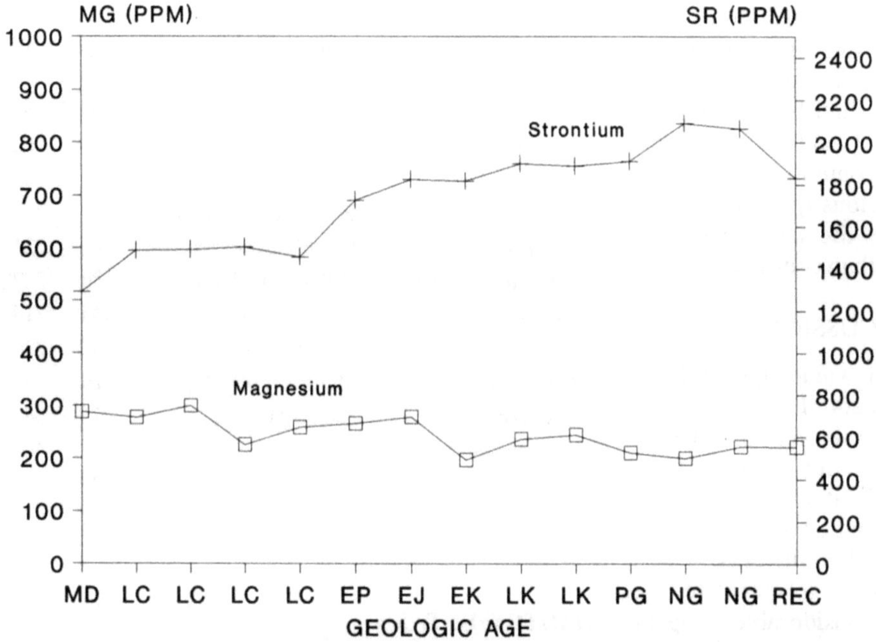

FIGURE 2. Sr and Mg contents of nautiloid cephalopod shells as a function of time. Age abbreviations are: D, Devonian; C, Carboniferous; P, Permian; J, Jurassic; K, Cretaceous; PL, Paleogene; NG, Neogene; REC, Recent; L, Late; M, Middle; E, Early.

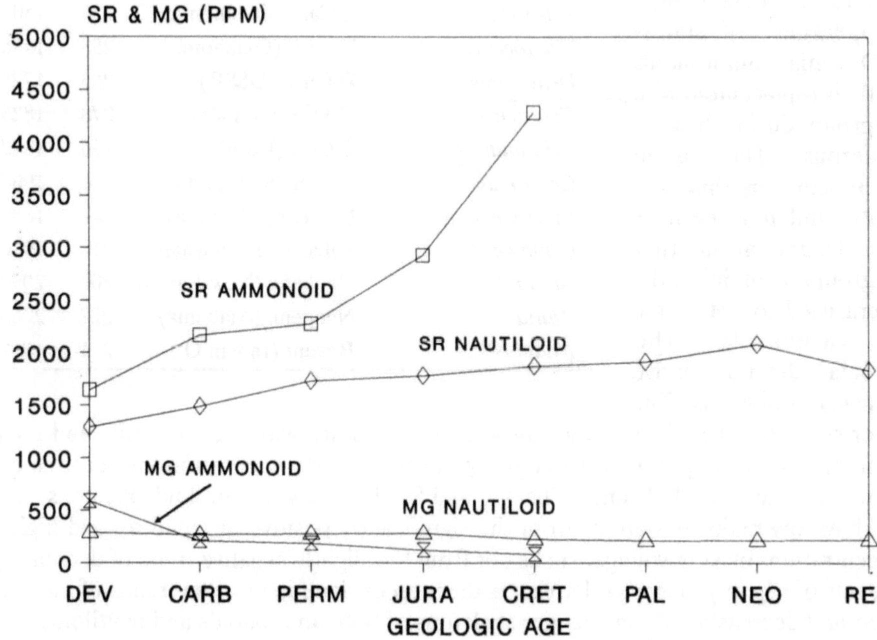

FIGURE 3. Comparison of the data of Figs. 1 & 2 at the same scale. SR, strontium; MG, magnesium. Age abbreviations are: Dev, Devonian; Carb, Carboniferous; Perm, Permian; Jura, Jurassic; Cret, Cretaceous; Pal, Paleogene; Neo, Neogene; Rec, Recent.

Figure 3 also illustrates that the Mg content of early ammonoid skeletons was slightly higher than coeval nautiloids, but that Mg content of ammonoid skeletons quickly decreased to the same or slightly less than coeval nautiloids.

TABLE 1. Chemical data used to construct Figs. 1-3. Age designations are those of Figs. 1-3. Concentrations of Mg and Sr in parts per million

AMMONOID TAXA	AGE (GEOG)	MG	SR
Sandbergeroceras	L.Dev (New York)	590	1600
Manticoceras	L.Dev (New York)	577	1680
Girtyoceras	E.Carb (Oklahoma)	215	2110
Gastrioceras	L.Carb (Arkansas)	200	2218
Neoshumardites	E.Perm (Oklahoma)	180	2270
Eleganticeras	E.Jura (Germany)	345	3148
Cadoceras	M.Jura (Poland)	95	2768
Quenstedtoceras	M.Jura (Poland)	126	2850
Aconeceras	E.Cret (USSR)	77	3834
Deshayesites	E.Cret (USSR)	71	4069
Mortoniceras	L.Cret (Texas)	82	4485
Barroisiceras	L.Cret (Texas)	74	4329
Sphenodiscus	L.Cret (Mexico)	92	4689
NAUTILOID TAXA			
Archiacoceras	M.Dev (Germany)	287	1290
Mitorthoceras	L.Carb (Oklahoma)	276	1487
Brachycycloceras	L.Carb (Oklahoma)	299	1491
Solenochilus	L.Carb (Oklahoma)	225	1503
Titanoceras	L.Carb (Oklahoma)	258	1456
Dentoceras	E.Perm (USSR)	265	1725
Cenoceras	E.Jura (Germany)	278	1825
Strionautilus	E.Cret (USSR)	196	1817
Cenoceras	L.Cret (So.Dakota)	236	1900
Eutrephoceras	L.Cret (So.Dakota)	244	1888
Cymatoceras	Paleocene (Louisiana)	210	1910
Aturia	Neogene (New Jersey)	200	2090
Aturia	Neogene (Alabama)	222	2065
Nautilus	Recent (Indian O.)	221	1831

DISCUSSION

It would appear from the data plotted in Figure 1, that the skeleton of early ammonoids contained approximately two times the Mg and one-half the Sr of coeval nautiloids. It is tempting to speculate that the considerable drop in Mg concentration in ammonoid skeletons at the end of the Devonian is related to the documented disappearance of almost all Devonian ammonoids and their replacement with new groups during the Carboniferous. There is no way to confirm this possibility, and it is perhaps only circumstantial that the groups containing the genera used to obtain the Mg values for the Devonian did not survive to the Carboniferous. The rather small perturbations in the Mg and Sr curves are within error limits and most probably represent specimens from groups with Sr or Mg concentrations near the extremes for that period of time. The Mg and Sr chemistry of nautiloid skeletons does not show any major deviations from the significantly positive Sr trend toward higher concentrations of Sr in younger shells nor from the slightly negative trend of decreasing amounts of Mg in younger skeletons. Is there an explanation for the trends of increasing Sr and decreasing Mg in younger skeletons of both ammonoids and nautiloids?

Crick and Mann (1987) and Crick *et al.* (1987) offered an explanation based upon the kinetics of crystal chemistry and structure. The major inorganic constituents of the cephalopod shell, in order of relative abundance, are Ca, Sr and Mg, with ionic radii of 1.0, 1.13 and 0.72, respectively. Although aragonite is polymorphus with calcite, the aragonite structure is stable for ions larger than Ca (>1.0 radius), whereas the calcitic structure is stable for ions smaller than Ca (<1.0 radius). This stability factor for aragonite is conferred by the aragonitic structure, which consists of planar trigonal CO_3 groups with Ca ions in positions of hexagonal close packing; this situation gives aragonite its pseudo-hexagonal character. It is important to note that each CO_3 group lies between six Ca atoms and is arranged such that each O is linked to three Ca ions. This produces a radius ration (radius cation:radius anion) of Ca:O of approximately 0.71, which is intermediate between the coordination numbers 6 and 8. Similar radius ratios of Sr and Mg to O are 0.81 and 0.51, with corresponding coordination numbers of 8 and 6 (Mason & Moore, 1982). Thus, the Sr ion substitutes for the Ca ion in the aragonite structure more easily than the Mg ion. Further, the inclusion of Mg in the crystal lattice at Ca sites distorts the orthorhomic structure as the lattice "adjusts" to the smaller Mg ion; consequently, high-Mg aragonite will be more unstable and less mechanically viable than low-Mg aragonite.

An important consideration in this reasoning is a comparison between the relative ratios of Ca, Mg and Sr in world oceans (1.0:3.2:0.02) and the ratios of average Ca, Mg and Sr concentrations for *Nautilus* (1.99:0.0003:0.002). These ratios are the best of all extinct or extant nautiloid cephalopods. The body fluids of mollusks, and presumably extinct cephalopods, are or were essentially isosmotic with sea water, and for *Nautilus* as well as most extinct nautiloid cephalopods to produce aragonite with these or similar ratios, quantities of Mg on the order of 100 times the amount of Ca required for aragonite production are or would have had to be excluded from the composition of extrapallial fluids. Why cephalopods discriminate against Sr is unknown, although the larger ionic diameter of Sr would tend to slightly distort the aragonitic lattice and make Sr less desirable as a skeletal component. Excluding as much Mg as possible, and using Sr when necessary, provided the most efficient means of producing a skeleton composed predominantly of nacre (pseudo-hexagonal crystals). Organisms of lower evolutionary grade and skeletons of aragonite do not discriminate against Sr relative to Ca, e.g., taxa as diverse as planktonic foraminifera (Graham *et al.*, 1982) and coelenterates (Amiel *et al.*, 1973), and commonly have a Sr concentration of an order of magnitude greater than cephalopods. The combination of the near exclusion of Mg and the increase in Sr content in cephalopod skeletons with time is seen as evidence of the evolution of the physiochemical system toward the efficient production of a more stable and perhaps more mechanically strong skeleton.

Why ammonoids incorporated even more Sr and slightly less Mg in their skeletons from the Carboniferous through to the end of the Cretaceous is perhaps related to their different style of shell construction. For the most part, ammonoid shells tend to have thinner shell walls and septa than coeval nautiloids. This was particularly pronounced during their final 140 million years of evolution (Jurassic and Cretaceous). Did the combination of Sr rich aragonite and organic matrix produce a shell wall light enough for enhanced mobility yet strong enough to withstand several bars of pressure? Ammonoid shells generally have a higher ratio of nacre aragonite to prismatic aragonite than coeval nautiloids which suggests that the production of larger volumes of nacre may have been their goal. A shell designed for greater mobility may indicate that ammonoids evolved away from the scavenger mode of nautiloids toward more active predators. The higher concentration of Sr and lower concentration of Mg in ammonoid shells relative to coeval nautiloids may reflect the need for a more efficient or perhaps a faster means of growing the shell. For whatever reasons, the ammonoid shell was lighter than the coeval nautiloid shell and the trend toward increasing Sr content in ammonoid shells was continuing when the group disappeared.

REFERENCES

AMIEL, A.J., FRIEDMAN, G.M., & MILLER, D.S., 1973. Distribution and nature of incorporation of trace elements in modern aragonitic corals. Sedimentology, 20:47-64.

BRAND, E., & VEIZER, J., 1980, Chemical diagenesis of a multicomponent carbonate system, 1. Trace elements. Journal Sedimentology, 50:1219-1236.

CRENSHAW, M.A., 1982. Mechanisms of normal biological mineralization of calcium carbonate. *In* Biological Mineralization and Demineralization (ed., G.H. Nancollas), Springer-Verlag, Berlin:243-257.

CRICK, R.E., 1988. Buoyancy regulation and macroevolution in nautiloid cephalopods. Senckenbergiana Lethaea, 69:13-42.

CRICK, R.E., & MANN, K.O., 1987. Biomineralization and systematic implications. *In Nautilus*, The Biology and Paleobiology of a Living Fossil (eds., W.B. Saunders, & N.H. Landman), Plenum Press, New York:115-134.

CRICK, R.E., & OTTENSMAN, V.M., 1983. Sr, Mg, Ca and Mn chemistry of skeletal components of a Pennsylvanian and Recent nautiloid. Chemical Geology, 39:147-163.

CRICK, R.E., MANN, K.O., & CHAMBERLAIN, J.A., JR., 1987. Skeletal chemistry of *Nautilus* and its taxonomic significance. Biochemical Systematics and Ecology, 15:461-474.

CRICK, R.E., BURKART, B., CHAMBERLAIN, J.A., JR., & MANN, K.O., 1985. Chemistry of calcified portions of *Nautilus pompilius*. Journal of the Marine Biological Association, United Kingdom, 65:415-420.

GRAHAM, D.W., BENDER, M.L., WILLIAMS, D.F., & KEIGWIN, L.D., JR., 1982. Strontium-calcium ratios in Cenozoic planktonic foraminifera. Geochimica et Cosmochimica Acta, 46:1281-1292.

GRANDJEAN, J., GRÉGOIRE, C. and LUTTS, A., 1964. On the mineral components and the remnants of organic structures in shells of fossil molluscs. Bull. Acad. R. Belg., Classe Sci., Sér. 5, 50:526-595.

MASON, B., & MOORE, C.B., 1982. Principles of Geochemistry. John Wiley, New York, 344 p.

SAUNDERS, W.B., 1983. Natural rates of growth and longevity of *Nautilus belauensis*. Paleobiology, 9:280-288.

WILBUR, K.M., 1972. Shell formation in molluscs. *In* Chemical Zoology, Vol. VII, Mollusca (eds., M. Florkin, & B.T. Schneer), Academic Press, New York:103-145.

WILBUR, K.M., 1976. Recent studies of invertebrate mineralization. *In* The mechanisms of Mineralization in the Invertebrates and Plants (eds., N. Watabe, & K.M. Wilbur), University of South Carolina Press, Columbia:79-108.

WILBUR, K.M., & OWEN, G., 1964. Growth. *In* Physiology of Mollusca, Vol. 1 (eds., K.M. Wilbur, & C.M.Yonge), Academic Press, New York:211-242.

TIDALLY DEPOSITED BANDS IN SHELLS OF BARNACLES AND MOLLUSCS

Dennis John Crisp

Animal Biology Group, Marine Science Laboratories, Menai Bridge, Gwynedd, United Kingdom

INTRODUCTION

At its normal pH of 8.3, sea water is virtually saturated with calcium carbonate (Revelle & Fleming, 1934; Wattenberg & Timmermann, 1936). Little energy is therefore needed to form skeletons of calcite or aragonite. Almost all of the invertebrate phyla and a number of marine plants have evolved skeletons of calcium carbonate. Vertebrates have acquired calcium phosphate skeletons. However, calcareous skeletons are not always laid down continuously and uniformly, despite the ubiquity of calcium and the relevant anions in the surrounding seawater. Yearly growth interruptions, ascribed to annual temperature changes or other climatic variables, have long been observed and understood (Orton, 1923).

In 1963 Wells first drew attention to the possibility that minor microgrowth interruptions in calcification might represent astronomical events such as days or lunar months. His particular contribution was to suggest that the minute ridges in corals were due to daily cycles of deposition. Hence, were it possible accurately to count the number of minor ridges between major annulations, the changes in the earth's rotation might be estimated.

Paleontologists were not slow to recognize similar evidence of microgrowth interruptions in other fossils, notably in bivalve shells and stromatolite deposits (Rosenberg & Runcorn, 1975). These marks within the substance of the fossil could be seen in polished, etched sections from which "peels" were prepared and viewed microscopically. In this paper I shall refer to the regular, darker, and usually thinner moieties as "bands". These are separated by more transparent and usually thicker intermediate layers, here called "increments".

It soon became usual to call these microgrowth bands "daily bands", though with little evidence in support of the hypothesis that they had been deposited every day. Perhaps the term "daily bands" rolled off the tongue more readily than the more correct, but cacophonous, description "microgrowth bands"! Pannella *et al.* (1968) recognized this as an unproven assumption which they countered by the argument: "There is a large enough body of information on growth bandings in many taxons to justify the conclusion

Origin, Evolution, and Modern Aspects of Biomineralization in Plants and Animals
Edited by Rex E. Crick
Plenum Press, New York

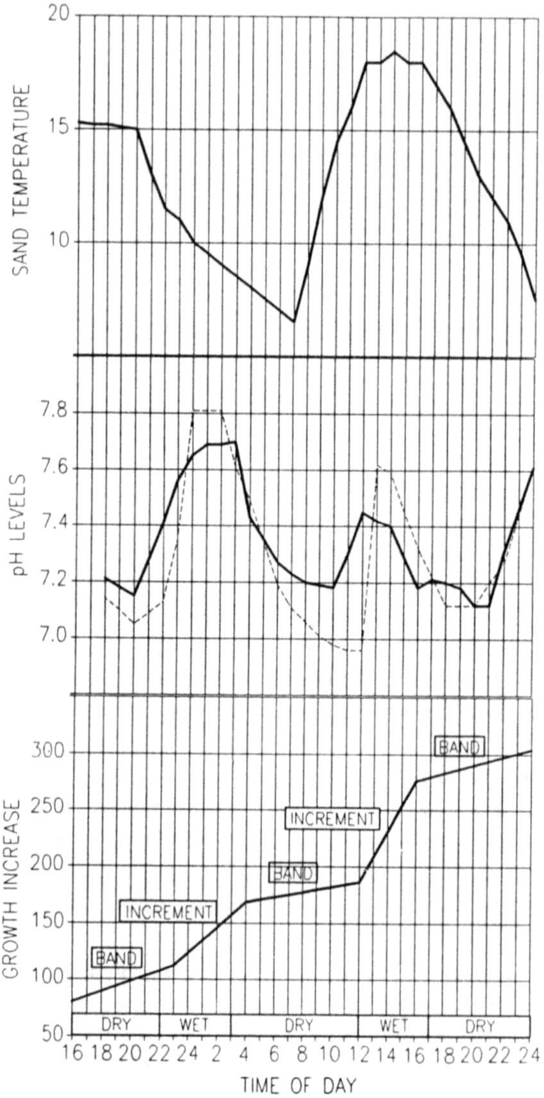

FIGURE 1. Changes observed over a 24 h period in a population of *Cerastoderma edule* on May 21-22, 1977. Samples were taken from just above mean tide level on a sandbank near the western entrance to the Menai Strait (Abermenai Point). *Upper Curve*: Sand temperature °C. *Middle Curves*: pH of mantle fluid (broken line) and of extra pallial fluid (continuous line). *Lower curve*: Percentage increase in growth at shell edge. Note that the fall in pH of body fluids when the bivalve is emersed is greater at daytime emersion than at night. The growth rate is greater during daytime immersion. The shell continues to grow slowly during emersion, but accelerates shortly after re-immersion. (Redrawn from Richardson *et al.*, 1981)

that daily growth increments are a widespread feature in the organic world." Nevertheless it was obvious that experimental support from living organisms was needed.

When are bands and increments formed?

Much evidence on the diel nature of shell banding rests on the belief that bands form at certain times in the 24 h cycle. MacClintock & Pannella (1969) did a short term experiment on *Mercenaria mercenaria* to answer this question. They concluded that the addition to the shell took place during the night, and therefore deposition of shell increments occurred every 24 hours. Whyte (1975) made a similar investigation, comparing the appearance under the electron microscope of shell increments. From the "lineated texture" of the surface membrane and the often porous structure of the underlying cross-lamellar shell laid down at night, he also concluded that calcification occurred at night. Moreover his measurements of the pH of the extrapallial fluid (between the shell and mantle) gave higher values at night which would have aided calcification. However, though he noted "interruptions" in the daily increments, which he ascribed rightly to tidal emersion, he did not demur from the views of his predecessors MacClintock & Pannella (1969). Pannella (1975) regarded such tidal effects as a minor influence on the daily rhythm: "Without the effect of tidal inundations the bivalves would deposit calcium carbonate only once a day following environmental temperature, light and food supply fluctuations which quite generally have a 24 h frequence."

A more detailed series of experiments on a population of cockles, *Cerastoderma edule*, growing near mean tide level on a sandbank, were described by Richardson *et al.* (1981). Measurements of growth, and of the pH of mantle and extrapallial fluid were made simultaneously, together with environmental temperature and tidal inundation records over a 24 h period. The results, shown in Figure 1, demonstrate two periods of falling pH accompanied by slower growth and coinciding with emersion. Growth during the warmer daylight period exceeded that during the night. In a longer series of observation during periods of good weather the width of the bands was observed to alternate; alternations occurring fortnightly a little before the maximum of spring tides.

These patterns of alternating strong and weak bands were observed also by Dolman (1975) using a photocell probe but, though he realized they were due to tidal emersions, he did not interpret them as purely tidal effects. He thought, probably incorrectly, that bands formed at night were more distinct than those formed during the day. The fortnightly switch to uniformity of banding whenever low tides occurred at twilight was almost certainly correct. He referred to the pattern as "Diurnal tidal" growth lines. More importantly, he applied Fourier analysis to obtain power spectra with periods of semi-annual, monthly and lunar fortnightly periodicity just as one would expect for purely tidal influences. Dolman's results for *Cardium edule* (1975, Fig. 1) show very clear tidal periodicity.

Fortnightly periodicities

Pannella (1975) provides data from living and fossil shells of an interaction between a solar daily and lunar tidal periodicity which results in a regularly changing pattern of fortnightly periods. Pannella & MacClintock (1968) had carried out experiments on the shells of samples of *Mercenaria mercenaria* and *Tridacna squamosa* growing at low water. These experiments revealed "tidal modulations" of lunar periodicity in addition to the supposed "daily lines". An explanation of these fortnightly periodicities was given by Richardson *et al.* (1981). In the early afternoon emersion period, when air temperature rose to 14–18° C and the animals are thereby more severely stressed, broader bands were formed, whereas at night when low water temperatures fell to 12–13° C, narrower bands were formed. During neap tide lows, which at this site occur in early morning and late evening, conditions are similar for the formation of both sets of tidal bands, hence differences in width are then small and irregular (Fig. 2).

These fortnightly alternations in the shell pattern are important geophysically. Whichever of the interpretations are put upon them, they represent a rhythm of lunar periodicity derived from the moon's orbital motion and therefore of use in geochronology in whichever fossils they occur.

Are the bands daily or tidal?

The initial assumption that there was a band formed at daily intervals appeared at first to be borne out by experiments on living bivalves (MacClintock & Pannella, 1969)

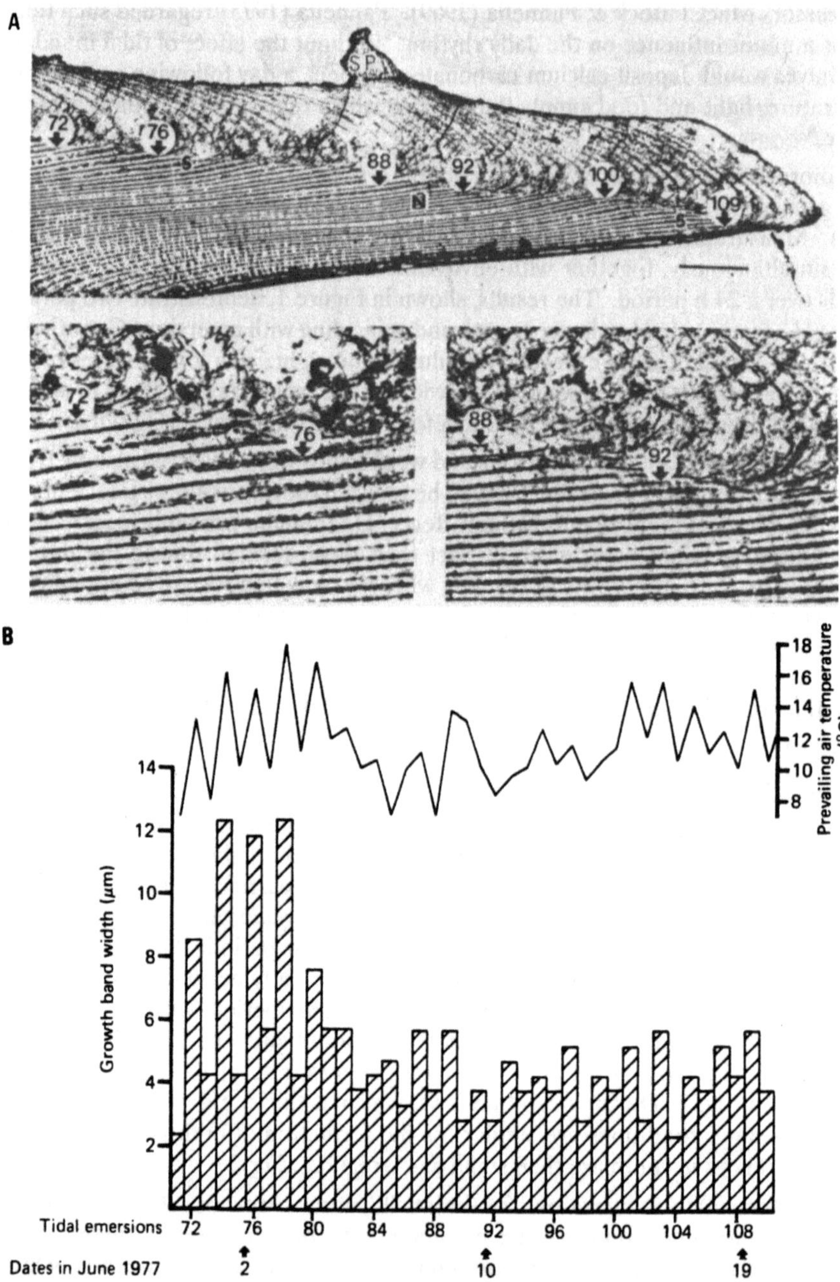

FIGURE 2. Upper figure.- Peel of *Cerastoderman edule* growing at Abermenai Point showing alternating bands during the spring tide period, **S**, (around band No. 76) and non-alternating bands during the neaps period, **N**, (around band 92), **S.P.**, Spine. Bar = 100 μm. *Lower figure.* - Measurement of band widths showing their correspondence with diurnal alterations of temperature. The alternating wider bands (strictly increments) are formed during the heat of the afternoon. (From Richardson *et al.*, 1981)

though at least some "bidaily" patterns were admitted. This belief in daily bands was so strongly held that even though Pannella (1975) correctly recognized tidal lines ("organic lines") as having been formed by closing the shell and withdrawing the mantle during air exposure, he remained convinced of an underlying diel calcification rhythm.

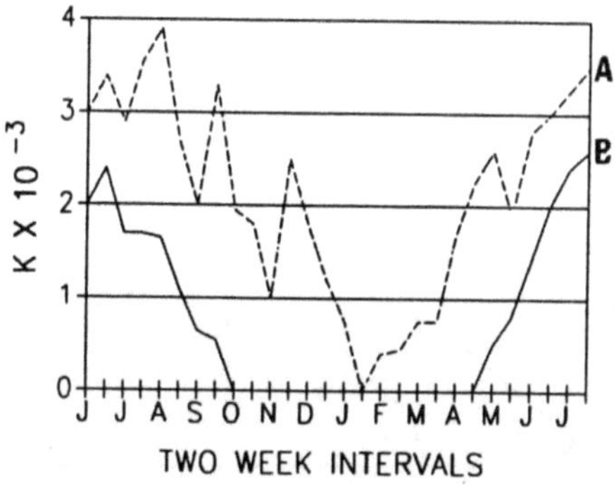

FIGURE 3. Rates of growth of *Cerastomderma edule* measured over a full year from tidal band distances. Averaged values for 14 day sample were converted to the Bertalanffy exponent *K*, which corrects for the influence of size. ▪A. (broken line) From Menai Strait, northern Wales with mild winters. Tidal bands formed almost continuously. ▪B. (full line) From Sobotn, Ramfjord northern Norway (approx. 50 km south of Tromsö) with severe winters. Note approximately 6 months of negligible growth without recognizable tidal increments. (Redrawn from Richardson *et al.*, (1980a).

The issue had been confused by House & Farrow's (1968) and Farrow's (1971) reports of some 360 microgrowth bands lying between the annual growth checks. Cockles from Southend on Sea, southeast England were used by Farrow (1971). They concluded that the calcification cycle was daily. However, we know now that visible bands are laid down regularly about twice a day but only so long as the animal is growing at a satisfactory rate. During late autumn and winter, as growth slows down, the bands become thinner and eventually become indistinguishable within the winter growth check. The effect of such winter deceleration was reported by Pannella & MacClintock (1968), when shells of *Mercenaria* grown at Woods Hole, Massachusetts, were examined. Richardson *et al.* (1980a) showed that few bands were lost in winter when *Cerastoderma edule* was grown in the milder climate of North Wales compared with the large number lost in Northern Norway (Fig. 3). Hence, counting a series of bands is not a sure method of estimating periodicity. The sharp winters of the more continental climate of southeast England probably reduced the number of identifiable bands to values close to those observed.

Another source of confusion arose from ridges on the shell of several species of scallop, first studied as fossils (Davenport, 1935). Clarke (1968) announced that *Pecten* diagensis formed fine external ridges once a day when grown in aquaria. However, the agreement between ridges and days elapsed applied only to 25% of specimens! Those that did not fit were dismissed because "environmental stress" commonly causes halts in growth. Other species were later found to produce ridges at a maximum but not an average of one per day (Clarke, 1975). Clearly the results as a whole do not bear out the assertion. Similarly Broom & Mason (1978) concurred with the prevailing orthodoxy, though frankly stating that in *Chlamys* opercularis "one lamella is probably laid down every day, but a lamella is not necessarily laid down every day."

Eventually Gruffydd (1981) demolished the myth of daily growth ridges in pectinids. He showed that these were simply shell decorations belonging to the animals morphology and uninfluenced by astronomical events. There was no evidence of any diurnal rhythm in their onset. Furthermore, at lower ambient temperature ridges formed more

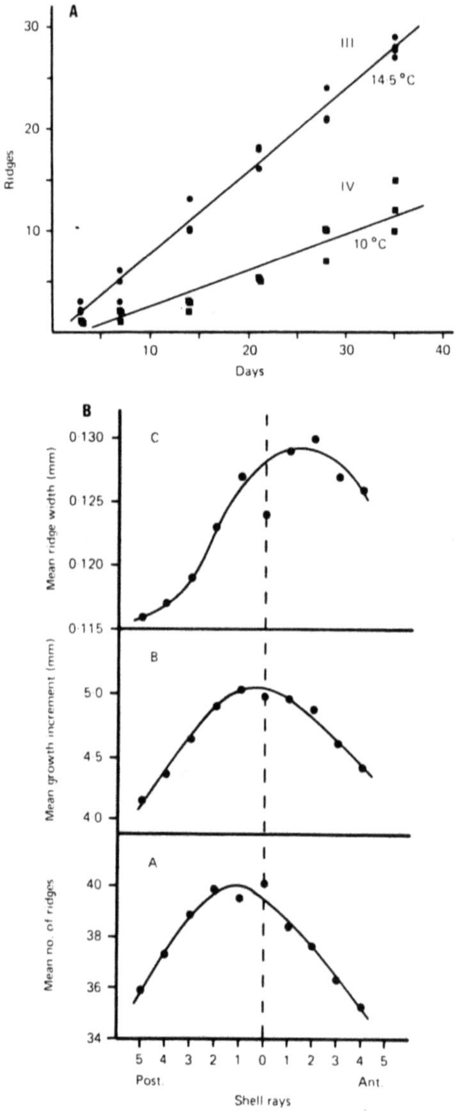

FIGURE 4. Ridges or laminae on *Pecten maximus*. ▪A. The rate of ridge production at 10° and 14.5° C. Note that ridge production is a linear function of time and that the rate rises ×2.6 for a 4.5 deg. C increase, implying a very high Q10 in the region of 6. ▪B. The asymmetric distribution of shell ridges around the median ray (broken line). There are always more ridges posterior to the central ray (A), but growth is nearly symmetrical (B) because the shortage of anterior ridges is compensated by greater width (C). Thus the ridges do not form a continuous series, and this is incompatible with regular periodicity of initiation of growth. (From Gruffydd, 1981)

slowly yet kept their distance of separation virtually unaltered (Fig. 4a). Moreover each ridge was not a continuous boundary with every part generated simultaneously as with a time-dependent growth band. The number of ridges varied with the angle about the median ray (Fig. 4b).

Other examples of suggested daily bands in molluscs have been put forward by Wilbur (1972), Evans & Le Messurier (1972), Koike (1973, 1975), Hall *et al.* (1974), and Kenny (1977). Some of these claims were rejected by Ekaratne & Crisp (1982) on re-examination of the data. Le Gall (1970) claimed, without numerical support, that *Mytilus edulis* had tidal microgrowth bands, but then denied any influence of tidal emersion: "Leur formation ne semble cependent pas être en relation directe avec le phénomène des marées."

It is to Evans (1972, 1975) that we owe the first correct and unequivocal interpretation of microgrowth bands in bivalves. In a delightfully straightforward paper he brought order out of chaos. He demonstrated how the Pacific cockle, *Clinocardium nuttalli*, which lives in a mixed diurnal-semidiurnal tidal regime, produces one microgrowth band on each emersion, independently of the solar daylight cycle. He rendered the "daily increment" view unsustainable for this animal at least. The pattern of shell bands exactly replicated the pattern of emersions as shown in Fig. 5, and the agreement needs no statistics to prove it! Figure 5 incidentally, also shows that when the animal was not quite emmersed (*e.g.* from 28 June to 8 July) it formed a series of faint alternating bands indicative of the endogenous

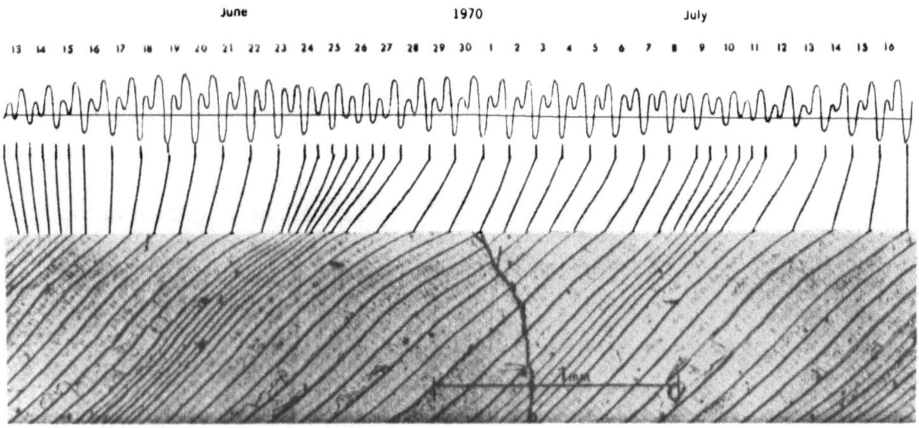

FIGURE 5. Internal growth lines in *Clinocardium nuttalli* collected at Charleston, Oregon, compared with tidal predictions for the same period. The horizontal line drawn through the tidal curves marks the intertidal position (+0.6 m) at which the specimen was found. Lines are formed when the water level falls below the +0.6 m tide level. (Reproduced with permission from Evans, J.W., copyright 1972 by the American Association for the Advancement of Science).

rhythm demonstrated later by Richardson *et al.* (1980b). In 1976 Pannella published almost identical banding patterns showing the coincidence of tidal fluctuations and micro growth bands, but did not refer to Evans (1972) pioneer work.

Entirely independent evidence of a similar phenomenon in barnacles arose from a detailed study of barnacle growth. Bourget & Crisp (1975a) observed microscopically the growth in diameter of a juvenile barnacle settled on the underside of a coverglass and exposed to a simulated tidal cycle. Shell growth was not continuous; indeed the authors showed that when emersed there would be insufficient sea water retained in the mantle space to provide the calcium needed to sustain the high rate of growth that took place when it was immersed. Sections of the shell, and especially of the part known as the "sheath" (Darwin, 1854), were found to have micro-growth bands which corresponded to the periods of emersion separated by increments of shell growth. The thickness of these increments could be increased by allowing the barnacle to remain immersed longer or by increasing the calcium content of the seawater. Conversely more frequent emersions of less than 12 h resulted in more frequent banding and thinner increments separating the bands. Date marking, for example by prolonged feeding in calcium enhanced seawater, allowed the effects of variable tidal emersion and insolation on the appearance of the shell banding to be studied both in *Balanus balanoides* (Bourget & Crisp, 1975a,b) and *Elminius modestus* (Crisp & Richardson, 1975). Bands were occasionally missed if barnacles living at high water level failed to be immersed for a sufficient time or were subjected to prolonged sunlight in summer.

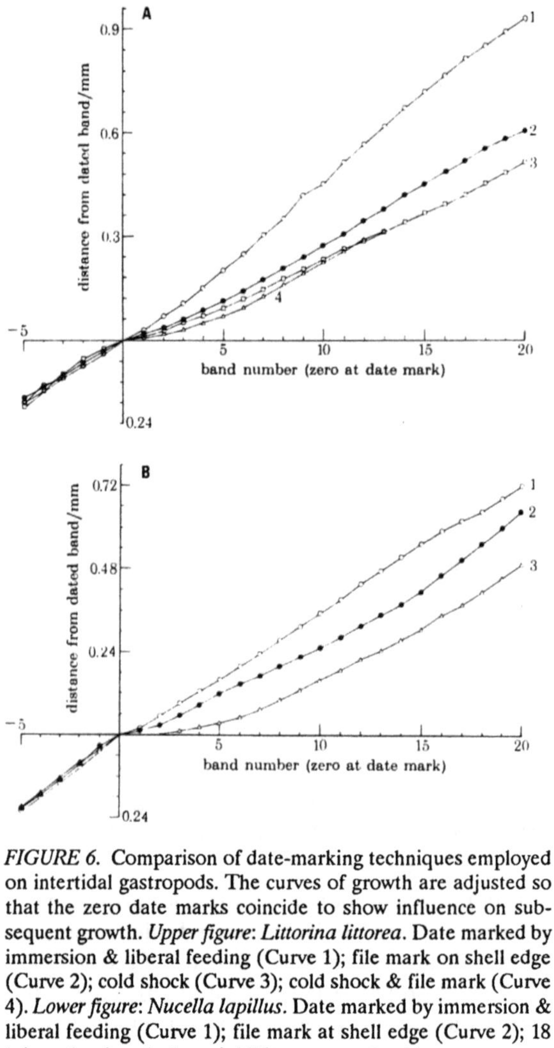

FIGURE 6. Comparison of date-marking techniques employed on intertidal gastropods. The curves of growth are adjusted so that the zero date marks coincide to show influence on subsequent growth. *Upper figure*: *Littorina littorea*. Date marked by immersion & liberal feeding (Curve 1); file mark on shell edge (Curve 2); cold shock (Curve 3); cold shock & file mark (Curve 4). *Lower figure*: *Nucella lapillus*. Date marked by immersion & liberal feeding (Curve 1); file mark at shell edge (Curve 2); 18 h in carbonic anhydrase inhibitor, acetazolamide at concentration 2.25×10^{-4} molar (Curve 3). (From Ekeratne & Crisp 1982)

Gastropods also were found to have microgrowth bands laid down at tidal intervals, each band conforming to the shape of the shell aperture. The shell needed to be sectioned along the plane of maximum growth following the helical line of growth of the shell whorl if clearly visible peels were to be obtained (Antoine & Quemerais-Pencreac'h, 1980, Ekaratne & Crisp, 1982). Polyplacophoran shells also exhibit recognizable periodicities at intervals of approximately 28 bands (Jones & Crisp,M., 1985), suggesting tidal and lunar cycles of deposition.

To determine the periodicity of microgrowth bands, a time scale over which the bands are laid down is essential. The winter growth checks are unsatisfactory because bands within them become indistinguishable. The most satisfactory rationale is to mark the shell at the beginning of an experiment and then carry it out under good conditions of growth in the field. Date-marking techniques include the following: file marks at the growing tip, cold shock at near zero temperature for 2–4 days, suppression of growth by carbonic anhydrase inhibitors, feeding *ad lib* under continuous immersion to produce an exceptionally wide increment, and staining with alizarin or tetracycline. Ekaratne & Crisp (1982) found that, in gastropods, all but the immersion mark method tended sometimes to interfere with shell growth. Acetazolamide treatment was the least desirable (Fig. 6; Table 1). Alizarin dye similarly upsets shell growth. Nevertheless, even with acetazolamide growth recovered in due course and few bands were lost. Hence experiments of 20 tides or more can establish an identity between growth increments and tidal emersions, usually within a few percentage of expectation. Table 1 gives some examples to show the good agreement obtained with three groups of organisms.

TABLE 1. Agreement between number of immersions and number of increments of shells of barnacles, bivalves and gastropods, date marked by the method specified

OBSERVER	SPECIES	METHOD	CONDITIONS	INCREMENTS		
				1	2	3
BARNACLES						
Bourget	*Balanus balanoides*	Immersion	L.W.N.T.L	53	49.0 ± 2.1	0.92
& Crisp (1975b)	" "	"	H.W.N.T.L.	53	49.8 ± 2.2	0.94
	" "	"	Laboratory	45	48.0 ± 2.0	1.04
	" "	"	Simulated semi-diurnal	52	52.0 ± 2.0	1.00
Crisp	*Elminius modestus*	Stress band	M.T.L.	24	23.86	0.99
& Richardson (1975)	" "	Immersion band	M.T.L.	24	23.30	0.97
	" "	Enhanced calcium	M.T.L.	24	23.60	0.98
BIVALVES						
Richardson,	*Cerastoderma edule*	Cold shock	M.T.L.	42	41.7	0.99
et al. (1979)	" "	"	Cont. 12L/12D	61	61.2 ± 12.8	1.003
	" "	"	Immersion 24D	61	57.3 ± 10.3	0.94
			Immersion 24L	43	39.7 ± 5.4	0.92
GASTROPODS		2.25×10^{-4}M Acetazolamide				
Ekaratne	*Patella vulgata*	28 h	M.T.L.	38	37.0 ± 0.45	0.97
& Crisp (1982)	*Nucella lapillus*	28 h	M.T.L.	38	33.4 ± 1.01	0.88
	Littorina littorea	18 h	M.T.L.	38	36.5 ± 0.31	0.96
	" "	28 h	M.T.L.	38	33.4 ± 0.68	0.88
	" "	72 h	M.T.L.	39	32.1 ± 0.84	0.82
	" "	File mark	M.T.L.	69	69.3 ± 0.13	1.004
	" "	Immersion mark	M.T.L.	38	37.9 ± 0.15	0.997
	" "	Cold shock, 4 d	M.T.L.	55	55.1 ± 0.10	1.002

LWNTL: Low water neap tide level
MTL: Mean tide level
HWNTL: High water neap tide level
[1]Number expected (n_e) from immersions
[2]Number observed (n_o) from increments
[3]n_o/n_e

12L/12D: 12 hours light; 12 hours dark
24D: 24 hours dark 0 hours light
24L: 24 hours light 0 hours dark

ENDOGENOUS TIDAL RHYTHMS IN BIVALVES AND BARNACLES

There was good evidence in bivalves of an endogenous circa-tidal rhythm (Richardson *et al.*, 1980b). Cockles growing well up the intertidal zone, having characteristically clear microgrowth bands (*cf.* Rhoads & Pannella, 1970) were transferred to a continuously immersed situation in a live box suspended from a raft. The previously well marked bands ceased to be laid down and a layer of continuous and presumably rapid growth followed (Fig. 7). However after a time faint bands began to reappear. These were formed at intervals comparable with those of the tides (see Table 1) but not in as exact a correspondence as those of animals exposed to the rise and fall of tides on the shore. These endogenous rhythms are therefore "circa-tidal". Normally, to cor-

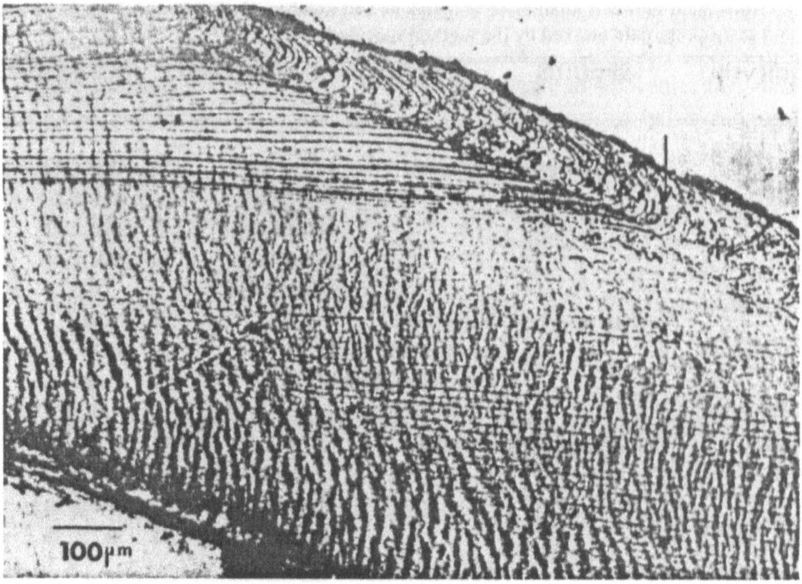

FIGURE 7. Phase contrast micrograph of an acetate peel of a cockle that had been transplanted from the intertidal zone to a continuously submerged live box on a raft and exposed to natural photoperiod. Arrow marks point of transfer. Note the well marked, narrow growth bands and increments laid down intertidally to the left of the arrow, present in **P**, the outer reflected layer (curved bands) and in **C**, the cross-lamella layer (parallel bands). After transfer, the microgrowth bands disappear at first, but weak circatidal banding can be seen to reappear at the lower half of the figure. This is considered to be a result of the endogenous rhythm. Irregular crystallites can be seen running orthogonally to the bands. (From Richardson *et al.*, 1979).

respond strictly to the external tides they must be entrained by actual tidal immersions and emersions. Under carefully controlled laboratory conditions, in which all "Zeitgebers" caused by periodic changes in temperature, light or food supply were eliminated, faint circa-tidal bands were again formed. Bourget & Crisp (1975b) reported similar weak banding in barnacles kept continuously immersed, as evidence of an endogenous tidal rhythm, but it seemed less well developed than in bivalves. In *Venerupis semidecussata* (now changed to *Tapes philippinarum*) which had been given a simulated daily (24 h) tidal emersion, strong bands are laid down only at the time of emersion. However, Richardson (1987) found that these clearly marked bands were sometimes accompanied by weak bands at intermediate 12 h intervals, corresponding to the time of expected low water though unaccompanied by actual emersion. Cockles similarly produce such additional bands when subjected to an artificial 24 h tide. The endogenous rhythm explains why bivalves living between the low water of neap and spring tides or even below the lowest springs continue to form tidal bands, albeit faint ones (*cf.* Evans, 1972).

A noteworthy feature of the many laboratory experiments (Richardson *et al.*, 1979, 1980b; Richardson, 1987) in which tidal emersions and periods of illumination were modified, was the total absence of any influence of light on the banding patterns formed in the shell. This is a very strong argument against the existence of any daily or circadian component to the banding pattern (see Table 1).

ARE THERE CALCAREOUS SKELETONS WITH DAILY PERIODICITY?

Species of the two groups found to display tidal bands, barnacles and molluscs are all from shallow water or the intertidal habitat. I have examined shells of species such as *Nautilus* sp. (*contra* Pannella & MacClintock, 1968), *Balanus hameri* and *Bathylasma corroliforme* which are not intertidal and none of these had clear microgrowth bands. Similarly, not all shallow water or intertidal forms display microgrowth bands by the techniques used for barnacles and bivalves. I have examined several species of calcareous algae, corals, and foraminifera which showed no trace of microgrowth within the skeletons, either of tidal or diurnal periodicity. All the microgrowth bands of proven tidal origin have in common structural features within the calcite or aragonite skeleton, whereas all the surface decorations, spines, ridges or lamellae, as yet investigated adequately, have been shown to be related solely to the animals' morphology and not to be laid down at a predetermined periodicity. The minute ridges on the epitheca of hermatypic corals, to which Wells (1963) first drew to the attention of geophysicists, would be an exception if they were formed at daily or tidal intervals. Since corals depend on photosynthetic symbionts for energy accumulation during the day, it would not be surprising if they formed shell during daylight when the rise in pH would encourage calcification. Barnes (1972) found that the coral lappets took up an actively secreting posture at the growing edge during the day and withdrew at night, suggesting that they formed the epithecal ridges once a day. He also date-marked the epitheca with alizarin dye on the scleractinian coral *Manicina areolata* to show daily growth and counted about 360 epithecal ridges between the larger yearly increments. Nevertheless he doubted their reliability as a geochronological clock. Scrutton (1965, 1970, 1978) found in Devonian corals 27–34 small, supposedly daily, increments grouped into 13 supposedly lunar periods. Buddemerier & Kinzie (1975) reviewed the earlier work on corals and their own X-ray density photographs of *Porites lobata*. They identified annual density changes between which smaller bands were laid down at monthly or bimonthly (probably lunar) intervals, but they expressed doubts about the use of corals as paleoecological clocks. The position regarding corals has still to be resolved.

Stromatolites are an ancient group of organisms that are thought to have been photosynthetic and therefore candidates for a diurnal banding pattern. Modern equivalents are to be found in high saliinity pools in which supersaturation levels of calcium has been reached. Here robust blue green algae raise the pH by day through photosynthesis and lower it by night through respiration. Thus layers of calcium carbonate precipitate formed by day are sandwiched between dark matted algal filaments remaining there at night. In theory therefore the stratifications typical of stromatolites represent daily events. Unfortunately experimental observations of such deposits have reached differing conclusions. Monty (1965), Gebelein & Hoffman (1968) and Hardie & Ginsberg (1971) ascribed the periodicites, respectively, to diurnal, tidal and meteorological changes.

Another category that would not be expected to display tidal rhythms of growth are the pelagic groups, notably cephalopods and fish. These organisms cannot sense the tides directly, but they could appreciate diurnal rhythms from light intensity changes and associated effects such as vertical plankton migrations. Little definite is known regarding cephalopods; Choe (1963) suggested that marks on cuttle shell were diurnal, but reported no experiments. Pannella (1975) figures major and minor annulations in a Devonian cephalopod suggesting 26.5 fortnightly tidal cycles per year.

Fish otoliths have microgrowth bands in addition to the well established annual changes in appearance, namely, Hickling's "concentric lamellae". Since the first reports by Pannella (1971, 1974) the evidence that these finer bands are formed daily has been widely used by fishery scientists, perhaps without sufficient scrutiny (Beamish & McFarlane, 1983). Pannella (1971) studied otoliths from the red hake. From repetitive cycles of bands of 14, 29, 100, 200 and 360 frequencies he concluded that the bands were clearly lunar and annual. However, frequencies were inconsistent and grew smaller with increasing age. Brothers *et al.* (1976) and Brothers (1978) studied fish otoliths from larval stocks, assuming the initial point in time when bands began to form. The results were however compromised by the admission of the existence of "sub-daily bands".

Schmidt & Fabrizio (1980) did a similar experiment by successively sampling a wild population of *Micropteras salmoides* over an interval of 32 days. The range of counts between individuals of different age was very high (*c.* 25 bands) but the average increase in bands was 33.9 $\times$ 1.7 (S.E.). Essig & Cole (1986) obtained good correspondence between daily sampling intervals and growth band counts in cohorts of larvae of *Alosa pseudoharengus* (alewife) but the "two-day" time interval was too short for accuracy.

Date marking with Tetracycline (Schmidt, P. 1984) has been widely used to test fish otolith periodicity. Wild & Forman (1980) recaptured date-marked tuna after a time interval in the wild, but though one species had laid down approximately one line per day the other had not. Campana & Neilson (1982) obtained values of 0.9 to 1.06 for the regression of counts on days elapsed after date marking *Platichthys stellatus*, the starry flounder. As with other observers, they mention subdaily lines. This problem may have been already resolved by Taubert & Coble (1977) whose laboratory study of *Lepomis* species and the mouth brooder, *Tilipia mossambica*, gives the most convincing demonstration of daily bands. In *L. cyanellus* excellent agreement was obtained between days elapsed after "swim up" and numbers of bands, with coefficients of variation of observed/expected bands as low as 0.03. The photomicrographs however showed many small bands between the distinct major daily bands. These authors were able to produce a broad "cold shock" band and to show that, in an experiment without a 24 h cycle of light and dark, the well marked daily bands were not produced. We may therefore conclude that some fish and fish larvae display increments in the otoliths that are formed at daily, not tidal, intervals under the influence of the solar day as the zeitgeber. Further work on fish living in the intertidal habitat should be done, since such fish might show tidal and 14 day lunar periodicities.

ORIGIN AND NATURE OF TIDAL BANDS

The bands separating growth increments are formed when the shell is closed or partially closed and the body fluids have become more acidic (Richardson *et al.* 1981). As pH falls, the rate of calcification diminishes or may even be reversed. Virtually all investigators, including Wilbur (1972) and Pannella (1975) accept that the change in pH in the extrapallial fluid is the underlying cause of interruptions in calcification rate. Pannella (1975) and Lutz & Rhoads (1977) refer to the organic matrix that is thought to precede calcification and consider that the bands are "organic lines" or at least organically rich layers where calcification is reduced or interrupted. Presumably "organic rich" must imply "mineral poor", especially if resorption is believed to have occurred (Dugal, 1939; Crenshaw & Neff, 1969). Pannella considered these areas of suspended calcification are followed by a "rebound" of mineralization when the tide returns, assisted by an independent diel rhythm controlled by light and temperature.

FIGURE 8. Heavily etched shell of *Cerastoderma edule* as seen under SEM. The narrower bands stand out clearly from the broader increments which have been preferentially dissolved by the etching agent. Bar = 100 µm. (Courtesy of Dr. C.A. Richardson)

If the bands were organically rich, they could conceivably be more resistant to acid. They would then give rise to raised organic bands on acid etching, but if treated with alkaline oxidizing agents, such as NaOCl, to depressions. Deith (1985) showed that, whether acid etched or treated with NaOCl, the bands were left as ridges of calcium carbonate raised proud of the general surface. Koike (1980), however, was the first to show that the bands were raised, mineral rich, organically deficient regions of the shell. Moreover, C.A. Richardson of U.C.N.W. (private comm.) applied protein stains and found, that the stronger colour representing organic matter developed in the increments thus reinforcing the view that the increments are organically richer than the bands.

Some time ago J.A. Nott at U.C.N.W. (private comm.) photographed under the scanning electron microscope "stereo pairs" of lightly etched sections of cockle shell. These had the appearance of slopes and scarps, the high points corresponding to bands with the steep descent corresponding to the calcification "rebound" at the end of band formation. Richardson's (unpublished) EM photograph of a more heavily etched shell (Fig. 8) shows the band as an outstanding wall of aragonite crystals that appear larger than those of the adjacent increments. Shell sections when undergoing decalcification can be seen under the light microscope to have protruding elements that resist solution opposite the positions where the bands reach the surface. Rosenberg (1973) and Rosenberg & Jones (1975) examined by electron microprobe elemental analysis shells of *Clione* and *Cardum* (cockle) which had been ground polished and cleaned. Their traverses indicated slightly higher calcium levels in the narrower increments, which were presumably the bands. A similar result was obtained by Nott (personal communication) and by Deith (1985) but the number of bands and specimens used were not sufficient to be conclusive.

Since the mineral component of the band is more resistant and probably higher in calcium than that of the the increment, we may put forward the view (*cf.* Deith, 1985) that this is simply the result of slower precipitation at reduced pH. Such an hypothesis would explain the widespread incidence of tidal bands in shells, whether of calcite or aragonite, in so many unrelated animals living in the intertidal zone. These creatures have in common only the crystalline nature of the shell and the exposure of their body fluids to the influence of desiccation stress at low water, resulting in lowered pH.

It is a well established principle, due to von Weimarn (Philbrick & Holmyard, 1932), that the dispersion coefficient , which is a measure of the number of units into which a precipitate is divided, depends on the supersaturation level at the moment of formation. The crystal size is proportional to δ^{-1}, and

$$\delta^{-1} = \eta^{-1} S/(C\text{-}S)$$

where η is the viscosity of the medium, S is the equilibrium solubility and C the supersaturation concentration of the precipitate. It predicts that, as pH falls, the value of C and the rate of mineralization will also fall, but crystals will increase in size and approximate more closely to perfection. There is much further evidence in support of this hypothesis provided by other properties of the shell banding patterns. Richardson (unpubl.) demonstrated that bands are mechanically harder than increments. The more amorphous increments reflect light at their many boundaries while the more crystalline bands transmit light more readily. Hence in sections as seen by reflected light the increments appear light and the bands dark, and vice versa by transmitted light. Further, the many less perfect crystallites of the increments will have a larger surface area to volume ratio. If each crystal were enclosed in an organic sheath there would be more stainable tissue in the increments as indeed Richardson (see above) observed. Furthermore their greater surface area would increase the attraction by capillarity of fluids of low surface energy, while imperfections would render the increments more porous. A simple experiment can demonstrate that these differences exist. If a felt tip pen is gently brushed along a polished, but not etched, section of shell, ink rapidly penetrates the increments which, as a result, stain darker than the bands.

After many millions of years of fossilization, when the organic matrix may have disappeared, the crystal pattern remains remarkably intact. Similarly if fresh shells are placed in boiling sulphur, the organic matter is charred or destroyed and the shells become brittle; yet on sectioning the banding pattern is still visible (Crisp, unpubl.). After decalcification by ethylene diamine tetraacetic acid, the organic remains can still be seen with the characteristic banding pattern intact within. We therefore conclude that the organic moiety does not itself determine the shell bands and increments. It simply models them in compliance with the crystallite pattern. Its function may be to improve the resilience of the shell and hold the crystallites in place.

CHANGES IN ROTATION OF THE EARTH AND OF THE MOON'S ORBIT

In this secton the "dot" convention above a symbol represents the differential with time of angular rotation, n is orbit velocity, Ω is rotational velocity and e, m, subscripts are for earth and moon respectively.

There are two changes in the primeval calendar that result from the transfer of momentum, due chiefly to the tidal friction of the earth's oceans and seas. Momentum is transferred from earth to moon because the earth's rotation rate ($\Omega_e = 7.27 \times 10^{-5}$ rad sec^{-1}) exceeds that of the moon's orbit about the earth ($n_m = 0.247 \times 10^{-5}$ rad sec^{-1}).

The loss of kinetic energy reduces both angular velocities, but since angular momentum is conserved, the moment of inertia of the earth-moon system must be increased by the moon's moving away from the earth. Though the transfer of momentum accelerates the moon's orbit relative to the earth, the absolute acceleration, $\dot{n}_m$, is negative (in the order of -0.135×10^{-22} rad sec^{-2}, Lambeck, 1980) as is that of the earth's rotation, $\dot{\Omega}_e$, in the order of -5.5×10^{-22} rad sec^{-2}.

It may be assumed that the annual circuit of the earth round the Sun does not change hence, as the earth's rotation slows down, there will be fewer days per year (N_1) and the length of day (L.O.D.) will increase over the eons. Similarly the number of days per synodic (lunar) month (N_2) will also decrease. However the number of synodic months per year (N_3) must rise since, relative to the earth, the moon's orbital velocity is accelerating and the month is therefore getting shorter. Clearly these calendar changes are connected by $N_1 = N_2 \cdot N_3$.

Though tidal friction in seas and oceans is now considered to have changed little over the period during which the ocean topography has been roughly constant, the measurements of Ω and n show large short-term fluctuations. These are caused by such factors as regular changes in wind stress, precipitation and polar wobble, and irregular changes such as seismic events, mantle and core changes, glaciations, mountain building and solar events. Hence the earth and moon decelerations, as measured by the more accurate modern methods using atomic clocks, may not be at all representative of the past. According to O'Hora (1975), recent lunar observations of the earth's retardation greatly exceeded the long term average. Calculations based on telescope observations over the past 300 years give rise to quite discordant estimates for $-\dot{n}_m$ and values of $-\dot{\Omega}$ of 4-5×10^{-22} rad sec^{-2} (see Lambeck, 1980, Table 10.4). A third approach is based on a comparison of current positions of heavenly bodies with ancient records of astronomical events. The periods are sufficiently long to be representative of tidal deceleration, but many ancient recorded dates require speculative antiquarian research and may not be reliable.

From a summary of all such physical data, Lambeck (1980) recommends the use of $\dot{\Omega}_e = -5.5 \times 10^{-22}$ rad sec^{-2} and $\dot{n}_m = -0.135 \times 10^{-22}$ rad sec^{-2}. These values are close to those of Munk & MacDonald (1960) which were used to estimate tidal torque, but there are in the literature considerably higher values.

Of great interest to geophysicists is the evolution of the earth-moon system. If rotational and orbital motions could be determined sufficiently far back in geological time, the vexed question of the moon's origin might be solved. Unfortunately the earth's shape and mass, its moment of inertia, its ocean topography and even the gravitational constant itself, cannot be assumed constant over a period of some 3×10^9 years. Back extrapolation of the decelerations of the moon's and earth's motions are therefore invalid. However, astronomers believe that at a very early stage the moon approached the earth, creating very large tides. Since then it has moved away (Gerstenkorn, 1969).

Paleontologists have offered the possibility of deriving information concerning the primeval calendar from evidence of periodic structures of daily tidal and annual formations in fossils – the so called "Paleontological Clocks". These go back 3×10^9 years in the Rhodesian Bulawayan Stromatolites, 4×10^8 years in the Devonian Corals and bivalves and 1×10^8 years in the Cretaceous barnacles. What does this evidence tell us, and with what certainty?

PALEONTOLOGICAL CLOCKS

Essentially the paleontological evidence for changes in the primeval calendar is bedevilled by two defects. Both stem from having made assumptions regarding observations on fossils without having studied the same features in living organisms. The criticisms are (a) wrong or highly dubious interpretations of periodic structures and (b) insufficient attention being paid to the probable errors in counting. As has been shown above, firm evidence of tidal periodicity in skeletal structures has been obtained, though only for molluscs and barnacles living in shallow water or intertidally. Slightly less strong evidence exists for daily structures in the ridges of corals and the otoliths of fish. We may therefore assume that the same applies to fossils of these groups.

The major error made in the past was the interpretation of tidally formed bands in bivalves, of which currently *c.* 706 are formed per year. As daily bands, only *c.* 365 would be formed. Despite this error, the remarkably consistent increase in the past of supposed days per year published by Pannella (1972) strongly suggests that, in some way, weaker bands were omitted, perhaps alternately, so that the final results did not differ from expectation of days/year (see Table 2). Moreover Pannella *et al.* (1968) and Pannella (1972) gave measurements for lesser annulations, assumed to be months, in the order of 29, precisely the number that one would obtain by counting the bands between one set of alternations to the next in a spring to spring tidal series (Fig. 2). Berry & Barker (1975), however, obtained almost exactly half that number! No such criticism can be levied against the data of Wells (1970), Scrutton (1970) or Mazzullo (1971) if epithecal ridges in corals do indeed represent daily events.

The second criticism rests on the apparent precision embodied in these results. The convenient summaries in Lambeck (1980) allows the coral data to be regressed as days per year on geological age. One then obtains $-6.75 \pm 1.13 \times 10^{-22}$ rad sec^{-2} for the earth's acceleration, $\dot{\Omega}_e$, implying only ± 7 days per year as the standard deviation about the regression. Similarly for Pannella's bivalve data, $\dot{\Omega}_e = -6.6 \pm 1.67 \times 10^{-22}$ rad sec^{-2} with a standard deviation of ± 8.8 days per yearly count.

It must be remembered that the accurate identity of bands laid down to tides elapsed given in Table 1 depended on accurate dating of the initial and final bands. Evans' (1972) original display of bands and tides (Fig. 5) depended equally on the alignment of bands to particular dates in the tidal progression. Such accuracy would not be possible where the dates can only be surmised by the cessation of bands during winter months, or from the approximation of ever-narrower bands in winter. This is especially so in tropical organisms, such as corals, where winter checks are likely to be weakly imprinted. The same problem applies to the recognition of neap and spring tides from banding patterns (Fig. 2). As Evans (1975) remarked "The trouble is that counting growth lines is not easy. It is constantly necessary to make subjective decisions about whether a line is really

TABLE 2. Summary of total annual and monthly increments from fossil bivalves observed from (Pannella, 1972)

PERIOD	Ma[1]	$\bar{X}$ INCREMENTS/A[2] $\overline{N_1}$	RANGE	N SPEC[3]	N COUNTS[4]
Recent	0	359	353-366	9	32
U. Cretaceous	70	375	371-379	3	16
M. Triassic	220	372	365-375	3	7
U. Carboniferous	290	383	380-389	3	11
L. Carboniferous	340	398	397-399	2	9
M. Devonian	380	406	-	1	6

PERIOD	Ma[1]	$\bar{X}$ INCREMENTS/M[2] $\overline{N_1}$	RANGE	N SPEC[3]	N COUNTS[4]
Recent	0	29.2	29.0-29.6	7	186
U. Tertiary	14	29.4	29.2-29.8	5	197
M. Tertiary	38	29.8	29.6-29.9	2	40
L. Tertiary	54	29.6	29.4-30.0	3	141
U. Cretaceous	70	29.9	29.6-30.2	7	159
M. Triassic	220	29.7	29.4-30.0	3	77
U. Carboniferous	300	30.2	29.9-30.7	4	59
U. Devonian	350	30.4	30.2-30.5	4	168
U. Ordovician	445	30.3	29.8-30.7	2	84

[1]Millions of years
[2]A = years; M = months
[3]Number of specimens
[4]Number of counts

a line or where an annual or monthly series begins or ends. As a result it is not surprising to find that counts often come out close to hypothesized values."

The evidence based on changes in the days per lunar month (N_2) is less easy to interpret in terms of the earths rotation, than the changes in number of days per year (N_1). For N_1 the calculations is simple

$$dN_1/dt = \dot{\Omega}_e/n_e.$$

For N_2, both earth's and moon's decelerations are involved

$$dN_2/dt = \dot{\Omega}_e(n_m-n_e)-\dot{n}_e(\Omega_e-n_e)/(n_m-n_e)^2.$$

Lambeck (1980) solved the latter equation for both coral and bivalve data. In the latter, Lambeck (1980) assumed that Berry & Barker's (1975) values for N_2 should be doubled, so that they become consistent with Pannella & MacClintock (1968). Table 3 shows the remarkable agreement between the bivalve data, the coral data, and the data

TABLE 3. Computation of Earth and Moon deceleration from fossil evidence (from Lambeck, 1980)

SOURCE	Ma	$\dot{\Omega}_e$ rad sec^{-2}	$\dot{n}_m$ rad sec^{-2}
Coral epitheca	-440 Ma	$-6.3 \pm 0.7 \times 10^{-22}$	$-1.5 \pm 0.4 \times 10^{-22}$
Bivalve shell bands	-360 Ma	$-5.9 \pm 0.6 \times 10^{-22}$	$-1.3 \pm 0.3 \times 10^{-22}$
Modern astronomy	1700 AD	$-5.5 \pm 0.5 \times 10^{-22}$	$-1.35 \pm 0.1 \times 10^{-22}$

of modern physics. Bearing in mind the difficulties in counting, the doubts concerning periodicites, and the deviations in laying down bands due to such contingencies as cold snaps, storms, and predators, they look too good to be true!

Insofar as these results are valid, they indicate that no appreciable change has taken place in tidal friction over the past 400 Ma, which includes the breaking up of Pangaea and the reductions in sea level during severe glaciations. It is generally believed that energy dissipation is likely to have been less in the past, but paleontological evidence does little to support this view.

The interpretation of stromatolites is even more speculative. There is no agreement between Pannella's (1972) and Mohr's (1975) fossil evidence, nor between recent interpretations of experiments on modern stromatolites. It is most unfortunate, because the period of the first fossil stromatolites (-2000 to -3000 Ma) was a highly significant period in the evolution of the moon's orbit. Then it was much closer to the earth and may have generated large tidal forces when the seas, but not the ocean basins, had just formed.

THE VALUE OF PALEONTOLOGICAL EVIDENCE FROM CALCAREOUS SKELETONS

Recent research outlined in this paper shows how an understanding of processes in living forms may cast suspicion on the testimony of fossils in regard to the primeval calendar. The message, however, is not that we should disregard this evidence, but that it should be reinterpreted on the better basis that we now have, and re-examined using more objective criteria (*e.g.*, Dolman, 1975). Indeed, far more is required to be done on living forms, especially on fish, corals and stromatolites, before we can take full advantage of the very important geophysical information that may be locked away in fossils. The relationship in extant species between shell banding patterns and depth, climate, ecological factors, and tidal regimes requires further study.

Already one can see the possibilities of additional kinds of information, especially in the realm of paleoecology, arising from such work. Those bivalves with prominent banding patterns were probably living intertidally, and the details of such patterns could reveal the tidal regime and so the strengths of the component harmonics as suggested by Pannella (1976). This in turn would allow speculation on the topography and the oscillation of tides in the primeval oceans. For a given species, seasonal changes in breadth of increments, indicative of growth rate, could reveal latitudinal, climatic, and annual temperature changes of the deposit in which it was found.

ACKNOWLEDGEMENTS

Participation in The Fifth International Symposium on Biomineralization was possible by funds provided by The Organized Research Fund of The University of Texas at Arlington and The Sea Grant Program administered by Texas A & M University.

REFERENCES

ANTOINE, L. & QUEMERAIS-PENCREAC'H, D., 1980. Stries et rythmes de croissance chez la *Patella vulgata* L. C.r. hebd. Seanc. Acad. Sci. Paris D 290:1127-1130.

BARNES, D.J., 1972. The structure and formation of growth ridges in scleractinian coral skeletons. Proc. Roy. Soc. Lond. B 182:331-350.

BEAMISH, R.J. & MCFARLANE, G.A., 1983. The forgotten requirement for age validation in fisheries biology. Trans. Amer. Fisheries Soc., 112:735-743.

BERRY, W.B.N. & BARKER, R.M., 1975. Growth increments in fossil and modern bivalves. *In* Growth rhythms and the history of the earth's rotation. Rosenberg, G.D. and Runcorn, S.K.(Ed.) John Wiley and Sons pp. 9-25.

BOURGET, E. & CRISP, D.J., 1975a. Factors affecting deposition of the shell in *Balanus balanoides*. J. mar. Biol. Ass. U.K., 55:231-249.

BOURGET, E. & CRISP, D.J., 1975b. An analysis of the growth bands and ridges of barnacle shell plates. J. mar. Biol. Ass. U.K., 55:439-461.

BROOM, M.J. & MASON, J., 1978. Growth and spawning in the pectinid *Chlamys opercularis* in relation to temperature and phytoplankton concentration. Marine Biology, 47:277-285.

BROTHERS, E.B., 1978. Exogenous factors and the formation of daily and subdaily growth increments in fish otoliths. Amer. Zool., 18:631.

BROTHERS, E.B., MATTHEWS, C.P. & LASKER, R. 1976. Daily growth increments in otoliths from larval and adult fishes. Fish. Bull., 74:1-8.

BUDDEMEIER, R.W. & KINZIE III, R.A., 1975. The chronometric reliability of contemporary corals. *In* Rosenberg, G.D. and Runcorn, S.K. (Ed.) Growth rhythms and the history of the Earth's rotation. John Wiley and Sons., pp. 135148

CAMPANA, S.E. & NEILSON, J.D., 1982. Daily growth increments in otoliths of starry flounder *Platichthys stellatus* and the influence of some environmental variables on their production. Can. J. Fish. Aquat. Sci., 39:937-942.

CHOE, S. 1963. Daily age markings on the shell of cuttle fishes. Nature, Lond. 197:306-307.

CLARKE, G.R. II, 1968. Mollusk shell: Daily growth lines. Science, 161:800-802.

CLARKE, G.R. II, 1975. Periodic growth and biological rhythms on experimentally grown bivalves. *In* Growth Rhythms and the history of the Earth's rotation. Ed. Rosenberg, D. and Runcorn, S.K. (Ed.) John Wiley and Sons., pp. 103-117.

CRENSHAW, M.A. & NEFF, J.M., 1969. Decalcification at the mantle-shell interface in molluscs. Amer. Zool., 9:881-885.

CRISP, D.J. & RICHARDSON, C.A., 1975. Tidally produced internal bands in the shells of *Elminius modestus* Darwin. Mar. Biol., 33:155-160.

DARWIN, C., 1854. A monograph of the subclass Cirripedia. II The Balanidae, the Verrucidae etc. Ray Soc. London, 684pp.

DAVENPORT, C.B., 1935. Growth lines in fossil pectens as indicators of past climates. J. Palaeont. 12:514-515.

DEITH, M.R., 1985. The composition of tidally deposited growth lines in the shell of the edible cockle *Cerastoderma edule*. J. mar. Biol. Ass. U.K., 65:573-581.

DOLMAN, J., 1975. A technique for the extraction of environmental and geophysical information from growth records in invertebrates and stromatolites. *In* Growth rhythms and the history of the Earth's rotation. Rosenberg, G.D. and Runcorn, S.K. (Eds) John Wiley and Sons, pp. 191-221.

DUGAL, L.P., 1939. The use of calcareous shell to buffer the product of anaerobic glycolysis in *Venus mercenaria*. Cellul. Comp. Physiol., 13:235-251.

EKARATNE, S.U.K. & CRISP, D.J., 1982. Seasonal growth studies of intertidal gastropods from shell microgrowth band measurements, including a comparison with alternative methods. J. mar. Biol. Ass. U.K., 64:183-210.

ESSIG, R.J. & COLE, C.F., 1986. Methods of estimating larval fish mortality from daily increments in otoliths. Trans. Amer. Fish. Soc., 115:34-40.

EVANS, J.W., 1972. Tidal growth increments in *Clinocardium nuttalli*. Science, N.Y. 176:416-417.

EVANS, J.W., 1975. Growth micromorphology of two bivalves exhibiting non-daily growth lines. *In* Growth rhythms and the history of the Earth's rotation (Ed. Rosenberg, G.D. and Runcorn, S.K. (Eds.) John Wiley and Sons, pp. 119-134.

EVANS, J.W. & LE MESURIER, M.H., 1972. Functional micromorphology and circadian growth of the rock boring clam *Penitella penita*. Can. J. Zool., 50:1251-1258.

FARROW, G.E., 1971. Periodicity structures in the bivalve shell: experiments to establish growth controls in *Cerastoderma edule* from the Thames estuary. Palaeontology 14:571-588.

GEBELEIN, C.D. & HOFFMAN, P., 1968. Intertidal stromatolites from Cape Sable Florida. Geol. Soc. Am. Spec. Pap., 121:109 (Abstract).

GERSTENKORN, H., 1969. The earliest past of the Earth Moon system. Icarus 11:189-207.

GRUFFYDD, LL.D., 1981. Observations on the rate of production of external ridges on the shell of *Pecten maximus* in the laboratory. J. mar. Biol. Ass. U.K., 61:401-411.

HALL, C.A., DOLLASE, W.A. & CORBATO, C.E., 1974. Shell growth in *Tivela stultorum* (Maure, 1823) and *Callista chione* (Linnaeus, 1758) (Bivalvia): Annual periodicity, latitudinal differences and diminution with age. Palaeogr. Palaeoclim. Palaeoecol., 15:33-61.

HARDIE, L.A. & GINSBERG, R.N., 1971. The sedimentary record of a tidal flat lamination. Geol. Soc. Am. Abst. Prog., 3:591.

HOUSE, M.R. & FARROW, G.E., 1968. Daily growth banding in the shell of the cockle, *Cardium edule*. Nature, 219:1384-1386.

JONES, P. & CRISP, M., 1985. Microgrowth bands in chitons: evidence of tidal periodicity. J. mollusc. Studies, 51:133-137.

KENNY, R., 1977. Growth studies of the tropical intertidal limpet *Acmaea antillarum*. Marine Biology, 39:161-170.

KOIKE, H., 1973. Daily growth lines of the clam *Meretrix lusoria* - a basic study for the estimation of prehistoric seasonal gathering. J. Antrop. Soc. Nippon, 81:122-138.

KOIKE, H., 1975. The use of daily and annual growth lines of the clam *Meretrix lusoria* in estimating seasons of Jornon Pemod shell gathering. Bull. R. Soc. N.Z., 18:189-193.

KOIKE, H., 1980 Microstructure of the growth increment in the shell of *Meretrix lusoria*. *In* The mechanics of biomineralisation in animals and plants. Proceedings of the Third International Biomineralisation Symposium, 1977. (Omori, M, and Watabe, N., Eds) Tokai University Press, pp. 93-97.

LAMBECK, K., 1980. The Earth's variable rotation: geophysical causes and consequences. Cambridge University Press xi + 449 pp.

LE GALL, M.P., 1970. Méthode d'étude des stries de croissance de *Mytilus edulis* L. Mise en évidence du rhythme et des modalités de leur formation. C.r. hebd. Seanc. Acad. Sci. Paris D., 270:590-511.

LUTZ, R.A. & RHOADS, D.C., 1977. Anaerobiosis and a theory of growth line formation. Science, N.Y., 198:1222-1227.

MACCLINTOCK, C. & PANNELLA, G., 1969. Time of calcification in the bivalve mollusk *M. mercenaria* (L.) during the 24 h period. Ann. meeting Geol. Soc. Amer., p140.

MAZZULLO, S.J., 1971. Length of the year during Silurian and Devonian periods: New Values. Geol. Soc. Am. Bull., 82:1085-1086.

MOHR, R.E., 1975. Measured Periodicities of the Biwabik (Precambrian) Stromatolites and their geophysical significance. *In* Growth rhythms and the history of the Earth's rotation. Rosenberg, G.D. and Runcorn, S.K. (Eds) John Wiley and Sons, pp. 43-56.

MONTY, C.L.V., 1965. Recent algal stromatolites in the Windward lagoon, Andros Island, Bahamas. Ann. Soc. Geol. Belg., 88:296-376.

MUNK, W.H. & MACDONALD, G.J. 1960. The rotation of the Earth. Cambridge University Press. London & NY, 313 pp.

O'HORA, N.P.J., 1975. The detection of recent changes in the Earth's rotation. *In* Growth rhythms and the history of the Earth's rotation. Rosenberg, G.D. and Runcorn, S.J.(Eds) John Wiley and Sons, 427-444.

ORTON, J.H., 1923. On the significance of "rings" on the shells of Cardium and other molluscs. Nature, London, 112:10.

PANNELLA, G., 1971. Fish otoliths: daily growth layers and periodical patterns. Science (Washington D.C.), 173:1127-1127.

PANNELLA, G., 1972. Palaeontological evidence on the Earth's rotational history since early precambrian. Astrophys. Space Sci., 16:212-237.

PANNELLA, G., 1974. Otolith growth patterns: an aid in age determination in temperate and tropical fishes. *In* Bagenal, T.B. (Ed.) The aging of fish. Unwin Bros, Surrey, England. pp.28-29.

PANNELLA, G., 1975. Palaeontological clocks and the history of the Earth's rotation. *In* Growth rhythms and the history of the Earth's rotation. Rosenberg, G.D. and Runcorn, S.K. (Eds) John Wiley and Sons, pp. 253-287.

PANNELLA, G., 1976. Tidal growth patterns in recent and fossil mollusc bivalves: a tool for the reconstruction of Paleotides. Die Naturwissenschaften, 63:539-543.

PANNELLA, G. & MACCLINTOCK, C., 1968. Biological and environmental rhythms reflected in molluscan shell growth. J. palaeontol., 42(5) Suppl. 64-80.

PANNELLA, G., MACCLINTOCK, C. & THOMPSON, M.M., 1968. Palaeontologic evidence of variations in length of synodic month since late Cambrian. Science, 162:792-796.

PHILBRICK, F.A. & HOLMYARD, E.J., 1932. A textbook of theoretical and inorganic chemistry. Temple Press, Letchworth., vii - 803pp.

REVELLE, R. & FLEMING, R.H., 1934. The solubility product constant of calcium carbonate in seawater. Proc. Fifth Pacific Cong. Canada, 1933, Vol. 3:2089-2092.

RHOADS, D.C. & PANNELLA, G., 1970. The use of molluscan shell patterns in ecology and palaeoecology. Lethaia 3:143-161.

RICHARDSON, C.A., 1987. Tidal bands in the shell of the clam *Tapes philippinarum* (Adams & Reeve, 1950). Proc. R. Soc. Lond. B. 230:367-387.

RICHARDSON, C.A., CRISP, D.J. & RUNHAM, N.W., 1979. Tidally deposited growth bands in the shell of the common cockle, *Cerastoderma edulis* (L.). Malacologia 5:277-290.

RICHARDSON, C.A., CRISP, D.J., RUNHAM, N.W., & GRUFFYDD, LL..D., 1980a. The use of tidal bands in the shell of *Cerastoderma edule* to measure seasonal growth rates under cool temperate and sub-arctic conditions. J. mar. Biol. Ass. U.K., 60:977-990.

RICHARDSON, C.A., CRISP, D.J. & RUNHAM, N.W., 1980b. An endogenous rhythm in shell deposition in *Cerastoderma edule*. J. mar. Biol. Ass. U.K., 60:991-1004.

RICHARDSON, C.A., CRISP, D.J. & RUNHAM, N.W., 1981. Factors influencing shell deposition during a tidal cycle in the intertidal bivalve, *Cerastoderma edule*. J. mar. Biol. Ass. U.K., 61:465-476.

ROSENBERG, G.D., 1973. Calcium concentration in the bivalve *Clione andatella* Sowerby. Nature, London, 244:155-156.

ROSENBERG, G.D. & JONES, C.B., 1975. Approaches to chemical periodicities in molluscs and stromatolites. *In* Rosenberg, G.D. and Runcorn, S.K. (Ed.) Growth rhythms and the history of the Earth's rotation. John Wiley and Sons, pp. 223-241.

ROSENBERG, G.D. & RUNCORN, S.K., 1975 (Eds.) Growth rhythms and the history of the Earth's rotation, xvi+559pp. John Wiley and Sons, London, New York, Sydney, Toronto.

SCHMIDT, P.O., 1984. Marking growth increments in otoliths of larval and juvenile fish by immersion in tetracycline to examine the rate of increment formation. US Nat. Fish Services, Fish Bull., 82:237-242.

SCHMIDT, R.E. & FABRIZIO, M.C., 1980. Daily growth increments on otoliths for aging young of the year largemouth bass from a wild population. Progressive Fish Culturist, 42:78-80.

SCRUTTON, C.T., 1965. Periodicity in Devonian coral growth. Palaeontology 7:552-558.

SCRUTTON, C.T., 1970. Evidence for a monthly periodicity in the growth of some corals. *In* Palaeogeophysics. (Ed. Runcorn, S.K.) Academic Press, London and New York.

SCRUTTON, C.T., 1978. Periodic growth features in fossil organisms and the length of day and month. *In* Tidal friction and the Earth's rotation (Ed. Broche, P. and Sunderland, J.) Springer-Verlag, Berlin, pp. 154-196.

TAUBERT, B.D. & COBLE, D.W., 1977. Daily increments in otoliths of three species of *Lepomis* and *Tilapia mossambica*. J. Res. Board Canada, 332-340.

WATTENBURG, H. & TIMMERMANN, E., 1936. Uber die Sattingung des Seewassers an CaCO3, und die anorganogene Bilding von Kalksedimenten. Ann. d. Hydro. u. Mar. Meteor, 23-31:1936.

WELLS, J.W., 1963. Coral growth & geochronometry. Nature, Lond,. 197:948-950.

WELLS, J.W., 1970. Problems of annual and daily growth rings in corals. Photogrammetry. U.S. Coast and Geodetic Survey, Washington.

WHYTE, M.A., 1975. Time, tide and the cockle. *In* Rosenberg, G.D. and Runcorn, S.K. (Eds) Growth rhythms and the history of the Earth's rotation. John Wiley and Sons, pp. 177-189.

WILBUR, K.M., 1972. Shell formation in molluscs. *In* Chemical Zoology Vol. 7 *Mollusca* (Florkin, M. and Scheer, B.T. Eds.) Academic Press, N.Y., pp. 103-145.

WILD, A., & FORMAN, T.J., 1980. The relationship between otolith increments and time for Yellow fin and Skipjack Tuna marked with tetracycline. Inter America Tropical Tuna Commission, La Jolla, Ca. USA, Bull. 17:509-560.

SILICIFICATION PROCESSES IN SPONGES: *GEODIA* ASTERS AND THE PROBLEM OF MORPHOGENESIS OF SPICULE SHAPE

Tracy L. Simpson

Department of Biology, University of Hartford, West Hartford, CT 06117

ABSTRACT. The deposition of silica by sponges involves a special membrane and an internal organic matrix, the axial filament. A hexagonal cross sectional shape of the filament may act as a basis, through differential growth, of the formation of a trigonal shape. Dynamic processes involved in silica deposition include membrane and organic filament syntheses and assembly; transport, accumulation and polymerization of silicic acid; and termination of these processes. The theoretical details of these processes are discussed. The pattern of development of complex siliceous structures (asters) in *Geodia cydonium* involve the formation of short, terminal processes which then become fused together and finally form a star-shaped pattern. It is difficult to invoke the presence of a centrally located silicification center as a basis for this sequence of morphogenesis. Such centers appear to be present during the development of other morphological types of siliceous sponge structures and such a center may be present in the early stages of development of *Geodia* asters before the maturation of the tips of the structures. Both sodium and calcium ions appear to interact with the sponge silicification system as determined by their effects on spicules formed in the presence of germanic acid. The absence of central bulbs in marine spicules grown in germanic acid may be partly due to the high concentration of these (and other) ions in sea water.

INTRODUCTION

The question of how siliceous sponge spicules come to possess a specific and predictable shape involves two separate aspects: 1) the biological components which are essential for such morphological determination, and 2) the morphogenetic (*i.e.*, dynamic) processes which result from the activity of these components. This paper will examine, particularly, the latter aspect of sponge silicification emphasizing recent results which significantly complicate our present views.

DETERMINANTS OF SHAPE

Accumulated data (as reviewed by Garrone *et al.*, 1981; Simpson, 1984) demonstrate the involvement of two known morphological entities that are necessary for silicification in sponges. These are an intracellular, surrounding, unit-type membrane now referred to as the silicalemma and secondly, a central organic core, the axial filament, which develops *within* the space bounded by the silicalemma. Opaline silica is deposited around

Origin, Evolution, and Modern Aspects of Biomineralization in Plants and Animals
Edited by Rex E. Crick
Plenum Press, New York

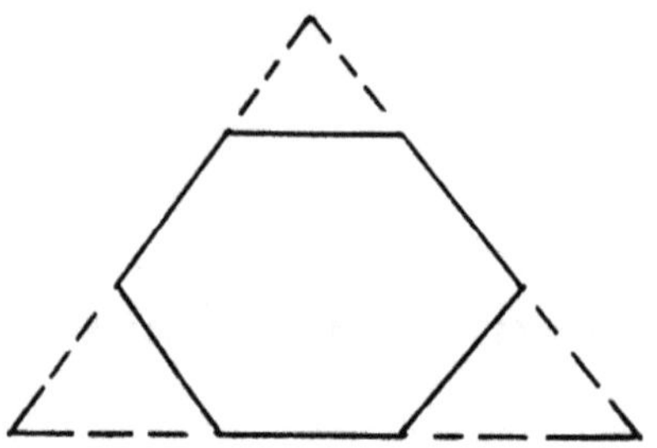

FIGURE 1. Hypothetical mechanism for the transformation of an axial filament with a hexagonal cross sectional shape to one with a trigonal shape.

the filament and within the silicalemma. The view has emerged that the shape of the filament and the surrounding silicalemma act as the determinants of the final spicule shape, since the axes of the mature opaline silica are cored by the filament; *i.e.*, when the filament is branched so is the resulting siliceous structure. Sponges represent the best known case of the involvement of an organic matrix in a biological silicification system. However, all filaments are not identical in their dimensions, namely, in their width; in some sponge species filaments are considerably larger and apparently display internal heterogeneity (Simpson *et al.*, 1985a; Mackie & Singula, 1983) while others are thinner and homogeneous (see Garrone *et al.*, 1981). Even the smallest known sponge spicules with widths of 0.5 μm or less possess organic filaments (Donadey, Station maritime d'Endoume, Marsaille, France, pers. comm.). Very large sponge spicules and fused silica networks possibly present unique features relative to axial filaments but little is known for sure. The cross-sectional shape of filaments has been reported to be triangular in some sponge groups and it has been hypothesized that this confers upon them the potential to branch in four different planes (Reiswig, 1971). Other workers have found filaments in these groups to be hexagonal (Garrone, 1969; Simpson & Vaccaro, 1974) and a recent report suggests cross-sectional filament shape may be labile (Simpson *et al.*, 1985a). Garrone (pers. comm.) has suggested that a triangular cross-sectional shape may develop directly by differential growth of three of the faces from an original hexagonal one (Fig. 1). Such a mechanism could account for reports of both hexagonal and trigonal filaments within the same sponge groups. Some available data (Garrone, 1969) establish that some filaments contain ordered elements in three different planes further supporting such a hypothesis. This important aspect of the relationship of the organic matrix to shape determination of the silica requires further investigation.

DYNAMIC ASPECTS OF SPICULE FORMATION

Although the processes which are involved in initiating and completing the final siliceous structure have not been fully investigated it is possible to outline their major features. The following chemical and morphogenetic functions can be recognized: 1) initial synthesis and assembly of the silicalemma, 2) initial synthesis and assembly of the axial filament, 3) transport, entrapment and polymerization of silicic acid, 4) growth of the silicalemma, 5) growth of the axial filament, 6) termination of growth of the silicalemma, 7) termination of growth of the axial filament, and 8) termination of silicic acid transport, entrapment and polymerization. Although speculative, it is highly likely that these processes are coupled since the geometric shape of the silica structures is highly defined and reproduced in each type of structure and, importantly, is highly precise. The initiation of the organic filament, and also the silicalemma, could involve an *in toto* assembly of all or an initial portion of the structure or their initiation could

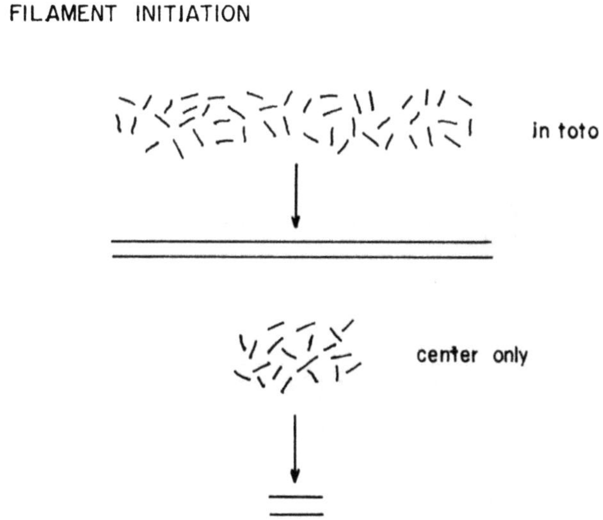

FIGURE 2. Two different mechanisms for the assembly of axial filaments. **Upper**–filament formed simultaneously along its length through the assembly of subunits. **Bottom**–filament forming at its center thus requiring further assembly during growth.

involve the assembly of only a small central portion (Fig. 2). Under the latter conditions all filaments, and membranes would have to then undergo at least some growth (elongation). A dramatic demonstration (Simpson & Langenbruch, 1984) that the latter is probably the case is a recent report that even very small structures undergo growth at their extremities. However, because this process is normally so rapid, such growth is difficult to discern. During growth, newly assembled filament or membrane could be added through elongation or, less likely, through central deposition and "pushing" of the filament (and membrane) outward (Figs. 3 & 4). The actual mechanism of growth is unknown but it is known that most filaments undergo an increase in length which is sometimes very rapid and thus difficult to observe. The pattern of silicic acid transport and processing could be diffuse silicic acid being pumped along the length of the membrane or it could occur at specific loci central or terminal (Fig. 5). Each pattern can also act as a model for the assembly and addition of new silicalemma. Interestingly, the latter possibility implies the movement, between the silicalemma and filament of a trapped form of silicic acid and as discussed below appears to occur at least in some cases. Since the silicalemma is assumed to be the site of silicic acid transport (Sullivan & Volcani, 1981), this could be coupled to membrane assembly and both could also be coupled to filament growth thus giving a tripartite, coupled system. Or, any combination of coupling and pattern of activity might be expressed in *differently* shaped structures thereby determining to some extent the final morphology; or, such a mixture of non-coupled and coupled activities could be expressed at different times in the development of the *same* spicule.

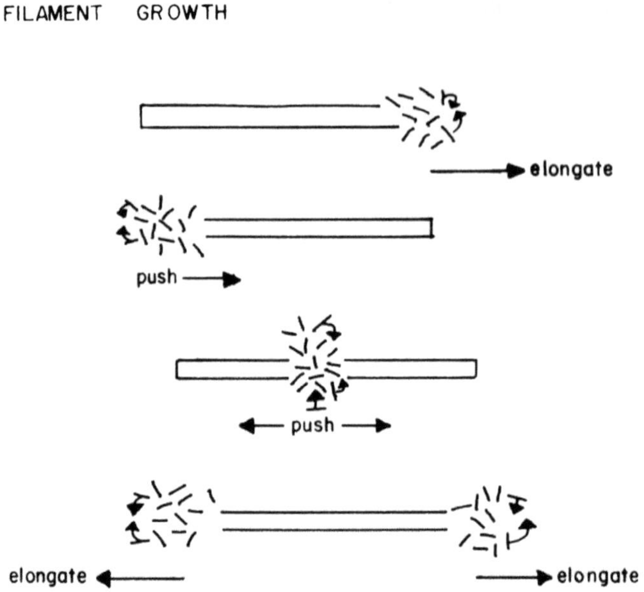

FIGURE 3. Possible mechanism for the growth in length of axial filaments. Filament is pictured as assembling and elongating only at one extremity. The same growth pattern could also result if the filament assembly occurs at one extremity and pushes the preformed filament outward in one direction. Or, central assembly and pushing could result in bidirectional growth which could also occur as the result of assembly and elongation at both extremeties.

RECENT RESULTS ON DYNAMICS

The use of germanium and its isotope ^{68}Ge as an analog of silicon has recently provided interesting, although limited, data on certain dynamic aspects of opaline silica deposition in freshwater sponges (Simpson *et al.*, 1979, 1983; Davie *et al.*, 1983; Simpson & Langenbruch, 1984). These results support the views that there are localized regions of the silicalemma which are more active and thus act as centers for silicic acid and silica accumulation, and that silicic acid or its derivatives (dimers, trimers, etc.) are probably mobile within the silicalemma and thus are translocated from the center to other areas of the developing structure. Two arrangements of active centers have been identified in two different spicule types (see Fig. 5). In the *first* a center occurs at approximately the midpoint of the axis of the developing structure and in the *second* there appears to be two centers at the extremities of the structures. The data also strongly indicate that there are secondary centers for accumulation particularly in the former system. In all cases, germanium is directly incorporated into the silica (Davie *et al.*, 1983; Simpson *et al.*, 1983).

These results are unexpected since the prevalent view of silicic acid transport and accumulation is that both processes occur along the length of the silicalemma (and thus the length of the axial filament) as depicted in Figure 5. Interestingly, silicification patterns in diatoms are viewed as beginning in localized areas of the complexly arranged

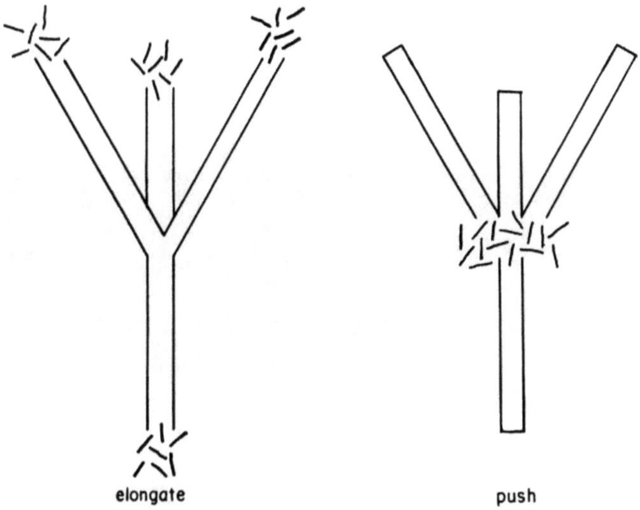

FIGURE 4. Two possible mechanisms for the growth of the axial filament in silica structures which have four separate axes, each in a different plane. Such spicules occur in so-called primitive demosponges. **Left**–filament elongating at its extremities. **Right**–filament assembling at the center of four axes and pushing outward.

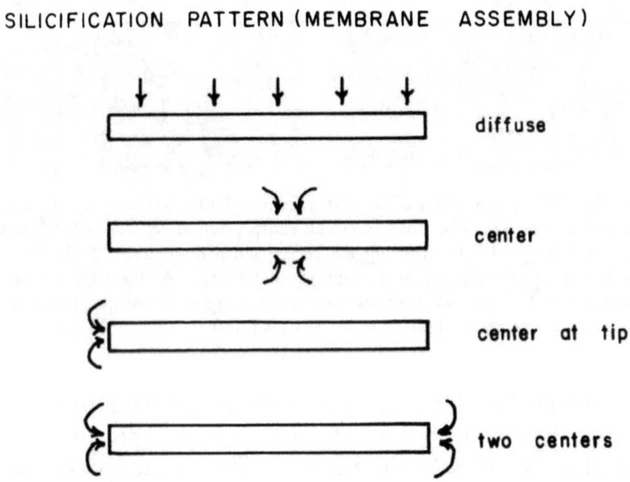

FIGURE 5. Possible patterns of silicic acid transport and accumulation. *Upper.* Transport/accumulation pictured as occurring along whole length of the structure. Available data support two and possibly three patterns (*bottom*)(nondiffuse) in two different spicule types.

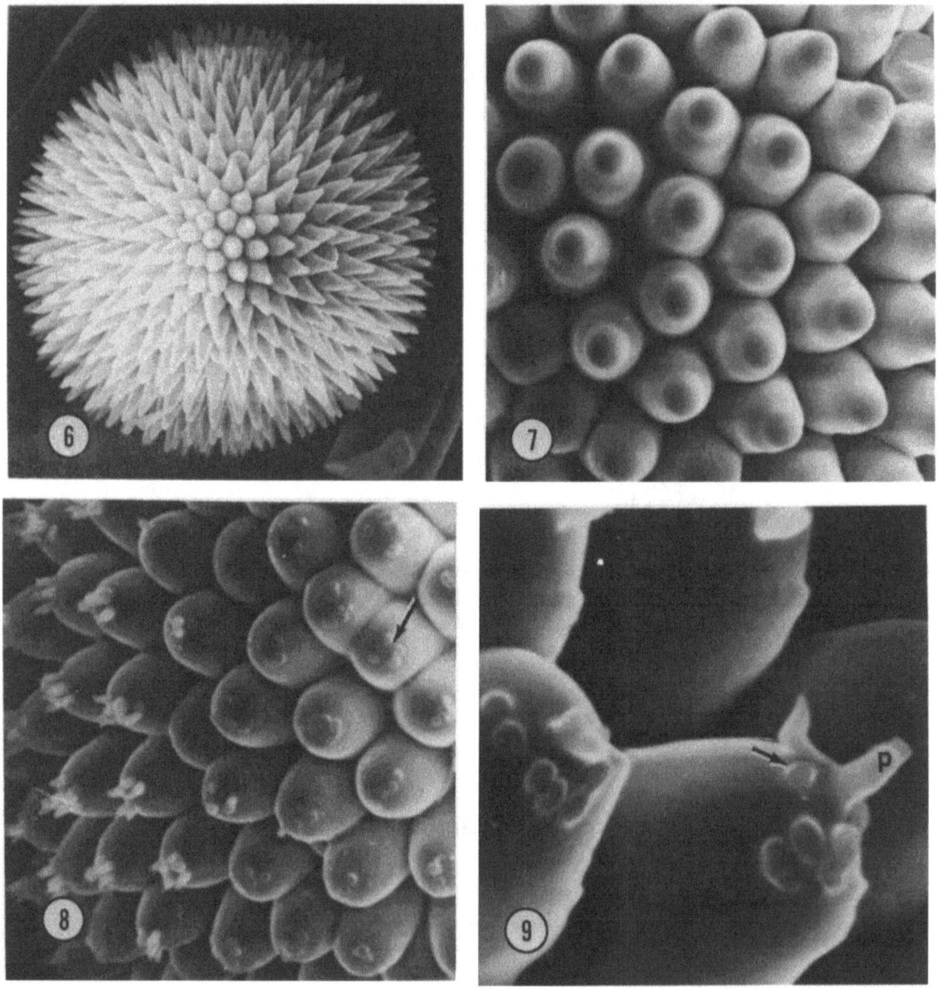

FIGURES 6–9. ▪*6.* SEM of a young aster of *Geodia cydonium*. Note that each ray emerges from the spicule center which is solid (see Fig. 12). The ray tips are all sharp-pointed. × 900. ▪*7.* Rays tips of older aster. Ends are rounded and "rings" below tips suggest recent silica deposition. × 3,400. ▪*8.* Short, upright processes develop at the tip of each ray at a more advanced stage. At the right (arrow), processes were broken during preparation. × 2,800. ▪*9.* Ends of two rays as in Fig. 8. A few upright processes (**p**) are intact, others are broken and in some an axial structure (arrow) is faintly visible. × 11,000.

silicalemma and the silicification process then spreads outward to other areas (Volcani, 1981). Based upon the sponge system at least some of this spreading may be due to the movement of silicic acid within the silicalemma rather than to the activation of transport at more distant sites.

For the most part, the contours of the surfaces of sponge spicules are notably regular; that is, they are not "lumpy" or otherwise misshapen. We can ask how they come to possess such uniformity and regularity. The absence of discrete, surface growth incre-

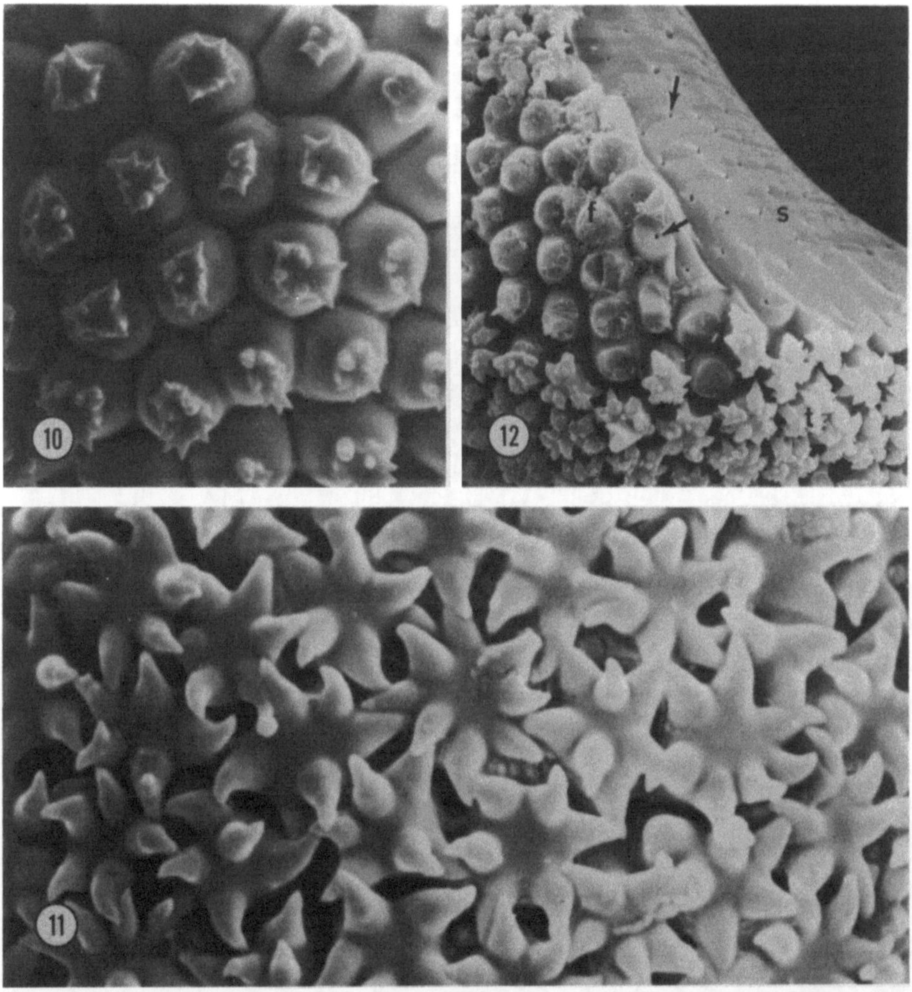

FIGURES 10–12. ▪*10.* Later stage than Fig. 9 showing filling of silica between previously isolated, upright, short processes at ends of rays. × 3,300. ▪*11.* Mature ray tips. Each is a star-shaped structure in which the points began as short processes as in Figs. 9 & 10. × 4,900. ▪*12.* Portion of a fractured, mature aster showing intact star-shaped tips (t). fractured ray tips (f), the solid interior of the structure (s), and the axial filament (arrows) which form the core of each ray and traverses the center of the spicule. × 11,800.

ments suggests that polymerization is a continuous, highly uniform process. If, indeed, silicic acid is continuously moved (diffused or otherwise translocated) within the silicalemma this may account for this uniformity of surface contours. Interestingly, in young, *normally* developing structures, transient swellings in the silica can occur approximately at the midpoint of an axis (Simpson, 1981). Such transient accumulations could be a result of the initial turning on of the silicification center. The control of the succeeding, uniform deposition of silica may be greater through movement and then immobilization of silicic acid within the silicalemma, than it could be by direct control of the rate of transport across the whole surface of the silicalemma. Under the latter

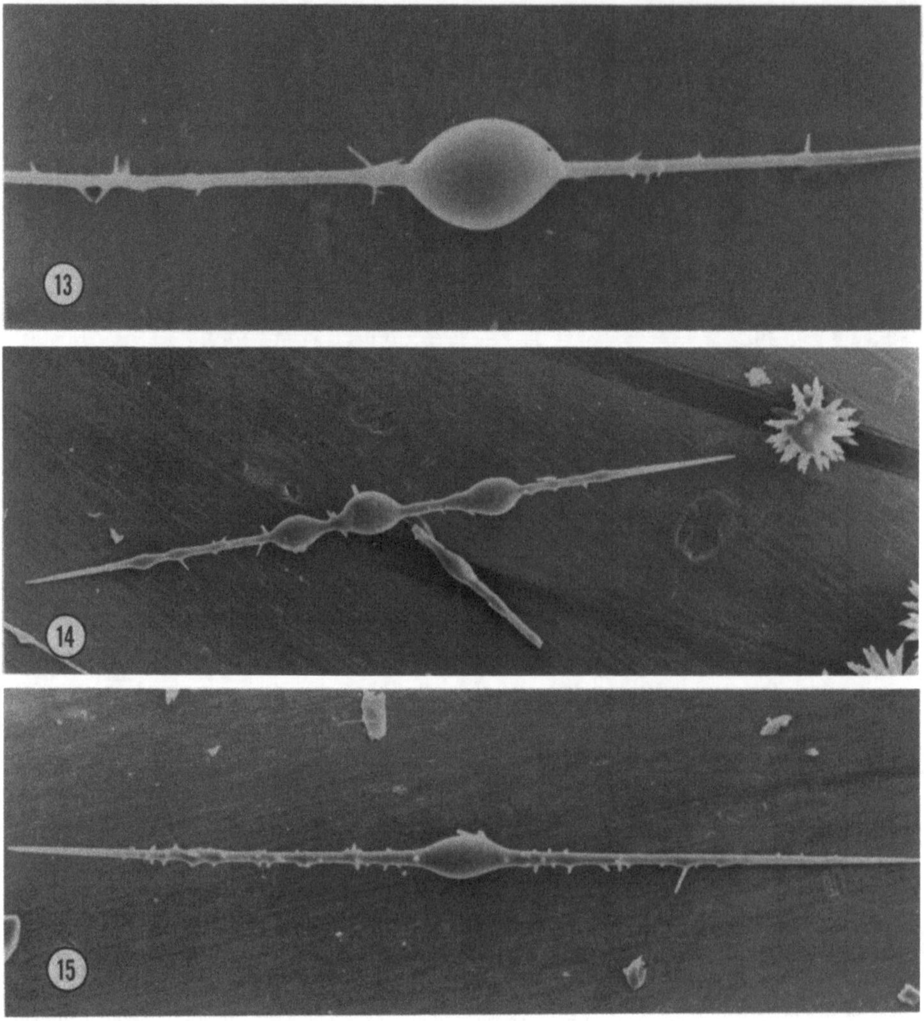

FIGURES 13–15. ▪*13.* Spicule of *Ephydatia muelleri* formed in control medium to which was added sufficient germanic acid to give Ge/Si molar ratio of 0.1. Note prominent, central bulb of accumulated silica. Short spines are also present along the shaft. × 1,100. ▪*14.* Spicule formed in control media with Ge/Si of 0.1 plus 10 mM NaCl. Note secondary bulbs of accumulated silica lateral to the central bulb. × 1,100. ▪*15.* Spicule formed in control media with Ge/Si of 0.1 plus 6.7 mM $CaCl_2$. Note the numerous spines lateral to the central bulb and the thicker shaft. × 1,100.

conditions literally all pumping sites would have to be functioning at precisely the same level. Stated in other terms, uniformity of morphology may be more readily achieved through localized silicic acid accumulation followed by translocation and polymerization than by global, equal transport with little translocation. However, as presented in the following section, the pattern of development of at least some morphological types of structures makes it difficult to visualize the presence of silicification centers and mobility as the only basis for silica deposition.

TABLE 1. Effects of sodium and calcium on the response togermanium in the fresh-water sponge *Ephydatia muelleri*

EXPERIMENTAL CONDITIONS	N	PERCENT SPICULES WITH SHAFT SPINES	CENTRAL BULBS	2° BULBS	NO BULBS
Control	88	6.8	77.3	14.8	7.9
Control + 10 mM NaCl	86	18.6	13.9	72.1	13.9
Control + 6.7 mM CaCl2	59	83.1	83.1	13.6	3.4

THE PATTERN OF SILICIFICATION OF ASTERS IN *GEODIA*

Hartman (1981) originally presented preliminary data on the development of the radial silica aster of a species of *Geodia*. The data presented here is considerably more comprehensive and deals with *Geodia cydonium* collected from the Adriatic Sea. The data demonstrate the complex nature of silica deposition and the difficulty of invoking the involvement of silicification centers in all processes. Methods for preparation and examination of sponge spicules with scanning electron microscopy (SEM) have been previously described in detail (Simpson *et al.*, 1983). The youngest spicules observed consist of numerous, highly pointed rays all of which radiate out from a smaller center (Fig. 6). Based upon data from another species (Simpson *et al.*, 1985a) these structures are each formed within a single cell. This early stage in development could involve a silicification center at the spicule center but the pathway of silicic acid transport *to* the center from the extracellular space is difficult to envisage because the spicule fills the cell as it is deposited and the spicule center is positioned farther and farther from the cell surface. Furthermore, as the spicules grow the central region becomes more and more silicified. As this occurs the rays grow outward and become blunt ended (Fig. 7). Then, at the end of each ray a cluster of small upright processes develops (Fig. 8). Within each of the processes an axial filament appears to be present (Fig. 9). The space between them then becomes silicified and with continued filling and lateral growth of the processes (Fig. 10) the end of each ray finally forms a star-shaped structure (Fig. 11). Such a process of infilling likely involves the expansion of the silicalemma. As shown in fractured structures an axial filament is present coring each major ray (Fig. 12). All of the filaments come together (not shown) at the spicule center and thus are branches of the same organic core.

The initial deposition of silica in these radial structures is indeed likely to begin at their center as appears to be the case in a previously studied spicule in which the *ends* of the structure are radial and in which silica accumulation occurs at the extremities in the presence of germanium (Simpson & Langenbruch, 1984). However, as the tips of the rays of *Geodia* asters mature (Figs. 7-12), morphogenesis of the silica takes place at points far removed from the spicule center. In this case it is difficult to consider that silicic acid transport/accumulation occurs centrally and is followed by translocation because the ray tips lie peripherally at a point closest to the extracellular environment from which silicic acid is secured. Thus, although not constituting proof of a diffuse mechanism for transport (see Fig. 5), this appears likely at least during the maturation of the ray tips. The final development of the star-shaped ray tips must involve the silicalemma since these terminal structures begin as discrete, upright processes. The surrounding membrane must expand in order for the areas between the processes to become filled with silica. This sequence emphasizes that in this case, and possibly others, there is coordinated (coupled) activity between the filament coring the small upright processes and the surrounding silicalemma as previously suggested (Simpson, 1981).

SILICIFICATION CENTERS AND THE PROBLEM OF MARINE SPICULES

Although germanium and ^{68}Ge have so far proven very useful probes for examining various aspects of sponge silicification in fresh-water species, these silicon analogs do not have the same morphological effect upon siliceous structures deposited by marine forms (Simpson *et al.*, 1985b). Specifically, no centers of silica accumulation develop. Germanium is, as determined by SEM-coupled, wavelength dispersive, X-ray analyses, evenly incorporated into the silica which lacks bulbous regions. On the other hand, germanium inhibits the growth in length of these structures as in fresh-water forms (Simpson *et al.*, 1979). In some cases when growth is severely affected, a single spherical structure is formed. A possible basis for this dramatic difference is the very significant difference in ion concentration in sea water versus fresh water (Simpson *et al.*, 1985b). We have initiated a first series of experiments to test the possible role of ions in the lack of response of marine spicules to germanium. The experimental conditions employed are the same as those previously used (Simpson *et al.*, 1983) except that in one series of experiments the sodium concentration (as NaCl) was elevated to 10 mM and in a second series the calcium (as $CaCl_2$) concentration was brought to 6.7 mM. Gemmules of the fresh-water sponge *Ephydatia muelleri* were grown in these media all of which contained a Ge/Si molar ratio of 0.1 (Simpson *et al.*, 1983). In the control medium containing Ge/Si 0.1, the spicules develop large central bulbs as usual (Fig. 13) and *some* short spines on the shaft of some spicules. In 10 mM sodium there is an obvious increase in the incidence of structures which possess secondary bulbs, located lateral to the central one (Fig. 14) and in 6.7 mM calcium there is a dramatic increase in the incidence of spicules which possess spines along the shaft (Fig. 15). The results are summarized in Table 1 and while there remain many more experiments to be carried out, there is evidence that both of these ions can affect the usual germanium response of the system. The increase in incidence of secondary bulbs suggests a stimulation of secondary centers for accumulation and as documented in diatoms this is possibly due to a Na-ATPase pump which transports silicic acid (Bhattacharyya & Volcani, 1980). The increase in spination in elevated calcium suggests either a direct or indirect role of this divalent cation in silicification. Each of these ions partially antagonizes the germanium response suggesting that, indeed, their elevated concentration in sea water may contribute to the absence of a morphological effect of germanium in marine spicules. However, it may also be that there are other significant differences between the fresh-water and marine silicification systems which lead to an absence of a typical germanium response in the latter. For example, the transport and/or accumulation mechanism in them may be much more permissive (less specific) thus allowing free translocation of the Ge-Si hybrid product and its intermediate forms. Much future work remains to be undertaken in order to fully appreciate the differences.

CONCLUSIONS

The effects of germanium on the freshwater sponge silicification system suggests the presence of silicification centers. The morphogenesis of sponge spicules must now include consideration of these centers; namely, the silicalemma-axial filament basis for sponge silicification appears to be overlain by the development of localized regions of activity. Consequently, the morphogenesis of spicule shape appears to involve a dynamic component in addition to the morphological components. In the case of *Geodia* asters and possibly in some other morphological types of sponge spicules (see Hartman, 1981) some aspects of morphogenesis may not be heavily influenced by localized centers. The absence of accumulated silica in response to germanium in marine spicules may be related, at least in part, to the ionic strength of the medium indicating preliminarily some relationship between both sodium and calcium and the silicification process.

ACKNOWLEDGEMENTS

Part of the work reported here was carried out during the tenure of a Foreign Exchange Scientist position in the U.S. France Cooperation in Science Program (NSF-CNRS). Thanks go to the following laboratories for their assistance: Laboratoire d'Histologie Experimentale, Universite Claude Bernard, Villeurbanne, France; Istituto di Zoologia Ed Anatomia Comparata, Universita di Bari, Bari, Italy; Zoologisches Institut, Entwicklungsgeschichtliche Abteilung, Universitat Bonn, Bonn, West Germany. Participation in the Symposium was made possible by grants from The Organized Research Fund of The University of Texas at Arlington and The Sea Grant Program administered by Texas A & M University.

REFERENCES

BHATTACHARYYA, P. & VOLCANI, B.E., 1980. Sodium dependent silicate transport in the apochlorotic marine diatom *Nitzschia alba*. Proc. Nat. Acad. Sci. U.S.A., 77:6386-6390.

DAVIE, E.I., SIMPSON, T.L., & GARRONE, R., 1983. Experimental germanium incorporation into siliceous sponge spicules. Biol. Cell., 48:191-201.

GARRONE, R., 1961. Collagene, spongine et squellette mineral chez l'eponge *Haliclona rosea* (O. S.) (Demosponge Haploscleride). J. Microsc., 8:581-598.

GARRONE, R., SIMPSON, T.L., & POTTU-BOUMENDIL, J., 1981. Ultrastructure and deposition of silica in sponges. *In* Silicon and Siliceous Structures in Biological Systems (ed., T.L. Simpson & B.E. Volcani), pp. 495-525. New York: Springer Verlag.

HARTMAN, W.D., 1981. Form and distribution of silica in sponges. *In* Silicon and Siliceous Structures in Biological Systems (ed. T.L. Simpson & B.E. Volcani), pp. 453-493. New York: Springer Verlag.

MACKIE, G.O. & SINGULA, C.L., 1983. Studies on hexactinellid sponges. I. Histology of *Rhabdocalyptus dawsoni* (Lambe). Phil. Trans. Roy. Soc. Lond., 301:365-400.

REISWIG, H.M., 1971. The axial symmetry of sponge spicules and its phylogenetic significance. Cahiers Biol. mar., 12:505-514.

SIMPSON, T.L., 1981. Effects of germanium on silica deposition in sponges. *In* Silicon and Siliceous Structures in Biological Systems (ed., T.L. Simpson & B.E. Volcani), pp. 529-550. New York: Springer Verlag.

SIMPSON, T.L., 1984. The Cell Biology of Sponges. New York: Springer Verlag.

SIMPSON, T.L. & LANGENBRUCH, P-F., 1984. Effects of germanium (Ge) on the morphogenesis of a complex silica structure and on the assembly of the collagenous gemmule coat in a freshwater sponge. Biol. Cell., 50:181-190.

SIMPSON, T.L. & VACCARO, C.A., 1974. An ultrastructural study of silica deposition in the fresh water sponge *Spongilla lacustris*. J. Ultrastr. Res., 47:296-309.

SIMPSON, T.L., REFOLO, L.M., & KABY, M.E., 1979. Effects of germanium on the morphology of silica deposition in a fresh water sponge. J. Morph., 159:343-354.

SIMPSON, T.L., GARRONE, R., & MAZZORANA, M., 1983. Interaction of germanium (Ge) with biosilicification in the freshwater sponge *Ephydatia muelleri*: Evidence of localized membrane domains in the silicalemma. J. Ultrastr. Res., 85:159-174.

SIMPSON, T.L., LANGENBRUCH, P-F., & SCALERA-LIACI, L., 1985a. Silica spicules and axial filaments of the marine sponge *Stelletta grubbi* (Porifera, Demospongiae). Zoomorphology, 105:375-382.

SIMPSON, T.L., GIL, M., CONNES, R., DIAZ, J-P., & PARIS, J., 1985b. Effects of germanium (Ge) on the silica spicules of the marine sponge *Suberites domuncula*: Transformation of spicule type. J. Morph., 183:117-128.

SULLIVAN, C.W. & VOLCANI, B.E., 1981. Silicon in the cellular metabolism in diatoms. *In* Silicon and Siliceous Structures in Biological Systems (ed. T.L. Simpson & B.E. Volcani), pp. 15-42. New York: Springer-Verlag.

VOLCANI, B.E., 1981. Cell wall formation in diatoms: Morphogenesis and biochemistry. *In* Silicon and Siliceous Structures in Biological Systems (ed. T.L. Simpson & B.E. Volcani), pp., 157-200. New York: Springer Verlag.

STRUCTURE OF MOLLUSCAN PRISMATIC SHELL LAYERS

Harry Mutvei

Swedish Museum of Natural History, Department of Palaeozoology, S-10405 Stockholm, Sweden

ABSTRACT. Structure of the calcitic megaprisms is studied with SEM in the outer shell layer of two archaeogastropod species: *Haliotis rufescens* Swainson, from the Pacific at San Francisco, and *Haliotis tuberculata* L., from the Atlantic, Dinard, France. In *H. rufescens* where the calcitic megaprisms have a regular size, shape and orientation, the prisms are traversed by horizontal organic intracrystalline sheets. These sheets are regularly spaced (about 1 micron apart) and have a parallel fibrous structure. The occurrence of well defined intracrystalline sheets in the prismatic structures has not been reported previously. In *H. tuberculata* where the calcitic megaprisms occur together with aragonitic granules, and are precipitated at irregular intervals, the prisms have a highly varying size, shape and orientation. The intracrystalline organic matrix seems to form also here horizontal sheets but these are fragile and composed of very thin fibres with reticulate arrangement. Consequently, there exists a close correlation between the organization of the calcitic megaprisms and the intracrystalline organic matrix.

Extremely thin, horizontal, organic sheets are described in the semi-prismatic shell layer in the cephalopod *Nautilus pompilius*. The spacing of these sheets is similar to that of the interlamellar organic sheets in the adjacent nacreous layer. The following structural changes take place when the semi-prismatic layer is transformed into the nacreous layer: (1) the acicular crystallites acquire a strictly vertical orientation and form nacreous tablets which become twinned, and (2) the simple, horizontal, soluble organic sheets increase in thickness, become three-layered and partially insoluble.

INTRODUCTION

Specimens of the recent archaeogastropod genus *Haliotis* show a highly varying mineralogical composition of the outer shell layer. This layer may be aragonitic, calcitic, or mixed aragonitic/calcitic. The occurrence of two $CaCO_3$ polymorphs in one and same shell layer is extremely rare in invertebrate exoskeletons. In addition to *Haliotis* (Mutvei *et al.*, 1985), it has been described in a single trochid archaeogastropod species: *Cittarium pica* (L.) (Wise & Hay, 1968; Erben, 1971), and in shell regenerates of the freshwater Pulmonate snails (Watabe & Dunkelberger, 1979).

The present paper deals with structural differences in calcite megaprisms in two *Haliotis* species: *H. rufescens* and *H. tuberculata*. In *H. rufescens* the outer shell layer consists exclusively of calcitic megaprisms, whereas in *H. tuberculata* the calcitic megaprisms occur together with aragonitic granules of spherulitic/prismatic structure.

Origin, Evolution, and Modern Aspects of Biomineralization in Plants and Animals
Edited by Rex E. Crick
Plenum Press, New York

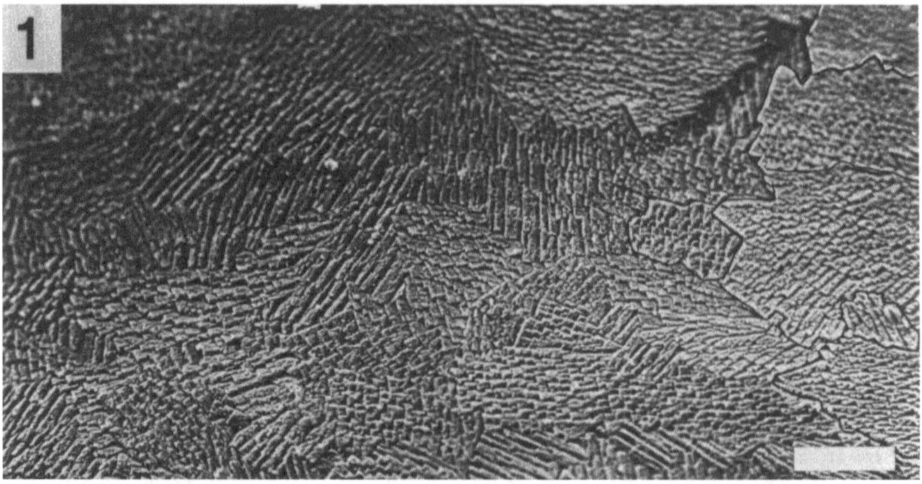

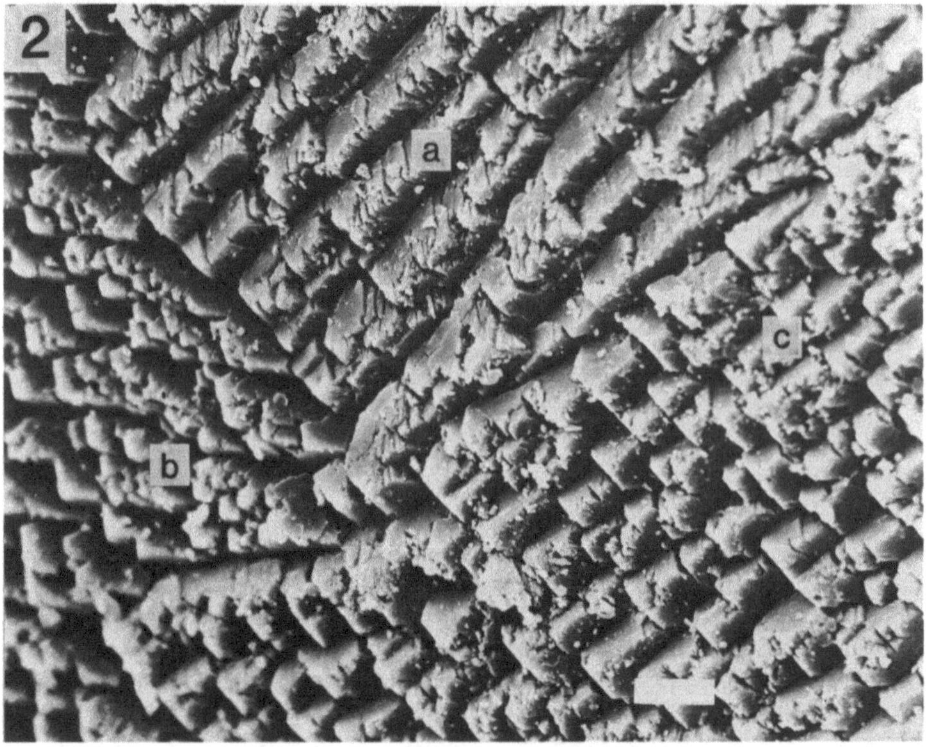

FIGURE 1 & 2. ▪*1 - Haliotis rufescens.* Growth surface of calcitic megaprisms at shell margin; outlines of megaprisms indicated on right hand side. Bar = 100 μ. ▪ *2 - Haliotis rufescens.* Detail of growth surface of three adjacent megaprisms (a,b,c); note the different orientation of the minor prismatic units in each megaprism. Bar = 10 μ.

Particular attention was paid on the structure and arrangement of the intracrystalline organic matrix in the megaprisms. This matrix was made insoluble by using various fixatives. As generally agreed, the soluble organic fraction plays a dominant role in shell mineralization by binding Calcium and initiating nucleation (for the review see Crenshaw, 1982). Studies of the spatial relationship between the soluble organic fraction and the crystalline components are therefore of great importance for understanding the shell formation.

In addition to the calcitic megaprisms in *Haliotis*, the semi-prismatic shell layer in the cephalopod *Nautilus pompilius* L. was investigated with respect to its soluble organic matrix. This layer can be easily transformed into the structurally more complex nacreous layer, and vice versa.

MATERIALS AND METHODS

Specimens

Archaeogastropods: *Haliotis rufescens* Swainson (Red Abalone), Pacific, north of San Francisco. *Haliotis tuberculata* L. Atlantic, Dinard, France.

Cephalopod: *Nautilus pompilius* L. Pacific, Solomon Islands.

Preparation Methods

The following methods of preparation were used.

In order to study the growth surfaces of the megaprisms in the outer shell layer of *Haliotis* these surfaces were cleaned from organic matrix by a treatment with sodium hypochlorite (8-12% active chlorine) for 20 minutes.

For studies of the soluble intracrystalline organic matrix, numerous horizontal and vertical shell sections were polished with aluminum oxide and etched either with chromium sulphate for 30 seconds (Mutvei, 1972a,b), or with 25% glutaraldehyde at +40°C for 20 minutes. The preparations etched with glutaraldehyde were dehydrated in alcohol and acetone, and dried in critical point apparatus.

In some polished sections of *Haliotis* shell the intracrystalline organic matrix was dissolved by a treatment with the proteolytic enzyme trypsin, dissolved in 0.4 M cacodylate buffer, at pH 8.2 and +40°C for 5 days.

RESULTS

Haliotis rufescens Swainson

Calcitic megaprisms. - Nakahara *et al.* (1982) described the outer shell layer in this species as composed of "calcite crystallites". According to the terminology proposed by Suzuki & Uozumi (1981), this layer in our adult specimens is composed of calcitic megaprisms. In horizontal thin sections studied in polarized light with crossed nicols, each prism appears as a single crystallographic unit. The crystallographic orientation of the adjacent prisms is different. The diameter of the megaprisms ranges from 100 to 200 microns (Fig. 1). The growth surface of the megaprisms, from which the organic matrix was removed by sodium hypochlorite treatment, shows in SEM preparations numerous (30-100) minor prismatic units. These units are about 10 microns in diameter

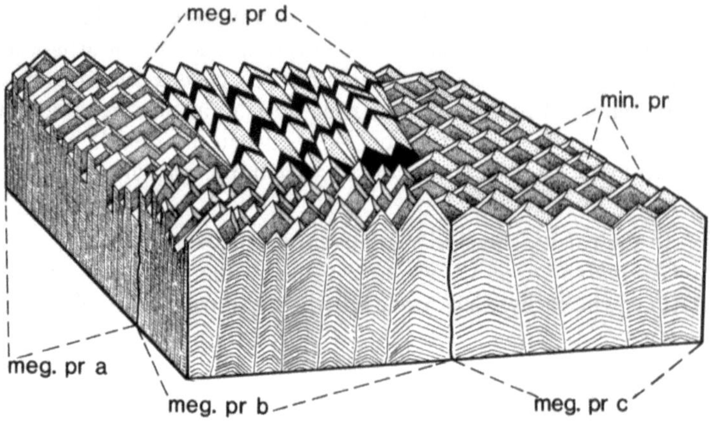

FIGURE 3. Haliotis rufescens. Block diagram of growth surface of four adjacent megaprism (**meg.pr**) (a,b,c,d): note different orientation of minor prismatic units (**min.pr**) in each megaprism.

and have a uniform shape and orientation in each megaprism (Figs. 1, 2, 3). In adjacent megaprisms, the crystal faces of the minor prismatic units are somewhat differently oriented (Figs. 1, 2, 3). The interfacial angles could not be measured in our SEM photographs.

According to Nakahara *et al.* (1982:35), the main portion of the growth surface of the outer shell layer "is composed of somewhat rhomboid-shaped minerals that generally exhibit an imbricated arrangement". These writers studied very young shells. It is therefore probable that the shape of the minor prismatic units (= "rhomboid-shaped minerals") changes with increasing age, and that these prismatic units become arranged into well defined megaprisms.

Vertical polished and etched sections reveal that each minor prismatic unit is composed of numerous vertical acicular crystallites, about 0.1-0.2 μ in diameter. These crystallites are oriented parallel to the long axis (c-axis) of each megaprism (indicated by arrows in Figs. 4, 5). However, this orientation is slightly different in adjacent megaprisms (indicated by arrows in Fig. 8). The latter condition explains the different crystallographic orientation of the cross sections of the megaprisms in polarized light.

Organic sheets. - As pointed out by Nakahara *et al.* (1982:35), the growing "crystallites" in the outer layer are surrounded by an electron-dense envelope. The results of the present study show that the megaprisms are traversed by numerous intracrystalline organic sheets, not reported previously in the molluscan prismatic layers. The horizontal sheets are regularly spaced, one micron or somewhat less apart. The regular spacing of these sheets is best seen in the middle part of the Figure 4. They seem to traverse the acicular crystallites without interrupting their vertical continuity (Fig. 5). They have a zigzag course which reflects the shape of the angular surfaces of the minor prismatic units in the megaprisms (compare Fig. 2 and Fig. 8). It can therefore be concluded that each organic sheet in question originally covers the growth surface of the minor prismatic units, and then becomes embedded into the prisms when these continue to grow in vertical direction.

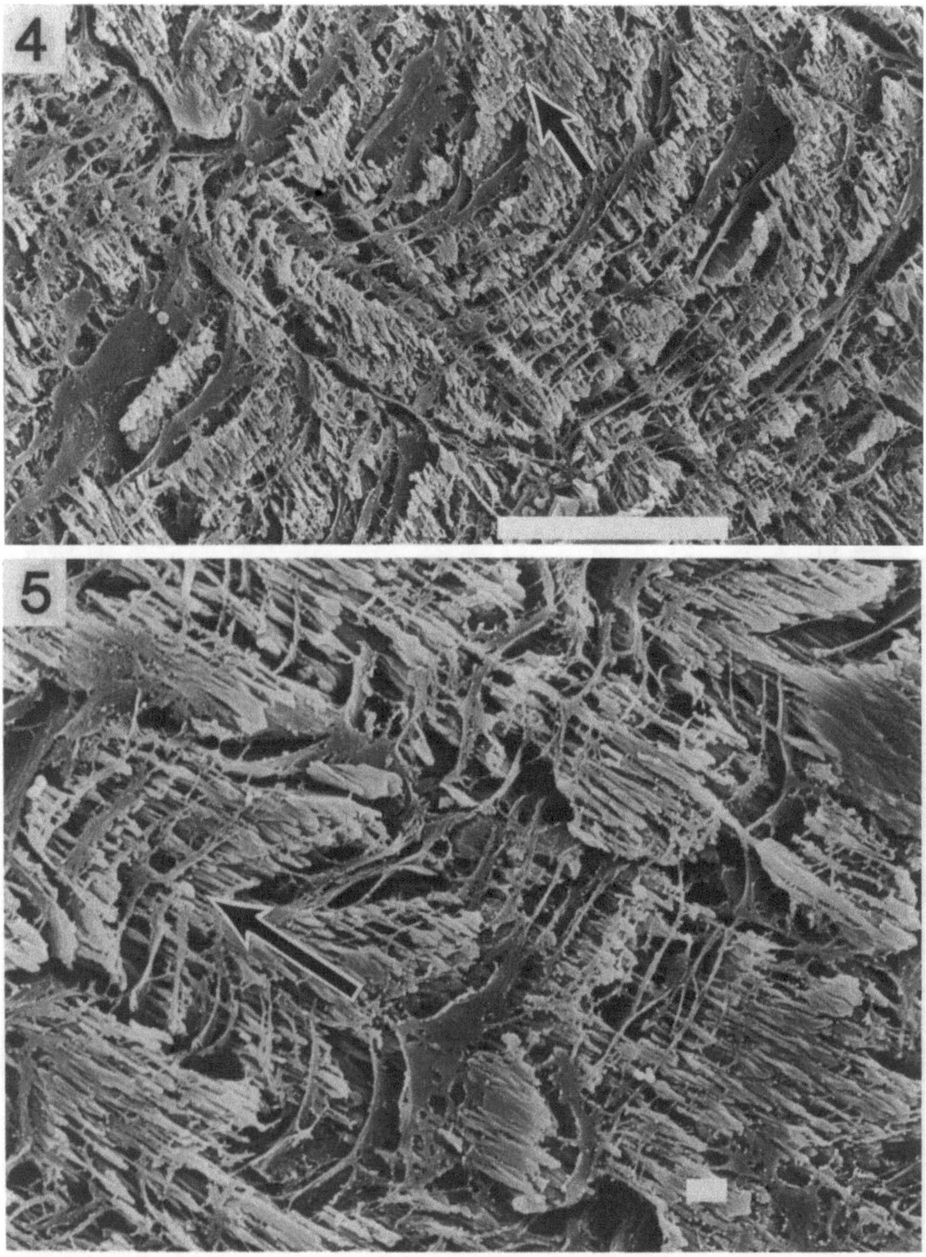

FIGURE 4 & 5. Haliotis rufescens. Vertical polished sections of calcitic megaprisms etched with glutaraldehyde solution to show acicular crystalline components and regularly spaced, horizontal organic sheets; directions of c-axis in prisms indicated by arrows. Bar (Fig. 4) = 10 μ; bar (Fig. 5) = 1 μ.

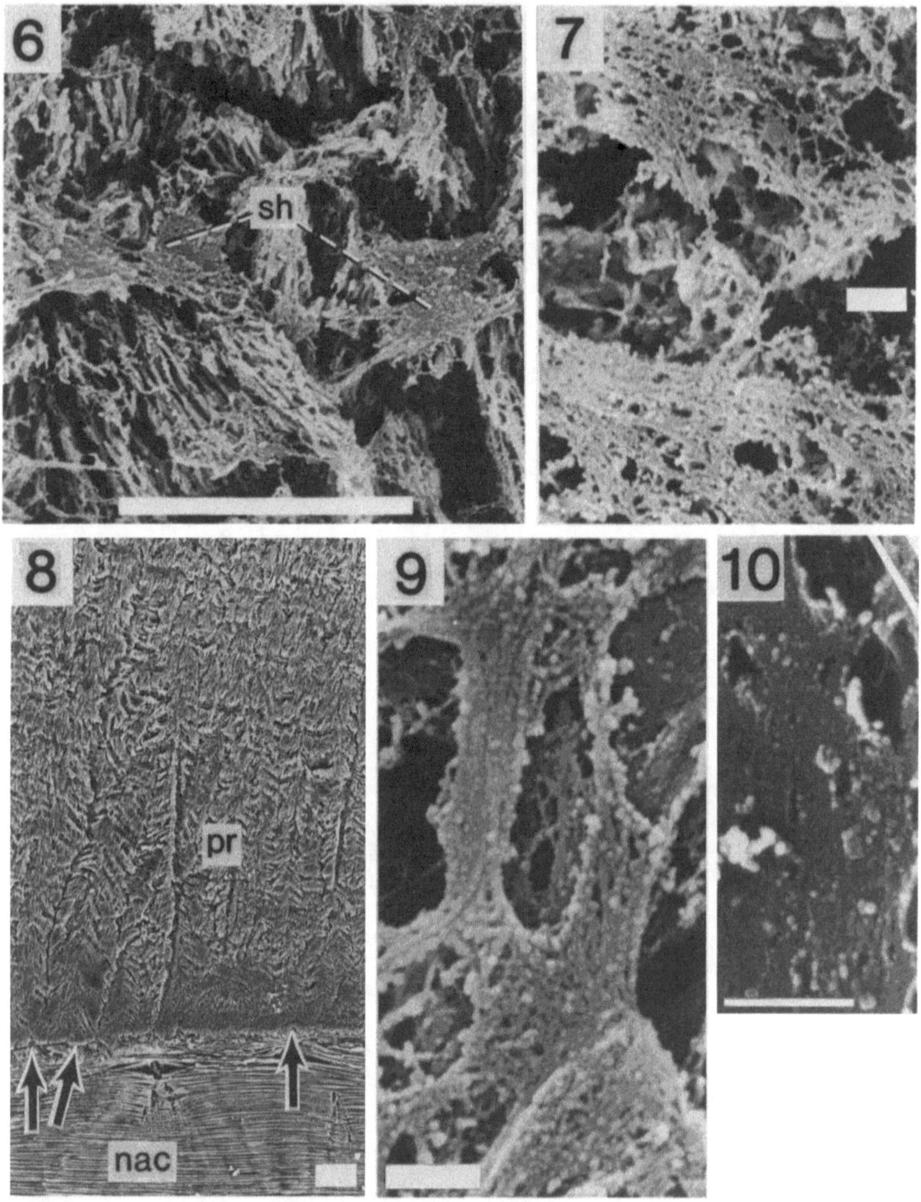

FIGURE 6–10. Haliotis tuberculata. ▪*6.* Remnants of horizontal organic sheets (**sh**) in spherulitic/prismatic (aragonitic) portion of outer shell layer. Bar = 10 μ. ▪*7.* Organic sheet in higher magnification to show fibres and their reticulate arrangement. Bar = 1 μ. ▪*8. Haliotis rufescens.* Vertical sections of the nacreous (**nac**) and the calcitic prismatic (**pr**) layer etched with glutaraldehyde; directions of *c*-axis in prisms indicated by arrows. Bar = 10 μ. ▪*9, 10.* Parallel fibrous structure of horizontal organic sheets. Bar = 1 μ.

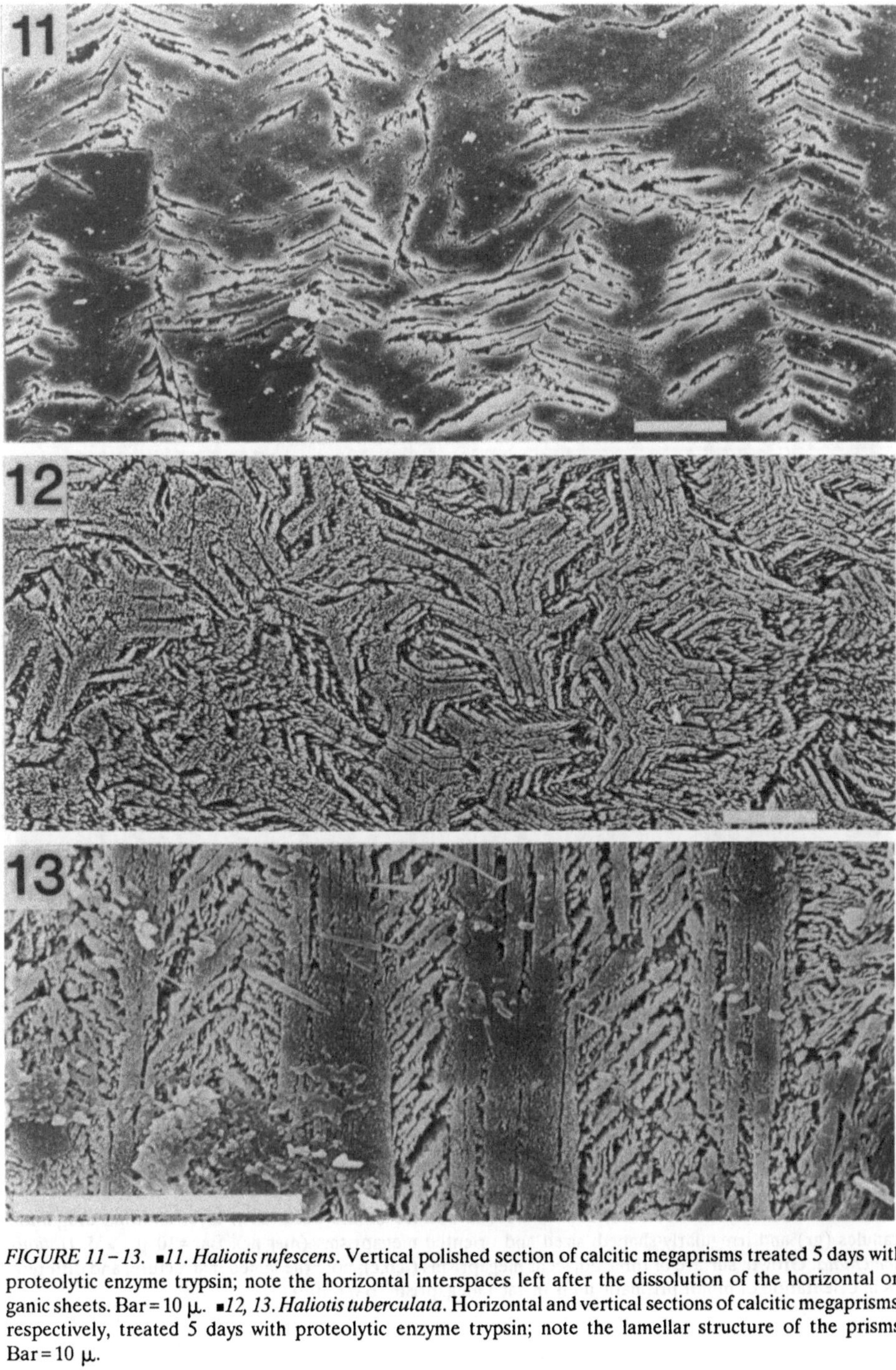

FIGURE 11–13. ▪*11. Haliotis rufescens.* Vertical polished section of calcitic megaprisms treated 5 days with proteolytic enzyme trypsin; note the horizontal interspaces left after the dissolution of the horizontal organic sheets. Bar = 10 μ. ▪*12, 13. Haliotis tuberculata.* Horizontal and vertical sections of calcitic megaprisms, respectively, treated 5 days with proteolytic enzyme trypsin; note the lamellar structure of the prisms. Bar = 10 μ.

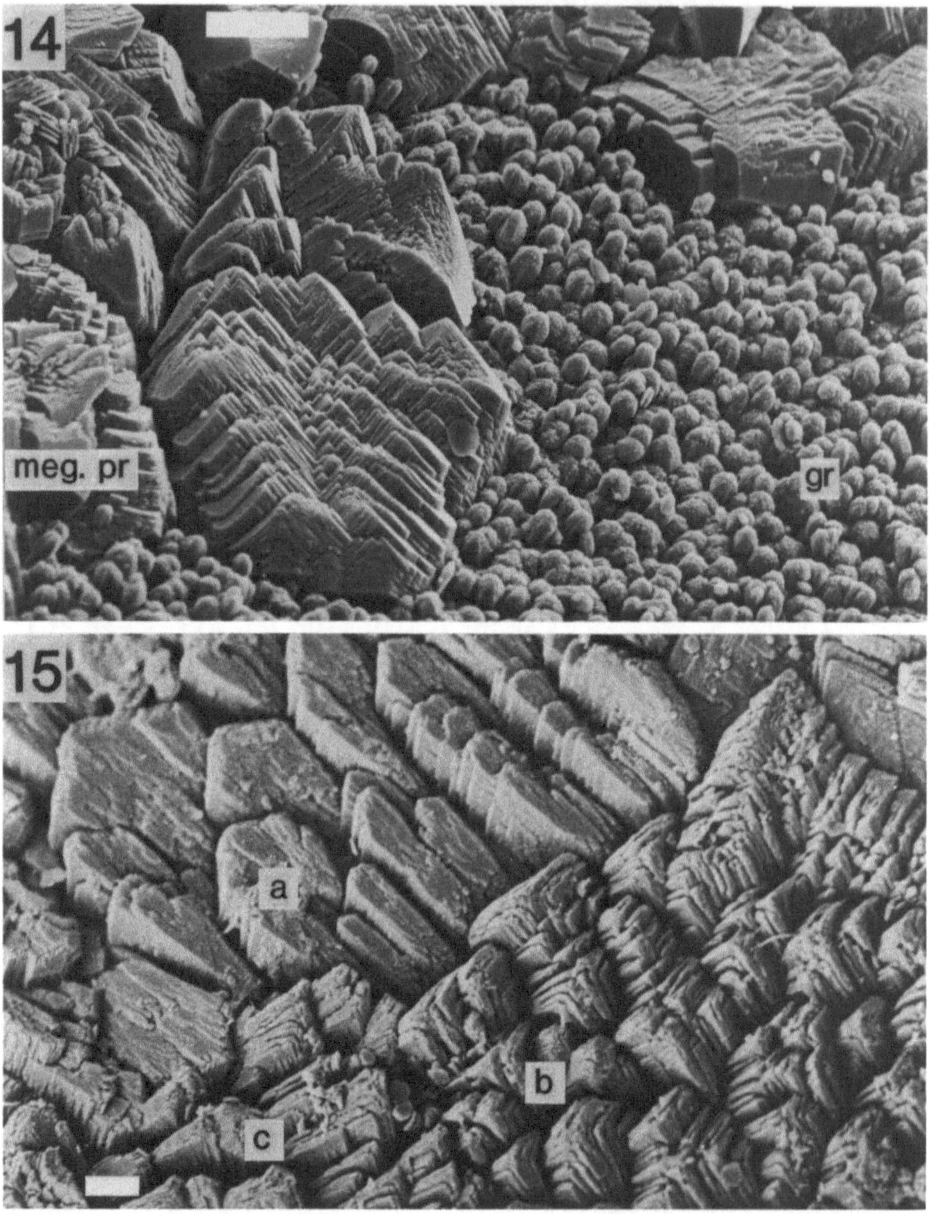

FIGURE 14–15. ▪*14. Haliotis tuberculata.* Growth surface of the outer shell layer showing aragonitic granules (**gr**) and irregularly shaped, sized and oriented megaprisms (**meg.pr**). Bar = 10 μ. ▪*15. Haliotis tuberculata.* Growth surface of three adjacent megaprisms (**a**,**b**,**c**); note the lamellar structure, and differences in orientation of minor prismatic units in each megaprism. Bar = 10 μ.

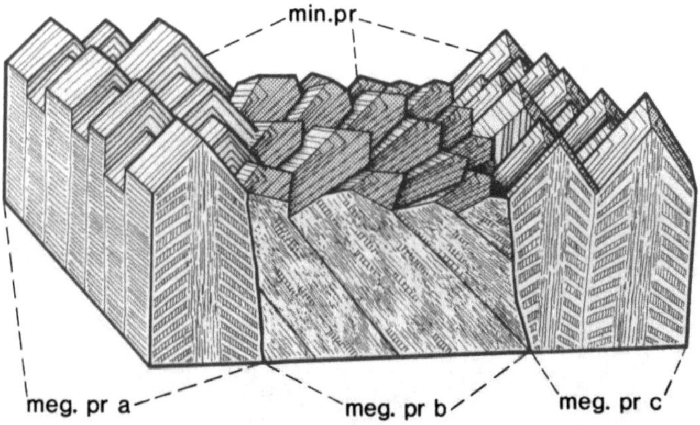

FIGURE 16. Haliotis tuberculata. Block diagram of growth surface of three adjacent megaprisms (**meg.pr, a,b,c**); note different orientation of minor prismatic units (**min.pr**) in each megaprism.

The horizontal intracrystalline sheets are fully calcified and cannot therefore be dissolved in sodium hypochlorite, not even after a prolonged treatment. However, they dissolve slowly in the proteolytic enzyme trypsin. After the latter treatment, narrow horizontal and regularly spaced interspaces appear at the sites of the horizontal intracrystalline organic sheets (Fig. 11).

The horizontal organic sheets are comparatively sturdy and therefore clearly visible after glutaraldehyde etching and critical point drying. In high magnification they show a composition of parallel fibres (Figs. 9, 10). These fibres have a diameter of about 0.1 m and tuberculated surfaces.

Haliotis tuberculata L.

Calcitic megaprisms. - As pointed out (Mutvei *et al.*, 1985), the outer shell layer in this species has a mixed aragonitic/calcitic composition. The aragonite forms granules of spherulitic/prismatic structure. Heavily etched spherulitic granules are shown in Figure 6. The calcitic megaprisms vary considerably in size, shape and orientation. They are precipitated at irregular intervals in the basal portion of the outer shell layer. In several places, numerous aragonitic granules are trapped into the calcitic megaprisms (see Mutvei *et al.*, 1985, Pl. 2, Figs. 1, 2, 3, 6; Pl. 3, Figs. 1, 3, 6).

The growth surface of the outer shell layer, cleaned from the organic matrix by sodium hypochlorite treatment, shows in Figure 14 aragonitic granules (gr) and large calcitic megaprisms (meg.pr). The megaprisms have a highly varying shape and orientation. In Figures 15 and 16 the growth surface of three adjacent megaprisms is shown (meg.pr a, b, c). Each megaprism is composed of numerous minor prismatic units (min.pr) which have a distinct lamellar structure and a different orientation in each megaprism. In comparison with those in *Haliotis rufescens*, the megaprisms in *H. tuberculata* are (a) considerably more irregular in size, shape and orientation, and (b) their minor prismatic units are generally much larger and have a characteristically lamellar structure (compare Figs. 1, 2, 3, and Figs. 14, 15, 16).

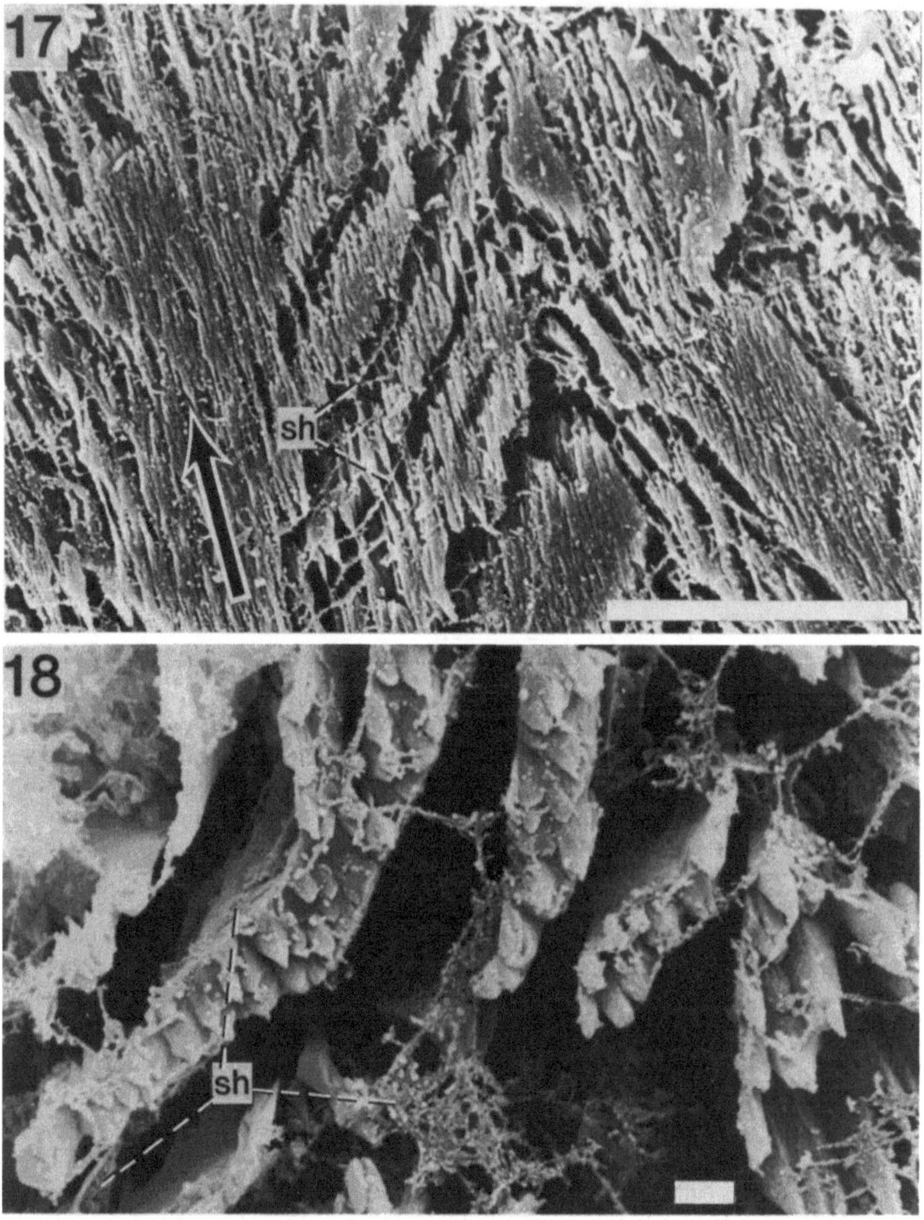

FIGURE 17–18. ▪*17. Haliotis tuberculata.* Vertical polished section of calcitic megaprisms, etched with glutaraldehyde showing acicular crystalline components and horizontal organic sheets (**sh**); direction of the *c*-axis of the megaprism indicated by arrow. Bar = 10 μ. ▪*18. Haliotis tuberculata.* Same as Fig. 17 except scale = 1 μ.

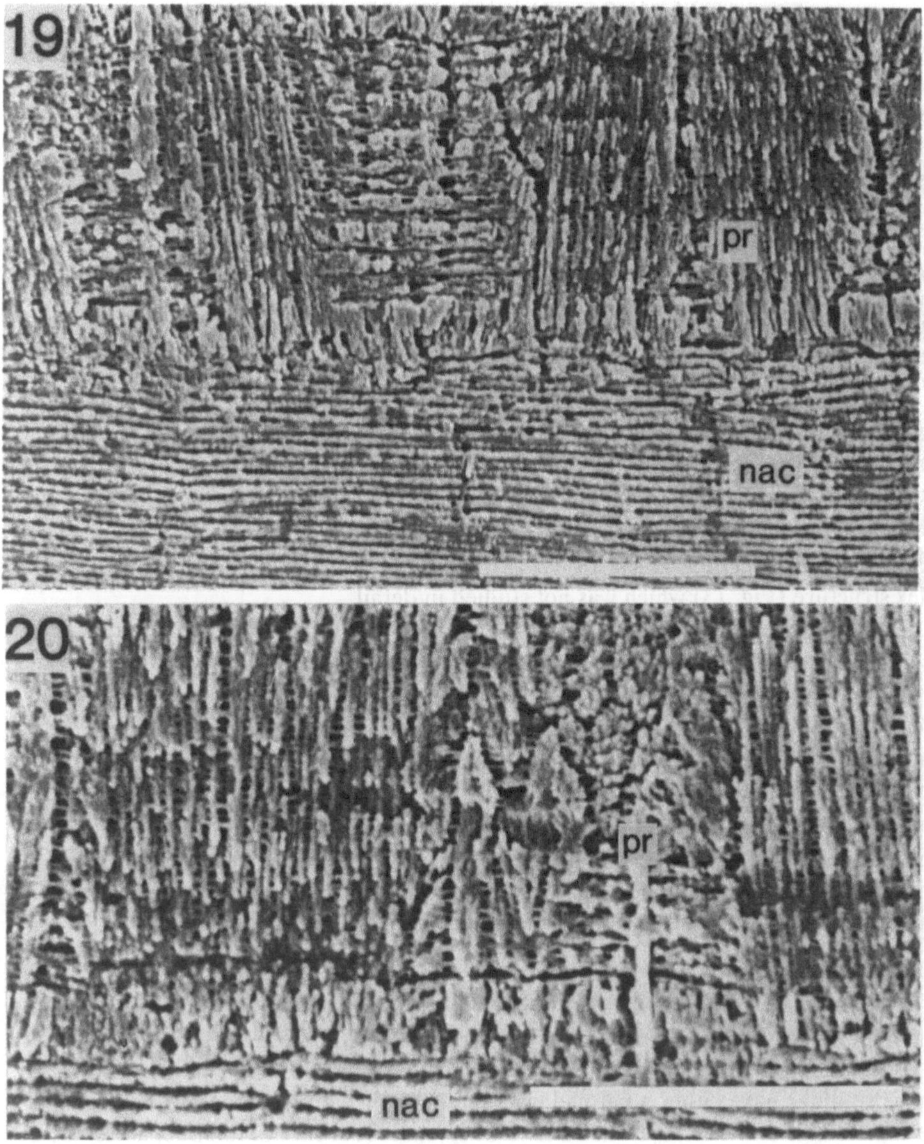

FIGURE 19–20. Nautilus pompilius. Vertical polished sections of the semi-prismatic myostracal layer (**pr**) and the adjacent nacreous layer (**nac**) etched with chromium sulphate; note the occurrence of regularly spaced, thin, horizontal organic sheets in the myostracal layer. Bar = 10 μ.

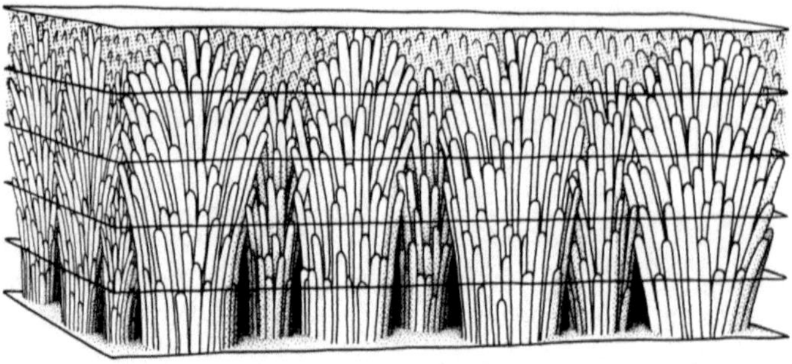

FIGURE 21. Nautilus pompilius. Block diagram of myostracal layer, to show bundles of acicular crystallites, traversed by thin, horizontal, organic sheets.

Horizontal and vertical polished and etched sections reveal that each minor prismatic unit is composed of acicular crystalline components, oriented parallel to the long axis of the prism (indicated by an arrow in Fig. 17). The shape and size of these acicular crystallites agree to those in *Haliotis rufescens* (compare with Figs. 4, 5). The lamellar structure of the minor prismatic units is elucidated in a horizontal section (Fig. 12) and in three vertical sections (Figs. 13, 17, 18). Some of the lamellae seem to have a complex structure (Fig. 13) which was not studied in detail.

Organic sheets. - Vertical polished sections of the outer shell layer, etched with glutaraldehyde, show remnants of extremely thin organic sheets. Figure 6 elucidates such remnants (sh) in the region composed of aragonitic granules. As indicated in Figures 17 and 18, the organic sheets (sh) seem to occur between the oblique horizontal calcitic lamellae of the minor prismatic units. These sheets seem to be composed of thin fibres with reticulate arrangement (sh, Figs. 6, 7, 18).

The intracrystalline organic sheets in *Haliotis tuberculata* differ considerably from those in *H. rufescens*: (a) they are much thinner, more fragile, and their spatial relationship to the mineral components is therefore more difficult to study, and (b) the fibres in the sheets seem to have a smaller diameter and reticulate arrangement.

Nautilus pompilius L.

Aragonitic myostracal layer. - As pointed out (Mutvei, 1964), the myostracal layer in the shell wall of *Nautilus* has a semi-prismatic structure. It is composed of acicular crystallites which are often arranged in bundles (pr, Figs. 19, 20, Fig. 21). In each bundle the crystallites have either a parallel orientation or they diverge more or less distinctly in distal direction. The adjacent bundles of crystallites seem not to be separated by interprismatic organic sheets as is the case in typical aragonitic and calcitic prisms.

Organic sheets. - In polished vertical sections of the myostracal layer, etched with chromium sulphate, numerous organic sheets are visible in the interspaces between the adjacent acicular crystallites (pr, Figs. 19, 20, Fig. 21). These sheets are extremely thin and seem to be horizontally continuous. The distances between the consecutive sheets vary from 0.2 m to 0.5 μ. The sheets seem to traverse the acicular crystallites without interrupting their vertical continuity.

In comparison with the interlamellar organic sheets in the adjacent nacreous layer (nac, Figs. 19, 20.), the organic sheets in the myostracal layer are (1) considerably thinner, (2) their spacing is somewhat smaller and less regular, and (3) they seem to consist entirely of soluble organic matrix, whereas the interlamellar sheets are partially insoluble (*e.g.* Grégoire, 1962; Mutvei; 1969,Bevelander & Nakahara, 1980).

Similar thin, horizontal, organic sheets were also found in the inner semi-prismatic portion of the outer shell layer of *Nautilus*.

DISCUSSION

The outer calcified shell layer in the species of the archaeogastropod genus *Haliotis* shows unique mineralogical and structural variations: (1) in one group of species this layer consists entirely of calcitic megaprisms, (2) in second group of species it has a mixed mineralogy in being composed of aragonitic granules and calcitic megaprisms, and (3) in third group of species it consists entirely of aragonitic granules with spherulitic/prismatic structure (Mutvei *et al.*, 1985).

In the present paper calcitic megaprisms were described and compared in two species: *Haliotis rufescens* (group 1) and *H. tuberculata* (group 2). In both species the megaprisms are composed of acicular crystallites, oriented parallel to the long axis (c-axis) of the prisms; the growth surface of each megaprism shows numerous minor prismatic units. However, in the following features the megaprisms exhibit considerable differences between these two species: (1) in *H. rufescens* the outer shell layer consists exclusively of calcitic megaprisms, where as in *H. tuberculata* such megaprisms are precipitated at irregular intervals in the outer shell layer which consists predominantly of aragonitic granules. (2) In *H. rufescens* the variations in size, shape and orientation of the megaprisms are considerably smaller than those in *H. tuberculata*. (3) In *H. rufescens* the minor prismatic units are horizontally traversed by regularly spaced, thin but sturdy, intracrystalline organic sheets with parallel fibres, whereas in *H. tuberculata* such sheets are much more fragile and seem to consist of thinner fibres with reticulate arrangement.

In conclusion, the differences in the structure of the mineral components are significantly correlated with the differences in the structure of the sheets of the intracrystalline organic matrix: well defined organic sheets occur together with regularly precipitated calcitic megaprisms, whereas poorly defined organic sheets occur together with irregularly precipitated calcitic megaprisms, mixed with granular aragonite. It is therefore justified to assume that the nature of these sheets excerts some influence on the precipitation of the calcitic megaprisms.

Nakahara & Bevelander (1971), Cuif *et al.* (1981) and Suzuki & Uozumi (1981) found that the intracrystalline organic matrix in the bivalve calcitic prisms does not form sheets but occurs as spongy-like horizons. The well developed horizontal organic sheets in the calcitic megaprisms in *Haliotis rufescens* differ therefore considerably from the spongy-like horizons in bivalves.

In addition to the regularly spaced, intracrystalline organic sheets in the calcitic megaprisms of *Haliotis*, organic sheets occur in the aragonitic semi-prismatic myostracal layer in *Nautilus*. Also in the latter layer the sheets show a regular spacing. In a previous study (Mutvei *et al.*, 1985), parallel organic filaments were described in the aragonitic-spherulitic outer shell layer in some *Haliotis* species. It is probable that these filaments are remnants of continuous organic sheets, but a conclusive evidence for such an interpretation is still missing.

The discovery of the regularly spaced, horizontal organic sheets in the simple aragonitic semi-prismatic shell layer helps us to understand better the structural differences between this layer and the nacreous layer. As pointed out (*e.g.* Mutvei, 1964, 1972a, b; Mutvei *et al.*, 1985; Bandel, 1977), the semi-prismatic/spherulitic layer and the nacreous layer are easily interchangeable in the molluscan shells. In *Nautilus* this takes place, for example: (a) in the septal necks where the spetal nacre acquires distally a more or less distinct prismatic structure, (b) at the boundary between the outer semi-prismatic layer and the nacreous layer of the shell wall, and (c) at the apertural margin of adult shells where the nacreous layer is successively transformed into the semi-prismatic layer. A similar structural transformation is also known in the shell septa of a Jurassic ammonoid (Kulicki & Mutvei, 1982).

It is still difficult to understand all the changes which have to take place during the transformation from the structurally simple spherulitic/semi-prismatic layer to the structurally complex nacreous layer, and vice versa. This transformation seems to require that the acicular crystallites, which form the spherulites and simple prisms (= sectors of spherulites), acquire a strictly vertical orientation and form the nacreous tablets which become twinned in a more or less complex manner. At the same time, the simple, thin, organic sheets of the semi-prismatic layer probably increase in thickness and form the interlamellar organic sheets which are composed of three layers and which are situated between the consecutive mineral lamellae of nacreous tablets. The interlamellar sheets are composed of both soluble and insoluble fraction (*e.g.* Grégoire, 1962; Mutvei, 1969; Bevelander & Nakahara, 1980), whereas the simple organic sheets in the semi-prismatic layer seem to be composed of the soluble fraction only.

ACKNOWLEDGEMENTS

Many of the results forwarded in the present paper have been obtained in co-operation with Dr. Y. Dauphin, Université Paris VI, and Drs. J.-P. Cuif and A. Denis, Université Paris XI (Orsay). This study was financially supported by Grant 287-115 of the Swedish Natural Science Research Council. Participation in the Sixth International Symposium on Biomineralization at The University of Texas at Arlington was made possible by grants from The Organized Research Fund of The University of Texas at Arlington and The Sea Grant Program administered by Texas A & M University.

REFERENCES

BANDEL, K., 1977. Ubergänge von der Perlmutter-Schicht zu prismatischen Schichttypen bei Mollusken. Biomineralization, 9:28-47.

BEVELANDER, G. & NAKAHARA, H., 1980. Compartment and envelope formation in the process of biological mineralization. *In* The mechanism of biomineralization in animals and plants (ed. M. Omori & N. Watabe), pp. 19-27. Tokai Univ. Press.

CRENSHAW, M.A., 1982. Mechanism of normal biological mineralization and demineralization. Dahlem Konferenzen 1982 (ed. G.H. Nancollas), pp. 243-257. Berlin, Heidelberg, New York: Springer-Verlag.

CUIF, J.-P., DENIS, A. & GASPARD, D., 1981. Recherche d'une méthode d'analyse ultrastructurale des tests carbonatés d'Invertébrés. Bull. Soc. geol. France, 23:525-534.

ERBEN, H.K., 1971. Anorganische und organische Schalenkomponenten bei *Cittarium pica* (L.) (Archaeogastropoda). Biomineralization, 3:51-64.

GRÉGOIRE, C., 1962. On submicroscopic structure of *Nautilus* shell. Bull. Inst. Roy. Soc. Natur. Belgique, 38:1-71.

KULICKI, C. & MUTVEI, H., 1982. Ultrastructure of the siphonal tube in *Quenstedtoceras* (Ammonitina). Stockholm Contr. Geol., 37:129-138.

MUTVEI, H., 1964. On the shells of *Nautilus* and *Spirula* with notes on the shell secretion in non-cephalopod molluscs. Arkiv fr Zoologi, 16:221-278.

MUTVEI, H., 1969. On the micro- and ultrastructure of the conchiolin in the nacreous layer of some recent and fossil molluscs. Stockholm Contr. Geol., 22:1-17.

MUTVEI, H., 1972a. Ultrastructural relationships between the prismatic and nacreous layers in *Nautilus* (Cephalopoda). Biomineralization, 4, 81-86.

MUTVEI, H., 1972b. Ultrastructural studies on cephalopod shells. Bull. Geol. Instn. Univ. Uppsala N.S. 3:237-261.

MUTVEI, H., DAUPHIN, Y., & CUIF, J.-P., 1985. Observations sur l'organisation de la couche externe du test des *Haliotis* (Gastropoda): un cas exceptionnel de variabilité mineralogique et microstructurale. Bull. Mus. Nat. Hist., Paris, 4e sér., 7:73-91.

NAKAHARA, H., 1983. Calcification of gastropod nacre. *In* Biomineralization and biological metal accumulation (ed. P. Westbroek & E.W. DeJong), pp. 225-230. D. Reidel Publ. Co.

NAKAHARA, H., BEVELANDER, G. & KAKEI, M., 1982. Electron microscopic and amino acid studies on the outer and inner shell layers of *Haliotis rufescens*. Venus (Jap. J. Malac.), 41:33-46.

SUZUKI, S. & UOZUMI, S., 1981. Organic components of prismatic layers in molluscan shells. J. Fac. Sci., Hokkaido Univ., Ser.4, 20:7-20.

WATABE, N. & DUNKELBERGER, D.G., 1979. Ultrastructural studies on calcification in various organisms. Scan. Elec. Microsc. 1979/II:403-416.

WISE, JR., S.W. & HAY, W.W., 1968. Scanning electron microscopy of molluscan shell ultrastructures. Trans. Amer. Microsc. Soc., 87:419.

REFERENCES

ARCHITECTURE AND CHEMICAL COMPOSITION OF THE MAGNETITE-BEARING LAYER IN THE RADULA TEETH OF *CHITON OLIVACEUS* (POLYPLACOPHORA)

Paul van der Wal[1], John J. Videler[1], Piet Havinga[2] and Roel Pel[3]

[1]*Department of Marine Biology, University of Groningen, P.O. Box 14, 9750 AA Haren, The Netherlands;* [2]*Centre for Medical Electron Microscopy, University of Groningen, Oostersingel 69/2, 9713 EZ Groningen, The Netherlands;* [3]*Department of Microbiology, University of Groningen, P.O. Box 14, 9750 AA Haren, The Netherlands*

ABSTRACT. The radula apparatus in chitons (Polyplacophora) is a toothed ribbon used to excavate algae living on and in rocky substrates. Magnetite (Fe3O4) is a component of mineralized radula teeth. It is contained in a layer beneath the surface that leads in the cutting direction. The architecture and composition of this layer in fully mineralized teeth of *Chiton olivaceus* were studied by scanning electron microscopy. Worn, broken, sectioned, and etched teeth were examined. Etching was done with 4N HCl or by incubating teeth in cultures of bacteria that exclusively used chitin as a source of carbon and energy.

The magnetite-bearing layer is composed of closely packed rod-shaped and elongated concavo-convex (trough-shaped) units oriented at right angles to the cutting edge making an angle with the leading surface of about 60°. Near the leading surface extensions of these units curve baseward over an angle of about 90°. During the excavating action of the radula wear predominantly occurs through fracturing along the boundaries between the elongated units. This implies that the angle between the surface of wear and the leading surface, the wedge angle, is determined by the internal structure of the tooth and is fixed at about 60°. It is discussed that the microanatomical design of the magnetite-bearing layer is such that during wear material losses are minimized.

Magnetite occurs in granular form and approximately comprises 97% by weight of the magnetite-bearing layer. Chitin is a component of the organic matrix surrounding the mineral grains and forming the boundary layer of the elongated units.

INTRODUCTION

Chitons (Polyplacophora) are marine molluscs. Most species live on hard substrates and feed on endo- and epilithic algae. This food source is scraped and excavated from the substrate with the radula, a toothed ribbon in the buccal cavity of these animals. The radula is composed of a membrane on which are implanted numerous transverse rows of teeth. During radular growth the teeth are conveyed onto the back of the membrane towards more advanced positions, and meanwhile become mature. Eventually they

Origin, Evolution, and Modern Aspects of Biomineralization in Plants and Animals
Edited by Rex E. Crick
Plenum Press, New York

enter that part of the radula that is actively involved in the scraping process. There they wear down and finally break loose from the ribbon.

The radula owes most of its erosive capability to the presence in each transverse row of two mineralized teeth. In species of the family Chitonidae these teeth are mineralized by three minerals, each in a separate compartment: 1) magnetite (Fe_3O_4) in a layer beneath the surface that leads in the cutting direction, 2) lepidocrocite (γ-FeOOH) in a layer behind this one, and 3) a carbonate apatite mineral [$Ca_5(PO_4,CO_3,OH)_3(F,OH)$] in the space between the lepidocrocite-bearing layer and the trailing surface of the tooth (Lowenstam & Weiner, 1984). Mineralized teeth also contain proteinaceous material and chitin (Lowenstam & Weiner, 1984; Runham, 1963).

A research project has begun at the University of Groningen in which hard-rock excavating radulae are being used as model systems for the development of alternative dredging machinery. The aim is to look for dynamic, structural, physical, and chemical properties that might be translated into industrial designs. In this contribution we report on a scanning electron microscopical (SEM) study of the magnetite-bearing layer of *Chiton olivaceus* Spengler, a species belonging to the family of the Chitonidae. Its architecture was studied in worn, broken, sectioned, and acid etched teeth. With quantitative X-ray microanalysis the relative amounts of mineral and organic matter were estimated. Also an attempt has been made to localize chitin by etching teeth in cultures of bacteria, belonging to the genus *Clostridium*, that exclusively use chitin as a source of carbon and energy.

MATERIAL AND METHODS

Specimens of *Chiton olivaceus* were collected near Calvi on Corsica at STARESO, the marine biological station of the University of Liège, in waters varying in depth from 0 to 10 m. They were fixed in either 70% ethanol or in seawater containing 4% formaline. Radulae were dissected and, unless stated otherwise, were cleaned in 1N NaOH at 100° C during 1 to 4 hours. They were then washed in distilled water and stored in 70% ethanol.

Some radulae were cleaned ultrasonically, dried in air, and mounted on aluminum stubs with a carbon-containing adhesive. Some teeth were deliberately broken by pressing the point of a needle against their surface. The preparations were then coated with a 10 nm thin layer of gold in a diode sputtering device and viewed in an ISI SEM type DS 130 at 10 kV.

Other radulae and also polished sections of radulae embedded in Technovit or EPON were incubated in an aqueous solution of 4N HCl during a period of up to 24 hours. After that they were cleaned in distilled water, dried at the critical point of carbon dioxide, mounted on aluminum stubs, coated with gold, and viewed in an ISI SEM also at 10 kV. Embedment in Technovit occurred after the radulae had been dried in air; prior to embedment in EPON the radulae were dehydrated in a graded series of ethanol and then infiltrated with propylene oxide.

Radulae fixed in 4% formaline or 70% ethanol, cleaned with 1N NaOH or not cleaned in this way were incubated in growth media inoculated with strictly anaerobic bacteria, belonging to the genus *Clostridium*, that exclusively use chitin as a source of

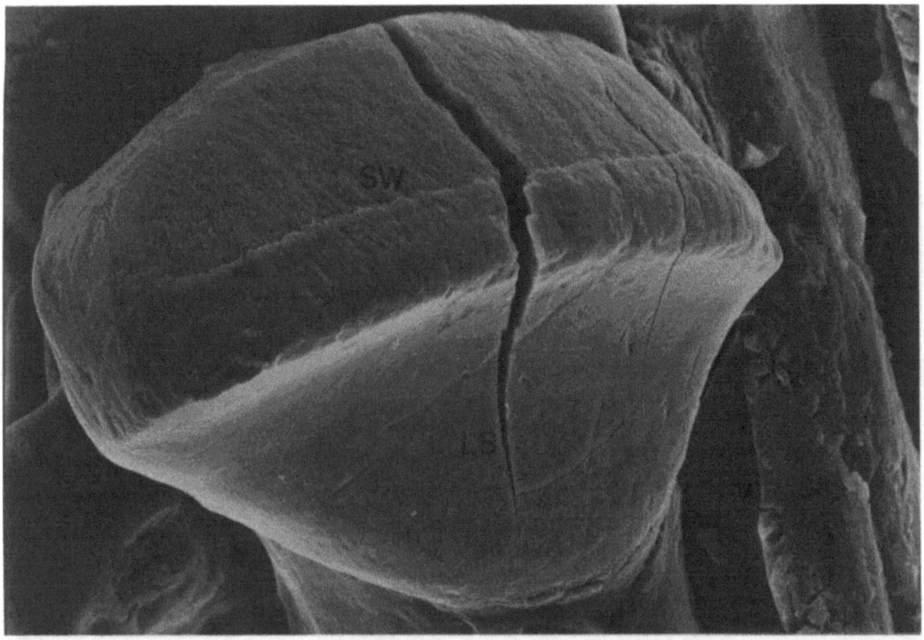

FIGURE 1. Micrograph of a worn tooth slightly etched with 0.01 N HCl during 10 sec. **SW**, surface of wear. **LS**, leading surface. The boundary between the iron-containing layer and the apatite-containing layer is visible in the surface of wear as a line parallel to the cutting edge. The magnetite-bearing layer is positioned directly behind the leading surface. Fissures splitting the tooth longitudinally are caused by dessication during preparation. × 6,750.

carbon and energy. The bacteria were isolated from anaerobic sediments of the intertidal mudflats of the Waddenzee (The Netherlands) and were extensively tested on their substrate specificity (Pel & Gottschal, 1986). Prior to inoculation the radulae were either autoclaved together with the growth medium, or were, after a formaldehyde sterilization, aseptically transferred to autoclaved growth medium. For the composition of this medium one is referred to the work of Pel & Gottschal (1986). After 3 to 8 weeks of incubation the radulae were washed in distilled water, cleaned ultrasonically, dried, mounted on aluminum stubs, coated with gold, and viewed in a ISI SEM at 10 kV. As a check on these experiments, some radulae were processed in the same way, except that the media were not inoculated with bacteria.

X-ray microanalysis was done on polished sections of a radula embedded in Technovit. The specimen was coated with palladium and viewed in an ISI SEM type DS 130, at 25 kV with the specimen holder tilted 35° toward the X-ray detector. For quantitative analysis a polished piece of iron and a magnetite crystal were analyzed under the same circumstances as the radula.

RESULTS

Architecture

There is a distinct structural difference between the compartments that contain magnetite and lepidocrocite and the layer containing apatite. Mild acid treatment, which

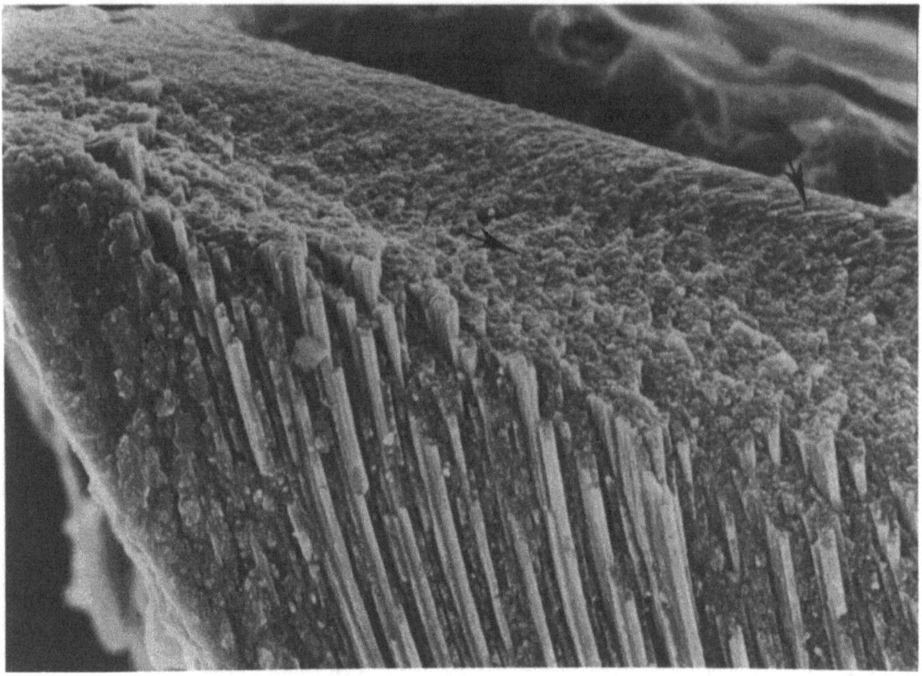

FIGURE 2. Micrograph of longitudinal fracture plane showing part of the magnetite-bearing layer near the top of the tooth (at left). The layer consists of sheets of elongated units that make an angle with the leading surface (at the top of the micrograph) of about 60°. Near this surface they pass into less distinct structures that curve toward the base (arrows). × 6,750.

only affects the apatite, accentuates the discontinuity (Fig. 1). The magnetite-bearing layer is about 15 μm thick. From fractographs (Figs. 2 & 3) it is shown to consist of closely packed elongated units extending over the total width of the layer, except for the zone immediately bordering the leading surface. These elongated units are oriented at right angles to the cutting edge and make an angle with the leading surface of about 60° (Fig. 2). The form of the units in transverse or oblique section are well revealed in polished sections etched with 4N HCl. Two types exist, a trough-shaped elongated unit and, enclosed between four of these, a rod-shaped unit (Fig. 4). The trough-shaped units are oriented with their concave sides toward the top of the tooth. Figure 5 is a schematic drawing of these units showing their arrangement, shape, and size.

The surface of the elongated units is formed by closely spaced smooth ridges laying parallel to the elongation of the structures (Fig. 3). The interiors of the units are stuffed with equidimensional granules smaller than 60 nm in diameter (Fig. 4). In fracture planes the individual granules are hardly or not distinguishable (Figs. 2 & 3). The boundaries between the units often form the most predominant sites along which fracturing occurs. This is especially the case where fracturing results from applying a pressure directly onto the leading surface. On the surface of wear the structural units are seen

FIGURE 3. Fractograph showing part of the magnetite-bearing layer. Fracturing occurred either parallel or transverse to the elongation of the units, respectively showing the ridged outer faces and the granular interiors. × 11,475.

to be exposed over large parts of their length clearly displaying their ridged surfaces. In other words the outer faces of the units predominantly form the surface of wear (Fig. 6).

In the zone immediately bordering the leading surface the units with their characteristic shapes are absent (Figs. 2 & 4). This zone is narrow at the top of the tooth and broadens toward the base to a width of about 5 μm. Its granular nature is most conspicuous. The granules have the same size as those constituting the trough- and rod-shaped units. They are arranged in lamellar and rod-like structures, which in both etched sections and fracture planes are faintly exposed or obscure. These structures are continuous with the trough- and rod-shaped units and curve toward the base over an angle of about 90°.

Structure of the organic matrix

When whole teeth are incubated in an aqueous solution with 4N HCl for 24 hours, all mineral matter dissolves. What remains is an organic skeleton with an outer shape identical to that of the tooth prior to demineralization. The exterior of this skeleton is smooth; its interior is filled with densely packed parallel tubular and filamentous elements. At a distance of about 15 μm from the leading surface a structural discontinuity exists (Fig. 7), which is thought to be homologous to the discontinuity observed in the

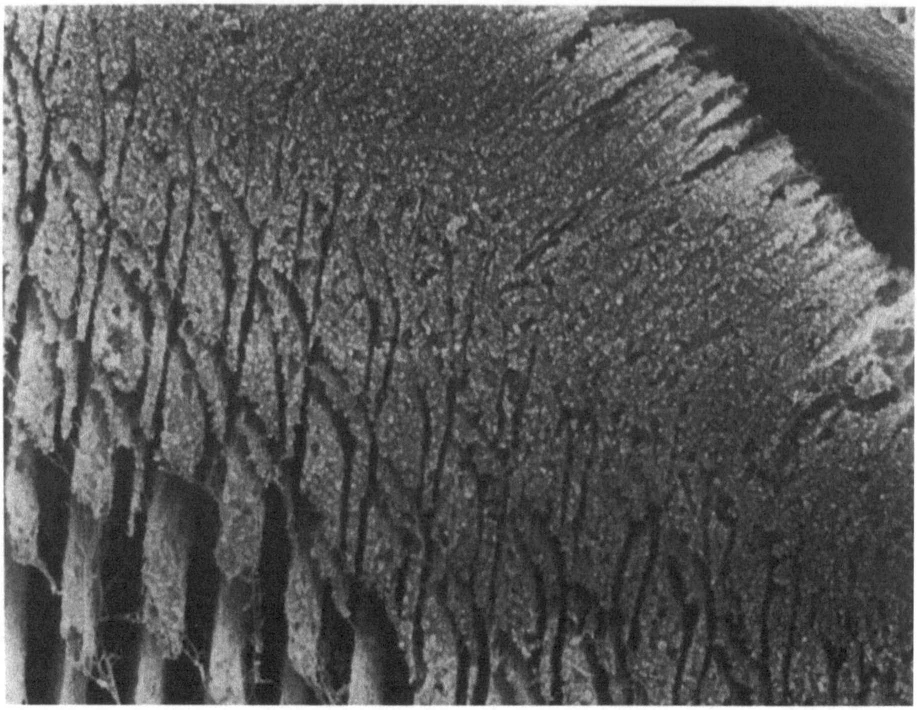

FIGURE 4. Micrograph of a polished section through a mineralized tooth showing part of the magnetite-bearing layer near the leading surface (upper right corner). The section has been etched with 4N HCl and reveals the concavo-convex outlines of the trough-shaped units. At the bottom left, these units are absent and only the rod-shaped units withstood the acid action. The units pass into less distinct structures that curve toward the bse. × 17,000.

untreated teeth dividing the iron-containing and the apatite-containing compartments. Due to the preparation procedure the two parts are often seen to be separated from each other by a wide gap.

The structural elements of the magnetite-bearing layer as observed in demineralized teeth and those of the same layer as observed in non-demineralized teeth are identical in their orientation. There are also some similarities in shape. At the magnetite-bearing side of the discontinuity the organic material consists of about 0.5 μm wide filaments and flattened tubes oriented perpendicularly to the cutting edge making an angle of about 60° with the leading surface (Figs. 7 & 8). They are composed of a monolayer of fibers smaller than 30 nm in diameter laying parallel to the gross structure (Fig. 9).

Near the leading surface the structural elements pass into a multitude of narrow tubules of about 130 nm in diameter that curve toward the base over an angle of about 90° (Fig. 8). Close to the leading surface the tubules pass into fibers laying parallel to this surface (Fig. 10). A smooth layer forms the outer extension of the organic framework. It should be noted here that due to the critical point drying technique the material might have been shrunken by 20-25%.

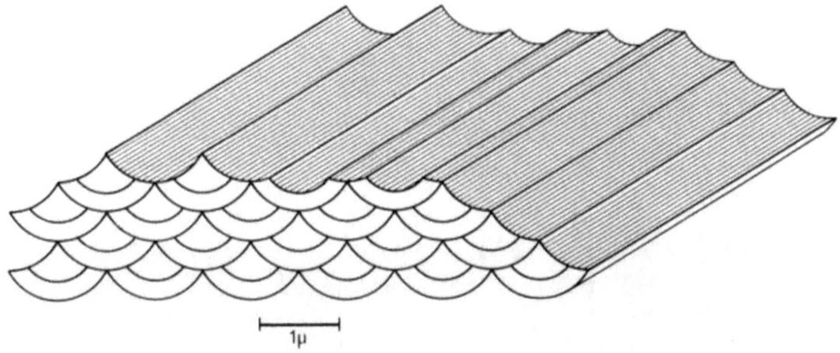

FIGURE 5. Schematic drawing of the rod-shaped and elongated trough-shaped units.

Etching with chitin-degrading bacteria

When whole radulae were subjected to mild ultrasonic treatment after incubation in media containing chitin-degrading bacteria, the membrane, non-mineralized teeth, and the immature teeth in the developmental stage prior to magnetite mineralization disintegrated. Only the base and cusp of the mineralized teeth remained intact. Close examination revealed that these teeth showed only signs of etching at the surface of wear, and at other places where the interior had been exposed to the bacteria. Only the iron mineral-containing layer appeared to be affected (Fig. 11). The ridged boundary layer

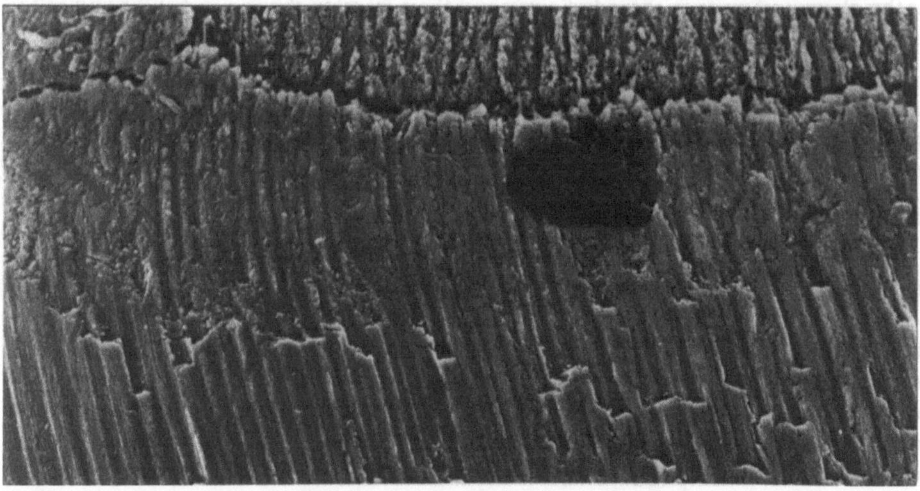

FIGURE 6. Detail of Figure 1. Elongated units are exposed over large parts of their length clearly displaying their ridged surface. The discontinuity at the top of the micrograph represents the boundary between the iron-containing and the apatite-containing layer. Between the lepidocrocite- and magnetite-bearing layer there is no discontinuity. × 3,400

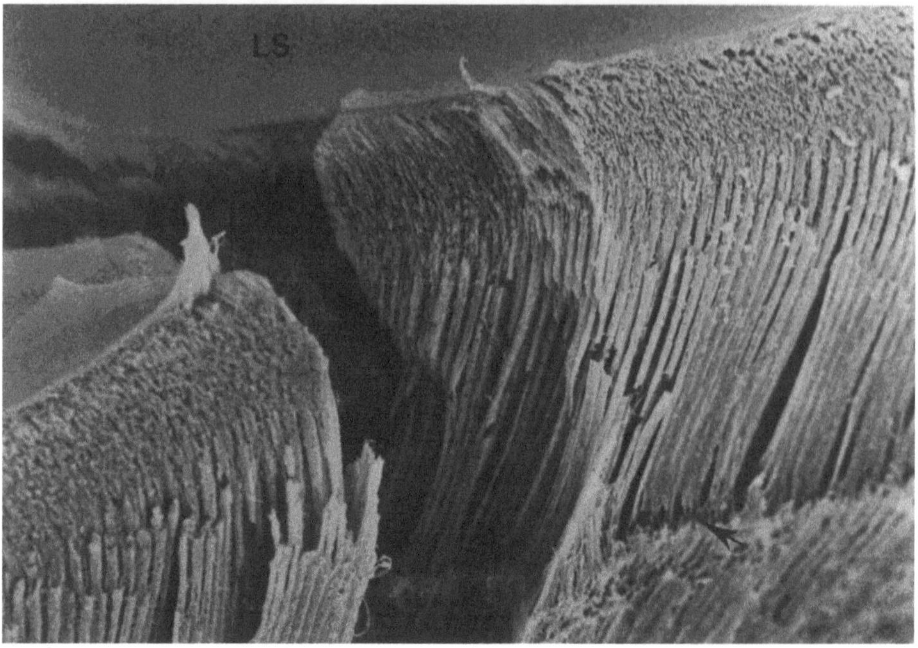

FIGURE 7. Micrograph of tooth etched with 4N HCl for 24 hours showing the shape of the organic framework of the magnetite-bearing layer. A wide longitudinal fissure permits a 3-dimensional view of this layer. The leading surface (LS) is covered with a smooth sheet of organic material. The arrow points to a discontinuity in the organic framework that probably represents the boundary between the iron-containing and the apatite-containing compartments. × 3,400.

of the elongated units is corroded away and their interior shows a granular and pitted appearance. The granules are well outlined and measure about 40 nm.

To verify the corrosive action of the bacteria some radulae were incubated in the same media, but without the chitin-degrading bacteria. After an incubation period of 8 weeks the radulae did not disintegrate upon ultrasonic treatment, nor showed any sign of etching.

X-ray microanalysis

Table 1 lists the results of the quantitative X-ray microanalysis of the magnetite-bearing layer. Spread over the total width of this layer and at distances of 3 μm, six point analyses were performed. The small deviation from the mean intensity of *K*-lines of iron indicates that the iron is homogeneously distributed. Two standards were used to determine the iron content: one containing 100% magnetite and the other containing 100% iron. An ideal standard should resemble the object to be quantified in mean atomic number and in density. Therefore, the better standard is the one containing 100% magnetite. In fact, the intensity of the *K*-lines of iron measured in the magnetite-bearing layer differed only by a small amount from that measured in the magnetite standard. This means that the mean iron content of 70.24% (w/w) is very close to the real value.

The magnetite-bearing layer contains about 97% (w/w) magnetite, assuming that all iron is in the form of this mineral. Most of the remaining 3% of this layer is probab-

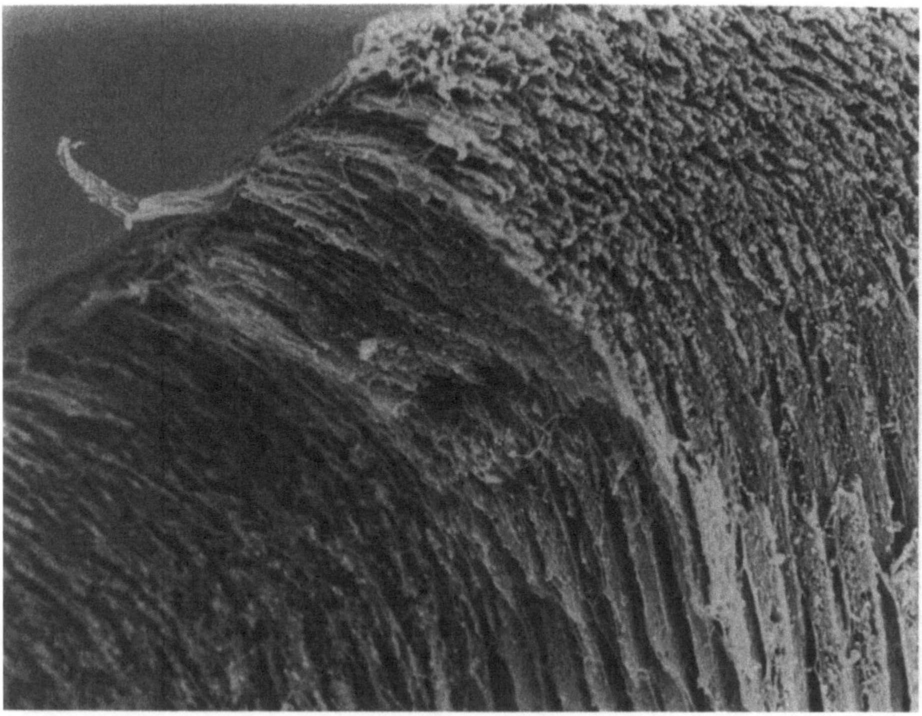

FIGURE 8. Detail of Figure 7. × 11,475.

ly composed of organic material. Elements other than iron were not detected in the magnetite-bearing layer.

TABLE 1. Quantitative X-ray microanalysis of the magnetite-bearing layer using standards of 100% iron and 100% magnetite

	Mn – layer $\mu \pm \sigma$ n = 6	100% Mn n = 1	100% Fe n = 1
Intensity of K-lines of Fe (counts/sec.nA)	419.13 ± 6.14	431.76	617.01
Fe-content (weight %) standard: 100% Fe	67.93 ± 0.99	69.98	-
Fe-content (weight %) standard: 100% magnetite	70.24 ± 1.03	72.36[2]	-
Fe_3O_4-content (weight %) standard: 100% magnetite	97.07 ± 1.42[1]	-	-

[1]Assumes all Fe in the form of magnetite.

[2]Calculated on basis of mean atomic weights of Fe and O.

DISCUSSION

Form and function: A model

We have made a detailed description of the architecture of the magnetite-bearing layer. In the following discussion an attempt is made to relate these results to the function of the tooth as a cutting device.

We have shown that the trough-shaped and rod-shaped structures are oriented at right angles to the cutting edge at an angle with the leading surface of about 60° and that in the surface of wear these structures are exposed over almost their

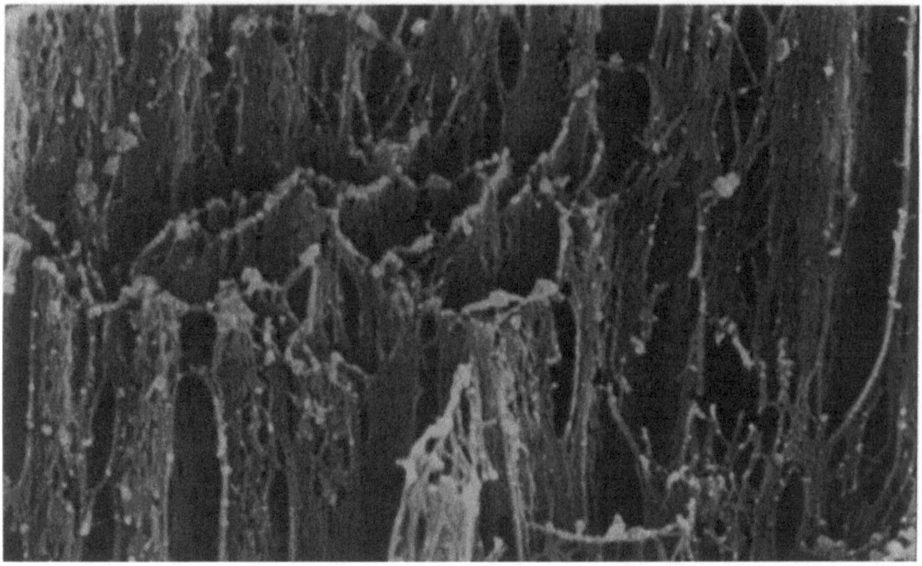

FIGURE 9. Organic matrix of elongated units of the magnetite-bearing layer revealed after etching with 4N HCl for 24 h. × 22,9500.

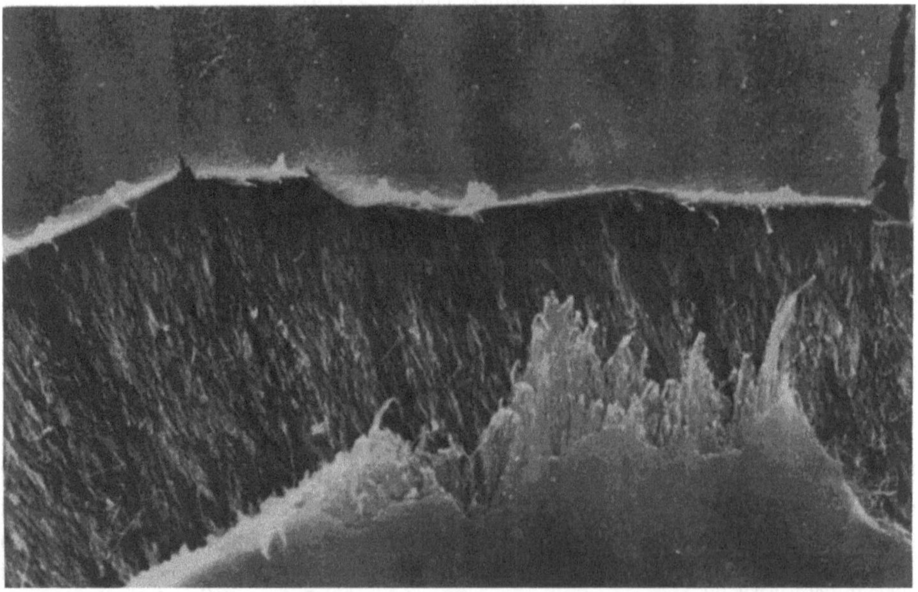

FIGURE 10. Frontal view of leading surface of tooth etched with 4N HCl for 24 h. Directly behind the smooth outer layer, bundles of fibers lay parallel to the leading surface. × 4,300.

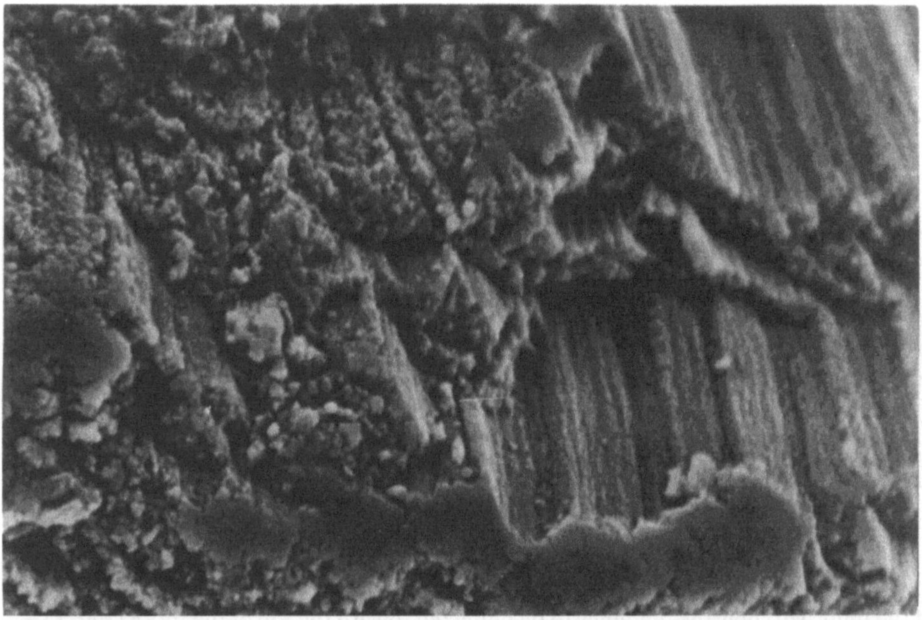

FIGURE 11. Micrograph of tooth incubated in a culture of chitin-degrading bacteria showing part of the magnetite-bearing layer. Elongated units on the left are strongly etched and sharply contrast with the units on the right, which were exposed after the incubation period through fracturing due to ultrasonic treatment. ×22,950.

full length showing their ridged outer side. This implies that the angle between the surface of wear and the leading surface, the wedge angle, is fixed at about 60°. By this property the tooth is a very efficient cutting device, because it is never blunt and never over-sharp. The latter condition which would make it fragile (Fig. 12). Furthermore, wearing is identical to fracturing on micrometer scale and fracturing predominantly occurs between the elongated units.

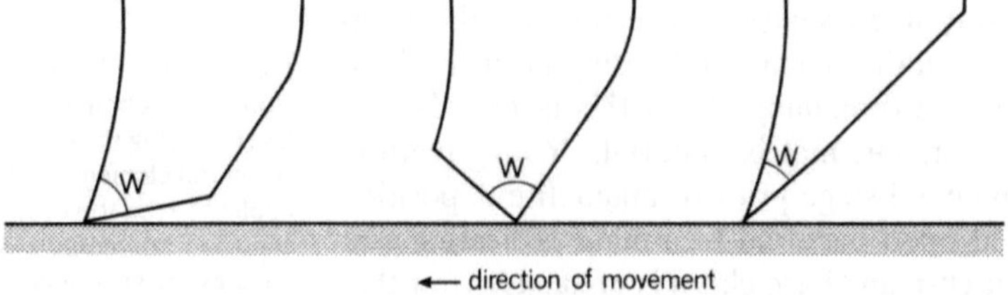

FIGURE 12. Schematic drawing showing side view of teeth pressed against substrate. The wedge angle (W) is fixed at about 60° (A), thus the tooth is never blunt (B) and never over-sharp (C).

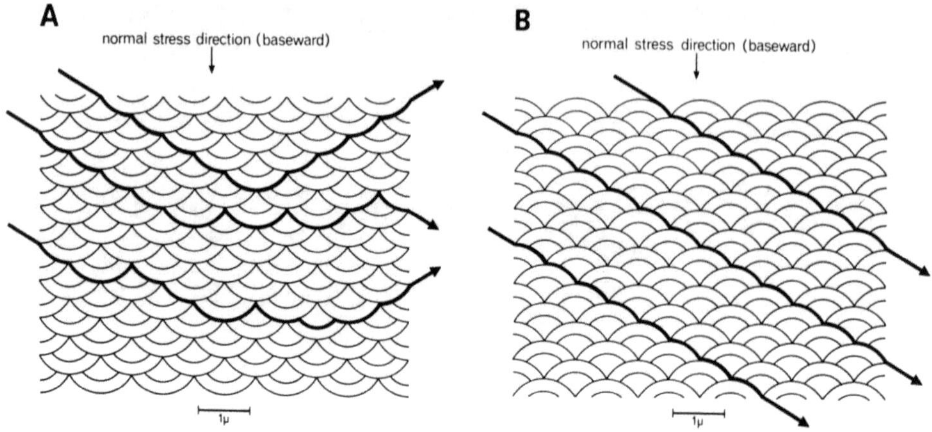

FIGURE 13. Schematic drawing of section transverse to the elongation of the rod-shaped and trough-shaped units. ▪A. Normal orientation. ▪B. Reversed (hypothetical) orientation. Arrows, possible fracture paths.

The concave sides of the elongated units were shown to be oriented toward the top of the tooth (Fig. 5). Due to this orientation a fracture front following the outer faces of the units will show the tendency to be deflected toward the top of the tooth (Fig. 13A), which greatly restricts material losses. For comparison the reversed (hypothetical) orientation of the units is shown in Figure 13B. In that case fracture fronts would penetrate deeply into the tooth's interior without having a chance to rise again.

We have shown that near the leading surface the trough- and rod-shaped units pass into less well-structured elements that curve baseward and eventually end in fibers laying parallel and in close proximity to this surface. We think that these strongly curved extensions of the structural units serve as anchors preventing shear between the units and that their direction fits to sustain the tensile stress generated in this zone (Fig. 14).

In the future the validity of these ideas will be tested by means of a mathematical modelling technique known as finite element analysis.

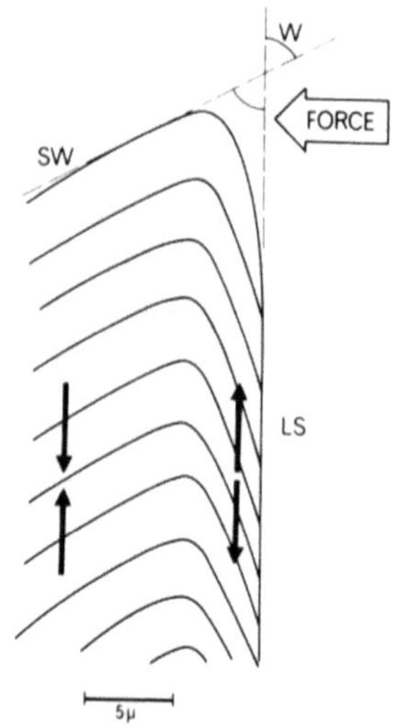

FIGURE 14. Schematic cross-section through magnetite-bearing layer showing the orientation of the structural elements. When a force is applied to the top of the tooth, tensile (↔) and compressive (→ ←) stresses are generated in this layer.

Composition of the Organic Matrix

With X-ray microanalysis we estimated the magnetite content in the magnetite-bearing layer at about 97% (w/w). The remaining 3% of this layer is largely composed of organic matrix material. Young teeth in the developmental stage prior to magnetite deposition disintegrated upon bacterial treatment indicating that in both their cusp and base chitin is present. After that stage mineralized teeth were not affected by the bacteria, except the organic matrix positioned within the magnetite-bearing layer. We think that during or prior to magnetite deposition the chitin undergoes a chemical change making it inaccessible for the chitinases produced by the bacteria. The chemical change involved might be the linking of chitin end groups with aspartyl and/or glutamyl residues of the radula protein, as suggested by Runham's (1963) histochemical studies on radulae of six different species. The chitin present in the magnetite-bearing layer is not or only in part changed in this way. It is possible that this material was already enclosed by magnetite and thereby shielded against any chemical interaction prior to completion of the chemical change.

Chitin is also present in non-mineralized teeth and in the radula membrane. The chemical change mentioned above does not take place in these parts of the radula or remains incomplete. The presence of chitin in the radula membrane and the immature to be mineralized teeth of chitons is consistent with the findings of Runham (1963) and Lowenstam and Weiner (1984). However, Runham (1963) was not able to detect chitin histochemically in mineralized teeth. It is possible that the number of stainable chitin end groups are so diminished in these parts of the radula that the staining technique used was not sensitive enough to give a detectable reaction.

The chitin in the magnetite-bearing layer is present in material surrounding the mineral grains and in material forming the boundary layer of the structural units. Proteins also form part of the organic matrix of chiton teeth (Runham, 1963). In the future we shall make an attempt to localize these matrix components by etching radulae with proteinases.

Evaluation of previous results

Kirschvink & Lowenstam (1979) have made a SEM study of mineralized teeth of several chiton species. They reported that the surface of the magnetite-bearing layer was subdivided into thick-walled hexagonal polygons, about 5 – 7 μm in diameter, which were filled with a thin-walled honeycomb-like network with mesh openings of about 100 nm wide. They claimed that these structures form the organic matrix of the magnetite-bearing layer, that they have a continuation to within the tooth's interior, and that within the confinement of each cell of the honeycomb a single equidimensional crystal of magnetite is formed during mineralization. We have never found such structures, neither on the outside of the tooth, nor in the tooth's interior, and we are strongly of the opinion that these authors reached erroneous conclusions due to misinterpretation of their SEM-data as explained below.

In the zone where the radula is formed and mineralization takes place, narrow epithelial cells, the odontoblasts, extend to the surface of the teeth where they terminate in a bundle of 500 to 1000 closely packed microvilli surrounded by a thin mitochondrion-filled shell (Nesson & Lowenstam, 1985). In ultrathin sections this structure bears a striking resemblance in both shape and size to the polygon and honeycomb structure shown in the work of Kirschvink & Lowenstam (1979). We think that these structures are identical. The polygons and honeycomb-like networks do not form part of the organic matrix, but are the remnants of the apical parts of odontoblast cells still adhering to the surface of the tooth. Obviously, Kirschvink & Lowenstam (1979) viewed improperly cleaned material.

These authors also obtained size distributions of magnetite grains of several chiton species by erroneously taking the mesh openings of the honeycomb-like network as the diameter of single monocrystalline magnetite grains. Grain size determines in large part the magnetic behavior of monocrystalline magnetite, and they related their findings to magnetic properties of whole teeth, which were also studied. It is clear that the results of this work must be reinterpreted after a proper assessment of the size distribution of magnetite grains.

In a preliminary ultrathin section electron microscope study of mature teeth of *Chiton olivaceus* we found that magnetite occurs in monocrystalline equidimensional grains of about 20 nm in diameter. In our SEM images we observed granules varying in width between 40 and 60 nm. As a rule, however, the single monocrystalline grains would always show larger in these images, because the grains are covered with a 10 nm thick layer of gold and are wrapped in organic material which adds to their thickness. Furthermore, the granules may represent aggregates of monocrystalline grains. In the near future we shall perform a transmission electron microscope study in an attempt to determine the size distribution of magnetite grains in a number of chiton species.

ACKNOWLEDGEMENTS

We are indebted to Dr. G. Boom for performing the quantitative X-ray microanalysis. Thanks are due to Mr. D. Huizinga and Mr. H.R.A. Meiborg for photographic assistance and to Mr. D. Visser for preparing the drawings in Figs. 5, 12, 13, and 14. This research was subsidized by the Stichting voor de Technische Wetenschappen. Participation in the Symposium was made possible in part by grants from The Organized Research Fund of The University of Texas at Arlington and The Sea Grant Program administered by Texas A & M University.

REFERENCES

KIRSCHVINK, J.L., & LOWENSTAM, H.A., 1979. Mineralization and magnetization of chiton teeth: paleomagnetic, sedimentologic, and biologic implications of organic magnetite. Ear. Planet. Sci. Lett., 44:193-204.

LOWENSTAM, H.A., & WEINER, S, 1985. Transformation of amorphous calcium phosphate to crystalline dahllite in the radular teeth of chitons. Science, 227:51-53.

NESSON, M.H., & LOWENSTAM, H.A., 1985. Biomineralization processes of the radula teeth of chitons. *In* Magnetite Biomineralization and Magnetoreception in Organisms (ed. Kirschvink, J.L., Jones, D.S., & MacFadden, B.J.), pp. 333-363. Plenum Publishing.

PEL, R., & GOTTSCHAL, J.C., 1986. Mesophilic chitin-degrading anaerobes isolated from an estuarine environment. FEMS Microbiol. Ecol. 38:39-49.

RUNHAM, N.W., 1963. The histochemistry of the radulas of *Acanthochitona communis*, *Lymnaea stagnalis*, *Helix pomatia*, *Scaphander lignarius* and *Archidoris pseudoargus*. Ann. Histochim., 8:433-442.

CALCIUM BINDING SUBSTANCE IN THE HERMATYPIC CORAL, *ACROPORA HEBES* (DANA)

Yeishin Isa

Department of Biology, University of the Ryukyus, Nishihara, Okinawa 903-01, Japan

ABSTRACT. A characteristic of skeleton formation in the hermatypic coral *Acropora hebes* has been investigated with special reference to the calcium-binding substance assay. The skeletal soluble organic matrix such as protein and carbohydrate did not bind calcium ions in the Sephadex G-75 chromatography. On the other hand, the insoluble organic matrix of the skeleton showed the calcium-binding ability, which was reduced by treatment with the organic solvent. Presumably, the calcium-binding substances in the skeletal insoluble matrix are firmly attached to the calcium carbonate crystals and will not be extracted by the aqueous or organic extracting solutions. The major component of the calcium-binding substances in the skeleton is phospholipid, but not the other lipids such as neutral lipid or glycolipid. A high-performance TLC revealed that the skeletal calcium-binding phospholipids consisted mainly of acidic phosphatidylserine.

INTRODUCTION

Recent studies on the mechanisms of skeleton formation in scleractinian corals (see review by Johnston, 1980) have suggested that the skeletal organic substance are intimately related to the calcium carbonate crystal formation, *i.e.*, nucleation, growth, orientation and type of crystals. The presence of an organic matrix seems to be necessary for the formation and growth of calcified structures in most organisms (Watabe, 1981). The organic matrix is believed to be involved in the skeleton formation process as an activator or inhibitor of crystal growth by binding calcium. Although invertebrate calcified tissues contain much protein and carbohydrate which bind calcium (Crenshaw, 1972; Abantangelo *et al.*, 1978; Samata & Krampitz, 1982; Isa & Ikeda, 1986), they also contain calcium-binding lipid compositions (Rabinowitz, 1976; Abolins-Krogis, 1979). Some evidence has been provided that hard tissue lipid components may play an important role in the mineralization of vertebrate cartilages and bones (Cotmore *et al.*, 1971; Wutheir, 1975; Boskey *et al.*, 1978; Majeska *et al.*, 1979).

Biochemical and electron microscope studies (Wainwright, 1963; Young, 1971a; Johnston, 1979; Isa & Yamazato, 1981) have revealed that scleractinian skeleton consists of aragonite crystals and a small amount of organic matrix. The skeletal organic matrix has been shown to be incorporated extracellularly into skeleton of the hermatypic coral, *Acropora hebes* (Isa, 1986). According to the classic study of Goreau (1956) mucopolysaccharide-like substances are responsible for the epitaxial growth of crystal

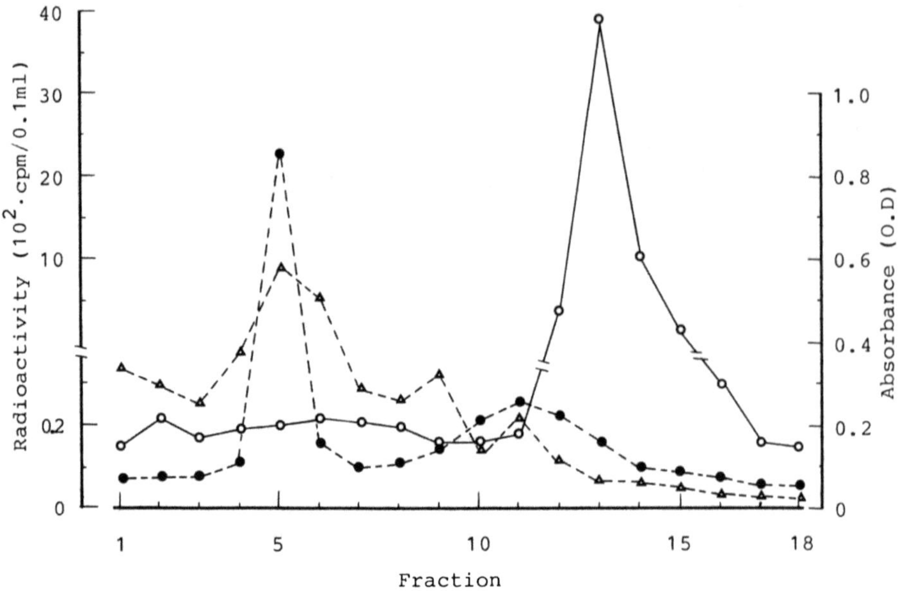

FIGURE 1. Sephadex G-75 chromatography of soluble organic matrix. Fractions were assayed for Ca-binding activity (open circle), and for protein concentration (dot) by absorbance at 280 nm or for carbohydrate concentration (triangle) at 490 nm.

in corals. On the other hand, Mitterer (1978) suggested that the aspartic acid-rich protein might be involved in the formation of calcium carbonate crystals. Lipids were also discovered in the organic matrix of scleractinian corals (Young *et al.*, 1971; Isa & Okazaki, 1987). However, no information is available concerning calcium-binding properties of these lipids, except that some shallow-water corals are found to contain calcium-binding lipids in the skeleton (Isa & Okazaki, 1986). This paper reports the results of biochemical investigations of calcium-binding capacities of different components of organic matrices of the skeleton of *Acropora hebes*.

MATERIALS AND METHODS

Material

The hermatypic coral *Acropora hebes* (Dana) was collected alive from a tide pool (1-3 m in depth) in the southern part of Okinawa, Japan, and stored at -20°C until use.

Preparation of calcium-binding substance

Powders of skeletons were prepared according to the methods by Isa and Okazaki (1987). The skeletons (10g dry weight) were completely decalcified with 10% EDTA (pH 8.0) or with 1N HCl at room temperature, and remaining organic materials were subsequently centrifuged at 15,000 × G for 20 min. The resulting pellet was used as the insoluble organic matrix of skeleton. The supernatant was dialysed five times against 5 l of distilled water, concentrated to 10 ml volume by a rotary evaporator (Eyela N-21) and centrifuged at 15,000 × G for 20 min. The resulting supernatant was used as the soluble organic matrix of skeleton. Extraction and separation of lipids from skeletal and soft tissue fractions of *Acropora hebes* were carried out according to the method by Isa and Okazaki (1987).

Assay of calcium-binding capacity

calcium-binding capacity of the soluble matrix was assayed by gel-filtration chromatography at room temperature. A mixture of 1 ml of the matrix with 2 ml of 0.05 M Tris-HCl (pH 7.5) and 1 mM $^{45}CaCl_2$, containing about 3 μCi of ^{45}Ca, was loaded on a Sephadex G-75 column (1.5 × 40 cm) equilibrated and then eluted with 0.01 M Tris-HCl solution (pH 7.5). The flow rate was 20 ml/hr and 5 ml fractions were collected. Each of the eluates was scanned by 280 nm for determination of concentration of proteins using a spectrophotometer (Hitachi 101). The concentration of carbohydrate of each eluates was determined by the method of Hodge and Hofreiter (1962). Subsequently, the radioactivity of the aliquote of each fraction was counted with a gas-flow counter (Aloka, FC-1E).

For the determinations of calcium-binding capacity of insoluble matrix, six mg of the matrix were suspended in 2 ml of 0.02 M Tris HCl solution (pH 7.5), and then mixed with 1 ml of $^{45}CaCl_2$ solution (containing about 3 μCi of ^{45}Ca) for 10 min. and centrifuged at 15,000 × G for 20 min. The pellet was dissolved in 1 ml of 1N HCl solution and the radioactivity of the aliquote of the solution was determined by a gas-flow counter as above. In some experiments, the insoluble matrix used was treated for 10 min. with 10 volume of chloroform/methanol (V/V, 2:1) solution or the final concentration of 1 μM of $LaCl_3$ was added to the reaction mixture.

Calcium-binding capacity of lipid fraction was assayed by the procedure of Isa and Okazaki (1987).

Thin layer chromatography

Extraction and separation of phospholipids from skeletal and soft tissue fractions were performed according to Isa and Okazaki (1986). Standard phospholipids (phosphatidylcholine, phosphatidylethanolamine, phosphatidylserine, phosphatidylinositol, and phosphatidic acid) were obtained from Serdary Research Laboratories, London. About ten ul drops of phospholipids were spotted with a micropippete onto a high-performance TLC plate (10 × 10 cm, E. Merk, Darmatsdt, Unit, Germany), which had been previously activated for 5 min. at 110°C. Spots were confined to an area of about 5-7 mm diameter. The plate was developed at room temperature up to 8 cm from the bottom with the solvent methyl-acetate- n-propanol- chloroform- methanol- 0.25% aqueous potasium chloride (25:25:25:10:9) and evaporated for 10 min under a hood, and allowed to dry at room temperature. For the visualization of phospholipids, the plates were evenly sprayed with molybdate reagent (Ditter & Lester, 1964).

RESULTS

Figure 1 shows an elution profile of the soluble organic matrix of skeleton by Sephadex G-75 chromatography. Although peaks of both protein and carbohydrate were clearly obtained in the void volume (fraction no. 4-6), they did not bind calcium

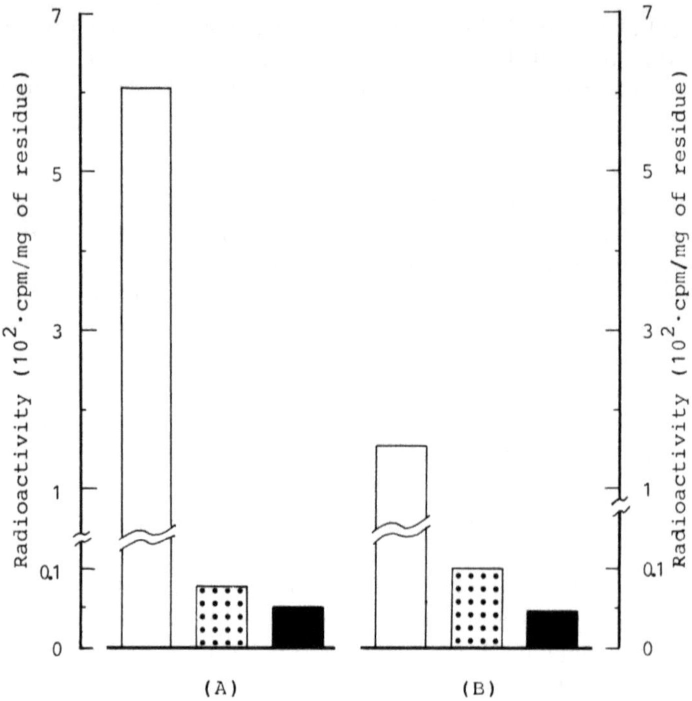

FIGURE 2. Ca-binding activity of insoluble organic matrices of skeleton decalcified with EDTA (▪A) and with 1 N HCl (▪B). The matrix was untreated (open bar) or treated (stippled bar) with chloroform/methanol (2:1) solution. Addition of 1 mM La^{3+} ions to the reaction mixture containing the untreated matrix reduced its calcium-binding activities (solid bar).

ions. Free calcium ions in the reaction medium were almost eluted near the total volume (fraction no. 12-14) of the column used.

Figure 2 shows the calcium-binding capacity of the insoluble organic matrix of skeleton which was decalcified with EDTA or with HCl solutions. The EDTA residue bound more calcium ions than that of HCl residue. The calcium-binding capacity of both of these matrices decreased significantly by a treatment with chloroform/methanol (V/V, 2:1) solution.

The calcium-binding of lipid fractions from skeletal and soft tissue is shown in Figure 3. Total lipids from decalcified skeleton showed a higher binding capacity than those extracted from the soft tissue. The lipids from skeleton sample prior to decalcification were hardly able to bind calcium ions. In both skeletal and soft tissue lipids, the calcium-binding capacity of the phospholipid fractions was the highest, while it was negligible or relatively low in the neutral lipid and glycolipid fractions. The capacity of calcium-binding was higher in skeletal than in soft tissue phospholipids.

Figure 4 shows a chromatographic separation by a high-performance TLC of phospholipids from the skeletal and soft tissue of the coral and of standard phospholipids from the commercial source. As indicated by the clearly well-defined spots, the phos-

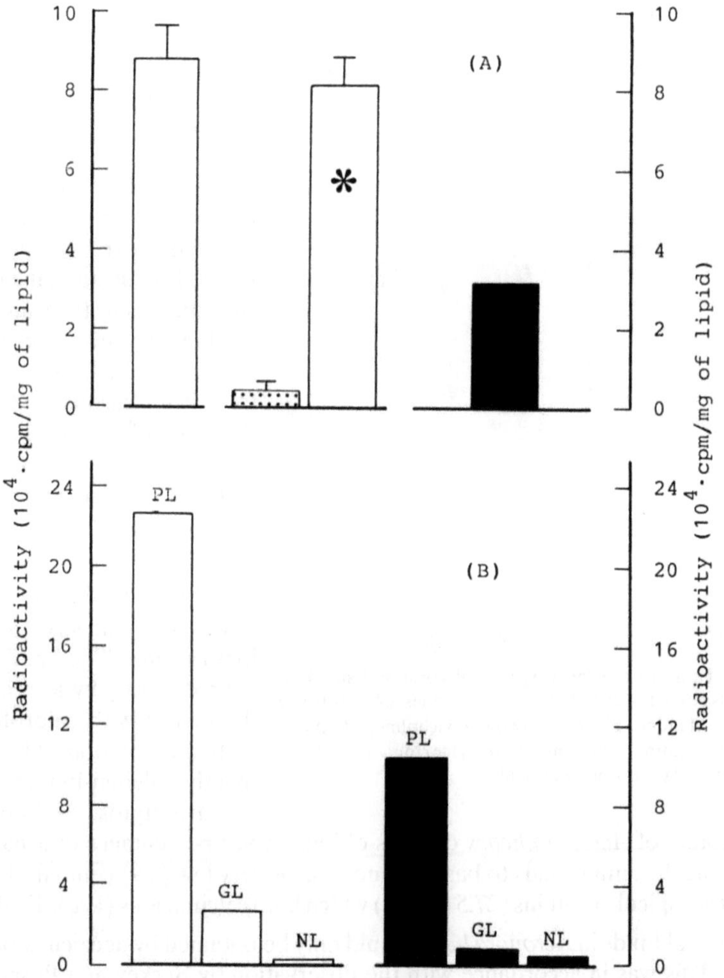

FIGURE 3. ▪A. Ca-binding activity of total lipids extracted from the skeleton decalcified (open bat) and undecalcified (stippled bar), and from soft tissue (solid bar). * = the skeleton treated with the lipid extracting organic solution prior to decalcification. The values of skeleton samples represent a mean ± standard deviation (n = 6). ▪B. Ca-binding activity of lipid compositions extracted from the skeleton decalcified (open bar) and from the soft tissue (solid bar). PL, phospholipid; GL, glycolipid; NL, neutral lipid.

pholipids of soft tissue were consisted of several kinds of phospholipids components. However, the skeletal phospholipid mostly consisted of phosphatidylserine.

DISCUSSION

This study reveals for the first time the calcium-binding substance, phosphatidylserine, in the organic matrix of scleractinian skeleton and substantiates the hypothesis previously proposed by Isa and Okazaki (1987).

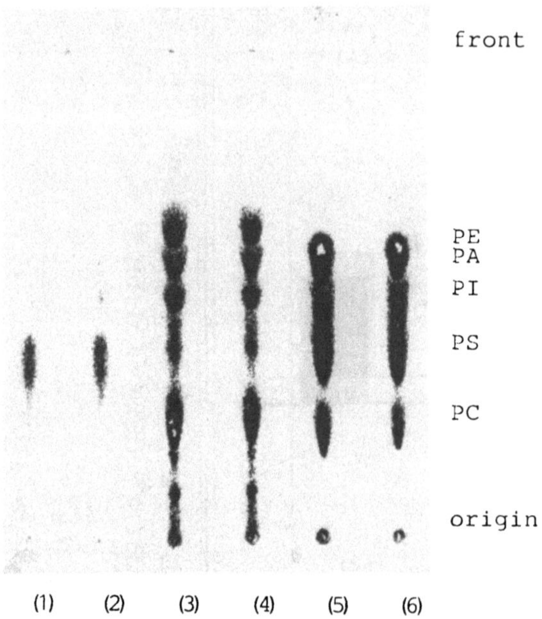

FIGURE 4. Thin layer chromatograms of coral and standard phospholipids (**PL**). Lanes 1, 2, skeletal PL; Lanes 3,4, soft tissue PL; Lanes 5,6, standard PL. **PC**, phosphatidylcholine; **PE**, phosphatidylethanolamine; **PS**, phosphatidylserine; **PI**, phosphatidylinositol; **PA**, phosphatidic acid.

The skeletal soluble matrix, extracted by EDTA, did not show calcium-binding capacity by Sephadex G-75 chromatography, neither did the HCl extract (not shown). The EDTA extract of coccolith from *Cricosphaera roscoffensis* var. *haptonemofera* showed a substantial binding of calcium ions by the same method used in this study (personal communication by Okazaki). Therefore, it is reasonable to assume that both of skeletal protein and carbohydrate, detected in the soluble organic matrix of the EDTA extract, are not the calcium-binding substances.

The insoluble matrix, however, bound strongly calcium ions *in vitro*. The binding disappeared mostly by a treatment of the matrix with chloroform and methanol solution. This suggests that the calcium-binding substances are lipids. Although the skeletal protein of *Acropora hebes* consists of high molar percentages of aspartic acid, the ratio of acidic amino acids to basic amino acids is very low (2.9) compared with that of alcyonarian spicule proteins (27.5 to 37.9) which bind calcium ions (Isa & Ikeda, 1986).

The skeletal lipids in *Acropora hebes* could only be obtained by decalcification of the skeletons. This was in accordance with the observation by Boskey and Posner (1976) that phospholipids could be extracted only from demineralized bone. This indicates that the lipids are firmly attached to the calcium carbonate crystals and can not be extracted by the washing medium used for the preparation of the skeleton samples. These observations support the proposed hypothesis that initial crystal formation occurs in the organic substance in *A. hebes* (Isa, 1986).

The marked difference in calcium-binding capacity between skeletal and soft tissue lipids indicate that the skeletal lipid extract is not contaminated with that of soft tissue lipid. Phospholipids of vertebrate and invertebrates play roles in the hard tissue formation (Rabinowitz *et al.*, 1976; Boskey *et al.*, 1978). In this connection, it is interesting that the calcium-binding capacity in the scleractinian skeleton are mainly due to the phospholipids. Though calcium-binding by the phospholipids of soft tissue was higher than the neutral lipid and glycolipid, the binding was lower than that of skeletal phospholipid. This suggests that skeletal phospholipids are composed of different components from soft tissue phospholipids. As shown by the high-performance TLC, the skeletal phospholipids were mostly consisted of acidic phosphatidylserine, in contrast to those of soft tissue being composed of several kinds of phospholipids. This was also the case in *Porites nigrescens* (Isa, unpublished).

Hsu *et al.* (1978) suggest that the acidic phospholipids in isolated, membrane-bound matrix vesicles are possibly initial calcification sites of vertebrate bone, teeth and cartilage, and play an important part in the induction of $CaPO_4$ deposition *in vitro*. It has also been reported that phospholipids are responsible for cartilage tissue mineralization in invertebrates (Rabinowitz *et al.*, 1976), and that lipofucsin-type pigment granules, contained in the shell-repair membrane of snails, serve as an initial site for membrane calcification (Abolins-Krogis, 1979). Isa and Okazaki (1987) showed that the calcium-binding capacity of the skeletal phospholipids increased markedly above pH 9.0, especially in the presence of HCO_3 ions. However, this does not seem to be the characteristic of the skeletal phospholipids in that Cotmore *et al.* (1971) described a similar effect of PO_4^{3-} ions on *in vitro* binding of calcium by commercial phosphatidylserine. The acidic phosphatidylserine in skeletal organic matrix of scleractinians may concentrate calcium ions and may serve as a seeding site or nucleator in calcium carbonate depositions.

ACKNOWLEDGEMENTS

I express my gratitude to Dr. M. Okazaki, Tokyo Gakugei University, for arranging my visit to his laboratory and for his advice and helps. I deeply appreciate Professor N. Watabe, University of South Carolina, for reviewing the manuscript. Financial support for this study was provided in part by the Ministry of Science and Culture, Japan (No. 57740402). Participation in the Symposium was made possible by grants from The Organized Research Fund of The University of Texas at Arlington and The Sea Grant Program administered by Texas A & M University.

REFERENCES

ABANTANGELO, G., DAGA-GORDINI, D., CASTELLANI, I. & CORTIVO, R., 1978. Some observations on the calcium ion binding to the eggshell matrix. Calcif. Tiss. Res., 26:247-252.

ABOLINS-KROGIS, A., 1979. The effect of adenosine triphosphate, magnesium chloride and phospholipids on crystal formation in the mineralized shell-repair membrane of the snail, *Helix pomatia* L. Cell Tissue Res., 204:497-505.

BOSKEY, A.L. & POSNER, A.S., 1976. Extraction of a calcium-phospholipid-phosphate complex from bone. Calcif. Tissue Res., 19:273-283.

BOSKEY, A.L., GOLDBERG, M.R. & POSNER, A.S., 1978. Calcium-phospholipid-phosphate complexes in mineralized tissue. Proc. Soc. Exp. Biol. Med., 157:590-593.

COTMORE, J.M., NICHOLS, G. & WUTHEIR, R.E., 1971. Phospholipid-calcium migration in the presence of phosphate. Science, 172:1339-1341.

CRENSHAW, M.A., 1972. The soluble matrix from *Mercenaria mercenaria* shell. Biomineralization, 6:6-11.

DITTER, J.D. & LESTER, R.L., 1964. A simple specific spray for the determination of phospholipids on thin-layer chromatography. J. Lipid Res., 5:126-127.

GOREAU, T.F., 1956. Histochemical detection of mucopolysaccharide-like substances and alkaline phosphatase in madreporaria. Nature, 177:1020.

HODGE, J.E. & HOFREITER, B.T., 1962. Determination of reducing sugars and carbohydrates. *In* Methods in Carbohydrate Chemistry. pp. 388. Ed. by R.L. Whistler & M.L. Wolfrom. Academic Press.

HSU, H.H.T., CECIL, R.N.A. & ANDERSON, H.C., 1978. Role of adenosine triphosphatase, phospholipid and vesicular structure in the calcification of isolated and reconstituted matrix vesicles. Metabl Bone Dis. Rel. Res., 1:169-172.

ISA, Y. & YAMAZATO, K., 1981. The ultrastructure of calicoblast and related tissues in *Acropora hebes* (Dana). Proc. 4th Int. Coral Reef Symp., 2:99-105.

ISA, Y., 1986. An electron microscope study on the mineralization of the skeleton of the staghorn coral, *Acropora hebes*. Mar. Biol., 93:91-101.

ISA, Y. & OKAZAKI, M., 1987. Some observations on the Ca binding phospholipid from scleractinian coral skeletons. Comp. Biochem. Physiol., 87b:507-512.

ISA, Y. & IKEDA, K., 1986. Some properties of the Ca binding substance of spicule from the alcyonarian corals. Ryudai RI Lab. Reports., 2:1-4.

JOHNSTON, I.S., 1979. The organization of a structural organic matrix within the skeleton of a reef-building coral. Scann. Electron Microsc., 2:421-431.

JOHNSTON, I.S., 1980. The ultrastructure of skeletogenesis in hermatypic corals. *In* International review of cytology, vol. 67, pp. 71-214. Ed. by H. Bourne & J.F. Danielle, New York Academic Press.

MITTERER, R.M., 1978. Amino acid composition and metal binding capacity of the skeletal protein of corals. Bull. Mar. Sci., 28:173-180.

MAJESKA, R.J., HOLWERDA, D.L. & WUTHEIR, R.E., 1979. Localization of phosphatidylserine in isolated chick epiphyseal cartilage matrix vesicles with trinitrobenzene-sulfonate. Calcif. Tissue Intl., 141:41-46.

RABINOWITZ, J.L., TRAVARES, C.J., LIPSON, R. & PERSON, P., 1976. Lipid components and *in vitro* mineralization of some invertebrate cartilages. Biol. Bull., 150:69-79.

SAMATA, T. & KRAMPITz, G., 1982. Calcium-binding polypeptides in oyster shells. Malacologia, 22:225-233.

WAINWRIGHT, S.A., 1963. Skeletal organization in the coral, *Pocillopora damicornis*. Quart. J. Micr. Sci., 104:169-183.

WUTHEIR, R.E., 1975. Lipid composition of isolated epiphyseal cartilage cells, membranes and matrix vesicles. Biochim. Biophys. Acta, 409:128-143.

YOUNG, S.D., 1971a. Organic material from scleractinian coral skeletons --- I. Variation in composition between several species. Comp. Biochem. Physiol., 40B:113-120.

YOUNG, S.D., O'CONNER, J.D. & MUSCATINE, L., 1971. Organic material from scleractinian coral skeletons -- II. Incorporation of ^{14}C into protein, chitin and lipid. Comp. Biochem. Physiol., 40B:945-958.

SKELETAL ORGANIZATION IN CARIBBEAN *ACROPORA* SPP. (LAMARCK)

Brent R. Constantz[1]

Earth Sciences Board, University of California, Santa Cruz, CA 95064; [1]Present address: NORIAN, 1025 Terra Bella, Mt. View, CA 94043

"Although the zooxanthellae seem to play an important role in determining calcification rates in reef-building corals, certain, as yet unknown, physiological factors operate to control the basic mineralization process in a manner which bears no obvious relationship to the number of algae present in a given species. This is illustrated by the fact that large apical polyps of some of the branching acroporid corals contain few zooxanthellae but calcify several times faster per unit of tissue nitrogen than the yellowish brown lateral polyps which are literally stuffed with algae." – T.F. Goreau (1959:67)

ABSTRACT. The size and morphology of calcium carbonate crystals of the exoskeletons of the Caribbean Scleractinian coral species of the genus *Acoropora* are described. The fundamental units of the skeleton are the trabeculae, which are linearly aggrading spherulitic fans of polycrystalline aragonite fiber bundles. Each spherulitic fan originates from a center of calcification (Ogilvie, 1896; Wells, 1956) that is composed of packets of submicron calcium carbonate crystals in an amorphous matrix. The nucleating packets appear to have an intracellular origin and their production stimulated by zooxanthellate photosynthesis.

INTRODUCTION

Goreau (1959:66) made the significant observation that even in complete darkness, zooxanthellate specimens of *Oculina diffussa* and *Manicina areolata* calcified more rapidly than specimens of the same species with the zooxanthellae removed. This, and Goreau's observation that branching acroporids are non-zooxanthellate at the branch tips, where the rates of linear growth were highest, both indicated that algal enhancement of calcification was not driven simply by photosynthetic removal of CO_2 alone (Goreau 1959). Goreau (1959, 1961, 1963) and Wainwright (1963) believed that skeletal growth in Scleractinians was mediated by an organic matrix. The organic matrix was proposed to nucleate crystals and guide crystal growth (Johnston 1980). Wainwright (1963), working with *Pocillopora damicornis*, proposed that zooxanthellae provided corals with the metabolic energy required to synthesize this structural organic matrix. Evidence for the excretion of glycerol from zooxanthellae to cnidarian and molluscan hosts (Muscatine, 1967) suggested that calcification was driven by zooxanthellae-derived food. Pearse & Muscatine (1971) demonstrated that labeled organic products from zooxanthellae on proximal portions of a ramose acroporid branch were translocated to

Edited by Rex E. Crick
Plenum Press, New York

the non-zooxanthellate branch tips. They suggested that organic matrix synthesis, and thus calcification, at the non-zooxanthellate branch tips were fueled by the translocated metabolic products. Vandermeulen *et al.*, (1972) demonstrated that zooxanthellate photosynthesis had an important effect on calcification in *Fungia scutaria*, *Montipora verrucosa* and, *Pocillopora damicornis*. Inhibition of photosynthesis reduced the rate of $^{45}Ca^{2+}$ incorporation to dark levels, even in the presence of light. These studies, mentioned above, corroborated a general model of algal-enhanced calcification whereby translocated algal metabolites supplied the metabolic energy required for organic matrix synthesis.

Barnes (1972) observed the daily growth patterns in the epitheca of *Manicinia areolata*. He noticed that one portion of the epitheca grew during the day and another portion at night. A diel variation was subsequently demonstrated by Chalker (1976) in the rate of calcification at saturating light intensities in *Acropora cervicornis*, with maxima at sunrise and sunset; this pattern persisted even under constant light and was correlated with a similar rhythm in the photosynthetic capacity of the zooxanthellae. Barnes and Crossland (1980) showed that branch extension in *Acropora acuminata* was more rapid at night than the day, but that calcification was 2 to 3 times greater in the day than the night. Based on this finding, in a diel study of the axial corallite of *Acropora cervicornis*, Gladfelter (1982, 1983) found that a framework of "1−3 μm fusiform crystals" are errected at the tip of the axial corallite during the night, setting up a scaffolding for subsequent epitaxial growth of aragonite fibers during the day. Gladfelter (1982, 1983) proposed a general model for skeletal growth whereby 1−3 μm fusiform crystals were formed at the tips of the axial corallites during the night. Jackson and Gladfelter (1985) described an organic matrix of "crystal bearing envelopes" in which these crystals form. It was proposed that this was followed by a second phase during the day where aragonite fibers grow from the "fusiform crystals" in an epitaxial manner. The model of Gladfelter (1982, 1983) explains the pattern of axial corallite extension and calcification observed by Barnes & Crossland (1980).

Gladfelter's (1982, 1983) model requires that the initial sites of mineralization be extracellular because of the 1−3 μm size of the "fusiform" crystals. The cells of the calcicoblastic ectoderm usually form a 2−3 μm thick sheet. Even in another branching Acroporid, *Acropora hebes*, on the tips of axial corallites where the cell sheet is unusually thick, there is no evidence of large intracellular crystals (Isa & Yamzato, 1981; Isa, 1988−this volume). The difficulty with Gladfelter's model is that it sidesteps the petrographic and transmission electron microscopic evidence of Wainwright (1964) on *Pocillopora damicornis* (zooxanthellate) and *Lophelia pertusa* (non-zooxanthellate). Wainwright (1964) observed that the centers of calcification are surrounded by a brown zone that formed a broad zone of unoriented aragonite crystals that had wide intercrystalline boundaries. This crystalline brown zone became progressively oriented and less porous moving out from the center of calcification to the aragonite fiber bundles. Based on his X-ray microdiffraction results, Wainwright (1963) determined that the crystals in the centers of calcification are randomly oriented and less than 1 μm across. Using the same criteria, he determined that the brown zone was grannier than the centers of calcification (*i.e.* > 1 μm crystals). In a petrographic and scanning electron microscopic study of *Acropora palmata*, James (1974) also observed the "brown zone" described by

Wainwright (1963). James (1974) identified: 1) dark to opaque centers of calcification composed of small (1 μm or less) randomly oriented aragonite crystals, 2) a brown zone of randomly oriented equant, micron sized crystallites and oriented elongate needles of aragonite. I observed 1 – 4 μm equant crystals in several coral species (Constantz 1986). I too found them different from the fine material at the centers of calcification, which are composed of crystals smaller than 0.1 μm across. In other words, Wainwright (1963), James (1974), and Constantz (1986) observed the micron plus equant crystals (Gladfelter's "fusiform crystals") that form a band around the centers of calcification, but collaborated with the long-standing views of Ogilvie (1896) and Wells (1956), that the aragonite fibers originated from the centers of calcification. Gladfelter (1982, 1983) and Jackson & Gladfelter (1985) assumed that the "fusiform crystals" were the crystals at the centers of calcification. If the "fusiform" crystals seen in *A. cervicornis* by Gladfelter (1982, 1983) nucleate the aragonite fiber bundles then how are the centers of calcification related to skeletogenesis? Alternatively, since all Scleractinians have centers of calcification (Wells 1956), but micron plus, equant crystals have been reported only in a few taxa, how important are they?

This study describes the skeletal organization of the Caribbean *Acropora* species: *A. palmata*, *A. prolifera*, and *A. cervicornis*. The three species have very similar skeletons that differ only in their respective styles of axial corallite branching (Randal, 1981; Constantz, 1984). In fact, a graded morphological range through the three species occurs in the Caribbean and exact species definitions are difficult at best. Gel electrophoretic data indicates that all three species may represent one highly variable species (Ohlhurst, 1984). With the morphological and mineralogical data presented here, plausible explanations for mechanisms of calcification will be evaluated.

METHODS

Sample collection and preparation

Living samples of the arborescent corals *Acropora palmata*, *A. prolifera*, and *A. cervicornis* were taken at four locations in the Caribbean. Samples from the US Virgin Islands, were collected at Water Island, off St. Thomas, at 4 m depth at 0800 hours during April 1983. Samples from Barbados were collected from the fringing reef off Jamestown at 5 – 15 m depth at 0600 hours, 1200 hours, 1800 hours, and 2400 hours during July 1983 and February 1984. In the Grenedines, samples were collected in the Tobago Cays, off Jamesby, at 2 m depth at 0600 hours, 1200 hours, and 1800 hours. Jamaican samples were collected in Discovery Bay at 2 m depth at 0600 hours and at Rio Bueno at 2 m depth at 1200 hours during August 1983 and at 10 – 20 m depth at 0600 hours, 1200 hour, 1800 hours, and 2400 hours during August 1984. *A. prolifera* was only present at the Jamaican site.

At every collection site, several samples were taken from two or more individuals, including two or more branches from each individual. After collection, samples were air-dried in the sun for a day or two. In Barbados, one group of samples was transferred live to a running sea water system at Bellairs Research Institute. The following morning, at 0600 hours, the live samples were fixed by slow infiltration of the fixation solution introduced to the surrounding sea water with a hypodermic needle for about 20 seconds, then by total immersion of the sample in the fixation solution. The fixation solution was 0.1 M Na coccoadylate, 0.25 M sucrose, 3% glutaraldehyde, and 0.5% tannic acid (Johnston, 1980) adjusted to pH 8.2. Fixation was 1 hour at 25° C and then 24 hours at 10° C.

Optical petrography

Air-dried samples were cut on a diamond saw. The cuts were oriented either parallel or perpendicular to the axis of the branch. Each specimen was mounted on a slide and ground to approximately 20 μm thickness. Specimens were investigated on a Zeiss photomicroscope (#62846) with 31x, 79x, 200x, and 500x objectives.

Bulk mineralogy

The upper 2–4 cm of air-dried branch tips were ground to a fine powder. Samples were scanned on a Philips 2500D X-ray diffractometer calibrated with a silica standard. Scanning was from 20–40 2θ, to include all major peaks for aragonite, calcite, and dolomite. The lower limit of detection using the technique is 2–3 modal percent of the sample (Milliman, 1974). Any crystalline phase present in volumes lower than 2–3% of the bulk sample would not have been detected.

Scanning electron microscopy

Tissue was removed from specimens of *Acropora palmata*, *A. prolifera*, and A. *cervicornis* using a dilute bleach solution of 1:1 deionized water and bleach (5% NaOCl) or by ultrasonic cleaning in dichloromethane. Specimens were washed in 100% acetone for 1 min and air-dried.

Fixed specimens were cut into 1 cm sized cubes. Care was taken to protect the tissue from mechanical damage. The diamond-bladed saw was cooled with wash solution of the same composition as the vehicle in primary fixation. Specimens were post-fixed with the same vehicle containing 1.0% osmium tetraoxide for 1 hour. These specimens were stored in a final wash, then wrapped in a cryoprotectant and plunged into a liquid nitrogen slush and immediatedly fractured. Specimens were then infiltrated with a series of absolute ethanol, followed by freon 113, and critical-point dried with CO_2.

For analysis of the ultrastructure of the skeleton, air-dried samples were sectioned with a diamond-bladed saw and diamond paste-polished to a mirror finish. These specimens were etched with 1.0% formic acid for 1 min. All specimens for SEM were set on aluminum stubs with silver paint, evacuated to 10^{-5} torr, and sputter coated with gold-paladium or coated with an evaporated carbon coat. Finished specimens were investigated in a Semco Nanolab 7 scanning electron microscope at accelerating voltages ranging from 5 kV to 25 kV. A combined Princeton Gamma Tech energy dispersive spectrometer allowed qualitative elemental analyses of selected areas. The sensitivity of detection of calcium, magnesium, and strontium with this instrument was 10,000 ppm (1%).

Transmission electron microscopy

For transmission electron microscopy of skeletal ultrastructure, 1 cm cubes were cut from air-dried specimens with a diamond-bladed saw. Specimens were impregnated into epoxy resin cylinders. The cylinders were cut at planes perpendicular or parallel to the axis of the branches, making two specimens from each original. These specimens were ground and polished to a mirror finish using diamond paste. The polished surface was then etched with 1% formic acid for about 20 sec. Acetone-softened, acetate peels were pressed on to the etched surfaces, then peeled off. The acetate peels were coated with a carbon film on the impression side. This surface was counter-shaded with germanium, evaporated at a 20° angle to the plane of the replica. The shaded carbon peels were immersed in acetone, dissolving the acetate, and the replicas were picked up on copper grids. These specimens were investigated on a Siemens 100B transmission electron microscope at accelerating voltages ranging from 40 kV to 100 kV.

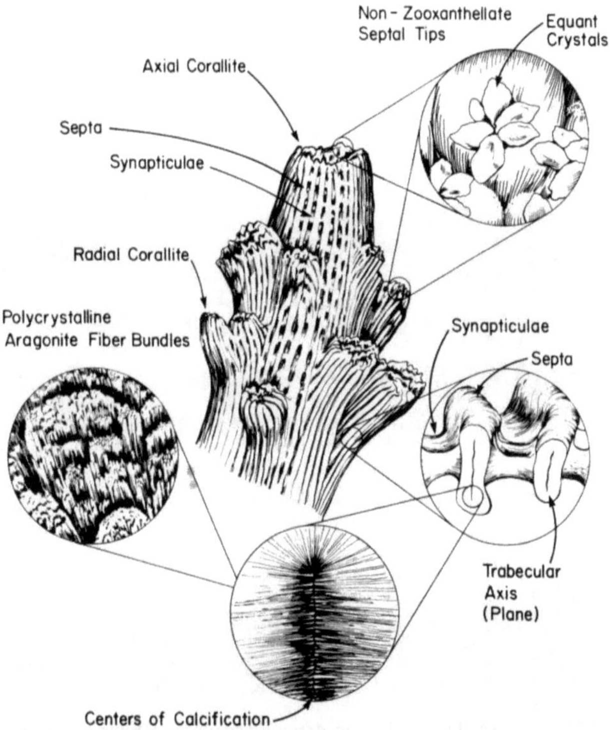

FIGURE 1. Illustration of an *Acropora* spp. axial corallite. Radial corallites branch from the axial corallite. Vertical rods are septa. Horizontal rods are synapticulae. Centers of calcification make-up the trabecular axe that run through the septae and synapticulae. Equant crystals nucleate and grow on the non-zooxanthellate septal tips.

For TEM analysis of skeletal ultrastructure, directly, uncovered ultra-thin sections (< 1 μm) were ground from air-dried specimens. Copper grids were mounted at specific points (100 μm dia.) on the section. Specimens were then removed from the glass slide and areas protected by the copper grids broken away from the main specimen. Specimens held in grids were argon beam ion-thinned in a Technics Auto-Min ion thinner, until several beveled holes, 100 μm diameter each, were milled in portions of the specimen. These specimens were investigated with a Siemens 100B transmission electron microscope at an accelerating voltage of 100 kV.

RESULTS

Petrographic and electron microscopic results are reported as several good examples of representative observations. Images are grouped so that several aspects of an individual sample are together for comparison. All features reported were observed in all three species. Where a particular feature was observed in only a limited number of

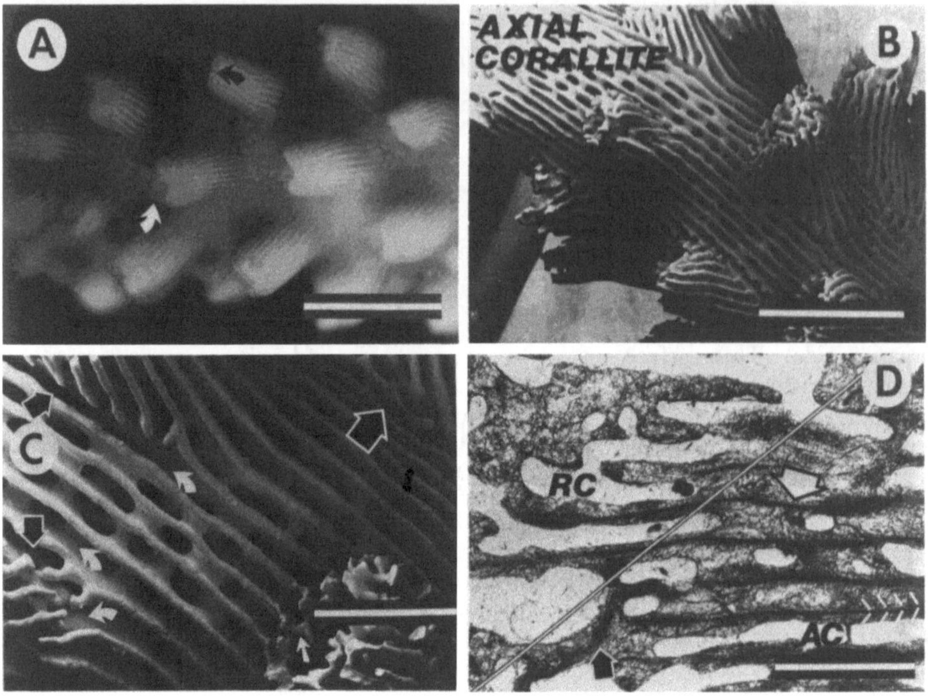

FIGURE 2. ▪A. Photo of living *Acropora cervicornis*. Bar = 6 μm. ▪B. *Acropora* spp. axial corallite. Bar = 4 mm. ▪C. *Acropora* spp. axial corallite. Bar = 2 μm. ▪D. Septa on axial corallite. Bar = 2 μm.

instances, specific information on the number of times observed is provided. Mineralogic and elemental data is presented at the end of the results section.

Figure 1 is a schematic diagram of an *Acropora* spp. axial corallite. The purpose of this schematic is to place all structures described below in a proper geometrical framework. Several radial corallites bud off the axial corallite. The vertical rods are septa. The horizontal bars are synapticulae. Running the length of every septum and synapticulum are trabecular centers which consist of centers of calcification. The centers of calcification are composed of small crystals less than 1 μm in diameter. The crystals radiating from the centers of calcification are polycrystalline aragonites fiber bundles. At the tips of the axial corallite and to a lesser degree, on the tips of the radial corallites, equant aragonite crystals greater than 1 μm in diameter first form. These crystals later form a distinct brown zone in the skeleton around the centers of calcification.

Figure 2a is a photograph of a living *Acropora cervicornis* colony from Barbados. The field of view of this photograph is 1.5 cm (1.5×10^4 μm). This photograph illustrates the general morphology of the axial and radial corallites of all three species and the position of the tissue overlying the corallites. The polyps are partially withdrawn and only parts of the tentacles are visible (white arrows). The main shaft running the length of the photograph is the axial corallite of a branch, 2 cm proximal to the tip. The smaller lateral shafts are the radial corallites. Note that the coenosteal tissue covers the entire corallum and has a rich, dark color, indicating the presence of pigmented endosymbiotic zooxanthellae. Only approaching the tips of the radial corallites does the rich, dark color of the tissue fade. Unlike the well known transparent tissue at the growth tips of the axial corallites (Goreau, 1959; Pearse & Muscatine, 1971; Barnes & Crossland, 1980), which almost completely lacks pigmented zooxanthellae, the tissue at the tips of the radial corallites is only slightly depleted in pigmented zooxanthellae relative to the coenosteal tissue. At the extreme distal tips of the radial corallites the tissue covering the extending septa is nearly transparent and has the color of the white skeleton beneath (black arrows). These characteristics were also observed at the tips of several axial corallites of both *A. palmata* and *A. prolifera*.

Figure 2b is a SEM of an *Acropora cervicornis* branch tip. Notice that the non-zooxanthellate parts noted in Figure 2a are the extending septal tips. This specimen originated from the U.S. Virgin Islands. The horizontal field width of this micrograph is 12 mm (12 x 10^3 μm). The corallite wall or theca is principally synapticulothecate (Wells, 1956:345) and partially septotheca (Wells, 1956:345; Richart y Menendez & Friedman, 1977). It consists of simple trabecular septa (white arrows; Wells, 1956:337) that run its length and simple synapticulae that connect the septa (black arrows; Wells, 1956:343). The extending corallite wall is led by the extending septa which occur at the parts of the skeleton where the concentration of pigmented zooxanthellae are the lowest, most notably on the axial corallite and to a lesser degree on the radial corallites.

In *Acropora cervicornis* the branch has a single axial corallite that runs its entire length. The branches of *A. palmata* have several axial corallites that have a common coenosteum connecting them structurally (Constantz, 1984). The growing tips of the axial corallites in *A. palmata* protrude ahead of the distal margin of a frond and are characteristically non-zooxanthellate. *A. prolifera* is intermediate between *A. cervicornis* and *A. palmata*. There is one axial corallite at proximal portions of *A. prolifera* branches which branch and form a discontinuous frond. The axial corallites of *A. prolifera* have isolated non-zooxanthellate tips like both *A. palmata* and *A. cervicornis*, but unlike *A. cervicornis* they do not remain isolated, and unlike *A. palmata*, they do not become totally fused. Radial corallites in all three species develop from axial corallites by extratentacular budding (Wells, 1956:348) and have a nariform shape (Wallace, 1978:275). The microarchitecture of *A. palmata*, *A. prolifera* and, *A. cervicornis*, as just described, was observed in all specimens in the study and is consistent in Caribbean *Acropora* spp. (Constantz, 1984).

Figure 2c is a SEM at higher magnification of the axial corallite wall of the specimen in Figure 2b. The horizontal field width of this figure is 9 mm (9 x 10^3 μm). The septa are spaced apart 0.25 mm (2.5 x 10^2 μm) to 0.50 mm (5.0 x 10^2 μm). The synapticulae are spaced apart 0.50 mm (5 x 10^2 μm) to 2 mm (2 x 10^3 μm). Two white arrows indicate surfaces on the septa and synapticulae where growing fasciculi are visible. A few of the septal tips of the radial corallite in the lower right corner of the micrograph appear broken. The fasciculi, or aragonite fiber bundles, radiate away from trabecular axes which are at the centers of the septae and synapticulae. Note the spatial relationship between the septa on the axial corallites and the radial corallites (black arrow). Septa that originate proximal to radial corallites continue up the radial corallite's outer wall, whereas septa on the lateral and inner walls of radial corallites originate at the base of the radial corallites. This general pattern was observed in every specimen in the study.

Figure 2d is a polarized light micrograph of a region corresponding to the section of an axial corallite in Figure 2c. This specimen is another individual of *Acropora cervicornis* collected in the Grenedines. The most conspicuous feature of the axial corallite (AC) is the prominent dark streaks (white arrow) that run the lengths of the septa discontinuously. These streaks have a central dark line and a diffuse fringe. The width of the dark streaks ranges from 50–200 μm. The dark streaks comprise 10% to 60% of the cross-sectional area of the axial corallite septa. The distinct dark line at the center of the dark streaks appears less than 50 μm in width, if present at all. The dark line is commonly absent and only a diffuse region less intense than the dark streaks is present (black arrow). The aragonite fiber bundles appear to radiate away from the dark streak forming a pattern indicated by the white chevrons on the micrograph. The dark streak is finely crystalline and appears dark because of enhanced back reflection of the light on the many crystalline interfaces. Note that at this magnification it is impossible to discern the point from which the aragonitic fiber bundles radiate because the dark streak often obscures the point of radiation.

The axial corallite in Figure 2d (separated from an adjoining radial corallite by a line) buds a radial corallite (RC) in the left portion of the micrograph. The septa in this portion of the micrograph do not have prominent dark streaks. Although the dark streaks run up septa at the base of radial corallites (RC), dark streaks are sporatic, diffuse, or entirely absent in the septa of radial corallites. This characteristic appeared in every petrographic thin section of all three species in which the section included the septa of radial corallites.

Figure 3a is a polarized light micrograph of a petrographic thin section taken parallel to the outer surface of an axial corallite of *Acropora cervicornis*. This section is from a set of serial petrographic thin sections taken from the specimen in Figure 1d. This section cuts through both the synapticulae and the septa. Two black arrows indicate a gap in the dark streak. The dark streak on the septa to the right of the gapped septa is quite bold. Note that the axis of the surface from which the aragonite fibers nucleate (Fig. 3b), and the dark streaks of the axial corallite septa are not always coincident. The dark streaks fluctuate in both intensity and distribution, even while the radiating aragonite fiber bundles are a consistent component of the septa.

Figure 3b is a SEM of the septa of an axial corallite of *Acropora cervicornis* 2 cm proximal of the distal tip collected in the U.S. Virgin Islands. The field width of this micrograph is 805 μm. The upper part of this micrograph is the outer surface of the axial corallite. The three fractured faces in the micrograph are cross-sections through the septa of the axial corallite. Black arrows indicate the width (taken in a plane through the branch axis) of the axial surface from which the aragonite fiber bundles radiate. Note that this surface is nearly as wide as the width of the septa. For instance, in the fractured septum in the middle of the micrograph, the width of the septum is approximately 400 μm and the radial width of the plane from which the aragonite fiber bundles nucleate is approximately 300 μm. Note that the synapticulae (white arrows) connect to the sides of the septa near the middle of the septa. Petrographic thin sections taken parallel to the outer surface of the axial corallites that cut through both the

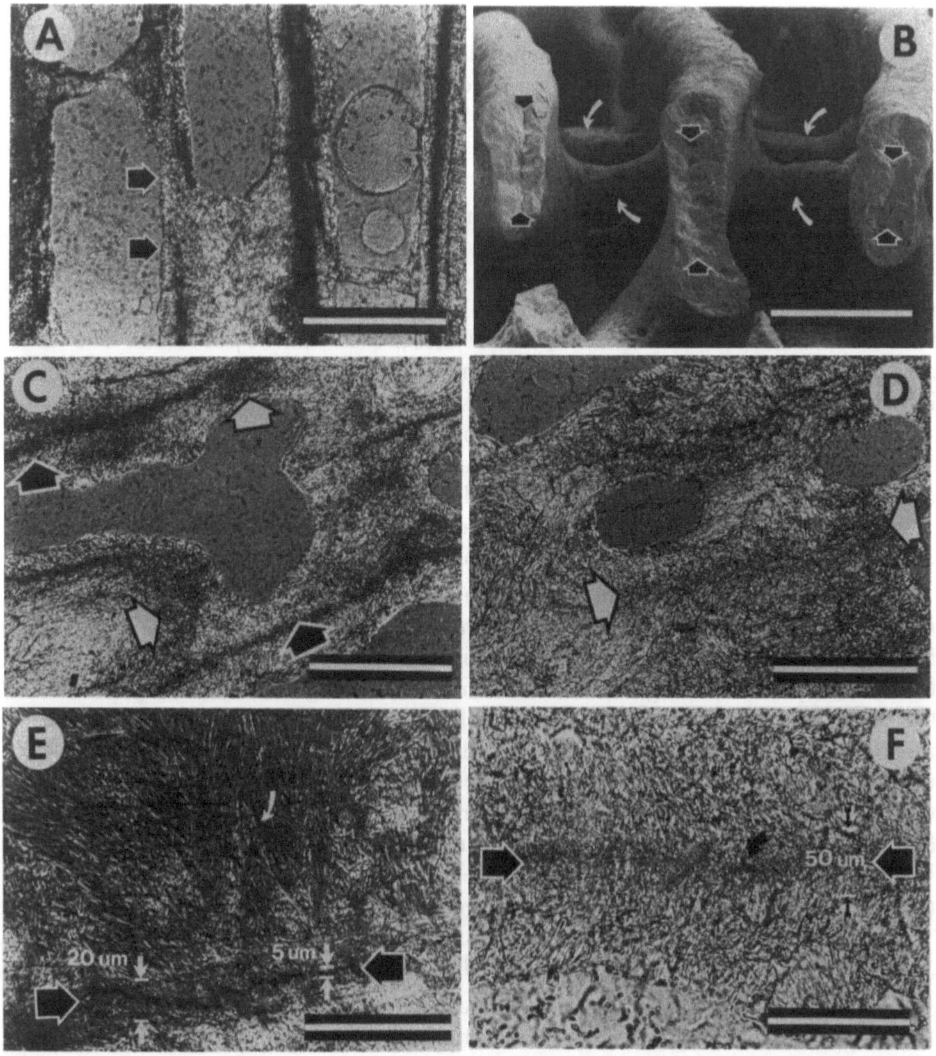

Figure 3. ▪A. Gap in dark streaks in septa of axial corallite (black arrows). Bar = 0.5 μm. ▪B. Fracture through septa perpendicular to axial corallite. Trabecular plane denoted by black arrows. Synapticuae denoted by white arrows. Bar = 0.25 μm. ▪C. Diffuse region of the dark streak (white arrows) and resrticted regions of the dark streak (black arrows). Bar = 0.5 μm. ▪D. White arrows denote trabecular axis on radial corallite. Archers arrows point to sparse equant crystals. Bar = 0.5 μm. ▪E. Black arrows denote trabecular axis. Irregular line if chain of nuclei packets. White arrow points to an aragonite fiber bundle. Bar = 50μm. ▪F. Large black arrows denote trabecular axis. Small curved black arrow denotes visible nuclei packets. Bar = 100 μm.

septa and the synapticulae must cut through the axial surface from which the aragonite fibers radiate, forming a line. This general geometry, that synapticulae are connected to the septa adjacent to the plane from which the aragonite fiber bundles nucleate, was apparent in all specimens included in this investigation, and is not a configuration peculiar to this micrograph. Thus, in Figure 3a, where both the septa and the synapticulae are sectioned, the gap in the dark streak must not be an artifact caused by two dimensional sectioning of a three dimensional object. Rather the gap in the dark streak observed in Figure 3a, and many other specimens, must be a true discontinuity in the dark streak. This observation was easily repeatable in all petrographic thin sections containing portions parallel to the outer wall of the corallite which were sectioned through both septa and synapticulae.

Figure 3c is a polarized light micrograph of a petrographic thin section of a septa at the base of a radial corallite. This specimen is from a set of serial petrographic thin sections from the specimen in Figure 2d. Black arrows point to a the dominant dark streaks running the length of the septa. The central lines of these streaks are very distinct and the broad, diffuse zone is absent. The central lines are less than 50 μm across. The white arrows point to local areas of diffuse darkening. Close observation of both the central lines (black arrows) and the diffuse regions shows that they contain or are composed of small (*c.* 1–4 μm), equant crystals. It is possible to observe in petrographic thin-section that these crystals are less than 10 μm across but larger than 1 or 2 μm. Note that the diffuse regions are not extensive. Instead, each is separated from the others by areas of aragonite fiber bundles. By using high power it was possible to observe the equant crystals in the dark streaks and diffuse regions in virtually every specimen.

Figure 3d is a polarized light micrograph of a petrographic thin section of the septa in a distal portion of a radial corallite in *Acropora cervicornis*. This specimen is from a set of serial sections taken from another individual collected at the same time and location as the specimen in Figure 2d. This section cuts through both the synapticulae and the septa in a plane parallel to the outer wall of the corallite, thus sampling the axis of aragonite fiber bundle radiation. The two white arrows delineate the surface from which the aragonite fiber bundles radiate. Relatively few equant crystals (archer's arrows) are visible in the septal axis here. This micrograph is representative of the general trend that the dark streaks and diffuse regions of non-fibrous crystals become sparser in the septa of the radial corallites, moving progressively away from the axial corallite. This trend, as just described, was observed in every petrographic thin section in the study. The nucleation axis on the upper septa of this specimen contains a characteristic *Cliona* sp. boring of connected chambers that parallels the trace of the surface.

Figure 3e is a SEM of a polished, etched section taken through the plane of a surface from which the aragonite fiber bundles radiate. This specimen was collected at Rio Bueno, off Jamaica's north coast. The field width of this micrograph is 164 μm. The axis of a surface from which aragonite fiber bundles nucleate is indicated by two black arrows. A broad band of enhanced dissolution about 20–30 μm wide is visible within which a very discrete irregular line of very enhanced dissolution is planely visible. This very distinct, irregular line varies between 2–7 μm in width. The aragonite fiber bundles that radiate from this axis appear to be about 5–10 μm in width (white arrow). Finding the precise points where the aragonite fiber bundles radiate from was very difficult and time consuming. Unlike light microscopy, SEM only gives an image of secondary electrons from the surface of a section. Even the dark streaks were not obvious. Twelve good examples of this thin irregular line at the point of aragonite fiber radiation were found.

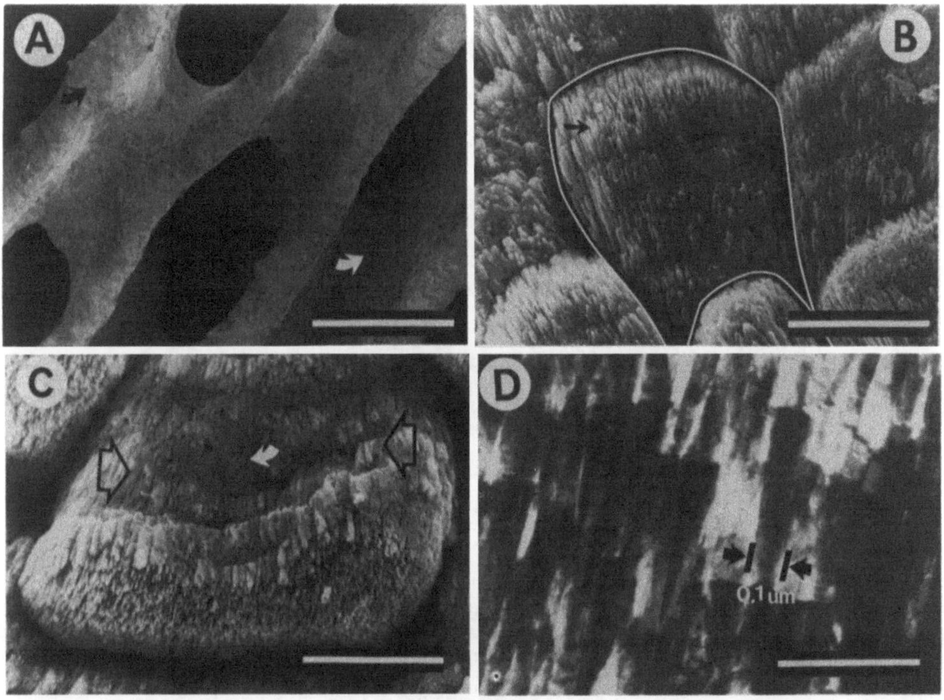

FIGURE 4. ▪A. Detail of septa on axial corallite. Black arrow points to smooth region. White arrow points to growing fasciculi. Bar = 2 μm. ▪B. Individual polycrystalline aragonite fiber bundle is outlined at the surface. Black arrow points to an individual fiber. Bar = 10 μm. ▪C. Black arrows denote region of smoothing. White arrow points to region of notable truncation of aragonite fibers. Bar = 5 μm. ▪D. Indiviual aragonite fibers. Note width. Bar = 0.5 μm.

Figure 3f is a polarized light micrograph of an axis of aragonite fiber radiation in a radial corallite septum. This specimen is a serial-section from the specimen in Figure 2d. A band of non-fibrous, equant crystals delineate the axis indicated by two black arrows. This band is less than 50 μm in width. Within this band are a discontinuous chain of localized dark areas (black arrow) that are roughly 10 μm across and occupy central portions of the axis.

Figure 4a is a SEM of the axial corallite wall of the same specimen in Figure 3b. At the outer edge of the septa, the surface of the skeleton is smooth, having no indication of the internal ultrastructure (black arrow). A white arrow indicates an area on a synapticulae, internal to the corallite wall, where the growing tips of internal fiber bundles are just visible at the surface.

Figure 4b is a SEM of an individual polycrystalline aragonite fiber bundle (outlined). This specimen was collected and fixed in Barbados. The field width of this micrograph is 30 μm. Each fiber bundle is in direct contact with its neighboring fiber bundles. The entire fiber bundle is approximately 14 μm across at its widest point. The black arrow points to one of the aragonite fibers that comprise the fiber bundle. This acicular morphology is characteristic to the growing tips of the aragonite fiber bundles in the skeleton

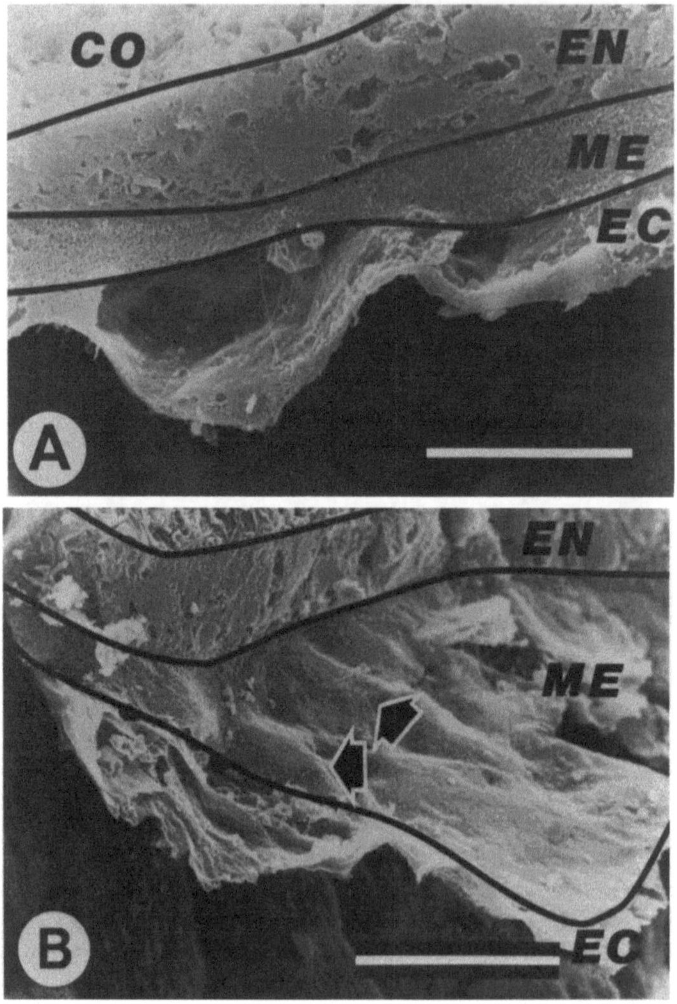

FIGURE 5. ▪A. Cryofracture across aboral tissue layers. Note coelenteron (CO), gastrodermis (EN), mesoglea (ME), and calicoblastic ectoderm (EC). Bar = 8 μm. ▪B. Cryofracture across aboral tissue layers. Note undulations in mesoglea. Bar = 16 μm.

of *Acropora cervicornis*, except at the non-zooxanthellate tips of growing septa. Notice that the shape of the fibers changes between the leeding end of the fiber bundle and the base. At the tip the fibers are acicular, whereas at the base the fibers become less pointed or more angular. There are distinct troughs perpendicular to the direction of fiber growth at a 5 – 10 μm interval on the fiber bundle. Fibers in this region have irregular, angular morphologies. This pattern of troughing and deterioration of the characteristic acicular fiber morphologies moving away from the most extended parts of the fiber bundle was commonly observed in about half of the fiber bundles studied.

FIGURE 6. ▪A. Center of calcification (**NP**) is composed of nuclei packets from which fasciculi (or polycrystalline aragonite fiber bundles) radiate. Black arrow denote direction of radiation. Bar = 30 μm. ▪B. Close-up of above. Black arrows point to individual nuclei packets. White arrows denote region were two packets are molded together. Bar = 10 μm. ▪C. Close-up of above. Black arrows indicate individual subpackets. White arrows point to fiber which have nucleated. [Bar = 4 μm]

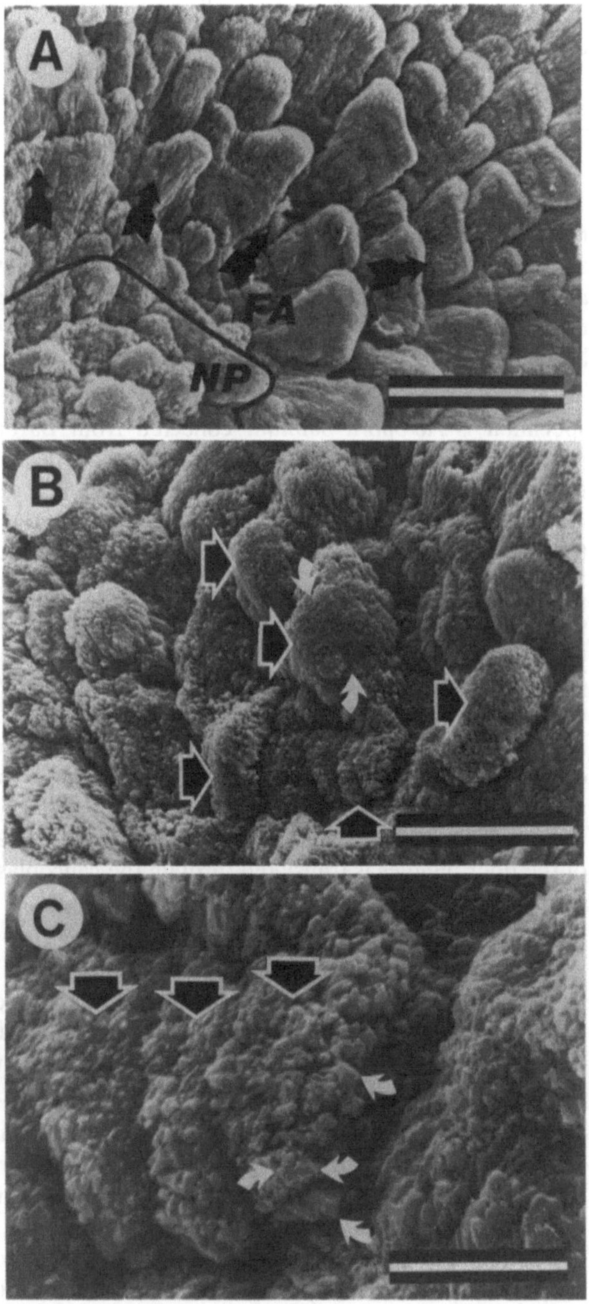

Figure 4c is a SEM of an individual fiber bundle about 1 cm proximal to the axial corallite. This bundle is noticably troughed in linear regions moving backward from the growing tip (black arrows). A white arrow indicates an area where the truncated outlines of the former crystals are observable. This pattern is characteristic to large areas of many fiber bundles 100 μm or more across, merging into smooth skeletal surfaces at the periphery.

Figure 4d is a transmission electron micrograph of a germanium-shaded, carbon replica taken from a polished and etched sample of *Acropora cervicornis*; individual aragonite fibers are embedded in the replica. At least three layers of these fibers are overlapping in the section. Since the fibers each measure 0.1 μm in thickness the section is approximately 0.3–0.4 μm thick. Each individual fiber has been broken across its length in the replication process, however the width of the fibers is primary. This dimension was further corroborated on ion-thinned sections and it was consistently found in all transmission electron microscopic samples. In over 25 different specimens investigated, the limits of fiber diameter were 0.05 – 0.15 μm. Because of beam drift, power fluctuations, and variability of focus, these dimension were in error by at least 0.05 μm.

Selected area diffraction of individual aragonite fibers within fiber bundles and of non-fibrous, equant crystals demonstrated that both crystal types are single crystals of aragonite. Oriented selected area diffraction patterns with bright and dark field imaging indicated that the individual aragonite fibers elongated in the *c*-axis direction, as noted in *Picukkioira damicornis* by Wainwright (1963). The non-fibrous, equant aragonite crystals had distinct diffraction patterns indicating they are monocry stalline.

Figure 5a is a SEM of the lower tissue layers immediately over the skeleton. This specimen is *Acropora palmata* collected from Barbados. The field width of this micrograph is 24 μm. The area above the tissue layers is the coelenteron (CO). The upper tissue layer is the aboral gastroderm (EN). The middle tissue layer is the mesoglea (ME). The lower tissue layer is the calicoblastic ectoderm (EC). The calicoblastic ectoderm is characteristically out-pocketed from the mesoglea in areas over fiber bundles.

Figure 5b is a SEM of an area about 25 μm to the right of Figure 5a. The field width of this micrograph is 48 μm. In this area the freeze-fracture has traveled across the aboral gastrodermis (EN) perpendicular to the tissue layer, but has moved along the contact between the gastrodermis and the mesoglea (ME), and then cut down through the calicoblastic ectoderm at an oblique angle. Note the undulations in the mesoglea (black arrows). Note that these undulations correspond in spacing to the fiber bundles on the skeleton below. The rounded tops of the fiber bundles are not visible in this micrograph. This corresponding pocketing is a characteristic of the calicoblastic ectoderm (EC) in places adjacent to fiber bundles. Note that here the irregular surface of the fiber bundles is translated through the calicoblastic ectoderm to the mesoglea.

Figure 6a is a SEM of a center of calcification and radiating fan of aragonite fiber bundles directly adjacent to Figures 5a & b. The field width of this micrograph is 135 μm. Black arrows indicate the direction of radiation of the aragonite fiber bundles (FA). The center of calcification (NP) is outlined. (This zone of transition between the fiber bundles and the center of calcification is wider than the fine black line used here to avoid covering-up too much of the micrograph; the width of this zone is approximately 5 μm.).

Figure 6b is a SEM of the center of calcification in Figure 5a. The field width of this micrograph is 45 μm. The center of calcification is composed of round packets (black arrows) of very fine granular crystals. Each granular packet is approximately 8 μm across. Each packet appears to be composed of subpackets, each consisting of the same granular material. These subpackets appear as though they had been accreted together in a softened or gelatinous state. This is indicated by rounded suture lines between the subpackets (white arrows). The dimensions of these irregular subpackets is less than 4 μm.

Figure 6c is a SEM of the three distinct granular subpackets at the lower, middle black arrow of Figure 6b. The field width of this micrograph is 8 μm. Each sub pack-

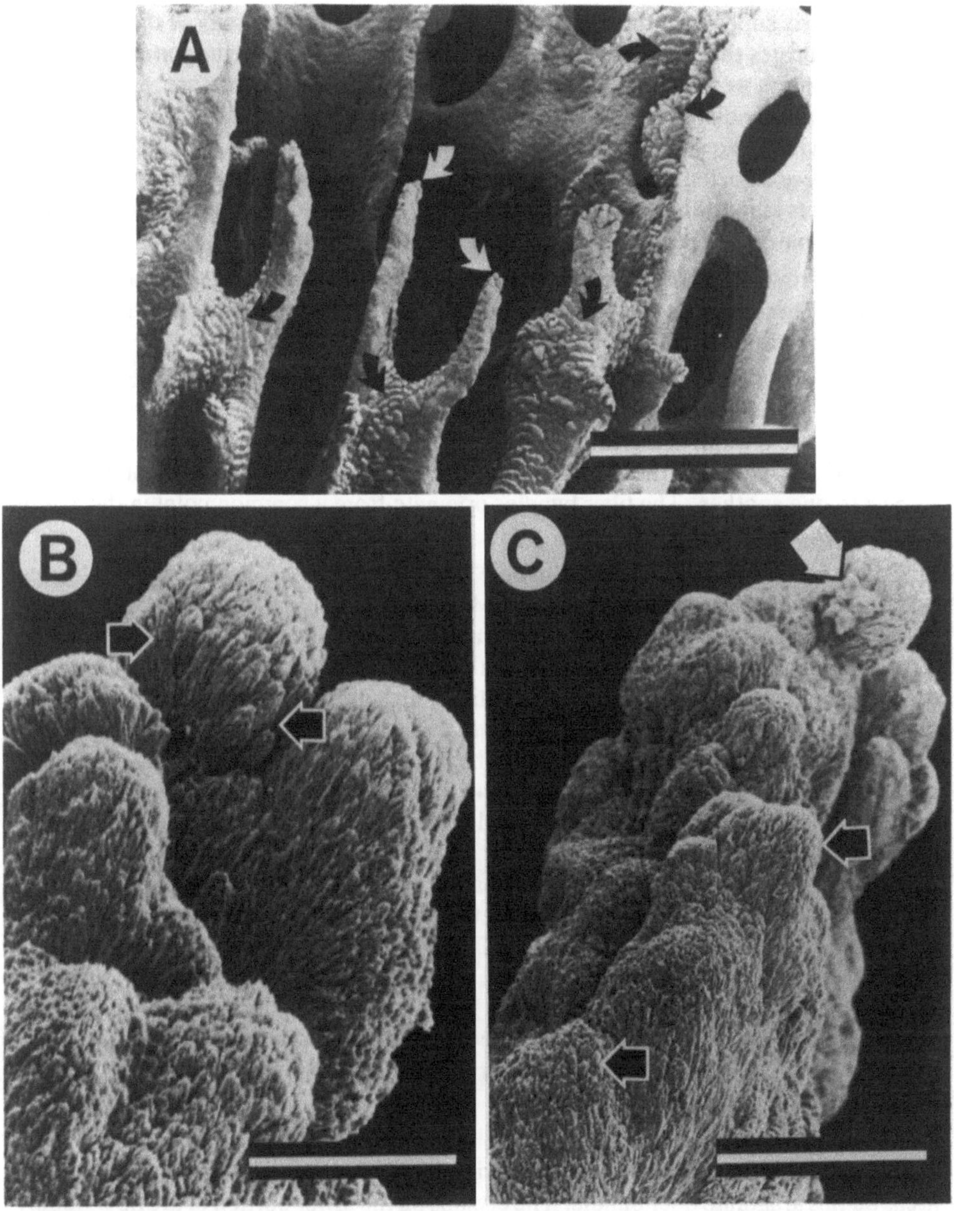

FIGURE 7. ▪A. Extending radial corallite. Black arrows denote growing fasciculi. White arrows point to septal tip seen in 6B and C. Bar = 200 μm. ▪B. Septal tip at right-hand white arrow. Black arrows denote region of fasciculate divergence. Bar = 10 μm. ▪C. Septal tip on left-hand. Black arrows denote fasciuli. White arrow denotes equant crystals. Bar = 25 μm.

et is indicated by black arrows. The individual grains making-up the granular subpackets are approximately 0.05 μm to 0.25 μm across. There appears to be an even finer grained matrix supporting these granules. The irregular morphology of the individual granules indicates that each granule may be composed of smaller components. Six calcification centers were observed using surface scanning electron microscopy and all six contained this granular material. Energy dispersive spectral analysis of these centers of calcification have shown only calcium peaks above background. Neither magnesium or strontium were detected at levels above background. Attempts to localize these extremely small points in ion-thinned specimens or on CO_3 replicas using transmission electron microscopy have not yet met with success in any *Acropora* spp.

White arrows designate larger crystals growing off the right-hand granular packet in Figure 6c. These crystals are approximately 0.1 μm in diameter and oriented perpendicular to the granular packet.

Figure 7a is a SEM of the extending septa in a radial corallite. This specimen was collected at the same time (0800 hours) and locality as the specimens in Figures 2b & c and 3b. The field width of this micrograph is 1.28 μm. Black arrows designate aragonite fiber bundles. Note that the aragonite fiber bundles have subparallel orientation with respect to one another. The fiber bundles originate from a common point in the axis of the synapticulae or septa.

The white arrows in Figure 7a designate extending septa that have been investigated at higher magnification. Figure 7b is a SEM of the septa designated by the white arrow to the right. The field width of this micrograph is 32 μm. The black arrows indicate the most distal aragonite fiber bundle. Notice that by tracing back the fibers on either side, they culminate at a common point of radiation. This indicates that the fiber bundles in the group nucleated at the same site. Note that this is not the base of a single aragonite fiber bundle, but a site of nucleation from which several fiber bundles radiate. The width of the base of the group of fibers is less than 5 μm. Scanning around this septal point, no non-fibrous, equant crystals were found. This characteristic was observed in some 45 radial corallites investigated. Note that each aragonite fiber between the black arrows diverges rapidly from the site of nucleation. Yet net direction of growth of the entire fiber bundle is not so radial. Instead the fibers oriented in the upward direction have extended the greatest amount. Note on the group of aragonite fiber bundles to the right, the same trend is observable. Here, however, the fiber bundle has grown a further distance from the common point of radiation. Note that the fiber bundle has preferentially elongated in the vertical direction and fiber bundles on the flanks of the group have noticably deteriorated fiber morphologies compared to the very fine acicular fibers at the top of the fiber bundle.

Figure 7c is a SEM at the white arrow on the left septa in Figure 6a. The field width of this micrograph is 60 μm. Note the variation in fiber morphology between the two areas on this aragonite fiber bundle. The crystals on the lower portion of the fiber bundle has crystals lackng acicular morphologies, whereas the fibers on the top of thefiber bundle have the characteristic acicular morphologies. This non-uniformity in aragonite fiber morphology, which is characteristic on the tips of the septa of radial corallites, is very uncommon for the aragonite fiber bundles of the skeleton.

In Figure 7c, growing off one side of the most distal fiber bundle, are a cluster of about 5 non-fibrous, equant crystals that form a distinct rosette (white arrow). Each non-fibrous crystal is approximately 5 μm across. Note that this crystal cluster is not positioned toward the direction of septal extension. Instead, these crystals have nucleated at a common point on the side of the most distal fiber bundle. These crystals were seen only at septal tips. They are particularly abundant at the septal tips of the axial corallites were they have been reported before (Gladfelter, 1982, 1983; Le Tissier & Brown, 1985). These crystals are positioned in a way to later form the dark streaks of fine crystals that are characteristic of the axial corallite septa (Figure 2d), as noted by Gladfelter (1983). They are relatively uncommon on the tips of the radial corallite septa and dark streaks are not prominent in these septa (Figure 3c). This case is one of only four instances where I observed the nonfibrous, equant crystals at the extending tips of septa on the radial corallites.

X-ray diffraction of ground samples of the upper 2–4 cm of axial corallites of all three species showed only aragonite peaks above background. If calcite or dolomite were present in the samples they represented less than 2–3 modal percent of the samples.

DISCUSSION

The findings reported here demonstrate that the aragonite fibers bundles in the skeletons of *Acropora palmata*. *A. prolifera* and, *A. cervicornis* nucleate on granular packets of submicron crystals. These nuclear packets (Constantz, 1986) form an irregular chain of centers of calcification (Figure 7a, b, c) along the axes of aragonite fiber nucleation which are termed the trabecular axes (Ogilvie, 1896). Constantz (1986:153, Fig. 1a) presented corroborating evidence of similar nuclear packages at trabecular axes in Porites porites and has observed nuclear packages composing the centers of calcification in a broader range of Scleractinian taxa.

SEMs from this study demonstrate the finely crystalline texture of the nuclear packets in the *Acropora* spp. studied (Figures 7a, b, c). The crystals in the nuclear packets are typically less than 0.1 μm across. The centers of calcification are composed of many individual nuclear packets (Figure 7b). Individual nuclear packets have a globular shape and a granular surface texture (Figure 7c). The irregular contacts between adjoining nuclear packets indicates that they were not rigid when incorporated into the skeleton.

Attempts to localize these centers in *Acropora palmata*, *A. prolifera* and, *A. cervicornis* for ion-thinning have been unsuccessful because of their extremely small size, so no selected area electron diffraction information about their mineralogy is available. The energy dispersive spectra demonstrate that granular packages in the centers of calcification contain calcium, but magnesium and strontium are both below the detection level of 1%, so mineralogies of dolomite and magnesian calcite are unlikely. Based on the data presented here it is plausible that the nuclear crystals have an aragonite or low magnesian-calcite mineralogy. Even though calcite was not detected in the skeleton with powder X-ray diffraction, since the centers of calcification appear to comprise less than 1% of the volume of the skeleton, it would not be possible to detect it even if the nuclear packets were calcite. It is also unknown whether each crystal nucleous, seen thus far with scanning electron microscopy only, is monocrystalline, or an aggregate of even smaller mineral grains, as the irregular crystal shapes and sizes would indicate. Constantz & Meike (1988) have reported that in *Mussa angulosa* the granular contents of the centers of calcification are submicron crystals of the $CaCO_3$ polymorph: calcite and are polysynthetically growth-twinned. The possible inclusion of organic material in the matrix of the packets is also a possibility that has not been evaluated.

The position of the centers of calcification suggests that the nuclear packets are the initial mineralogic phase in Scleractinian calcification (Ogilvie, 1896; Bryan & Hill, 1941; Wells, 1956; Constantz, 1986). The small grain size of each nucleous renews the possibility that the initial site of mineralization in scleractinians is intracellular. The 0.1 μm nuclei are small enough to have formed within the 2–3 μm thick squamous cells of the calicoblastic ectoderm (Hayasi, 1937; Isa & Yamazato, 1981; this volume). The geometry of the nuclear packets restricts the number of plausible explanations for their formation. Since they were incorporated into the skeleton in a non-rigid state, one plausible hypothesis for their formation could be that the individual 0.1 μm nuclei are added to the skeleton one by one at a specific site for some restricted interval, such as an afternoon, thus forming each distinct nuclear packet. Another plausible hypothesis could be that the entire nuclear packets are added to the skeleton at once. Because whole nuclei packets are 2–4 μm across, this later hypothesis would argue against an intracellular origin for the nuclei.

There are no examples of extracellular processes that could mediate either of these alternatives. By contrast, the exocytosis of granular material is already well documented in scleractinians.

Johnston (1980), working with *Pocillopora damicornis* described calicoblastic ectodermal cells which contained secretory vesicles, usually concentrated along the distal cell boundary. Johnston (1980) estimated that these vesicles are about 250 nm (0.25 μm) in diameter. Many of these vesicles appear empty, but others were filled with a discrete granular material. Johnston (1980:195, Fig. 11) shows a micrograph of the granular contents of one of the secretory vesicles being exocytosed from the calicoblastic ectoderm to the skeleton. Although Johnston (1980) interpreted these vesicles as " organic matrix precursor vesicles", he did not provided any evidence to determine whether the 10 nm (0.01 μm) to 20 nm (0.02 μm) diameter granular contents of these vesicles were organic or a very fine mineral material.

Kinchington (1980) working with the fully developed larvae of the non-zooxanthellate solitary coral Caryophyllia smithi, described several distinct vesicles in the aboral epidermal cells (Kinchington, 1980:146, Fig. 2). Each vesicle appears to be about 0.5 μm across. Kinchington's (1982) X-ray microanalysis of these vesicles demonstrated that the vesicles contained calcium. It will be impossible to evaluate the origin of the granular packets at the centers of calcification until we have selected area electron diffraction from both the centers of calcification and the granular contents of the calcium-bearing secretory vesicles. Even if this formidable research objective is accomplished, it must must be considered that the granular contents may not become truly crystalline until they are exocytosed and interact with the solution in the extracellular milieu. Further analysis of both the granular packets in the skeleton and the the secretory vesicles in the calicoblastic epithelium may prove a fruitful line for future studies.

Evidence from polished, etched specimens indicates that the irregular, dark line (*c.* 5 – 10 μm in width, Fig. 3e) at trabecular axes corresponds to centers of calcification seen with scanning electron microscopy (Fig. 7a, b, c). Centers of calcification have been described in the entire scleractinian order of stony corals by Ogilvie (1896) and by Wells (1956). Ogilvie (1896) and Wells (1956) found that the distribution of centers of calcification is very specific to every scleractinian taxon, and thus have used this distribution to construct the systematic classification of the *Scleractinia* (Wells, 1956, Fig. 237). If all scleractinians have similar mechanisms of skeletal growth, *i.e.* initiated by nuclear packets at the centers of calcification (Constantz, 1986), this indicates that the geometrical relationships whereby the nuclear packets are included into the skeleton form the basis for our present systematic methods in the *Scleractinia* (Ogilvie, 1896; Wells, 1956).

In *Acropora palmata*, *A. prolifera* and, *A. cervicornis* the trabecular axes are irregular chains of centers of calcification (*c.* 5 – 10 μm in width) running the lengths of the septa and synapticulae. In petrographic thin sections (two-dimensional) the trabecular axes appear as a single chain of calcification centers that are difficult to distinguish because of their small size and because they are largely obscured by dark streaks of 1 – 4 μm sized equant crystals, particularly in the axial corallites (Figs. 2d, 3a & c). In petrographic thin sections of the radial corallites of all three species, the centers of calcification are marginally visible at the trabecular axes where 1 – 4 μm equant crystals are sparse (Fig. 3f, black arrow). Most Scleractinians do not have dark streaks of equant crystals at their trabecular axes and the points of trabecular radiation are visible as distinct centers of calcification (Ogilvie, 1896; Wells, 1956). When the septa and synapticulae are viewed in cross section with scanning electron microscopy (Fig. 3b), it is clear that the trabecular axes are in fact a planar surface or trabecular plane (*c.* 300 – 400 μm wide and 5 μm thick) of centers of calcification running the length of the septa. This gives the septa an oval cross section (Fig. 3b) instead of a round cross section that a single trabecular axis would produce.

The aragonite fiber bundles (Figs. 4b, c, and 6a) that nucleate on the centers of calcification in *Acropora palmata*, *A. prolifera* and, *A. cervicornis* are termed the fasciculi (Ogilvie, 1896). The fasciculi are composed of acicular fibers that are about 0.1 μm in diameter (Fig. 4d) Using scanning electron microscopy, Constantz (1984) estimated the diameter of these fibers to be 0.5 μm; the estimate made here with transmission electron microscopy is the more accurate because of the higher resolving power of the transmission microscope. These aragonite fibers are very pointed in the long direction; this crystallographic configuration is termed acicular. Petrographic thin sections (Fig. 3f) and SEMs (Fig. 6a) across trabecular axes indicate that all fasciculi originate at centers of calcification. It is possible to trace every fasciculus back to the trabecular axis (Fig. 3b, 6a). New individual aragonite fibers nucleate in parallel alignment with the basal fibers within the fasciculi (Fig. 4b). Since fasciculi were not observed having any other origin it is plausible that the nuclear packets of the centers of calcification provide a nucleation surface that determines the unique, polycrystalline growth habit of the fasciculi. Constantz (1986, Fig. 2F) has presented scanning electron microscopic evidence from Diploria labyrinthiformis of abiotic marine cements nucleated on septal granulations in vacant portions of a corallum that have organized into fasciculus-like bundles; nuclear packets are shown on septal granulations in living parts of the corallum, indicating that some aspect of the nuclear packets are responsible for the fasciculus-like organization of the cements. The nuclear packets appear to have a high potential to nucleate very high densities of aragonite fibers both in scleractinian skeleto-genesis and abiotically.

The fasciculi nucleate on, and grow from, the nuclear packets. Yet based on the data presented here it is not possible to identify the mechanism whereby they subsequently grow. Fasciculi differ fundamentally from common abiotic marine cements (Bathurst, 1973) in that they are polycrystalline (Wainwright, 1963, Constantz, 1986). They are also very different from other biominerals in that they grow competitively, filling voids, until they are out of space. By contrast, molluscan crystals nucleate on an organic matrix (Crenshaw, 1984), grow to a very uniform size and shape (Addadi & Weiner, 1985), and then stop growing. Ogilvie (1896) noted that the undersurface of the calicoblastic epithelium is pocketed with impressions that match, exactly, the surfaces of growing fasciculi on the adjacent skeleton. She posed the question of whether the fasciculi were making impressions in the undersurface of the tissue as they grew or whether the impression-like structures formed the shape of the crystals by guiding or directing crystal growth. Even with the aid of SEM, we have still not devised a way of answering Ogilvie's question!

If the fasciculi were forcing their way up into the calicoblastic ectoderm, the distension of the calicoblastic ectoderm may be expected to translate through to the mesoglea (Fig. 5b), distending it too. Alternatively, if the fascicular impressions were, in fact, prefabricated crystalline envelopes that actively direct or passively guide fasciculate growth, one would not expect the overlying mesoglea to be distended by the growth of the fasciculi because the pocketed calicoblastic tissue would be in its proper configuration. Cryofracture data presented here indicates that the mesoglea is distended at a scale and regularity that reflects the surface of the underlying calicoblastic epithelium (Fig. 5c).

Despite this observation direct evidence to answer Ogilvie's (1896) question is still lacking. The documentation of abiotic fasciculate formation on nuclear packages in *Diploria labyrinthiformis* supports the interpretation that the fasciculi grow in a physiochemical manner akin to marine cements (Constantz, 1986). However, the uniformity of the individual aragonite fibers within the fasciculi of *Acropora palmata* and *A. cervicornis* can not be explained by simple physiochemical crystal growth kinetics alone. Constantz (1986) has shown evidence that a broad range of fiber morphologies exist in the fasciculi of scleractinians. Data on the species studied here indicates that the morphologies of these fibers are highly uniform over most of the corallum, baring the non-zooxanthellate septa tips, of these related species. Given & Wilkinson (1985) have shown that the morphology of aragonite fibers growing abiotically may be related to the saturation state of the parent solution for $CaCO_3$. Such an explanation might be plausible if the fascicular fibers were growing from a solution that was kept at a constant physiological pH. Recent studies by Addadi & Weiner (1985) indicate that soluble organic constituents of a parent solution, particularly amino acids, may control the uniform shapes of calcite crystals growing from solution. Information on both the exact saturation state and the solution composition of the fluid below the calicoblastic ectoderm in any scleractinian species is presently lacking, but future studies may provide valuable insights into this question.

On growing skeletal surfaces, fasciculi that are parallel to the surface of the skeleton (Figs. 4b,c, 6a) are at a low angle to the surface and have the apparence of shingles on a house (Figs. 4b,c, 6a, 7a). The surfaces of fasciculi there are noticably furrowed perpendicular to the direction of fiber growth (Fig. 4b). This furrowing of the fasciculi could be a result of diurnal growth cycles in the fasciculi, having a periodicity of several micrometers. Each day would be represented by a new fasciculate tuft of acicular fibers. In some areas were the surface of these fibers are noticeably smooth, they appear to have been demineralized. Constantz (1986, Figs. 1E,F) has reported similar demineralization surface in *Colpophyllia natans* and has speculated that this process is probably chemical, involving a weak acid, rather than mechanical. These smoothened surfaces apparently stop fasciculate growth thereby forming the finished corallum. Subsequent thickening of the corallum, termed "secondary thickening" by Gladfelter (1982), is, in fact, common marine cement (Lightly, 1985). Marine cement should not be considered a unique skeletal element.

SEM (Figs. 3e, 7a,b,c) indicates that the nuclear packages at the centers of calcification make-up the trabeculae axes described by Ogilvie (1896), Wells (1956), Wainwright (1963, 1964), and James (1974). Fasciculi nucleate here and radiate out in a spherulitic manner forming the trabeculae. Petrographic thin sections indicate that 1–4 μm equant crystals (Wainwright, 1963; James, 1974; Gladfelter, 1982, 1983; Constantz, 1986) form a broad zone (*c.* 50+ μm wide) in the axial corallites that appear as dark streaks (Fig. 2d) that obscure polarized light viewing of the trabecular axes. These dark streaks are analogous to the "brown zones" of Wainwright (1963) and James (1974). This broader zone is very irregular and often discontinuous (Fig. 3a,b). At intervals spaced roughly 500–1000 μm apart this broad zone of equant crystals becomes very diffuse (Fig. 3c). By contrast, on the radial corallites this broad zone of equant crystals is almost non-existing (Fig. 3d). Some equant crystals can be seen associated with the trabecular axis, but these occurrences are very sparse. The pattern of diffuse spots on the dark streak in axial coralites is particularly interesting because the spacing overlaps with the average daily linear growth rates of axial corallites. For instance, published linear growth rates for *Acropora palmata* are 25–40 cm/yr (Vaughan, 1915), 88 cm/yr (Bak, 1976), and 47–83 cm/yr (Gladfelter *et al.*, 1978); these annual rates translate to daily rates of 680–110 μm/day, 2411 μm/day, and 1297–2274 μm/day respectively. This indicates that the nightly accumulation of equant crystals at the tips of the axial corallite demonstrated by Gladfelter (1982, 1983) may be recorded in the skeleton as distinct bands of dispersion in the dark streak. This pattern, that the zone of equant crystals is very concentrated and periodically diffuse in the axial corallites but almost nonexistent in the radial corallites, corroborates with SEM information on the surfaces at the tips of growing septa.

Skeletal elements observed at the tips of septa were 1–4 μm equant crystals (Fig. 7c, white arrow; Wainwright, 1963; James, 1974; Gladfelter, 1982, 1983; Le Tissier & Brown, 1985) and fasciculi with morphologies non-characteristic of *Acropora palmata*, *A. prolifera* and *A. cervicornis*. (Plate 7b, Ogilvie 1896; Wells 1956). Gladfelter (1983), working with the axial corallite of *A. cervicornis*, has demonstrated that these equant crystals nucleate and grow during the night and are most abundant in the early morning at the axial corallite septal tips. Le Tissier & Brown (1985) have noted a similar pattern in *Porites damicornis*, were Wainwright (1963) first described the "brown zone".

Constantz (1986) noted the occurrence of equant crystals in Porites porites. These equant crystals have the morphology and size of inorganically precipitated marine cements (Towe & Malone, 1970). Because these equant crystals clearly make-up greater than 2–3% of the skeletons volume and X-ray diffraction detected no calcite above background in the any of the samples, it is most likely that the equant cements are aragonite. This finding is corroborated with several selected area electron diffraction patterns of these crystals indicating they are aragonite. Gladfelter (1982, 1983) found that during the daylight, fully developed fasciculi nucleate on these equant aragonite cements. Based on the microscopic evidence presented by Gladfelter (1982, 1983), it is not clear what the fasciculi nucleate on. Moreover, the "fully developed fasciculi" at the tips of the axial corallite presented by Gladfelter (1982, 1983) do not resemble the fasciculi of the proximal skeleton, both reported here and in Gladfelter (1983, Fig. 8B). Gladfelter's (1982, 1983) study concentrated on the non-zooxanthellate septal tips of the axial corallites, where the equant cements are very abundant. Gladfelter (1982, 1983) did not recognize nuclear packages at the axial corallites [see Gladfelter (1983) Figure 3B several globular submicron bodies, are visible studding the surfaces of the equant crystals at the tips of an axial corallite collected at 1800 hours; at this low magnification these globular bodies appear very similar to nuclear packets]. Because no equant cements were found on living portions of the corallum proximal to the non-zooxanthellate tips of axial corallites or at zooxanthellae-depleted septal tips of radial corallites, its appear that the precipitation of these equant cements is related to a lack of zooxanthellae in the overlying tissue. Moreover, since no equant crystals were found on zooxanthellate parts of the corallum at any time of collection, night or day, this suggests that these equant crystals are not related simply to a lack of zooxanthellate photosynthesis, but to total zooxanthellae absence.

The morphology of the aragonite fibers in the fasciculi at the extreme tips of septa in *Acropora palmata*, *A. prolifera* and, *A. cervicornis* are non-characteristic of the fasciculate morphologies of proximal portions of the skeleton. The developing aragonite fiber bundles on costa of the extending spine of the axial corallite reported by Gladfelter (1982, Fig. 3E) do not have the characteristic acicular fiber morphologies found on the remainder of the corallum of *A. cervicornis* (Gladfelter, 1983, Fig. 8B). There are, in fact, no examples from this study, or in Gladfelter (1982, 1983) of fully developed fasciculi at the tips of the axial corallites. Non-characteristic morphologies in the fasciculi at the tips of radial corallites were always observed in this study (Figs. 7b, c). Comparing Figures 7b, c from this study and Figure 3E from Gladfelter (1982) indicates that the fasciculi at the tips of the non-zooxanthellate axial corallites have non-characteristic morphologies. The aragonite fiber bundles at the tips of the radial corallite lack the characteristic acicular fiber morphologies found on the remainder of the corallum (Figs. 4b,c, 6a). Because the characteristic acicular fasciculi are only found on zooxanthellate portions of the corallum, but have non-characteristic low-angle points on non-zooxanthellate axial corallite tips and septal tips on zooxanthellae-depleted radial corallites, both day and night, the uniform acicular morphologies of fasciculi characteristic to *A. palmata*, *A. prolifera* and, *A. cervicornis* appear to be related to the presence of zooxanthellae in the overlying tissue. Again, there is no evidence for diurnal changes in the morphology of fasciculate fibers; it appears to be the simple presence or absence of zooxanthellae alone, not the action of zooxanthellate photosynthesis, that is responsible for this pattern.

CONCLUSIONS

Because all fasciculi nucleate at centers of calcification and their polycrystalline, fibrous habit appears to be related to some characteristic of the nuclear packets, whereas the 1–4 μm equant aragonite cements nucleate at random (Gladfelter, 1982, 1983) and only during the night on non-zooxanthellate parts of the skeleton, the equant, monocrystalline, non-fibrous, aragonite cements may have such an organization because they nucleate in the absence of the nuclear packets. This interpretation is corroborated by the work of Towe & Malone (1971), who precipitated almost identical equant aragonite cements in glass beakers with no nuclei. Assuming this interpretation is valid, nuclear packets must be deposited only on non-zooxanthellate parts of the skeleton during daylight. At night, nuclear packets become covered with fasciculate overgrowth and new crystals nucleate at random as 1–4 μm equant aragonite cements. On zooxanthellate parts of the corallum, equant aragonite cements never or seldom grow, indicating that nuclear packets are always available. This might indicate that the production of nuclear packets is related the presence of zooxanthellae in the adjacent tissue.

It follows from this that the increased rate of daytime net calcification (Barnes & Crossland, 1980) may be due to increased nuclear packet production during hours of zooxanthellate photosynthesis. During the day, photosynthesis on proximal portions of the corallum may translocate (Pearse & Muscatine, 1971), or even leak, some metabolic factor to the distal portion of branches to have a potentiating effect on nuclear package production. The net effect of this process would be that nuclear package production and fasciculate nucleation would occur on non-zooxanthellate portions of the corallum principally during the day. Translocation of potentiating factors of nuclear production to distal, non-zooxanthellate tips of the corallum would only be important during the day. The potentiating factors may be the local removal of intracellular CO2 by the photosynthesizing zooxanthellae, which would increase intracellular pH, perhaps faster than the cell could regulate pH. This would form CaCO3 by combining with free Ca2+ in the cytoplasm (Goreau & Bowen, 1956). The intracellular CaCO3 would be localized in intracellular vesicles and exocytosed. By sequestering the free Ca2+ in a mineralogic form, it could not re-diffuse back into the cells and need to be re-regulated.

At the distal tips of axial corallites nuclear package production would take place only during times of zooxanthellate photosynthesis on proximal portions of the corallum, depositing nuclear packets during the day. At the non-zooxanthellate distal tips of the corallum, only randomly oriented equant cements would nucleate during the evening because of the absence of nuclear packets. This would explain the pattern of growth at the axial corallite first mentioned by Goreau (1959), who noted that zooxanthellae have a potentiating effect on calcification in scleractinians, even in the dark.

ACKNOWLEDGEMENTS

Funded by National Science Foundation Grant EAR-8313414, a doctoral dissertation grant to the author sponsored by Professor L.F. Laporte. This paper was a chapter in the authors dissertation. Critical comments on the manuscript are greatly appreciated from N.P. James, L.F. Laporte, M. Moore, D.C. Potts, and R.K. Trench. Participation in the Symposium was made possible by grants from The Organized Research Fund of The University of Texas at Arlington and The Sea Grant Program administered by Texas A & M University.

REFERENCES

ADDADI, L. & WEINER, S., 1985. Interactions between acidic proteins and crystals: Steriochemical requirements in biomineralization. Proc. Nat. Acad. Sci., USA, 82:4110-4114

BARNES, D.J., 1970. Coral skeletons: an explanation of their growth and structure. Science, 170:1305-1308.

BARNES, D.J., 1972. The structure and function of growth ridges in scleractinian corals. Proc. Roy. Soc. London B., 182:331-350.

BARNES, D.J., & CROSSLAND, C.J., 1980. Diurnal and seasonal variations in the growth of a staghorn coral measured by time-lapse photography. Limn. Ocean., 25:1113-1117.

BRYAN, W.H., & HILL, D., 1941. Spherulitic crystallization as a mechanism of skeletal growth in the hexacorals. Proc. Royal Soc., Queensl., 52:78-91.

CHALKER, B.E., 1976. Calcium transport during skeletogenesis in hermatypic corals. Comp. Biochem. and Physiol. A., 54:455-459.

CRENSHAW, M.A., 1984. Mechanisms of normal biological mineralization of calcium carbonates, in Nancollas, G.H., ed. Biological Mineralization and Demineralization, Dahlem Konferenzen 1982. New York, Springer. 243-257.

CONSTANTZ, B.R., 1984. Comparative function of microarchitecture in *Acropora cerviornis* and *Acropora palmata*. Palaeontographica Americana, 54:48-52.

CONSTANTZ, B.R., 1986. Coral skeleton construction: a physiochemically dominated process. Paliaos, 1:152-157.

GLADFELTER, E.H., 1982. Skeletal development in *Acropora cervicornis*: I. Patterns of calcium carbonate accretion in the axial corallite. Coral Reefs, 1:45-51.

GLADFELTER, E.H., 1983. Skeletal development in *Acropora cervicornis*: II. Diel patterns of calcium carbonate acretion. Coral Reefs, 2:91-100.

GOREAU, T.F., 1959. The physiology of skeletal formation in corals. I. A method for measuring the rate of calcium deposition by corals under different conditions. Biol. Bull., 116:59-75.

GOREAU, T.F., 1961. Problems of growth and calcium deposition in reef corals. Endeavor, 20:32-39.

GOREAU, T.F., 1963. Calcium carbonate deposition by coralline algae and by corals in relation to their roles as reef builders. Ann. New York Acad. Sci., 190:127-167.

GOREAU, T.F., & BOWEN, B.T., 1955. Calcium uptake by a coral. Science, 122:1188-1189.

HAYASI, K., 1937. On the Detection of calcium in the calicoblasts of some reef corals. Palaos Trop. Biol. Sta. Stud., no. 2(8):169-179.

ISA, Y., & YAMAZATO, K., 1981. The ultrastructure of the calicoblast and related tissues in *Actopora hebes* (Dana). Proc. Fourth Inter. Coral Reef Sym., 2:99-105.

JAMES, N.P., 1974. Diagenesis of scleractinial corals in the subaerial vadose environment. J. of Sed. Pet., 48:785-799.

JACKSON, J., & GLADFELTER, E.H., 1985. Ultrastructure of the tissue/skeleton interface of *Actopora cervicornis* over a diel cycle. Proc. Fifth Inter. Coral Reef Cong., Tahiti, UNESCO, in press.

JOHNSTON, I.S., 1980. The ulrastructure of skeletogenesis in hermatypic corals. Inter. Rev. of Cytology, 67:171-214.

KINCHINGTON, D., 1980. Localisation of intracellular calcium within the epidermis of a cool temperate coral. *In* Tardent, P. & Tardent, R. (eds.), Developmental and cellular biology of coelenterates. Elsevier., 143-147.

LIGHTLY, R.G., 1985. Preservation of internal reef porosity and diagenetic sealing of submerged early Holocene Barrier Reef, Southeast Florida Shelf. *In* Schneiderman, N. & Harris, P.M. (eds.) SEPM Spec. Pub. 36, Carbonate Cements., 123-151.

LE TISSIER, M., & BROWN, B.E., 1985. Diurnal patterns of skeletal formation in *Pocillopora damicornis* (*Linnaeus*): Proc. Fifth Inter. Coral Reef Cong., Tahiti, UNESCO, in press.

MUSCATINE, L., 1967. Gycerol extretion by symbiotic algae from carals and *Tridacna* and its control by the host. Science, 156:516-519.

PEARSE, V.B., & MUSCATINE, L., 1971. Role of symbiotic algae (zooxanthellae) in coral calcification. Biol. Bull., Woods Hole., 141:350-363.

OGILVIE, M., 1896. Microscopic and systematic study Madreporarian types of corals. Phil. Trans. Roy. Soc. London B., 187:83-345.

OHLHORST, S.L., 1984. The use of poolyacrylamide gel electrophoresis in coral taxonomua. Paleontographica Americana, 54:45-48.

RICHART Y MENENDEZ, F., & FRIEDMAN, G.M., 1977. Submarine diagenesis of the axial corralite of *Actopora cervicornis.* Proc. Third Inter. Coral Reef Sym., Miami, 163-166.

TOWE, K.M.,& MALONE, P.G., 1970. Precipitation of metastable carbonate phases from sea water Nature, 226:348-349.

VANDERMEULEN, H.J., DAVIS, N.D. & MUSCATINE, L., 1972. The effects of inhibitors of photoynthesis on zooxanghellae in corals and other invertebrates. Mar. Biol., 16:185-191

WAINWRIGHT, S.E., 1963. Skeletal organizationin the coral, *Pocillopora damicornis*. Quar. J. Microsc. Sci., 104:169-193.

Wainwright, S.A., 1964. Studies of the mineral phase of coral skeleton. Exp. Cell Res., 32:213-230.

WALLACE, C.C., 1978. The coral genus *Actopora* (Scleractinia: Astrocoeniina: Acroporidae) in the central and southern Great Barrier Reef Province. Mem. Queensl. Mus., 18:273-319.

WELLS, J.A., 1956. Scleratinia. *In* Moore, R.C., (ed) Treatise on Invertebrate Paleontology: Geol. Soc. Amer. & Univ. Kans., Lawrence. F328-444.

CALCITE CENTERS OF CALCIFICATION IN *MUSSA ANGULOSA* (SCLERACTINIA)

Brent Constantz[1,3] and Annemarie Meike[2]

[1]*Earth Sciences Board, University of California, Santa Cruz, CA 95064;* [2]*Department of Geology and Geophysics, University of California, Berkeley, CA 94720;* [3]*Present address: NORIAN, 1030 Hamilton Court, Menlo Park, CA 94025*

ABSTRACT. Transmission electron microscope (TEM) analysis demonstrates that centers of calcification in the exoskeleton of the scleractinian coral *Mussa angulosa* are composed of submicron sized crystals of calcite, the rhombohedral polymorth of calcium carbonate.

INTRODUCTION

More than a century ago Sorby (1879) found that the specific gravity of both perforate and non-perforate coral skeletons was about 2.75. He concluded that scleractinian coral exoskeletons must be wholly aragonite, although he was not certain that calcite was entirely absent (Sorby, 1879). Low-Mg calcite has more recently been reported as an important, primary constituent of coral skeleton based on X-ray diffraction and atomic adsorptive spectroscopy (Houck *et al.*, 1975). This finding has been criticized because petrographic and morphological evidence, combined with wave-length dispersive spectral data, indicate extensive alteration of skeletal aragonite by microborings filled with secondary calcite (Macintyre & Towe, 1976). We found substantial evidence of microborings in over one-half the specimens used in our study. We eliminated all but about 10% of the initial specimens, and used only those that contained no calcite filled endolithic borings or any other skeletal alteration.

All corals were collected live from the west coast of Barbados and transferred to a running sea water system. Specimens were never exposed to fresh water. Extreme caution was exercised at every step of sample preparation to assure that no secondarily precipitated calcite was included. Part of each sample was immediately fixed over 24 hours in a solution of sodium cacodylate, sucrose, glutaraldehyde, and tannic acid at pH 8.2. The remainder of the sample was air dried. Thin sections (approximately 10 μm thick) of the youngest centimeter of skeletal growth from both the fixed and the air-dried samples were cut with a diamond saw and ground using kerosene. Uncovered thin sections were mounted with crystal bond for later removal. Copper hex-grids were mounted on centers of calcification (Ogilvie, 1896; Wells, 1956; Constantz, 1986) with epoxy resin. The coral wafers were then removed from the glass slide and the hex-grid

Origin, Evolution, and Modern Aspects of Biomineralization in Plants and Animals
Edited by Rex E. Crick
Plenum Press, New York

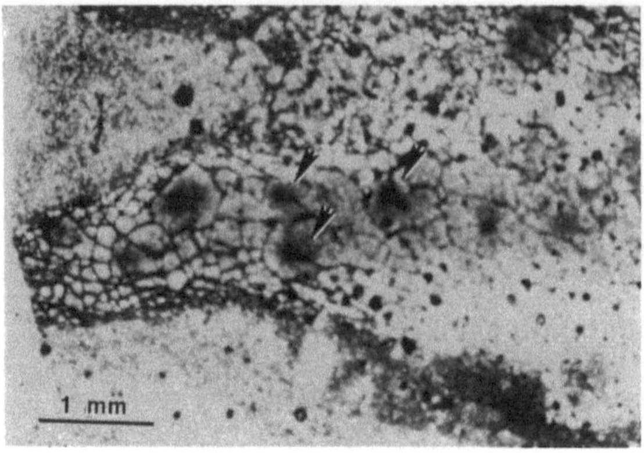

FIGURE 1. Thin section photomicrograph of *Mussa angulosa* septal spine. Arrows indicate cloudy centers of calcification. Fractures originated from sample preparation.

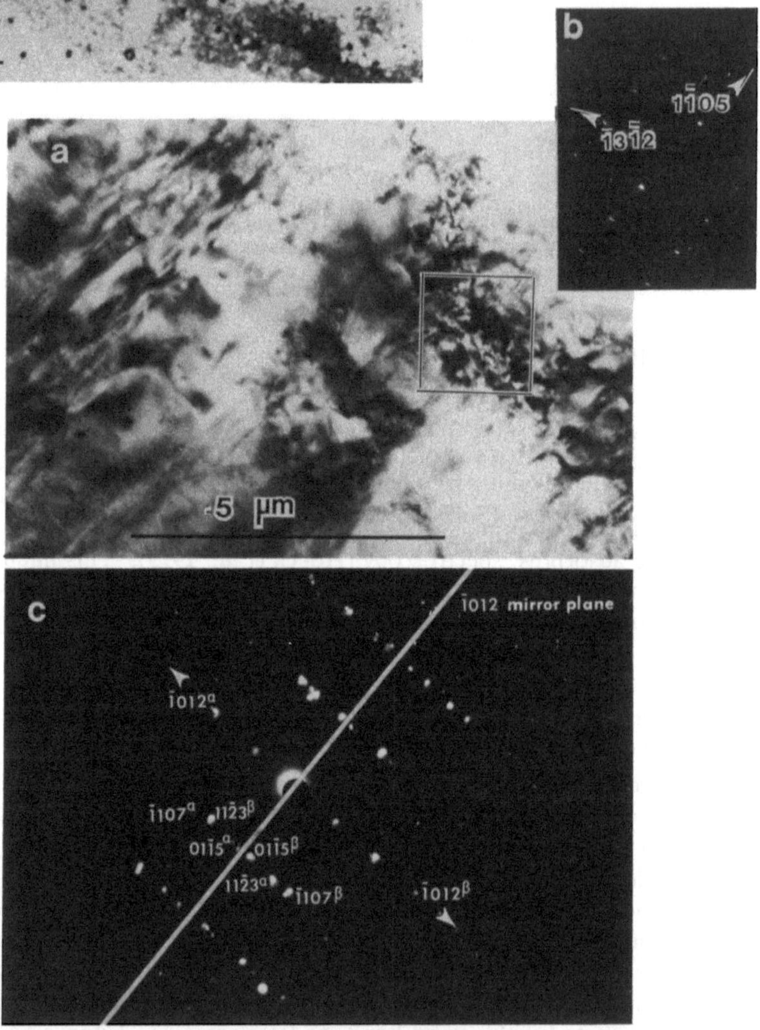

FIGURE 2. – a. Transmission electron photomicrograph (1.5 MeV) showing the tiny crystallite microstructure of the center of calcification. – b. Selected area diffraction of – a shows calcite pattern. – c. Selected area diffraction pattern of – 2a in a different orientation shows the presence of calcite twins on the {$\bar{1}$012} plane. α and β are arbitrary designations for the two twin orientations.

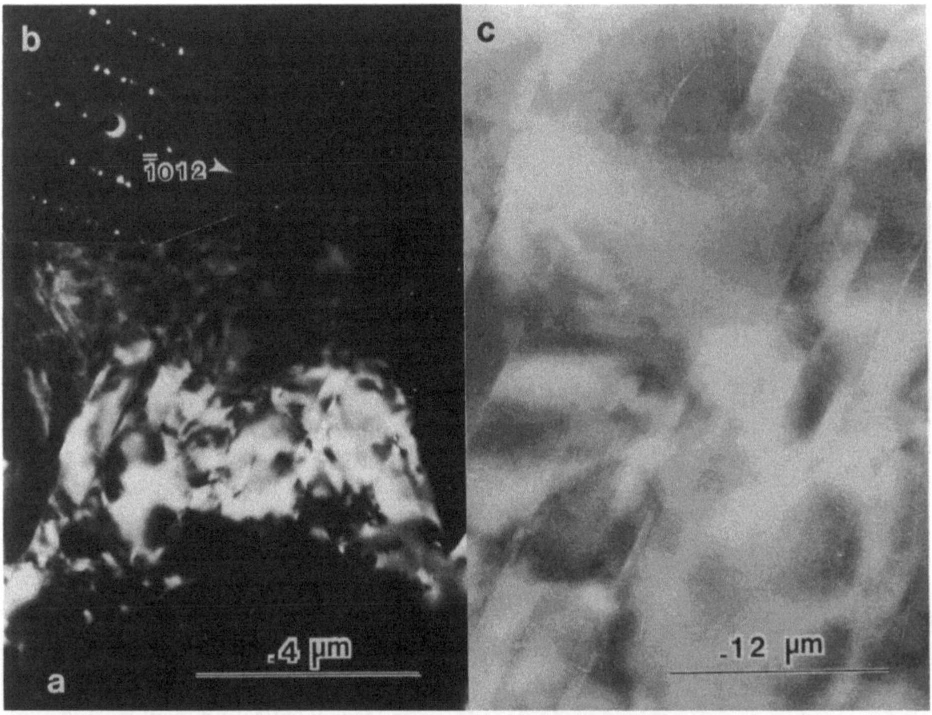

FIGURE 3. ▪a. Dark field image of the boxed region in Fig. 2a using the $(10\bar{1}5)\alpha$ diffraction point. The boundaries of twinned crystals can be seen. ▪b. Selected area diffraction pattern showing the mirror plane symmetry between the α and β twin crystals' diffraction patterns. ▪c. Bright field image of ▪a at a greater magnification.

portion excised. We prepared ultra-thin foils of these specimens using argon ion-thinning procedures (Thomas & Goringe, 1979). Analysis included both standard transmission electron microscopy (100 Kv) combined with an energy dispersive X-ray spectrometer, and high voltage electron microscopy (1.5 MeV). With TEM we have not only resolved ultrafine microstructure, but also easily substantiated, by electron diffraction, the crystal symmetry of the calcium carbonate polymorph within very small regions of the sample.

The polymorphs of calcium carbonate, such as aragonite, calcite and vaterite, cannot be distinguished by chemical analysis. However, they can be identified with symmetry information from a diffraction pattern. Each spot in a diffraction pattern represents a set of lattice planes in the crystal. The magnitude of the vector from the central transmitted beam to a diffraction spot is inversely proportional to the lattice spacing of the plane it represents. The angles between two vectors is determined by the angles between the respective crystal lattice planes. We used hexagonal indices and the X-ray unit cell to represent the reciprocal lattice of calcite. The X-ray unit cell indices should not be confused with those of the morphological unit cell.

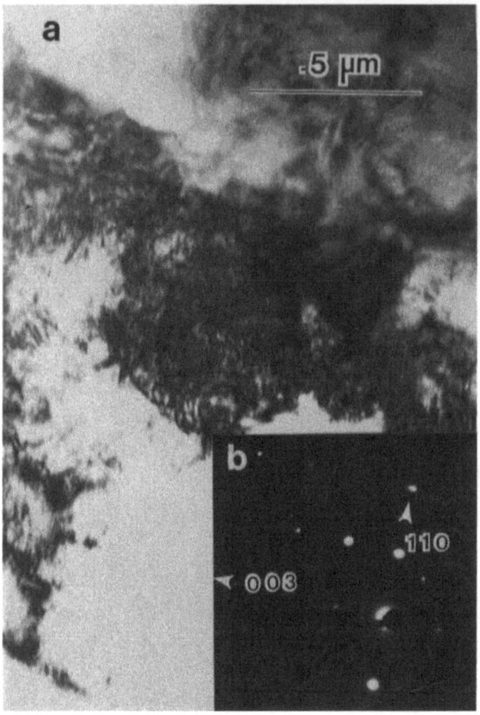

FIGURE 4. ▪a. Bright field image of calcification center. ▪b. Selected area diffraction pattern of ▪a showing a low order calcite pattern.

The scleractinian species that we have investigated to date are *Porites porites*, *Siderastrea radians*, *Monastrea annularis*, *Diploria labyrinthiformis*, and *Mussa angulosa*. The size and distribution of the centers of calcification varies considerably among these species. In most scleractinian species, each calcification center is a crystallite cluster only several microns to a few tens of microns across. In most species, the grid placement and ion-thinning necessary to observe these minute areas was extremely difficult. However, one species studied, *M. angulosa*, has large centers of calcification, several hundred microns across (Fig. 1). Consistent success in grid placement and ion-thinning of *M. angulosa* samples allowed repeatable observations in different specimens.

The TEM of a portion of a *M.angulosa* calcification center (Fig. 2a) shows that it is composed of tiny crystallites. The dark field image (Fig. 3a), explained below, more explicitly defines the boundaries of each crystallite in the boxed region of Figure 2a. Figures 2b & c, and several other diffraction patterns demonstrate unequivocally that the centers of calcification studied here are calcite. Diffraction patterns of single grains in several orientations were indexed. Twinning and sample tilting relationships were verified with a stereonet to insure the uniqueness of the indentification. The mineral does not have the orthorhombic symmetry of aragonite. In addition, calcite (whose symmetry group is $R\bar{3}c$) was distinguished from vaterite (whose symmetry group is $P6_3/mmc$) through space group symmetry extinctions in the diffraction patterns. A low index orientation (Fig. 4a & b) as well as another image in which two orientations of the same grain are shown (Fig. 5a, b & c) are also consistent with calcite.

Figures 2b & c are selected area electron diffraction patterns of the same area in two orientations. The twin patterns $\{\bar{1}012\}$ as in Figure 2c are frequently observed. This twin has equivalent morphological unit cell indices $\{\bar{2}021\}$ and is a common growth twin in calcite (Parker & Diehl, 1945). The extremely well-defined diffraction patterns for such tiny crystallites indicate that many submicron sized domains are in near-coincident orientation and the twinning may be polysynthetic. Randomly oriented crystallites would generate a diffraction pattern of concentric rings. Radially oriented crystallites

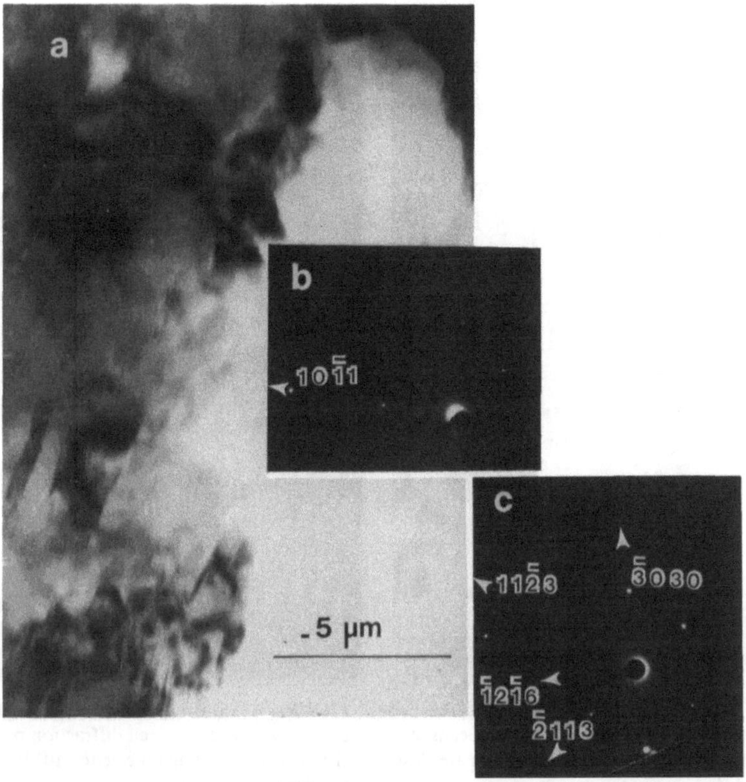

FIGURE 5. ▪a. Bright field image of calcification center. ▪b. Selected area diffraction pattern in the same orientation as Fig. 2a. ▪c. Selected area diffraction pattern of the region shown in Fig. 2a at a different orientation.

would also generate a concentric ring pattern in all orientations except those that include the axis of symmetry.

The "dark field" TEM technique isolates a single diffraction spot to illuminate the areas of the image for which it is responsible. In this case, we isolated a diffraction spot unique to one twin in order to contrast one set of crystals from their mirror twins that remained dark. Figure 3a is a dark field image illuminating one set of twins. In Figure 3b, the diffraction pattern and the diffracting beam in the image are shown. In two other orientations (Figs. 6a & b and 7a & b) the diffraction pattern is consistent with twinning on the $\{\bar{1}012\}$ of calcite.

Our instrument detected no Mg in the energy dispersive analyses of calcification centers, and have no evidence to suggest that this is a magnesian calcite. Centers of calcification constitute only a small percentage of the total exoskeleton volume, less that 2% in most species. Although calcite much more readily accepts Mg into its lattice structure than aragonite, the calcitic calcification centers could not account for the values of up to 46% low Mg calcite reported earlier (Houck *et al.*, 1975). Their analyses must have included algal microborings filled with secondary calcite (Macintyre & Towe, 1976).

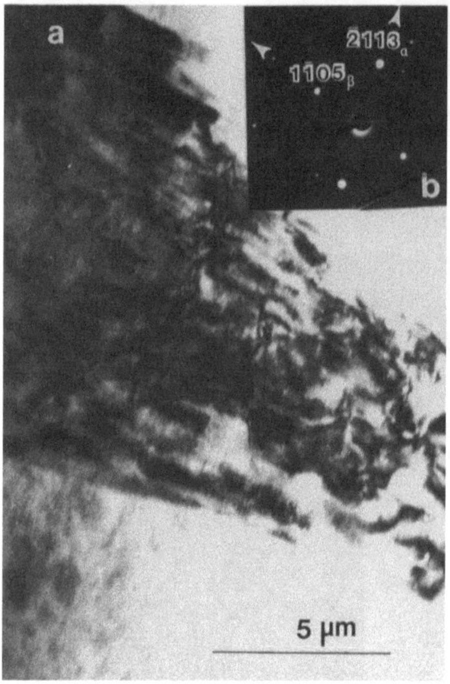

FIGURE 6. ▪a. Bright field image of calcification center. ▪b. Selected area diffraction pattern consistent with twinning on calcite $\{\bar{1}012\}$. The twins are designated α and β.

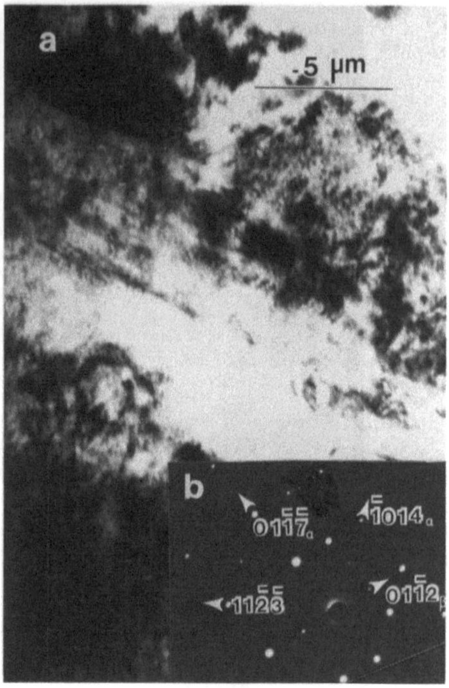

FIGURE 7. ▪a. Bright field image of calcification center. ▪b. Selected area diffraction pattern consistent with twinning on calcite $\{\bar{1}012\}$. The twins are designated α and β.

Our selected area electron diffraction data from carefully screened and prepared scleractinian exoskeletons demonstrates that centers of calcification in *Mussa angulosa* are composed of submicron-sized calcite crystals. Because these calcite crystals nucleate spherulitic fans of aragonite fibers they have an important role in the scleractinian mineralization. The origin of these crystal nuclei and their role in polycrystalline aragonite fiber bundle nucleation are important questions for future studies of skeletogenesis.

ACKNOWLEDGEMENTS

We thank Hans-Rudolf Wenk for initiating this collaborative project and for the use of the 100 kV transmission electron microscope at the Department of Geology and Geophysics, University of California, Berkeley. We also thank the Lawrence Berkeley Laboratory International Center for Electron Microscopy for permission to use the 1.5 MeV electron microscope. We appreciate comments on the manuscript, especially from I.G. Macintyre, K.M. Towe, and P.A. Sandberg. Funded by NSF EAR-8313414 and EAR-8305865. Participation in the Symposium was made possible by grants from The

Organized Research Fund of The University of Texas at Arlington and The Sea Grant Program administered by Texas A & M University.

REFERENCES

CONSTANTZ, B.R., 1986. The Skeletal Ultrastructure of Scleractinian Corals. Ph.D. Thesis. 172 pp., University of California, Santa Cruz.

HOUCK, J.E., BUDDIMEIER, R.W., & CHAVE, K.E., 1975. Skeletal low-magnesium calcite in living scleractinian corals. Science, 189:997-999.

MACINTYRE, I.G. & TOWE, K.M., 1976. Skeletal calcite in living Scleractinian corals: microboring fillings, not primary skeletal deposits. Science, 193:701-702.

OGILVIE, M.M., 1896. Systematic study of Madreporan corals. Phil. Trans. Roy. Soc. Lond., 187:83-345.

PARKER, R.L. & DIEHL, E., 1945. Uber Calcitvierlinge von Robat-i-Karim (Iran). Schw. Min. Petr. Mittr., 25:341-348.

SORBY, H.C. Anniversary address of the president: structure and origin of limestones. Quarter. J. Geol. Soc. Lond., 35:56-95.

THOMAS, G. & GORINGE, M.J., 1979. Transmission Electron Microscopy of Materials, 388 pp., New York:John Wiley.

WELLS, J.W., 1956. Scleractinia. *In* Treatise of Invertebrate Paleontology, Part F, (ed. R.C. Moore). pp. 328-444. Geol. Soc. Amer., Lawrence, Kansas.

EXTRA-, INTER-, AND INTRACELLULAR MINERALIZATION IN INVERTEBRATES AND ALGAE

Norimitsu Watabe[1] and Roni J. Kingsley[2]

[1]*Electron Microscopy Center and Department of Biology, University of South Carolina, Columbia, SC 29208;* [2]*Dental Research Center, University of North Carolina, Chapel Hill, NC 27514*

ABSTRACT. Biomineralization occurs extra-, inter- or intracellularly. Almost all structures of extracellular formation are secretory products of multicellular epithelial tissues. Intracellular mineralization is most commonly seen within vacuoles and vesicles. Many of the structures formed intracellularly eventually become extracellular entities, and two methods are known for the intra-extracellular transition. Intercellular mineralization is not common.

There does not seem to be standard methods of calcium transport characterizing the extra- and intracellular mineralization. However, much work is needed before conclusions may be drawn concerning the ion transport for mineralization.

The majority of structures originated in vacuoles and vesicles exhibit species-specific, almost symmetrical, and highly intricate morphology. Discussion is given on several factors which have been considered to affect the morphogenesis of these structures. The presence of regularly spaced inhibitors of crystal growth in the vacuoles/vesicles is suggested as one of the possible factors controlling the morphogenesis. The intracellularly formed structures attain a uniform size characteristic for each structure at its maturity, but not much is known concerning the mechanisms to regulate the final size. Some taxa-specific neurosecretory products or analogies and/or precursors of hormonal substances could regulate the mineralization.

INTRODUCTION

Biomineralization occurs at different sites in different organisms. It was first reported that these sites were either intracellular or extracellular (see Watabe, 1974, 1981). However, it has become clear that this classification is an over-simplification. In this paper, mineral formation at various sites will be examined in invertebrates and algae, and its significance relative to the mechanisms of biomineralization will be discussed.

SITES OF MINERAL FORMATION

Biomineralization is a cell mediated process, therefore, there exist three possible sites of the mineral formation, *i.e.*, extracellular, intercellular and intracellular. In fact, all three of these sites are present in the lower organisms.

Origin, Evolution, and Modern Aspects of Biomineralization in Plants and Animals
Edited by Rex E. Crick
Plenum Press, New York

Extracellular mineral formation

There are two types in this category, *i.e.*, outside and inside of the organism's body. Most of extracellular formation takes place completely outside of body coverings, such as in the extracellular spaces adjacent to the outer surface of the epidermis, or at the outer cell wall. The mineral formation occurring directly on the cell wall such as seen in bacteria is termed "epicellular mineralization" by Mann (1987, this volume). Examples are tests of foraminiferans, exoskeletons of scleractinian corals, serpulid worm tubes, external shells of molluscs, carapaces of crustaceans, exoskeletons of bryozoans, cones of sipunculids, brachiopod shells, and outer crusts of calcareous algae, etc. (see Watabe, 1981 for references). Most of these structures are calcareous. However, brachiopod shells are either calcareous or made of calcium phosphate.

In some organisms, mineral formation occurs extracellularly but within the organisms' tissues. Structures such as aragonite spicules of the turbellarian *Florianella bipolaris* are within the basement membrane (basal lamina) of the epidermis (Rieger & Sterrer, 1975a). In octocorals, structure such as the calcite nodules (*Veretillum cynomorium*) (Ledger & Franc, 1978) and amorphous calcareous granules (*Leptogorgia virgulata*) (Kingsley & Watabe, 1982a) are present within the central axis, but outside of the axial epithelium.

Regardless of whether the sites of mineralization are outside or inside the organisms' soft tissue, almost all structures of extracellular formation are secretory products of multicellular epithelial tissues. Those formed outside are usually "massive" skeletons overlaying the organisms' body surface as protective covers. Secretion from the multicellular epithelial tissue onto the body surface is a direct and therefore, perhaps energy-saving process, evolved to achieve this protective function. The large surface areas of the epithelium facilitates formation of such "massive" skeletons.

Intracellular mineral formation

With few exceptions, the mineralized structures originated intracellularly exhibit species-specific, almost symmetrical and highly intricate morphology. Various subcellular sites of mineral formation are known in this category. The most commonly seen is within vacuoles, many of which are probably derived from Golgi complexes. Examples are calcite spicules in the scleroblasts of the octocorals *Leptogorgia* (Kingsley & Watabe, 1982b) and *Renilla* (Dunkelberger & Watabe, 1974); calcareous spicules in epidermal spicule-forming cells of the acochilidacean molluscs (Rieger & Sterrer, 1975b); larval spines (see Okazaki *et al.*, 1980), ossicles (Stricker, 1985), teeth (Kniprath, 1974), regenerated spines and tests (Shimizu & Yamada, 1980) in sclerocytes of echinoderms; aragonite spicules in scleroblasts of the didemnids (compound ascidians) (Kniprath & Lafargue, 1980); calcium phosphate stylets in styletocytes of nemertean worms (Stricker, 1984), and calcite coccoliths of unicellular algae (Wilbur & Watabe, 1963; see also Blackwelder *et al.*, 1979). Silica structures are formed within vesicles of rhizopod amebae (Harrison *et al.*, 1976), siliceous sponges (see Garrone *et al.*, 1981), diatoms (see Volcani, 1981), radiolarians (Riedel & Sanfilippo., 1981), chrysophytes (McGrory & Leadbeater, 1981) and choanoflagellates (Leadbeater, 1981).

Other sites of mineralization are cisternae of the endoplasmic reticulum as observed in calcareous granule formation in the protozoan *Poronodon* (André & Fauré-Fremiet,

1967) and the homopteran *Philaenus squmaris* (Gourantan, 1968). Mitochondria of the calciferous gland of earthworms (Crang *et al.*, 1968), and multivesicular bodies of the calcium secreting gland of tube regenerating serpulid worms (Neff, 1971) contain rhombohedral calcite crystals.

As discussed by Watabe (1981) and Kingsley (1984) many of the structures formed intracellularly eventually become extracellular entities. The intra-extracellular transition appears to follow either one of the two methods. As is seen in the formation of gorgonian spicules (Kingsley & Watabe, 1982b), and ossicles (Stricker, 1985; Donachy, 1984), larval spicules (Uemura & Watabe, unpublished data), spines and tests (Shimizu & Yamada, 1980) of echinoderms, the membranes of vacuoles in which mineralized structures grow come to fuse with the plasma membranes and the structures are exposed to the extracellular environment by the formation of a breach in the membranes. Further growth of the mineralized structure occurs in echinoderms in the extracellular environment. This is accomplished by secondary sclerocytes which migrate to the immediate vicinity of the mineralized structure (Shimizu & Yamada, 1980). It is not known whether or not spicules will still grow extracellularly in *Leptogorgia*.

The other known method of transition is by migration of vacuoles and vesicles, containing mature mineralized structures, to the cell surface where their contents are extruded extracellularly via exocytosis. This is seen in coccolith formation and siliceous structure formation in chrysophytes, choanoflagellates, and rhizopod amebae (see Simpson & Volcani, 1981). In coccoliths (de Jong; Watabe, pers. obs.), no further mineralization occurs once they become extracellular. This may also hold true in chrysophytes, and other unicellular organisms, although assemblage of mature extruded units into sheaths takes place extracellularly (Leadbeater, 1981).

Not all of the mineralized structures of intracellular origins become extracellular. Examples are silica spicules of sponges and silica skeletons of radiolarians and walls of diatoms (see Simpson & Volcani, 1981) in which the final skeletal structures are completed intracellularly.

Intercellular mineral formation

Only a few examples are known. One is the calcite spicules of calcareous sponges. The monaxon is formed within a narrow space between a pair of sclerocytes, and the triax arises in a intercellular space formed between each two of six aggregated sclerocytes (Jones, 1970; Ledger & Jones, 1977). One of the sites of spherule formation in the amphipod *Orchestia* is the intercellular space. However, spherule formation will be treated separately and will not be discussed in this section.

The intercellular formation in calcareous sponges is actually an extracellular event. However, the intercellular spaces where spicule formation occurs are obliterated by septate junctions and are almost isolated from the "extracellular" environment. Therefore, the formation can be assumed to be of an intracellular, or pseudo-intracellular nature. The fact that the spicules show symmetrical morphology, characteristic of intracellular mineralization, further supports the view.

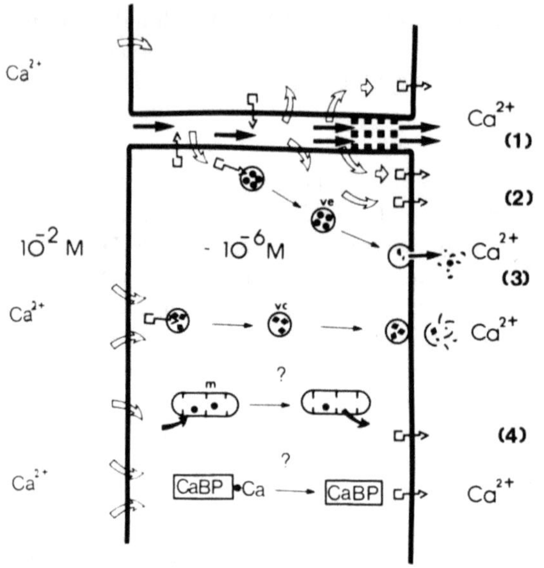

FIGURE 1. Transepithelial calcium transport for extracellular mineral formation. ▪1. Intercellular transport through septate junction (black squares). ▪2. Intracellular transport via gradient. ▪3. Intracellular transport by electron dense bodies (**ve**), or as amorphous and/or crystalline granules in vacuoles (**vc**). ▪4. Intracellular transport by mictochondria (**m**) or calcium binding proteins (**CaBP**). Open arrows indicate calcium transport via gradient; arrows with boxes indicate transport by pumps.

Calcareous spherule formation

The soft tissues of many invertebrates contain numerous calcareous spherules, approximately 0.5 to 200 μm in diameter (Watabe *et al.*, 1976). These spherules (or granules by Simkiss & Mason, 1983; Fournié & Chétail, 1984; Wilbur, 1985) are of calcium carbonate or phosphate and often contain other cations such as magnesium, zinc, and manganese (Simkiss & Mason, 1983).

All three sites of formation have been described for calcareous spherules. Extracellular formation has been reported in earthworms in which the spherules are produced in the lumen of the calciferous gland epithelium (Morgan, 1981). In the amphipod *Orchestia* at the pre-exuvial period during molting, epithelial cells of the posterior caeca secrete ca ions into the posterior caeca lumen and form spherules extracellularly as a calcium storage (Graf, 1968). Immediately after exuviation, the spherules in the posterior caeca lumen dissolve, and calcium is liberated into the network of channels between cells in the epithelium, and new spherules are formed intercellularly (Graf, 1971; Graf & Meyran, 1983). Examples of intracellular spherule formation are numerous (see Watabe *et al.*, 1976). This type of formation occurs within vacuoles of calcium cells in various connective tissues.

Many of those calcareous spherules do not remain at the sites of formation, but dissolve, mobilizing the calcium into circulatory fluid in response to various physiological stimuli (see Watabe *et al.*, 1976; Simkiss & Mason, 1983; Fournié & Chétail, 1984). It is understandable, therefore, that they exist in the readily soluble form of amorphous calcium compounds or vaterite, and do not exhibit elaborate morphology regardless of their site of formation.

METHODS OF CALCIUM TRANSPORT TO SITES OF MINERAL FORMATION.

Calcium transport processes have not been investigated extensively in invertebrates or algae. However, the following schemes can be derived from existing reports.

Transport for extracellular mineralization

A majority of the extracellularly formed minerals are secretory products of epithelial cells. This implies that calcium ions taken up by the organism from the environment or food, and entering the hemolymph or body fluid, must move across the epithelium to reach the site of mineralization. Two routes are conceivable for transepithelial calcium transport, *i.e.*, the intercellular and intracellular routes. The intercellular route has been suggested by Graf (1971), Neff (1972), Simkiss (1974, 1976a, b), Simkiss & Wilbur (1977), Watabe & Blackwelder (1980), and Meyran *et al.* (1984). Although invertebrate epithelial cells are joined together by septate junctions near the apices, these junctions are considered permeable to Ca^{2+} (Meyran *et al.*, 1984). Therefore, when the concentration is high enough, Ca^{2+} would pass through the intercellular space into the outer surface of the epithelium via septate junctions (Fig. 1-1). One such example of this type of intercellular calcium transport has been reported in calcareous spherule formation in the posterior caeca lumen of *Orchestia* (Meyran *et al.*, 1984).

The other method of intercellular calcium transport has been reported in the mollusc *Mercenaria mercenaria* by Neff (1972), in which calcium precipitated in the form of minute (100 – 2000 Å) granules apparently passes through the intercellular space by a leaky junction into the extrapallial space. A similar method has been proposed by Johnston (1980) in hermatypic corals in which vesicles, containing matrix and calcium ions, are discharged onto the calcification site through spaces between calicoblasts. However, Wilbur & Saleuddin (1984) cautioned about attributing great importance to intercellular transport as a factor in molluscan shell formation. That hemolymph fluid passing through intercellular channels cannot be the sole source of calcium and extrapallial fluid is shown by differences in composition of the hemolymph and extrapallial fluid. Other evidence also demonstrates that the mantle cells contribute to the composition of extrapallial fluid (see Wilbur & Saleuddin, 1984).

Several methods have been suggested for the transepithelial transport of calcium through the cytoplasm. Since the ionic concentration of calcium in hemolymph or body fluid and cytosol can be assumed to be approximately $10^{-2} - 10^{-3}$M, and $10^{-6} - 10^{-7}$ M, respectively, Ca^{2+} would diffuse into the epithelial cytosol down the concentration gradient. Cytosolic Ca^{2+} could then be actively transported out of the cell and concentrated into the extraepithelial space (Fig. 1-2). This transport is possibly mediated by pumps such as Ca^{2+}-ATPase or $Na^{+}K^{+}$-ATPase, or by antiports such as Na^{+}-Ca^{2+} exchange (Wood, 1973; Roer & Dillaman, 1984; Meyran *et al.*, 1984) localized at the apical plasma membranes. Portions of calcium ions moving through the intercellular space in *Orchestia*, mentioned previously, are derived from the cytosolic calcium that moved into this space by these pumps localized at the lateral plasma membranes (Meyran *et al.*, 1984). However, calcium transport down an electrochemical gradient across the membranes of mantle epithelium has also been shown in several species of molluscs (see Simkiss, 1976b; Wilbur & Saleuddin, 1983 for references).

Another method which has been considered for transepithelial calcium transport is via vesicles and vacuoles (Fig. 1-3). The atremate brachiopod *Glottidia pyramidata* possess shells of (CO_3F) containing apatite (LeGeros *et al.*, 1985). Calcium (and phosphate), which possibly diffused into the mantle cytoplasm from the hemolymph, appear to be sequestered into vacuoles within the mantle epithelium and deposited as

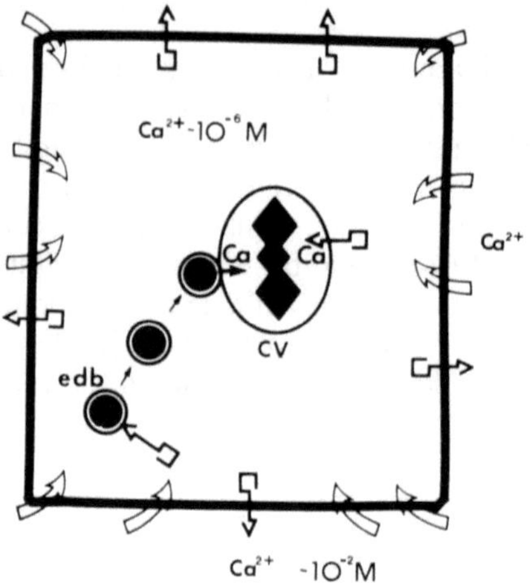

FIGURE 2. Calcium transport for intracellular mineral formation. **CV**, calcifying vacuole; **edb**, electron-dense body; open arrows, calcium transport via gradient; arrows with boxes, calcium transport by pumps.

amorphous and/or crystalline granules (Watabe & Pan, 1984). When shell mineralization is occurring, these vacuoles are observed to come into close contact with the apical plasma membranes of the epithelium, move out into the extrapallial space, and release their content into the space (Pan, 1985). Calcium transport via vesicles has also been shown in operculum formation in gastropods (Vovelle, 1973). Calcium ions are also transported in bound forms as seen in a hermatypic coral in which calcium is contained in electron dense bodies in the epidermis (Kinchington, 1980).

Mitochondria have been implicated in the transport of calcium (and phosphorus) in granular forms (see Lehninger, 1983) (Fig. 1-4). However, they are also suggested to serve as repositories for excess cytosolic calcium (Simkiss, 1976b). Transport of calcium by calcium-binding proteins is another possibility (Fig. 1-4). Although evidence has not been shown in invertebrates, transport by calcium-binding proteins has been suggested to occur through fish gills, (Flik *et al.*, 1985), and the presence of this protein has been reported in the mantle epithelium of the freshwater gastropod *Lymnaea stagnalis* (Dogterom & Doderer, 1981). However, the reality of this trans-epithelial transport method needs to be critically evaluated because of its slow rate of transport (M.A. Crenshaw, personal communication).

Transport for intracellular mineralization

Transport of calcium into cells appears to be through diffusion as has been suggested in spicule formation in the scleroblasts of *Leptogorgia virgulata* (Kingsley & Watabe, 1985a,b). A potential difference (inside negative) has been found in the calcareous alga *Emiliania huxleyi* which favors the diffusion together with the low cytosolic Ca^{2+} concentration (Sikes & Wilbur, 1982). In embryonic spicule formation in the sea urchin *Hemicentrotus pulcherrimus*, Ca^{2+} uptake is mainly coupled with influx of chloride into the cells (Yasumasu *et al.*, 1985). However, silicate uptake in diatoms is energy dependent (Sullivan & Volcani, 1981).

Calcium ions which enter the cells are pumped into vacuoles and vesicles where mineral formation takes place, and also pumped out of the cell to maintain low cytosolic calcium levels (Fig. 2) (see Wilbur, 1980). In diatoms, silicate uptake by vesicles is Na^+-dependent and Na^+K^+-ATPase has been strongly suggested to be involved (Sullivan & Volcani, 1981). Calcium transport into mineralizing vacuoles is also carried out by electron-dense bodies as reported in *Leptogorgia* (Kingsley & Watabe, 1985b). The presence of calcium-ATPase in plasma membranes to pump Ca^{2+} out into the extracellular space is shown in this organism as well (Kingsley & Watabe, 1985b).

In summary, there does not seem to be standard methods of calcium transport characterizing the extra- and intracellular mineralization. No information is available on calcium transport in the intercellular spicule formation of calcareous sponges. However, much work is needed before conclusions may be drawn concerning ion transport.

FACTORS CONTROLLING THE MORPHOLOGY OF MINERALIZED STRUCTURES

Regardless of whether they are of extra-, inter-, or intracellular origin, the gross morphology of most mineralized structures or skeletons are elaborate, very well regulated and characteristic of each taxa. Calcareous spherules are exceptions in that they are somewhat spherical to oval in shape without species-specific morphological elaborations. However, the individual crystals and their aggregates which comprise extracellularly-formed structures (*e.g.*, mollusc shells) are not necessarily regularly shaped and their size and morphology may vary greatly depending on physiological and environmental conditions. In contrast, the majority of structures originating in vacuoles and vesicles exhibit species-specific, almost symmetrical, and highly intricate morphology (Figs. 3–6). The size is also generally uniform in each structure. With the exception of calcareous sponge spicules and echinoderm test plates, spines and spicules, these intracellularly-formed structures are polycrystalline aggregates or aggregates of amorphous units. Their individual crystals or amorphous units are also usually uniform in size and shape. Intracel-

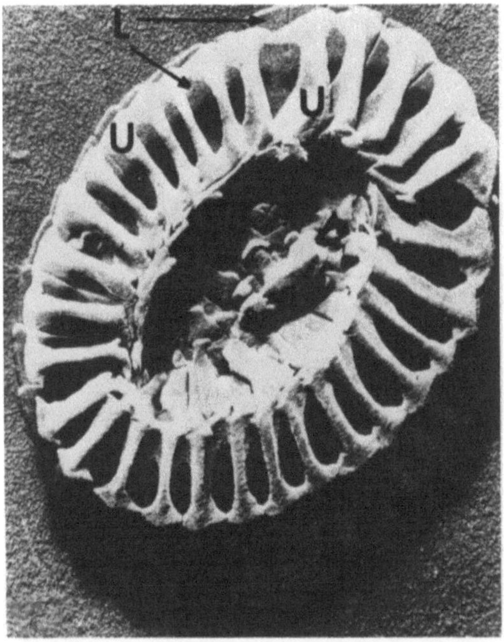

FIGURE 3. Coccolith of the alga *Emiliania huxleyi*. U, upper elements; L, lower elements. × 24,800.

FIGURE 4. Sterraster of the demosponge *Geodia gibberosa* (Hartman, 1981). × 1,300.

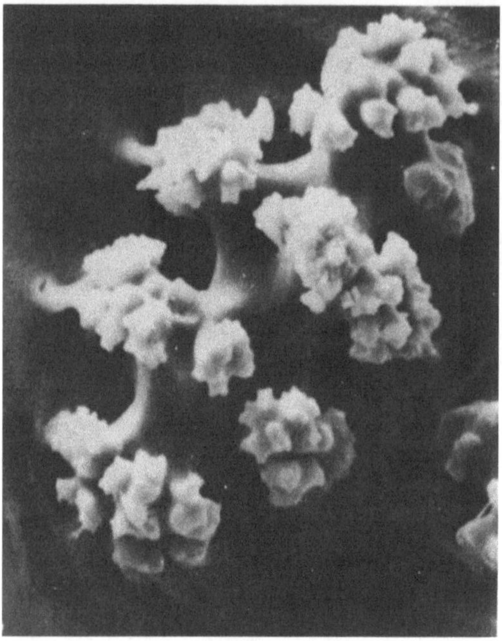

FIGURE 5. Spicule of the gorgonian *Leptogorgia virgulata*. × 1,200.

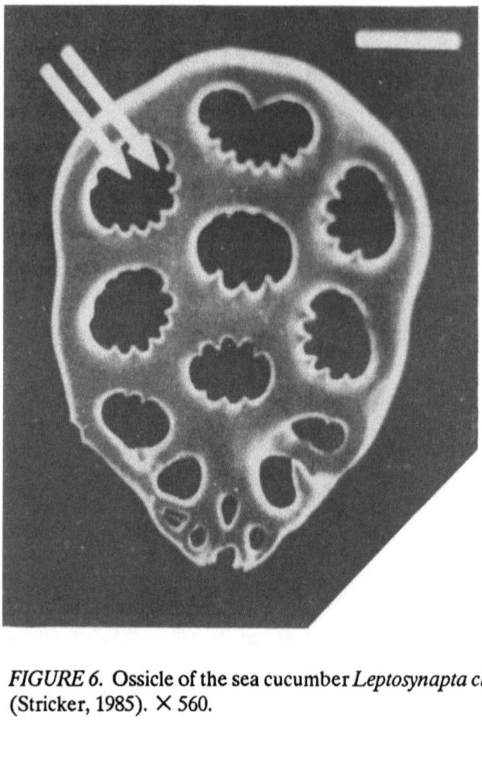

FIGURE 6. Ossicle of the sea cucumber *Leptosynapta clarki* (Stricker, 1985). × 560.

lularly formed structures can be considered to be under a more direct and higher degree of cellular control than in the extracellular counterparts. This is indicated by the more precise architectural forms, and their sites of formation which are within membrane-bounded organelles. In this section, we will discuss the possible factors controlling the morphology of intracellularly-formed mineralized structures. In spite of recent developments in electron microscopy, very few papers exist concerning the morphogenesis of mineralized structures except for siliceous skeletons (Simpsons *et al.*, 1985).

Several factors have been considered to affect the morphogenesis of structures formed within vacuoles and vesicles. Environmental factors such as calcium concentration, temperature, and salinity certainly affect the morphology and dimensions of calcified structures (see Watabe & Wilbur, 1966; Levi, 1972; Blackwelder, 1976; Jones, 1978). However, these are factors modifying the basic morphology of mineralized structures, and will not be discussed further. It is apparent that physical and/or biochemical components of membranes and contents of vacuoles/vesicles are the determinants of morphogenesis. It has been suggested that the coccolith vacuoles function as a mold during the morphogenesis of coccoliths (Wilbur & Watabe, 1963; Klaveness, 1972; Mann, 1983). However, these vacuoles do not show the same level of morphological intricacy as the coccoliths themselves (see also Van der Wal., 1984). Similarly, vacuoles of *Leptogorgia* scleroblasts do not exhibit the elaborate morphology of the spicules (see Kingsley & Watabe, 1982b). The overall shape of mineralized structure may be determined partly by the physical constraints of the vacuole membranes in some organisms, but it is unlikely that it exerts effects on detailed morphology. An exception may be the organic sheath's control over calcareous sponge spicule formation (Ledger & Jones, 1977).

FIGURE 7. Diagram depicting a vacuole which contains a spicule (*cf.* Fig. 5) and possible crystal growth inhibitors (6 dotted areas) controlling spicule morphology (*Leptogorgia*).

Direction of morphogenesis is probably accompanied by the organic and inorganic materials within the vacuoles. These materials' chemical components, concentrations, and local variations in distribution within the vacuole, all would affect crystal orientation, crystal habit, and crystal growth directions once nucleated. As for the organic materials, two types can be envisaged, *i.e.*, inductive and inhibitory, and the insoluble and soluble fractions of the organic matrix of mineralized structures have been shown to carry such functions (see Watabe *et al.*, 1986 for review). In the coccolithophorids, the oval shaped organic base plate, on which rhombohedral crystals develop peripherally, determines the oval-disc shaped morphology of coccoliths (Wilbur & Watabe, 1963; Van der Wal, 1984). Each siliceous sponge spicule contains an organic axial filament in its core. The overall contour of the filament, as well as its shape in cross section, may be involved in the morphogenesis of the final spicule shape (Simpson *et al.*, 1985). The base plate and the filament core are presumably the insoluble organic matrices and serve as nucleators of mineral formation (see Watabe, 1981). Succeeding nucleation, growth of crystal (or amorphous silica) aggregates may proceed in definite directions as determined by the matrix molecular arrangement. Or, they grow in all directions if the matrix exerts no directional effect and/or the growth conditions are homogeneous within the vacuole.

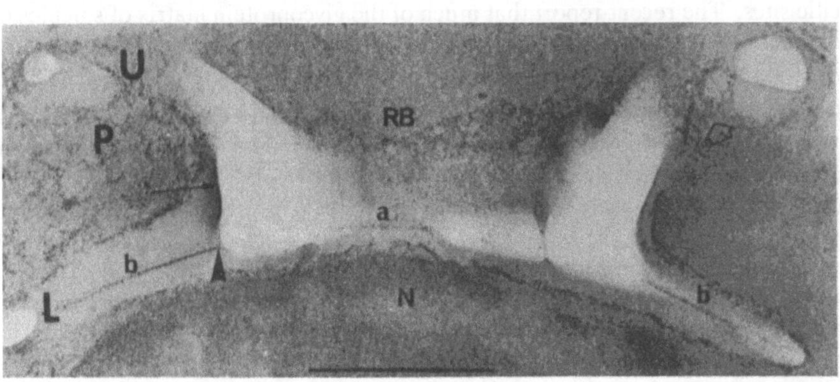

FIGURE 8. Electron micrograph of a vertical section of a coccolith (*cf.* Fig. 3) within a cell of *Emiliania huxleyi*, depicting the presence of polysaccharides (P) between the upper (U) and lower (L) elements (after Van der Wal, 1984). ▪a. Central portion of the organic plate. ▪b. Laterally extended portion of the plate; **RB**, reticular body; **N**, nucleus. × 50,000.

The spherulitic silica sterrasters of the demosponge *Geodia gibberosa* (Fig. 4) could be a result of this latter type of growth. On the other hand, if some growth inhibitors are present with a certain regular distribution within the vacuoles, especially along the inner vacuolar membranes, growth in those regions would be inhibited. Thus, regularly spaced indentations in the mineralized structures would result. Because of such growth inhibition, the composition and/or concentration of the solution adjacent to the inhibited regions may change and growth in those regions may be modified to show morphology which is different from the rest of the mineralized structure. This may explain the intricate form of gorgonian spicules (Fig. 5, 7). The presence of such inhibitors has been suggested for *Emiliania huxleyi* (Watabe, 1967). It is interesting to note that the soluble fraction of coccolith matrix, consisting of polysaccharides (de Jong *et al.*, 1976, 1979), has been shown to inhibit crystallization of calcium carbonate (Borman *et al.*, 1983) and that the polysaccharides have been localized in the indented regions of coccoliths (Fig. 8) (Van der Wal, 1984). The soluble organic matrix fraction may play inhibitory roles in other organisms as well. The soluble matrix of *Leptogorgia* has been found to consist mainly of calcium-binding proteins (Samata, Kingsley, & Watabe, unpublished), and calcium-binding proteins of the soluble matrix from mollusc shells have been reported to inhibit nucleation of calcium carbonate crystals (Wheeler *et al.*, 1981; Sikes & Wheeler, 1983, 1986).

The scheme discussed above mandates the presence of inhibitors along the vacuole membranes in a symmetrical fashion with regular spacings in order to form these intricate structures with regular geometrical order. At present, there is no direct evidence to support this. We are currently carrying out immunocytochemical investigations on the spatial orientation of the soluble and insoluble matrix and their calcium-binding proteins fractions within the spicule-forming vacuoles of *Leptogorgia*. Even if we obtain the evidence for inhibitors localized in a regular fashion along the membranes, the sites of formation and attachment of the inhibitors on the membranes must be characterized. Inhibitors may be synthesized outside of the calcifying vacuole and subsequently brought to the mineralizing site via organelles such as vesicles (Kingsley & Watabe, 1984). Alternatively, the vacuoles may synthesize specific inhibitors at specific sites. The recent report that much of the glycoprotein matrix of spicules of sea urchin embryos is formed in the intracellular vacuole of the mesenchyme cells (Benson *et al.*, 1986) may support the latter view.

Almost all of the intracellularly-formed structures attain a uniform size characteristic for each structure, which then becomes extracellular. This implies that mechanisms exist to regulate the final size of these structures. In coccolithophorids, the mechanisms may be within the vacuoles since coccoliths are formed in succession in a single cell, Or they may be outside of the vacuoles in a system like *Leptogorgia* since one spicule is formed in one vacuole in a single scleroblast (Kingsley & Watabe, 1982b). These ideas, however, are speculative. We have no experimental data concerning such control mechanisms. In vertebrates, several hormones such as PTH and calcitonin, etc., control calcification. There may be some taxa-specific neurosecretory materials or analogues and/or precursors of vertebrate endocrine products present in lower organisms, which could regulate mineralization. In fact, many substances which cross-react with vertebrate hormones such as calcitonin, thyroxine and its precursors, insulin, etc., have been found in lower organisms (Spangenberg, 1971, 1974; LeRoith *et al.*, 1981; Deftos, *et al.*, 1985; Watabe & Kingsley, unpublished).

We have suggested several possible factors which may affect the morphogenesis and size of some mineralized structures, but other possibilities may exist. Biochemical, physiological, cytochemical, and ultrastructural investigations of mineralizing vacuoles and vesicles will be essential to elucidate the problems.

ACKNOWLEDGEMENTS

Several studies reported in this paper have been supported by National Science Foundation grants #PCM8201389 and DCB8502698, and funds from Electron Microscopy Center, University of South Carolina. We thank Mrs. Betty Bynum for typing and editing the manuscript. Participation in the Symposium was made possible by grants from The Organized Research Fund of The University of Texas at Arlington and The Sea Grant Program administered by Texas A & M University.

REFERENCES

ANDRÉ, J., & FAURÉ-FREMIET, E., 1967. Formation et structure des concretions calcaires chez *Poronodon margani* Khal. Journal Microscopie, 6:391-398.

BENSON, S.C., BENSON, N.C., & WILT, F., 1986. The organic matrix of the skeletal spicule of sea urchin embryos. Journal Cell Biology, 102:1878-1886.

BLACKWELDER, P.L., 1976. Temperature relationships in coccolith morphology and dimension in fossil and living *Emiliania huxleyi* (Chrysophyta: Haptophyceae) Ph.D. thesis, 106 pp. Columbia: University of South Carolina.

BLACKWELDER, P.L., BRAND, L.E., & GUILLARD, R.L., 1979. Coccolith morphology and paleoclimatology. II. Cell ultrastructure and formation of coccoliths in *Cycloccolithina leptopora* (Murray and Blackman) Wilcoxon and *Gephyrocapsa oceanica* Kamptner. Scanning Electron Microscopy, 1979, II:417-420.

BORMAN, A.H., DE JONG, E.W., HUIZINGA, M., & WESTBROEK, P., 1983. Inhibition of CaCO3 precipitation by a polysaccharide associated with coccoliths of *Emiliania huxleyi*. *In* Biomineralization and Biological Metal Accumulation (ed. P. Westbroek & E.W. de Jong), pp. 303-305. Dordrecht: D. Reidel Publishing Co.

CRANG, R.E., HOLSEN, R.C., & HITT, J.B., 1968. Calcite production in mitochondria of earthworm calciferous glands. Biological Sciences, 18:299-301.

DE JONG, E.W., BOSCH, L., & WESTBROEK, P., 1976. Isolation and characterization of a Ca^{2+}-binding polysaccharide associated with coccoliths of *Emiliania huxleyi* (Lohmann) Hay & Mohler. European Journal Biochemistry, 70:611-621.

DE JONG, E.W., VAN RENS, L., WESTBROEK, P., & BOSCH. L., 1979. Biocalcification by the marine alga *Emiliania huxleyi* (Lohmann) Hay & Mohler. European Journal Biochemistry, 99:559-567.

DEFTOS, L.J., LEROITH, D., SHILOACH, J., & ROTH, J., 1985. Salmon calcitonin-like immunoactivity in extracts of *Tetrahymena pyriformis*. Hormone Metabolism Research, 17:82-85.

DOGTEROM, A.A., & DODERER, A., 1981. A hormone dependent calcium-binding protein in the mantle edge of the freshwater snail *Lymnaea stagnalis*. Calcified Tissue International, 33:505-508.

DONACHY, J.E., 1984. Arm regeneration in the brittle star *Ophiothrix angulata* (Say): Salinity and calcium concentration effects and ultrastructure of ossicle formation. M.S. thesis, 90 pp. Columbia: University of South Carolina.

DUNKELBERGER, D.G., & WATABE, N., 1974. An ultrastructural study on spicule formation in the pennnatulid colony *Renilla reniformis*. Tissue and Cell, 6:573-586.

FLIK, G., VANRIJS, J.H., & WENDELAAR BONGA, S.E., 1985. Evidence for high-affinity Ca^{2+}-ATPase activity and ATP-driven Ca^{2+}-transport in membrane preparations of the gill epithelium of the cichlid fish *Oreochromis mossambicus*. Journal Experimental Biology, 119:335-347.

FOURNIÉ, J., & CHÉTAIL, M., 1984. Calcium dynamics in land gastropods. American Zoologists, 24:857-870.

GARRONE, R., SIMPSON, T.L., & POTT-BOUMENDIL, J., 1981. Ultrastructure and deposition of silica in sponges. *In* Silicon and Siliceous Structures in Biological Systems (ed. T.L. Simpson & B.E. Volcani), pp. 495-525. New York: Springer-Verlag.

GOURANTAN, J., 1968. Composition, structure et mode de formation des concretions minerales dans l'intestin moyen des homopteres Cercopides. Journal Cell Biology, 37:316-328.

GRAF, F., 1968. Le stockage de calcium avant la mue chez les Crustacés Amphipodes *Orchestia* (talitridé) et *Niphargus* (Gammaridé hypogé). Doctoral Thesis, Dijon, CNRS no. 2690:1-216.

GRAF, F., 1971. Dynamique du calcium dans l'epithélium des caecum postérieurs *d'Orchestia cavimana* Heller (Crustacé, Amphipode). Rôle de l'espace intercellulaire. Comptes rendus des séances de l'Académie des Sciences, Paris, 273:1828-1831.

GRAF, F., & MEYRAN, J-C., 1983. Premolt calcium secretion in midgut posterior caeca of the Crustacean *Orchestia*: ultrastructure of the epithelium. Journal Morphology, 177:1-23.

HARRISON, F.W., DUNKELBERGER, D.G., WATABE, N., & STUMP, A.B., 1976. The cytology of the testaceous rhizopod *Lesquereusia spiralis* (Ehrenberg) Penard. II. Ultrastructure and shell formation. Journal Morphology, 150:343-358.

HARTMAN, W.D., 1981. Form and distribution of silica in sponges. *In* Silicon and Siliceous Structure in Biological Systems (ed. T.L. Simpson & B.E. Volcani), pp. 453-493. New York: Springer-Verlag.

JOHNSON, I.S., 1980. The ultrastructure of skeletogenesis in hermatypic corals. International Review Cytology, 67:171-214.

JONES, W.C., 1970. The composition, development, form and orientation of calcareous sponge spicules. Symposia Zoological Society London, 25:91-121.

JONES, C.W., 1978. Spicule growth and production in juvenile *Sycon ciliatum*. Colloques internationaux du CNRS, No. 291:67-77.

KINGSLEY, R.J., 1984. Spicule formation in the invertebrates with special reference to the gorgonian *Leptogorgia virgulata*. American Zoologist, 24:883-891.

KINGSLEY, R.J., & WATABE, N., 1982a. Ultrastructure of the axial region in *Leptogorgia virgulata* (Cnidaria: Goronaceae). Transactions of the American Microscopical Society, 101:325-339.

KINGSLEY, R.J., & WATABE, N., 1982b. Ultrastructural investigation of spicule formation in the gorgonian *Leptogorgia virgulata* (Lamarck) (Coelenterata: Gorgonacea). Cell and Tissue Research, 22:325-334.

KINGSLEY, R.J., & WATABE, N., 1984. Synthesis and transport of the organic matrix of the spicules in the gorgonian *Leptogorgia virgulata* (Lamarck) (Coelenterata: Gorgonacea). An autoradiographic investigation. Cell and Tissue Research, 235:533-538.

KINGSLEY, R.J., & WATABE, N., 1985a. An autoradiographic study of calcium transport in spicule formation in the gorgonian *Leptogorgia virgulata* (Lamarck) (Coelenterata: Gorgonacea). Cell Tissue Research, 239:305-310.

KINGSLEY, R.J., & WATABE, N., 1985b. Ca-ATPase localization and inhibition in the gorgonian *Leptogorgia virgulata* (Lamarck) (Coelenterata: Gorgonacea). Journal Experimental Marine Biology and Ecology, 93:157-167.

KINCHINGTON, D., 1980. Localization of intracellular calcium within the epidermis of a cool temperate coral. *In* Development and Cellular Biology of Coelenterates (ed. P. Trardent & R. Tardent), pp. 143-148. Amsterdam: North-Holland Biomedical Press.

KLAVENESS, D., 1972. *Coccolithus huxleyi* (Lohmann) Kamptner. I. Morphological investigations on the vegetable cell and the process of coccolith formation. Protistologica, 8:335-346.

KNIPRATH, E., 1974. Ultrastructure and growth of the sea urchin tooth. Calcified Tissue Research, 14:211-228.

KNIPRATH, E., & LAFARGUE, F., 1980. Spicule formation in the Didemnidae (Compound Ascidians). *In* The Mechanisms of Biomineralization in the Animals and Plants (ed. M. Omori & N. Watabe), pp. 31-36. Tokyo: Tokai University Press.

LEADBEATER, B.S.C., 1981. Ultrastructure and deposition of silica in loricate Choanoflagellates. Silicon and Siliceous Structures in Biological Systems (ed. T.L. Simpson & B.E. Volcani), pp. 295-322. New York: Springer-Verlag.

LEDGER, P.W., & FRANC, S., 1978. Calcification of the collagenous axial skeleton of *Veretillium cynomorium* Pall (Cnidaria: Pennatulacea). Cell and Tissue Research, 192:249-266.

LEDGER, P.W., & JONES, W.C., 1977. Spicule formation in the calcareous sponge *Sycon ciliatum*. Cell and Tissue Research, 181:553-567.

LEGEROS, R.Z., PAN, C-M., SUGA, S., & WATABE, N., 1985. Crystallo-chemical properties of apatite in atremate brachiopod shells. Calcified Tissue International, 37:98-100.

LEHNINGER, A.L., 1983. The possible role of mitochondria and phosphocitrate in biological calcification. *In* Biomineralization and Biological Metal Accumulation (ed. P. Westbroek & E.W. de Jong), pp. 107-121. Dordrecht: D. Riedel Publishing Co.

LEROITH, D., LESNIAK, M.A., & ROTH, J., 1981. Insulin in insects and annelids. Diabetes, 30:70-75.

LEVI, S.F., 1972. Effects of salinity on growth, calcification, and ultrastructure in a calcareous alga, *Hymenomonas carterae*. M.S. thesis, pp. 148, Columbia: University of South Carolina.

MANN, S., 1983. Mineralization in biological systems. Structure and Bonding, 54:125-174.

MANN, S., 1987. The role of inorganic phosphate in iron oxide biomineralization. *In* Origin, Evolution, and Modern Aspects of Biomineralization in Plants and Animals (ed. R.E. Crick), pp. ___. New York: Plenum Press.

MCGRORY, C.E., & LEADBEATER, B.S.C., 1981. Ultrastructure and deposition of silica in the Chrysophyceae. *In* Silicon and Siliceous Structure in Biological Systems (ed. T.L. Simpson & B.E. Volcani), pp. 201-230. New York: Springer-Verlag.

MORGAN, A.J., 1981. A morphological and electron-microbe study of the inorganic composition of the mineralized secretory products of the calciferous gland and chloragogenous tissue of the earthworm, *Lumbricus terrestris* L. The distribution of injected strontium. Cell and Tissue Research, 220:829-844.

MYRAN, J.C., GRAF, F., & NICAISE, G., 1984. Calcium pathways through a mineralizing epithelium in the crustacean *Orchestia* in premolt: Ultrastructural cytochemistry and x-ray microanalysis. Tissue and Cell, 16:269-286.

NEFF, J.M., 1971. Ultrastructural studies of the secretion of calcium carbonate by the serpulid polychaete worm, *Pomatoceros caerulens*. Zeitschrift für Zellforschung, 120:160-186.

NEFF, J.M., 1972. Ultrastructure of the outer epithelium of the mantle in the clam *Mercenaria mercenaria* in relation to calcification of the shell. Tissue and Cell, 4:591-600.

OKAZAKI, K., MCDONALD, K., & INOUE, S., 1980. Sea urchin larval spicule observed with scanning electron microscope. *In* The Mechanisms of Biomineralization in the Animals and Plants (ed. M. Omori & N. Watabe), pp. 159-168. Tokyo: Tokai University Press.

PAN, C.-H., 1985. Ultrastructural and physiological investigations of the mineralization of the inarticulate brachiopod, *Glottidia pyramidata* (Simpson). Ph.D. thesis, pp. 165. Columbia: University of South Carolina.

RIEDEL, W.R., & SANFILIPPO. A., 1981. Evolution and diversity of form in Radiolaria. *In* Silicon and Siliceous Structures in Biological Systems (ed. T.L. Simpson & B.E. Volcani), pp. 323-343. New York: Springer-Verlag.

RIEGER, R.M., & STERRER, M., 1975a. New spicular skeletons in Turbellaria and the occurrence of spicules in marine meiofauna. I. Zietschrift für Zoologische Systematik und Evolutionsforschung, 13:207-248.

RIEGER, R.M., & STERRER, M., 1975b. New spicular skeletons in Turbellaria and the occurrence of spicules in marine meiofauna. II. Zietschrift für Zoologische Systematik und Evolutionsforschung, 13:249-278.

ROER, R., DILLAMAN, R., 1984. The structure and calcification of the crustacean cuticle. American Zoologist, 24:893-909.

SHIMIZU, M., & YAMADA, J., 1980. Sclerocytes and crystal growth in the regeneration of sea urchin test and spines. *In* The Mechanisms of Biomineralization in the Animals and Plants (ed. M. Omori & N. Watabe), pp. 169-178. Tokyo: Tokai University Press.

SIKES, C.S., & WHEELER, A.P., 1983. A systematic approach to some fundamental questions of carbonate calcification. *In* Biomineralization and Biological Metal Accumulation (ed. P. Westbroek & E.W. de Jong), pp. 285-289. Dordrecht: D. Reidel Publishing Co.

SIKES, C.S., & WHEELER, A.P., 1986. The organic matrix from oyster shell as a regulator of calcification *in vivo*. Biological Bulletin, 170:494-505.

SIKES, C.S., & WILBUR, K.M., 1982. Functions of coccolith formation. Limnology and Oceanography, 27:18-26.

SIMKISS, K., 1974. Calcium translocation by cells. Endeavour, 33:119-124.

SIMKISS, K., 1976a. Intracellular and extracellular routes in biomineralization. Symposia Society Experimental Biology, 30:423-444.

SIMKISS, K., 1976b. Cellular aspects of calcification. *In* Mechanisms of Mineralization in the Invertebrates and Plants (ed. N. Watabe & K.M. Wilbur), pp. 1-32. Columbia: University of South Carolina Press.

SIMKISS, K., & MASON, A.Z., 1983. Metal ions: metabolic and toxic effects. *In* The Mollusca, Vol. 2, (ed. P.W. Hochachka), pp. 101-164. New York: Academic Press.

SIMKISS, K., & WILBUR, K.M., 1977. The molluscan epidermis and its secretions. Symposia of the Zoological Society of London, 39:35-76.

SIMPSON, T.L., LANGENBRUCH, P-F., & SCALERA-LIACI, L., 1985. Silica spicules and axial filaments of the marine sponge *Stelletta grubii* (Porifera, Demospongiae). Zoomorphology, 105:375-382.

SIMPSON. T.L., & VOLCANI, B.E., 1981. Introduction. *In* Silicon and Siliceous Structures in Biological Systems (ed. T.L. Simpson & B.E. Volcani), pp. 1-12. New York: Springer-Verlag.

SPANGENBERG, D.B., 1971. Thyroxine induced metamorphosis in *Aurelia*. Journal Experimental Zoology, 178:183-194.

SPANGENBERG, D.B., 1974. Thyroxine in early strobilation in *Aurelia aurita*. American Zoologist, 14:825-834.

STRICKER, S.A., 1984. Styletogenesis in nemertean worms: The ultrastructure of organelles involved in intracellular calcification. Journal Morphology, 179:119-134.

STRICKER, S.A., 1985. The ultrastructure and formation of the calcareous ossicles in the body wall of the sea cucumber *Leptosynapta clarki* (Echinodermata, Holothuroida). Zoomorphology, 105:209-222.

SULLIVAN, C.W., & VOLCANI, B.E., 1981. Silicon in the cellular metabolism of diatoms. *In* Silicon and Siliceous Structures in Biological Systems (ed. T.L. Simpson & B.E. Volcani), pp. 15-42. New York: Springer-Verlag.

VAN DER WAL, P., 1984. Calcification in two species of coccolithophorid algae. Gus Papers Geology, Series 1, No. 20-1984:1-112.

VOLCANI, B.E., 1981. Cell wall formation in diatoms: morphogenesis and biochemistry. *In* Silicon and Siliceous Structures in Biological Systems (ed. T.L. Simpson & B.E. Volcani), pp. 157-200. New York: Springer-Verlag.

VOVELLE, J., 1973. Transfert du calcium à travers l'epithélium du repli opercularie chez *Astrea rugosa* L. (Turbinidae). Malacologia, 14:47-51.

WATABE, N., 1967. Crystallographyic analysis of the coccolith of *Coccolithus huxleyi*. Calcified Tissue Research, 1:114-121.

WATABE, N., 1974. Crystal growth of calcium carbonate in biological systems. Journal Crystal Growth, 24/25:116-122.

WATABE, N., 1981. Crystal growth of calcium carbonate in the invertebrates. Journal Crystal Growth and Characterization, 4:99-147.

WATABE, N., BERNHARDT, A.M., KINGSLEY, R.J., & WILBUR, K.M., 1986. Recalcification of decalcified spicule matrices of the gorgonian *Leptogorgia virgulata*. Transactions of the American Microscopical Society (In press).

WATABE, N., & BLACKWELDER, P.L., 1980. Ultrastructure and calcium localization in the mantle epithelium of the freshwater gastropod *Pomacea paludosa* during shell regeneration. *In* The Mechanism of Biomineralization in Animals and Plants (ed. M. Omori & N. Watabe), pp. 131-144. Tokyo: Tokai University Press.

WATABE, N., MEENASKSHI, V.R., BLACKWELDER, P.L., KURTZ, E.M. & DUNKELBERGER, D.G., 1976. Calcareous spherules in the gastropod, *Pomacea paludosa*. *In* The Mechanisms of Mineralization in the Invertebrates and Plants (ed. N. Watabe & K.M. Wilbur), pp. 283-308. Columbia: University of South Carolina Press.

WATABE, N., & PAN, C-M., 1984. Phosphatic shell formation in atremate brachiopods. American Zoologist, 24:977-985.

WATABE, N., & WILBUR, K.M., 1966. Effects of temperature on growth, calcification and coccolith form in *Coccolithus huxleyi*. Limnology and Oceanography, 11:567-575.

WHEELER, A.P., GEORGE, J.W., & EVANS, C.A., 1981. Control of calcium carbonate nucleation and crystal growth by soluble matrix of oyster shell. Science, 212:1397-1398.

WILBUR, K.M., 1980. Cells, crystals and skeletons. *In* The Mechanisms of Biomineralization in Animals and Plants (ed. M. Omori & N. Watabe), pp.3-11. Tokyo: Tokai University Press.

WILBUR, K.M., 1985. Topics in molluscan mineralization: Present status, future directions. American Malacological Bulletin, Special Edition, 1:51-58.

WILBUR, K.M., & SALEUDDIN, A.S.M., 1984. Shell formation. *In* the Mollusca, Vol. 4 (ed. A.S.M. Saleuddin & K.M. Wilbur), pp. 235-284. New York: Academic Press.

WILBUR, K.M., & WATABE, N., 1963. Experimental studies on calcification in molluscs and the alga *Coccolithus huxleyi*. Annals New York Academic Sciences, 109:82-112.

WOOD, D., 1973. Calcium movement and electrogenesis across the isolated clam mantle. Ph.D.thesis, 122 pp. Pullman: Washington State University, 122 pp.

YASUMASU, I., MITSUNAGA, K., & FUJINO, Y., 1985. Mechanism for electrosilent Ca2+ transport to cause calcification of spicules in sea urchin embryos. Experimental Cell Research, 159:80-90.

ULTRASTRUCTURAL AND PROTEIN ASPECTS OF APATITE FORMATION IN VERTEBRATE HARD TISSUES

Hiroshi Nakahara and Mitsuo Kakei

Department of Oral Anatomy, Josai Dental University, Sakado, Saitama, Japan

ABSTRACT. The central dark line (*CDL*) which is one of the 100 planes of hydroxyapatite and represents the site of initiation of crystal growth, was observed not only in the crystallites of enamel but in dentin, bone and baleen as well. The area of *CDL* probably contains high concentration of carbonate ions. It was demonstrated in all the hard tissues tested that the earliest appearance of mineral structure is composed of a ribbon- or disc-shaped precursor mineral plate intimately surrounded by an organic envelope. The *CDL* appears at the center of the mineral portion as the first step in crystal growth.

Carbonic anhydrase activity was observed in the area where crystal nucleation occurs in developing enamel, dentin, cartilage and bone. The enzyme probably contributes the carbonate ion to the crystal nucleation site where the carbonate-rich *CDL* is being formed. By means of immunoblotting method it has been shown that carbonic anhydrase is one of the major protein constituents of first formed enamel.

The similarities in the mineralization process among the apatite forming tissues suggest that the matrices of these hard tissues contain similar groups of proteins, probably derived from basic cytoplasmic proteins. Some of these proteins such as troponin T and I, actin and myosin, were in fact identified in developing enamel matrix by means of the immunoblotting technique. The presence of troponins and actin were also demonstrated in calcifying dentin. Calmodulin was shown to be present only in the later (maturing) stage of enamel. These cytoplasmic proteins contained in the calcifying matrix are in a somewhat altered condition compared with their original form in the cytoplasm. The alteration is probably due to the results of the complicated process of degradation and recombination that occurs in these proteins.

INTRODUCTION

The majority of vertebrate hard tissues produce hydroxyapatite [$Ca_{10}(PO_4)_6(OH)_2$] as a final form of mineral which always contain certain amounts of carbonate. Growing mineral crystallites are always surrounded by an organic envelope which is composed mainly of protein and may be responsible for promoting and regulating crystal growth. In order to clarify the essential problem of mineral formation in vertebrates, we have attempted for several years to study the initial and growing crystallite, and associated organic envelope by means of relatively high magnification transmission electron microscope (TEM) observations. Recently we have also identified a number of protein

Origin, Evolution, and Modern Aspects of Biomineralization in Plants and Animals
Edited by Rex E. Crick
Plenum Press, New York

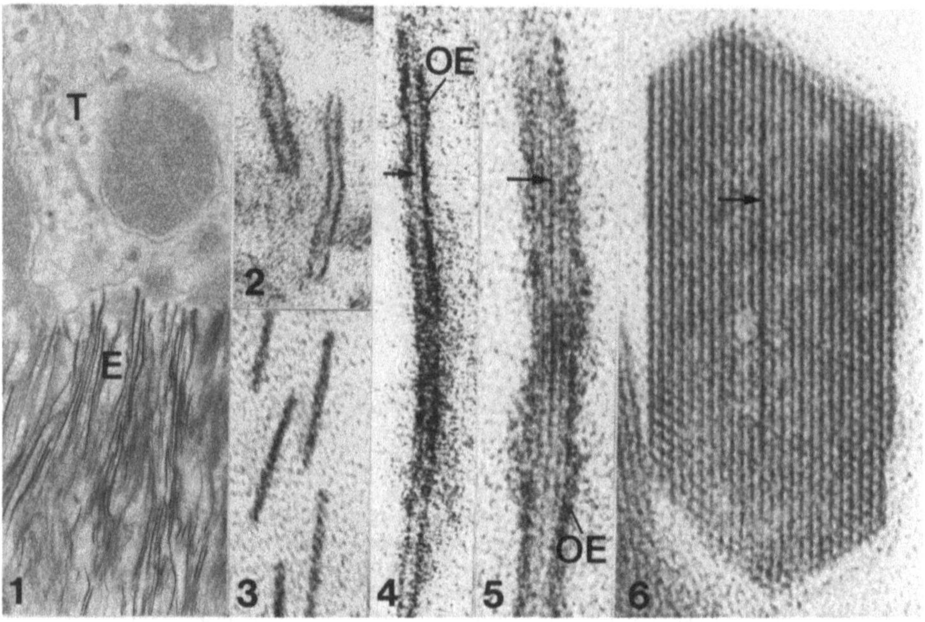

FIGURES 1–6. ▪*1.* Section of early enamel (matrix formation stage) of a monkey (*Macaca fascicularis*), showing ribbon-shaped early minerals (E) and Tomes process of ameloblast (T). Uranyl acetate-lead citrate double stain. × 20,000. ▪*2.* Early enamel of rat incisior, higher magnification of doubly stained section, showing cross cut of ribbon-shaped elements that consist of an electron dense organic envelope and a relatively electron lucent inner zone (precursor mineral). × 800,000. ▪*3.* Unstained section of early enamel of rat lower incisor. The organic envelope is not clearly seen (*cf.* Fig. 2). Early mineral portion appear relatively electron dense. × 700,000. ▪*4.* Double stained section of cat enamel showing a longitudinal cut of ribbon-shaped structure. Arrow indicates sharp central line which is the first appearance of *CDL*. OE: organic envelope. × 1,100,000. ▪*5.* Longitudinal cut of a crystal in early maturation stage enamel of cat. Double stain. 100 fringes of 0.82 nm periodicity are arranged parallel to the *CDL* (arrow). This micrograph also indicate that double staining does not cause dissolution of the apatite crystals. OE: organic envelope. × 1,700,000. (Figures 3–5 from Nakahara, 1983). ▪*6.* Nearly mature crystallite of rat enamel, cut transversely in reference to the longitudinal axis of crystal. *CDL* (arrow) does not reach the periphery of the crystallite. Unstained. × 1,900,000.

constituents of the cytoplasmic origin in an attempt to relate them to the possible role they may play in the process of mineralization.

CRYSTALLITES AND THE ORGANIC ENVELOPES OF DEVELOPING ENAMEL

The important features we have observed in the developing enamel are (1) the central dark linecentral dark line (*CDL*), and (2) the organic envelope. The *CDL* has long been a controversial feature in as much as high-magnification observations failed to reveal this structure. It has been shown however in mature human enamel (Marshall & Lawless, 1981) and in developing rat and kitten enamel (Nakahara & Kakei, 1983, 1984b) at relatively high magnification. We have suggested that electron beam damage after prolonged beam exposure may be the main reason why some studies failed to demonstrate the *CDL* (Nakahara & Kakei, 1983).

TEM observations of early developing enamel indicates the ribbon-shaped structure (Fig. 1) that appears in the earliest mineralized enamel is composed of an inner mineral portion devoid of a crystalline lattice, and surrounding organic envelope (Fig. 2). Unstained sections of the same materials show moderately electron dense rod-shaped structures that apparently represent the inner mineral portion appearing as a somewhat fuzzy outline; the organic envelopes are not clearly shown (Fig. 3). These observations indicate that the earliest mineral laid down in developing enamel is not crystalline, further, these non-crystalline mineral and organic envelope are probably forming a molecular complex.

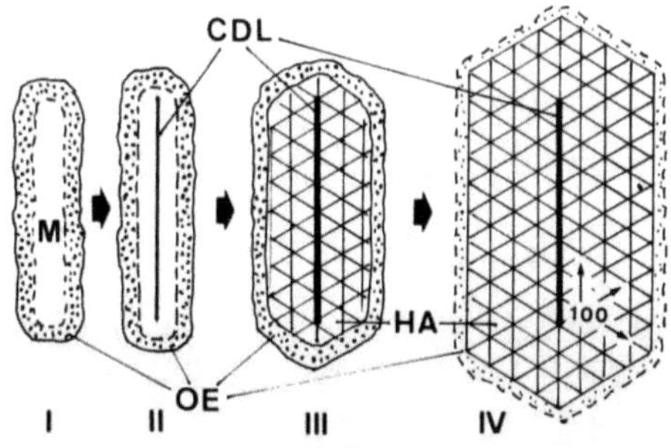

FIGURE 7. Schematic illustration of the formation of enamel crystallite drawn from the sections cut perpendicular to the *c*-axis of apatite crystals. These models, especially step I to III, also represent the crystal formation of dentin, bone or baleen. **OE**: organic envelope. **M**: mineral portion of the early organic-mineral complex which is ribbon-shaped in enamel and disc-shaped in dentin, bone or baleen. *CDL*: Central dark line. **HA**: hydroxyapatite crystal. 100: (100) fringes of hydroxyapatite.

The first sign of crystal formation in the ribbon-like "envelope-mineral complex" is the appearance of sharp line located at the center of the inner zone composed of mineral (Fig. 4). Later, as the crystals grow, the line apparently become the *CDL* (Figs. 5 & 6). In the nearly mature enamel, apatite crystals are slender rod-shaped structures, with their *c*-axis parallel to the long axis; in cross section they appear roughly as an elongated hexagon. The high magnification image of enamel apatite in cross-section to the long axis of the crystallite exhibits 3 sets of 100 fringes (0.82 nm spacing) arranged at an angle of 60° to each other. Among these 100 fringes, a line located at the center which always appears as the longest diagonal line in the elongated hexagon is more apparent than the other lines and represents the *CDL* (Figs. 6 & 7). The *CDL* does not reach periphery of the crystal, it ends within crystallites; this, because the *CDL* is the initiation site of crystal formation (Marshall & Lawless, 1981) and is always positioned at the center of the crystallites during their growth (Nakahara & Kakei, 1983, 1984b). This also explains why the cross section of enamel crystal (cross to *c*-axis) always appears as an elongated hexagon instead of an equilateral hexagon.

There are some definite indications that the area of the *CDL* contains a higher concentration of carbonate ion than the other parts of the crystallite (Marshall & Lawless, 1981; Nakahara & Kakei, 1984b), this problem will be discussed later in connection with the formation of other hard tissues.

CENTRAL DARK LINE AND ORGANIC ENVELOPE OF THE CRYSTALLITE OF DENTIN AND BONE

The features observed in enamel crystallites previously referred to were also demonstrated in developing and mature apatite crystallites of dentin and bone (Nakahara, 1982; Nakahara & Kakei, 1984a). The matrix of these tissues contain numerous collagen fiber bundles, a high magnification resolution of these tissues is not

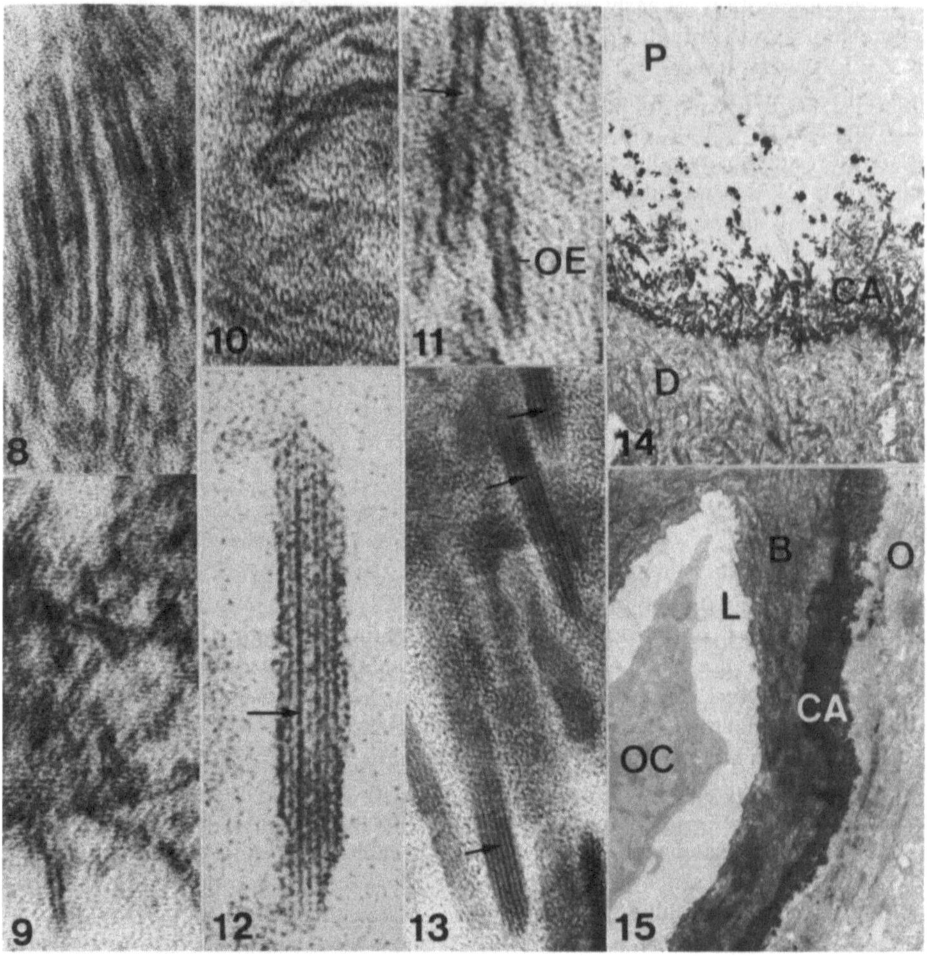

FIGURES 8–15. ▪*8.* Early calcified area of dentin; double stain. Thin plates consist of relatively electron lucent inner plate (precursor mineral) and surrounding electron dense organic envelope. × 700,000. ▪*9.* Calcifying bone (calvaria) of 6 day-old rat; double stain, showing plate-shaped elements consisting of an electron dense organic envelope and a relatively electron lucent inner mineral layer. × 900,000. ▪*10.* Unstained rat calvaria showing moderately electron dense mineral plates. Organic envelope is not clearly shown. × 900,000. ▪*11.* Dentin crystallites of rat approximately 5 μm removed from the calcification front. A dark line (arrow) appears at the center of inner mineral plate surrounded by an envelope (OE). Double stain. × 1,500,000. (Nakahara & Kakei, 1984a). ▪*12.* Side view of a plate-shaped mature human dentin crystallite, showing *CDL* (arrow) which does not reach the periphery of crystal. Lines parallel to *CDL* are ordinary 100 fringes. × 1,700,000. (Nakahara, 1982). ▪*13.* Calvaria of mature rat, showing *CDL* (arrows) and ordinary 100 fringes (parallel to *CDL*) in the side view of crystallites. Unstained. × 1,000,000. (Nakahara & Kakei, 1984). ▪14. Area of dentin (D) and predentin (P) of rat incisor, showing carbonic anhydrase activity (CA) as dark deposit. The enzyme activity is localized only at the very early mineralized zone of dentin where the crystal nucleation is supposed to be taking place. Slightly stained with double stain. × 8,000. ▪*15.* Calvaria of 6 day-old rat, showing carbonic anhydrase activity (CA) as dark deposit at the growth surface (calcification front) of calcified bone. Note that the inner surface of lacuna (L) does not show any enzyme activity. **B**: calcified bone, **O**: osteoid, **OC**: osteocyte. Double stain. × 7,500.

as clear as that of enamel. The shape of the crystals in the mature state of dentin and bone are reported to be an irregularly contoured plate of diverse sizes (5 to 70 nm in the longest dimension). Accordingly the side view shows a rod-like appearance of varying lengths (Johansen & Parks, 1960; Bocciarelli, 1970; Nakahara, 1982). The *CDL* of dentin and bone crystallites is clearly observed in the rod-shaped side view, and appears as one of the 100 fringes at the center, or in some cases off the center (Figs. 12 & 13). As in the case of enamel, the *CDL* does not reach the peripheral edge of the crystallite; it ends within crystallite, leaving a line-free area at the periphery. The actual shape of the *CDL* of dentin or bone therefore is disk-like, and is confined within the crystallite (Nakahara, 1982).

The organic envelope similar to that observed in developing enamel also encloses the dentin and bone crystallites (Nakahara & Kakei, 1984a). In high magnification electron micrographs of developing dentin or bone, particularly at the growing surface (calcification front), we can observe the clusters of the earliest mineral deposit containing the profile of early envelope-mineral complex similar to that observed in enamel. In the doubly stained sections stained with uranyl acetate-lead citrate, the complex is composed of an inner precursor mineral plate (relatively electron lucent), measuring 0.5 to 0.8 nm in thickness and a surrounding organic envelope 0.7 to 1 nm in thickness (Figs. 8 & 9). Unstained sections do not show the envelope; they show however moderately electron-dense rod-shaped structures of inner mineral portion which always show fuzzy outline (Fig. 10).

The appearance of the *CDL* within the mineral portion of the envelope-mineral complex (Fig. 11) occurs at some distance from the calcification front; 3 to 5 μm in rat incisor dentin, 1 to 2 μm in calvaria of 6-day-old rat. Subsequently, ordinary 100 fringe lines of apatite appear successively parallel to the initial central line. In mature dentin or bone 6 to 8 lines are observed most frequently in crystallites viewed from the side (Figs. 12 & 13).

CENTRAL DARK LINE AND CARBONATE

There are some indications that the area of *CDL* contains higher concentration of the carbonate ion than the other areas of the apatite crystallite (Marshall & Lawless, 1981; Nakahara, 1982; Nakahara & Kakei, 1984 a,b). The reasons for this are as follows:

1. During development, the carbonate concentration in the mineral portion of enamel (Hiller *et al.*, 1975; Robinson *et al.*, 1979) or dentin (Quint *et al.*, 1980) decreases as maturation proceeds. This means that younger crystals which contain a higher proportion of *CDL* area, contain higher percentage of the carbonate ion.

2. The central portion (area of *CDL*) is reported to be the initial site of dissolution during carious alteration in enamel (Vogel & Frank, 1977a; Arends, 1982) as well as in dentin (Takuma, 1971; Vogel & Frank, 1977b; Nakahara, 1982). The carbonate concentration decreases sharply in the early stage of carious alteration in enamel (Hallsworth *et al.*, 1972) and dentin (Johansen, 1963).

3. It should also be noted that the inorganic portion of enamel contains less carbonate (2 to 3%) than that of dentin or bone (about 3.5%). Since enamel crystals are larger than the others, enamel should have a lesser *CDL* area.

4. Carbonic anhydrase which accelerates the formation of carbonate or bicarbonate ion from carbon dioxide, is found in the area where crystal nucleation is apparently taking place in enamel, dentin and bone. This topic will be elaborated in the following paragraphs.

CARBONIC ANHYDRASE

Since the *CDL* serves as the site of crystal initiation and the area of the *CDL* contains a higher concentration of the carbonate ion, relatively large quantities of carbonate ion should be produced in the very limited area of calcification front zone of enamel, dentin and bone. To explain the cause of sudden increase in the supply of the carbonate ion, we suspected that the presence of carbonic anhydrase activity in the calcification front area might account for this phenomenon. Indeed, histochemical tests, revealed the enzyme activity in enamel, dentin, bone and calcifying cartilage as follows: in growing dentin and bone, the narrow surface zone at calcification front shows highly positive reaction of carbonic anhydrase (Fig. 14 & 15) (Kakei & Nakahara, 1983a,b, 1984; Takeyama & Kiyomura, 1984). In enamel, only the surface zone (near Tomes' process of ameloblast) of early stage (matrix formation stage) shows a positive reaction (Kakei & Nakahara, 1983a,b; Nakahara & Kakei, 1984b). All the enzyme positive areas correspond to the site where the initiation of crystal growth occurs. The immunoblotting method has shown that carbonic anhydrase is also one of the major protein components of early enamel (matrix formation stage) (Fig. 20 & 21) and calcifying dentin (Fig. 20) (Kakei & Nakahara, 1985 a,b).

According to the results obtained in the study referred to above we believe that the carbonic anhydrase activity is involved in the initiation of the crystal growth of apatite. Further the function of the enzyme is to supply carbonate ion to the early envelope-mineral complex which subsequently give rise to initiate crystal formation.

BALEEN

The organic envelope and the *CDL* are found in an unusual type of hard tissue, baleen, which is located in the oral cavity of whalebone baleen. Calcification occurs entirely within keratinized epidermal cells (Pautard, 1963; Halstead, 1974). We have obtained a specimen of baleen from sei whale; *Baraenoptera borealis* which was sectioned by routine methods for TEM observation. Low magnification electron micrographs show the epidermal (heavily hornified) cells probably do not have an active metabolizing function. they do however contain various concentrations of mineral deposits (Figs. 16-19) which are reported to be composed of hydroxyapatite crystallites (Pautard, 1963). Our observations (unpublished) show that the mineral crystallites of baleen are essentially similar to that of dentin or bone, they are platelet-shaped and surrounded by an organic envelope (Fig. 17). Unstained sections show mineral portions similar shape to those observed in other hard tissues (Fig. 18). A side view of these crystallites show varying thicknesses indicating of diverse stages of mineral growth in the cell. In some of the thinnest platelets the central line is lacking. Thicker platelets show a *CDL (Fig. 17), and the more developed crystallites show several 100 fringes of 0.82 nm intervals parallel to the CDL* (Fig. 19).

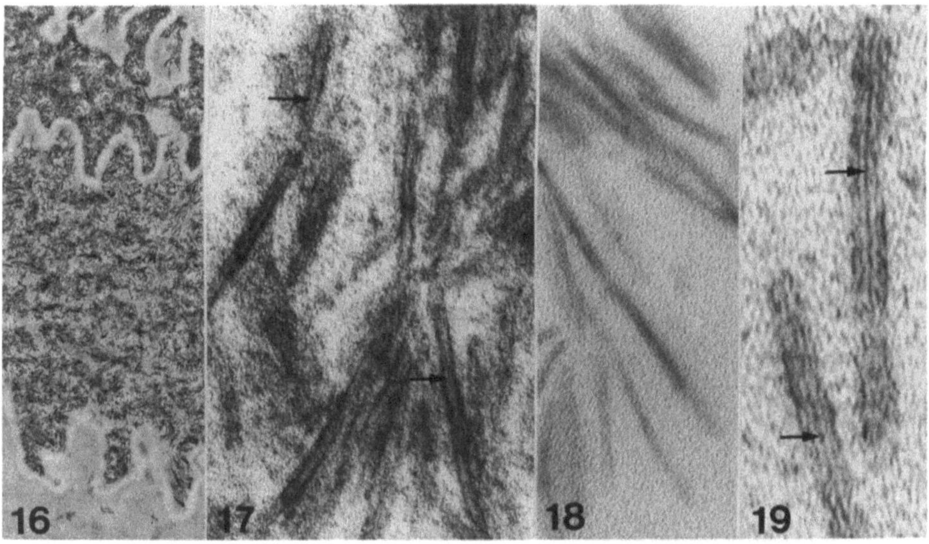

FIGURES 16–19. ▪*16.* Showing keratinized epidermal cells containing mineral deposits. Double stain. × 20,000. ▪*17.* Plate-shaped early minerals (envelope-mineral complex) composed of electron dense organic envelope and inner mineral plate. Some mineral plates show *CDL* (arrows). Double stain. × 600,000. ▪*18.* Unstained section showing mineral plates only. Organic envelopes are not clearly seen. × 600,000. ▪*19.* High magnification of mature crystals, showing *CDL* (arrows) and ordinary 100 planes (parallel to *CDL*) of hydroxyapatite. Double stain. × 1,500.000.

Though the apatite crystals of baleen are formed entirely within the cell boundary, our TEM observation reveal that the organic envelope and the *CDL* are present, similar to the situation in early enamel, dentin and bone which form crystallites outside the cell. Since intercellular substance is almost nil in baleen tissue (Fig. 16), the organic substances associate with or responsible for mineral formation could not be supplied from the outside of cell, they could only be derived from cytoplasmic components. This problem will be discussed in following paragraphs.

PROTEINS

It has been clearly shown in the preceding description that all the apatite forming tissues we have examined, *i.e.*, enamel, dentin, bone, calcifying cartilage and baleen, all show essentially a similar process of crystal nucleation and growth. The points of similarities are: 1) presence of the *CDL* which represents the initiation site of crystal growth and probably contains a high concentration of carbonate, 2) presence of the carbonic anhydrase activity in the area where crystal nucleation is taking place as observed in enamel, dentin, bone and calcifying cartilage and the enzyme appears to be responsible for supplying the carbonate ion to the *CDL* area of the earliest mineral phase, and 3) presence of the organic envelope which, together with the precursor mineral, form an envelope-mineral complex in the early calcifying stage.

These basic similarities suggest that the initiation and subsequent early growth of crystals are governed by a similar mechanism and mediated by similar organic substan-

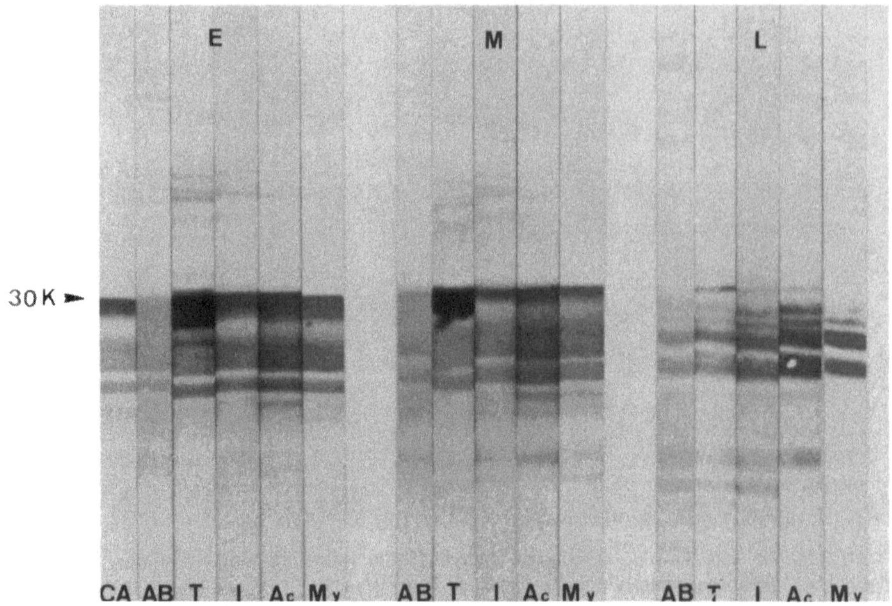

FIGURE 20. Electrophoretic blotting of one dimensional gel using antibodies specific for carbonic anhydrase (**CA**), troponin-T (**T**), troponin-I (**I**), actin (**Ac**) and myosin (**My**), respectively. Developing enamel matrix proteins of each stage were excised from root apex (E), middle (M) and white opaque (L) regions, represent early, middle and late stage respectively. **AB**: Amido black staining. (Kakei & Nakahara, 1985b).

ces; this also suggests that the hard tissues should have utilized similar substances to form organic envelopes and other organic structures associated with mineral formation during their phylogenetic and ontogenetic development despite the difference in embryological origin. We have also predicted that the proteins essential for mineralization are probably derived from widely distributed substances in the biological system; the most likely candidates are basic cytoplasmic components (cytoskeletal proteins).

It was previously suggested by Lowenstam & Margulis (1980a,b) that some intracellular proteins responsible for calcium regulation such as troponin may be involved in establishing the biomineralization mechanism during the very early stage of organic evolution. Although the vertebrate appeared in later evolutional period, it may be more reasonable to consider that the mineral formation of each hard tissue such as enamel, dentin, bone or baleen evolved independently of one another.

Based on the concept referred to above we have attempted to ascertain whether the basic cytoplasmic proteins (actin, myosin, troponin T, troponin I and calmodulin tested to-date) are contained in the organic matrix of developing enamel and calcifying dentin. Enamel and dentin was utilized in as much as these tissues are relatively easy to obtain from laboratory animals (rat incisor), and easy to avoid contamination. To identify each cytoplasmic protein we utilized an immunoblotting method (Kakei & Nakahara, 1985b).

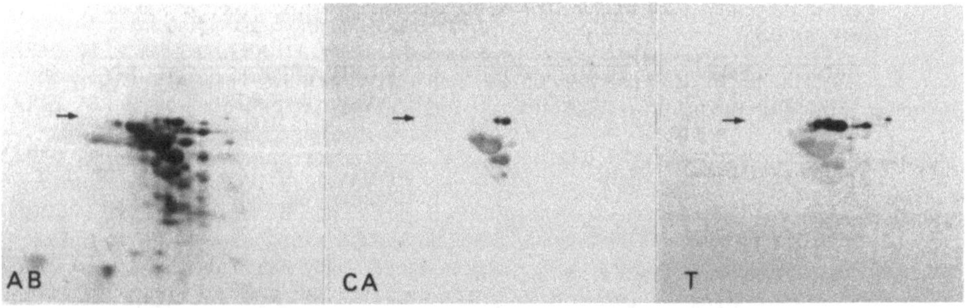

FIGURE 21. Electrophoretic blotting of two-dimensional gel of the rat immature enamel matrix proteins extracted with dissociation buffer containing 1% SDS, 1% β-mercaptoethanol, 10% glycerol, 0.01M sodium phosphate, pH 7.1. Isoelectric focusing was performed according to O'Farrell (1975) with some modification. The second dimension was performed in 12% slab gel according to Laemmli (1970). Electrophoretic blotting was performed by the method of Towbin *et al.*, (1979). After electrophoretic transfer of proteins to nitrocellulose membrane, the membranes were incubated with a 1:1,000 dilution of 1st antibodies for 2 h individually. The membranes were washed, and then incubated with a 1:2,000 dilution of 2nd antibody for 1 h. Next these were washed, incubated with a 1:1,000 dilution of rabbit PAP complex reagent and the color developed as described previously (Kakei & Nakahara, 1985a, b). **AB**: Amido black staining showing whole protein, **CA**: Antibody against carbonic anhydrase, **T**: Antibody against troponin-T. Arrows indicate 30K daltons molecular weight.

Enamel proteins are generally classified into two groups: 1) amelogenin which dominates the earlier stage of developing enamel, and 2) enamelin which is prominent in the later (maturing) stage of enamel (Eastoe, 1979; Fincham, 1984; Deutsch *et al*, 1984). However, these proteins appear to be composed of several different sub-groups, further, their activity in connection with the mineral formation is virtually unknown.

The results of our immunoblotting test (Fig. 20 & 21) indicates that actin, myosin, troponin T and troponin I make up part of the matrix of calcifying enamel (Kakei & Nakahara, 1985b); actin and troponins are also components of calcifying dentin matrix (unpublished). These results also indicate that later stages of enamel matrix contain a higher content of EDTA insoluble proteins (Table 1) which probably are the components of the scaffolding of the organic matrix. Another protein we have detected is calmodulin, an acidic calcium binding protein. This protein occurs only in the later (maturing) stage of enamel formation indicating that the crystal growth of the later enamel formation may be controlled by rather different mechanism from that of early enamel or calcifying dentin (Kakei & Nakahara, 1986). The results of immunoblotting are summarized in Table I. The immunoblotting patterns of each protein show multiple bands in one dimensional strips (Fig. 20) or multiple spots in two dimensional sheets (Fig. 21) instead showing single proper bands: this indicate these cytoplasmic proteins contained in the calcifying matrix are probably in a more or less altered state, differing the original intra-cytoplasmic condition. The alteration may be due to the complicated process of degradation and recombination among the proteins. It is well known that some of these proteins have a calcium-binding or phosphate-modulating function which may be related to apatite formation. However, much more studies need to be done in order to obtain full understanding of the process of mineralization.

TABLE 1 - Summarized Immunoblotting Results. Adapted from Kakei & Nakahara (1985a,b; 1986)

RAT INCISOR MATERIALS:	DEVELOPING ENAMEL		DEVELOPING ENAMEL (WHOLE)		CALCIFY-ING DENTIN
Anti-proteins tested	Early	Late	EDTA soluble	EDTA insoluble	
Carbonic anhydrase	+ + +	--	+ + +	--	+ +
Troponin T	+ + +	+	+	+ + +	(+)
Actin	+ +	+	+	+ +	(+)
Myosin	+	+	+	+ +	(-)
Calmodulin	--	+ +	--	+ +	(-)

CONCLUSIONS

The *CDL*, the organic envelope, carbonic anhydrase and some proteins of apparent cytoplasmic origin are common features in apatite forming tissues of vertebrates. The *CDL* which represents the site of crystal initiation contains high concentrations of the carbonate ion. Carbonic anhydrase probably supplies carbonate for initiating crystals. Carbonic anhydrase and cytoplasm-originated proteins are the major components of the calcifying matrix.

ACKNOWLEDGEMENTS

This research was supported in part by a grant from the Ministry of Education of Japan; No 60304014. We are grateful to Dr. Gerrit Bevelander for his valuable advice. Participation in the Symposium was made possible by grants from The Organized Research Fund of The University of Texas at Arlington and The Sea Grant Program administered by Texas A & M University.

REFERENCES

ARENDS, J., 1982. Mechanism of dental caries. *In* Biological Mineralization and Demineralization (ed. G.H. Nancollas), pp. 303-324. Berlin: Springer-Verlag.

BOCCIARELLI, D.S., 1970. Morphology of crystallites in bone. Calc. Tiss. Res., 5:261-269.

DEUTSCH, D., SHAPIRA, L., ALAYOFF, A., LEVIEL, D., YOELI, Z. & ARAD, A., 1984. Protein and mineral changes during prenatal and postnatal development and mineralization of human deciduous enamel. *In* Tooth Enamel IV (ed. R.W. Fearnhead & S. Suga). pp. 234-239, Amsterdam: Elsevier Science Publishers.

EASTOE, J.E., 1979. Enamel protein chemistry-past, present and future. J. Den. Res., 58 (B):753-763.

FINCHAM, A.G., 1984. Amelognins: Progress and problems. *In* Tooth Enamel IV (ed. R.W. Fearnhead & S. Suga), pp. 114-119, Amsterdam: Elsevier Science Publishers.

HALLSWORTH, A.S., WEATHERELL, J.A. & ROBINSON, C., 1973. Loss of carbonate during the first stage of enamel caries. Caries Res., 7:345-348.

HALSTEAD, L.B., 1974. Vertebrate Hard Tissues. pp. 100-102. London: Wykeham Publications.

HILLER, C.R., ROBINSON, C. & WEATHERALL, J.A., 1975. Variations in the composition of developing rat incisor enamel. Calc. Tiss. Res., 18:1-12.

JOHANSEN, E., 1963. Ultrastructural and chemical observations on dental caries. *In* Mechanisms of Hard Tissue Destruction (ed. R.F. Sognnaes), pp. 187-211, Washington, D.C., Amer. Assoc. Adv. Sci.

JOHANSEN, E. & PARKS, H.F., 1960. Electron microscopic observations on the three-dimensional morphology of apatite crystallites of human dentin and bone. J. Biophys. Biochem. Cytol., 7:743-746.

KAKEI, M. & NAKAHARA, H., 1983a. A light microscopic study of the localization of carbonic anhydrase activity in the developing dentin and enamel of the rat lower incisor. Jap. J. Oral Biol., 25:374-377.

KAKEI, M. & NAKAHARA, H. 1983b. Ultrastructural localization of carbonic anhydrase activity in developing enamel and dentin of the rat incisor. Jap. J. Oral Biol., 25:1129-1133.

KAKEI, M., & NAKAHARA, H., 1984. Histochemical localization of carbonic anhydrase activity in epiphyseal growth cartilage and calvaria of rat. Jap. J. Oral Biol., 26:554-558.

KAKEI, M., & NAKAHARA, H., 1985a. Electroimmunoblotting study of carbonic anhydrase in developing enamel and dentin of the rat lower incisor. Jap. J. Oral Biol., 27:357-361.

KAKEI, M., & NAKAHARA, H., 1985b. Demonstration of cytoplasmic structural proteins in developing enamel matrix of the rat incisor. Jap. J. Oral Biol., 27:1001-1005.

KAKEI, M. & NAKAHARA, H., 1986. Identification of calmodulin in developing enamel matrix by the immunoblotting technique. Bull. Josai Dent. Univ., 15:309-312.

LAEMMLI, U.K., 1970. Cleavage of structural proteins during the assembly of the head of bacteriophage T4. Nature, 227:680-685.

LOWENSTAM, H.A. & MARGULIS, L., 1980a. Calcium regulation and the appearance of calcareous skeleton in the fossil record. *In* The Mechanisms of Biomineralization in Animals and Plants (ed. M. Omori & N. Watabe), pp. 289-300, Tokyo: Tokai University Press.

LOWENSTAM, H.A. & MARGULIS, L., 1980b. Evolutionary prerequisites for early phanerozoic calcareous skeletons. BioSystems, 12:27-41.

MARSHALL, A.F. & LAWLESS, K.R., 1981. TEM study of the central dark line in enamel crystallites. J. Den. Res., 60:1773-1782.

NAKAHARA, H., 1982. Electron microscopic studies of the lattice image & "central dark line" of crystallites in sound and carious human dentin. Bull. Josai Dent. Univ., 11:209-215.

NAKAHARA, H. & KAKEI, M., 1983. The central dark line in developing enamel crystallite: An electron microscopic study. Bull. Josai Dent. Univ., 12:1-7.

NAKAHARA, H., & KAKEI, M., 1984a. TEM observations on the crystallites of dentin and bone. Bull. Josai Dent. Univ., 13:259-263.

NAKAHARA, H. & KAKEI, M., 1984b. Central dark line and carbonic anhydrase: Problems relating to crystal nucleation in enamel. *In* Tooth Enamel IV. (ed. R.W. Fearnhead, & S. Suga) pp. 42-46, Amsterdam: Elsevier Science Publishers.

O'FARRELL, P.H., 1975. High resolution two-dimensional electrophoresis of proteins. J. Biol. Chem., 250:4007-4021.

PAUTARD, F.G.E., 1963. Mineralization of keratin and its comparison with the enamel matrix. Nature, 199:531-535.

QUINT, P., ALTHOFF, J., HÖHLING, H.J., BOYDE, A. & LAABS, W.A., 1980. Characteristic molar ratios of magnesium, carbon dioxide, calcium and phosphorus in the mineralizing fracture callus and predentin. Calc. Tiss. Intern., 32:257-261.

ROBINSON, C., BRIGGS, H.D., ATKINSON, P.J. & WEATHERELL, J.A., 1979. Matrix and mineral changes in developing enamel. J. Den. Res., 58(b):871-880.

TAKEYAMA, H. & KIYOMURA, H., 1984. A histochemical study of carbonic anhydrase activity in alveolar bone of the rat. J. Jap. Orthod. Soc., 43:356-360.

TAKUMA, S., 1971. Dentin caries. *In* A New Atlas of Dental Pathology, 2nd Edition (ed. S. Takuma), pp. 138-183. Tokyo: Ishiyaku Shuppan. (In Japanese).

TOWBIN, H., STAEHELIN, T. & GOROLON, J., 1979. Electrophoretic transfer of proteins from polyacrylamide gels to nitrocellulose sheets: Procedure and some applications. Proc. Nat. Acad. Sci. U.S.A., 79:4350-4354.

VOGEL, J.C. & FRANK, R.M., 1977a. Stages in the dissolution of human enamel crystals in dental caries. Calc. Tiss. Res., 24:19-27.

VOGEL, J.C. & FRANK, R.M., 1977b. Ultrastructural study of apatite crystal dissolution in human dentin and bone. J. Biol. Buccale, 5:181-194.

THE PHOSPHATIC MODE OF CALCIFICATION IN ONTOGENY AND PHYLOGENY OF THE INTEGUMENT AND SKELETON OF VERTEBRATES

Norman E. Kemp

Department of Biology, The University of Michigan, Ann Arbor, Michigan 48109

ABSTRACT. The earliest known fossil vertebrates, the ostracoderms, had already evolved the principal vertebrate hard tissues of calcified cartilage, bone, dentine and enamel (enameloid). These tissues develop with the calcium phosphate mineral hydroxyapatite as the inorganic constituent of their matrices. As suggested by Gans & Northcutt (1983), adoption of this phosphatic mode of mineralization may have resulted from evolution of a new type of head featuring the emergence of neural crest tissue and epidermal placodes as the vertebrate progenitors diverged from protochordate ancestors. Mineral crystallites in the hard tissues undergo nucleation and growth in close association with protein fibrils or tubular sheaths polymerized either from collagen as in the dermal or endoskeletal matrix of scleroblasts in calcified cartilage, bone, dentine and tooth cementum, or from the amelogenins and enamelins of the enameline matrix secreted by ameloblastic epithelial cells.

Development of mineral crystallites has been investigated in electron microscopic studies of the hard tissues of frogs and sharks, aquatic vertebrates considered representative of the vertebrate patterns of mineralization. Thyroxine-treated frog tadpoles were utilized for tracing the progress of deposition of hydroxyapatite crystallites in association with collagen fibrils during development of bone in the femur. Shark jaw endoskeleton and teeth were investigated to elucidate differentiation of calcified cartilage, dentine and enamel. Crystallites of calcified cartilage are small and needle-like similar to those in bone and dentine. The crystallites of enamel develop within tubular sheaths, grow much larger than those of the other hard tissues, and assume the hexagonal shape considered to be characteristic of apatite crystals.

INTRODUCTION

The earliest fossils we recognize as "vertebrate" are those of the agnathan ostracoderms first appearing in strata of the Late Cambrian Period more than 500 million years ago (Kemp, 1984). One of the great mysteries surrounding the divergence of vertebrates from their invertebrate ancestors is why the vertebrates adopted the calcium phosphate mineral hydroxyapatite for hardening their exoskeletons and thereby abandoned the typically invertebrate exoskeletal minerals, calcite, aragonite and vaterite. Some invertebrates in fact do utilize calcium phosphate in their skeletons, including certain species of protozoans, corals, inarticulate brachiopods (*e.g.*, *Lingula*) and the protostomian phyla annelids, molluscs and arthropods (Pautard, 1976), but these

Origin, Evolution, and Modern Aspects of Biomineralization in Plants and Animals
Edited by Rex E. Crick
Plenum Press, New York

are outside of the deuterostomian line of evolution leading to the vertebrates (Lehman, 1977).

Early deuterostomes, the echinoderms, have dermal ossicles of calcium carbonate enriched with magnesium (Barnes, 1974). Other invertebrate deuterostomes, chaetognaths, pogonophorans and protochordates, are soft-bodied, lacking calcified exoskeletons. Ancestral protochordates had already evolved the notochord, dorsal nerve cord and gill slits characteristic of the phylum Chordata, but vertebrates represented a significant departure from the invertebrate way of life. What started the ball rolling in our direction, according to the innovative speculation of Gans & Northcutt (1983) was the evolution of a new type of head enabling a changeover from passive filter feeding to active, predatory suction feeding. Sense organs, muscles and connective and supportive tissues were all involved in the transition. Neural crest tissue and epidermal placodes, which gave rise to head mesenchyme and to sensory epithelium and ganglia were key inventions of the vertebrate lineage. Both neural crest and placodes produce cells which may secrete mineralizing matrices. Gans & Northcutt (1983) speculate that the earliest step in vertebrate mineralization may have been the calcification of the matrix around electroreceptor cells of the head epithelium. This hard material would have had dielectric properties serving to enhance the electrical gradient between internal and external portions of the cells. As they express it (Northcutt & Gans, 1983: 17), "In this view, crystallization produced or induced by these sensory cells represents the earliest calcification of the integument." Once under way, dermal mineralization would have conferred the additional benefits of support and protection, resistance to osmotic water intake, and storage of phosphate ions (Berrill, 1955).

Whatever the pattern of origin, four types of mineralized tissue were present in ostracoderms. Their integumental armor consisted of bone in their dermal plates, scutes and scales, which often bore rounded protuberances known as odontodes. The latter were composed of dentine-like tissue capped with a layer of enamel (enameloid); thus they were primitive forerunners for the evolution of vertebrate scales and teeth (Ørvig, 1977). Ostracoderms also had mineralized plaques of calcified cartilage, known as endoskeletal tesserae, in their head skeletons (Denison, 1967; Kemp, 1985).

Bone, dentine and calcified cartilage are collagen-associated hard tissues, *i.e.*, their mineral component is associated with collagen in the matrix secreted by their scleroblasts the osteoblasts, odontoblasts, and chondroblasts respectively. Enamel is a markedly different kind of hard tissue, which may be described as amelogenin-enamelin-associated. The glycoproteins known as amelogenins and enamelins, which are associated with enamel crystallites, are secreted by ameloblasts derived from the inner dental epithelium. In mammalian teeth the enamel crystallites become hexagonal and grow to be very much larger than the crystallites in other hard tissues (Arends *et al.*, 1983).

Despite obvious differences in the size and shape of mineral crystallites in enamel and the collagen-associated tissues, it is remarkable that the chemical composition of the crystallites is the same in all of the hard tissues. Mineral crystallites in all of the principal hard tissues of vertebrates are composed of calcium phosphate, which may go through an amorphous phase of deposition but which is eventually deposited as hydroxyapatite. This mineral type also characterizes the otoliths (statoconia) in the ear labyrinths of cyclostomes, the most primitive living vertebrates (Hardisty, 1979). Beginning with elasmobranch fishes, however, otoliths are an exception to the phosphatic mode of mineralization in vertebrates. Instead of apatite the mineral comprising the otoliths of most vertebrates is composed of calcium carbonate. Evidently the fluid milieu of the ear labyrinth lacks the necessary matrix components for inducing calcium phosphate crystallization.

In addition to the principal proteins (collagen or amelogenins and enamelins) hard tissue matrix components include carbohydrates complexed with proteins as proteoglycans and glycoproteins. At least in the collagen-associated tissues these complexes are probably degraded as a prelude to mineralization. Characteristic phosphoproteins have been identified among the non-collagenous proteins of bone and dentine matrix, and are also found in enamel matrix. Somehow these organic constituents of the matrix facilitate the union of Ca^{2+} ions and orthophosphate ions ($H_2PO_4^-$, HPO_4^{2-}, and PO_4^{3-}). At physiological pH, these ions interact and bind OH^- ions to produce hydroxyapatite, $Ca_{10}(PO_4)_6(OH)_2$, which may also attract substitution ions such as Mg^{2+}, Fl^-, and CO_3^{2-} (Neuman & Neuman, 1958; Kemp, 1984).

In order to understand the mechanism of mineralization in tissues, we must elucidate what causes the nucleation of mineral crystallites in their extracellular matrices, and also what regulates crystal growth. If mineral turnover occurs, we need to know how the crystallites are resorbed and their ions recycled. Probably the basic mechanisms of tissue mineralization are homologous in all vertebrates; hence I will consider these mechanisms from the standpoint of previous investigations on mineralizing systems in representative aquatic vertebrates, namely, frogs and sharks (Kemp & Hoyt, 1969; Kemp & Park, 1974; Kemp & Westrin, 1979; Kemp, 1984, 1985).

MATERIALS AND METHODS

Development of hydroxyapatite crystals has been studied in thyroxine-treated frog larvae. Tadpoles of *Rana pipiens* were induced to undergo precocious metamorphosis by treatment with 1-thyroxine solution (6.25×10^{-8} M) beginning at Taylor Kollros stage IX prior to the onset of ossification in the femur (details in Kemp & Hoyt, 1969). Limbs fixed in glutaraldehyde and OsO_4 were harvested daily for a period up to 10 days following initial exposure to thyroxine. Tissue was processed for both light and electron microscopy. Epon-embedded tissue was sectioned through the femur with a diamond knife mounted on an LKB ultratome. Thin sections were stained with uranyl acetate and examined with an RCA EMU 3E electron microscope operating at 50 kV.

Structure and development of calcified cartilage were investigated in endoskeletal tissues of three shark species, *Carcharhinus menisorrah*, *Triaenodon obesus* and *Negaprion brevirostris* (Kemp & Westrin, 1979). Skeletal tissues from the jaws, fins, gill arches and vertebrae were fixed and processed for light microscopy and for both scanning electron microscopy (SEM) and transmission electron microscopy (TEM).

Development of dentine and enamel was investigated in teeth of the same shark species (Kemp & Park, 1974; Kemp, 1985). The principal emphasis in these studies was on enamel differentiation because of the long-standing controversy over the homology of enamel in elasmobranch and mammalian teeth. According to some authors, enamel of both shark and mammalian teeth is derived from epithelial ameloblasts; hence shark

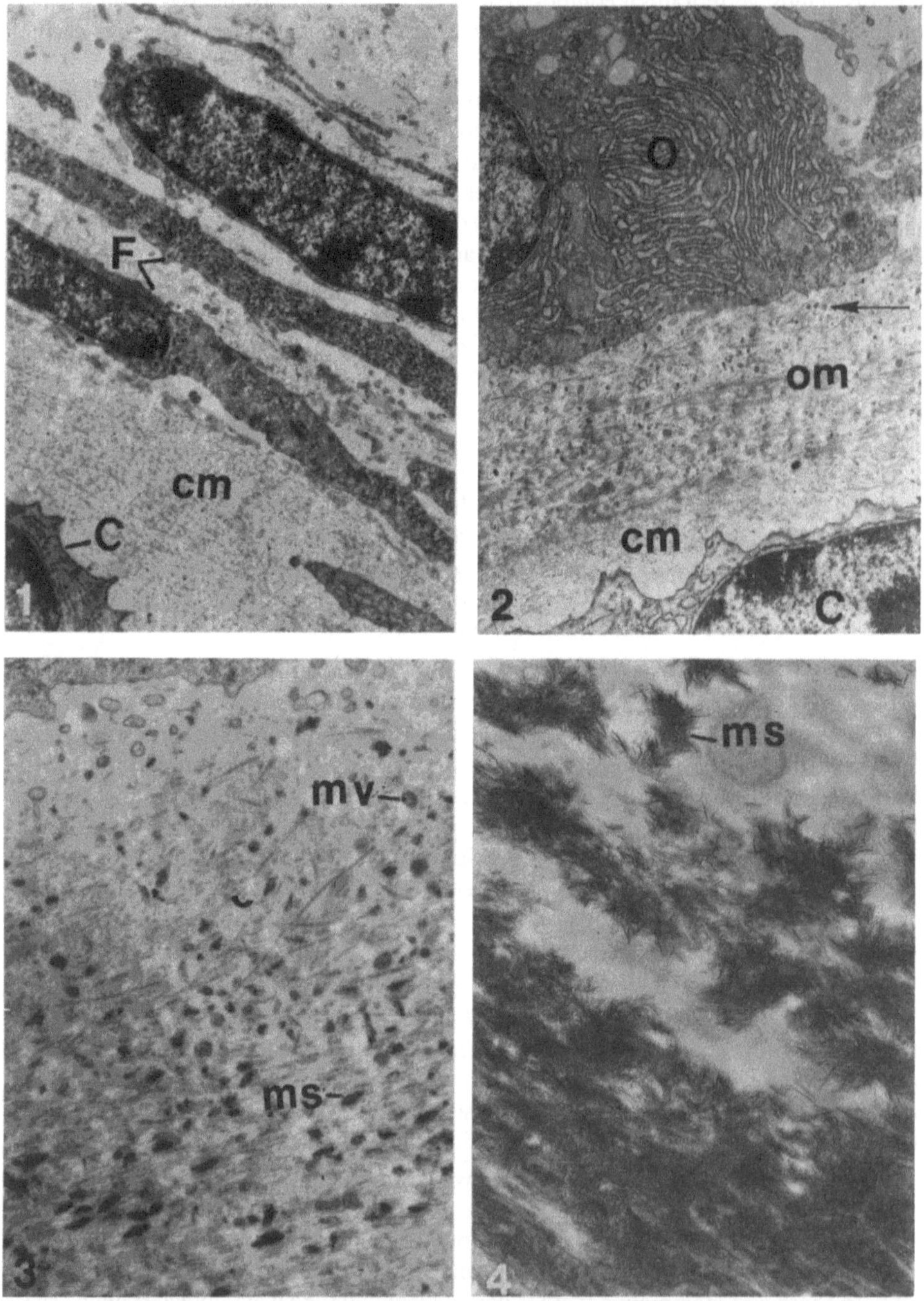
F
cm
C
1
O
om
cm
C
2
mv
ms
3
ms
4

enamel is considered to be homologous with mammalian enamel. Other authors, however, have contended that shark tooth enamel is not "true" enamel but instead is a modified type of dentine derived from odontoblasts. Because of the uncertainty the term enameloid is commonly used in reference to fish tooth enamel. Blocks of shark jaw tissue with attached teeth were fixed and processed for light and electron microscopy. Thin sections of both unmineralized and mineralizing teeth were cut with a diamond knife and examined by TEM.

RESULTS

Histogenesis of bone in frog femur

The frog femur first develops as a cartilaginous model upon which periosteal bone is subsequently deposited (Kemp & Hoyt, 1969). This chondrogenic phase is illustrated in Figure 1, which shows a portion of a chondrocyte and its matrix surrounded by perichondrial fibroblasts. Although this specimen had been treated with thyroxine for 1 day, its morphology is like that of untreated controls.

A specimen treated for 5 days (Fig. 2) shows that cells of the inner of fibroblasts in the perichondrium have become transformed into osteoblasts; hence the perichondrium has become a periosteum. Cytoplasm of the osteoblasts is engorged with cisternae of rough endoplasmic reticulum and elongated mitochondria. A prominent layer of osteoid matrix has been deposited over the cartilage matrix. Collagen fibrils in osteoid are generally about twice as large in diameter as those in cartilage. Scattered among the collagen fibrils in the osteoid are many matrix vesicles derived by pinching off from protoplasmic processes of the osteoblasts.

Mineralization has begun in a specimen treated with thyroxine for 6 days (Fig. 3). Mineralizing sites are electron-dense and scattered throughout the osteoid. Apparently these sites coincide with the distribution of matrix vesicles, for the vesicles disappear as mineralization advances. At the stage illustrated in Figure 3 there is a gradient in the distribution of matrix vesicles and mineralizing sites. Toward the osteoblast there are more vesicles and fewer mineralizing sites than in the osteoid toward the underlying cartilage, where the ratio is reversed. The precise relationship between matrix vesicles, collagen fibrils and nucleation sites for crystallites has not been determined. Whether mineral deposition begins within intact matrix vesicles, or not until after they have ruptured, is an important but unanswered question.

Further progress in mineralization is seen in a specimen treated with thyroxine for 9 days (Fig. 4). Mineralizing sites at this time have expanded and consist of closely massed collagen fibrils and needle-like crystallites. Hydrolysis of the ground substance which earlier kept collagen fibrils well separated probably accounts for the clumping of

◄

FIGURE 1-4. Sections of femurs of frog tadpoles treated with thyroxine to hasten limb development. ▪*1.* Cartilaginous femur after 1 day of exposure to thyroxine beginning at Taylor-Kollros stage IX. C, chondrocyte; **cm**, cartilage matrix; **F**, fibroblasts of perichondrial sheath. × 9,210. ▪*2.* Osteoid matrix in femur of tadpole exposed to thyroxine for 5 days. **O**, osteoblast; **om**, osteoid matrix with coarse collagen fibrils and matrix vesicles (arrow); **cm**, cartilage matrix with fine collagen fibrils; × 9,420. ▪*3.* Early mineralization in osteoid matrix of tadpole after 6 days of exposure to thyroxine. Electron-dense masses scattered throughout the osteoid are mineralizing sites. **ms**, mineralizing site; **mv**, matrix vesicle. × 33,510. ▪*4.* Mineralizing matrix of femur in tadpole exposed to thyroxine for 9 days. × 69,740.

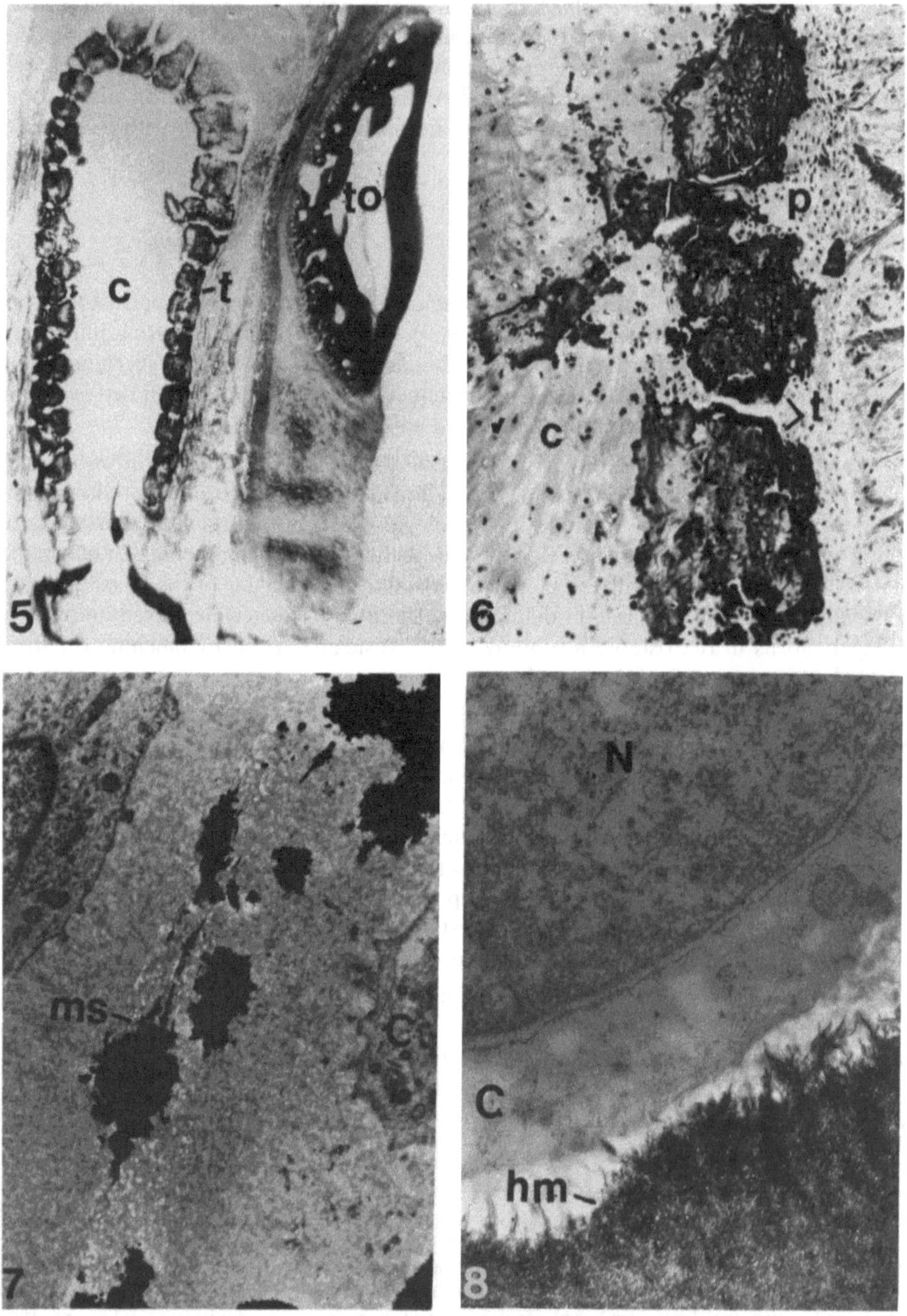
5
c
t
to
6
p
c
t
7
ms
C
8
N
C
hm

fibrils in these mineralizing sites. With continued expansion the sites coalesce and form the heavily mineralized bone matrix which consists of a conglomerate of collagen fibrils and hydroxyapatite crystallites with their long axes aligned.

Histogenesis of calcified cartilage in shart jaw

Endoskeletons of chondrichthyan fish, Elasmobranchii and Holocephali, are cartilaginous, as the name implies. Yet the periphery of the skeletal cartilages in these fish becomes calcified by the development of a mosaic of surface plaques called tesserae (details in Kemp & Westrin, 1979). A longitudinal section of the jaw of the shark *Carcharhinus menisorrah* (Fig. 5) illustrates this tesserate pattern. Individual tesserae are 3-dimensional blocks bordered by perichondrial connective tissue on their outer surfaces and by hyaline cartilage on their inner and lateral surfaces (Fig. 6).

Within the tesserae the matrix of entrapped chondrocytes or perichondrial fibroblasts is calcified by deposition of hydroxyapatite crystallites, as in bone of the frog femur. Crystallites first form in discrete mineralizing masses of collagen fibrils at some distance from chondrocytes (Fig. 7). As mineralization proceeds, the volume of hyaline matrix is progressively reduced and the mineralizing masses enlarge and coalesce until chondrocytes become surrounded by a tight jacket of heavily mineralized matrix (Fig. 8). The crystallites in this matrix of calcified cartilage are small and needle-like, resembling those of bone. In contrast to the pattern of mineralization in bone, however, mineralization of tesseral calcified cartilage does not require deposition of an osteoid-like matrix as a prelude to calcification. Whatever modification of the cartilage matrix precedes calcification appears to be mediated by chondrocytes or perichondrial fibroblasts directly without their transformation into secretory cells like osteoblasts. Chondrocytes encased within the tesserae remain viable.

Histogenesis of shark teeth

The teeth of cacharhinid sharks are highly efficient incisors with triangular cusps, sharpened and in some species serrated along their lateral borders to facilitate cutting. Behind the row of upright functional teeth are several rows of developing teeth at progressively younger stages of differentiation from outer (anterior) rows to inner (posterior) rows. As the shark grows, its rows of teeth move forward so that old functional teeth may be removed from the battle line and replaced by a new row of able recruits. Thus throughout its life the shark retains teeth at various stages of development from uncalcified primordia through the stages of calcification to the fully calcified functional dentition. This report is based on studies of teeth at early stages of calcification (Kemp & Park, 1974; Kemp, 1985).

FIGURE 5-8. Sections of shark jaws showing calcification of tesserae. ▪*5.* Photomicrograph of jaw of *Carcharhinus menisorrah* showing plaques of calcified cartilage (tesserae) surrounding hyaline cartilage of the interior of the jaw. **c**, region of hyaline cartilage; **t**, tesserae; **to**, base of large tooth embedded in oral mucosa. × 11. ▪*6.* Enlarged view of tesserae, showing relation to hyaline cartilage in the interior of the jaw and connective tissue of the perichondrium externally. **p**, perichondrium. × 55. ▪*7.* View of internal border of a tessera, showing mineralizing masses of calcified cartilage at some distance from a tesseral chondrocyte. **ms**, mineralizing site; **C**, tesseral chondrocyte. × 7,570. ▪*8.* View of tesseral chondrocyte which has become almost completely surrounded by heavily mineralized matrix. **N**, nucleus of chondrocyte; **C**, cytoplasm of chondrocyte; **hm**, heavily mineralized matrix. × 20,590.

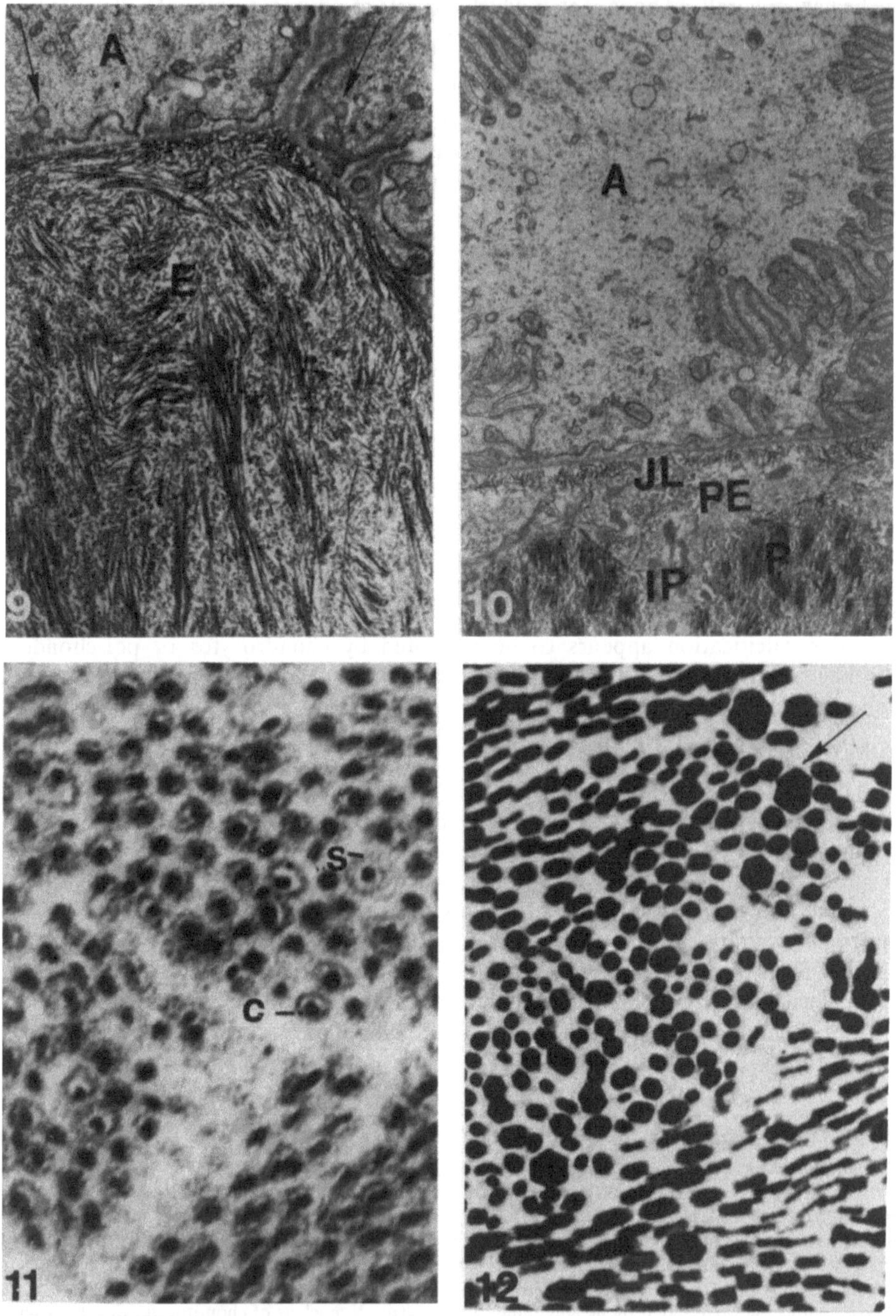
A
E
9
A
JL
PE
IP
P
10
S
C
11
12

From an evolutionary standpoint, the teeth of sharks are modified scales. Both the placoid scales and teeth of elasmobranchs develop with an interior calcified zone of dentine capped with a layer of enamel (enameloid). Shark dentine is a product of odontoblastic secretion and contains small hydroxyapatite crystallites like those of dentine in other vertebrates. Its development resembles that of bone.

Development of enamel in the shark *Negaprion brevirostris* begins with elongation of inner dental epithelial cells to become columnar ameloblasts, as in the teeth of higher vertebrates. Ameloblastic vesicles develop within the ameloblasts and secrete their contents at their inner ends bordering the enameline matrix (Fig. 9). It appears that the ameloblastic secretions contribute to the basal lamina and also pass through it to contribute constituents of the enameline matrix.

Crystallization of hydroxyapatite crystallites begins immediately beneath the basal lamina in a juxtalaminar zone (Fig. 10). Underlying this region is a zone of relatively unmineralized fibrils, which I have termed the pre-enamel zone. More deeply the region of heavy mineralization of the enameline matrix is subdivided into mineralized zones called palisades (Garant, 1970) separated by unmineralized inter-palisadal zones.

The earliest observed crystallites are separated from one another within individual tubular sheaths (Fig. 11). Groups of ensheathed crystallites are organized into bundles within which tubules and crystallites are oriented in the same direction. Substructural units in the tubular sheaths indicate that the sheaths are polymerized from protein filaments. After their initial deposition the crystallites grow enormously compared with the tiny crystallites of the collagen-associated tissues. Sections across the c-axis of older crystallites show the hexagonal shape of apatite crystals (Fig. 12).

DISCUSSION

Mineralization of collagenous hard tissues

The precursor matrices for mineralization in bone, dentine and calcified cartilage are osteoid, predentine and the hyaline matrix of chondrocytes. Fibrillar collagen is the principal protein constituent of the matrix in these hard tissues; hence it is appropriate to refer to their mineralization as collagen-associated (Kemp, 1984, 1985). Proteoglycans and glycoproteins constitute most of the ground substance around collagen fibrils. Matrix vesicles which bud from scleroblasts may be present in the matrix of these tissues at the onset of mineralization.

FIGURE 9-12. Sections of shark teeth showing development of enameline matrix and hydroxyapatite crystallites. ▪*9.* Inner ends of ameloblasts of *Negaprion brevirostris* and underlying matrix already mineralizing. Ameloblastic vesicles (arrows) in ameloblasts secrete contents which contribute to basal lamina and enameline matrix. **A**, ameloblast; **E**, mineralizing enameline matrix. × 16,530. ▪*10.* Inner end of an ameloblast of *Negaprion brevirostris* and four zones of underlying enameline matrix. Ameloblast shows lateral processes interdigitating with those of neighboring ameloblasts. **JL**, juxtalaminar zone of fine crystallites; **PE**, pre-enamel zone of unmineralized fibrils; **P**, palisadal zone of mineralization; **IP**, interpalsisadal zone of unmineralized fibrils. × 8,520. ▪*11.* Crystallites of enamel layer in tooth of *Carcharhinus menisorrah* showing development within tubular sheaths sectioned in cross- section (above) or obliquely (lower right). **cr**, crystallite; **s**, tubular sheath, × 183,330. ▪*12.* Large crystallites in enamel layer of tooth of *Triaenodon obesus*. The largest crystallites typically are hexagonal in cross section (arrow). × 59,330.

The scenario for mineralization is still speculative, but degradation of constituents of the ground substance in order to expose collagen fibrils to the mineralizing ions appears to be a necessary preliminary step (Dziewiatkowski *et al.*, 1968; Kemp & Westrin, 1979; Kemp, 1984). It is not known whether nucleation of hydroxyapatite crystallites begins on the collagen fibrils or on some other constituent of the matrix. According to the theory of Glimcher & Krane (1968), nucleation begins within "hole zones" between the tropocollagen molecules of collagen fibrils. Growth of crystallites might be restricted initially by limited intrafibrillar space, and later by availability of interfibrillar space (Glimcher, 1981). According to Bonar *et al.* (1985), there is not enough intrafibrillar space in fully mineralized bovine bone to accommodate its mineral content; thus much of the mineral must be between fibrils.

Nucleation sites for hydroxyapatite may be primarily phosphate acceptors which attract calcium ions, as theorized by Glimcher (1981), or they may be calcium acceptors which bind phosphate secondarily. Glycoproteins such as osteocalcin or osteonectin, phosphate carrier proteins known as phosphophoryns, and phospholipids have been implicated in the chemistry of nucleation (review in Kemp, 1984).

The role of matrix vesicles in mineralization has been of great theoretical interest since they were first described by Anderson (1969) and Bonucci (1970). According to one idea (Ornoy, 1983), calcium phosphate first accumulates within the matrix vesicles as amorphous deposits, which subsequently are converted to hydroxyapatite crystallites. The vesicles then rupture and liberate the crystallites for alignment along collagen fibrils. This theory awards matrix vesicles a primary role in mineralization and relegates collagen to a passive, mechanical function. Since mineralization of collagen fibrils may proceed in vitro, however (Glimcher, 1959), it is logical to infer that collagen is actively involved in nucleation in vivo. The principal role of matrix vesicles may be to liberate hydrolytic enzymes or phospholipids facilitating modification of the matrix so as to permit nucleation on or within the collagen fibrils. Neuman & Neuman (1958) have suggested that collagen may act as an organic crystalline substance which induces mineral crystallization by epitaxy. An alternative explanation for the initiation of crystallization (Glimcher, 1981) is that collagen fibrils serve to concentrate phosphate and calcium ions in restricted space, thereby causing salt precipitation without the constraints of epitaxy.

Mineralization of enamel

Two obvious features distinguish tooth enamel from the collagen-associated hard tissues, despite the fact that the mineral crystallites are hydroxyapatite in both types of tissue: 1) enamel crystallites are much larger than those of bone, dentine and calcified cartilage, and 2) the principal proteins are amelogenin and enameline rather than collagen. Assuming that close packing of collagen fibrils is a major factor in restricting growth of crystallites in the collagen-associated hard tissues, it appears that the absence of these fibrils in the enameline matrix renders the conditions for crystalline growth more favorable there. Loose packing of the enameline proteins may permit early robust growth of crystallites, and removal of much of the protein from the matrix during tooth maturation (Reith, 1967) probably provides space for additional growth.

Amelogenins are hydrophobic glycoproteins with a molecular weight of 5,000-40,000 daltons, which account for most of the enameline protein in mammalian teeth during the period of ameloblastic secretion (Fincham *et al.*, 1982a). Enamelins are acidic glycoproteins with a molecular weight of 8,000-72,000 daltons. During tooth maturation the enamelins are conserved while amelogenins are declining in amount (Fincham *et al.*, 1982b). Using fluorescent antibodies to mammalian amelogenin, Herold *et al.* (1980) and Slavkin *et al.* (1983) have demonstrated that shark tooth ameloblasts synthesize and secrete cross-reactive proteins homologous to amelogenin. Such proteins have indeed been extracted from shark teeth (Graham, 1985). Biochemical evidence thus has substantiated the similarity of the organic matrix of enamel in sharks and mammals.

Morphological evidence likewise supports the concept of homology of the enamel layer in these widely disparate groups. Development of enamel crystallites in shark teeth is remarkably similar to development of their mammalian counterparts. Shark tooth crystallites develop within tubular sheaths, as demonstrated here and in earlier papers (Kemp & Park, 1974; Kemp, 1985). A similar pattern was proposed by Travis & Glimcher (1964) and Jessen (1968), but the morphological relationship between sheath protein and crystallites in mammalian teeth is still being investigated (Warshawsky *et al.*, 1984). Shark tooth ameloblasts are secretory, and they are the source of the sheath protein as in mammalian teeth (Kemp & Park, 1974; Kemp, 1985; Kerebel *et al.*, 1977). Another similarity is that the mineral crystallites of shark tooth enamel become large and exhibit hexagonal symmetry like those of mammalian teeth (Kemp, 1974, 1984, 1985; Daculsi & Kerebel, 1980), in sharp contrast to the small, needle-like crystallites in collagen-associated hard tissues. The possibility that hexagonal electron microscopic images of rat enamel crystallites may be shadow images of rectangular crystallites has been advanced by Warshawsky *et al.* (1984).

The concept that the patterns of mineralization exemplified by the collagen-associated and amelogenin-enameline-associated hard tissues are homologous throughout the vertebrate kingdom bolsters the faith that what we learn about the mechanisms of mineralization in fish and frogs is relevant to similar mechanisms in mice and men. Moreover, the same concept enables us to project our comprehension backward in time to the dawn of vertebrate evolution and to visualize how our Cambrian forebears, the ostracoderms, first began to practice the extraordinary alchemy of transforming their soft connective tissues and dermal epithelium into the treasure of mineralized armor.

ACKNOWLEDGMENTS

The original research serving as the basis for this contribution was supported by NSF Grant 4317, USPHS Grant AM13745, and grants from the University of Michigan Cancer Research Institute. I thank the organizing committee for the 5th International Symposium on Biomineralization for travel support enabling me to participate in the symposium.

REFERENCES

ANDERSON, H.C., 1969. Vesicles associated with calcification in the matrix of epiphyseal cartilage. J. Cell Biol., 41:59-72.

ARENDS, J., JONGEBLOED, W.L. & SCHUTHOF, J., 1983. The ultrastructure of surface enamel in relation to de- and remineralization. *In* Demineralization and Remineralization of the Teeth (eds. S.A. Leach and W.M. Edgar). pp. 155-163. Oxford: IRL Press.

BARNES, R.D., 1974. Invertebrate Zoology, Third Ed. 870 pp. Philadelphia: W.B. Saunders.

BERRILL, N.J., 1955. The Origin of Vertebrates. 257 pp. Oxford: Clarendon Press.

BONAR, L.C., LEES, S., & MOOK, H.A., 1985. Neutron diffraction studies of collagen in fully mineralized bone. J. Mole. Biol., 181:265-270.

BONUCCI, E., 1970. Fine structure and histochemistry of "calcifying globules" in epiphyseal cartilage. Zeits. Zellfor. mikrosk. Anat., 103:192-217.

DACULSI, G. & KEREBEL, L.M., 1980. Ultrastructural study and comparative analysis of fluoride content of enameloid in sea-water and fresh-water sharks. Arch. Oral Biol., 25:145-151.

DENISON, R.H., 1967. Ordovician vertebrates from western United States. Fieldiana (Geology), 16:131-192.

DZIEWIATKOWSKI, D.D., TOURTELLOTTE, C.D., & CAMPO, R.D., 1968. Degradation of protein-polysaccharide (chondromucoprotein) by an enzyme extracted from cartilage. *In* The Chemical Physiology of Mucopolysaccharides (ed. G. Quintarelli). pp. 63-79. Boston: Little, Brown & Co..

FINCHAM, A.G., BELCOURT, A.B., LYARUU, D.M., & TERMINE, J.D., 1982a. Comparative protein biochemistry of developing dental enamel matrix from five mammalian species. Calc. Tiss. Intern., 34:182-189.

FINCHAM, A.G., BELCOURT, A.B., & TERMINE, J.D., 1982b. Changing patterns of enamel matrix proteins in the developing bovine tooth. Caries Res., 16:64-71.

GANS, C. & NORTHCUTT, R.G., 1983. Neural crest and the origin of vertebrates: a new head. Science, 220:268-274.

GARANT, P.R., 1970. An electron microscopic study of the crystal-matrix relationship in the teeth of the dogfish *Squalus acanthias*. J. Ultrastr. Res., 30:441-449.

GLIMCHER, M.J., 1959. Molecular biology of mineralized tissues with particular reference to bone. *In* Biophysical Science -- A Study Program (eds. J.L. Oncley, F.O. Schmitt, R.C. Williams, M.D. Rosenberg and R.H. Bolt). pp. 359-393. New York: John Wiley.

GLIMCHER, M.J., 1981. On the form and function of bone; from molecules to organs. Wolff's law revisited. *In* The Chemistry and Biology of Mineralized Connective Tissues (ed. A. Veis). pp. 617-675. New York: Elsevier-North Holland.

GLIMCHER, M.J. & KRANE, S.M., 1968. The organization and structure of bone and the mechanism of calcification. *In* Treatise on Collagen, vol. 2, part B, Biology of Collagen (gen. ed. G.N. Ramachandran, ed. B.S. Gould). pp. 67-251. New York: Academic Press.

GRAHAM, E.E., 1985. Isolation of enamel-like proteins from blue shark (*Prionace glauca*) enameloid. J. Exp. Zool., 234:185-192.

HARDISTY, M.W., 1979. Biology of the Cyclostomes. 428 pp. London: Chapman and Hall.

HEROLD, R.C., GRAVER, H.T., & CHRISTNER, P., 1980. Immunohistochemical localization of amelogenins in enameloid of lower vertebrate teeth. Science, 207:1357-1358.

JESSEN, H., 1968. Elliptical tubules as unit structure of forming enamel matrix in the rat. Arch. Oral Biol., 13:351-352.

KEMP, N.E., 1984. Organic matrices and mineral crystallites in vertebrate scales, teeth and skeletons. Amer. Zool., 24:965-976.

KEMP, N.E., 1985. Ameloblastic secretion and calcification of the enamel layer in shark teeth. J. Morph., 184:215-230.

KEMP, N.E. & HOYT, J.A., 1969. Ossification of the femur in thyroxine-treated tadpoles of *Rana pipiens*. Devel. Biol., 20:387-410.

KEMP, N.E. & PARK, J.H., 1974. Ultrastructure of the enamel layer in developing teeth of the shark *Carcharhinus menisorrah*. Arch. Oral Bio., 19:633-644.

KEMP, N.E. & WESTRIN, S.K., 1979. Ultrastructure of calcified cartilage in the endoskeletal tesserae of sharks. J. Morph., 160:75-102.

KEREBEL, B., DACULSI, G., & RENAUDIN, S., 1977. Ultrastructure des améloblasts au cours de la formation de l'émailoide des Sélaciens. Biol. Cell., 28:125-130.

LEHMAN, H.E., 1977. Chordate Development. 369 pp. Winston-Salem, N.C.: Hunter.

NEUMAN, W.F. & NEUMAN, M.W., 1958. The Chemical Dynamics of Bone Mineral. 209 pp. Chicago: Univ Chicago Press.

NORTHCUTT, R.G. & GANS, C., 1983. The genesis of neural crest and epidermal placodes; a reinterpretation of vertebrate origins. Quar. Rev. Biol., 58:1-28.

ORNOY, A., BEN-HUR, H., & LANCET, M., 1983. Matrix vesicles and mineralization of cartilage in young human fetuses. Calc. Tiss. Intern., 35 (Supplement), A15 (Abstract).

ØRVIG, T., 1977. A survey of odontodes ("dermal teeth") from developmental, structural, functional, and phyletic points of view. *In* Problems in Vertebrate Evolution. (eds. S.M. Andrews, R.S. Miles and A.D. Walker). pp. 53-75. New York: Academic Press.

PAUTARD, F.G.E., 1976. Calcification in single cells: with an appraisal of the relationship between *Spirostomum ambiguum* and the osteocyte. *In* Mechanisms of Mineralization in the Invertebrates and Plants (eds. N. Watabe and K.M. Wilbur). pp. 33-53. Columbia, S.C.: Univ. S. Carolina Press.

REITH, E.J., 1970. The stages of amelogenesis as observed in molar teeth of young rats. J. Ultrastr. Res., 30:111-151.

SLAVKIN, H.C., SAMUEL, N., BRINGAS, JR., P., NANCI, A. & SANTOS, V., 1983. Selachian tooth development. 2. Immunolocalization of amelogenin polypeptides in epithelium during secretory amelogenesis in *Squalus acanthias*. J. Craniofacial Gene. Devel. Biol., 3:43-52.

TRAVIS, D.F. & GLIMCHER, M.J., 1964. The structure and organization of, and the relationship between the organic matrix and the inorganic crystals of embryonic bovine enamel. J. Cell Biol., 23:447-497.

WARSHAWSKY, H., BAI, P., NANCI, A., & JOSEPHSEN, K., 1984. Morphological visualization of two categories of enamel proteins in relation to the crystals of rat incisor enamel. *In* Tooth Enamel IV. (eds. R.W. Fearnhead and S. Suga). pp. 177-182. New York: Elsevier Sci. Publ..

KEMP N.E. & PARK J.H. 1974. Ultrastructure of the enamel layer in developing teeth of the shark Carcharhinus menisorrah. Arch. Oral Biol. 19: 633-644.

LOW TEMPERATURE CARBONATE PHOSPHATE MATERIALS OR THE CARBONATE–APATITE PROBLEM: A REVIEW

H. Catherine W. Skinner

Department of Geology and Geophysics, Yale University, New Haven, CT 06511

ABSTRACT. Hydroxyapatite and fluorapatite are well known and documented crystalline mineral phases. Apatite containing carbonate, the usual species present in vertebrate tissues and in phosphorites, however, remains an enigma. It has been impossible to accurately determine the chemical composition and the precise crystalline configuration of the mineral apatitic phase formed in biologic environments.

Experimental studies in the $CaO-P_2O_5-H_2O-CO_2$ system at high temperature and elevated CO_2 pressures have led to suggestions of two possible lattice sites for carbonate substitution in the Ca-apatites [$Ca_5(PO_4)_3(OH)$], the (F,OH) site and for (PO_4). Several structural configurations of such substitutions have been presented as models for carbonate apatite.

Extrapolation of these results, to the low temperatures and pressure conditions appropriate for phosphorites and vertebrate tissues, is not straightforward. The mineral in these localities is extremely fine grained. It may be composed of more than one mineral species, intimately associated with bio-organic compounds, and usually undergoes alteration post initial (primary) precipitation. Uptake of ions and/or resorbtion and redeposition are characteristic of the biologically mediated mineral(s).

During the formation, accumulation and maturation of sedimentary phosphorites, and/or vertebrate tissues, local chemical and physical factors, such as the concentration of fluorine or sites for nucleation, play significant roles in the composition and structure of the apatitic mineral phase.

INTRODUCTION

There is a long history of the study of apatitic minerals and many people have contributed to the problems presented by the carbonate apatites. As the main constituent of the sedimentary rocks known as phosphorites, the composition range of carbonate apatite has been outlined and discussed by Gulbrandson (1969), McClellan & Lehr (1969), McConnell (1973), Burnett (1977), and more recently by Thompson *et al.* (1984), and McArthur (1985). The vertebrate mineralizing system also produces apatites. In common association with the CO_2 of cell metabolism the composition suggested for these biominerals approximates a carbonate apatite (Zapanta-LeGeros, 1965; Neuman & Mulryan, 1967; McConnell, 1970; Simpson, 1972; Driessens, 1982).

Origin, Evolution, and Modern Aspects of Biomineralization in Plants and Animals
Edited by Rex E. Crick
Plenum Press, New York

Some general background on apatites will serve to introduce the carbonate apatites. Apatite is a name applied to a series of phosphate minerals with a common crystal structure type. The Ca-apatites [$Ca_5(PO_4)_3(OH,F,Cl)$] with small compositional variations occur in all the many different rocktypes that cover the surface of the earth. Crystals of Ca-apatites are common as accessory minerals in igneous and metamorphic rocks. Occasionally extremely large single crystals (up to 6 ft in circumference) may be found in some pegmatites. These occurrences are usually the result of crystallization at high temperatures (greater than 300° C) and elevated pressures (greater than 1 Kbar). Apatites are also the mineral species that distinguishes a class of sedimentary rocks known as phosphorites. Marine deposits with wide geographic and geologic occurrence, phosphorites contain at least 10% by volume apatite and have formed in many localities over the globe from the Precambrian to Recent times. Phosphorites may be precipitating under restricted marine conditions today, but not in the volume typical of the past. The formation of fine disseminated particles of apatites in soils, muds and other hydrous surficial environments, however, is consistently observed.

Perhaps the most familiar bio-apatites are those that form the mineral portion of vertebrate bones and teeth and the shell parts of a few invertebrates, *i.e. Lingula* (Tuross & Fisher, 1988).

Both the phosphorites and the endo- and exoskeletal materials are low temperature apatites and associated with biologic activity, but the precise conditions and mechanisms which dictate the location and chemical composition of the mineral remain under study. It may, however, be possible to distinguish between *primary* or *secondary* apatite. *Primary* means the nucleation and direct precipitation of an apatitic solid from solution with or without the aid of a template (epitaxy). *Secondary*, means the chemical alteration of an already formed solid. The alteration may be simple such as F^- ions substituting or OH^- ions in the Ca-apatite lattice producing a slightly different, but related species. A *secondary* mineral phase may also result from molecule by molecule replacement of part or all of the original mineral in response to a changed environment, *e.g.*, $CaCO_3$ replaced by $Ca_5(PO_4)_3(OH)$.

The distinction between *primary* or *secondary* mineral phases calls attention to the variety and reactivity of low temperature apatitic minerals. For example, sequential mineralization is typical of vertebrate bones and teeth (Albright & Skinner, 1987). Osteons, distinctive apatitic mineralized structures in bone tissue, are formed, resorbed and reformed over the lifetime of an individual. The viability of bone during life depends on the delicate balance and deposition of the mineral phase. Post-mortem, under certain conditions of burial, bone tissues may take up fluorine.

Alternately, limestones (calcium carbonate) may become phosphatized. Substitution of phosphate for carbonate in or on the carbonate minerals in sediments is not unusual. There may also be coprecipitation of phosphates and carbonates. Nodules which show concentric bands of accretion incorporate fluorine over time.

Both examples suggest that several different apatite species may co-exist within a particular site or sample. A single grain of Ca-apatite within the skeleton or in a sedimentary horizon probably has had a complex history.

Apatites produced in association with biological activity have the following characteristics:

1. They are fine grained mixtures of several minerals with organic or bio-organic compounds. Phosphorites may contain the minerals quartz, chalcedony, feldspar, clay, calcite, dolomite, glauconite, and goethite as well as shell fragments, bacterial remains and fecal pellets in addition to the mineral apatites. Along with apatite the minor constituents found in bones and teeth include brushite, ($CaHPO_4 \cdot 2H_2O$), monetite, ($CaHPO_4$), "amorphous" Ca-phosphate, (resembles apatite), whitlockite, (Ca_3PO_4), octacalcium phosphate, ($Ca_8H_2(PO_4)_6 \cdot 5H_2O$) and calcium carbonate. The latter, in the form of the mineral calcite, has been identified in the medulary cavity of the chick femur (Schraer & Schraer, 1970).

2. The mineral materials are very fine grained. The grain size of the apatite crystals in bone, in teeth and in phosphorites is in the range of tens to a few thousand angstroms (Å, 10^{-10}m).

3. Within the apatites there is a wide range of chemical substitution. Ca-apatites, whose ideal chemical composition has been expressed as $Ca_5(PO_4)_3(F,OH,Cl)$, usually take many other elements into the crystal lattice; over 25 different elements have been detected (Cheney *et al.*, 1974). Phosphorite apatite may contain over 1% sodium and up to 6% carbonate, as well as measurable quantities of strontium, magnesium, or other elements and several to tens of parts per million uranium or thorium. Bone mineral has been called the storehouse of the elements required to satisfy the metabolic needs of the body. It is not surprising that analyses of tissues from different bone or animal species show a range of constituents in addition to the major elements, calcium and phosphorus (Armstrong & Singer, 1965; Zipkin, 1970).

TABLE 1. Comparison of the composition of bone and phosphorites. (in wt %, unless otherwise specified)

	BONE		PHOSPHORITE	
	Bovine[1]	Human[2]	Namibia[3]	Peru[4]
Ca	>10.0	22.5	33.01	28.76
Na	1.0 – 10.0	0.52	–	0.67
K	<1.0	0.089	0.075	0.67
Mg	0.1 – 1.0	0.26	0.426	0.44
Mn	0.001	–	–	–
Fe	0.01 – 0.1	–	0.014	–
Pb	0.001 – 0.01	–	–	–
Cu	0.01 – 0.1	–	7.0*	–
Sr	0.01 – 0.1	–	1650.0*	–
Sn	0.001 – 0.01	–	–	–
Ni	<0.001	–	<3.0*	–
Si	0.001	–	0.188	6.26
Al	0.001 – 0.01	–	0.002	1.80
P	>10.0	10.3	13.40	12.09
Ti	0.0	–	0.012	–
CO_2	–	3.5	1.1@	–
F	–	0.054	2.74	2.29

[1]Bovine (Corital) bone from Skinner *et al.* (1972) Tables 1 & 2. Elements looked for but not detected: Ag, As, B, Be, Bi, Cd, Ce, Co, Ga, Ge, Hf, Hg, In, Ir, La, Mo, Nb, Os, Pd, Pt, Rh, Ru, Sb, Sc, Ta, Te, Th, Tl, U, V, W, Y, Zn, Zr.

[2]Human (adult) bone (dry fat free wt. basis) from Zipkin (1970) Table 1, sample labelled "Mean". Sample also contains 0.11 Cl.

[3]Phosphorite, Namibia from Thompson *et al.* (1984) Table 2, sample 5800 recalculated. Sample contains 275 ppm Ba, <4 ppm Pb, <2 ppm Rb, 0.3 ppm Th, 58 ppm U, 4 ppm Y, 20 ppm An, 335 ppm Zr.

[4]Phosphorite, Peru/Chile from Burnett (1977) Table 1, sample PD 21-25 recalculated. Sample contains 61% Apatite, 20% Plagioclase, 8% K-feldspar, 8% Quartz.

– = Not determined

* = ppm

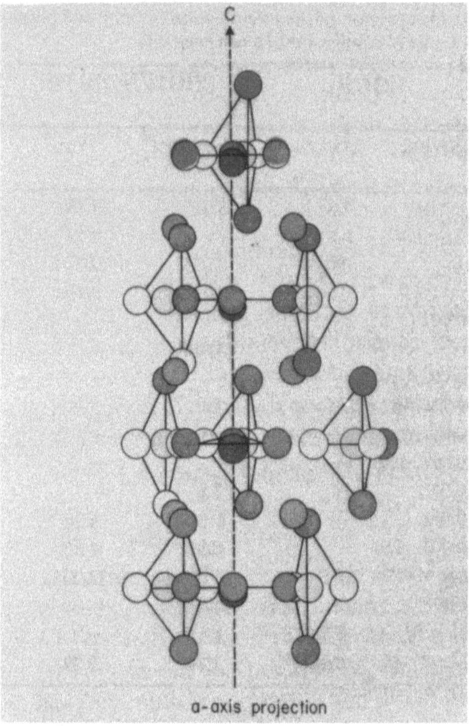

FIGURE 1. Projection of the Ca-apatite crystal structure down the *c*-axis showing the rhombohedral shape of the unit cell (enclosed in dark lines). The *c*-axis is perpendicular to the plane of the paper. Numbers on arrows show exact position of two superposed elements along *c*-axis. Modified after Young & Elliot (1966).

In Table 1 the elements and molecular species that have been identified in the apatitic phase of phosphorites are compared with an analyses of bones. Note that the marine precipitates are fluorine containing, a fluorapatite, while hydroxyapatite is the characteristic form found in mineralized tissues of living vertebrates.

CRYSTAL STRUCTURE OF CALCIUM APATITES

Figures 1–3 provide several views of the three dimensional crystal structure of an ideal Ca-apatite, $Ca_{10}(PO_4)_6$-$(OH,F)_2$. The illustration demonstrates that apatite is a phosphate mineral in which the tetrahedral phosphate groups are coordinated to two crystallographically distinct types of calcium ions, labelled Ca_1 and Ca_2. Ca_1 ions are located in the center of the unit cell marked off by black lines connecting the fluorine (or OH) position at the corners of the rhombohedron in the *c*-axis projection of the structure (Fig. 1). Ca_2 ions are located in what appears to be hexagonal array around the fluorine site. The hexagonal pattern shown in this projection is misleading as can be seen from

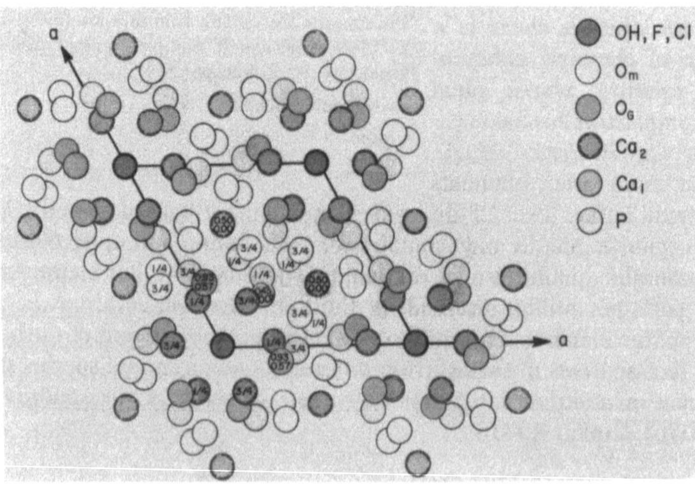

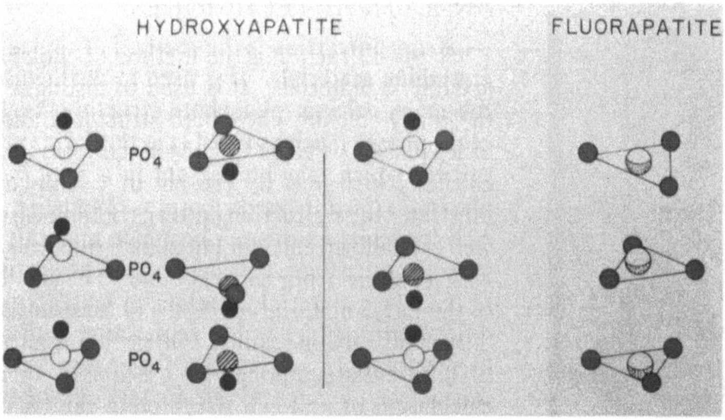

FIGURE 3. The possible relationships of the OH^- to the Ca triangle in hydroxyapatite showing the directionality of H^+ and the position of F relative to the Ca triangle in fluorapatite. Modified after Young & Elliot (1966)

another view of the structure (Fig. 2). Three Ca_2 atoms form a trigonal plane and the other three calcium ions are displaced along the *c*-axis and rotated 60°. The trigonal association is reinforced by the F or OH atom which is centrally located relative to the plane of the three calcium atoms.

In fluorapatite, the fluorine ion is ideally situated exactly in the plane of the Ca_2 triangle (Fig. 3), whereas the OH of hydroxyapatite will be somewhat out of the plane, and the proton, H^+, may point in alternate directions (Young & Elliott, 1966).

The substitution of elements for calcium in the two crystallographic sites has not been systematically investigated so it is not possible to say whether Ca_1 is the preferential site for the sodium, magnesium, strontium, and other elements known to occur in apatites (Table 1). However, one could calculate the expected maximum substitution at this site assuming an ideal apatite crystal structure and based on the contents of the unit cell:

Total calcium in $Ca_{10}(PO_4)_6(F)_2$ = 400/1008 = 39.7 wt%, of which 4 out of 10 calcium sites in the structure are Ca_1. If all substitutions were at this site up to 16 wt% foreign atoms could be accommodated. A similar calculation shows that an ideal fluorapatite contains 3.77 wt% fluorine.

As early as 1940, McConnell & Gruner identified two types of mineral apatite containing CO_3^{3-}. It has since been documented and accepted that carbonate is ubiquitously associated with the sedimentary precipitates (McClellan & Lehr, 1969) and with bone mineral (Biltz & Pelligrino, 1969). The association may be detected when a separate calcium carbonate phase is identified but often the carbonate is intimately associated with the apatite and assumed to be intracrystalline, that is, part of the crystal structure.

FIGURE 2. The *a*-axis projection of the Ca-apatite, $Ca_{10}(PO_4)_6(OH,F)_2$, crystal structure showing the tetrahedral configuration of the phosphate groups that dominate the structure. The *c*-axis of the unit cell is vertical. Modified after Young & Elliot (1966).

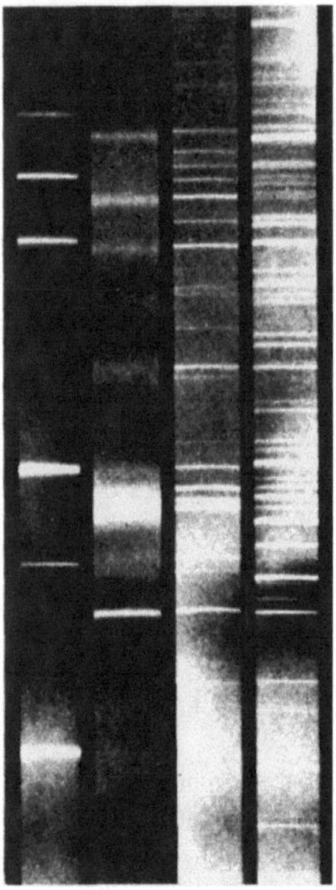

FIGURE 4. X-ray diffraction patterns of Ca(OH)2, Cortical Bone (CORT.), hydroxyapatite (HA) and whitlockite, Ca3(PO4)2 (WHIT.) showing the relationships between different composition solid materials and the relative diffusness of the typical bone pattern.

ANALYTICAL TECHNIQUES

X-ray diffraction is the method of choice to identify crystalline materials. It is used to distinguish between the many calcium phosphate structures and to detect other phases (such as $CaCO_3$ in the forms aragonite or calcite) which may be present in a sample. Figure 4 illustrates the diffraction patterns obtained from several well crystallized calcium phosphate minerals compared with a pattern from cortical bone. The small grain size of the latter material accounts, at least in part, for the diffuse diffraction results. Diffraction is also employed to differentiate among the Ca-apatite species. The calculation of unit cell parameters can be used to estimate compositional variations (Skinner, 1968). Table 2 lists the unit cell parameters (*a* & *c*) for Ca-apatites of different compositions from several sources, both high and low temperature materials have been included. The data show a remarkable consistency for *c* and a small range for *a* with the fluorine containing species exhibiting the smallest *a* values. There are overlaps between high and low temperature, natural and synthetic samples.

Gulbrandson (1970) suggested a peak-pair X-ray diffraction method for estimating the chemical composition of apatites. Using the diffraction maxima predominantly contributing to the *a* and *c* dimensions shifts related to the level of substitution can be measured. The peak-pairs (300) – (002) or (410) – (004) were selected by Gulbrandson to calculate the percentage of carbonate in marine phosphorites, mostly of Miocene and older ages. A range of 1.8 – 4.2% carbonate was found using this method (Gulbrandson, 1970). LeGeros (1969) prepared carbonate apatite by precipitation from solutions at 100° C and estimated the carbonate content of the products using the peak-pair method. It should be noted that above 22 wt% carbonate a separate and second solid phase was detected, $CaCO_3$ as aragonite.

The sensitivity of the X-ray techniques for determining composition and the results are directly related to the quality of the diffraction maxima used for the measurements. Low temperature precipitates (below 50° C), synthetic or natural, usually give extremely diffuse patterns (*i.e.*, bone, Fig. 4) making accurate estimate of carbonate content by the diffraction method difficult, if not impossible. Synthetic studies have demonstrated that carbonate intra-apatite is disruptive to the crystal structure resulting in poor "crystallinity", and probably contributions to small crystallite size.

Carbonate determinations on carbonate-apatites using other techniques, such as Infrared or Raman spectroscopy or direct chemical detection of evolved CO_2 for

TABLE 2. Unit cell parameters of mineral and synthetic apatites. *a* and *c* in nm

APATITES		a	c	SOURCE
HIGH TEMPERATURE				
Fluorapatite				
	Natural	0.9391	0.6878	Young *et al.*, 1969
	Synthetic	0.9372	0.6886	Bonel, 1972
Hydroxyapatite				
	Natural	0.9432	0.6881	Posner et al., 1958
	Synthetic	0.9414	0.6873	Skinner, 1980
Carbonate Apatite				
	Natural	0.9436	0.6883	Simpson, 1965
	Synthetic	0.9399 – 0.9414	0.6874 – 0.6909	Nelson & Featherstone, 1982
		0.9367 – 0.9537	0.6865 – 0.6934	Driessens *et al.*, 1983
LOW TEMPERATURE				
Francolite				
	Staffel, Ger.	0.9346	0.6887	Brophy & Nash, 1968
	Corneillue, Fr.	0.9361	0.6893	Bonel, 1972
Dahllite				
	Saxony	0.9419	0.6886	Brophy & Nash, 1968
Bone				
	Living	0.942	0.688	McConnell, 1970
	Fossil	0.9357	0.6888	Brophy & Nash, 1968
Phosphorites (extracted)		0.9322 – 0.9385	0.6883 – 0.6900	McClellan & Lehr, 1969

example, may be hindered by the presence of other carbon containing compounds in the sample. Biologic apatitic materials contain bio-molecules and probably carbonate, if not $CaCO_3$, associated with the apatitic minerals. For example the biological remains in phosphorite samples are difficult to separate from the fine grained apatitic mineral phase. In bone the crystallites are intimately intermingled with fibrous proteins, predominantly collagen, and other cell products. Estimation of the portion of carbonate within the apatite phase in mineralized tissues is exceedingly difficult and probably not accurate in spite of the diligence of investigators, delicate techniques and extraction procedures.

CARBONATE APATITES: EXPERIMENTAL INVESTIGATIONS

In experimental systems, control over the partial pressures of CO_2 and H_2O and cation composition allows substitution within the apatite structure (Bonel, 1972). Two sites have been suggested: *Type A*, carbonate substituting for the F/OH), and *Type B*, carbonate substituting for phosphate (Labarthe *et al.*, 1971). Bonel (1972) produced a *Type A* carbonated apatite (carbonate in place of OH) by treating hydroxyapatite with a stream of CO_2 gas at 900° C for about 5 h. There is an increase in the *a*-axis parameter (Table 1). The carbonate for phosphate substitution was investigated by Zapanta-Le-Geros (1965) who suggested that additional and coordinate cation substitution, *i.e.*,

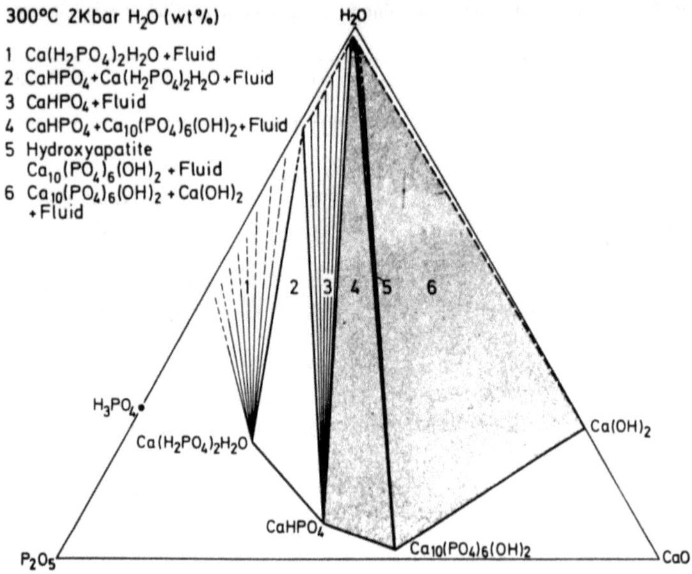

FIGURE 5. Equilibrium phase diagram of the $CaO-P_2O_5-H_2O$ system at 300° C and 2 Kbars H_2O pressure. The several solid phases present in the different phase fields are listed. Note that hydroxyapatite, in association with another solid phase and fluid occurs over much of the diagram but only in phase field #5 is it the only solid phase present with the fluid phase (Skinner, 1973, mod.).

sodium displacing some calcium, would be required to maintain charge balance. Other researchers seeking to accommodate carbonate substitution in calcium-only apatites at the phosphate site have suggested that vacancies might occur in Ca and OH sites (Labarthe *et al.*, 1973).

In addition to maintaining charge balance, another problem associated with the substitution of carbonate for phosphate is the molecular configuration. The tetrahedral motif of phosphate is the major contributor to the apatite crystal structure (Figs. 1 & 2). It has been suggested that the planar trigonal carbonate ion be coupled with F or OH to form a tetrahedral group, $[F+(CO_3)]$ or $[(OH)+(CO_3)]$, as a substitute for phosphate; that combinations of Na or K + CO_3 replace the tetrahedral ion; or that four carbonate groups might substitute for three phosphate groups to satisfy the structural considerations.

The fact that carbonated apatitic materials can be experimentally produced either at high temperatures and high CO_2 pressures or precipitated in the temperature range 50-100° C. from solutions that contain sodium or other ions is not in doubt. However, sophisticated techniques such as infrared and Raman spectroscopy, X-ray diffraction and chemical analyses, used to determine the concentrations and locations of carbonate have not produced a consensus on the carbonate configuration or substitution mechanisms within the apatite structure.

The most recent experimental investigations have been focused on mixed *A* and *B* type substitutions. Driessens *et al.* (1983a,b), using high temperature/pressure data, favored an interpretation where vacancies occur in the Ca and OH sites when no other cations, *i.e.*, sodium are available or coupled substitution of Na + CO_3 for Ca + PO_4 with 40% of the carbonate at OH sites (*Type A* substitution). Nelson & Featherstone (1982), compared high temperature (>900° C) and low temperature (85 – 100° C) synthetic data, and suggested that four carbonate groups replace three phosphate groups with Na and OH groups involved in the substitution. Santos & Gonzalez-Diaz (1977) interpreted data on solution synthesized *Type A B* material as carbonate substituting for HPO_4^{2-}-groups. An initial non-stoichiometric hydrated structure precipitates, and during maturation all possible HPO_4 groups are substituted by carbonate with the final solid as a mixture of calcium carbonate, hydroxyapatite and water. These investigators asserted that "For stoichiometric hydroxyapatite substitution of CO_3^{2-} is obviously impossible." (Santos & Gonzalez-Diaz, 1977:2133). In summary, all researchers, including earlier experimenters (Simpson, 1964; Bonel, 1972) propose crystal structural ramifications (vacancies or different configurations) to allow carbonate incorporation into an apatitic phase.

Experimental investigations characterize the carbonate containing apatite as a single phase although obtaining such samples require a great deal of effort. Driessens (1983b:20) states that attaining equilibrium to assure the synthesis of ideal composition Ca-apatites took up to six months with a high temperature (870° C), low humidity, CO_2 gas stream system. The addition of sodium hastened the attainment of equilibrium; *i.e.*, when the apatite phase produced in the gas stream at elevated temperatures exhibited consistent X-ray parameters after continued exposure to the conditions. Similar difficulty in obtaining equilibrium was expresed by Skinner (1973:566) using hydrothermal synthesis and elevated pressures during the construction of the equilibrium phase diagram for the system CaO-P_2O_5-H_2O. Crystalline stoichiometric hydroxyapatite crystals in association with fluid grew in a very limited phase field (#5 in Fig. 5). Over most of the composition range of the diagram, two solid phases (one of which is hyroxyapatite), are in equilibrium with a fluid phase. Driessens single phase apatite field is also very restricted (Driessens *et al.*, 1983b:21, Table 1) but it is not clear from the data presented how much OH might be present in a particular carbonate apatite sample since the unit cell parameters range from those typical of fluorapatite to *Type A* carbonate apatite (Table 2). Note that no fluorine was used in the preparation of these materials. The difficulty of obtaining a unique, single phase, apatite, even under controlled conditions in a simplified experimental system, cannot be overstated.

CARBONATE APATITES: NATURAL OCCURRENCES

The several alternative models of carbonate in Ca-apatites based on recent experimental investigations reiterate earlier suggestions of Gruner & McConnell (1937), McConnell & Gruner (1940) and Gruner, McConnell & Armstrong (1937) who had examined carbonated apatite mineral samples, dental enamel and dentine. They proposed inclusion of carbonate in the place of phosphate, with or without the addition of other cations. The name dahllite was applied to the low fluorine hydrated biomaterials and formulae were presented to account for differences in the stoichiometry of these apatitic materials (McConnell, 1971).

McClellan & Lehr (1969) studied the sedimentary carbonate fluorapatites known as francolites. To such fluorapatites, they assigned the general formula $Ca_{10-a-b}Na_a Mg_b(PO_4)_{6-x}(CO_3)_xF_y$, where the value of x ranged from near zero to 1.5, (or 25% carbonate) based on chemical and X-ray analyses for over 100 commercial samples from many localities and ages (McClellan & Lehr, 1969:1379). These natural materials contain many elements in addition to calcium, phosphorus and carbonate (see Table 1) and are relatively high in F, H_2O and OH. The difficulty of assessing and assigning H_2O in apatitic materials has been discussed by McConnell (1973).

The high concentration of fluorine (Table 1), often above that required for satisfying the site in the apatite structure (Figs. 1−3) has led to the suggestion that carbonate in these samples is probably located in the phosphate sites and a $F+CO_3$ substitution 1:1 for phosphate was proposed (McClellan & Lehr, 1969). In a recent paper based on considerations of cationic size and coordination ratios for a single phase apatite, McClellan (1980) predicted a maximum substitution of 1.33 carbonate for the 6.00 phosphate groups per unit cell. The calculated limiting value (approximately 6.5 wt%) was in good agreement with the average carbonate determined by direct analyses, 6.3 wt% (McClellan, 1980:678, Table 3). The calculated maximum cation composition (assuming that the carbonate is associated with F or OH for these francolite samples was ($Ca_{9.51}$, $Na_{0.35}$, $Mg_{0.14}$) also in good agreement with the observed maximum amount of Na=0.51 and Mg=0.25. Major chemical substitutions, such as suggested by these samples, within the apatite structure, would cause a shift in the *a*-axial unit cell parameter. Using X-ray data on phosphate rock from Tunisia, 5.70% carbonate and 1.20% Na_2O, were estimated to be in samples containing 92.2% apatite, on the assumption that all P_2O_5 in the rock analysis was as apatite (McClellan, 1980:678-679, Tables 4 & 5). Note that these samples exhibit a small *a* unit cell dimension (Table 2) commensurate with their high fluorine content.

The fluorine level is used as an indicator of the age of the samples. There is evidence that the initial precipitate is more hydrated and fluorine uptake is a post depositional phenomenon (secondary). Modern samples from Namibia show lower fluorine content than older samples (Thomson, 1983). The fluorine content of phosphorites vary, perhaps with age, but not directly with P_2O_5 (Thomson *et al.*, 1984). Any fluorine uptake will be influenced not only by the concentration of fluorine in sea or pore water but by the basic composition (*i.e.*, carbonate) content of the solid undergoing transformation.

DISCUSSION

One of the major drawbacks in defining carbonate apatites has been the impossibility of obtaining well crystallized material. In fact, it is difficult to experimentally produce large single crystals of stoichiometric hydroxyapatite [$Ca_{10}(PO_4)_6(OH)_2$] much less carbonate containing apatite which appears to require vacancies or extra cations (*i.e.*, sodium) to maintain electric neutrality and structural integrity. There remains differences of opinions on the precise location and configuration of carbonate in the apatite lattice in spite of investigations spanning 45 years. In view of the difficulties, it might be appropriate to revive an alternate suggestion: that of the two phase system in which calcium carbonate and apatite are separate solid phases but in intimate association at the atomic level.

McConnell states that there is no natural carbonated apatite where carbonate substitutes for OH (McConnell, 1971:1667). If one agrees and assumes a single phase apatitic material, carbonate must occupy a position similar to that of phosphate and coordinate with Ca, OH, F, or H_2O. The incorporation will require structural changes to accommodate the relatively large amounts of carbonate obtained on analysis of biologically mediated apatites. Further, it is likely that incorporation of carbonate within the Ca-apatite structure is probably primary, *i.e.*, precipitation occurs when local conditions provide the high concentration of the appropriate carbonate species (HCO_3 or CO_3). Carbonated apatites will be metastable, especially with respect to fluorapatite (McClellan, 1980) and over time a stoichiometric calcium fluor-apatite will develop. Because of the nature of the ionic apatite structure, natural phosphate-carbonate mixed materials will always be in a state of flux as they respond to the environment.

Phosphorites are sedimentary rocks containing a calcium (sodium and magnesium) carbonate fluorapatite. The model of formation presently favored (Thomson *et al.*, 1984) is precipitation within the pore solutions of diatom oozes with phosphorus derived from degeneration of biological materials, a primary or authogenic precipitation of apatite. This environment is also appropriate for the generation of CO_2, and hence the of carbonate to be included in the solid.

The growth rates of peletal phosphate and nodules have been shown to be much slower than surrounding sediment accumulation (Burnett *et al*, 1982). The source of phosphorus is believed to be bacteria and moved upward with the flux of pore waters rather than derived from seawater. The carbonate included in the apatite of phosphate nodules could be from the same source. A relatively closed environment might produce elevated carbonate in the sediment pore water and lead to local precipitation of carbonate or phosphate minerals or a mixed carbonate-phosphate phase. With continual variations in the concentrations of phosphate, or carbonate, the phases precipitating might be different mineral mixtures or a family of apatitic compounds with a range of compositions. The assumption underlying these proposals is that the intracrystalline carbonate or a separate phase in a mixed crystal, is part of the of the initial precipitate rather than a secondary alteration of the apatite in which carbonate replaces phosphate groups. And, as McArthur (1985) concluded from examination of the chemistry of several samples of unaltered francolites, the rate of francolite formation is "not an important factor in controlling the composition" (McArthur, 1985:29). He states that the predominant control on carbonate substitution is pH.

Examination of the carbon isotopic signature of modern phosphorites has suggested that there are different zones of formation related to the metabolic status of the bacteria aiding precipitation (McArthur *et al.*, 1980). Post depositional diagenesis would alter the composition of the apatitic mineral in phosphorite from a francolite to the more stable fluorapatite. Expulsion of intra-lattice carbonate could aid the nucleation of separate mineral phases such as calcite or dolomite (McClellan, 1980).

The carbonate content of Ca-apatites in vertebrate tissues may also vary. By analogy, the compositional differences of the initial bone mineral precipitate should reflect the metabolic rate and pH generated and maintained by the local cell population. Since the fine-grained mineral is not usually bathed in solutions containing high concentrations of fluorine, but does act as an exchange system, the solids may take up other cations (Na, K, Sr) in time. However the carbonate content and the crystallinity may be relatively unaffected resulting in Ca-apatites that show few differences whether they are newly deposited or have existed over the lifetime of the individual. Since the viability of vertebrae hard tissues depends, to some extent, on the ease of resorption and redeposition of the mineral phase, the fact that the Ca-apatites retain carbonate and low crystallinity may be important factors in the processes of biological selection. The calcium phosphate system, rather than other compositions with different crystal chemistries, may have been selected for vertebrate mineralization because of the requirement of intraskeletal mineral turnover.

SUMMARY

Natural low temperature apatite samples may be primary precipitates but they continually respond to changes in the environment. Small grainsize and poor crystallinity, possibly the result of the initial incorporation of carbonate into the Ca-apatite structure, favors such dynamic reactivity. With their large surface area these fine apatitic materials offer advantages to the living systems but make it difficult to identify the specific mineral species involved.

Analysis of the mineral, beyond identification as "apatitic", has, and should continue to provide, useful information on the mechanisms of formation of sediments and mineralized tissues and their post-depositional reactions. Further experimental investigations are needed to isolate and characterize the possible carbonate apatite phase and/or other minerals that might be present in the tissues and sedimentary rocks.

ACKNOWLEDGEMENT

Participation in the Symposium was made possible by grants from The Organized Research Fund of The University of Texas at Arlington and The Sea Grant Program administered by Texas A & M University.

REFERENCES

ALBRIGHT, J.A., & SKINNER, H.C.W., 1987. Bone: Structural organization and remodeling dynamics. *In* Scientific Basis of Orthopaedics, 2 ed. (J.A. Albright & R.A. Brand) 161-198. New York: Appleton-Century-Crofts.

ARMSTRONG, W.D., & SINGER, L., 1965. Composition and constitution of the mineral phase of bone. Clin. Orthop., 38:179-190.

BILTZ, R.M., & PELLIGRINO, E.D., 1969. The chemical anatomy of bone. I. A comparative study of bone composition. J. Bone Joint Surg., 51 A:456-466.

BONEL, G., 1972. Contribution a l'être de le carbonatation des apatites. I. Synthese et etude des proprieties physico-chemiques des apatites carbonatees du type A. Ann. chem. (Paris), 7:65-78.

BROPHY, G.P. & NASH, T.J., 1968. Composition, infrared and x-ray analysis of fossil bone. Amer. Mineral., 53:445-454.

BURNETT, W.C., 1977. Geochemistry and origin of phosphorite deposits off Chile and Peru. Bull. Geol. Soc. Amer., 88:813-823.

BURNETT, W.C., BEERS, M.J., & ROE, K.K., 1982. Growth rates of phosphate nodules from the continental margin off Peru. Science, 215:1616-1618.

CHENEY, T.M., MCCLELLAN, G.H. & MONTGOMERY, E.S., 1979. Sechura phosphate deposits, their stratigraphy, origin and composition. Econ. Geol., 74:232-259.

DRIESSENS, F.C.M., 1982. Mineral Aspects of Dentistry. Basil, Switzerland, S. Karger.

DRIESSENS, F.C.M., VERBEECK, R.M.H. AND KIEHENS, P., 1983a. Mechanism of substitution in carbonated apatites. Zeit. Anorganic Alloganic Chem., 504:195-200.

DRIESSENS, F.C.M., VERBEECK, R.M.H., & HEIJLIGERS, H.J.M., 1983b. Some physical properties of Na-and CO3-containing apatites synthesized at high temperatures. Inorgan. Chim. Acta, 80:19-23.

GRUNER, J.W., & MCCONNELL, D., 1937. The problem of carbonate apatites. The structure of francolite. Zeit. Kristallogr., 97A:208-215.

GRUNER, J.D., MCCONNELL, D., & ARMSTRONG, W.D., 1937. The relationship between crystal structure and chemical composition of enamel and dentine. J. Biol. Chem., 121:771-781.

GULBRANDSON, R.A., 1966. Chemical composition of phosphorites of the Phosphoria formation. Geochim. Cosmochim. Acta, 30:769-778.

GULBRANDSON, R.A., 1969. Physical and chemical Factors in the formation of marine apatite. Econ. Geol., 64:365-382.

GULBRANDSON, R.A., 1970. Relation of carbon dioxide content of apatite of the Phosphoria Formation to regional facies. U.S. Geol. Surv. Prof. Pap., 700B:9-13.

LABARTHE, J.C., BONEL, G., & MONTEL, G., 1971. Sur la localization des ions carbonate dans le reseau des apatites calciques. Comp. Ren. Acad. Paris, 273C:349-355.

LABARTHE, J.C., THERASSE, M., BONEL, G. & MONTEL, G., 1973. Sur le structure des apatites phosphocalcique carbonees de type B. Compte Rendue Academie de Paris, 276C, 1175-1178.

LEGEROS, R.Z., TRAUTZ, O.R., KLEIN, E. & LEGEROS, J.P., 1969. Two types of carbonate substitution in the apatite structure. Experimentia, 25:5-7.

MCARTHUR, J.M., 1980. Post depositional alteration of the carbonate-fluorapatite phase of Moroccan phosphates. *In* Marine-Phosphorites-Geochemistry, Occurrence Genesis (Y.E. Bentor, ed.), p. 53-60. Soc. Econ. Paleontol. Mineral. Spec. Publ. 29:53-60.

MCARTHUR, J.M., 1985. Francolite geochemistry-compositional controls during formation, diagenesis, metamorphism and weathering. Geochim. Cosmochim. Acta, 49:23-35.

MCCLELLAN, G.H., 1980. Mineralogy of carbonate fluorapatite. Geol. Soc. London, 137:675-681.

MCCLELLAN, G.H., & Lehr, J.R., 1969. Crystal chemical investigations of natural apatites. Amer. Mineral., 54:1374-1391.

MCCONNELL, D., 1970. Crystal chemistry of bone mineral: Hydrated carbonate apatites. Amer. Mineral., 55:1659-1669.

MCCONNELL, D., 1971. Crystal chemistry of phosphorite. Econ. Geol., 66:1085-1086.

MCCONNELL, D., 1973. Apatite. Its Crystal Chemistry, Mineralogy, Utilization, and Geologic and Biologic Occurrences. New York, Springer Verlag.

MCCONNELL., D. & GRUNER, J.W., 1940. The problem of carbonate apatites. III Carbonate apatite from Magnet Cove, Arkansas. Amer. Mineral., 25:157-167.

NELSON, D.G.A., & FEATHERSTONE, J.D.B., 1982. Preparation, analysis and characterization of carbonated apatite. Calc. Tiss. Inter., 34:S69-S81.

NEUMAN, W.F., & MULRYAN, B.J., 1967. Synthetic Hydroxyapatite Crystals. III The carbonate System. Calc. Tiss. Res., 1:94-104.

SANTOS, M. & GONZALEZ-DIAZ, P.F., 1977. A model for B Carbonate Apatite. Inorgan. Chem., 16:2131-2134.

SCHRAER, R., & SCHRAER, H., 1970. The avian shell gland: A study in calcium translocation. *In* Biological Calcification: Cellular and Molecular Aspects (H. Schraer, ed.), pp. 347-373. New York: Appleton-Century-Crofts.

SIMPSON, D., 1964. The nature of alkali carbonate apatites. Amer. Mineral., 49:363-376.

SIMPSON, D., 1965. Carbonate in hydroxlapatite. Science, 147:501-502.

SIMPSON. D., 1972. Problems of the composition and structure of bone minerals. Clin. Orthop. Rel. Res., 86:260-286.

SKINNER, H.C.W., 1968. X-ray diffraction analysis technique to monitor composition fluctuation within the mineral group-apatite. Appl. Spectros., 22:412-414.

SKINNER, H.C.W., 1973. Phase relations in the CaO-P2O5-H2O system from 300° C. to 600° C. at 2 kb H2O pressure. Amer. J. Sci., 273:545-560.

SKINNER, H.C.W., KEMPNER, E.S., & PAK, C.Y.C., 1972. Preparation of the mineral phase of bone using ethylenediamine extraction. Calc. Tiss. Res., 10:257-268.

THOMSON, J.M., 1983. Offshore Peruvian phosphorite: a reappraisal of its age and genesis. Chem. Geol., 38:93-105.

THOMSON, J., CALVER, S.E., MUKERJEE, S., BURNETT, W.C. & BREMNER, J.M., 1984. Further studies of the nature, composition and ages of contemporary phosphorite from the Namibian Shelf. Ear. Planet. Sci. Let., 69:341-353.

YOUNG, E.J., MYERS, A.T., MUNSON, E.L., & CONKLIN, N., 1969. Mineralogy and geochemistry of fluorapatite from Cerro de Mercado, Durango, Mexico. U.S. Geol. Surv. Prof. Pap., 650:84-89.

YOUNG, R.A., & ELLIOTT, J.C., 1966. Atomic-scale bases for several properties of apatites. Arch. Oral Biol., 11:699-707.

ZAPANTA-LEGEROS, R., 1965. Effect of carbonate on the lattice parameters of apatite. Nature, 206:403-404.

ZIPKIN, I., 1970. The inorganic composition of bones and teeth. *In* Biological Calcification: Cellular and Molecular Aspects (H. Schraer, ed.), pp. 69-103. New York: Appleton-Century-Crofts.

PHOSPHORUS AND THE FERRITIN IRON CORE: FUNCTION-BALANCED BIOMINERALIZATION

Kenneth M. Towe

Department of Paleobiology, Smithsonian Institution, Washington, DC 20560

ABSTRACT. In the biomineralization of storage iron, the proteins ferritin and hemosiderin sequester iron in a mineral form of colloidal dimensions (<80 Å). This mineral is the metastable hydrous ferric oxide *ferrihydrite* ($5Fe_2O_3 \cdot 9H_2O$). Ferrihydrite represents an appropriate mineralogical choice to balance the important but opposing physiological functions of iron storage; and iron mobility. The ferrihydrite in both ferritin and hemosiderin contains important amounts of non-structural phosphorus adsorbed to the crystallite surfaces. Experimental data support the suggestion that this phosphorus can act to preserve the role of iron mobility by inhibiting the potential during iron storage for inversion of the more readily soluble, metastable ferrihydrite to one of the less soluble, more stable phases such as goethite ($\alpha FeOOH$) or hematite (αFe_2O_3).

INTRODUCTION

In biological systems, the processes of iron storage and transport can be viewed as a matter of achieving and maintaining an appropriate balance in the stability/instability of the storage compound(s) involved. A highly soluble compound could result in iron toxicity or iron storage deficit while a highly stable, insoluble form could impair iron mobility with the potential for an iron storage overload. Thus, in biological systems iron must be bound in such a way as to keep it non-toxically sequestered, yet it must be readily available when required for transport and use. The purpose of this paper is to provide support for the view that in size, chemistry and mineralogy the ferritin iron core, has evolved toward a functionally-balanced condition ideally suited to its dual roles.

Ferritin is the principal iron-protein complex; involved in biological iron storage (Harrison, 1969; Ford *et al.*, 1984). The apoprotein is subspherical with an outside diameter of 125 Å (Fig. 1). Surrounded by the protein, the iron "core" or "micelle" of ferritin is made up of one or more crystalline particles of hydrous ferric oxide; that form within a cavity of about 80 Å diameter. Hemosiderin, a related but less well-defined protein, occurs as insoluble granules with crystalline iron particles that are closely similar to those found in ferritin (Fischbach *et al.*, 1971).

Initial X-ray and electron diffraction data (Haggis, 1965; Harrison *et al.*, 1967; Towe & Bradley, 1967) indicated that the crystal structure of the ferritin core differed from that of other known iron oxide minerals. Although it was similar to the mineral hematite (αFe_2O_3), there were distinct differences in the diffraction patterns. The similar

Origin, Evolution, and Modern Aspects of Biomineralization in Plants and Animals
Edited by Rex E. Crick
Plenum Press, New York

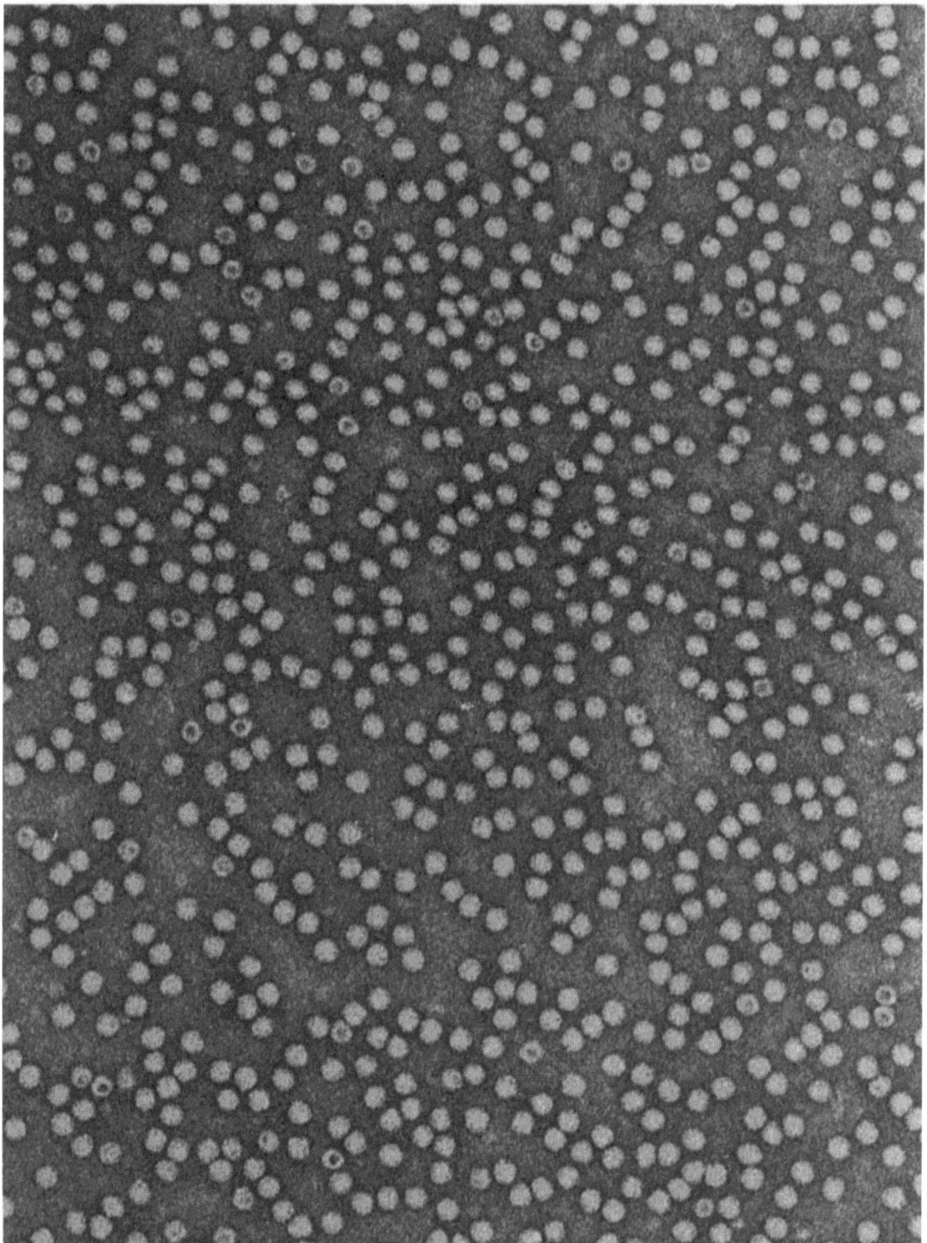

FIGURE 1. TEM of horse spleen apoferritin molecules negatively-stained with 2% Na-phosphotungstate at pH 7.0 (× 258,000). A few ferritin molecules (with dark cores) can be seen. The material is a *native* apoferritin preparation (=not "synthesized") from ferritin by using dithionite treatment to remove iron. Sucrose density gradient ultracentrifugation and Cd(SO4) crystallization were used to concentrate and purify the lighter fraction (predominantly apoferritin) from the heavier iron-containing ferritins.

diffraction data on hemosiderin, as well as data gathered from a number of geological and other natural occurrences (van der Giessen, 1966; Towe & Lowenstam, 1967; Jackson & Keller, 1970; Chukrov *et al.*, 1971) were all sufficiently clear and distinct that in 1973 the International Mineralogical Association's Commission on New Minerals and Mineral Names approved the new mineral *ferrihydrite* (Chukrov *et al.*, 1973) with an ideal formula composition of $5Fe_2O_3{\cdot}9H_2O$.

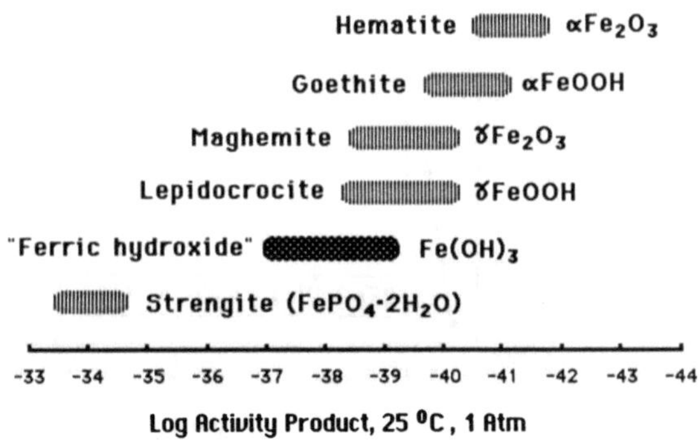

FIGURE 2. Comparative solubility data for a variety of iron minerals. The stippled "ferric hydroxide" field encompasses the range of activity products expected for ferrihydrites.

RATIONALE

Ferrihydrite is metastable with a marked tendency on aging to form increased particle sizes and more stable minerals, usually hematite (αFe_2O_3) or goethite ($\alpha FeOOH$) (Towe & Bradley, 1967; Chukrov *et al.*, 1973; Schwertmann & Fischer, 1973). Although there are no published experimental data on the thermodynamic stability of ferrihydrite (identified as such), many so-called "amorphous ferric hydroxides" are now known to be ferrihydrite (Towe & Bradley, 1967; Schwertmann & Fischer, 1973). Thus, the solubility of ferrihydrite should bracket the various solubility measurements reported from the many mineralogically uncharacterized precipitates with activity products in the range of 10^{-37} to 10^{-39} (stippled area, Fig.2). The activity product of ferrihydrite can reasonably be estimated, therefore, from the compiled sources of Langmuir (1969, 1971) to be approximately 10^{-38}. The formation of hematite or goethite from metastable compounds within the stippled area of Figure 2 would result in substantial, order of magnitude increases in stability with the resulting activity products in the shaded areas ranging from 10^{-40} to 10^{-42}.

This has relevance to biological systems because, regardless of what factors control biological iron metabolism (*i.e.*, enzymes, chelation, redox reactions), these mechanisms must operate within the constraints imposed by the physical/crystal chemistry of the compounds involved. Therefore, from the point of view of biological iron *storage* (or iron biomineralization) an increase in stability might be considered desirable. On the other hand, any increase in stability must also reduce the solubility and hence the ease of mobility and *transport* of iron.

Because of the small size of their ferritin cores the large surface areas favor a ready solubility for iron transport. The available space within the average apoprotein molecule constrains most cores to a diameter of about 80 Å (Banyard *et al.*, 1978). Although exceptional iron cores as large as 98 Å have been reported (Massover & Cowley, 1973), the majority of the ferrihydrite cores in most ferritins are closer to 60-70 Å in diameter (Harrison & Hoy, 1973). To assist in the iron transport function this size limitation

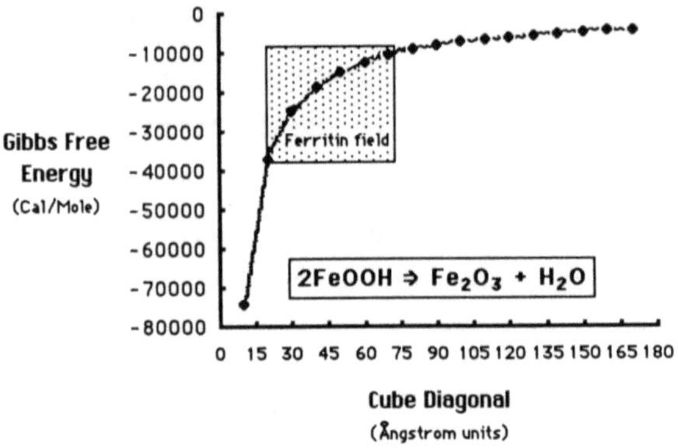

FIGURE 3. Illustration of the marked change in stability of goethite with respect to hematite at very small crystallite sizes. Size range of most ferritin iron cores superimposed in stippled area. Data replotted from Langmuir (1971).

serves to prevent the automatic reduction in solubility that would accompany increasingly larger cores with their correspondingly reduced surface areas.

From the point of view of mineralogy and crystallite size, ferritin appears "well-designed" for the purposes of iron storage and transport. On the other hand, in the normal ferritin particle size range (stippled area, Fig. 3) kinetics and thermodynamics markedly favor the formation of hematite by inversion from less stable compounds (Langmuir, 1971). In the case of ferritin, this reaction would lead to a marked decrease in solubility for the iron core unrelated to surface area considerations. Thus, from the point of view of iron metabolism the formation of hematite with a much greater insolubility compared with ferrihydrite, would create an imbalance in favor of storage and against transport. How has this metabolic dilemma been solved, or at least brought into balance?

RESULTS AND DISCUSSION

In contrast to ferrihydrites synthesized in the laboratory, analyses of both ferritin; and hemosiderin micelles always show small, but significant amounts of phosphorus. Because changes in the phosphorus (and loosely-bound water) content of the cores do not affect the diffraction data (Harrison *et al.*, 1967) neither is considered an integral component of the crystal structure of the core mineral. That most of the phosphorus is adsorbed to the surface of the iron cores is also supported by ferritin core reconstitution experiments (Treffry & Harrison, 1978) and by the fact that, for the most part, the iron occurs in ratios with the phosphorus that are consistent with the volume-to-surface area and diameter relationships expected of an ideal sphere. The iron compound in ferritin (and hemosiderin) may therefore be classified as ferrihydrite with variable adsorbed water

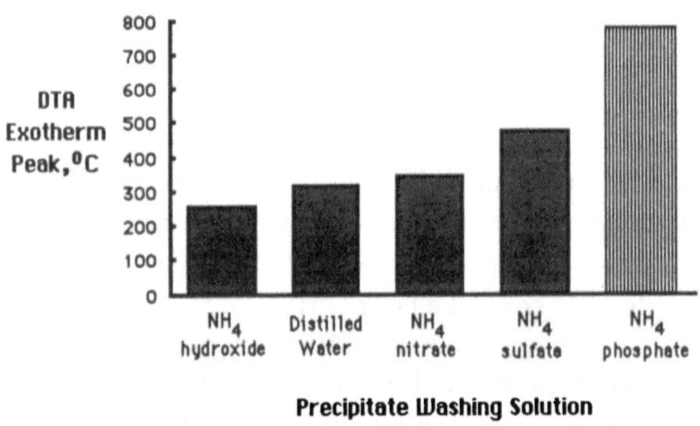

FIGURE 4. Influence on thermal stability of freshly-prepared, cold-precipitated hydrous iron oxides by various anions surface-adsorbed during washing process. Data from R.C. Mackenzie (1952).

and phosphate. It is suggested here that this phosphorus serves to even out the imbalance described above for the storage and transport functions in ferritin.

It is well-known to soil chemists that surface-adsorbed anions can influence the stability of colloidal iron compounds. As illustrated by the data of Mackenzie (1952) plotted in Figure 4, phosphate adsorbed to freshly-precipitated colloidal hydrous ferric oxides (so-called "ferric hydroxides") markedly inhibits their complete transformation to more stable phases. Specifically, the differential thermal analysis (DTA) exotherm peak signals the formation of hematite (αFe_2O_3) by recrystallization from the "$Fe(OH)_3$" starting material.

TABLE 1. COMPARATIVE DIFFRACTION DATA. X-ray and electron diffraction data on 10 different variously treated ferritin iron cores and ferrihydrites. There is a general increase in stability toward thermal treatment from left to right. (observed d-spacings given in Ångstrom units)

	A	B	C	D	E	F	G	H	I	J	
					3.70				3.67	3.67	
				2.71	**2.70**		**2.70**	**2.70**	**2.69**	**2.69**	⇐
	2.54	2.54	2.54	2.54	2.52	2.54	2.52	2.52	2.52	2.51	
	2.24	2.24	2.24	2.24	2.22	2.24	2.23	2.23	2.20	2.20	
⇒	**1.98**	**1.98**	**1.98**	**1.98**	**1.98**	**1.98**	**1.97**	**1.98**			
					1.84			1.85	1.84	1.84	
	1.72	1.72	1.72	1.72	1.70	1.72	1.72	1.70	1.69	1.69	
										1.64	
					1.60				1.60	1.60	
	1.51	1.51	1.51	1.51		1.51	1.52	1.52			
	1.47	1.47	1.47	1.47	1.48	1.47	1.47	1.46	1.49	1.48	
									1.46	1.45	

A Reference ferrihydrite
B Ferritin, untreated, wet
C Ferritin, 5% NaOCl, dried 50ºC
D Ferritin, water, 100ºC, 2 hours
E Ferrihydrite, water, 100ºC
F Ferritin, oven-dried, 110ºC, 1 hour
G Ferritin, oven-dried, 400ºC, 1 hour
H Ferritin, oven-dried, 500ºC, 1 hour
I Ferritin, oven-dried, 600ºC, 1 hour
J Reference hematite

⇐ Diagnostic hematite peak
⇒ Diagnostic ferrihydrite peak

To test whether this effect could have any significance for iron biomineralization, the iron cores of ferritin (isolated without heat treatment from a fresh horse liver) were prepared free of their protein shell by treatment with commercial sodium hypochlorite bleach (5.25% NaOCl). Synthetic ferrihydrites (free of phosphorus) were prepared according to the method of Towe and Bradley (1967). Both preparations were heated to various temperatures and monitored by means of X-ray and/or electron diffraction methods for their inversion to hematite. Although more time-consuming and difficult, the diffraction method provides a more sensitive monitoring of the mineral transformations than DTA analyses.

Because ferrihydrite and hematite have closely similar crystal structures identification of these minerals from their diffraction patterns requires care. A comparison of their observed *d*-spacings is given in Table 1: columns B and J. As shown, most of the diffraction peaks overlap one another and are of little value in permitting the detection of one of these minerals in the presence of the other. However, two peaks are the most diagnostic: the very strong hematite (104) peak at *d*(calc.) 2.70 Å, $I=100$ and the moderately weak ferrihydrite (113) peak at *d*(calc.) 1.97 Å, $I=20$. Thus, in evaluating the results of the heating experiments the presence or absence of either of these two diagnostic peaks is significant. The appearance of the (104) hematite peak, because of its strong diffraction intensity, can mean the appearance of very small amounts of hematite mixed in with ferrihydrite. On the other hand, the presence of the (113) ferrihydrite peak, because of its relatively weak diffraction intensity, means that significant amounts of ferrihydrite remain in the sample, unrecrystallized. The appearance of the moderately weak (108) hematite peak at *d*(calc.) 1.60 Å, $I = 30$ signals substantial hematite in the sample.

Diffraction patterns of untreated, wet ferritin as well as of the NaOCl-treated ferritin confirmed that neither air-drying nor removal of the protein shell affects the mineralogy of the iron cores. Movement of adsorbed phosphorus from ferritin during alkali treatment was not monitored, but as much as 80% could have been lost in the process (Granick, 1942). Transmission electron microscopy indicates that although there is a strong tendency during hydrolysis for the ferrihydrite preparations to form 300 Å diameter spherical-to subspherical aggregates, the sizes of the individual crystallites are closely similar to those of ferritin (Towe & Bradley, 1967).

The results of the heating experiments are summarized in Table 1. It is clear from this table that ferritin iron cores resist thermal alteration much more strongly than do phosphorus-free ferrihydrite crystallites. Column E shows that ferrihydrite simply brought to a boil in distilled water has been substantially converted to hematite. On the other hand, ferritin cores held at boiling for two hours (Column D) or oven-heated in air to 400° C for one hour (Column G) show only limited evidence for hematite formation. Even at 500° C (Column H) the ferritin cores, while mostly hematite, still show a ferrihydrite component (peak at *d*(obs.) 1.98 Å) which is finally eliminated at 600° C (column I). Ferrihydrite was completely converted to hematite at 425 °C in one hour (unshown). Chukrov *et al*., (1972) report that ferrihydrite ages to αFe_2O_3 on heating to 40° C for 10 to 14 days.

The above results fit well with the DTA transformations of Mackenzie (1952) illustrated in Figure 4. In his experiments the hydrous iron oxide precipitates washed with NH_4-phosphate resisted complete recrystallization until 780°C while those washed in distilled water (or NH_4OH) were completely transformed by 345° C. All of Mackenzie's preparations (air-dried at 105° C prior to DTA) lost their loosely-bound water similarly at 150° C.

CONCLUSIONS

The heating experiments described here provide strong support (but not proof) for the suggestion that the adsorbed phosphorus in mammalian ferritins (and hemosiderins) serves to inhibit the otherwise thermodynamically-spontaneous transformation of the ferrihydrite core to the most stable, insoluble of all ferric oxides αFe_2O_3 (hematite). Such kinetically-favored transformations, if allowed to proceed could be metabolically devastating, perhaps leading to iron overload. Ferritin iron cores therefore appear to be ideally suited to their opposed biological functions. They exist as an easily-formed compound that is mineralogically stable enough for adequate storage of iron, are small enough for easy solubility and ready transport, yet are chemically balanced to resist the inevitable transformation to the more insoluble compounds that their small size otherwise dictates.

In order to mimic the native iron cores of either mammalian ferritin or hemosiderin more closely, ferritin reconstitution or iron metabolism studies should give crystal-chemical considerations a high priority. Some of the models for such studies have wisely incorporated phosphorus into the iron compounds involved (*e.g.*, Treffry & Harrison,

1978; Mansouri *et al.*, 1985). But many of these preparations have been of uncertain mineralogical composition, or of a crystal chemistry quite different from the native condition. Similarly, iron complexes used regularly in the treatment of iron deficiencies have not been ideal mineralogically (Towe, 1981). Ferrihydrites with adsorbed phosphate would seem the best choice as a ferritin iron core analog.

ACKNOWLEDGEMENTS

Special thanks are due Dr. Albert Kahn (University of Copenhagen) for his advice and assistance regarding the isolation of the native apoferritin fraction. Participation in the Symposium was made possible by grants from The Organized Research Fund of The University of Texas at Arlington and The Sea Grant Program administered by Texas A & M University.

REFERENCES

BANYARD, S.H., STAMMERS, D.K., & HARRISON, P.M., 1978. Electron density map of apoferritin at 2.8 Å resolution. Nature, London, 271, 282-284.

CHUKROV, F.V., ZVYAGIN, B.B., GORSCHKOV, A, YERMILOVA, L.P., & RUDNITSKAYA, E.S., 1971. The Towe-Bradley phase - a product of the supergene alteration of ores. Izvestiya Akademii Nauk SSSR, seriya geologicheskaya, no. 1.

CHUKROV, F.V., ZVYAGIN, B.B., GORSCHKOV, A, YERMILOVA, L.P., & BALASHOVA, V.V., 1973. O ferrigidrite. Izvestiya Akademii Nauk SSSR, seriya geologicheskaya, no. 4, 23-33. English translation: Ferrihydrite. International Geology Review, 16, 1131-1143.

FISCHBACH, F.A., GREGORY, D.W., HARRISON, P.M., HOY, T.G., & WILLIAMS, J.M., 1971. On the structure of hemosiderin and its relationship to ferritin. Journal of Ultrastructure Research, 37, 495-503.

FORD, G.C., HARRISON, P.M., RICE, D.W., SMITH, J.M.A., TREFFRY, A., WHITE, J.L., & YARIV, J., 1984. Ferritin: design and formation of an iron-storage molecule. Philosophical Transactions Royal Society of London, B304, 551-565.

GRANICK, S., 1942. Ferritin. I. Physiological and chemical properties of horse spleen ferritin. Journal of Biological Chemistry, 146, 451-461.

HAGGIS, G.H., 1965. The iron oxide core of the ferritin molecule. Journal of Molecular Biology, 14, 598-602.

HARRISON, P.M., 1969. The biochemistry of iron. *In* Iron Metabolism and Anemia, Pan American Health Organization, Scientific Publication No. 184, 2-20.

HARRISON, P.M. & HOY, T.G., 1973. Ferritin. *In* Inorganic Biochemistry, vol. 1 (ed. G.L. Eichhorn), pp. 253-279. Amsterdam: Elsevier.

HARRISON, P.M., FISCHBACH, F.A., HOY, T.G., & HAGGIS, G.H., 1967. Ferric oxyhydroxide core of ferritin. Nature, London, 216, 1188-1190.

JACKSON, T.A. & KELLER, W.D., 1970. Evidence for biogenic synthesis of an unusual ferric oxide mineral during alteration of basalt by a tropical lichen. Nature, London, 227, 522-523.

LANGMUIR, D., 1969. The Gibbs free energies of substances in the system Fe-O2-H2O-CO2 at 25°C. U.S. Geological Survey Professional Paper 650-B, B180-B184.

LANGMUIR, D., 1971. Particle size effect on the reaction goethite = hematite + water. American Journal of Science, 271, 147-156.

MACKENZIE, R.C., 1952. Investigations on cold-precipitated hydrated ferric oxide and its origin in clays. *In* Problems of Clay and Laterite Genesis, pp. 65-75, New York: American Institute of Mining Engineers.

MANSOURI, A.N., THOMPSON, C., THEIL, E.C., CHASTEEN, N.D., & SAYERS, D.E., 1985. Fe(III) ATP complexes: models for ferritin and other polynuclear iron complexes with phosphate. Journal of Biological Chemistry, 260, 7975-7979.

MASSOVER, W.H. & COWLEY, J.M., 1973. The ultrastructure of ferritin macromolecules. The lattice structure of the core crystallites. Proceedings of the National Academy of Sciences, U.S.A., 70, 3847-3851.

SCHWERTMANN, U. & FISCHER, W.R., 1973. Natural "amorphous" ferric hydroxide. Geoderma, 10, 237-247.

TOWE, K.M., 1981. Structural distinction between ferritin and iron dextran (Imferon). Journal of Biological Chemistry, 256, 9377-9378.

TOWE, K.M. & BRADLEY, W.H., 1967. Mineralogical constitution of colloidal "hydrous ferric oxides". Journal of Colloid and Interface Science, 24, 384-392.

TOWE, K.M. & LOWENSTAM, H.A., 1967. Ultrastructure and development of iron mineralization in the radular teeth of *Cryptochiton stelleri* (Mollusca). Journal of Ultrastructure Research, 17, 1-13.

TREFFRY, A. & HARRISON, P.M., 1978. Incorporation and release of inorganic phosphate in horse spleen ferritin. Biochemical Journal, 171, 313-320.

VAN DER GIESSEN, A.A., 1966. The structure of iron(III) oxide-hydrate gels. Journal of Inorganic and Nuclear Chemistry, 28, 2155-2159.

THE ROLE OF INORGANIC PHOSPHATE IN IRON OXIDE BIOMINERALIZATION

Stephen Mann

School of Chemistry, University of Bath, Bath BA2 7AY, United Kingdom

ABSTRACT. The involvement of inorganic phosphate in the formation of biogenic iron-oxides such as ferrihydrite ($5Fe_2O_3 \cdot 9H_2O$) in the iron-storage protein ferritin and goethite (α-FeOOH) in the radula teeth of the limpet *Patella vulgata* is discussed from a chemical perspective. The structural characteristics and spatial distribution of biominerals in these systems has been studied by high resolution transmission electron microscopy (HRTEM), energy dispersive X-ray analysis (EDXA), proton-induced X-ray emission (PIXE) analysis and Mössbauer spectroscopy. The results indicate that phosphate is an important component in the mediation of these solid state processes, for example, in the regulation of structural and redox states, and nucleation and growth phenomena. Phosphate-rich domains in limpet teeth and bacterial ferritin are non-crystalline whereas the absence or low level of phosphate results in the deposition of crystalline phases. Phosphate interactions also influence the morphology, magnetic properties and composition of these biominerals. A general rationalization of phosphate-iron-oxide interactions is presented based on the results of chemical studies *in vitro*. Several possible roles for phosphate are identified; (a) crystal growth inhibition, (b) crystal growth mediation (morphological enhancement, microcrystalline states), (c) kinetic stabilization with respect to phase transformation (dynamic behaviour in homeostasis), (d) solubility control via complexation, and (e) redox control over Fe(II) oxidation and reductive dissolution of Fe(III) oxides.

INTRODUCTION

Biomineralization is the biological regulation and tailoring of structural inorganic chemistry for functional potential. Although much attention has been directed towards the formation of calcium salts, recent work has shown that iron-oxide biominerals are widespread in distribution (Lowenstam, 1981; Mann, 1983; Webb, 1983). These materials have important industrial properties. Our interests lie in the understanding of the mechanisms of controlled crystallization and in this respect the role of inorganic phosphate in iron-oxide biomineralization may be paramount. The objective of this paper is therefore to discuss, on the basis of recent inorganic and biological studies, the role of phosphate in the mediation of iron-oxide formation and to present a rationalization of these interactions for further development.

Table 1 lists several iron-oxide/phosphate biominerals present in extant organisms. Note that these phases have phosphate as an integral component of the final mineral state and that phosphate is involved in both the mineralization of crystalline and

Edited by Rex E. Crick
Plenum Press, New York

TABLE 1. Biogenic Fe-oxides containing phosphate

Mineral	Occurrence	Function	References
Fe_3O_4 (magnetite)	Mollusca	Teeth	Towe & Lowenstam,1967
$5Fe_2O_3 \cdot 9H_2O$ (ferrihydrite)	Bacteria	Fe storage and transport	Mann *et al.*,1986a Watt *et al.*,1986
	Animalia	Fe storage and transport	Michaelis *et al.*,1943
αFeOOH (goethite)	Mollusca	Teeth	Grime *et al.*,1985 Mann *et al.*,1986b
Amorphous	Annelida	Gizzard plate sternal shield	Lowenstam,1972
	Mollusca	Chiton teeth	Lowenstam,1972
	Echinodermata	Dermal granules	Lowenstam & Rossman,1975

non-crystalline biominerals. This paper will discuss two biological systems in detail, (a) goethite formation in limpet teeth and (b) ferrihydrite deposition in mammalian and bacterial ferritin in order to highlight the important roles of phosphate in iron-oxide biomineralization.

BIOLOGICAL SYSTEMS

We have used structure-determination techniques such as high resolution transmission electron microscopy (HRTEM) and ^{57}Fe Mössbauer spectroscopy combined with high resolution analytical procedures (energy dispersive X-ray analysis (EDXA) and proton-induced X-ray emission (PIXE) spectroscopy) in the study of the following systems involving biomineralization.

Limpet teeth

The common limpet, *Patella vulgata* inhabits the intertidal zone where it grazes on rock-encrusting organisms such as minute algae using a specially developed tongue or radula. The radula is 7 cm long, 2 mm wide and consists of 200 transverse rows of teeth. The mineral composition of the teeth is 12% ferric oxide and 10% silica by weight. The iron-oxide is present in the form of goethite (α-FeOOH) (Lowenstam, 1962).

We have used PIXE analysis to determine the spatial localization of elements during maturation of limpet teeth (Grime *et al.*, 1985). This technique has considerable advantages over EDXA. The use of a proton beam combined with a scanning system enables regions in the range of 1 μm diameter to be probed with an analytical sensitivity of 10 ppm. This enhancement in sensitivity is due to the high probability of characteristic X-ray production and low-continuum background radiation from the interactions of protons, rather than electrons, with the sample.

At early stages of mineralization the PIXE elemental maps showed a preferential localization of iron and phosphorus in the base of the teeth (Fig. 1a and b). Both elements are highly concentrated within the same region of the base just below the junction zone (Grime *et al.*, 1985). On maturation, iron-oxide mineralization develops

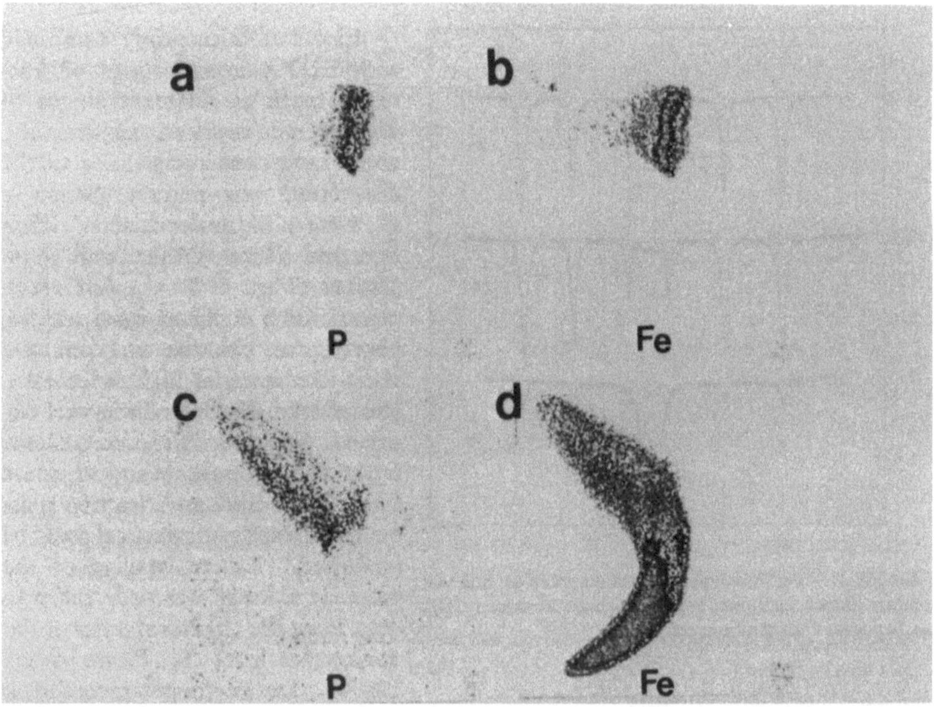

FIGURE 1. Scanning proton microprobe elemental maps of limpet teeth. ▪a. P map of early-maturing teeth (row 28). ▪b. Fe map of early-maturing tooth (row 28). ▪c. P map of mature tooth (row 120). ▪d. Fe map of mature tooth (row 120). In all cases the morphology of the teeth approximates that shown by the Fe map in ▪d. Highest concentrations are identified by the darkest grey shading. At row 28 only the base is mineralized. At row 120 Fe mineralization is in both the base and cusp but P is essentially localized in the base. From Grime *et al.*, 1985 (originals are in color).

in the tooth cusps whilst phosphate deposition remains essentially located within the bases (Fig. 1c and d).

The consequence of this precise spatial distribution on the structural properties of the goethite phase was studied by HRTEM (Mann *et al.*, 1986b) and ^{57}Fe Mössbauer spectroscopy (St. Pierre *et al.*, 1986b). Figure 2 shows Mössbauer spectra collected for (a) mature teeth, (b) separated cusps of mature teeth and (c) separated bases of mature teeth at 78 K. In spectrum (a), both a magnetically split sextet and a superparamagnetic doublet can be observed. The six line component shows a temperature dependence and magnetic hyperfine field (37.9T at 293 K) characteristic for well-ordered stoichiometric goethite crystals above 100 nm in dimension. In contrast, the doublet component shows a temperature dependence characteristic of a microcrystalline (<100 Å) or poorly-ordered material with a goethite-like environment around each Fe(III) ion. Figures 2b and 2c clearly show that the disordered-component is essentially located within the phosphate-rich region of the tooth bases. Analysis of spectra recorded from early-maturing teeth (base mineralization only) agreed with these results on the mature teeth. In the early-maturing teeth, 65% of the total spectral intensity was derived from the disordered component as compared with 16% in the fully mature teeth.

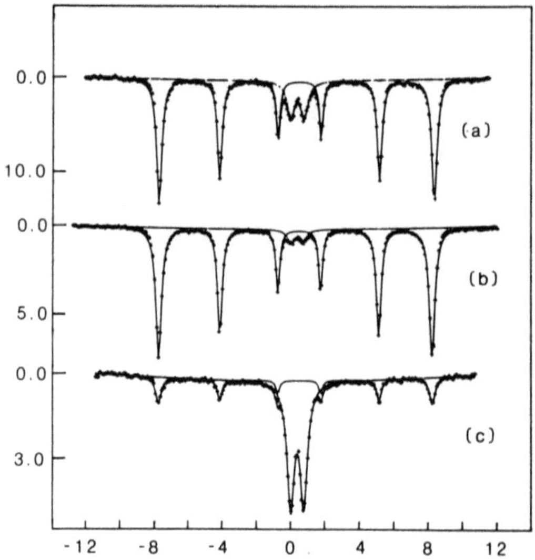

FIGURE 2. ^{57}Fe Mössbauer spectra recorded at 78 K. ▪a. Mature limpet teeth. ▪b. Separated cusps of mature teeth. ▪c. Separated bases of mature teeth.

Electron microscopy combined with EDXA investigations of fractured teeth at different stages of development revealed the structure, morphology and composition of the disordered base-material (Mann *et al.*, 1986b). Granular material rich in iron and silicon, calcium and phosphorus (Figs. 3 & 4), particulate material rich in silicon, iron, calcium, phosphorus, chlorine and zinc and sheet-like material high in calcium, phosphorus, silicon, and iron were observed. Electron diffraction patterns from these phases showed some evidence of random diffraction spots but no characteristic mineral could be identified. Lattice imaging on the granular material was undertaken to determine the structural order at the nanometre level (St. Pierre *et al.*, 1986b). Lattice fringes recorded at thin edges of the material clearly revealed the micro-crystalline and polycrystalline nature of this phase (Fig. 5). In many cases the lattice spacings were consistent with goethite interplanar *d* spacings and domain sizes were of the order of 5 nm. These results were in direct contrast to studies on material isolated from the tooth cusps which showed well-defined acicular crystals of pure goethite (Fig. 6).

Ferritin

The storage and mobilization of surplus iron in eukaryotes and some prokaryotes is regulated by the protein ferritin. Ferritin isolated from horse spleen consists of a hollow spherical shell of 24 symmetrically related protein subunits (Mr approx. 18,000 per subunit) surrounding a core of inorganic hydrated Fe(III) oxide. The volume of the cavity set by the protein shell is of the order of 70 to 80 Å resulting in an upper limit of 4500 iron atoms (approx. 30% (w/w) Fe) that can be stored within the molecule. Typically, iron levels are high (10 to 20%) but variable.

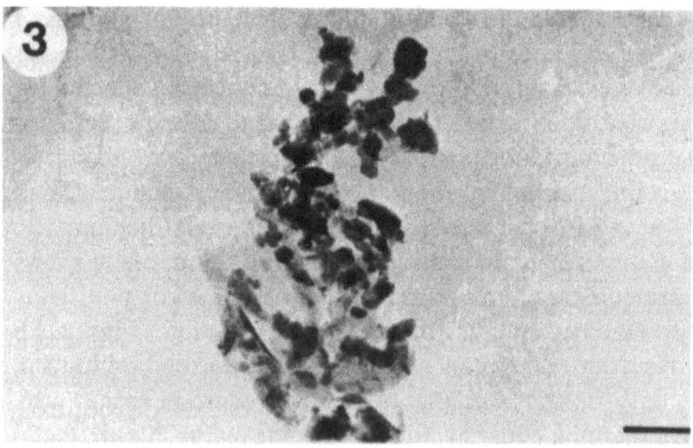

FIGURE 3. Electron micrograph of granular material extracted from limpet tooth bases. Bar = 200 nm.

The uniqueness of the ferritin structure arises from the metabolic requirement to organize and utilize dissolved iron at concentrations and pH levels that induce precipitation of potentially toxic solid phases. Not only is iron solubilized by micelle encapsulation but homeostatic control of iron is maintained. Thus the inorganic core must be rapidly precipitated and dissolved to match fluctuations in iron demand.

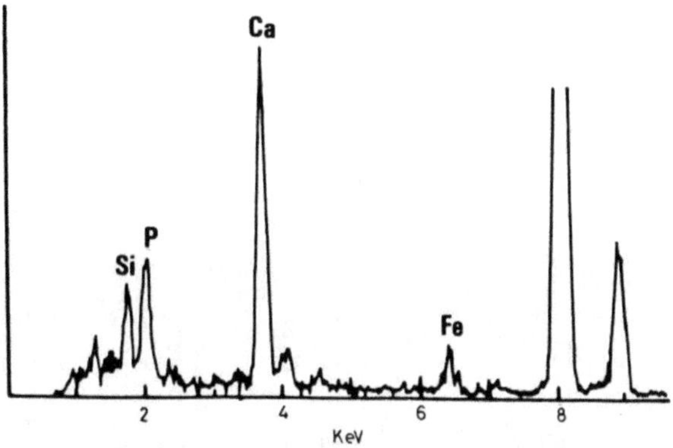

FIGURE 4. Corresponding EDXA spectra for the material shown in Figure 3.

The core material in mammalian ferritin has been shown, on the basis of electron diffraction (Towe & Bradley, 1967), and high resolution electron microscopy (Mann *et al.*, 1986a; Massover & Cowley, 1973; Isaacson & Ohtsuki, 1980) to be microcrystalline, non-stoichiometric, ferrihydrite ($5Fe_2O_3 \cdot 9H_2O$). Chemical analysis indicates that inorganic phosphate may be associated with the core. Levels are low in mammalian ferritins (Fe/P = 9 to 21, Mann *et al.*, 1986a; Michaelis *et al.*, 1943) whereas bacterial ferritins have much higher phosphate content (Fe/P = 1.4 to 1.7, Mann *et al.*, 1986a; Watt *et al.*, 1986).

The spatial distribution of phosphate within the ferritin cores is uncertain. The Fe/P ratio is rather lower than average in cores with low iron content and higher than average in cores of high iron content (Fischbach *et al.*, 1969). This observation together with the finding that approximately 77% of the phosphate is lost from the cores when separated from their protein shells by alkali treatment (Granick & Hahn, 1944) suggests that much of the phosphate is weakly adsorbed at surface sites on the ferrihydrite core. Other evidence (Van Kreel *et al.*, 1972) showed that the specific radioactivities of ^{32}P in ferritin molecules of different iron contents, isolated from livers perfused with media containing radioactive-labelled phosphate, were constant and equal to the specific radioactivity of phosphate remaining in the supernatents of liver homogenates. The dynamic equilibrium established between ferritin and liver phosphate supports the evidence for the phosphate surface adsorption hypothesis.

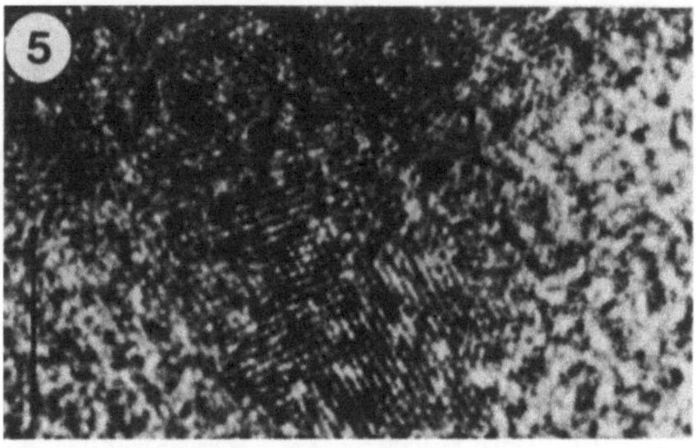

FIGURE 5. Lattice images of granular material from limpet teeth bases. Bar=5 nm.

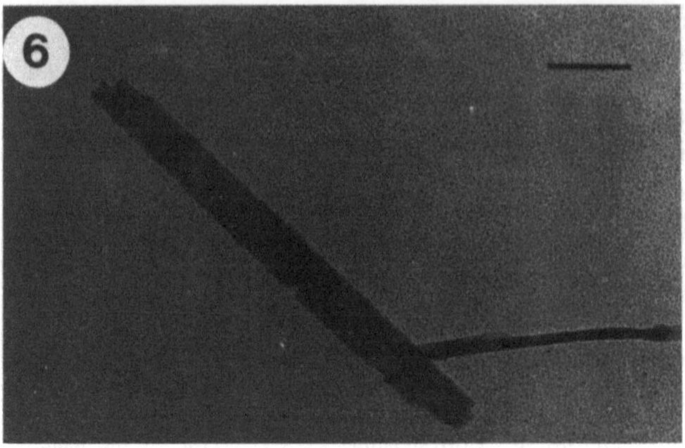

FIGURE 6. Well-defined acicular goethite crystal (phosphate-free) extracted from a limpet tooth cusp. Bar=200 nm.

The incorporation and release of inorganic phosphate in horse spleen ferritin has been studied by Treffry and Harrison (1978). These workers observed that phosphate was not a critical factor in ferritin core reconstitution from Fe(II), apoprotein and $KIO_3/Na_2S_2O_3$ (oxidant). Also, iron and phosphate release experiments showed that some phosphate was released somewhat ahead of iron suggesting a surface adsorption model. However, phosphate was continually released during the dissolution of the micelle suggesting that some phosphate is incorporated within the interior of the ferrihydrite core.

Although the above results suggest a mechanism of core formation *in vivo* in which phosphate is not present at intracellular sites of iron incorporation but is added later, the high phosphate content of bacterial ferritins suggest that both iron and phosphate are deposited concurrently.

We have used Mössbauer spectroscopy and HRTEM to study the magnetic and structural properties of mammalian, invertebrate and bacterial ferritin (Mann *et al.*, 1986a; St. Pierre *et al.*, 1986a). Our results implicate phosphate as an important factor in the modification of ferritin core structure.

Figure 7 shows ^{57}Fe Mössbauer spectra recorded at 4.2 K for (a) human and (b) bacterial ferritin (St. Pierre *et al.*, 1986a). The mammalian protein exhibits a broadened six-line magnetically split spectrum with a magnetic hyperfine field (49.4 T at 1.3 K) characteristic for the mineral ferrihydrite. The temperature dependence of this protein (data not shown) indicates that the cores show superparamagnetic behaviour (*i.e.*, antiferromagnetically ordered particles of small dimensions) with a blocking temperature of about 40 K corresponding to a range of core particle sizes with a mean value of 60-65 Å.

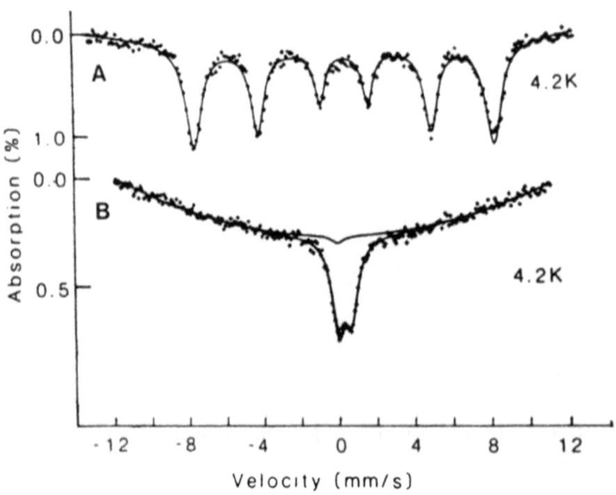

FIGURE 7. ^{57}Fe Mössbauer spectra recorded at 4.2 K. ▪a. Human ferritin. ▪b. bacterial ferritin.

In contrast the phosphate rich bacterial protein exhibits only a quadrupole split doublet at 4.2 K corresponding to a magnetically disordered (paramagnetic) Fe(III) phase above this temperature. Below this temperature the bacterial ferritin spectrum shows magnetic ordering but the hyperfine field is significantly smaller (42.0 T at 1.3 K) than for the human ferritin sample.

These results indicate that there is a fundamental difference in the nature of the magnetic interactions between neighbouring iron atoms in the human and bacterial ferritins. These differences which must arise from structural modifications (degree of crystallinity, density) were ascertained by lattice imaging of unstained proteins using HRTEM (Mann *et al.*, 1986a).

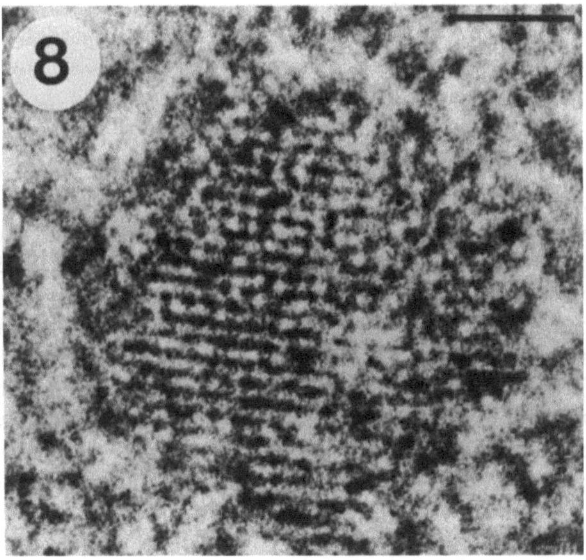

FIGURE 8. Lattice image of a human ferritin core. Bar = 2 nm.

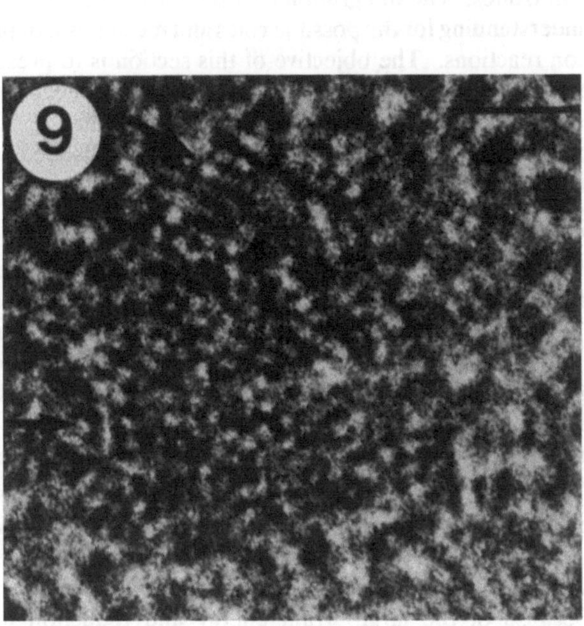

FIGURE 9. Lattice image of a bacterial ferritin core. Arrows indicate some short range order over 3 to 4 lattice spacings in localized regions of the amorphous particles. Bar = 2 nm.

Figure 8 shows a typical lattice image for a human ferritin core. Core sizes were in the mean range of 60 to 65 Å with a %(wt/wt) iron of 29.0 and Fe/P of 21.0. The fringes are well established although fluctuations in their intensity can be observed. The results show that in many instances the cores were single domain crystallites. Other crystalline cores showed evidence of lattice irregularities and disordered channels within the centre of the cores. The essential point is that many of the cores showed images consistent with well-crystallized ferrihydrite.

In contrast, lattice imaging of bacterial ferritin cores (%(wt/wt) iron of 8.72; Fe/P = 1.71, mean size range 60-65 Å) showed no evidence of long range crystal ordering. Figure 9 shows a "lattice image" of a bacterial core in which some short range order (10 to 20 Å) can be observed (arrows) above the incoherent pattern arising from the remaining areas of the iron-containing core.

Clearly, this difference in structural order within the two proteins results in the different magnetic properties determined by Mössbauer spectroscopy. A feasible hypothesis is that the high phosphate content of the bacterial core can be correlated with the non-crystallinity in this protein by inhibition of the crystal growth process. It seems most probable that there will be compositional as well as structural modifications in the presence of such high phosphate levels. Thus the bacterial core is more akin to a hydrated Fe(III) oxy-phosphate material than the ferrihydrite structure of mammalian ferritin. The presence of an amorphous hydrated iron-phosphate would also explain why we observed core sizes in the bacterial protein equivalent to those for human ferritin even though iron levels were 8.7% and 29% respectively. The more open, hydrated iron-phosphate structure results in large diffuse cores of low iron content.

IRON-OXIDE PHOSPHATE INTERACTIONS: AN OVERVIEW

The above results indicate the importance of phosphate in the biomineralization of iron-oxides. The integration of this data within a general chemical context provides an understanding for the possible roles and mechanism of phosphate in biological precipitation reactions. The objective of this section is to present a rationalization of possible phosphate interactions to enable a general conceptual framework to be developed. Figure 10 shows a scheme for the possible mediation of phosphate anions in iron-oxide biomineralization. Three areas of interaction need to be considered which involve solution, surface and redox chemistry.

Solution chemistry

Very little is known concerning the solution chemistry involved in iron-oxide biomineralization. At present, we have no information about concentrations, activities, the nature of phosphorus oxyanions involved, complexation reactions between Fe(II)/Fe(III) with phosphate or the possible solubility equilibria established. In general, we can assume that soluble iron is in the form of Fe(II) or low molecular weight Fe(III) complexes and that the low P(V)/P(III) redox couple favours P(V) oxyanions as being the more probable species involved in biomineralization.

Because of the strong affinity of phosphorus atoms for O atoms due to extensive $p\pi$-$d\pi$ bonding interactions (P-O bond length is short 1.5-1.6 Å), a range of P(V) oxyanions can be formed via the linking of phosphate tetrahedral units through corner sharing to give linear chains, rings and branched condensed phosphates (Table 2). Because of the large strain energy and reduction in P-O bond strength in these compounds, pyrophosphates, polyphosphates and metaphosphates are readily hydrolysed. Thus it is unresolved whether the inorganic phosphate involved in iron-oxide biomineralization derives from orthophosphate (via sugar phosphate, phosphate esters, phosphoproteins?) or via the polyphosphate pool.

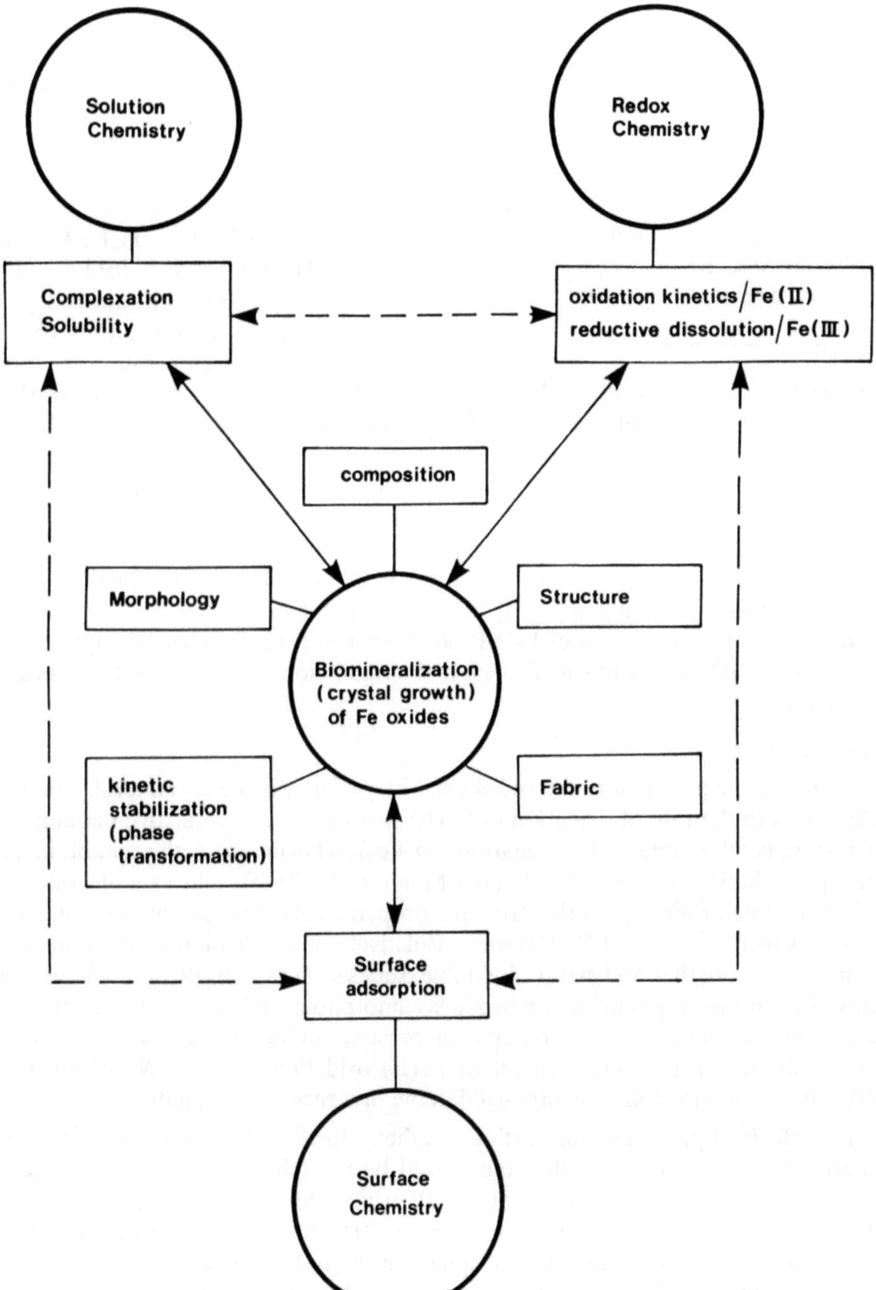

FIGURE 10. Generalized scheme for the interaction of phosphate in the chemistry of Fe-oxide biomineralization.

TABLE 2. Types of phosphorus (V) oxyanions

Form	Species
Orthophosphate	$H_2PO_4^-$, HPO_4^{2-}, PO_4^{3-}
Condensed phosphates	
Pyrophosphate	$H_2P_2O_7^{2-}$, $HP_2O_7^{3-}$, $P_2O_7^{4-}$
Tripolyphosphate (linear $[P_nO_{3n+1}]$)	$H_2P_3O_{10}^{3-}$, $HP_3O_{10}^{4-}$, $P_3O_{10}^{5-}$
Trimetaphosphate (cyclic $[P_nO_{3n}]^{n-}$)	$HP_3O_9^{2-}$, $P_3O_9^{3-}$

All P(V) oxyanions are acidic due to the easy ionization of the OH groups (Table 3). At pH 7 the predominant species which take part in solubility equilibria and complexation reactions are $H_2PO_4^-$ and HPO_4^{2-}, $H_2P_2O_7^{2-}$ and $HP_2O_7^{3-}$, $H_2P_3O_{10}^{3-}$ and $HP_3O_{10}^{4-}$ and $P_3O_9^{3-}$. Both ortho- and condensed phosphates form complexes, chelates and insoluble salts with Fe(II) and Fe(III) ions (Table 4). Although the equilibrium constants will be markedly influenced by activity products, pH and the presence of other ligands (SO_4^{2-}, CO_3^{2-}, F^-, organic species) the values for Fe(II) and Fe(III) phosphate complexes are considerable. At pH 7 the extent of complexation will be limited for Fe(III) due to the low solubility of Fe(III) hydrous oxides but, in contrast, Fe(II)-phosphate complexation can be significant. Thus the presence of phosphorous oxyanions at the sites of iron-oxide biomineralization will be an important factor in determining the distribution, nature, and concentration, of aqueous Fe(II) ions available for mineralization. In particular, the rate of hydrolytic scission of P-O-P bonds in iron-pyrophosphate complexes (and phosphate esters) may be of significance in the regulation of iron concentrations at the mineralization site.

Redox chemistry

Phosphate involvement in the redox chemistry of iron-oxide biomineralization may influence both the rate of oxidation of Fe(II) and the ease of reductive dissolution of solid phase Fe(III) oxides. The formation of α-FeOOH in limpet teeth probably involves direct precipitation from aqueous Fe(II) (Mann *et al.*, 1986b). The rate of oxidation is critical in the determination of the structural properties of the precipitate since it governs the local supersaturation of Fe(III) ions. Relatively slow oxidation kinetics favour the formation of more thermodynamically stable oxides such as goethite in the limpet tooth cusps. In contrast, rapid oxidation produces amorphous and pseudocrystalline phases such as those found in the tooth bases. The presence of impuritites such as silicate has been implicated in the enhancement of Fe(II) oxidation (Carlson & Schwertmann, 1981). A similar effect may be predicted in the presence of phosphate.

The effects of phosphate interactions on the reduction of Fe(III) oxides is of great importance in understanding the role played by phosphate in the iron storage and transport protein, ferritin. Phosphate levels within the inorganic core appear to be relatively low in mammalian ferritin (Fe/P = 9 to 21), but high in bacterial ferritin (Fe/P = 1.4 to 1.7). The role played by phosphate in the reductive dissolution of the cores during mobilization for metabolic use may be ultimately related to the functional adaption of these proteins in different biological systems. For example, structural, redox and compositional parameters will have a marked influence on the rates of storage and mobilization of iron by the ferritin molecule and hence on its role in iron homeostasis.

Experimental studies in simplified inorganic systems have shown that the stability of ferrihydrite under reducing conditions is enhanced in the presence of increased levels of phosphate adsorption. Thus lower redox (more negative) potentials are required to cause reductive dissolution of the oxide when phosphate is added (Willett & Cunningham, 1983). This fact accounts for the difference in reduction potentials for the bacterial ferritin from *Azotobacter vinelandii* (Steifel & Watt, 1979) of - 420 mV (pH 8) as against - 310 mV (pH 8) for mammalian ferritin (Watt *et al.*, 1985).

TABLE 3. Acidity and hydrolysis of phosphate (Sillen & Martell, 1964)

Equilibrium	Log Equilibrium Constant
$H_3PO_4 = H_2PO_4^- + H^+$	-2.2
$H_2PO_4^- = HPO_4^{2-} + H^+$	-7.0
$HPO_4^{2-} = PO_4^{3-} + H^+$	-12.0
$H_3P_2O_7^- = H_2P_2O_7^{2-} + H^+$	-2.4
$H_2P_2O_7^{2-} = HP_2O_7^{3-} + H^+$	-6.6
$HP_2O_7^{3-} = P_2O_7^{4-} + H^+$	-9.3
$H_3P_3O_{10}^{2-} = H_2P_3O_{10}^{3-} + H^+$	-2.3
$H_2P_3O_{10}^{2-} = HP_3O_{10}^{4-} + H^+$	-6.5
$HP_3O_{10}^{4-} = P_3O_{10}^{5-} + H^+$	-9.2
$HP_3O_9^{2-} = P_3O_9^{3-} + H^+$	-2.1

Surface chemistry

Inorganic solid surfaces such as iron-oxides readily adsorb inorganic phosphorus oxyanions at their surfaces. The driving force for these interactions lies in the thermodynamic stability of iron-phosphate bonding at the surface. Depending on the pH and concentration phosphate may either remain surface adsorbed or become structurally incorporated within the bulk solid phase.

LATTICE INCORPORATION

Because Fe(III) has a high affinity for both OH^- (log K = 11.8) and HPO_4^{2-} ligands (log K = 8.4) and the solubility products for FeOOH (log K_s = -38) and $FePO_4$ (log K_s = -22) are both extremely negative the precipitation of iron-oxides in the presence of phosphate is a complex process which is highly pH dependent. Under slightly acid conditions and suitable concentrations, pure $FePO_4$ can be precipitated (minimum solubility of $FePO_4$ is at pH 5.5). In contrast, at high pH only ferric oxides and hydroxyoxides will be formed. In neutral and slightly alkaline pH mixed Fe(III) phases containing both PO_4^{3-} and OH^- in variable proportions (approximately proportional to $[PO_4^{3-}]/[OH^-]^3$) can be precipitated. This is the case for the mineralization of the bacterial ferritin cores discussed above.

Structural incorporation of phosphate into these phases will be in the form of discrete phosphate tetrahedra with the P-O bond length remaining close to 1.55 Å. Flexibility in the O-P-O bond angle allows a range of structures to be adopted. Cations such as Fe(III) can be in both octahedral (*e.g.* monoclinic $Fe_3(PO_4)_2 8H_2O$ (Mori & Ito, 1950)) or tetrahedral sites (*e.g.* $FePO_4$ (Huttenlocher, 1935)). Three dimensional networks (crystalline and non-crystalline) can be adopted in which the four oxygen shared $[PO_2]^+$ unit can be effectively neutralized by the all corner sharing $[FeO_2]^-$ group· Thus a wide range of amorphous/gel-like structures of variable density, (as for example, in the bacterial ferritins) can be formed.

TABLE 4. Complex formation by phosphate (see Sillen & Martell, 1964; and Högfeldt 1982 for details).

Equilibrium	log Equilibrium Constant
$Fe^{2+} + HPO_4^{2-} = FeHPO_4$	3.6
$Fe^{2+} + H_2PO_4^{-} = FeH_2PO_4^{+}$	2.7
$Fe^{3+} + HPO_4^{2-} = FeHPO_4^{+}$	8.3
$Fe^{3+} + H_2PO_4^{-} = FeH_2PO_4^{2+}$	3.61
$Fe^{3+} + 2HP_2O_7^{3-} = Fe(HP_2O_7)_2^{3-}$	22.0
$Fe^{3+} + H_2P_2O_7^{2-} = FeH_2P_2O_7^{+}$	6.97
$Fe^{3+} + H_3P_2O_7^{-} = FeH_3P_2O_7^{2+}$	6.43
$Fe^{2+} + H_2P_3O_{10}^{3-} = FeH_2P_3O_{10}^{-}$	2.38
$Fe^{2+} + H_3P_3O_{10}^{2-} = FeH_3P_3O_{10}$	2.12
$Fe^{3+} + 2HP_3O_{10}^{4-} = Fe(HP_3O_{10})_2^{5-}$	20.63
$Fe^{3+} + H_2P_3O_{10}^{3-} = FeH_2P_3O_{10}$	7.03
$Fe^{3+} + H_3P_3O_{10}^{2-} = FeH_3P_3O_{10}^{+}$	6.37
$Fe^{2+} + HP_3O_9^{2-} = FeHP_3O_9$	1.15

Surface adsorption

The free energy of adsorption of phosphate at iron-oxide surfaces, ΔG_{ads} is given by

$$\Delta G_{ads} = \Pi + 2F\Psi$$

where Ψ is the electrochemical potential at the surface and Π is the chemical energy of adsorption due to non-electrostatic forces. (Often Ψ is *ca* 100mV, $2F\Psi$ = 2.3 kcal mol^{-1}; Π is often -2 to -8 kcal mol^{-1}). Note that both Π and Ψ are crystal-surface dependent such that phosphate adsorption will be crystallographically specific.

In terms of chemical interactions, phosphate adsorption can be considered as a ligand exchange process, for example

$$-\overset{|}{\underset{|}{Fe}}-OH + HPO_4^{2-} = -\overset{|}{\underset{|}{Fe}}-OPO_3H + OH^-$$

It is difficult in many instances to distinguish the above complex formation interaction and an exclusively coulombic interaction because of the continuous transition between electrostatic ion-exchange reactions and coordination by covalent bond formations. The complex formation description may be more appropriate in that the above process involves displacement of water and dislocation of electrons.

The extent of phosphate binding at an iron-oxide surface can be semiquantitatively estimated by comparing the affinity of the coordinating ligand (HPO_4^{2-}) for Fe^{3+} (*aq*) with its aqueous counterpart, OH^-. However, the stronger affinity of OH^- for Fe^{3+} rather than HPO_4^{2-} (or PO_4^{3-}) can be offset depending on the ratio $[HPO_4^{2-}]/[OH^-]$. Thus decreasing pH favours PO_4 binding.

The surface charge of Fe(III) oxides is strongly pH dependent (Table 5). Around neutral pH, oxides such as αFeOOH, γFe_2O_3, and Fe_3O_4 will have net negatively charged surfaces whereas amorphous ferrihydrite has a net positive charge and γFeOOH is close to neutral. Thus phosphate adsorption on ferrihydrite will be favoured by both chemical and electrostatic interactions. For oxides with negatively charged surfaces chemical interactions can often override the electrostatic repulsion energy. Adsorption of phosphate at these surfaces through electrostatic or chemical interactions will significantly modify the surface charge and thereby surface properties such as ion-exchange and particle aggregation. In general the adsorption of polyphosphates at iron-oxide surfaces will be generally greater that for orthophosphates. The tendency of condensation products to adsorb more strongly than discrete units can be attributed to the increased number of OH groups, the greater size (lower hydration) and enhanced chemical interactions involving several atoms.

TABLE 5. Zero Point of Charge for Fe-oxides (Parks, 1965)

Oxide	pHZPC
α-Fe_2O_3	5.2-8.6*
α-FeOOH	6.7
γ-Fe_2O_3	6.7
γ-FeOOH	7.4
Fe_3O_4	6.5
$5Fe_2O_3 \cdot 9H_2O$ (amorph.)	8.5
$Fe(OH)_2$	12.0

* Value dependent on preparative procedure

CRYSTAL GROWTH

The results presented above (Biological Systems) highlight the important role of phosphate in the mediation of crystal growth processes of biogenic iron-oxides. The major effect is in the inhibition of crystallization resulting in amorphous and pseudocrystalline biominerals. We have confirmed these observations in model systems. The influence of phosphorus (and silicon) in the crystallization of goethite in inorganic reaction systems has been studied by Mössbauer spectroscopy and HRTEM to determine the role of these elements in the biomineralization of limpet tooth bases (Quin *et al.*, 1986). Our results showed that phosphate-doped systems with P/Fe ratios greater than 4% resulted in complete inhibition of goethite crystallization from ferrihydrite suspensions at pH 12.0 and 80° C. Presumably all the potential nucleation and growth centres are effectively blocked by phosphate binding at these high levels of dopant.

The substitution of OH by HPO_4^{2-} at the crystal surface changes the number and magnitude of negative charges at the solution-crystal interface which in turn modifies the binding and exchange characteristics of cations bound at the surface. This has important consequences in crystal growth processes in which the rate determining steps involve chemical reactions and rearrangement on the crystal surfaces. Such reactions are crystal face specific such that, in principle, phosphate binding should result in the selective enhancement/inhibition of growth processes on particular crystallographic planes. For example, in the inorganic goethite described above, at low phosphate levels (<4%) the normal needle-like crystal morphology was severely modified to stunted rectangular crystallites which showed many regions of disorder when studied by HRTEM. The acicular morphology normally predominates in goethite because the crystal structure is dominated by a framework of double chains of FeO_6 octahedra running along the [001] direction. The modification in morphology in the presence of phosphate probably arises from attachment of phosphate anions at active sites at the ends of the FeO_6 chains resulting in partial blockage of growth along [001].

Another example, which may be important in explaining the unique crystal morphologies observed in magnetotactic bacteria (Mann *et al.*, 1984a,b; Matsuda *et al.*, 1983) is the selective enhancement of magnetite (Fe_3O_4) morphology in the presence of inorganic phosphate. We have shown that the aqueous synthesis of Fe_3O_4 from nitrate oxidation of Fe(II) solutions at high pH is highly dependent on the level of phosphate in the reaction medium (Couling & Mann, 1985). At 0% phosphate, the Fe_3O_4 crystals are crystallographically well-ordered but irregular in morphology. In the presence of 5-10% orthophosphate crystallization is still complete but the morphology is selectively enhanced in the form of octahedral crystallites. At higher phosphate concentrations (20-30%) crystallization of Fe_3O_4 is severely inhibited. The enhancement of the octahedral form over the irregular habit can be rationalized on the basis of selective adsorption of phosphate onto the {111} faces of Fe_3O_4 due to the high density of iron-tetrahedral sites in these crystal planes. Thus crystal growth along the [111] direction will be slow due to the screening of binding sites by phosphate resulting in well defined (111) faces in the final morphology.

The adsorption of phosphate at iron-oxide surfaces may result in charge neutralization which may be accompanied by the aggregation of particles due to the reduction in electrostatic repulsion of the charged surfaces. This may result in the stabilization of surfaces with respect to phase transformation. For example, the precipitation of Fe_3O_4 in the presence of phosphate, as described above (Couling & Mann, 1985), proceeds via an amorphous intermediate which undergoes further oxidation, dehydration and structural modification prior to crystallization. A similar phase has been identified in magnetotactic bacteria (Mann *et al.*, 1984a). The binding of phosphate to this precursor surface results in an inert, immobile network which effectively retards the chemical transformation to Fe_3O_4 via a dissolution-renucleation pathway. Thus at high phosphate levels (20-30%) the increased kinetic stability of the phosphate-containing precursor inhibits Fe_3O_4 crystal growth and allows continued oxidation of the Fe(II) (no longer stabilized within the Fe_3O_4 lattice) such that the gel rapidly oxidises to Fe(III) hydrated oxides in contact with air.

The enhanced stability of amorphous precursors such as ferrihydrite in the presence of phosphate could be an important role for the known phosphate content of mammalian and bacterial ferritins. These iron storage and transport proteins utilize rapid uptake and release of iron in response to metabolic requirements as a means of maintaining iron homeostatis. The kinetic stability of the ferrihydrite-phosphate composite prevents phase transformation to more thermodynamically stable (*i.e.*, less soluble) phases such as αFe_2O_3, α-FeOOH and Fe_3O_4. In particular, transformation of ferrihydrite to hematite (αFe_2O_3) is favoured at pH 7 and 24°C (Schwertmann & Murad, 1983). At present, no αFe_2O_3 phases are known to be mineralized in biological systems. Such phases would be essentially inert with respect to the dynamic properties required in the solution-solid redox equilibria inherent in the optimised functioning of the ferritin protein.

CONCLUSIONS

This paper has discussed the role of phosphate in the biomineralization of iron-oxides. Two biological systems and several inorganic (model) systems have been discussed. Phosphate interactions influence the structure, morphology, magnetic properties and composition of iron-oxide biominerals in limpet teeth and bacterial ferritin. Possible functional roles for phosphate involvement include (a) crystal growth inhibition (limpet teeth, ferritin), (b) crystal growth mediation (morphological enhancement in magnetotactic bacteria?), (c) kinetic stabilization (*e.g.*, ferrihydrite in ferritin) with respect to phase transformation leading to dynamic behaviour for homeostasis, (d) solubility control via complexation and (e) redox control over oxidation of aqueous Fe(II) (limpet teeth) and reductive dissolution of ferrihydrite (e.g., in ferritin).

ACKNOWLEDGEMENTS

I wish to thank several colleagues for their assistance and skill in the development of the studies reported in this paper. Professor R.J.P. Williams, Drs. J. Webb, T.G. St. Pierre, J.V. Bannister, D.P.E. Dickson, C.C. Perry, G.W. Grime, F. Watt, N.W. Runham, T.G. Quin, G.J. Long and S.B. Couling have all been joint contributers in the study of iron-oxide biomineralization as reported in the references below. Participation in the Symposium was made possible by grants from The Organized Research Fund of The University of Texas at Arlington and The Sea Grant Program administered by Texas A & M University.

REFERENCES

CARLSOÑ, L. & SCHWERTMANN, U., 1981. Natural ferrihydrites in surface deposits from Finland and their association with silica. Geochim. Cosmochim. Acta., 45:421-429.

COULING, S.B. & MANN, S., 1985. The influence of inorganic phosphate on the crystallization of magnetite (Fe3O4) from aqueous solution. J. Chem. Soc., Chem. Comm.:1713-1715.

FISCHBACH, F.A., HARRISON, P.M. & HOY, T.G., 1969. The structural relationship between ferritin protein and its mineral core. J. Mole. Biol., 39:235-238.

GRANICK, S. & HAHN, P., 1944. Speed of uptake of iron by the liver and its conversion to ferritin iron. J. Biol. Chem., 155:661-669.

GRIME, G.W., WATT, F., MANN, S., PERRY, C.C., WEBB, J. & WILLIAMS, R.J.P., 1985. Biological applications of the Oxford scanning proton microprobe. Trends Biol. Sci., 10:6-10.

HÖGFELDT, E., 1982. Stability constants of metal-ion complexes. chemical data series, no. 21, IUPAC.

HUTTENLOCKER, H.F., 1935. The crystal structure of aluminium orthophosphate. Zeits. Kristallogr., 90A:508-516.

ISAACSON, M. & OHTSUKI, M., 1980. Scanning transmission electron microscopy of small inhomogeneous particles: applications to ferritin. Scan. Elec. Microsc., 73-80.

LOWENSTAM, H.A., 1962. Goethite in radular teeth of recent marine gastropods. Science, 137, 279-280.

LOWENSTAM, H.A., 1972. Phosphate hard tissues of marine invertebrates: their nature and mechanical function and some fossil implications. Chem. Geol., 9:153-166.

LOWENSTAM, H.A., 1981. Minerals formed by organisms. Science, 171:487-490.

LOWENSTAM, H.A. & ROSSMAN, G.R., 1975. Amorphous, hydrous, ferric phosphatic dermal granules in *Molpadia* (Holothuroidea): physical and chemical characterization and ecologic implications of the bioinorganic fraction. Chem. Geol., 15:15-51.

MACARA, I.G., HOY, T.G. & HARRISON, P.M., 1972. The formation of ferritin from apoferritin.Biochem. J., 126:151-162.

MANN, S., 1983. Mineralization in biological systems. Structure and Bonding, 54:125-174.

MANN, S., FRANKEL, R.B., & BLAKEMORE, R.P., 1984a. Structure, morphology and crystal growth of bacterial magnetite. Nature (London), 310, 405-407.

MANN, S., MOENCH, T.T. & WILLIAMS, R.J.P., 1984b. A high resolution electron microscopic investigation of bacterial magnetite; implications for crystal growth. Proc. Roy. Soc. Lond. B., 221:385-393.

MANN, S., BANNISTER, J.V. & WILLIAMS, R.J.P., 1986a. Structure and composition of ferritin cores isolated from human spleen, limpet (*Patella vulgata*) hemolymph and bacterial (*Pseudomonas aeruginosa*) cells. J. Mole. Biol., 188:225-232.

MANN, S., PERRY, C.C., WEBB, J., LUKE, B. & WILLIAMS, R.J.P., 1986b. Structure, morphology, composition and organization of biogenic minerals in limpet teeth. Proc. Roy. Soc. Lond. B., 227:179-190.

MASSOVER, W.H. & COWLEY, J.M., 1973. The ultrastructure of ferritin macromolecules. The lattice structure of the core crystallites. Proc. Nat. Acad. Sci. U.S.A., 70:3847-3851.

MATSUDA, T., ENDO, J., OSAKABE, N. & TONOMURA, A., 1983. Morphology and structure of biogenic magnetite particles. Nature (London), 302:411-412.

MICHAELIS, L, CORYELL, C.D. & GRANICK, S., 1953. The magnetic properties of ferritin and some other colloidal ferric compounds. J. Biol. Chem., 148:463-480.

MORI, H. & ITO, T., 1950. The structure of vivianite and symplesite. Acta Crystallogr., 3:1-6.

PARKS, G.A., 1965. The isoelectric points of solid oxides, solid hydroxides and aqueous hydroxo complex systems. Chem. Rev., 65:177-198.

QUIN, T.G., LONG, G.J., BENSON, C.G., MANN S. & WILLIAMS, R.J.P., 1987. Study of the influence of silicon and phosphorus dopants on the structural and magnetic properties of synthitic goethites. Clays Clay Miner. (in the press).

SCHWERTMANN, U. & MURAD, E., 1983. Effect of pH on the formation of goethite and hematite from ferrihydrite. Clays Clay Miner., 31 (4):227-284.

SILLEN, L.G. & MARTELL, A.E. 1964. Stability constants of metal-ion complexes. Spec. Publ. No. 17. Chem. Soc., Lond.

STIEFEL, E.I. & WATT, G.D., 1979. *Azotabacter* cytochrome b557.5 is a bacterioferritin. Nature (London). 279:81-83.

ST. PIERRE, T.G., BELL, S.H., DICKSON, D.P.E., MANN, S., WEBB, J., MOORE, G.R. & WILLIAMS, R.J.P., 1986a. Mössbauer spectroscopy studies of the cores of human, limpet and bacterial ferritins. Biochim. Biophys. Acta, 870:127-134.

ST. PIERRE, T.G., MANN S., WEBB, J., DICKSON, D.P.E., RUNHAM, N.W. & WILLIAMS R.J.P., 1986b. Iron oxide biomineralization in the radula teeth of the limpet *Patella vulgata*; Mössbauer spectroscopy and high resolution transmission electron microscopy studies. Proc. Roy. Soc. Lond. B, 228:31-42.

TOWE, K.M. & BRADLEY, W.P., 1967. Mineralogical constitution of colloidal 'hydrous ferric oxides'. J. Colloidal Interface Sci., 24:384-392.

TOWE, K.M. & LOWENSTAM, H.A., 1967. Ultrastructure and development of iron mineralization in the radular teeth of *Cryptochiton stelleri* (Mollusca). J. Ultrastr. Res., 17:1-13.

TREFFRY, A. & HARRISON, P.M., 1978. Incorporation and release of inorganic phosphate in horse spleen ferritin. Biochem. J., 171:313-320.

VAN KREEL, B.K., PIJNENBERG, A.M.C.M., VANN EIJK, H.G. & LEIJNSE, B., 1972. The incorporation of ^{59}Fe, [^{3}H]-leucine and [^{32}P] inorganic phosphate in ferritin during the perfusion of isolated rat livers. Biochim. Biophys. Acta, 273:243-247.

WATT, G.D., FRANKKEL, R.B. & PAPAEFTHYMIOU, G.C., 1985. Reduction of mammalian ferritin. Proc. Nat. Acad. Sci. U.S.A., 82:3640-3643.

WATT, G.D., FRANKEL, R.B., PAPAEFTHYMIOU, G.C., SPARTALIAN, K. & STIEFEL, E.I., 1986. Redox properties and Mössbauer spectroscopy of *Azotobacter vinelandii* bacterioferritin. Biochemistry, 25:4330-4336.

WILLETT, I.R. & CUNNINGHAM, R.B., 1983. Influence of sorbed phosphate on the stability of ferric hydrous oxide under controlled Eh and pH conditions. Austr. J. Soil Res., 21:301-308.

WEBB, J., 1983. A bioinorganic view of the biological mineralization of iron. *In* Biomineralization and Biological Metal Accumulation (ed. P. Westbroek & E.W. de Jong), pp. 413-422. Dordrecht: Reidel.

HIGH CAPACITY CALCIUM-BINDING PROTEINS AS INTERMEDIATE CALCIUM CARRIERS IN BIOLOGICAL MINERALIZATION

Mary E. Marsh

University of Texas Health Science Center, Dental Science Institute, Houston, Texas 77225

ABSTRACT. High-capacity calcium-binding macromolecules have been identified at mineralization fronts in heterodont bivalves, vertebrate tooth dentin and unicellular algae. They appear to be functionally similar to casein, the mineral ion carrier of milk. The bivalve phosphoprotein studied in this laboratory occurs as discrete high molecular weight (MW) particles in the hemolymph and extrapallial fluid and at the inner shell surface during mineral deposition. The phosphoprotein particles are characterized by a high content of aspartic acid, phosphoserine and histidine residues which comprise 80% or more of the peptide chain. The native particles contain a protected pool of calcium and phosphate ions and an exchangeable pool of calcium and magnesium ions. The phosphoprotein monomers are covalently cross-linked via histidinoalanine residues and ionically cross-linked via divalent cations into compact particles with a MW of about 50 million. High concentrations of phosphoprotein particles accumulate at the mineralization front during stimulated biomineralization, but neither inhibit mineral deposition nor influence the crystal habits. The particles are probably calcium ion transporters. However, the mechanism by which protein bound calcium is utilized to form calcium carbonate crystals is unknown.

The bivalve phosphoprotein particles share common characteristics with vertebrate tooth dentin and algal coccolithosomes which have been studied in other laboratories. All three are high-capacity calcium-binding macromolecules which have been localized at mineralization fronts. Both the bivalve phosphoprotein particles and dentin phosphophoryn are aspartic acid-rich, highly phosphosylated proteins which have a tendency to cross-link and fragment. Both coccolithosomes and the bivalve phosphoprotein occur as discrete high MW particles about 20-40 nm in diameter surrounding growing calcium carbonate crystals in vivo. It is postulated that intermediate calcium carriers in the form of high-capacity calcium-binding macromolecules may be a general phenomenon in the biological calcification of skeletal tissues.

INTRODUCTION

The ability to sequester a large number of calcium ions is a property common to several proteins occurring in diverse tissues and phyla. Of particular interest for this symposium are the high-capacity calcium-binding proteins which are localized at mineralization fronts and have an intermediate function in tissue calcification. The purpose of this review is to summarize and compare the properties of these proteins in order to provide a basis for postulating a generalized role for the high capacity calcium-binding proteins as intermediate calcium carriers. Although the milk protein

Edited by Rex E. Crick
Plenum Press, New York

casein is not directly related to skeletal mineralization, a brief review of this protein is included because it is also an intermediate carrier of mineral ions.

CASEIN

The caseins are the major protein component of milk. They exist in a stable colloidal suspension as high molecular aggregates called micelles or particles. In bovine milk, the particles which range from 20 nm to 300 nm in diameter contain four different polypeptides, called αS1-, αS2-, β-, and κ-casein, in the approximate molar ratio of 3:1:3:1 (Schmidt, 1980). The particles also contain sequestered mineral ions, principally calcium and inorganic phosphate.

All of the caseins are phosphoproteins; however, the number of phosphate groups is variable due to differing levels of phosphorylation (Swaisgood, 1982). αS2 is the most highly phosphorylated casein; maximally 13 of its 207 residues are ester phosphates of serine and threonine. Generally 2.3% and 4.5% of the residues are phophorylated in β- and αS1-casein, respectively. In κ-casein, only one or two of its 169 residues are phosphorylated. Each of the caseins has a different primary structure. The only homologous sequence is SerP-SerP-SerP-Glu-Glu which occurs once in αS1- and β-casein and twice in αS2-casein (Swaisgood, 1982). In general the phosphorylated residues are restricted to acidic sequences which are also rich in glutamic and aspartic acid residues. The caseins differ from the high-capacity calcium-binding proteins described below in that only about 15% of the amino acid residues are negatively charged in the milk phosphoproteins.

The major inorganic constituents of the casein particles are calcium, magnesium, inorganic phosphate and citrate ions which together represent about 7% of the particle mass (Schmidt, 1982). A better appreciation of the mineral-binding capacity of casein can be obtained by expressing the number of mineral ions bound per 100 amino acid residues. Using the data of Holt (1982) which gives the number of mineral ions per casein molecule and using 200 as the average number of amino acid residues per molecule, the casein particles contain an estimated 11.5 calcium, 5.5 phosphate, 0.7 magnesium and 0.7 citrate ions per 100 amino acid residues. Holt (1982) found that 2.0 of the calcium ions and 0.45 of the magnesium ions are bound directly to the protein. The remainder of the calcium and magnesium ions are sequestered as ion clusters with inorganic phosphate and citrate ions. The ratio of calcium:magnesium:citrate:phosphate in the ion clusters is 1.61:0.044:0.097:1.0. The clustered mineral has been characterized as amorphous calcium phosphate since the casein particles are amorphous by X-ray and electron diffraction analysis, and no periodic lattice spacings can be imaged by high resolution electron microscopy (McGann et al., 1983; Lyster et al., 1984). The direct metal binding sites and the calcium phosphate cluster binding domains are apparently distinct (Holt, 1982). The size of the clusters and the number of cluster-binding domains per casein particle are unknown.

Mineralized casein particles form in the Golgi vesicles of the mammary gland (see references in Schmidt, 1982; and Mepham et al., 1982). Extracellular formation of protein-mineral complexes is unlikely since demineralized casein will not resequester calcium phosphate when dialyzed against skim milk (Holt, 1982). However, the demineralized casein will reacquire the directly bound metal ions. Due to the presence of casein, milk contains a larger concentration of dispersed calcium and phosphate ions than the fluid would otherwise hold because of the low solubility of calcium phosphate. In addition to serving as a nutrient source of amino acids, the casein particles are also intermediate mineral ion carriers, i.e., calcium and phosphate ions are transferred from the mammary glands in the mother to the digestive tract of her suckling young via the casein particles.

BIVALVE PHOSPHOPROTEINS

The composition and colloidal properties of casein and bivalve phosphoprotein particles are similar. The bivalve particles are compact high molecular weight structures consisting of phosphoproteins and mineral ions (Marsh & Sass, 1983). They are about 40 nm in diameter and occur as a stable colloidal suspension in the hemolymph and extrapallial fluid of heterodont bivalves. The concentration of the particles ranges from about 0.5 to 8 mg/ml.

The phosphoprotein particles from nine different species have been characterized (Marsh & Sass, 1985). In all species the amino acid composition is dominated by phosphoserine, aspartic acid and histidine residues. The content of phosphorylated residues ranges from 25-30 mole % in eight species, but is only 18 mole % in Mya arenaria. Nearly all of the phosphate is esterified to serine residues except in two species Macrocallista nimbosa and Mercenaria mercenaria. In the the latter about one-third of the phosphoserine is replaced by phosphothreonine. The aspartic acid level ranges from 19-35 mole %, and the histidine content from 31-42 mole %. The phosphorylated residues plus aspartic acid and histidine represent 80% or more of the peptide chain in all species. The phosphoproteins are also glycoproteins: there are 5-10 carbohydrate residues per 100 amino acid residues.

The molecular weight of the protein has not been determined, but based on the amino acid composition the minimum molecular weight of the monomer is about 140,000 (Marsh, 1986a). The phosphoprotein monomers are covalently cross-linked into aggregates of varying sizes, ranging from dimers and trimers up to enormous aggregates which are too large to be resolved by either agarose gel electrophoresis or molecular seive chromatography. Histidinoalanine is the covalent cross-linker, and it is derived from phosphoserine and histidine residues (Sass & Marsh, 1982; Marsh, 1986a). The variable molecular weight aggregates are in turn ionically cross-linked via calcium and magnesium ions into discrete particles of uniform diameter, ranging from about 20 nm in Mya arenaria to about 40 nm in Rangia cuneata.

The native particles also contain fragmented phosphoprotein chains. Both fragmentation and cross-link formation are probably due to the inherent instability of the phosphoserine residues. Phosphoserine can undergo β-elimination of phosphate to yield dehydroalanine residues. Dehydroalanine in turn can either 1) decompose into pyruvate resulting in peptide chain fragmentation or 2) react with an available histidine residue to form the histidinoalanine cross-links.

The mineral component of the native phosphoprotein particles is distributed into two distinct ion pools (Marsh & Sass, 1984). One pool is sequestered into protected domains and consists principally of calcium and inorganic phosphate ions. The protected pool will not exchange with free ions in the suspending medium. In Rangia cuneata the protected pool contains about 15 calcium and 6 phosphate ions per 100 amino acid residues. The protected mineral is probably sequestered as calcium phosphate clusters since neither ion will bind in the absence of the other. The mineral may be sequestered in a manner similar to the calcium phosphate of the casein particles. However, the calcium/phosphate ratio is about 2.5 in the former but only 1.6 in the latter.

The other mineral pool in the native bivalve particles consists of calcium and magnesium ions which are freely exchangeable with unbound ions in the suspending medium (Marsh & Sass, 1984). The exchangeable pool can accommodate a maximum of about 30 calcium or magnesium ions per 100 amino acid residues. Since there is no inorganic phosphate in this pool, the exchangeable ions bind directly to the protein through negatively charged residues. The relative amounts of calcium and magnesium in this pool are dependent on the relative concentrations of the free ions in the medium. The overall mineral ion content of individual particles in Rangia cuneata is about 3.3 M calcium, 0.54 M magnesium, and 0.55 M phosphate.

The casein and bivalve phosphoprotein particles sequester about the same amount of calcium phosphate per 100 amino acid residues. However, the pool of directly bound metal ions is much larger in the bivalve particles than in casein, due to the high content of phosphorylated residues in the former. When the demineralized bivalve protein is re-equilibrated with the hemolymph and extrapallial fluid, the protein reacquires the exchangeable pool of directly bound metal ions, but not the protected pool of calcium phosphate (Marsh & Sass, 1984). As with casein, mineralization of the bivalve phosphoprotein probably occurs intracellularly. Phosphoprotein particles have been identified in secretory vesicles of hemolymph cells, and under some conditions the outer mantle epithelium may also manufacture and secrete the native protein (Marsh, unpublished data).

The phosphoprotein particles have also been localized in a condensed solid phase on the inner shell surface (Marsh, 1986b). In Rangia cuneata, the inner shell surfaces dorsal to the pallial attachments are covered with a thin unmineralized coat of variable composition. In some populations the coat is predominately a fibrous polysaccharide and is devoid of phosphoprotein. In other populations the coat contains phosphoprotein particles bound to the polysaccharide matrix. In the latter, the phosphoprotein represents 10-90% of the coat material. The mature mineralized shell layers contain no phosphoprotein. The inner shell surface is a dynamic site of mineral metabolism where calcium carbonate is deposited during shell thickening and removed by acid dissolution during anaerobic respiration (Crenshaw, 1969). The occasional presence of phosphoprotein particles on the inner shell surface indicates that the protein is an intermediate in either the mineralization or demineralization process.

When Rangia are mechanically stimulated by scratching the inner shell surfaces, the thin organic coat develops into a thick calcified plaque over a period of 2-3 weeks (Marsh, 1986b). Mineralization of the coat is accompanied by a large accumulation of phosphoprotein particles. The plaque mineral is the aragonite polymorph of calcium carbonate like the normal shell mineral. However, the plaque mineral grows in spherulitic and dumbbell-shape clusters characteristic of aragonite precipitated from strictly inorganic solutions, while in the normal mineralized shell, the aragonite crystals are aligned in a complex cross-lamellar array. Therefore, although phosphoprotein particles accumulate in the coat during stimulated biomineralization, they do not influence the crystal habits. Equally important, since the aragonite is deposited in the presence of the phosphoprotein particles, the protein is not a functional mineralization inhibitor in vivo. This is an important point since other polyanious such as polyaspartic and polyglutamic acid strongly inhibit the precipitation of calcium carbonate from slightly supersaturated solutions in vitro (Wilbur & Bernhardt, 1982; Sikes & Wheeler, 1983).

Since phosphoprotein particles are high-capacity calcium binding proteins, they may be the transport vehicle for accumulating calcium ions at the mineralization front. In this respect the bivalve and casein phosphoprotein particles are functionally similar, both serving as intermediate calcium carriers. The mechanism by which protein bound calcium ions are utilized to form calcium carbonate crystals in the bivalve is unknown. The issue is further confused because the phosphoprotein particles contain a lot of inorganic phosphate, but very little carbonate. If appreciable amounts of carbonate are present in the phosphoprotein particles, it is not in a form which can be measured manometrically via CO2 release after acidification (Marsh & Sass, 1983).

PHOSPHOPHORYNS

Phosphophoryns are major noncollagenous proteins of the vertebrate tooth dentin matrix. Compositionally they are similar to the bivalve phosphoprotein. Only two amino acids, phosphoserine and aspartic acid, are represented in about 80% of the dentin protein, while in the bivalve protein the same two amino acids plus histidine represent 80% of the peptide chain. Bovine dentin in unerupted molars of fetal and adult animals contains principally a single high molecular weight phosphophoryn (Lee et al., 1983a; Stetler-Stevenson & Veis, 1983), which is degraded into lower molecular weight phosphopeptides as the tissue ages (Lee et al., 1983a). The parent protein has a molecular weight of about 150,000. Masters (1985) and Fujisawa et al., (1985) have postulated that the degradation is due to the slow β-elimination of phosphate over the long period of dentin maturation. β-elimination from phosphoserine residues yields dehydroalanine which can either react with lysine or histidine to form cross-links (lysinoalanine and histidinoalanine) or further decompose to pyruvate causing chain fragmentation (see above). In support of the autofragmentation theory, Fujisawa et al., (1985) have shown that both the pyruvate and lysinoalanine content of bovine phosphophoryn increase with dentin maturation, and that the pyruvate residues are concentrated in the lower molecular weight proteins. In the continuously erupting rat incisors, the dentin contains three major phosphophoryns with similar molecular weights (about 95,000) which may be different gene products (Butler et al., 1983).

Although a portion of the phosphophoryn secreted by the odontoblasts is retained in the dentin, Dimuzio and Veis (1978) demonstrated that the majority of the protein is turned over at the mineralization front. They postulate that some of the phosphophoryn binds to the collagen fibrils forming nucleation sites for hydroxyapatite, and that the remainder of the protein is degraded at the mineralization front releasing bound ions for crystal growth. The mineral ion content of newly secreted phosphophoryn is

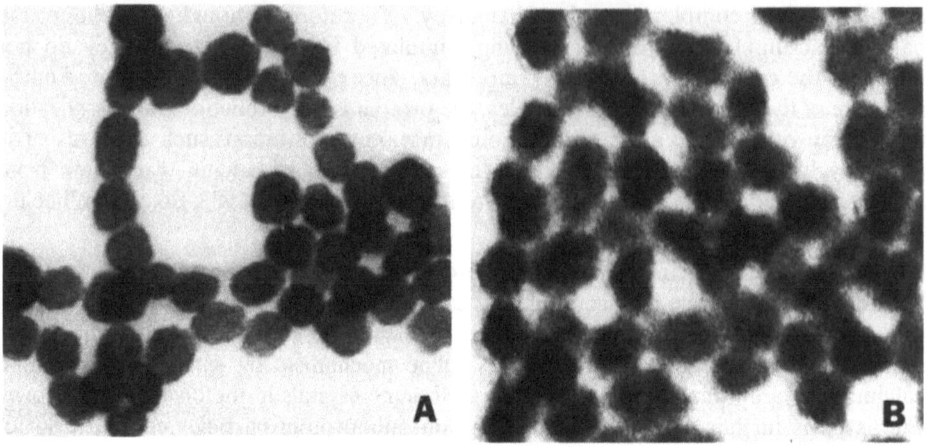

FIGURE 1. ▪A. Highly phosphorylated fraction of the rat dentin phosphorphoryns. The protein was equilibrated with 5 μM MgC12-20 μM BaC12 and evaporated onto a carbon coated grid. Electron micrograph, × 230,000. ▪B. Native phosphorprotein particles from the bivalve *Dinocardium robustum* equilibrated with 20 μM BaC12 and evaporated onto a carbon coated grid. Electron micrograph, × 230,000.

unknown. To date, the protein has been isolated only from the dentin matrix under demineralizing conditions.

In vitro at a low protein concentration, phosphophoryn will bind maximally about one calcium ion for each aspartic acid and phosphoserine residue which is equivalent to about 80 calcium ions per 100 amino acid residues (Lee et al., 1977). The protein will also bind inorganic phosphate ions in the presence of calcium (Lee et al., 1983b); however, the stoichiometry of the complex and the domains in which the mineral is sequestered are unknown. Kuboki et al., (1979) postulated that phosphophoryn may occur as mineral containing colloidial particles similar to casein during its transit from the odontoblasts to the mineralization front. This theory is consistent with the recent observation that rat phosphophoryns interact specifically in the presence of divalent cations to form high molecular weight particles about 40 nm in diameter (Marsh, unpublished data). Ultrastructurally the rat dentin and bivalve phosphoprotein particles are similar (Fig. 1); they are about the same size but the bivalve particles have a slightly more angular contour. In summary, phosphophoryn behaves like an intermediate calcium carrier, i.e. the protein self-associates, sequesters mineral ions and turns over at the mineralization front.

COCCOLITHISOMES

Coccolithisomes were discovered in the unicellular coccolithiphorid alga Hymenomas carterae by Outka and Williams (1971). To date these structures have not been isolated, and all information concerning their properties comes from ultrastructural and cytochemical studies. The coccolithisomes are electron dense particles about 25 nm in diameter. They are localized within Golgi saccules and in larger vesicles where they surround the calcite crystals on the rims of maturing coccoliths (Fig. 2a). The coccolithisomes preferentially stain with periodic acid-silver proteinate, demonstrating the presence of monosaccharide residues and indicating that the structures are either

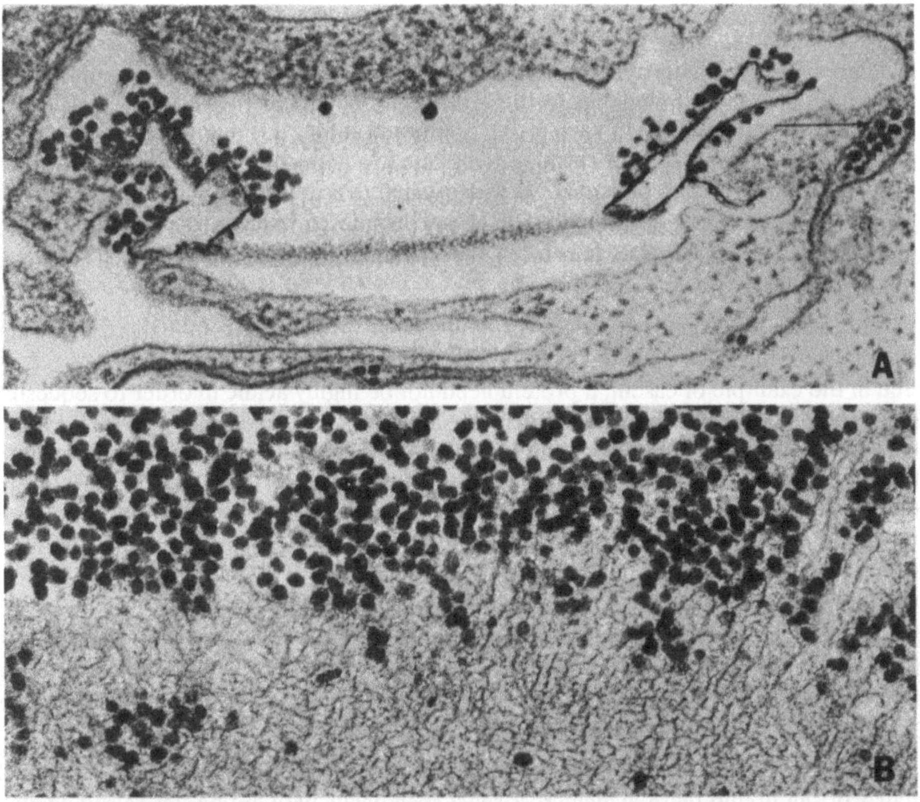

FIGURE 2. ▪A. Electron micrograph of a vesicle containing a maturing coccolith in the alga *Hymenomonas carterae*. The electron dense particles surrounding the calcite crystals are the coccolithisomes. Only the electron lucent profiles of the crystals are seen because the cells have been demineralized. Micrograph from Outka & Williams (1971). Reproduced with permission. Magnification, × 80,000. ▪B. Electron micrograph of an aragonite spherulite (bottom) surrounded by electron dense phosphorprotein particles (top) at the inner shell surface in the bivalve *Rangia cuneata*. Only the electron lucent profile of the spherulite is seen because the tissue has been demineralized. Magnification, × 60,000.

glycoproteins or polysaccharides (van der Wal et al., 1983a). By quantitative X-ray microanalysis, the individual coccolithisomes have an estimated calcium ion content of 6M (van der Wal et al., 1983b). This estimate is about twice the level of calcium in the bivalve phosphorprotein particles. Van der Wal (1983b) has postulated that the coccolithisomes function as intermediate calcium carriers for coccolith mineralization. The coccolithisomes surrounding the developing calcite crystals in the alga are ultrastructurally indistinguishable from the phosphoprotein particles surrounding the developing aragonite crystals in the bivalve (Fig. 2). The ultrastructure, calcium content and localization of the coccolithisomes strongly suggest that these particles are functionally and perhaps compositionally similar to the bivalve phosphoprotein.

DISCUSSION

High-capacity calcium-binding macromolecules have been localized at mineralization fronts in species belonging to three different groups of organisms (vertebrates, invertebrates and plants). In each group a different mineral is deposited. The calcitic and aragonitic polymorphs of calcium carbonate are deposited in the alga and bivalve respectively, and calcium phosphate as hydroxyapatite is deposited in vertebrate dentin. On the basis of these observations, it is worthwhile to explore the possibility that intermediate calcium carriers may be a general phenomenon in the biological calcification of skeletal tissues.

Very acidic highly phosphorylated proteins, which are obvious candidates for calcium carriers, have not been detected in vertebrate bone and calcified cartilage. However, as seen in the example of casein, a protein need not be highly acidic in order to sequester a large amount of calcium; very few acidic residues are required if the calcium is bound as relatively large clusters of calcium phosphate. Recently Butler's group has isolated a 44,000 Dalton phosphorylated glycoprotein from the rat bone matrix (Prince et al., 1987). About 4.2% of the residues are phosphoserine and phosphothreonine and maximally another 36.3% are aspartic and glutamic acid. Although the protein may bind little calcium directly, it is a possible candidate for a carrier of calcium ions via complexes with inorganic phosphate. In vertebrate growth plate cartilage, the acidic proteoglycans which are rich in uronic acid and sulfated hexosamine residues may be regarded as possible calcium carriers. In vitro these macromolecules are capable of suspending about 5g of CaHPO4 per 100g of proteoglycan (Pita et al., 1970), which is comparable to the mineral content of casein particles.

Recently Veis et al., (1986) have isolated phosphoproteins from the mineralized matrix of sea urchin teeth, which cross react specifically with antibodies raised against dentin phosphophoryns. Therefore, some degree of homology between these groups of proteins is indicated. Aspartic acid rich proteins have been identified in the shell matrices of several mollusks. Although these proteins are acidic, they contain no phosphorylated residues. The phosphoproteins of sea urchin teeth are possible candidates for calcium carriers, but the molluscan shell proteins are probably not. The latter are thought to influence the crystal habits of the shell mineral (Crenshaw, 1972; Sikes & Wheeler, 1983; Addadi & Weiner, 1985).

If intermediate calcium carriers are a general phenomenon, are the carriers which mediate the calcification of diverse skeletons in distantly related organisms derived from a common ancestral gene, or are the structural requirements for high capacity calcium carriers flexible enough to permit a parallel evolutionary development of carriers from diverse genetic elements? The primary structures of calcium carriers with a mineralizing function have not yet been determined. However, the similarity in amino acid composition suggests a sequence homology may exist between the dentin and bivalve phosphoproteins. It is difficult to even speculate on homology between these proteins and the algal coccolithisomes, since virtually nothing is known of the algal macromolecule. If future investigations demonstrate that the coccolithisomes are analogous to the bivalve and dentin phosphoproteins, a strong case can be made for a common ancestral gene. If, however, the coccolithisomes are principally acidic polysaccharides similar to cartilage proteoglycans, a parallel evolutionary development of calcium carriers will be indicated. Mineralized skeletons appeared abruptly in diverse metazoan organisms in the first 40-50 million years of the Cambrian period (Runnegar, 1987). Since plants and animals diverged long before Cambrian time, homology between algal and animal calcium carriers would indicate that pre-existing homologous proteins in distantly related organisms were later recruited for a common mineralizing function.

ACKNOWLEDGEMENTS

Helpful discussions with Dr. Bruce Runnegar on the evolution of mineralized structures are gratefully acknowledged. The author's contributions to the studies summarized in this review were supported by NIH Grant AM36239. Participation in the Symposium was made possible by grants from The Organized Research Fund of The University of Texas at Arlington and The Sea Grant Program administered by Texas A & M University.

REFERENCES

ADDADI, L., & WEINER, S., 1985. Interactions between acidic proteins and crystals: stereochemical requirements in biomineralization. Proc. Nat. Acad. Sci. U.S.A., 82:4110-4114.

BUTLER, W.T., BHOWN, M., DIMUZIO, M.T., COTHRAN, W.C., & LINDE, A., 1983. Multiple Forms of rat dentin phosphorproteins. Arch. Biochem. Biophys., 225:178-186.

CRENSHAW, M.A., 1972. The soluble matrix from *Mercenaria mercenaria* shell. Biomineralization, 6:6-11.

CRENSHAW, M.A., & NEFF, J.M. 1969. Decalcification at the mantle-shell interface in mollusks. Amer. Zool., 9:881-885.

DIMUZIO, M.T., & VEIS, A., 1978. The biosynthesis of phosphophoryns and dentin collagen in the continuously erupting rat incisor. J. Biol. Chem., 253:6845-6852.

FUJISAWA, R., KUBOKI, Y., & SASAKI, S., 1985. *In vivo* cleavage of dentin phosphophoryn following β-elimination of its phosphoserine residues. Arch. Biochem. Biophys., 243:619-623.

HOLT, C., 1982. Inorganic constituents of milk. III. The colloidial calcium phosphate of cow's milk. J. Dairy Res., 49:29-38.

KUBOKI, Y., FUJISAWA, R., AOYAMA, K., & SASAKI, S., 1979. Calcium-specific precipitation of dentin phosphoprotein: a new method of purification and the significance for the mechanism of calcification. J. Den. Res., 58:1926-1932.

LEE, S.L., KOSSIVA, D., & GLIMCHER, M.J., 1983a. Phosphoproteins of bovine dentin: evidence for polydispersity during tooth maturation. Biochemistry, 22:2596-2601.

LEE, S.L, GLONEK, T., & GLIMCHER, M.J., 1983b. ^{31}P Nuclear magnetic resonance spectroscopic evidence for ternary complex formation of fetal dentin phosphorprotein with calcium and inorganic orthophosphate ions. Calc. Tiss. Intern., 35:815-818.

LEE, S.L., VEIS, A., & GLONEK, T., 1977. Dentin phosphoprotein: an extracellular calcium-binding protein. Biochemistry, 16:2971-2979.

LYSTER, R.L.J., MANN, S., PARKER, S.B., & WILLIAMS, R.T.P., 1984. Nature of micellar calcium phosphate in cows' milk as studied by high-resolution electron microscopy. Biochim. Biophys. Acta, 801:315-317.

MARSH, M.E., 1986a. Histidinoalanine, a naturally occurring cross-link derived from phosphoserine and histidine residues in mineral-binding phosphoproteins. Biochemistry, 25:2392-2396.

MARSH, M.E., 1986b. Biomineralization in the presence of calcium-binding phosphoprotein particles. J. Exp. Zool., 239:207-220.

MARSH, M.E. & SASS, R.L., 1985. Distribution and characterization of mineral-binding phosphoprotein particles in Bivalvia. J. Exp. Zool., 234:237-242.

MARSH, M.S., & SASS, R.L.; 1984. Phosphoprotein particles: calcium and inorganic phosphate binding structures. Biochemistry, 23:1448-1456.

MARSH, M.E., & SASS, R.L., 1983. Calcium-binding phosphoprotein particles in the extrapallial fluid and innermost shell lamella of clams. J. Exp. Zool., 266:193-203.

MASTERS, P.M., 1985. *In vivo* decomposition of phosphoserine and serine in noncollagenous protein from human dentin. Calc. Tiss. Intern., 37:236-241.

MCGANN, T.C.A., KEARNEY, R.D., BUCHHEIM, W., POSNER, A.S., BETTS, F. & BLUMENTHAL, N.C., 1983. Amorphous calcium phosphate in casein micelles of bovine milk. Calc. Tiss. Intern., 35:821-823.

MEPHAM, T.B., GAYE, P., & MERCIER, J.C., 1982. Biosynthesis of milk proteins. *In* Developments in Dairy Chemistry-1 (ed. P. F. Fox), pp. 115-156. New York: Elsevier.

OUTKA, D.E., & WILLIAMS, D.C., 1971. Sequential coccolith morphogenesis in *Hymenomonas carterae*. J. Protozool., 18:285-297.

PITA, J.C., CUERVO, L.A., MADRUGA, J.E., MULLER, F.J., & HOWELL, D.S., 1970. Evidence for a role of protein polysaccharides in regulation of mineral phase separation in calcifing cartilage. J. Clin. Invest., 49:2188-2197.

PRINCE, C.W., OOSAWA, T., BUTLER, W.T., TOMANA, M., BHOWN, A.S., BHOWN, M., & SCHROHENLOHER, R.E., 1987. Isolation, characterization, and biosynthesis of a phosphorylated glycoprotein from rat bone. J. Biol. Chem., 262:2900-2907.

RUNNEGAR, B., 1989. This volume.

SASS, R.L. & MARSH, M.E., 1983. N τ- and Np π-Histidinoalanine: naturally occurring cross-linking amino acids in calcium-binding phosphoproteins. Biochem. Biophys. Res. Comm., 114:304-309.

SCHMIDT, D.G., 1982. Association of caseins and casein micelle structure. *In* Developments in Dairy Chemistry-1 (ed. P.F. Fox, pp. 61-86). New York, Elsevier.

SCHMIDT, D.G., 1980. Colloidial aspects of casein. Netherlands Milk Dairy J., 34:42-64.

SIKES, C.S., & WHEELER, A.P., 1983. A systematic approach to some fundamental questions of carbonate calcification. *In* Biomineralization and Biological Metal Ion Accumulation (ed. P. Westbroek & E.W. de Jong), pp. 285-289. Dordrecht: D. Reidel Publishing Co.

STETLER-STEVENSON, W.G., & VEIS, A., 1983. Bovine dentin phosphophryn: composition and molecular weight. Biochemistry, 22:4326-4335.

SWAISGOOD, H.E., 1982. Chemistry of milk protein. *In* Developments in Dairy Chemistry-1 (ed. P.F. Fox), pp. 1-59. New York: Elsevier.

VAN DER WAL, P., DE JONG, E.W., WESTBROEK, P., DE BRUIJN, W.C., & MULDER-STAPEL, A.A., 1983a. Polysaccharide localization, coccolith formation, and Golgi dynamics in the Coccolithiphorid *Hymenomonas carterae*. J. Ultrastr. Res., 85:139-158.

VAN DER WAL, P., DE JONG, E.W., WESTBROEK, P., & DE BRUIJN, W.C., 1983b. Calcification in the Coccolithophorid alga *Hymenomonas carterae*. *In* Environmental Biogeochemistry, Ecological Bulletin No. 35 (ed. R. Hallberg), pp. 251-258. Stockholm: Publishing House/FRN.

VEIS, D.J., ALBINDER, T.M., CLOHISY, J., RAHIMA, M., SABSAY, B., & VEIS, A., 1986. Matrix proteins of the teeth of the sea urchin *Lytechinus variegatus*. J. Exp. Zool., 240:35-46.

WILBUR, K.M., & BERNHARDT, A.M., 1982. Mineralization of molluscan shell: effects of free and polyamino acids on crystal growth rate *in vitro*. Amer. Zool., 22:952.

MORPHOLOGY AND BIOMINERALIZATION: A CARBONATE MODEL

Henri Petit[1]

Department of Orthodontics, Baylor College of Dentistry, 3302 Gaston Avenue, Dallas, TX. [1]Present address: 16-18 Rue du Cloître Notre-Dame 75004 Paris, France

ABSTRACT. The relationships between form and function in nature have been the focus of attention for centuries. Nevertheless, we still are in the dark in our understanding of the transmission of the genetic information related to the form of biological structures. Since 1974, *Amblema plicata perplicata* Conrad 1834, has provided us with a good model to study the role of the free periostracum. The periostracum carries out, without any contact with the cells of the mantle, a complex multi-structural organization at the edge of the forming shell. A dynamic and elaborate system of matrix/crystal relationships is described for the nacre formation as well as for the prism formation. This system can be disrupted and altered experimentally. Such temporary disorganization accordingly modifies the physical properties of the shell.

INTRODUCTION

The preparation for electron microscopy of frozen block specimens of mantle edge, free periostracum and attached forming edge of the shell was started as early as 1974 with *Amblema plicata perplicata* Conrad 1834 as an experimental animal. Later, the mineralization of the edge of the shell of a few other unionid mussels, as well as a series of Veneridae were investigated with a similar technique, which also was applied to the study of the periostracum of glutaraldehyde fixed samples of *Bathymodiolus thermophylus*. The close relationships of the periostracum and the adjacent epithelial cells within the periostracal groove was demonstrated and appeared in full agreement with the findings of previous researchers. These studies clearly show that layers of specific glycocalitic coatings were deposited on both sides of a pellicle secreted initially, by a specific row of cells at the bottom of the periostracal groove. Even if the initial formation of the pellicle differs markedly within species, coming either from "prepackaged periostracum" or from a cell membrane unbalance, the periostracum, within the periostracal groove and later between its exit of the mantle edge and its junction with the forming edge of the shell, appears as a conveyor belt capable of the transmission of various cellular information, dealing with biomineralization patterns.

Moreover, the periostracum can be seen as a multilaminar belt. This is obvious with the type of periostracal formation where the pellicle is coated on both sides thus forming at least three series of layers. This is also consistently observed in other systems for

Origin, Evolution, and Modern Aspects of Biomineralization in Plants and Animals
Edited by Rex E. Crick
Plenum Press, New York

other species and one can assess that the rythmicity of the glycocalitic deposition, associated with a slow extrusion of the whole periostracum makes a complex dynamic system which provides several temporal labels within the structure of the periostracum itself. It also assigns specifically some type of apposition material to each side of the pellicle. For example, one side will collect the glycocalytic deposition coming from the long microvillosities of the long epithelial cells, when the other side will be coated with the deposition coming from the short microvillosities of the short epithelial cells of the periostracal groove. The deposition may not be a perfectly continuous process, therefore, growth lines and irregular layering are obvious within each side of the coated pellicle. Finally, the periostracum matures and before the exit from the periostracal groove one can identify and dissect, depending on the species, two or three main layers: one outer layer, one pellicle and one inner layer (in unionids and Veneridaes), or one inner layer and one outer layer (in *Bathymodiolus* for an example). Each of these layers, but the pellicle are generally multilayered, and we have identified as many as 50 layers in the internal layer of *Amblema*.

Later, after the exit from the periostracal groove, where a tanning of the outer layer modified the physiology and chemical organization of the periostracum, the structural rearrangement of the periostracum is even more complicated by the addition of a new layer of glycocalytic material poured from the inner aspect of the mantle edge and added, internally onto the inner layer of the initial periostracum. The periostracum becomes identifiable again with three new layers that can be dissected. The innermost layer is made from this newly added material, the middle layer is made of the previous three layers, minus the tanned outer layer, the third layer is the newly tanned outer layer. This complex stratification is further connected to the mineral structures of the forming edge.

There is a continuum between the outer layer of the free periostracum (tanned) and the outer layer called also periostracum covering the outer aspect of the shell. The protective role of these layers seems essential for the biochemical processes of mineralization to happen within the underneath periostracal structures and within the palleal space. The middle layer, a complex layer with part of the three previous periostracal layers formed within the periostracal groove, is directly connected to the matrices of the forming prisms, at the edge of the forming shell. The innermost layer is directly connected with the matrices of a series of layers of "initial nacre" directly deposited under the primatic structures, and on which the mantle epithelial cells will later organize a continuous layering of the bulk of the nacre, through their glycocalytic coating.

More sophisticated studies have shown the formation of initial needles of aragonite within the periostracal layers in both the middle and the innermost layers. These needles are specifically associated into secondary types of structures which differs with the layers. An architectural organization of the biomineralization pattern for the prisms starts early in the middle layer of the periostracum with the association of the needles into little spherical structures. At the same time vacuolization can be observed, then the fusion of vacuoles within the middle layer of the periostracum helps organize some larger vacuoles called "antrum". In this process, partitions within the antrum delineate the future prismatic matrices. These matrices are later simply filled with the aragonite spherical structures. Therefore, the forming prisms appear as a tertiary architectural organization and not as a monocrystal to start with.

The innermost layer of the periostracum (also multilayered) is in contact with the pallial fluid and starts its journey toward the edge of the shell with the cleavage of some of its innermost layers making a fews loops inside the pallial fluid. At this time an amorphous association of Ca with the proteinaceous material of the periostracal layer can be seen. Later, more distally, the formation of aragonite needles is obvious. At the same time a geometric pattern of equally distant layers of proteinaceous material is formed to limit and contain the association of packages of parallel needles of aragonite into nacreous bricks. These nacreous bricks, also, are a tertiary structure during the formation of the initial nacre.

The prismatic layer as well as the initial nacreous layers seem to be formed and mineralized under the control of the periostracum. Some intricate networks of matrices connect together these various characteristic mineral structures and guide their architectural arrangements. They have been demonstrated to derive from the periostracum.

The morphogenetic role of the periostracum is not a simple one and in some species like the unionid mussels, we have been capable of following very precisely the sequences of events taking place during the formation of the shell, either on the surface or within the periostracal layers. From these observations one can draw several conclusions.

First, the layers of the periostracum are multilaminar and present at least three main cleavable groups of layers (inner layer, pellicle and associated innermost layers, and outer layer). They are proteinaceous in nature and vary in amino-acid composition and ratio from the bottom of the periostracal groove to the formed edge. They also vary within layers, the layers related with prism formation being different from the layers related to nacre formation. Moreover, slight variations in amino-acid content and ratio can be seen within different species.

Second, during the process of mineralization, the periostracal layers play a chemically active role. The combination of some amino-acids to the forming "crystal shell", the organization of the mineral/matrix relationships from amorphous material to tertiary structures, as well as the changes in the geometry of the matrices are each specific events which bring together some dramatic modifications in the amino-acid ratio of the periostracum. In other words, the periostracum before mineralzation is different in amino-acid ratio from the periostracum after one functional event which is associated with one biomineralization stage. Minute amino-acid analyses demonstrate clearly that each of the biomineralization steps are associated with some specific periostracal modification which alter the initial amino-acid ratio. As a consequence, the analyses of the amino-acid composition of the matrices of one structure does not reveal the composition in specific amino-acids related to the biomineralization mechanism of the associated structure. Instead, it expresses the "left over" or the "non-used" part of the periostracum for the construction of this given structure. A wise approach to the understanding of the amino-acid ratio and sequencing would therefore be to look at the biomineralization sequences "a retro", trying to identify what has been taken, and therefore missing, in the periostracal mass for a specific event.

Third, after the maturation of the structures some specific matrix-crystal relationships can be observed. They involve a certain energy level of cohesion. Ongoing experiments show that disruption of the matrix/crystal relationships can alter the physical property of the structure and help temporarily modify its plasticity.

CONCLUSION

The molluscan model helps understand the elaboration of specific morphological structures, shaped with the assistance of proteinaceous matrices, which dynamically change in their biochemical composition and organization at each stage of maturation of the forming shell.

MINERAL INDUCTION BY THE SOLUBLE MATRIX FROM MOLLUSCAN SHELLS

Ed M. Greenfield[1] and Miles A. Crenshaw

Marine Science Curriculum and Dental Research Center, University of North Carolina, Chapel Hill, NC 27514; [1]Present address: Pathology Department, Jewish Hospital, Washington University Medical Center, St. Louis, MO 63110

ABSTRACT. Whole matrix (a soluble/insoluble complex) was prepared from *Nautilus* septal nacre. Spherules formed on the whole matrix in solutions with activities resembling extrapallial fluid. Tripling the calcium in the incubation solution resulted in the formation of blade-like and plate-like crystals on the whole matrix. Mineral did not form on insoluble matrix even in the triple calcium incubation solution. Thus, the soluble matrix appears necessary to induce mineral. In contrast, a number of investigators have shown that the soluble matrix, when it is free in solution, binds to growing crystals and inhibits further growth. Accordingly, we conclude that the soluble matrix must be immobilized in order to induce mineralization.

INTRODUCTION

Molluscan shell formation is a well known example of "matrix mediated" mineralization (Wilbur, 1964; Lowenstam, 1981). The matrix can be separated into soluble and insoluble fractions during decalcification in EDTA (Crenshaw, 1972a). It has been suggested that the EDTA soluble fraction of the matrix is responsible for inducing mineral formation. This conclusion is primarily based on two lines of evidence: (1) the soluble matrix is localized at the site of initial mineral formation (Crenshaw & Ristedt, 1975, 1976; Greenfield *et al.*, 1984) and (2) the soluble matrix selectively binds calcium (Crenshaw, 1972a; Greenfield *et al.*, 1984; Worms & Weiner, 1986). The calcium binding induces anion binding, which induces further calcium binding (Greenfield *et al.*, 1984). This cooperative ion binding has been termed ionotropy and is believed to be the initial step in mineral formation (Crenshaw, 1982; Greenfield *et al.*, 1984).

When the soluble matrix is free in solution, it binds to growing crystals and inhibits further growth (Wheeler *et al.*, 1981; Sikes & Wheeler, 1983, 1986; Wheeler & Sikes, 1984; Addadi & Weiner, 1985). However, when the soluble matrix is immobilized, it appears to mediate crystal deposition. We have shown that mineral forms on whole matrix (a soluble/insoluble complex) incubated under conditions where the insoluble matrix by itself is inactive (Greenfield *et al.*, 1984). In addition, Addadi and Weiner (1985) have immobilized an "acidic" fraction of the soluble matrix by adsorption onto either polystyrene or glass. Calcite that is formed in the presence of this immobilized

Edited by Rex E. Crick
Plenum Press, New York

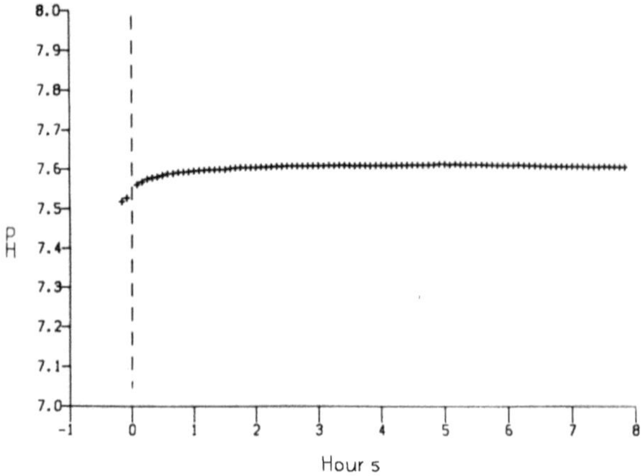

FIGURE 1. Stability of the pH during eight hour incubations. Each point represents a five minute average of 640 readings. Both calcium chloride and matrix were added at zero hours.

soluble matrix expresses an unusual 001 face. The 001 face is formed in contact with and appears to be induced by the immobilized soluble matrix. Thus, the soluble matrix may perform multiple functions: inducing mineral when immobilized on a support and inhibiting/regulating crystal growth when free in solution (Crenshaw, 1982; Greenfield *et al.*, 1984).

This study was undertaken to compare the mineral induction potential of whole matrix and insoluble matrix in solutions that do not spontaneously precipitate and that resemble the *in vivo* calcifying solution, the extrapallial fluid. However, the ionic strength of the incubation solutions had to be very low to prevent extraction of the soluble matrix from the whole matrix. Accordingly, the two types of matrix were incubated in solutions with ion activities resembling the extrapallial fluid, but with minimal ionic strength.

MATERIALS AND METHODS

The incubation solutions were based on the ion concentrations found in the *Mercenaria mercenaria* (Linne) extrapallial fluid (Crenshaw, 1972b) and have been described in detail (Greenfield, 1987). Briefly, the extrapallial fluid concentrations were corrected for calcium binding to macromolecules (Crenshaw, 1972; Misogianes & Chasteen, 1979) and for anaerobiosis (Crenshaw & Neff, 1969). The corrected extrapallial fluid concentrations are listed in Table 1. Activity coefficients (γ) were calculated from the concentrations with the Davies equation. The resultant $\gamma \times M_t$ values are also listed in Table 1. Incubation solution values of $\gamma \times M_t$ of calcium, carbonate and bicarbonate were set equal to the calculated extrapallial fluid values. The incubation solution concentrations (Table 1) were then back calculated iteratively with the Davies equation, while minimizing the ionic strength. In some incubations, the calcium was tripled to encourage mineral formation. In those cases, the calculations were performed in the manner described above, except that the calcium $\gamma \times M_t$ was set at 5.37 instead of 1.8. The average pH during the incubations was adjusted to 7.6 by trial and error additions

TABLE 1. Ion concentrations and $\gamma \times M_t$ in extrapallial fluid and the standard incubation solution

	EXTRAPALLIAL FLUID		INCUBATION SOLUTION	
	(mM)	$\gamma \times$ mM	$\gamma \times$ mM	mM
Calcium	7.9	1.79	1.79	2.8
HCO_3^-	4.16	2.87	2.87	3.21
$CO_3^=$	0.041	0.0093	0.0093	0.014
Sodium	444.0	306.0		3.3
Chloride	472.0	326.0		5.6
Magnesium	55.6	12.6		
Sulfate	23.0	5.23		
Potassium	9.6	6.63		
Ionic Strength	0.640		0.0117	
pH	7.6		7.6	
pCO_2			0.5%	
Saturation:				
Calcite	2.6		3.7	
Aragonite	1.9		2.8	

of sodium hydroxide. Dissolved inorganic carbon was supplied by equilibrating the incubation solution with mixtures of CO_2 and nitrogen. The Stumm and Morgan (1970) constant pCO_2 model was used to determine the appropriate CO_2 concentration. The degree of saturation in the incubation solutions and in extrapallial fluid was calculated with Garrels and Thompson's (1962) solubility constants after determining the extent of ion pairing. Ion pairing was determined using the computer program MINEQL (Westall *et al.*, 1976) after addition of a subroutine that iteratively corrects the ionic strength for the ion pairing (D.C. Wilson, Univ. N. Carolina-Chapel Hill, personal communication). As expected, the calculated degree of saturation in the incubation solution and in extrapallial fluid are similar (Table 1). Therefore, the errors due to the assumptions and approximations inherent in the design of the incubation system are small (for details see Greenfield, 1987).

Whole matrix was prepared by decalcifying *Nautilus pompilius* (Linne) septal nacre in 0.5% cetylpyridinium chloride (CPC), 0.1 M EDTA (pH 8) for 7 minutes (Crenshaw & Ristedt, 1975, 1976). The soluble matrix can be extracted by either increasing the length of the decalcification period or by using high ionic strength EDTA without CPC. Thus, insoluble matrix was prepared by decalcifying in 0.5% CPC, 0.1 M EDTA for 20 minutes or in 0.5 M EDTA (pH 8) for 7 minutes (Crenshaw & Ristedt, 1975, 1976). Matrix sheets were teased off of the side of the septum which had faced the mantle. The matrix was briefly rinsed in water and transferred to the incubation solutions, which had been equilibrated at 25°C with 0.5% CO_2. After addition of the matrix, the gas flow was shifted from bubbling through the incubation solution to flushing the head space of the reaction vial. This was necessary to protect the matrix from being physically disrupted by the bubbling. At the end of the incubation period, the matrix was briefly rinsed in dilute ammonium hydroxide and then in acetone (Blumenthal *et al.*, 1972).

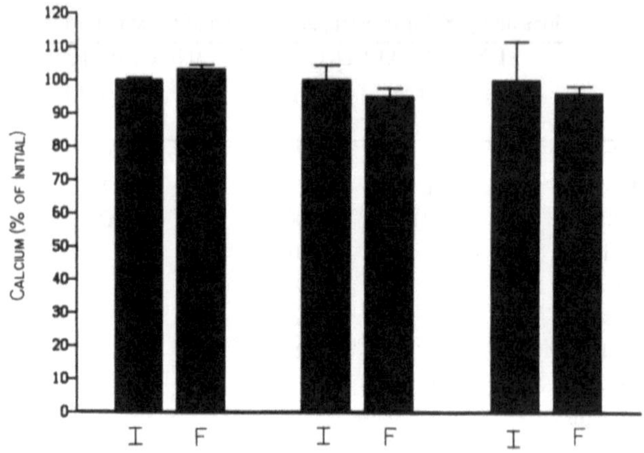

FIGURE 2. Stabililty of the calcium concentration during eight hour incubations. Calcium was determined by atomic absorption spectrophotometry. The three pairs of bars represent the initial (I) and final (F) calcium concentration (± standard deviation) of three separate experiments. In no case was there a significant ($P<0.05$) decrease in the calcium concentration. All three experiments included whole matrix in the triple calcium activity incubation solution.

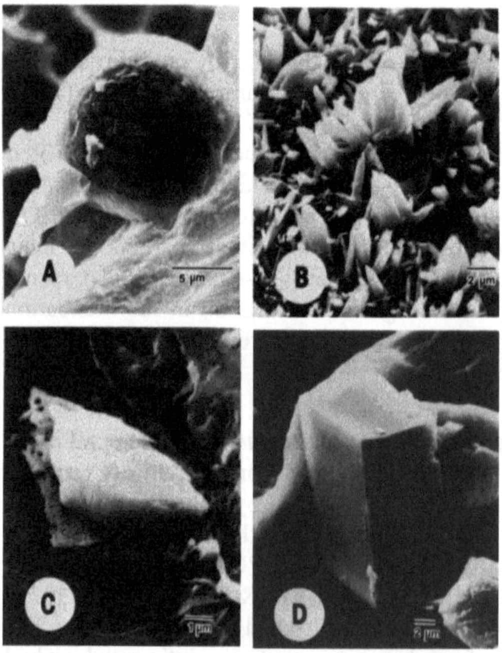

FIGURE 3. Scanning electron micrographs of the crystals formed on the whole matrix during the incubations. Spherules (▪A) formed in the standard incubation solution. Blade-like (▪B), plate-like (▪C), and rhombohedral (▪D) crystals formed in the triple calcium activity incubation solution. However, the rhombohedrons also formed in the absence of matrix.

The matrix was air dried, sputter coated with gold and examined with an ETEC Autoscan scanning electron microscope.

RESULTS AND DISCUSSION

It was difficult to obtain an average pH of precisely 7.60 during the incubations. Accordingly, all results were based on incubations with average pH's both slightly above and slightly below 7.6 (between 7.58 and 7.62). The pH of the incubation solution was very stable during the eight hour incubations (Fig. 1). This was true even when the incubation solution containing the triple calcium activity was used (data not shown). The constant pH and pCO_2 indicate that the CO_3 levels were also stable. The calcium concentration in the incubation solution was also constant during the incubations, even with the triple calcium activity solution (Fig. 2). Thus, the incubations were performed under essentially constant conditions.

$CaCO_3$ spherules formed on the whole matrix during the eight hour incubations (Fig. 3A). Mineral was not found, however, after six hour incubations. If the solution was incubated without matrix and then filtered through 0.2 mm polycarbonate filters (Nuclepore`), no mineral was found on the filters. Thus, mineral was only found in the presence of whole matrix. Of course, crystals smaller than 0.2 mm may have formed in the incubation solutions. However, the mineral that formed on the whole matrix was much larger than 0.2 mm. Therefore, the formation of these relatively large crystals was due to the whole matrix, probably acting either as a crystal nucleator or a crystal agglomerator.

When whole matrix was incubated in the triple calcium solution, many small blade-like crystals were found after four to eight hours (Fig. 3B). After eight hours, there were also some plate-like crystals on the whole matrix (Fig. 3C). If this solution was incubated without matrix, neither of these crystal types were found upon filtration. A few rhombohedral crystals were also found on the whole matrix after the eight hour incubations (Fig. 3D). However, these crystals also formed in the absence of matrix and, therefore, must be discounted.

To determine which fraction of the matrix was responsible for the mineral formation, insoluble matrix was tested in the incubation solutions. In contrast to whole matrix, insoluble matrix did not induce mineral formation even in the triple calcium solution. Thus, the soluble matrix is necessary for mineral formation in these solutions and, possibly, *in vivo*. This finding does not exclude a role for the insoluble matrix. It may still be involved in the regulation of crystal polymorph, orientation, growth rate and/or morphology. Recently, Bernhardt *et al.* (1985) suggested that the insoluble matrix may orient growing crystals *in vitro*.

CONCLUSIONS

Solutions were prepared with ion levels similar to extrapallial fluid, but with very low ionic strength. These solutions were stable during eight hour incubations.

Spherules formed on the whole matrix incubated in the above solutions. When the calcium was tripled, blade- and plate-like crystals formed on the whole matrix. Mineral did not form on the insoluble matrix in either of these solutions. Thus, the soluble matrix mediates mineral formation *in vitro* when it is immobilized on its *in vivo* support, the insoluble matrix.

ACKNOWLEDGEMENTS

The authors are indebted to Doug Wilson for providing the MINEQL computer calculations. This project was supported by the Office of Naval Research Contract NOOO14-84-C-0183 and the North Carolina Biotechnology Center. This is paper No. 18 from the North Carolina Biomolecular Engineering and Materials Application Center. Participation in the Symposium was made possible by grants from The Organized Research Fund of The University of Texas at Arlington and The Sea Grant Program administered by Texas A & M University.

REFERENCES

ADDADI, L. & WEINER, S., 1985. Interactions between acidic proteins and crystals: stereochemical requirements in biomineralization. Proc. Nat. Acad. Sci. U.S.A., 82:4110-4114.

BERNHARDT, A.M., MANYAK, D.M. & WILBUR, K.M., 1985. *In vitro* recalcification of organic matrix of scallop shell and serpulid tubes. J. Moll. Stud., 51:284-289.

BLUMENTHAL, N.C., POSNER, A.S. & HOLMES, J.M., 1972. Effect of preparation conditions on the properties and transformation of amorphous calcium phosphate. Mat. Res. Bull., 7:1181-1190.

CRENSHAW, M.A., 1972a. The soluble matrix from *Mercenaria mercenaria* shell. Biomineralization, 6:6-11.

CRENSHAW, M.A., 1972b. The inorganic composition of molluscan extrapallial fluid. Biol. Bull., Mar. Biol. Lab., Woods Hole, Mass., 143:506-512.

CRENSHAW, M.A., 1982. Mechanisms of normal biological mineralization of calcium carbonates. *In* Biological Mineralization and Demineralization, (ed. G.H. Nancollas), pp. 243-257. Berlin: Springer-Verlag.

CRENSHAW, M.A. & NEFF, J.M., 1969. Decalcification at the mantle-shell interface in molluscs. Amer. Zool., 9:881-885.

CRENSHAW, M.A. & RISTEDT, H., 1975. Histochemical and structural study of nautiloid septal nacre. Biomineralization, 8:1-8.

CRENSHAW, M.A. & RISTEDT, H., 1976. The histochemical localization of reactive groups in septal nacre from *Nautilus pompilius*. *In* The Mechanisms of Mineralization in the Invertebrates and Plants, (ed. N. Watabe & K.M. Wilbur), pp. 355-367. Univ. S. Carolina Press.

GARRELS, R.M. & THOMPSON, M.E., 1962. A chemical model for seawater at 25°C and one atmosphere total pressure. Amer. J. Sci., 260:57-66.

GREENFIELD, E.M., WILSON, D.C. & CRENSHAW, M.A., 1984. Ionotropic nucleation of calcium carbonate by molluscan matrix. Amer. Zool., 24:925-932.

LOWENSTAM, H.A., 1981. Minerals formed by organisms. Science, 211:1126-1131.

MISOGIANES, M.J. & CHASTEEN, N.D., 1979. A chemical and spectral characterization of the extrapallial fluid of *Mytilus edulis*. Anal. Biochem., 100:324-334.

SIKES, C.S. & WHEELER, A.P., 1983. A systematic approach to some fundamental questions of carbonate calcification. *In* Biomineralization and Biological Metal Accumulation, (ed. P. Westbroek & E.W. deJong), pp. 285-289. Holland: Reidel Publishing.

SIKES, C.S. & WHEELER, A.P., 1986. The organic matrix from oyster shell as a regulator of calcification *in vivo*. Biol. Bull., Mar. Biol. Lab., Woods Hole, Mass., 170:494-505.

STUMM, W. & MORGAN, J.J., 1970. Aquatic Chemistry, 583 pp. New York: John Wiley.

WESTALL, J.C., ZACHARY, J.L. & MOREL, F.M.M., 1976. MINEQL, a computer program for the calculation of chemical equilibrium composition of aqueous systems. Mass. Inst. Tech., Depart. Civil Eng. Tech. Note, 18:1-91.

WHEELER, A.P., GEORGE, J.W. & EVANS, C.A., 1981. Control of calcium carbonate nucleation and crystal growth by soluble matrix of oyster shell. Science, 212:1397-1398.

WHEELER, A.P. & SIKES, C.S., 1984. Regulation of carbonate calcification by organic matrix. Amer. Zool., 24:933-944.

WILBUR, K.M., 1964. Shell formation and regeneration. *In* Physiology of Mollusca, (ed. K.M. Wilbur & C.M. Yonge), pp. 243-282, New York: Academic Press.

WORMS, D. & WEINER, S., 1986. Mollusk shell organic matrix: fourier transform infrared study of the acidic macromolecules. J. Exp. Zool., 237:11-20.

COMPOSITION AND ASSOCIATION OF ORGANIC MATTER WITH CALCIUM CARBONATE AND THE ORIGIN OF CALCIFICATION

Richard M. Mitterer

Department of Geosciences, University of Texas at Dallas, PO Box 830688, Richardson, TX 75083-0688

ABSTRACT. Calcium carbonate is uniquely and characteristically associated with aspartic acid-rich (asp-rich) organic matter as an internal organic matrix and as an adsorbed phase. In addition to its adsorptive preference for carbonate surfaces, asp-rich organic matter also binds calcium ions in aqueous solution. Conversely, non-carbonate sediments are characterized by organic matter that is depleted in aspartic acid. Quartz preferentially adsorbs an asp-poor organic component. These properties indicate that asp-rich organic matter plays a significant role in controlling the nucleation and crystal growth of calcium carbonate in biological and abiological systems.

The development of calcification in the latest Precambrian-Early Cambrian represents a milestone evolutionary event. The subsequent fossil record, greatly enriched by the presence of these readily preserved calcareous skeletons, contains pulses of major extinctions. These two biological events - calcification and extinctions - are related here in a new hypothesis that provides a common explanation for both. The key to the hypothesis is a recent finding that calcification in modern marine organisms requires a minimum level of strontium in ambient seawater. Thus, the development of calcification may have been triggered when a critical level of strontium in the oceans was reached. Subsequent periodic decreases in the oceanic concentration of strontium may have been responsible for marine invertebrate extinctions.

INTRODUCTION

Calcium carbonate is one of two mineral phases (silica is the other) produced extensively by organisms. Accumulations of these biominerals account for much of the widespread carbonate rocks and cherts in the Phanerozoic geologic record. Both minerals also form abiogenically.

Calcified structures form through the interaction of inorganic ions and proteinaceous molecules in the body fluids of organisms. The organic substrate typically consists, at least in part, of an aspartic acid-rich (asp-rich) protein. Beyond these broad generalizations, major unanswered questions persist. Exactly how organisms calcify and how they achieve the wide variety of structures is not known. A common but very sketchy view is that asp-rich protein is an important agent of calcification through the interaction between carboxyl groups of the protein molecule and calcium ions of the fluids.

Origin, Evolution, and Modern Aspects of Biomineralization in Plants and Animals
Edited by Rex E. Crick
Plenum Press, New York

The association of calcium carbonate with asp-rich organic matter is not limited to biogenic carbonates. Abiogenic carbonates, such as ooids, also contain an "organic matrix" and this organic matter is also rich in aspartic acid. The association of asp-rich organic matter with biogenic and abiogenic carbonates highlights the potential importance of this molecular composition in the calcification process.

The purpose of this paper is several fold: (1) to emphasize the unique association between asp-rich organic matter and calcium carbonate; (2) to describe the adsorptive and Ca-binding properties of asp-rich organic matter; (3) to relate these results in a model of organic matter-calcium carbonate nucleation; and (4) to speculate on the origins of calcification.

COMPOSITIONAL PATTERN OF ORGANIC MATRIX

The organic matrix of invertebrate skeletons is proteinaceous with a characteristic predominance of aspartic acid. Weiner et al. (1983) have summarized the composition of the organic matrix in various phyla and have noted the soluble nature of asp-rich protein.

Figure 1a illustrates the amino acid composition of a representative molluscan example. Aspartic acid is the predominant component and comprises about 28% (280 residues/1000) of the total amino acids. The generality of this compositional pattern in carbonate-secreting organisms is demonstrated by a carbonate sediment containing a mixture of skeletal grain types from various contributing organisms in the environment (Fig. 1b). Glutamic acid usually comprises about 10% of these asp-rich proteins. Taken together the acidic amino acids, aspartic and glutamic acids, comprise a significant proportion of organic matrix protein. Due to the lesser abundance of glutamic acid, however, its role in calcification is not considered as important as that of aspartic acid.

The organic matrix of abiogenic carbonate sediments such as ooids also shares the same asp-rich compositional pattern. Ooids are sand-sized, spheroidal, concretionary grains that form in many shallow, high-energy, carbonate environments. Modern deposits occur, for example, in the Bahamas, Cuba, Red Sea and Persian Gulf. Ancient examples are known well into the Precambrian. The origin of ooids has been the subject of much speculation and controversy. Inorganic precipitation remains the popular theory, but other origins have been proposed. The presence of an asp-rich organic matrix has led to the suggestion that ooids may form through some form of organic-inorganic interaction in a manner analogous to calcification in organisms (Mitterer, 1968, 1971a, 1972a,b; Trichet, 1968; Carter, 1978; Mitterer & Cunningham, 1985). The various theories for the origin of ooids will not be discussed here. Rather, the intention is to highlight the compositional similarity of the organic matrices of biogenic and abiogenic carbonates. The amino acid composition of the organic matrix from an ooid sample is illustrated in Figure 1c. While the ooid organic matrix is not a protein, as in biogenic carbonates, and can more accurately be classified as a humic substance, hydrolyzable amino acids comprise a significant (~one-third) portion of the organic matrix. Similar to the calcified membranes of biogenic carbonates, aspartic acid is the dominant amino acid of the ooid organic matrix.

In summary, the compositional pattern obtained from amino acid analyses of a variety of biogenic and abiogenic carbonates is that aspartic acid is the prevalent amino acid. Analyses of non-carbonate sediments, soft tissues, and seawater are characterized by a predominance of amino acids other than aspartic acid (Mitterer & Cunningham, 1985). Asp-rich organic matter, therefore, is a unique feature of calcium carbonate in the marine environment.

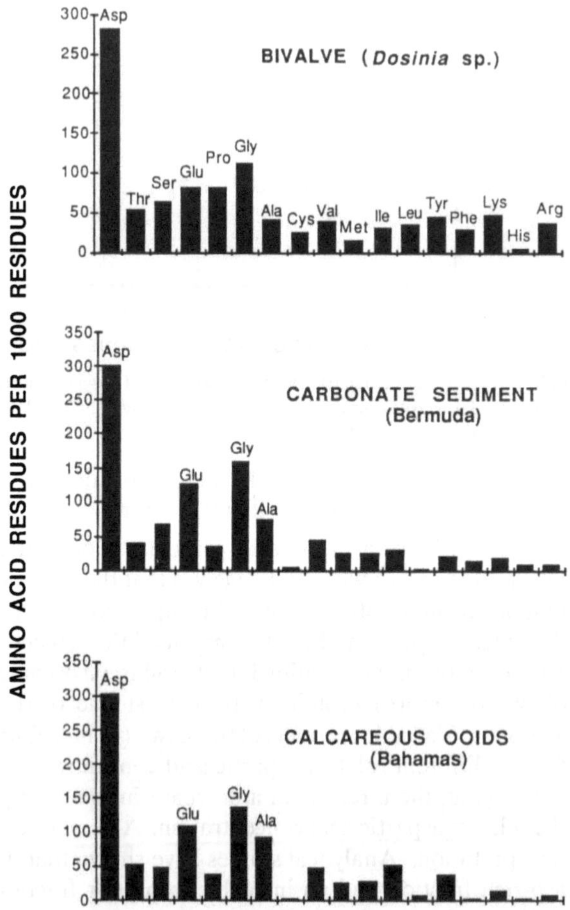

FIGURE 1. Amino acid composition, in residues per thousand, of organic matrix from mollusc shell, mixed skeletal sediment, and calcareous ooids. Amino acid abbreviations are: ASP, aspartic acid, THR, threonine, SER, serine, GLU, glutamic acid, PRO, proline, GLY, glycine, ALA, alanine, VAL, valine, MET, methionine, ILE, isoleucine, LEU, leucine, TYR, tyrosine, PHE, phenylalanine, LYS, lysine, HIS, histidine, ARG, arginine.

VARIABLE ASPARTIC ACID CONTENT OF MOLLUSCAN SHELL ORGANIC MATRIX

Not all skeletal organic matrices are asp-rich protein. The organic matrices of mollusc shells, which have been studied in greatest detail, have a range of protein contents and a variety of protein compositions. The protein content of bivalve shells, for example, varies by almost three orders of magnitude and ranges from about 3 to almost 900 micromoles (μM) of amino acid per gram of shell; aspartic acid in these shells ranges from a low of about 30 to a high of almost 350 residues/1000 (Fig. 2). Interestingly, the aspartic acid content is inversely proportional to the amount of protein present in the shells.

It has been suggested that the protein content of skeletal carbonates may be a measure of the efficiency of organisms in calcifying (Hare & Abelson, 1965). Organisms requiring a high protein content in their shells are relatively inefficient in calcifying, while organisms utilizing a lower protein content are able to calcify with greater efficiency. As aspartic acid concentration is inversely proportional to protein content, calcification efficiency can also be viewed in terms of the relative amount of aspartic acid in the

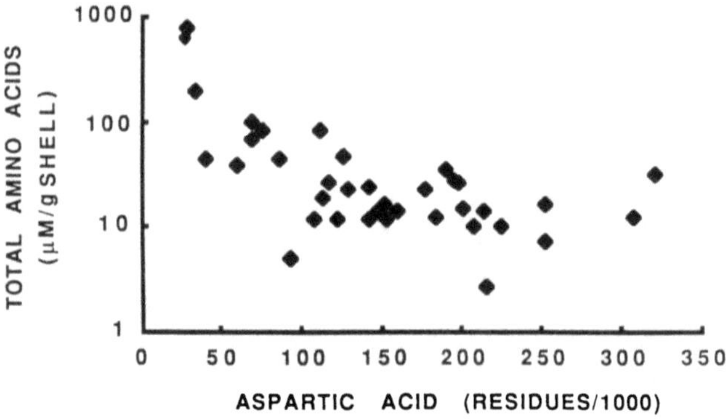

FIGURE 2. Relative aspartic acid composition of organic matrix as a function of the total amount of protein in bivalve shells. Data from Degens and Spencer (1966).

protein. Relatively high aspartic acid and low protein concentrations are inferred to represent efficient calcification, while relatively low aspartic acid and high protein concentrations signify less efficient calcification. This relative increase in aspartic acid concentration with decreasing protein content further demonstrates the importance of this amino acid in calcifying protein. Only amino acids that are critical to the calcification process will be evolutionarily favored; non-essential amino acids will decrease in relative abundance.

The range of aspartic acid contents of organic matrices in mollusc shells can be interpreted in two ways. One view is that the organic matrix is a homogeneous protein that has evolved into a variety of compositions with differing amounts of aspartic acid. Some loss of protein, then, accompanied the molecular evolution to an asp-rich nature. An alternative interpretation is that molluscan organic matrix is a heterogeneous mixture of two (or more) proteins with at least one of these being asp-rich and one being asp-poor. Variable amounts of these two (or more) proteins in bivalve shells can account for the different relative aspartic acid concentrations. Preferential loss of the asp-poor component, then, results in a decrease in the total protein content and an increase in the relative aspartic acid concentration. A number of lines of evidence support the latter interpretation. Analytical studies have shown that matrix protein consists of a soluble, asp-rich fraction and an insoluble, asp-poor fraction (Mitterer, 1972a; Weiner et al., 1983.

Further experimental evidence for the inhomogeneity of matrix protein is illustrated in Figure 3. The bivalve shell Pinna sp., with a protein content of about 80 mM per gram of shell and a relative aspartic acid concentration of about 10 percent (100 residues/1000), was treated with hot 2N NaOH for different periods of time. The hot base hydrolyzes any accessible protein within the shell. Analysis of the shell after treatment shows a fourfold decrease in protein content and a sixfold increase in relative aspartic acid concentration. Continued treatment with hot base does not significantly decrease the protein content or increase the relative

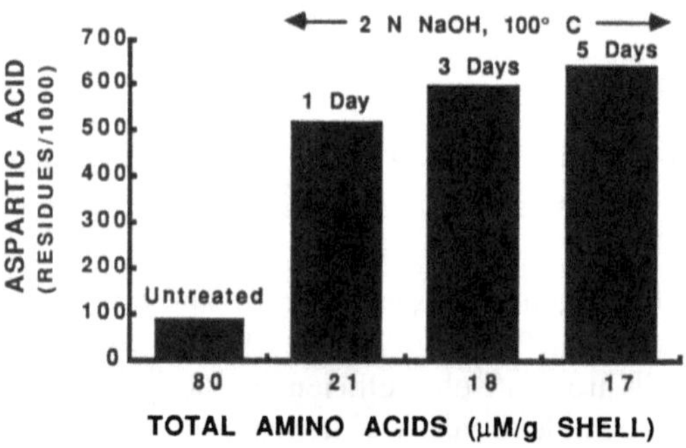

FIGURE 3. Relative aspartic acid composition of organic matrix of *Pinna* sp. before and after treatment with NaOH.

aspartic acid concentration beyond this limit. The interpretation of the results of this experiment is that the organic matrix consists of a mixture of proteins. A major portion of the organic matrix in Pinna is relatively "unprotected" and accessible to the hydrolytic alkaline solution while a minor portion evidently is "protected" by the shell and unaffected by the base. The hydrolyzed portion is asp-poor and the "protected" protein is asp-rich.

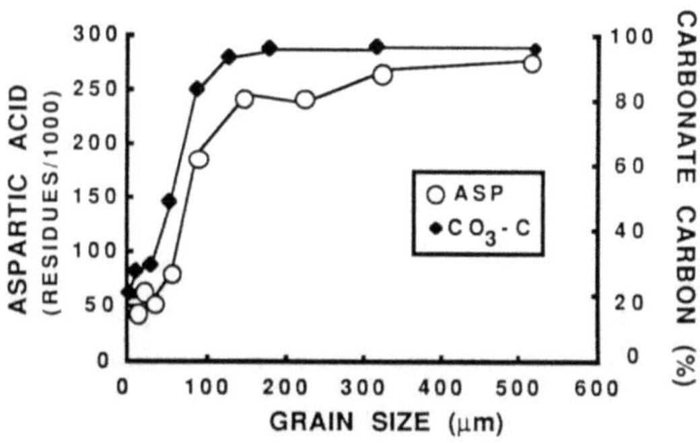

FIGURE 4. Relative aspartic acid composition and calcium carbonate content of sediments from the Flower Gardens Reef area, offshore Texas, as a function of sediment grain size. Both aspartic acid and calcium carbonate decrease significantly in the finer sediment fractions. Data from Carter and Mitterer (1978).

These results are in accord with the concept that the organic matrix consists of intercrystalline and intracrystalline moities (e.g., Watabe, 1965; Mutvei, 1970). Possibly the asp-poor intercrystalline organic matrix, apparently uncalcified and therefore "unprotected", is more easily removed by hydrolytic base. Asp-rich intracrystalline matrix, apparently well-calcified and more "protected", remains after alkaline treatment.

In summary, the evidence favors an interpretation that the organic matrices of mollusc shells are heterogeneous mixtures of proteins comprised in various proportions of non-calcified, asp-poor, intercrystalline matrix and calcified, asp-rich, intracrystalline matrix. Evolutionary loss of intercrystalline matrix in many organisms has led to shells with a low protein content and an asp-rich composition. The asp-rich, intracrystalline matrix may correspond to the asp-rich soluble organic matrix discussed by Crenshaw (1972), Mitterer (1972a), Weiner and Hood (1975), Weiner (1979) and Weiner et al. (1983), and the asp-poor intercrystalline matrix may represent insoluble matrix.

ASP-RICH ORGANIC MATTER PERFERENTIALLY ADSORBS TO CARBONATE SURFACES

Carbonate grains in the sedimentary environment have an organic coating of humic substances (Chave, 1965). Analyses of humic and fulvic acids extracted from carbonate surfaces show them to be asp-rich (Carter & Mitterer, 1978). Conversely, organic matter associated with non-carbonate sediments is depleted in aspartic acid. An illustration of this contrast is provided by sediments from Flower Gardens Reef on the Texas Shelf. Local supply of reef-derived detritus provides coarser carbonate grains to the immediately surrounding shelf sediment, while finer grain size siliciclastic fractions are transported from more distant continental sources. Amino acid analyses of individual size fractions of this sediment show that relative aspartic acid concentration correlates directly with calcium carbonate content (Fig. 4). Coarser carbonate grains contain organic matter with a predominance of aspartic acid, while organic matter adsorbed to finer siliciclastic grains is depleted in aspartic acid. Evidently, even within the same sedimentary environment there is a significant difference in the relative aspartic acid concentration of organic matter associated with carbonate and non-carbonate grains.

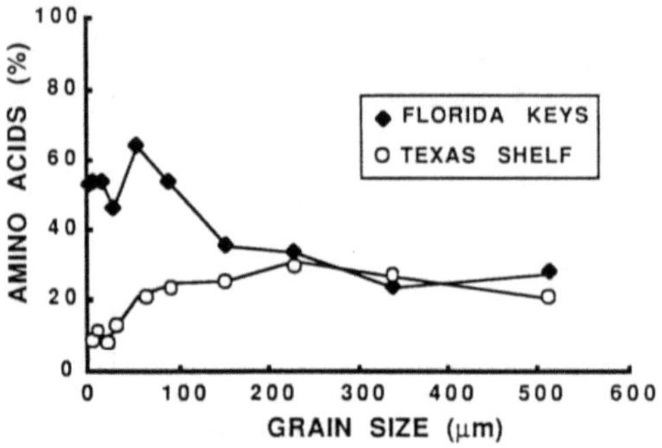

FIGURE 5. Percentage of organic carbon as amino acids in sediment samples from the Flower Gardens Reef of Fig. 4 and from the Florida Keys as a function of sediment grain size. The sediment sample from the Florida Keys is virtually all calcium carbonate while the Flower Gardens sample contains calcium carbonate primarily in the coarser sediment fraction. Data from Carter and Mitterer (1978).

The contrast between these different sediment types is not limited to relative aspartic acid content. Total concentration of hydrolyzable amino acids is also highly dependent on grain mineralogy. Amino acid concentration, as a percentage of organic carbon, decreases from about 30% in coarser carbonate sediment size fractions to about 10% of the organic carbon in finer siliciclastic fractions of Flower Garden sediments (Fig. 5). In contrast, carbonate sediments from the Florida Keys have comparable amino acid percentages in coarser carbonate size fractions, but the amino acid concentration increases to more than 50% of the organic carbon in finer carbonate fractions. As organic carbon content in various size fractions of the two sediments is virtually identical, the relative amino acid contents as a function of grain size reflect absolute increases or decreases in amino acid concentration. Thus, compared to siliciclastic sediments, not only do carbonate sediments have a different amino acid composition, but they also contain from three to six times as much organic matter in the form of hydrolyzable amino acids.

The attraction of asp-rich organic matter for carbonate surfaces is further elucidated by adsorption experiments. Figure 6 illustrates results from a series of adsorption experiments using asp-rich fulvic acid previously extracted from carbonate sediments (Carter, 1978). Experiments were conducted with calcite and quartz powders of identical size and with various ratios of fulvic acid and mineral. The series of experiments with calcite shows that, at all ratios of organic matter to mineral, a fraction richer in aspartic acid than the stock solution is adsorbed. The stock solution contains about 30% (300 residues/1000) aspartic acid, while the adsorbate contains from about 450 residues/1000 of aspartic acid at low organic matter/mineral ratios to about 650 residues/1000 of aspartic acid at high ratios. Two obvious interpretations of the results of these adsorption experiments are: (1) the fulvic acid stock solution is a heterogeneous mixture of compounds, and (2) calcite grains preferentially adsorb the molecular fractions richest in aspartic acid.

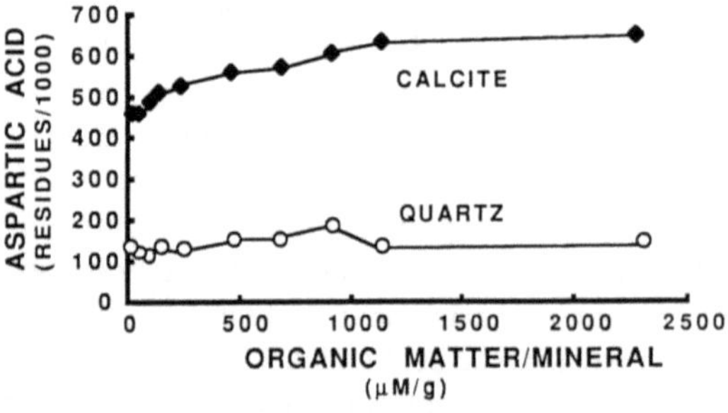

FIGURE 6. Relative aspartic acid content of adsorbed fulvic acid on calcite and quartz powders as a function of the amount of organic matter in solution. The aspartic acid composition of the solution is about 300 residues/1000. Data from Carter (1978).

An identical series of experiments but substituting quartz for calcite yields opposite results (Fig. 6). Asp-poor fulvic acid is preferentially adsorbed to quartz grain surfaces at all ratios of organic matter to mineral.

The results of these experiments, therefore, support the data obtained from sediment analyses. The major conclusion that can be drawn from these analytical and experimental investigations is that carbonate grains in the marine environment are preferentially associated with asp-rich organic matter, while non-carbonate grains are not.

ASP-RICH ORGANIC MATTER COMPLEXES CALCIUM

Metal binding studies, utilizing fulvic acids extracted from carbonate grains as well as soluble organic matrix from ooids and skeletal carbonates, have been performed to determine the degree of association of calcium and other cations and asp-rich organic matter. The details of this work have been presented previously, and only a brief summary is given here (Mitterer & Carter, 1977; Cunningham & Mitterer, 1980; Mitterer & Cunningham, 1985). The results indicate that both fulvic acid and soluble organic matrix bind metal ions to a significant degree, with the number of binding sites proportional to the aspartic acid concentration of the organic matter. Chemical blockage of carboxyl groups on the organic matter destroys binding ability. Thus, in addition to its adsorptive properties, asp-rich organic matter has the ability to bind calcium and other cations.

A MODEL FOR CALCIUM CARBONATE-ORGANIC MATTER INTERACTION

The picture that emerges from the analytical, adsorption and binding data can be summarized as follows:

1. The organic matrix of skeletal and non-skeletal carbonates contains asp-rich organic matter. Adsorbed humic substances on carbonate grain surfaces are also asp-rich. Conversely, the bulk organic matter of non-carbonate minerals, soft tissues and seawater is asp-poor. Asp-rich organic matter is uniquely associated with, and a characteristic feature of, all marine calcium carbonates.

2. In laboratory experiments carbonate grains preferentially adsorb asp-rich organic matter, while a non-carbonate mineral such as quartz preferentially adsorbs an asp-poor component.

3. Asp-rich organic matter in solution binds calcium and other metal ions. Carboxyl functions contributed by aspartic acid are the primary ligands involved in metal binding.

The potential significance of aspartic acid-containing protein to invertebrate calcification has been emphasized in a number of papers over the past 25 years (e.g., Hare, 1963; Degens et al., 1967; Degens, 1976; Towc & Hamilton, 1968; Mitterer, 1971, 1978; Weiner & Hood, 1975; Weiner, 1979; Weiner et al., 1983). This amino acid provides binding sites for Ca^{+2} via the negatively-charged beta carboxyl group. Because of a similar asp-rich composition, the organic matrix of abiogenic carbonates such as ooids has been linked to calcification in an analogous manner (Mitterer, 1968, 1971a, 1972a,b; Trichet, 1968; Carter, 1978; Mitterer & Cunningham, 1985).

Interaction of asp-rich organic matter with calcium carbonate minerals both in organisms and in sedimentary environments is highly specific and suggests an ordered

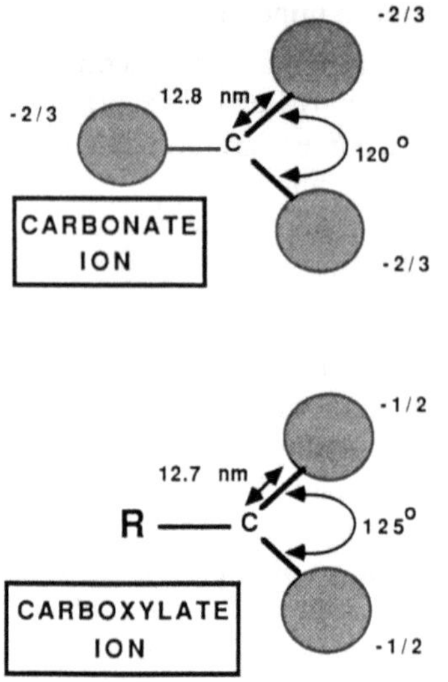

FIGURE 7. Geometrical comparison of carbonate anion with carboxyl functional group. C is carbon, large filled circles are oxygens, **R** is the organic moiety attached to the carboxyl group.

arrangement between functional groups on the organic matter and constituents of carbonate crystals. The beta carboxyl group of aspartic acid is the most probable interactive moiety based on the results described above. Confirming evidence is provided by experiments in which chemical blockage of carboxyl groups results in loss of both adsorptive and binding abilities of asp-rich organic matter (Mitterer & Cunningham, 1985). The nature of the interaction may be simple electrostatic attraction between negatively-charged carboxyl groups and positive charges on surface calcium ions in the calcite or aragonite lattice. A geometrical comparison between the carboxyl group and the carbonate anion, however, indicates an even more specific association (Fig. 7; Mitterer & Cunningham, 1985). Both groups are planar structures. The O-C-O bond angles are about 120-125o, and the C-O bond lengths are virtually identical. Finally, the negative charges are 2/3 for each carbonate oxygen and 1/2 for each carboxyl oxygen. Evidently, the similarities in geometry and charge are so close that the carboxyl group can substitute for carbonate ions in satisfying positive charges on surface calcium ions. Thus, the positioning of adsorbed asp-rich organic matter on a carbonate surface may be controlled by the spacing of surface calcium ions.

Adsorption of asp-rich organic matter to carbonate substrates does not complete the process of interaction. At this point, however, in the absence of a definitive understanding, a more speculative interpretation must follow. After adsorption of an asp-rich molecule, carboxyl groups not involved in surface interaction can bind dissolved calcium ions (Fig. 8). Then, either by epitaxy or by enhanced concentration of calcium due to the binding effect, nucleation of a new crystal of aragonite or calcite ensues (Mitterer, 1972a; Mitterer & Cunningham, 1985). After development of a new mineral layer, the carbonate surface again serves as a substrate for adsorption of another asp-rich membrane, and the process is repeated. At the molecular level, crystalline substrate and asp-rich membranes alternate as the struc-

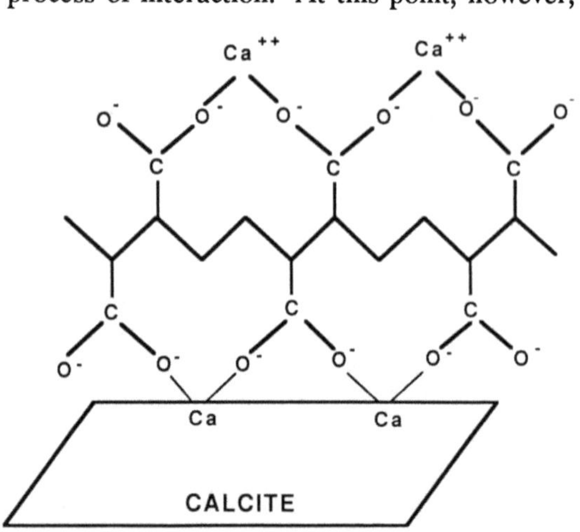

FIGURE *8.* Schematic model for the interaction of asp-rich organic matter with calcium carbonate surface and with calcium ions in solution.

ture grows. Asp-poor protein may serve to terminate or prevent crystal growth in certain directions or after the crystal has reached a certain size, thereby acting to control crystal morphology and shell structure. Viewed from the mineral perspective, the crystal structure, by selective and ordered adsorption of asp-rich membranes followed by Ca-binding and nucleation of a new layer, has managed to "replicate" itself. Because the adsorption-nucleation process may also operate outside of organisms, as in the case of ooids, the association of organic matter with calcium carbonate may be one of the most fundamental and primitive examples of highly specific mineral-organic matter interaction.

INVENTION OF CALCIFICATION

The previous discussion highlights the role of asp-rich protein in the modern invertebrate calcification process. Because of the close taxonomic affinity of many modern invertebrates to their earliest calcified ancestors, the basic calcification mechanism is undoubtedly little different, if at all, from that employed by the first calcifying organisms. Thus, an understanding of the calcification mechanism in modern invertebrates may elucidate the origin of calcification in the Early Cambrian. As Lowenstam and Weiner (1983) and Weiner et al. (1983) have noted, two choices exist for invention of calcification: (1) various organisms invented the calcification mechanism almost simultaneously and independently, or (2) organisms evolved the latent ability to calcify over a longer period of time with a subsequent triggering event specifically initiating calcification. The first possibility requires that the calcification formula spread to representatives of a number of invertebrate phyla within a geologically short span of time, a highly coincidental occurrence. The alternative scenario allows organisms to develop the latent ability to calcify over a longer period of time by gradual evolution of asp-rich protein and other requisites.

A change in ocean chemistry might be one important prelude to the development of invertebrate calcification. Because many of the first calcified organisms are inferred to have been phosphatic (Lowenstam, 1981; Lowenstam & Weiner, 1983), late Precambrian oceans might have been more conducive to phosphate mineralization. Subsequently, a change in ocean chemistry to that favoring carbonate mineralization would have redirected the efforts of organisms to form carbonate minerals instead. There is little indication, however, in the fragmentary late Precambrian-Early Paleozoic record that ocean chemistry changed significantly, although there is room for some deviations from modern element concentrations (Holland, 1984). Alternatively, a minor change in ocean chemistry might have had significant ramifications to calcification. A candidate for this role is strontium.

Bidwell et al. (1986) and Gallager et al. (this volume) have noted the important effect that strontium has on calcification in molluscs. They found that, depending on strontium concentration of ambient seawater in laboratory experiments, calcification in larval molluscs is either inhibited or permitted. At strontium concentrations from normal seawater values (8 ppm) to as low as 4 ppm, calcification proAplysiaceeds normally in the gastropod Aplysia californica. Between 4 and 0 ppm strontium, organisms develop abnormal shells. The bivalves, Mercenaria mercenaria and Bankia gouldi, require at least 6 ppm strontium for normal development, and identical deficits result when the

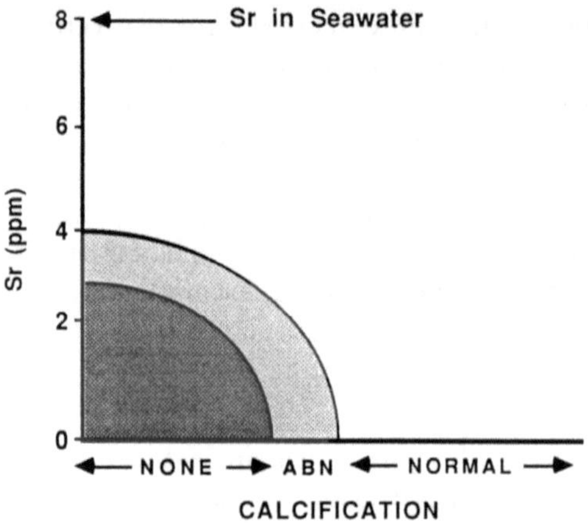

FIGURE 9. Schematic effect of variable Sr concentration of seawater medium on calcification in larval molluscs. Data from Bidwell *et al.* (1986) and Gallager *et al.*, (1988).

concentration is lowered (Gallager et al., this volume). At strontium concentrations below 3 ppm the organisms do not calcify although their soft tissues develop normally (Fig. 9). While the specific role that strontium plays in calcification is not known, and the number of organisms studied is small, if the experiments can be extrapolated from the laboratory to the "real world", the results suggest that variations in the concentration of strontium in seawater will have an important effect on the ability of organisms to calcify.

What evidence exists for possible variations in seawater strontium during the critical late Precambrian-Cambrian time period? Unfortunately, little if any direct evidence exists for variations in strontium concentration in seawater through time. Indirect evidence, however, based on 87Sr/86Sr in carbonate rocks, suggests that strontium concentration in the oceans may have varied over the past 700 million years. Figure 10 depicts very generalized variations in 87Sr/86Sr during this time interval (Burke et al., 1982; Veizer et al., 1983). A relationship between strontium concentration and its isotopic ratio in seawater is not established, and many explanations have been proposed to explain the generalized isotopic variations seen in Figure 10 and the more detailed isotopic variations shown in Figure 11. These include, but are not limited to, rate of seafloor spreading, volcanism, tectonic uplift, weathering and continental exposure, oceanic circulation, and climate (Peterman et al., 1970; Armstrong, 1971; Veizer & Compston, 1974; Brass, 1976; Burke et al., 1982; Veizer et al., 1983; Wadleigh et al., 1985). Recent modeling of the strontium cycle for the Cretaceous-Tertiary portion of the 87Sr/86Sr curve, the time period for which detailed isotopic data are available and for which spreading rates and other factors are more constrained, indicates that the increase in the ratio during this time interval is due either to an increase in river input of strontium at a constant isotopic ratio or to an increase in 87Sr/86Sr at a constant strontium concentration (Palmer & Elderfield, 1985; Hess et al., 1986). Hess et al. (1986) prefer the first interpretation while Palmer and Elderfield (1985) favor the second. The view presented here, in agreement with the interpretation of Hess et al. (1986) for the Cretaceous-Tertiary interval, is that variations in 87Sr/86Sr in seawater over the entire Phanerozoic are due primarily to changes in delivery of strontium to the oceans by rivers, thereby affect-

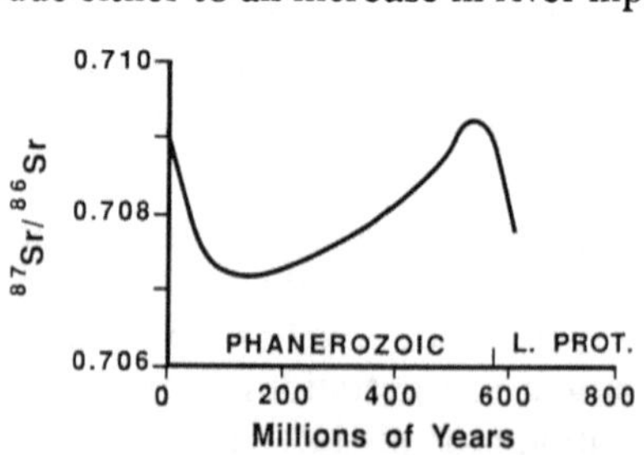

FIGURE 10. Generalized variation of $^{87}Sr/^{86}Sr$ in seawater for the last 700 Ma. Modified from Burke *et al.*, (1982) and Veizer *et al.* (1983).

ing both the strontium concentration and its isotopic ratio in seawater.

The 87Sr/86Sr of seawater (presently about 0.709) depends on input of strontium to the oceans from two major sources: (1) river input from continental runoff with a present 87Sr/86Sr value of about 0.711, and (2) hydrothermal alteration of oceanic basalt with a ratio of about 0.703 (Wadleigh et al., 1985). Although data indicate that hydrothermal reactions between seawater and basalt modify the isotopic composition of seawater strontium, this exchange may not significantly affect strontium concentration in the ocean (Wadleigh et al., 1985). Moreover, it has been postulated that changes in the rate of hydrothermal alteration of oceanic basalt have been small for about the last 750 m.y. so that input of 87Sr/86Sr to seawater from oceanic ridges during the Phanerozoic has been approximately constant (Keto & Jacobsen, 1985). If so, variations in strontium concentration and its isotopic ratio in the oceans during the Late Precambrian and Phanerozoic have been due primarily to variable input from continental sources.

Higher values of 87Sr/86Sr are interpreted to signify greater continental input and therefore higher strontium concentrations in the oceans. Conversely, low ratios denote lesser continental input and lower strontium concentrations. As seen in Figure 10, 87Sr/86Sr is at its highest value for the entire Phanerozoic (prior to the present) during the Early Cambrian. To reach the Early Cambrian value, the ratio rose from much lower levels in the Late Precambrian. Interpreted as changes in continental input, the increase in the 87Sr/86Sr value from its latest Precambrian low to its Early Cambrian high signifies a corresponding increase in the strontium concentration of the oceans. This interpreted increase in strontium occurs about the time when organisms evolved the calcification mechanism. In view of the experimental results of Bidwell et al. (1986) and Gallager et al. (this volume) on the importance of strontium to calcification in modern organisms, the inferred high level of strontium in the latest Precambrian-earliest Cambrian oceans may have been conducive to calcification. If so, organisms already poised to calcify, by the presence of asp-rich protein in their protective membranes and other requisites, were enabled to do so at the prevailing high strontium levels in the ambient sea water. In essence, then, organisms apparently evolved the ability to calcify in sea water containing a relatively high concentration of strontium and utilized this element in their calcification process.

It remains to be determined whether calcitic organisms are also sensitive to ambient strontium concentration. Although the experiments of Bidwell et al. (1986) and Gallager et al. (this volume) were performed on aragonitic organisms, the fundamental calcification process in calcitic and aragonitic organisms (some molluscs form both polymorphs) is undoubtedly identical. Strontium concentrations in calcareous shells and tests range from about 1000 to 10,000 ppm; the range in mollusc shells (calcite and aragonite) is considerably less, from 1000 to 4000 ppm (Milliman, 1974). Thus, the role of strontium in calcification may be unrelated to the mineralogy of the skeleton or its strontium content.

In summary, several steps are envisioned as requisites for invention of calcification. One of these is evolution of asp-rich proteins and another is utilization of strontium. That representatives of the various phyla evolved calcifying asp-rich protein and other requisites within a relatively short span of geological time is highly coincidental. Rather, this protein probably evolved over a period of time prior to onset of calcification, and its presence provided a latent potential to calcify pending subsequent critical steps.

IMPLICATION OF LOWER STRONTIUM CONCENTRATIONS

The foregoing discussion emphasizes the importance of high strontium concentration in ambient seawater to the calcification process. The opposite pattern of a lower strontium concentration, once calcification was established, has profound implications. Clearly, if a certain minimum strontium concentration is important to the calcifying ability of organisms, any decrease in strontium below the necessary threshold will have adverse effects on this ability. Decreases in oceanic strontium subsequent to the early Cambrian would have inhibited calcification in those organisms that had already evolved this process. Without the protection afforded by their exoskeletons the survival potential of these organisms would have been much lower. While some would undoubtedly survive because of other ecological factors such as a burrowing habit, many organisms experiencing calcification-inhibition would be easy prey for predators. Calcium toxicity induced by the inability to excrete calcium also may have been an important factor affecting survivability. Taken to the extreme, taxa whose representatives experienced calcification-inhibition may have become extinct as a consequence of the inability to calcify.

The importance of strontium to calcification therefore is two fold: (1) at high strontium levels calcification is initiated or maintained, and (2) at lower strontium levels calcification is inhibited. Various organisms may have different requirement levels for this element, so that all organisms would not necessarily cease calcifying at the same strontium concentration. If the strontium level fell sufficiently, however, calcification in many organisms would be inhibited, and major marine invertebrate extinctions may have ensued.

Figure 11 depicts the recognized extinction events of the Phanerozoic (Raup & Sepkoski, 1982, 1986) superimposed on the detailed strontium isotope curve of Burke et al. (1982). Major extinction events in the Paleozoic including those in the Late Ordovician, Middle to Late Devonian and Permian, follow times of lower 87Sr/86Sr. If lower ratios are considered to represent less continental input of strontium, then the concentration of strontium in the oceans at these times may have been sufficiently low to inhibit calcification and to trigger massive extinctions of calcareous marine invertebrate organisms. Oscillations of 87Sr/86Sr in Mesozoic oceans are less pronounced (as are extinction events except for terminal Cretaceous) than in Paleozoic oceans, and variations in oceanic strontium concentration may have been less extreme during the Mesozoic. Of interest, however, is that the minor, even upward, shift in the isotopic ratio at the Cretaceous-Tertiary boundary seemingly does not fit the expected isotopic pattern described above for other extinctions. The cause of this extinction event is highly controversial, and alternative causal mechanisms may indeed account for one or more extinction events.

The hypothesis outlined above is admittedly speculative. It is based on the as yet unexplained importance of strontium to calcification that has been reported by Bidwell et al. (1986) and Gallager et al. (this volume) and on the 87Sr/86Sr variations of seawater over the past 700 Ma documented by a number of authors. I have simply deduced logical

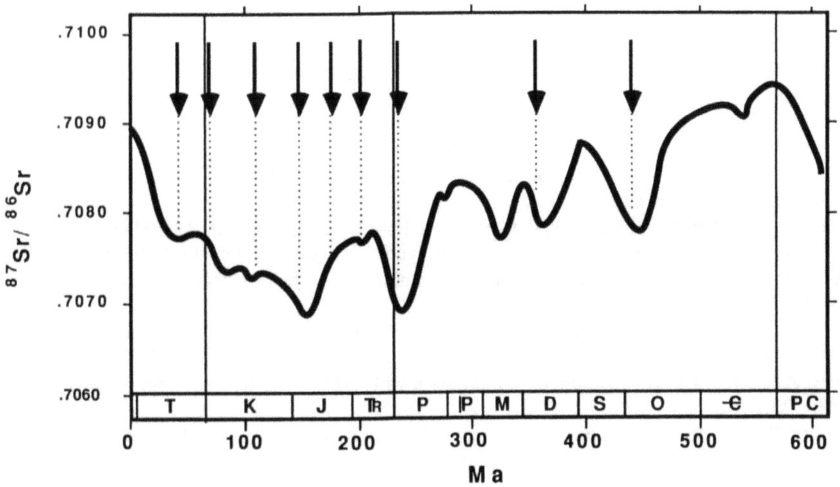

FIGURE 11. Detailed variation of $^{87}Sr/^{86}Sr$ in seawater during the Phanerozoic (from Burke *et al.* (1982) with extension from Veizer *et al.* (1983)) and marine invertebrate extinction events (indicated by arrows).

inferences from these data. The data certainly have much firmer independent foundations than the inferences I have drawn from them.

SUMMARY AND CONCLUSIONS

Biological and abiological calcification appears to be dependent on the presence of aspartic acid-rich organic matter, which is uniquely and characteristically associated with calcium carbonate. This amino acid contains a free carboxyl function that interacts with calcium ions in solution or with surface calcium ions on a carbonate crystal. Selective adsorption of asp-rich organic matter to a carbonate surface followed by nucleation of a new mineral layer on the organic membrane occurs repeatedly to build the mineral structure.

The invention of calcification was probably a multiple-step process that included evolution of asp-rich membranes and utilization of strontium. Reduced availability of this element may have also triggered subsequent extinctions of marine invertebrates.

ACKNOWLEDGEMENTS

Reviews of the manuscript by J. P. Bidwell, R. M. Garrels and R. B. Koepnick are gratefully acknowledged. This research was also greatly aided by the dedicated efforts of two former students, Robert Cunningham and Paul W. Carter. This work was supported by National Science Foundation Grants OCE-7925389 and EAR-8508298. Contribution No. 506 of the Department of Geosciences, University of Texas at Dallas. Participation in the Symposium was made possible by grants from The Organized Research Fund of The University of Texas at Arlington and The Sea Grant Program administered by Texas A & M University.

REFERENCES

ARMSTRONG, R.L., 1971. Glacial erosion and the variable isotopic composition of strontium in seawater. Nature, 230:132-133.

BIDWELL, J.P., PAIGE, J.A. & KUZIRIAN, A.M., 1986. Effects of strontium on the embryonic development of *Aplysia californica*. Biol. Bull., 170:75-90.

BRASS, G.W., 1976. The variation of marine $^{87}Sr/^{86}Sr$ ratio during Phanerozoic time: Interpretation using a fluxmodel. Geochim. Cosmochim. Acta, 40:721-730.

BURKE, W.H., DENISON, R.E., HETHERINGTON, E.A., KOEPNICK, R.B., NELSON, H.F., & OTTO, J.B., 1982 Variation of seawater $^{87}Sr/^{86}Sr$ throughout Phanerozoic time. Geology, 10:516-519.

CARTER, P.W., 1978. Adsorption of amino acid-containing organic matter by calcite and quartz. Geochim. Cosmochim. Acta, 42:1239-1242.

CARTER, P.W. & MITTERER, R.M., 1978. Amino acid composition of organic matter associated with carbonate and non-carbonate sediments. Geochim. Cosmochim. Acta, 42:1231-1238.

CHAVE, K.E., 1965. Calcium carbonate: Association with organic matter in surface seawater. Science. 148:1723-1724.

CRENSHAW, M.A., 1972. The soluble matrix from *Mercenaria mercenaria* shell. Biomineralization. 6:6-11.

CUNNINGHAM, R. & MITTERER, R.M., 1980. Metal binding study of fulvic acids from carbonate sediments using manganese as a magnetic resonance probe. *In* Biogeochemistry of Amino Acids (ed. P.E. Hare, T.C. Hoering & K. King, Jr.,), pp. 129-143. New York: Wiley.

DEGENS, E.T., 1976. Molecular mechanisms on carbonate, phosphate and silica deposition in the living cell. Top. Curr. Chem., 64:1-112.

DEGENS, E.T. & SPENCER, D.W., 1966. Data file on amino acid distribution in calcified and non-calcified tissues of shell-forming organisms. Tech. Rep. 66-27, Woods Hole Oceanogr. Inst., unpubl. manu.

DEGENS, E.T., SPENCER, D.W. & PARKER, R.H., 1967. Paleobiochemistry of molluscan shell proteins. Comp. Biochem. Physiol., 20:553-579.

GALLAGER, S.M., BIDWELL, J.P. & KUZIRIAN, A.M., 1988. Strontium is required in artificial seawater for embryonic shell formation in two species of bivalve molluscs. *In* The Origin, Evolution and Modern Aspects of Biomineralization in Plants and Animals (ed. R. E. Crick). New York: Plenum Press.

HARE, P.E., 1963. Amino acids in the proteins from aragonite and calcite in the shells of *Mytilus californianus*. Science, 139:216-217.

HARE, P.E. & ABELSON, P. H., 1964. Proteins in mollusk shells. Carnegie Inst. Washington Yearbook, 63:267-270.

HOLLAND, H.D., 1984. The Chemical Evolution of the Atmosphere and Oceans. 528 pp. Princeton: Princeton Univ. Press.

KETO, L.S. & JACOBSEN, S.B., 1985. The causes of $^{87}Sr/^{86}Sr$ variations in seawater of the past 750 million years. Abst. Prog., Geol. Soc. Amer. Ann. Meet., 628.

LOWENSTAM, H.A., 1981. Minerals formed by organisms. Science, 211:1126-1131.

LOWENSTAM, H.A. & Weiner, S., 1983. Mineralization by organisms and the evolution of biomineralization. *In* Biomineralization and Biological Metal Accumulation (ed. P. Westbrook & E. W. de Jong), pp. 191-203. Boston: D. Reidel.

MILLIMAN, J.D., 1974. Marine Carbonates, 375 p. Heidelberg: Springer-Verlag.

MITTERER, R.M., 1968. Amino acid composition of organic matrix in calcareous oolites. Science, 162:1498-1499.

MITTERER, R.M., 1971a. Influence of natural organic matter on CaCO3 precipitation. *In* Carbonate Cements (ed. O.P. Bricker), pp. 252-258. Baltimore: Johns Hopkins Univ. Press.

MITTERER, R.M., 1971b. Comparative amino acid composition of calcified and non-calcified polychaete worm tubes. Comp. Biochem. Physiol., 38b:405-509.

MITTERER, R.M., 1972a. Biogeochemistry of aragonite mud and oolites. Geochim. Cosmochim. Acta, 36:1407-1422.

MITTERER, R.M., 1972b. Calcified proteins in the sedimentary environment. *In* Advances in Organic Geochemistry 1971 (ed. H.R. von Gaertner & H. Wehner), pp. 441-451. Oxford: Pergamon.

MITTERER, R.M., 1978. Amino acid composition and metal binding capability of the skeletal protein of corals. Bull. Mar. Sci., 28:173-180.

MITTERER, R. M. & CARTER, P.W., 1977. Some analytical and experimental data on organic-carbonate interactions. Proc. 3rd Inter. Coral Reef Sym., 2:541-547.

MITTERER, R.M. & CUNNINGHAM, R., 1985. The interaction of natural organic matter with grain surfaces: Implications for calcium carbonate precipitation. *In* Carbonate cements, Spec. Publ. No. 36, (ed. N. Schneidermann & P.M. Harris), pp. 17-31. Tulsa: Soc. Econ. Paleontol. Mineral.

MUTVEI, H., 1970. Ultrastructure of the mineral and organic components of molluscan nacreous layers. Biomineralization, 2:48-72.

PETERMAN, Z.E., HEDGE, C.E. & TOURTELOT, H.A., 1970. Isotopic composition of strontium in seawater throughout Phanerozoic time. Geochim. Cosmochim. Acta. 34:105-120.

RAUP, D.M. & SEPKOSKI, J.J., Jr., 1982. Mass extinctions in the marine fossil record. Science, 215:1501-1503.

RAUP, D.M. & SEPKOSKI, J.J., Jr., 1986. Periodic extinction of families and genera. Science, 231:833-836.

TOWE, K.M., & HAMILTON, G.H., 1968. Ultrastructure and inferred calcification of the mature and developing nacre in bivalve mollusks. Calc. Tiss. Res., 1:306-318.

TRICHET, J., 1969. Etude de la composition de la fraction organique des oolites: Comparaison avec celle des membranes des bactéries et des cyanophycées. Comptes Rendus des seances de l'Academic des Sciences, Paris, 267:1492-1494.

VEIZER, J. & COMPSTON,W., 1974. $^{87}Sr/^{86}Sr$ composition of seawater during the Phanerozoic. Geochim. Cosmochim. Acta, 38:1461-1484.

VEIZER, J., COMPSTON, W., CLAUER, N., & SCHIDLOWSKI, M., 1983. $^{87}Sr/^{86}Sr$ in Late Proterozoic carbonates: evidence for a "mantle" event at ~900 Ma ago. Geochim. Cosmochim. Acta, 47:295-302.

WADLEIGH, M.A., VEIZER, J., & BROOKS, C., 1985. Strontium and its isotopes in Canadian rivers: Fluxes and global implications. Geochim. Cosmochim. Acta, 49:1727-1736.

WATABE, N., 1965. Studies on shell formation: Crystal-matrix relationships in the inner layers of mollusk shells. J. Ultrastr. Res., 12:351-370.

WEINER, S., 1979. Aspartic acid-rich proteins: Major components of the soluble organic matrix of mollusk shells. Calc. Tiss. Intern., 29:163-167.

WEINER, S. & HOOD, L., 1975. Soluble protein of the organic matrix of mollusk shells: A potential template for shell formation. Science, 190:987-989.

WEINER, S., TRAUB, W. & LOWENSTAM, H.A., 1983. Organic matrix in calcified exoskeletons. *In* Biomineralization and Biological Metal Accumulation (ed. P. Westbrook & E.W. de Jong), pp. 205-224. Boston: D. Reidel.

THE PROTEINS IN THE SHELL OF *LINGULA*

N. Tuross and L.W. Fisher

Geophysical Laboratory, Carnegie Institution of Washington, 2801 Upton Street, N.W., Washington D.C. 20008-3898; Bone Research Branch, National Institute of Dental Research, National Institute of Health, Bethesda, MD 20892

INTRODUCTION

The development of mineralized hard parts in animals is a fairly recent development in geologic time. Five hundred and fifty million years ago fossils with mineralized skeletons appear in the geologic record. Although the source and form of minerals used by organisms is diverse, the assumption that vertebrates used calcium phosphate ($CaPO_4$) and invertebrates use calcium carbonate ($CaCO_3$) to form hard parts is useful and widespread.

Phosphate mineralization does occur in the invertebrates, and this form skeletal formation is widely represented in early Cambrian fauna (Lowenstam, 1981; Runnegar, 1988). An interesting group of animals living today, the inarticulate brachiopods, contain several genera that mineralize their shells with $CaPO_4$. The genus *Lingula* has at least eleven species extant (Awati & Kshirsagar, 1903), and it has a continuous fossil record that reaches back to the Ordovician (c. 500 Ma).

The inorganic components of the *Lingula* shell have been analyzed by several investigators (Klement, R., 1938; McConnell, D., 1963), and are similar to the mineral fracolite, a carbonate containing calcium florapatite (LeGeros et al., 1985). Alternating organic/inorganic layers in the *Lingula* shell (Chapman, 1914) have recently been more completely described by Iwata (1981).

The organic fraction of the *Lingula* shell has been incompletely described, but one striking observation is the presence of the amino acid hydroxyproline, a hallmark of collagen in vertebrate bones and teeth. Proteins solubilized in 8 M urea have been described by Jope (1973, 1977), and represent some of the nonmineral-associated proteins found in the *Lingula* shell.

In the past fifteen years isolation and characterization of the noncollagenous proteins in vertebrate bones and teeth has progressed steadily (Veis *et al.*, 1972; Hauschka *et al.*, 1975; Price *et al.*, 1976; Termine *et al.*, 1981; Fisher *et al.*, 1983). Experience with the vertebrate systems may provide several lessons when extracting structural proteins in invertebrate systems (Fisher & Termine, 1985). Briefly, the rapid processing by grinding tissue under liquid nitrogen after removal from the animal is necessary to prevent loss

Origin, Evolution, and Modern Aspects of Biomineralization in Plants and Animals
Edited by Rex E. Crick
Plenum Press, New York

TABLE 1. Amino acid analysis of the nonmineral associated G extract

AMINO ACIDS RESIDUES/1000				
	G[1]	E[2]	PE[3]	ST[4]
OHP	17	0	21	4
ASP	101	129	93	124
THR	52	40	29	50
SER	59	70	36	61
GLU	118	138	61	179
PRO	54	68	46	45
GLY	121	128	148	66
ALA	130	111	303	109
VAL	39	40	44	29
ILE	29	23	12	29
LEU	51	55	30	78
TYR	23	20	28	17
PHE	52	46	22	70
HIS	47	40	32	33
OHL	-	-	2	2
LYS	57	43	22	70
ARG	53	48	71	45

[1]G extract
[2]E extract
[3]Residue after 2 extractions
[4]Soft tissue

and degradation of the proteins. The sequential dissociative methods developed by John Termine and coworkers at the Bone Research branch of the National Institute of Dental Research allow for the removal of proteins from the non-mineral compartment with 4 M Guanidine HCl plus protease inhibitors. This is followed by release of mineral associated proteins in 4 M Guanidine HCl 0.5M EDTA (ethylene diamine tetraacetic acid) plus protease inhibitors. In bone, the residue that remains after these two sequential extractions is collagen (Termine, 1983). Studies of vertebrate bones and teeth have also demonstrated that the age of an organism, and thus the degree of mineralization is important in studying the organic matrix associated with the hydroxyapatite crystals (Conn & Termine, 1985; Lee *et al.*, 1983; Fisher & Termine, 1985).

PROTEIN EXTRACTION AND CHARACTERIZATION

Lingula were collected live, and kept frozen until subjected to the protocols of sequential dissociative extraction and demineralization used in bones and teeth (Termine, 1983). The nonmineral associated "G" fraction and the mineral associated "E" fraction were rapidly desalted in order to prevent protein degradation (Fisher & Termine, 1985). The residue was then separated on gradient polyacrylamide 4-20% SDS gels by electrophoresis (Table 1). It is clear that in the G fraction, which is analogous to the urea extraction described by Jope (1973), contains more than the three molecular weight fractions seen in the earlier paper. The difference may be due to the preservation of the specimens or the increased release of protein components with grinding and longer extraction times. *Lingula* shells or extracts that were treated more casually (separated from soft tissue and allowed to air dry or dialysed as opposed to rapid desalting) showed an altered and degraded electrophoresis pattern (data not shown). The mineral compartment proteins, the "E" extract shows a predominant protein at 43,000 daltons (Fig. 1; arrow). Staining of the gel indicates that the 43,000 dalton mineral associated protein of the E extract has undergone post-translational modifications evidenced by an intense blue color and that the sharp band at 43,000 daltons in the G extract is a different unsubstituted protein. By the criteria used in assessing vertebrate mineralized tissue, the enrichment of a protein in the "E" extract, coupled by its exclusion in the "G" extract indicates that this protein may play a role in the nucleation or regulation of the mineral phase.

In order to test whether any of the mineral-associated E extract is collagen, the extract was incubated in the presence of bacterial collagenase for 10 h at 37° C. Rat-tail tendon collagen served as the control (Fig. 2). While the alpha bands of the collagen control were nearly completely digested by collagenase treatment, none of the bands from the E extract of *Lingula* were obviously digested by collagenase.

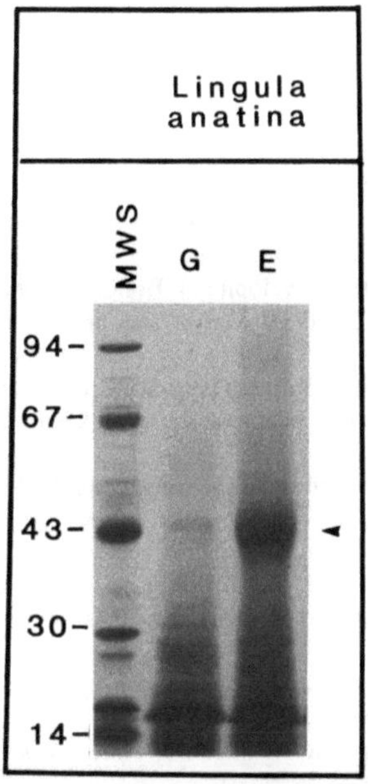

FIGURE 1. Coomassie stained SDS-gradient polyacrylamide (4-20%) gels of proteins from the nonmineral (G) compartment and the mineral (E) compartment of *Lingula* shell.

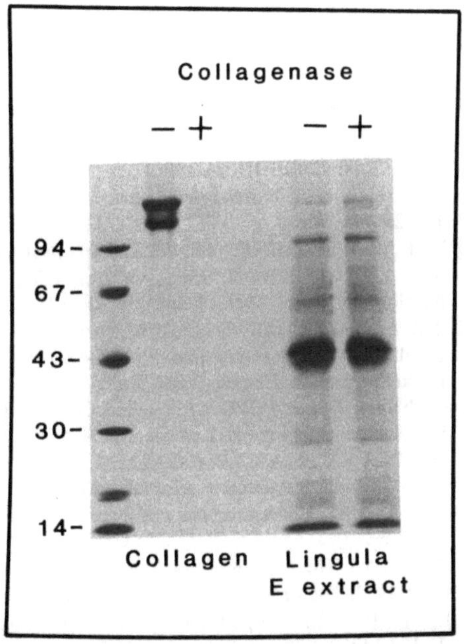

FIGURE 2. Coomassie stained SDS-gradient polyacrylamide (4-20%) gels of 200 μg of mineral compartment E protein and 30 μg of rat tail tendon collagen incubated for 10 h and 37C, pH 7.5, in the presence and absence of 100 units bacterial collagenase (Advanced Biofractures).

An amino acid analysis of the extracts supports the collagenase data. The E extract of the *Lingula* shell preferentially excludes the hydroxyproline containing proteins of *Lingula.* The presence of hydroxyproline in the G extract may arise from trace amount of soft body collagen due to muscle insertions into the shell. The hydroxyproline and hydroxylysine in the residue that remains after Guanidine and EDTA extraction is more interesting. Clearly the proteins of the post E residue are not a classic vertebrate fibrillar collagen: the large amount of alanine (300 residues/1000) militates against this. In general, the levels of hydroxyproline and the glycine/hydroxyproline ratios from the post E residue agree with those reported in whole shell by Iwata (1981). These data conflict with the 3/1 glycine/hydroxylproline ratio reported by Jope (1967). Thus, the possibility that a small amount of fibrillar collagen is present and tightly associated with the shell can not be ruled out and awaits further investigation.

What is clear from the electrophoretic and amino acid analysis of these fractions is that the isolation procedures used in vertebrate bones and teeth can be successfully applied to this class of invertebrates. Moreover, the segregation of heterogeneous acidic proteins in the G and E extracts coupled with the existence of a hydrophobic post E residue in the *Lingula* shell is reminiscent of other phosphate mineralizing systems-bones and teeth. Resolution of the nature of the organic substrate in the phosphatic inarticulate brachiopods and a comparison with the proteins in the CaCO3 inarticulate brachiopods is of interest to the field of biomineralization. The uneasy phylogenetic placement of these organisms may also be improved with this data (Gorjansky & Popov, 1986).

REFERENCES

AWATI, P.R. AND KSHIRSAGAR, G.R., 1903. *Lingula*. Zool. Mem., Univ. Bombay, 4:1-78.

CHAPMAN, F., 1914. Notes on shell structure in the genus *Lingula*, recent and fossil. Roy. Microsc. Soc. J., 5:28-31.

CONN, K.M. & TERMINE, J.D., 1985. Matrix protein profiles in calf bone development. Bone, 6:33-36.

FISHER, L.W., WHITSON, S.W., AVIOLI, L.V., & TERMINE, J.D., 1983. Matrix sialoprotein of developing bone. J. Biol. Chem., 258:12723-12727.

FISHER, L.W. & TERMINE, J.D., 1985. Purification of the noncollagenous proteins from bone: technical pitfalls and how to avoid them. Pp. 467-472. *In* A. Ornoy, A. Harell & J. Sela, eds. Current Advances in Skeletogenesis. Elsevier, New York.

GORJANSKY, W.J. & POPOV, L.Y., 1986. On the origin and systematic position of the calcareous shelled inarticulate brachiopods. Lethaia, 19:233-240.

HAUSCHKA, P.V., LIAN, J.B. & GALLOP, P.M., 1975. Direct identification of the calcium-binding amino acid θ-carboxyglutamate in mineralized tissue. Proc. Nat. Acad. Sci. USA, 72:3925-3929.

IWATA, K., 1981. Untrastructure and mineralization of the shell of *Lingula* unguis Linne, (inarticulate brachiopod). J. Fac. Sci. Hokkaido Univ., 20:3565.

JOPE, M., 1967. The protein of brachiopod shell I and II. Comp. Biochem. Physiol., 20:593-605.

JOPE, M.,1973. The protein of brachiopod shell V. Comp. Biochem. Physiol., 45B:17-24.

JOPE, M., 1977. Brachiopod shell proteins: their function and taxonomic significance. Amer. Zool., 17:133-140.

JOPE, M., 1979. The protein of brachiopod shell. Comp. Biochem. and Physiol., 63B:163-173.

KLEMENT, R., 1938. Die anorganische skeletsubstanz. Naturwiss., 26:145-152.

LEE, S.L., KOSSVA, D. & GLIMCHER, M.J., 1983. Phosphoproteins of bovine dentin: evidence polydispersity during tooth maturation. Biochem., 22:2596-2601.

LEGEROS, R.Z., PAN, C., SUGA, S., & WATABE, N., 1985. Crystallo-chemical properties of apatite in atremate brachiopods shells. Calc. Tiss. Intern., 37:98-100.

LOWENSTAM, H.A., 1981. Minerals formed by organisms. Science, 211:1126-1131.

MCCONNELL, D., 1963. Inorganic constituents in the shell of the living brachiopod *Lingula*. Geol. Soc. Amer. Bull., 74:363-364.

PRICE, P.A., OTSUKA, A.S., POSER, J.W., KRISTOPONIS, J. & RAMAN, N., 1976. Characterization of a θ-carboxyglutamic acid-containing protein from bone. Proc. Nat. Acad. Sci. USA, 73:1447-1451.

RUNNEGAR, B., 1988. The Evolution of Mineral Skeletons. *In* R.E. Crick, ed., Origin, Evolution and Modern Aspects of Biomineralization in Plants and Animals. Plenum, New York.

TERMINE, J.D., BELCOURT, A.B., CONN, K.M., & KLEINMAN, H.K., 1981. Mineral and collagen-binding proteins of fetal calf bone. J. Biol. Chem., 256:10403-10408.

TERMINE, J.D., 1983. Osteonectin and other newly described proteins of developing bone. Pp. 144-156. In W.A. Peck, ed., Bone and Mineral Research. Excerpta Medica, Amsterdam.

VEIS, A., SPECTOR, A.R. & ZAMOSCIANYK, H., 1972. Isolation of an EDTA-soluble phosphoprotein from mineralizing bovine dentin. Biochem. Biophys. Acta, 257:404-413.

AMINO ACIDS IN PLANKTONIC FORAMINIFERA: ARE THEY PHYLOGENETICALLY USEFUL?

Linda Stathoplos and P.E. Hare

Graduate School of Oceanography, University of Rhode Island, Narragansett, RI 02882-1197; Geophysical Laboratory, Carnegie Institution of Washington, 2801 Upton Street, Washington, DC 20008

ABSTRACT. Planktonic foraminiferal amino acid profiles from deep-sea core samples spanning over 3 million years and three ocean basins show that intra-specific variability, although increasing with sample age, is less than inter-specific variability. Profiles are not specific at the level of morphotype for the species examined. Thus, these profiles are useful for grouping morphotypes at a given time slice into species, and for tracing these species through time.

INTRODUCTION

Biomineralization in planktonic foraminifera is mediated by protein-containing organic matrix (Reiss, 1957, 1958; Towe & Cifelli, 1967; Banner *et al.*, 1973; Weiner & Erez, 1984). Because these proteins are encoded in the organism's DNA, the amino acid profiles of fossil planktonic foraminiferal tests will contain phylogenetic information if the following conditions are met: 1) differences in proteins are reflected in the relative abundances of their constituent amino acids; 2) diagenetic alteration does not obscure these differences; and 3) the protein content of organic matrices is not environmentally plastic. King & Hare's (1972a,b,c) studies in the early seventies of 16 species of planktonic foraminifera concluded that each species had a distinct amino acid profile, that profile differences agreed with phylogenies reconstructed from other evidence, and that these differences persisted as long as 18 million years. However, they did not consider intra-specific variability, nor did they compare this to inter-specific variability.

Despite their recommendation (King & Hare, 1972b,c) that amino acid profiles be exploited to infer phylogenies, subsequent work on amino acids preserved in fossil tests has focused on dating (reviewed in Wehmiller, 1986) and stratigraphic correlation (Miller, 1982; Miller *et al.*, 1983; Müller, 1984). Here, we examine the nature of the phylogenetic information retained in planktonic foraminiferal tests from two species to assess inter- and intra-specific variability.

Origin, Evolution, and Modern Aspects of Biomineralization in Plants and Animals
Edited by Rex E. Crick
Plenum Press, New York

MATERIALS AND METHODS

Core material

To look at spatial variability, samples were obtained from cores in three tropical regions: the eastern equatorial Atlantic, the western equatorial Pacific, and the Caribbean. The Atlantic and Pacific cores are Deep Sea Drilling Project (DSDP) sites; the Caribbean cores are trigger cores taken during "Trident" cruise 148. Near-surface samples from these cores represent ages from Holocene to late Pleistocene (Kennett, von der Borch *et al.*, 1986; Stein, 1985). To look at temporal variability, three stratigraphic levels were examined from the Pacific site, with ages ranging from late Pleistocene to early Pliocene.

Species analyzed

To compare intra-specific with inter-specific variability, two species, each with at least two distinct adult morphologies (morphotypes), were chosen: *Globorotalia tumida* (Brady) and *Globigerinoides sacculifer* (Brady). Most samples of *G. tumida* belonged to the *Globorotalia* (*Globorotalia*) *tumida tumida* (Brady) morphotype (Kennett & Srinivasan, 1983:159), abbreviated here as GT. Only one of the Atlantic samples is the *Globorotalia tumida flexuosa* (Koch) morphotype (Kennett & Srinivasan, 1983:159), abbreviated as GTF. Samples of *G. sacculifer* were separated into forms with and without a final sac-like chamber. The morphotype with the final sac-like chamber is abbreviated as GS. The morphotype without the final sac-like chamber has been called *Globigerinoides quadrilobatus* (d'Orbigny), and thus is abbreviated as GQ (the synonymy of forms of this species is discussed in Kennett & Srinivasan, 1983:66).

Table 1 summarizes location and stratigraphic data for the core samples used. Numbers of samples of each morphotype from each core level are also listed.

Amino acid analysis

The procedure for analysis was similar to that used by King & Hare (1972c). One milligram of foraminiferal calcite, representing 5-30 individual tests, was sufficient to yield enough protein for analysis. After sonicating and rinsing with Super-Q organic-free water, tests were air dried. The calcite was dissolved in fire-cleaned test-tubes by addition of an excess of 6 N HCl (100 ml/mg calcite). The reaction tube was then purged with N_2 to prevent oxidative loss of certain amino acids and capped tightly.

TABLE 1. Location and stratigraphic position of source material for samples of each morphotype: GT = *G. tumida tumida*; GTF = *G. tumida flexuosa*; GQ = *G. sacculifer* without sac; GS = G. *sacculifer* with sac.

OCEAN	SITE/CORE	LAT	LONG	AGE	Foram Zone	Morphotype GT	GTF	GQ	GS
CARIBBEAN	TR108-06(3-5cm)	13°22.4' N	75°45.8' W	Holocene	N23	2	-	1	-
CARIBBEAN	TR108-08(12-15cm)	13°00.9' N	76°23.9' W	Holocene	N23	2	-	1	-
E. EQ. ATLANTIC	366 1-1(0-5cm)	5°40.7' N	19°51.1' W	L Pleistocene	N23	2	1	1	2
W. EQ. PACIFIC	586 1cc	0°29.84' N	158°29.89' E	L Pleistocene	N23	4	-	3	6
W. EQ. PACIFIC	586A 1cc	0°29.84' S	158°29.89' E	L Pliocene	N21	6	-	4	1
W. EQ. PACIFIC	586A 5cc	0°29.84' S	158°29.89' E	E Pliocene	N19	3	-	1	3

The proteins released from the calcite matrix were hydrolyzed at 150° C for 20–25 min. This hydrolysis procedure has been found to be equivalent to a 22–24 h hydrolysis at 110° C for molluscs (Hare, 1975). The samples were centrifuge cooled and evaporated to dryness under N_2. Just before injection onto the high performance liquid chromatography (HPLC) column, they were diluted by a volume of pH2 buffer (1 ml of 12 N HCl in 200 ml dH_2O) equal to the amount of acid used for hydrolysis (100 ml/mg original calcite).

TABLE 2. Amino acids used in profile constructions, and their abbreviations

aspartate + asparagine	ASP
threonine	THR
serine	SER
glutamate + glutamine	GLU
glycine	GLY
alanine	ALA
valine	VAL
methionine	MET
isoleucine	ILE
alloisoleucine	AIL
total isoleucine	tILE (=AIL+ILE)
leucine	LEU
tyrosine	TYR
phenylalanine	PHE
histidine	HIS
lysine	LYS
arginine	ARG

Amino acids were separated using ion exchange HPLC (Hare, 1971) and quantified by fluorescence detection after post-column reaction with o-phthalaldehyde (OPA) (Roth, 1971; Benson & Hare, 1975). Proline and hydroxyproline are not detected by this method. The amino acids used to construct the profiles, along with their abbreviations, are listed in Table 2. Since isoleucine (ILE) epimerizes to alloisoleucine (AIL), which is resolved by this method, the two peaks are added to obtain total isoleucine (tILE).

Statistical analysis

Profiles for each sample express the abundance of fifteen amino acids as a percent of the total area represented by these amino acids. SAS (SAS Institute, Inc., 1982) routines are applied to these data to do a multivariate analysis of variance (MANOVA). SIMCA (Soft Independent Modelling of Class Analogy), a pattern recognition technique (Wold & Sjöström, 1977; Wold *et al.*, 1984), is used to construct a principal component (PC) model both of all the data and of each morphotype; each sample is then compared to these models. Samples can be assigned to clusters consisting of the samples closest to each model. The cluster significance test developed by McFarland & Gans (1986) is used to evaluate the significance of the clusters assigned from SIMCA results.

RESULTS

Reproducibility

The profiles used for analysis are shown in Table 3. Amino acids are listed in order of elution. Analytical reproducibility can be assessed by calculating the coefficient of variation (the standard deviation normalized by the mean) for multiple injections of a mixed sample (shown at the bottom of the Table 3). The coefficient of variation (cv) is less than 3.8% for the 7 amino acids that elute first: ASP, THR, SER, GLU, GLY, ALA, and VAL. Only LYS, of the subsequently eluting amino acid, is comparably reproducible. Subsequently eluting amino acids have cv's ranging from 5.3% to 17.4%.

The error introduced by peak integration is estimated by reintegrating four samples. These reintegrations follow their originals in Table 3, and are indicated by an (R). The coefficients of variation for these pairs are always less than 2% for the first seven amino

TABLE 3. Amino acid percentage data. MORPH = morphotype; see Tables 1 and 2 for other abbreviations. Outliers are indicated by (*). Repeated integrations (of sample directly above) are indicated by (R) after morphotype.

SITE/CORE	MORPH	ASP	THR	SER	GLU	GLY	ALA	VAL	MET	tILE	LEU	TYR
TR 148-06	GQ	24.79	7.12	9.61	13.17	14.14	11.40	5.18	0.75	2.50	2.88	0.00
TR 148-06	GT	19.22	6.35	9.89	12.17	17.42	12.96	5.47	0.43	3.20	4.31	0.00
TR 148-06	GT	21.55	6.82	8.79	11.70	15.32	13.24	5.93	0.91	3.00	4.13	0.00
TR 148-08	GQ	21.63	5.32	11.74	12.92	17.30	9.86	4.80	1.17	2.60	3.63	0.83
TR 148-08	GT	19.73	6.10	9.72	12.02	15.83	13.05	5.35	0.86	2.89	4.24	1.39
TR 148-08	GT*	17.82	5.50	10.02	11.58	15.76	11.66	5.24	7.66	2.64	3.71	0.91
366 1-1	GQ	27.49	4.78	7.25	13.54	12.97	11.16	5.85	1.08	3.14	3.32	0.81
366 1-1	GS	28.29	5.66	7.06	13.80	13.27	10.48	5.96	1.16	2.95	3.55	1.14
366 1-1	GS	29.38	5.40	6.34	13.42	12.37	10.80	5.77	1.12	2.88	2.88	0.84
366 1-1	GT	21.67	5.12	5.06	12.50	15.57	14.31	7.14	0.44	4.11	4.41	1.12
366 1-1	GT	21.72	5.03	5.25	12.43	15.76	13.77	7.05	0.55	4.03	4.39	1.30
366 1-1	GTF	20.43	5.05	6.63	12.74	16.29	14.07	6.94	0.82	3.99	4.30	0.88
586 1CC	GQ	28.97	5.14	6.65	13.11	13.82	10.79	5.47	1.25	2.82	3.82	0.99
-	GQ (R)	28.92	5.15	6.68	13.10	13.67	10.67	5.42	1.24	2.81	3.28	0.92
586 1CC	GQ	28.37	5.20	7.00	13.30	13.90	10.59	5.44	1.10	2.59	3.82	0.94
-	GQ (R)	27.90	5.19	6.97	13.17	13.76	10.53	5.44	1.21	2.78	3.52	1.08
586 1CC	GQ	29.62	5.79	6.40	13.61	12.34	10.66	5.96	1.08	2.91	3.44	0.97
586 1CC	GS	27.24	4.76	7.53	13.14	14.86	10.56	5.28	1.11	2.78	3.55	0.96
-	GS (R)	27.49	4.86	7.53	13.09	14.70	10.38	5.15	0.95	2.45	3.71	0.86
586 1CC	GS	28.08	5.36	6.16	13.01	13.60	11.10	5.60	1.31	2.90	3.57	0.80
-	GS (R)	28.07	5.22	5.99	12.80	13.72	11.19	5.65	1.32	2.97	3.44	0.82
586 1CC	GS	25.58	5.21	7.75	13.29	14.99	9.67	5.26	1.14	2.99	4.19	1.22
586 1CC	GS	27.31	5.41	6.90	13.19	13.61	10.16	5.54	1.00	2.78	4.09	1.28
586 1CC	GS	26.34	5.02	7.45	13.31	14.29	10.02	5.53	0.92	3.19	4.30	1.29
586 1CC	GS	28.29	5.61	6.79	13.12	13.39	10.42	5.82	0.96	3.12	3.60	0.94
586 1CC	GT	20.45	4.56	6.92	12.80	16.30	13.28	6.57	0.79	3.85	4.85	1.29
586 1CC	GT	20.98	4.81	5.63	12.53	15.63	15.71	6.70	0.60	3.99	4.82	1.40
586 1CC	GT	21.83	4.81	6.45	13.03	15.64	13.97	6.58	0.67	3.96	4.53	1.19
586 1CC	GT	21.74	5.17	5.79	12.89	15.74	13.92	6.78	0.21	3.82	4.62	1.21
586A 1CC	GQ	24.45	3.50	7.97	12.86	13.96	10.89	6.11	1.07	3.33	3.86	1.36
586A 1CC	GQ	23.36	3.53	9.35	13.03	15.61	10.50	5.91	0.97	3.22	3.93	0.97
586A 1CC	GQ	30.40	3.13	2.55	14.86	12.64	12.07	7.28	0.27	3.70	4.03	0.67
586A 1CC	GQ	28.46	3.03	5.79	14.54	14.39	11.05	6.70	0.54	3.61	4.08	0.95
586A 1CC	GS	23.58	3.19	4.17	15.64	17.19	10.48	5.71	0.76	3.26	5.10	2.08
586A 1CC	GT	21.86	2.77	1.10	14.81	15.22	16.90	8.63	0.18	4.88	4.83	1.20
586A 1CC	GT	19.67	2.45	5.05	15.12	18.93	14.01	6.43	0.27	3.82	4.78	1.67
586A 1CC	GT	19.14	2.88	4.98	14.75	16.87	14.15	7.47	0.16	4.24	4.75	0.97
586A 1CC	GT	21.20	2.88	3.28	14.92	15.31	14.83	7.81	0.18	4.31	4.79	0.82
586A 1CC	GT*	11.90	2.80	14.32	13.47	27.27	8.28	4.02	0.33	2.62	4.55	0.95
586A 1CC	GT*	15.54	2.92	11.96	15.12	21.88	10.65	5.49	0.41	3.13	4.32	0.98
586A 5CC	GQ	23.70	2.34	5.77	15.65	17.79	10.71	6.49	0.08	3.27	4.90	1.43
586A 5CC	GS	30.57	1.95	3.60	16.19	14.12	11.21	6.86	0.18	3.67	3.86	1.06
586A 5CC	GS	27.93	2.20	4.36	15.97	14.83	10.45	6.38	0.48	3.24	4.11	1.25
586A 5CC	GS*	18.26	2.54	8.80	14.67	20.19	8.81	5.29	0.62	3.68	5.60	1.63
586A 5CC	GT	17.91	1.80	3.91	15.29	18.34	15.09	7.34	0.20	4.89	5.33	1.21
586A 5CC	GT	20.67	3.30	7.40	13.50	13.86	11.82	6.69	0.22	4.00	5.56	1.49
586A 5CC	GT	19.44	2.37	3.88	16.00	15.47	15.45	7.71	0.15	4.97	5.49	1.48
586A 1CC	MIXED	21.38	2.71	2.71	14.60	15.52	15.59	7.88	0.72	4.37	4.56	0.94
586A 1CC	MIXED	21.22	2.78	2.77	14.95	15.25	15.56	8.26	0.68	4.19	4.52	1.17
586A 1CC	MIXED	21.78	2.74	2.57	14.83	15.00	15.48	8.11	0.52	4.72	5.07	0.98

TABLE 3. Continued

PHE	HIS	LYS	ARG%
2.05	1.58	2.53	2.31
2.70	1.67	1.81	2.40
2.54	2.17	1.77	2.12
2.24	1.23	2.44	2.28
3.06	1.84	1.84	2.07
2.50	1.51	1.69	1.81
1.97	3.31	1.70	1.63
1.75	1.92	1.66	1.36
2.21	3.60	1.67	1.33
2.84	2.69	1.64	1.38
2.38	3.58	1.33	1.41
3.02	1.11	1.82	1.93
2.13	2.03	1.68	1.31
2.11	2.28	2.00	1.74
1.89	2.48	1.69	1.67
2.03	2.73	2.04	1.65
2.00	1.94	1.89	1.42
2.23	1.80	2.16	2.04
2.42	2.09	2.18	2.13
2.32	2.34	2.05	1.81
2.19	2.61	2.12	1.88
1.97	2.31	2.54	1.90
2.00	2.81	1.91	2.03
1.89	3.16	2.03	1.26
1.92	2.51	1.94	1.56
2.70	1.59	1.93	2.13
2.80	0.94	1.70	1.77
2.64	1.26	1.66	1.76
2.53	1.88	1.88	1.82
2.47	4.10	2.52	1.55
2.15	3.44	2.62	1.42
2.19	3.42	2.17	0.63
2.19	1.79	1.72	1.15
1.77	1.47	2.99	2.61
2.88	2.11	1.74	0.89
2.68	1.72	2.57	0.83
2.65	3.12	2.60	1.27
2.84	3.16	2.74	0.93
2.27	2.08	2.64	2.51
2.45	1.21	2.17	1.77
2.50	1.78	2.58	1.00
2.22	1.87	2.21	0.45
2.57	2.95	2.55	0.73
2.82	1.88	3.34	1.88
3.02	1.94	2.38	1.34
3.21	2.94	3.91	1.43
3.64	1.39	2.13	0.43
3.33	2.74	2.13	0.81
3.18	2.73	2.11	0.62
3.00	2.41	2.19	0.6

acids. As with analytical reproducibility, the cv's of reintegration of the last eight peaks are infrequently as low as for the first seven, and can be as high as 13%. Thus integration variability, while usually less than analytical variability, is comparable in magnitude. Total experimental variability for these amino acids is thus approximately 5%.

The first seven eluting amino acids represent, on average, 82% (range 74–85%) of the total area of all 15 amino acids tabulated. These most reproducible amino acids are used to compare the varability of a population through space and time, and to do the cluster significance tests. However, despite their experimental variability, the last eight amino acids are included for MANOVA and SIMCA analyses to retain any information they might represent.

Variability

Experimental variability can be compared to the variability of profiles from populations, samples of the same species from the same stratigraphic level. For both species, the intra-specific variability in the late Pleistocene samples (from Sites 586 1cc and 366 1-1), is comparable to the experimental variability of 5% cv for the reproducible amino acids, with the exception of SER. Serine has a cv of 12% for *G. tumida* (7 samples) and 7% for *Globigerinoides sacculifer* (12 samples). The inter-specific variability, however, for all the late Pleistocene samples combined, is greater than experimental variability for all of the reproducible amino acids except GLU (cv's ranging from 6% to 16%; the cv for GLU=3%).

MANOVA allows testing of null hypotheses that the mean profiles of two groups are not significantly different. However, these tests require that the data are normally distributed and that the variances of the groups being compared are homogeneous (Natrella, 1963). Application of Bartlett's test for homogeneity of variance (Saila, *et al.*, unpub.) reveals that the raw percentage data do not meet the latter requirement; the distribution of most amino acid percentages is also found to be non-normal (Perillo & Marone, 1986). Therefore, the data are transformed by taking the arcsine of the square root of the percentage (Natrella, 1963:20-25) before multivariate testing with a SAS general linear model (GLM). Even after transformation, the data do not completely satisfy the test requirements; results of these tests must be interpreted with caution.

To test if the inter-specific variability is significant, only data from Site 586 (1cc) and Site 366 (1-1) are used (late

Pleistocene). The null hypothesis (H_O) is that the mean amino acid profile of *Globorotalia tumida* (GT + GTF) is not significantly different from the mean profile of *Globigerinoides sacculifer* (GS + GQ). The test is applied to the transformed data from late Pleistocene samples only. Four test statistics are calculated to estimate F: Wilk's criterion, Pillai's trace, the Hotelling-Lawley trace, and Roy's maximum root criterion (SAS Institute, Inc., 1982:11). The calculated F values can be compared to the F distribution to estimate the probability that the calculated difference in means would have arisen by chance. All four estimates give a value of F=78.85; the probability associated with this is 1.26%. We conclude that the mean profiles of the two species compared are significantly different.

Too few samples precludes calculating these F values for an intra-specific comparison of GQ (*Globigerinoides sacculifer* without a sac) to GS (*G. sacculifer* with a sac) at this same stratigraphic level. However, the mean profiles of the two morphotypes (GS *vs* GQ) are virtually identical by comparison with the differences observed between mean species profiles (GS + GQ *vs* GT + GTF).

The change in variability within a species at one location (Site 586) through time is also represented by too few samples to apply the MANOVA tests. Qualitatively, for the reproducible amino acids, we observe that the variability almost monotonically increases in both *Globorotalia tumida* and *Globigerinoides sacculifer* from cv's ranging from 1–7% (SER higher) at Site 586 1cc (late Pleistocene), to cv's ranging from 10–20% (except for THR and SER) by early Pliocene. (The cv of THR for *G. tumida* increases to 30%; for both species, the cv of SER increases to 40%.) Thus, by early Pliocene, variability is 2–4 times its level at late Pleistocene.

To investigate the robustness of these results, a different approach has also been taken to comparing inter- and intra-specific variability. SIMCA is a pattern recognition technique that can handle relatively small numbers of samples. Analysis consists of: 1) constructing a principal component (PC) model of all the data (see below); 2) comparing all the data to this model to detect "outliers" (samples which, for whatever reason, do not resemble the group as a whole and thus are distant from the PC model of all the data); 3) deciding whether to include or exclude outliers from subsequent analyses; 4) repeating steps 1–3 for each morphotype (using samples from the late Pleistocene time slice and each species, for all times) and finally 5) classifying all samples in terms of their distance from each of the calculated models (Wold, *et al.*, 1984). Models represent a multivariate data matrix X by a PC model with A components:

$$X = \underline{X} + TB + E$$

where $\underline{X}$ is the vector of variable means for the class, T is a matrix of the PC scores, B is the loading matrix, and E is the residual matrix not accounted for by the model. The significance of the calculated PC's is checked by cross-validation; if a PC is unstable within subsamples of a group, it is rejected as insignificant.

By comparing the distance of each sample to a variety of models, it can be determined which model it is closest to. Once groups are defined as all samples closest to a model, they can be tested for significance by applying the bootstrap cluster significance test (McFarland & Gans, 1986) to the original data. This test uses the data themselves to construct a probability distribution. For example, if a cluster of 5 samples from a data set of 50 samples is to be tested for significance, this technique randomly picks 5 samples many thousands of times from all the data and calculates the cluster size of each group as the mean squared distance (MSD) between all points of the group. The cluster size of the group to be tested for significance is then compared to the resulting MSD distribution; the probability of a cluster randomly having that size (MSD) can then be estimated. If the probability is small, it is likely that the cluster is significant.

The samples marked with an asterisk (*) in Table 3 are outliers compared to a model of all the data. Two of these (from site 586A 1cc) had already been identified as possibly contaminated because of the unusually large amount of total amino acids recovered compared to conspecific samples; both are excluded from further analysis. The other two outliers are not included in modelling because of the strong effect they can exert on principal component extraction. If a single point is distant from a tight group of points, it will determine where a "line" (PC) describing them will be placed. They are also excluded from the cluster significance tests because, if not in the cluster to be tested, they would artificially increase the chance that a cluster would be found to be significant by making all the data less grouped.

By combining data from Site 586 (1cc) and Site 366 (1-1), adequate models of two morphotypes, *Globorotalia tumida* (GT) and *Globigerinoides sacculifer* with a sac (GS), can be constructed for this time slice (late Pleistocene). The four samples of *G. sacculifer* without sacs (GQ) from this level do not allow for construction of a model with any significant principal components, but samples can still be compared to the weighted mean of this group as a model. Models are also constructed for each species (GS+GQ and GT+GTF) at this level.

Several questions are addressed: at the late Pleistocene level, do the *Globigerinoides sacculifer* samples without sacs (GQ) resemble the model of *G. sacculifer* samples with sacs (GS)? Does the *Globorotalia tumida flexuosa* (GTF) sample resemble the model of the *G. tumida tumida* samples (GT)? What is the discriminating power of the models for the two species? Can older and younger samples be identified as belonging to each species by these models?

To answer these questions, each sample is compared to each of the three morphotype models for GS, GQ (no significant PCs) and GT using SIMCA. Each is then assigned to a cluster based on which model it most resembles. On this basis all (except one) of the *Globigerinoides sacculifer* samples (with and without sacs) are closest to the model of GS (*G. sacculifer* with a sac), while all the *Globorotalia tumida* samples (including the *G. tumida flexuosa* sample) are closest to the model of GT (*G. tumida tumida*). The GS sample identified in Table 3 as an outlier more closely resembles the model of GT than the models of either GS or GQ. The apparently uncontaminated GT outlier from the Caribbean core is correctly assigned to the model of GT. This means that, with the exception of one sample, each species sample from other time-slices is correctly assigned using amino acid profiles (although samples from the modeled time-slice resemble the models much more closely).

We can also conclude that morphotypes of a species do not have unique amino acid profiles. All the samples of *Globigerinoides sacculifer* without sacs (GQ) from this time slice resemble the model of *G. sacculifer* with a sac (GS) more than their own model (a weighted average of the four GQ samples). Table 4a summarizes the results of cluster

TABLE 4. Results of cluster significance tests using percentages from the first seven (reproducible) amino acids (see Table 1 for morphotype abbreviations)

a. compared to all data from Sites 586 1cc and 366 (1-1), total number of samples = 17.

1	2	3	4	5	6
GS only	8	10000	0	0 .0	0.0
GS + GQ	12	10000	15	0.0015	0.00076
GQ only	4	10000	1548	0.1548	0.00709
GT only	6	10000	53	0.0053	0.00142
GT + GTF	7	10000	6	0.0006	0.00048

b. compared to all data from all cores samples, total number of samples = 39.

1	2	3	4	5	6
GS + GQ	22	10000	106	0.0106	0.00201
GT + GTF	17	10000	2920	0.292	0.00891

[1]Cluster to be tested
[2]Number of samples in cluster
[3]Subsets
[4]Subsets at least as tightly clustered
[5]Estimated P-value
[6]Uncertainity (95% confidence)

significance analyses using percentage data of the seven reproducible amino acids of samples from this late Pleistocene time slice. The group of only the GQ morphotype (*G. sacculifer* without a sac) is not significant because it is really a subset of the larger group of all *G. sacculifer*. Because only two species are represented in these data, random groups of four have a significant probability of being all the same species, and therefore being as tightly clustered as all GQ samples. Each species, however, forms a significantly clustered group compared to random groups of the same number of samples from all the data at this level.

A similar test of significance of the clusters of all *Globigerinoides sacculifer* and all *Globorotalia tumida* compared to all the data at all levels shows that the cluster of morphotypes GT and GTF has only a 29% probability of being significant (Table 4b). This may partly be due to the smaller number of samples in this cluster. A more definitive test of the coherence of amino acid profiles in planktonic foraminifera over millions of years will be possible once data from more species are available.

CONCLUSIONS

Pattern recognition (SIMCA) models of amino acid profiles and cluster significance tests allow us to conclude the following: 1) variability of a population at one location and stratigraphic level is only slightly greater than experimental variability; 2) this variability appears to increase with time (downcore); 3) the increase in variability does not prevent correct identification of a sample based on its amino acid profile in all but one sample; and finally 4) profiles are not specific at the level of morphotype, but are significantly different at the level of species for *Globigerinoides sacculifer* and *Globorotalia tumida*. This suggests that amino acid profiles will be useful for reconstructing phylogenies of planktonic foraminifera.

ACKNOWLEDGEMENTS

Generous support for travel to this symposium was provided to Linda Stathoplos by The Organized Research Fund of The University of Texas at Arlington and The Sea Grant Program administered by Texas A & M University. We thank J. P. Kennett and M. Bender for providing direction; M. Arthur, M. Leinen, and J. Quinn for sharing of space and equipment; finally, we thank K.-Y. Wei for providing helpful comments on the original draft of this manuscript.

REFERENCES

BANNER, F.T., SHEEHAN, R., & WILLIAMS, E., 1973. The organic skeletons of rotaline foraminifera: A review. J. Foram. Res., 3:30-42.

BENSON, J.R., & HARE, P.E., 1975. o-Phthalaldehyde: Fluorogenic detection of primary amines in the picomole range. Comparison with fluorescamine and ninhydrin. Proc. Nat. Acad. Sci., USA, 72:619-622.

HARE, P.E., 1971. Ultrasensitive amino acid analyzer. Carnegie Inst. Washington, Yearbook, 70:268-269.

HARE, P.E., 1975. Amino acid composition by column chromatography. *In* Protein Sequence Determination, 2nd edition (ed. S.B. Needleman), pp. 204-231. Berlin: Springer-Verlag.

KENNETT, J.P., & SRINIVASAN, M.S., 1983. Neogene Planktonic Foraminifera: A Phylogenetic Atlas. 265 pp. Stroudsburg, PA: Hutchinson Ross Publishing Co.

KENNETT, J.P., VON DER BORCH, C.C., *et al.*, 1986. Initial Reports of the Deep Sea Drilling Project, Volume 90. Washington: U.S. Govt. Printing Office.

KING, JR., K., & HARE, P.E., 1972a. Species effects in the epimerization of L-isoleucine in fossil planktonic foraminifera. Carnegie Inst. Washington Yearbook, 71:596-598.

KING, JR., K., & HARE, P.E., 1972b. Amino acid composition of planktonic foraminifera: A paleobiochemical approach to evolution. Science, 175:1461-1463.

KING, JR., K., & HARE, P.E., 1972c. Amino acid composition of the test as a taxonomic character for living and fossil planktonic foraminifera. Micropaleo., 18:285-293.

MCFARLAND, J.W., & GANS, D.J., 1986. On the significance of clusters in the graphical display of structure activity data. J. Medicinal Chem., 29:505-514.

MILLER, G.H., 1982. Quaternary depositional episodes, Western Spitsbergen, Norway; aminostratigraphy and glacial history. Arctic Alpine Res., 14:321-340.

MILLER, G.H., SEJRUP, H.P., MANGERUD, J., & ANDERSEN, B.G., 1983. Amino acid ratios in Quaternary molluscs and foraminifera from Western Norway: Correlation, geochronology and paleotemperature estimates. Boreas, 12:107-124.

MOORE, S., & STEIN, W.H., 1954. Procedures for the chromatographic determination of amino acids on four percent cross-linked sulfonated polystyrene resins. J. Biol. Chem., 211:893-906.

MÜLLER, P.J., 1984. Isoleucine epimerization in Quaternary planktonic foraminifera: Effects of diagenetic hydrolysis and leaching, and Atlantic-Pacific intercore correlations. "Meteor" Forsch.-Ergrebnisse (C), 38:25-47.

NATRELLA, M.G., 1963. Experimental statistics. National Bureau of Standards Handbook 91. 536 pp. Washington, D.C.: U.S. Govt. Printing Office.

PERILLO, G.M.E., & MARONE, E., 1986. Determination of optimal numbers of class intervals using maximum entropy. Math. Geol., 18:401-407.

REISS, Z., 1957. The Bilamellidea, nov. superfam., and remarks on Cretaceous globorotalids. Cushman Found. Foram. Res. Cont., 8:127-145.

REISS, Z., 1958. Classification of lamellar foraminifera. Micropaleo., 4:51-70.

ROTH, M., 1971. Fluorescence reaction foramino acids. Analyt. Chem., 43:880-882.

SAILA, S.B., RECKSIEK, C.W., PRAGER, M.H., & CHEN, X., unpub. BARTTR: Bartlett's test for homogeneity of variance (including four transformations). Fishery Sci. Appl. Prog..

SAS Institute, Inc., 1982. SAS User's Guide: Statistics, 1982 edition. 956 pp. Cary, NC: SAS Institute, Inc.

STEIN, R., 1985. The post-Eocene sediment record of DSDP Site 366: implications for African climate and plate tectonic drift. Geol. Soc. Amer. Mem., 163:305-315.

TOWE, K., & CIFELLI, R., 1967. Wall ultrastructure in the calcareous Foraminifera: crystallographic aspects and a model for calcification. J. Paleont., 41:742-762.

WEHMILLER, J.W., 1986. Amino acid racemization geochronology. *In* Dating Young Sediments (eds. A.J. Hurford, E. Jäger, & J.A.M. TenCate), pp. 139-158. Bangkok, Thailand: United Nations CCOP Technical Publication 16.

WEINER, S., & EREZ, J., 1984. Organic matrix of the shell of the foraminifer, *Heterostegina depressa*. J. Foram. Res., 14:206-212.

WOLD, S., ALBANO, C., DUNN III, W.J., EDLUND, U., ESBENSEN, K., GELADI, P., HELLBERG, S., JOHANSSON, E., LINDBERG, W., & SJÖSTRÖM, M., 1984. Multivariate data analysis in chemistry. *In* Chemometrics, Mathematics and Statistics in Chemistry (ed. B.R.Kowalski), pp. 17-95. Dordrecht, Holland: D.Reidel Publishing Co..

WOLD, S., & SJÖSTRÖM, M., 1977. SIMCA: a model for analyzing chemical data in terms of similarity and analogy. In Chemometrics, Theory and Application, (ed. B.R. Kowalski), pp. 243-252. Amer. Chem. Soc. Sym. Series No. 52.

ONTOGENETIC VARIATIONS IN THE DISTRIBUTION OF CA AND MG IN SKELETAL TISSUES OF VERTEBRATES AND INVERTEBRATES

Gary D. Rosenberg[1] and W. William Hughes[2]

[1]*Geology Department, Indiana/Purdue University, Indianapolis, IN 46202, USA;* [2]*Biology Department, Andrews University, Berrien Springs, MI 49104.*

ABSTRACT. There are important ontogenetic patterns in the distribution of Ca and Mg in both invertebrate and vertebrate skeletal tissues. These variations need to be defined before either the physiological or environmental significance of fluctuations in the inorganic skeletal components can be assessed fully. For example, biomineralogists have long known that organisms tend to exclude Mg from their skeletons because Mg inhibits calcification. However, we have found clearly defined, inverse Ca and Mg oscillations within skeletal tissues over a range in frequencies that are difficult to explain simply on the basis of random substitutions of Mg for Ca in the biomineral lattice. Moreover, our data suggest that there are ontogenetic trends in the development of the inverse oscillations. With mixed results, marine biologists have used Mg content in the skeletons of various invertebrates as an index of water temperature and of changing composition of seawater over the past 600 million years. Yet our analyses of the shells of invertebrates, such as brachiopods, reveal age-mediated concentrations of elements that complicate such generalizations.

We describe here the results of electron microprobe analyses of the skeletal tissues of two different animals to exemplify the above: the shells of the living brachiopod, *Terebratalia transversa*, and incisors of the rapidly aging Brookhaven National Laboratory Strain of Swiss Webster mice (BNLSW mice). The outer shell layer of the brachiopod was traversed across a series of structural growth increments that resemble bivalve growth patterns known to be deposited under tidal influence. The incisors were traversed from the enamel to the margin of the pulp cavity, across concentric growth increments representing about 20 - 30 days of growth. In both the brachiopod shell and the BNLSW mouse incisor, Ca and Mg were seen to oscillate inversely over a range of temporal and spatial periodicities. In cases where the structural growth patterns are a response to environmental changes (*e.g.* tidal oscillations in the brachiopod), the elemental patterns are a chemical response to the same parameter. There is some evidence that the inverse oscillations do not appear early in the ontogeny of the mouse incisor. It is not yet clear that the oscillations are confined to later ontogeny of the brachiopod shell. These results underscore our contention that an ontogenetic framework will help the physiological and environmental implications of changing skeletal composition appear comprehensible.

INTRODUCTION

We herein describe ontogenetic compositional variations in the outer shell layer of the living brachiopod, *Terebratalia transversa*, and in the incisor of the rapidly aging Brookhaven National Laboratory Strain of Swiss Webster mouse (BNLSW mouse). We

Edited by Rex E. Crick
Plenum Press, New York

have used the electron microprobe to map the distribution of Ca and Mg along the medial, anterior-posterior axis of the brachiopod shell and along radii of the medial section of the mouse incisor, from the enamel to the margin of the pulp cavity.

Very little is known about the ontogenetic changes in composition of either invertebrate or vertebrate skeletons. Because ontogenetic compositional patterns in skeletons have not been described systematically, it is very difficult to evaluate how skeletal composition responds to physiological and environmental factors. Rosenberg (1980) discussed this problem as it relates to bivalve molluscs, whose shells have been analyzed more extensively than those of any other invertebrate taxon. Few of these studies are ontogenetic. Most are whole-shell, mean value, analyses or are apparently random samples of parts of the shell. This problem exists even for mammals extensively studied for medical applications; gross abnormalities (*e.g.*, osteomalacia and osteoporosis) are known but minor, short-term changes in skeletal composition as superimposed on normal ontogenetic trends largely have been ignored. The lack of ontogenetic studies has prevented establishing consistent cause and effect relationships between changes in the environment and physiology on the one hand and changes in skeletal composition on the other.

A great deal of effort has been spent on trying to establish the environmental significance of Mg in the invertebrate skeleton, but with mixed results (Rosenberg, 1980). Most notably, the Mg content of the calcitic invertebrate skeleton has been advanced as being proportional to the temperature or inversely proportional to the salinity of the life habitat. But attempts to confirm these results have generated inconsistent data; generalizations have proven as elusive as will-o-the-wisps. Relevant to the title of this symposium and book, Wilkinson (1979) noted that the number of taxa possessing relatively high-Mg skeletons increased in importance after the Paleozoic and proposed that the change was due to an increase in the Mg/Ca ratio of seawater. However, Given and Wilkinson (1984) reconsidered Wilkinson's (1979) arguments and now the two researchers believe that carbon dioxide partial pressure in the atmosphere (and consequent carbonate concentration in seawater) is the driving force of changes in Mg/Ca ratios in carbonate.

Such erratic progress in establishing connections between skeletal composition and environmental and physiological changes is not surprising. Recently, MacKenzie *et al.* (1983) have shown that Mg-carbonates, of which the volumetrically most important are biogenic, are notoriously "sloppy" and inhomogeneous, preventing the establishment of thermodynamic and stoichiometric principles with which to explain their formation.

We submit our data to demonstrate that such compositional inhomogeneities are only apparently "sloppy". Ca and Mg concentrations may change regularly throughout invertebrate and vertebrate skeletons, with short-term oscillations superimposed upon long-term trends. Once the ontogenetic pattern of an organism's skeletal composition is known, it should be possible to test the hypothesis that skeletal composition responds to changes in particular environmental factors by simultaneously tracking skeletal chemistry and environmental change, as well as physiological parameters.

METHODS

Electron microprobe

Elemental distributions were mapped on polished sections with an electron microprobe scanning across growth increments, so that the patterns obtained were ontogenetic. Traverses were made in the outer shell layer of the brachiopod along the anterior-posterior axis and across medial sections of the incisor from the enamel to the margin of the pulp cavity. Standard microprobe techniques were used. The instrument was operated with a gun potential of 15 kV, and a specimen current of 0.015 mA on benitoite. The beam diameter was 1-5 μm and analyses were taken at intervals ranging between 1-10 μm, depending on the purpose of the traverse. Tests for gross repeatability sometimes used a relatively long measurement interval to save time. Apatite was used as a standard for Ca, and MgO was used as a standard for Mg. An LiF crystal was used to detect Cakα, and an RAP crystal was used to detect Mgkα X-rays. Additionally, Skα (using a pyrite standard and a PET crystal) and Pkα (using an apatite standard and an RAP crystal) were also analyzed, but only results for S distribution in the incisor are mentioned here in passing.

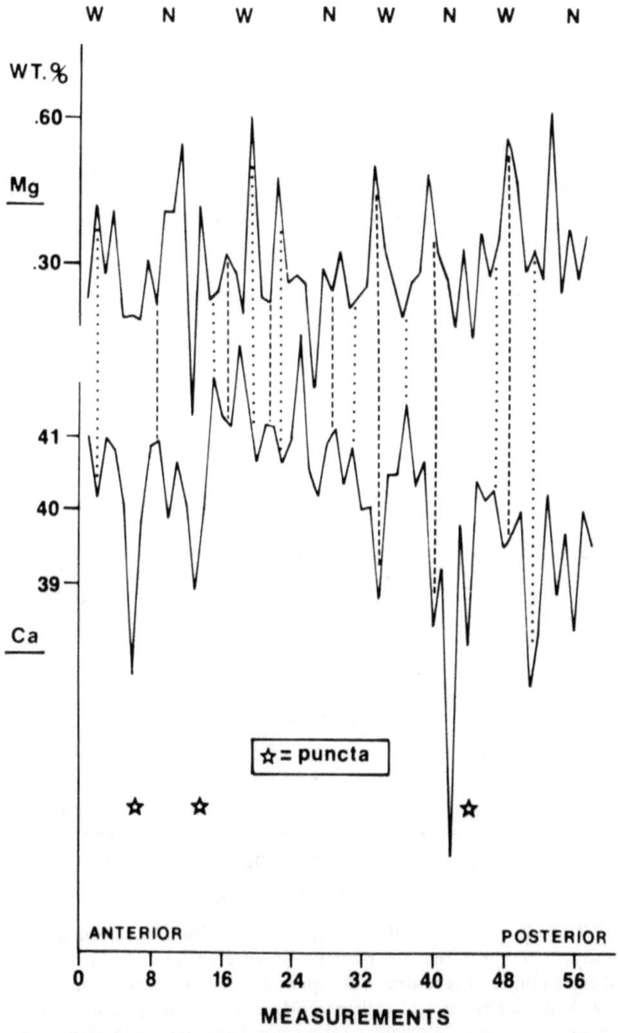

FIGURE 1. Distribution of Mg (top) and Ca (bottom) measured simultaneously with the electron microprobe in a portion of the outer shell layer of the living brachiopod *Terebratalia transversa*. Original (non-smoothed) data. Dashed correlation lines match inverse oscillations of the elements. Cross-correlation coefficient, $r = -0.384$ significant at $P <$, $n = 58$. Traverse proceeds from near the anterior margin at left, approximately 232 μm (58 measurements at 4 μm intervals) posteriorly, and crosses approximately 50 structural growth increments (seen faintly only in etched, but not polished sections). Structural growth increments widen and narrow (W & N) with a period of approximately 10 increments, tracking the tidal cycle as do bivalve growth increments. Scan intersected three punctae (*).

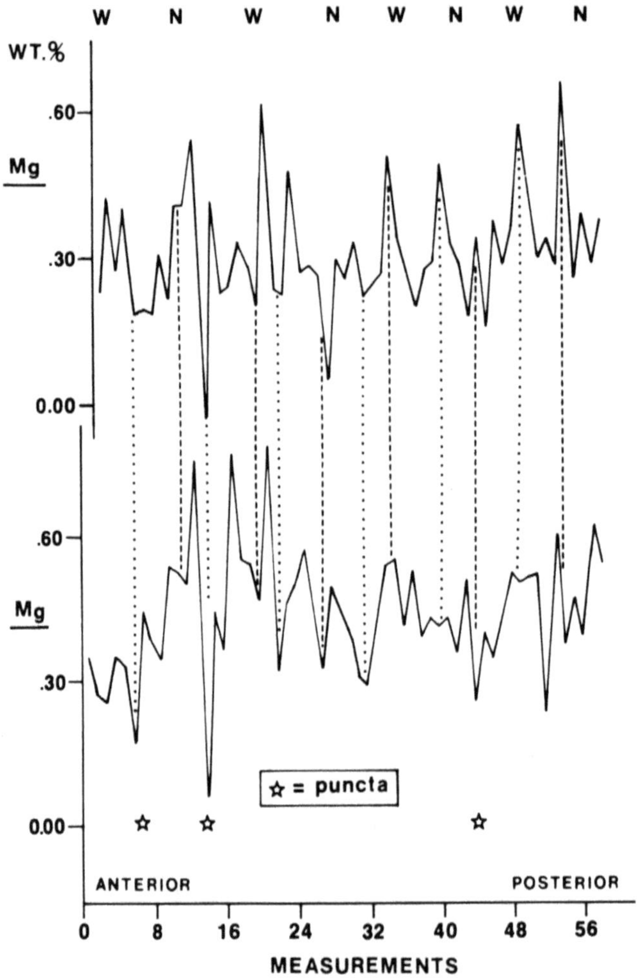

FIGURE 2. Demonstration of the repeatability of electron microprobe analyses of Mg in the outer shell layer of *Terebratalia transversa*. Graph shown in Figure 1 is re-drawn at top, with replicate below. Correlation lines match similar peaks and valleys of Mg distribution at offset of traverses shown. Both graphs original (non-smoothed) data with significant (+) cross-correlation coefficient, $r = +0.484$, $P < 0.01$, $n = 56$.

We determined empirically that ZAF corrections were not needed for our semiquantitative traverses because: 1) corrections were small relative to original values, 2) measurements retained their original relative differences after ZAF corrections, and 3) the smoothing process discussed below removed oscillations of even greater amplitude than those altered by ZAF corrections. Rationale (2) is especially important in rebutting the argument that Mg may only appear relatively low when Ca concentration is high because Ca absorbs Mg_{ka} X-rays, which are relatively low energy.

Drift and dead time corrections were made, however, and all traverses were repeated more than once to assess replicability.

Fourier techniques

Most of the data presented here are original (non-smoothed) data. But some traverses were smoothed using Fourier techniques (Rosenberg, *et al.*, 1980). A spectrum of cosine waves, each with a defined amplitude, frequency and phase was fitted to the data. Generally 12.5% to 50% of the cosine waves representing 90% of the variance of the traverses were added together (recombined) to generate a smoothed version of the elemental distribution. The highest frequencies were arbitrarily removed.

Cross correlations

Cross correlations of repeated analyses of a single element were made to test replicability. Cross correlations of Ca and Mg distributions, measured simultaneously (100% overlap), were made to determine phase differences in oscillation of the two elements.

RESULTS

Figure 1 shows the distribution of Mg and Ca in a portion of the outer shell layer of *Terebratalia transversa*. The traverse proceeds from near the anterior margin at left, posteriorly, for about 232 μm (there are 58 measurements taken at 4 μm intervals-- approximately one measurement per structural growth increment averaging 4 μm in thickness). The traverse crosses a series of poorly defined structural growth increments that oscillate in width (W and N at top = wide and narrow, respectively) over a period of about 10 increments, approximating the tidal cycle (an analysis of structural growth increment periodicities will be presented elsewhere). There is a slight indication of compositional cycles matching the (presumed) tidal structural cycles. There are approximately seven regions of relatively high-Mg/low-Ca content appearing at intervals of about 8-10 measurements (increments) and, although the

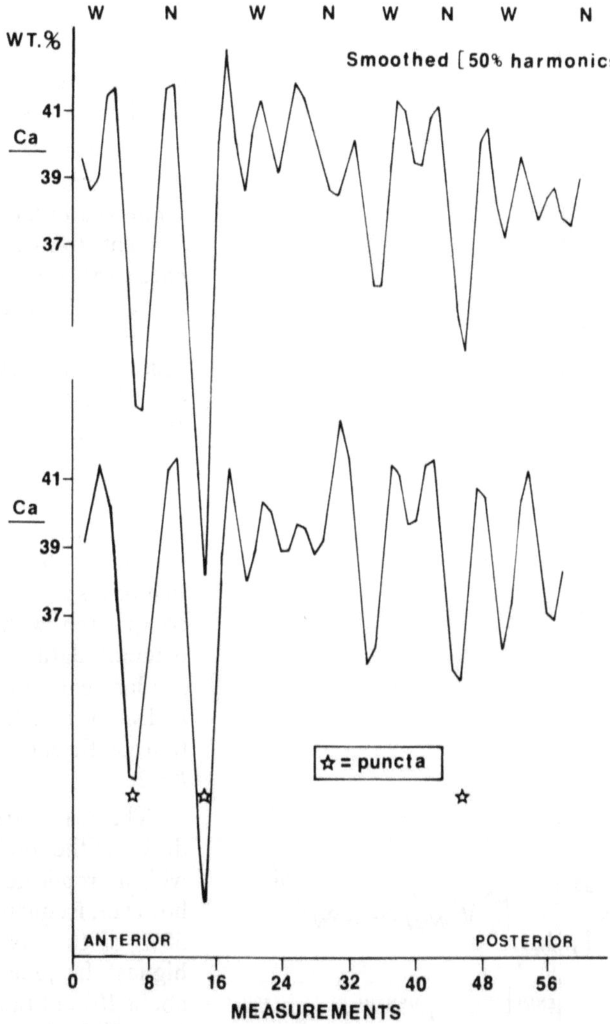

FIGURE 3. Demonstration of the repeatability of electron microprobe analyses of Ca distribution in the outer shell layer of *Terebratalia transversa*. Graph shown in Figure 1 is re-drawn at top, with replicate below. Both graphs smoothed using 50% of harmonics having about 90% of the variance in complete spectrum of cosine waves representing the data by Fourier transform.

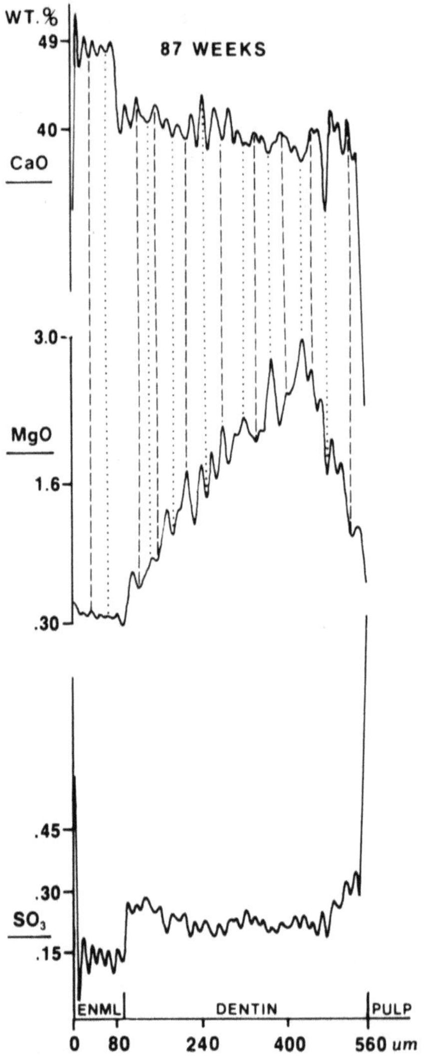

FIGURE 4. Distribution of Ca (top), Mg (middle), and S (bottom) in the BNLSW mouse incisor (87 weeks old). Data frequency smoothed using 12.5% of the harmonics (and approximately 90% of the variance for Ca and Mg) in the complete spectrum of cosine waves fitted to the data by Fourier transform. Traverse proceeds from enamel surface (left) to margin of pulp cavity (right). Correlation lines match simultaneous Ca and Mg measurements (fluctuating either inversely or directly, unlike correlation lines for the graphs of the brachiopod data). Most relatively high frequencies (circa one cycle per 16 μm) as well as long-term trends in means of Ca and Mg are inverse. S not correlated with Ca or Mg (see text).

presence of the punctae (marked with *) does present problems, we are currently trying to clarify phase relationships between the compositional and structural patterns.

However, the phase difference between the Mg and Ca oscillations is noteworthy. There is a pronounced tendency for the two elements to oscillate inversely, as the dashed lines correlating simultaneous measurements reveal. Both graphs are original (non-smoothed) data, so the peaks have not been shifted or distorted by Fourier recombination. The cross correlation coefficient, r, at 100% overlap is -.384, with $P<.05$, $n=58$, $K=2$ independent variables.

Repeatability of the Mg and Ca traverses in Figure 1 is depicted in Figure 2 and Figure 3, respectively. The entire Mg distribution in Figure 1 is re-drawn at the top in Figure 2 for comparison with its replicate below (also original data). Correlation lines match similar peaks and valleys of Mg distribution and, at the match position shown, the correlation coefficient, $r=+.484$, with $P<0.01$, $n=56$, $K=2$.

The Ca distribution in Figure 1 is re-drawn at the top in Figure 3, for comparison with its replicate below. These graphs are, however, frequency smoothed by eliminating 50% of the cosine waves (harmonics) of highest frequency (but representing only about 10% of the variance) in the spectrum. It is difficult to assess the statistical significance of cross correlations of smoothed series, but inspection does seem to suggest that the replications are very similar.

Figure 4 shows the distribution of Ca, Mg, and S in the incisor of the 87-week old BNLSW mouse incisor from the enamel (left) to the margin of the pulp cavity (right). Measurements were made at 4 μm intervals. Both graphs have been frequency smoothed by recombining 12.5% of the cosine waves of lowest frequency representing 90% of the variance of the original data. Figure 5 is a similar analysis in the incisor of a 117-week old mouse to demonstrate repeatability.

As expected, Ca concentration is high in the enamel relative to the dentine. Moving toward the pulp cavity (right), the Ca concentration reaches a nadir about 150-200 μm from the pulp cavity, and then increases slightly. Inversely, Mg content is low in the enamel, reaching a maximum within the dentine in the same area where Ca reaches a nadir.

Dashed correlation lines in these graphs match simultaneous measurements. But, unlike the brachiopod traverses, lines mark both direct as well as inverse oscillations in Ca and Mg. Nevertheless, in both the enamel and dentine, most of the relatively high frequency oscillations (approximating 1 cycle per 16 μm and representing daily frequency-- see Rosenberg & Simmons, 1980) are inverse and suggest that short-term inverse oscillations in Ca and Mg appear in vertebrate as well as invertebrate skeletal tissues. Correlation lines are not drawn between S and either of the other two elements as short term oscillations in S do not seem to maintain a constant phase relationship with either Ca or Mg. Cross correlations between Mg and Ca were not done because it is clear that the trends in the mean across the incisor are inversely proportional. These trends would obscure comparisons of the high frequency oscillations, further analyses of which will be reported elsewhere.

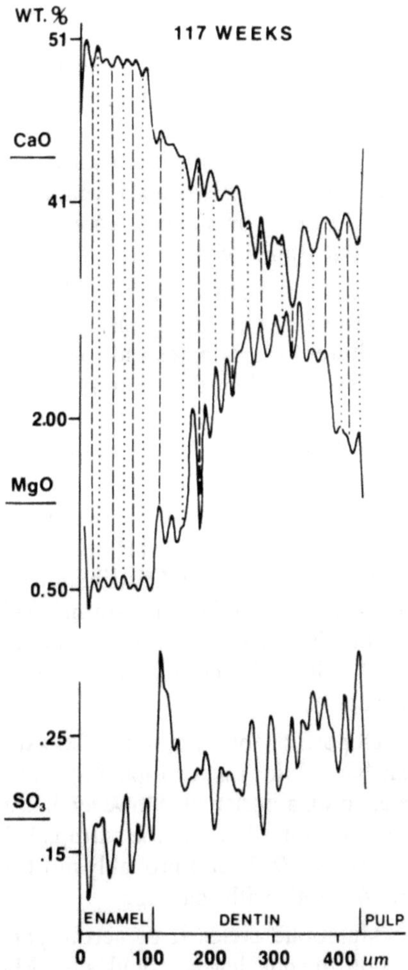

FIGURE 5. Distribution of Ca and Mg in the incisor of 117-week old BNLSW mouse, to show repeatability of patterns shown in Figure 4

Finally, we note that neither the inverse long-term (mean trends) nor the inverse short-term oscillations shown here appear in another, younger incisor. This suggests that the inverse fluctuations may develop only later in ontogeny of the incisor. These results will also be examined more fully elsewhere.

DISCUSSION

Our principal observation is that there are repeatable ontogenetic patterns in the distribution of Ca and Mg in the shell of the brachiopod, *Terebratalia transversa*, and in the incisor of the BNLSW mouse incisor. The two elements show a strong tendency to fluctuate inversely in concentration as skeletal growth proceeds in both animals. This suggests that Mg plays a more continuous, active role in calcification than previously appreciated. And its appearance in both the brachiopod shell and mouse incisor is consistent with the "Biochemical Principle": *A physiological process is fundamentally the same in whatever taxa it occurs* (Lehninger, 1983). Clearly, the biomineralogist can contribute to the understanding of the stoichiometry and dynamics of formation of Mg-containing carbonate and phosphate by demonstrating that compositional inhomogeneities in biominerals are not simply random "sloppiness". There really are (?genetic or ?physiological) "vital effects" (MacKenzie, *et al*, 1983) as well as environmental effects on mineral composition that can be appreciated only within an ontogenetic framework.

It has long been known that Mg inhibits calcification (see Wilbur & Bernhardt, 1984, for review). The Mg^{2+} cation is much smaller than the Ca^{2+} cation for which it substitutes in carbonates and apatites. Substitutions distort the crystal lattice, inhibiting calcification. Thus, the immediate mineralogic cause of inverse Ca and Mg fluctuations is not mysterious. But what, if any, are the physiological controls over the periodic inverse oscillations of Mg and Ca?

First, one could postulate that there is no direct physiological control. Random substitutions of Mg for Ca in the calcite or apatite crystal lattice could occasionally raise the Mg/Ca ratio in the skeleton. However, we would expect neither repeatable long term trends in elemental distribution in different individuals (cf. BNLSW mouse incisors, Figs. 4-5) nor regular, short-term oscillations (cf. both the brachiopod shell and mouse incisors, Figs. 1, 4-5). Moreover, it is not clear that all of the measured Mg is substituting for Ca within the crystal lattice (as the microprobe measures total Mg, regardless of phase).

Deposition of Mg within the skeleton could lower the Mg/Ca ratio within the calcifying fluid (*e.g.*, extrapallial), causing the next fraction of skeleton deposited to be relatively Ca-enriched. However, it is unlikely that Mg or Ca concentrations within such fluids are rate-limiting; the extrapallial fluid of the bivalve (Wilbur & Saleuddin, 1983; Crenshaw, 1972) and probably most other marine invertebrates such as brachiopods, are isosmotic with seawater.

Metabolic cycles (*e.g.*, aerobic/anaerobic), and the accompanying fluctuation in carbon dioxide levels, could alter Mg/Ca ratios in precipitated biominerals just as environmental carbon dioxide levels determine Mg/Ca ratios in inorganic carbonate (Given & Wilkinson, 1984). Such metabolic cycles are known, and links between them and calcification in both vertebrates and invertebrates have been proposed (Okada, 1943; Lutz & Rhoads, 1977), but the connections are not accepted by all biomineralogists (Wilkes & Crenshaw, 1979).

Finally, Mg or Ca transport through cells responsible for calcification could fluctuate periodically. At present, the complete sequence of such processes - let alone their temporal regularity - is known only poorly.

ACKNOWLEDGMENTS

The comments of several attendees, especially Miles Crenshaw, at this and previous symposia, have been of great benefit to our research. Research supported by NSF Grant

EAR 84-07133 and NASA Contract A21997C. Mudge Morris kindly prepared the figures. Rosenberg thanks the Organized Research Fund of the University of Texas at Arlington and the Sea Grant Program administered by Texas A&M University for a stipend that facilitated his attendance at the Symposium.

REFERENCES

CRENSHAW, M.A., 1972. The inorganic composition of molluscan extrapallial fluid. Biol. Bull., Mar. Biol. Lab., Woods Hole, Massachusetts, 143:506-512.

GIVEN, R.K. & WILKINSON, B.H., 1984. Kinetic control of morphology, composition, and mineralogy of abiotic sedimentary carbonates. J. Sed. Petrol., 55:109-119.

LEHNINGER, A.L., 1983. Principles of Biochemistry. Worth Publishers, New York: 1011 pp.

LUTZ, R.A. & RHOADS, D.C., 1977. Anaerobiosis and a theory of growth line formation. Science, 198:1222-1227.

MACKENZIE, F., BISCHOFF, W.D., BISHOP, F.C., LOIJENS, M., SCHOONMAKER, J., & WOLLAST, R., 1983. Magnesian calcites: low temperature occurrence, solubility and solid-solution behavior. *In* Reviews in Mineralogy, vol. 2 (ed. R.J. Reeder), Min. Soc. Amer.:97-144.

OKADA, M., 1943. Studies on the periodic pattern of hard tissues in animal body. Shanghai Evening Post, Medical Ed. September, 1943:26-31.

ROSENBERG, G.D., 1980. An ontogenetic approach to the environmental significance of bivalve shell chemistry. *In* Skeletal Growth of Aquatic Organisms (ed. D.C. Rhoads & R.A. Lutz), Plenum, New York: 133-168.

ROSENBERG, G.D., ASHTON, M., HEWITT, R., & SIMMONS, D.J., 1980. Application of normalized power spectra to the analysis of chemical and structural growth patterns. *In* Skeletal Growth of Aquatic Organisms (ed. D.C. Rhoads & R.A. Lutz), Plenum, New York: 675-686.

ROSENBERG, G.D. & SIMMONS, D.J., 1980. Rhythmic dentinogenesis in the rabbit incisor: circadian, ultradian and infradian periods. Calc. Tiss. Intern., 32:29-44.

WILBUR, K.M. & BERNHARDT, A.M., 1984. Effects of amino acids, magnesium, and molluscan extrapallial fluid on crystallization of calcium carbonate: *in vitro* experiments. Biol. Bull., Mar. Biol. Lab., Woods Hole, Massachusetts, 166:251-259.

WILBUR, K.M. & SALEUDDIN, A.S., 1983. Shell formation. *In* The Mollusca, vol. 4 (ed. A.S.M. Saleuddin & K.M. Wilbur), Academic Press, New York: 235-287.

WILKES, D.A. & CRENSHAW, M.A., 1979. Formation of a dissolution layer in molluscan shells. Scan. Elec. Micros., 2:469- 474.

WILKINSON, B.H., 1979. Biomineralization, paleoecology, and the evolution of calcareous marine organisms. Geology, 7:524-527.

STRONTIUM IS REQUIRED IN ARTIFICIAL SEAWATER FOR EMBRYONIC SHELL FORMATION IN TWO SPECIES OF BIVALVE MOLLUSCS

Scott M. Gallager[1], Joseph P. Bidwell[2,3] and Alan M. Kuzirian[4]

[1]*Department of Biology, Woods Hole Oceanographic Institution, Woods Hole, MA 02543;* [2]*Howard Hughes Medical Institute, Woods Hole Oceanographic Institution, Woods Hole, MA 02543;* [3]*present address: Endocrine Research Unit, Mayo Foundation, 5-164 West Joseph Building, Rochester, MN 55905;* [4]*Section on Neural Systems, Laboratory of Biophysics, NINCDS-NIH, Marine Biological Laboratory, Woods Hole, MA 02543*

ABSTRACT. Strontium is required in artificial seawater for the formation of the calcareous embryonic shell of the bivalves Mercenaria *mercenaria* (Linne, 1758) and *Bankia gouldi* (Bartsch, 1908). Embryos reared in defined media without strontium chloride become swimming larvae that appear normal except that the mineralized portion of the prodissoconch I shell is absent. Calcification is sensitive to small differences in strontium concentrations and a level of 6-8 ppm is required for normal development. Structural defects also result at high concentrations of this element, 3 to 10 times that of natural seawater, or when barium is substituted for strontium. Exposure to strontium is required only between hours 15-30 of the 40-45 hour embryonic period. Shell mineralization is initiated at hour 20 while the organic layer is observed with SEM at hour 18 and appears complete by hour 26. Larval shell growth (prodissoconch II) and statolith formation in *M. mercenaria* do not require the presence of strontium. These results are compared with those obtained earlier for the gastropod *Aplysia californica* (Cooper, 1863) and the implications are discussed.

INTRODUCTION

Recently we reported that strontium is required for the normal embryonic development of the marine gastropod *Aplysia californica* (Cooper, 1863)(Bidwell *et al.*, 1986). Embryos reared in artificial seawater which lacked strontium chloride became larvae with normally appearing soft tissue but with deformed shells and statocysts without the statolith granule. The development of mineralized tissue was sensitive to differences in strontium concentration as small as 1 ppm. Furthermore, preliminary experiments with another gastropod, a bivalve and two cephalopods provided evidence for similar reactions to the absence of this ion during embryonic development.

To date, this has been the most dramatic demonstration of strontium's biological activity and was the first indication of this element's widespread influence on calcification processes. The marine protozoa *Acantharia* spp. possess a test composed of strontium sulfate but these species have been viewed as singular curiosities and efforts to culture them for further study have failed (Anderson, 1981). The effects of strontium on calcification and cell division of the coccolithophorid *Cricosphaera* (*Hymenomonas*) *carterae* have been described (Blackwelder *et al.*, 1976; Weiss *et al.*, 1976). Although the presence of this element was effective in promoting both processes in calcium-deficient media, a requirement under normal conditions was not indicated.

Traditionally strontium has been cast into the role of contaminant to calcareous materials. This can be largely attributed to the following: 1) strontium is not required for the *in vitro* precipitation of these minerals, 2) its similar chemical and physical properties to calcium combined with its ubiquitous presence in both the bio- and geosphere facilitates substitution processes, and 3) until recently, there has been no compelling evidence for any metabolic role.

This view has been strengthened by the observation that a general correlation may exist between strontium levels in biogenic mineralized tissue and phyletic position (Lowenstam, 1964; Dodd, 1967; Gunatilaka, 1975). Lowenstam's classic survey of marine aragonitic organisms of Palau Island in the Pacific indicated that the more primitive evolutionary forms had Sr/Ca ratios near to or slightly greater than the seawater value whereas the more advanced organisms had lower levels of strontium (Lowenstam, 1964). Bivalves, gastropods and cephalopods are unique in having very low levels of this element present in their mineralized tissues (Lowenstam, 1964; Dodd, 1967; Gunatilaka, 1975). More recently, the latter author suggested that this may reflect an "evolutionary trend in the biochemical mechanisms of calcification and the concomitant development of the ability to reject certain ions from the growing skeleton" (Gunatilaka, 1981).

Paradoxically, although the higher molluscs seem particularly adept at maintaining low levels of this element in their shells and statoconia, strontium appears to have a profound affect on the initial mineralization of these tissues. Is strontium then a contaminant, a critical structural element or does its role change during ontogeny?

In this paper we report on our efforts to further characterize the effect of strontium on embryonic mineralization in molluscs. Experiments described here were conducted with the bivalves Mercenaria *mercenaria* (Linne, 1758)(Veneridae) and *Bankia gouldi* (Bartsch, 1908)(Teredinidae). Both species are oviparous broadcast spawners and development is external. During the embryonic period, which is non-feeding and complete within 40-45 hours (20° C), the animals proceed through a trochophore stage to the formation of the embryonic shell, prodissoconch I (Carriker, 1961; Culliney, 1975).

The effects of varying the strontium concentration on the embryonic development of Mercenaria *mercenaria* and *Bankia gouldi* are described and compared to those results reported for *Aplysia californica* (Bidwell *et al.*, 1986). The effects of high strontium concentrations on development are also included. Ion substitution studies with barium and a number of other metal ions are described. Experiments designed to characterize the critical period of exposure to strontium along with a discussion of some developmental events that occur during this critical window are presented.

MATERIALS AND METHODS

Adult Mercenaria *mercenaria* were collected locally between the months of May and July when the ambient seawater temperature was 18-20° C. *Bankia gouldi* was originally collected in wooden panels from Beaufort, N.C. and has been in laboratory culture for eight years. Both species were induced to spawn by exposure to elevated temperature (28° C). Once spawning had commenced, individual *M. mercenaria* were isolated in trays (20° C) to facilitate collection of gametes. Eggs and sperm released by *B. gouldi* were manually pipetted into separate containers. Fertilization was accomplished by addition of sperm to eggs (100:1) after which the zygotes were immediately rinsed and placed in basal medium (see below). Fertilized eggs were distributed into petri dishes in 50 ml of the test medium at a density of 100 animals/ml and incubated at 20± 0.5° C under continuous light.

The defined basal medium was prepared identically to that described by Bidwell *et al.* (1986) and contained the compounds given in Table I. The concentration of strontium was varied through additions from a stock solution of $SrCl_2 \cdot 6H_2O$.

Dose-response experiments were conducted to characterize the effect of varying the strontium concentration on embryonic development. Strontium is present in natural seawater at a concentration of 90 μmol/kg (approximately 8 ppm) at a salinity of 35 ppt (Bruland, 1983). Concentrations ranging from 0 to 9.07×10^{-5} M or approximately 0 to 8 ppm strontium in increments of 1 ppm were tested. The effects of greater amounts of this element were also observed at approximately 16, 24, 32, 40 and 80 ppm. Natural seawater (NSW) filtered to 1 μm was included in all experiments as a control and all treatments were replicated.

To test whether strontium may be substituting for another ion found in natural seawater, but not present in our artificial seawater, a more comprehensive formulation was devised. The enrichment compounds listed in Table 1 were added to the basal medium; concentrations reflect those reported for natural seawater (Bruland, 1983).

TABLE 1. Composition of artificial seawater media (salinity: 34 ppt, pH: 8.0-8.2)

COMPOUND	CONCENTRATION (g/L)
BASAL MEDIUM	
NaCl	23.88
$MgCl_2 \cdot 6H_2O$	10.68
Na_2SO_4	4.01
$CaCl_2 \cdot 2H_2O$	1.51
KCl	0.725
$NaHCO_3$	0.196
NaBr	0.086
[1]ENRICHMENT COMPOUNDS	
H_3BO_3	25.72 mg
$NaNO_3$	2.55 mg
LiCl	1.06 mg
Na4EDTA	0.35 mg
$NaH_2PO_4 \cdot H_2O$	0.32 mg
[2]NaF	0.28 mg
RbCl	0.17 mg
$NaIO_3$	79.16 μg
$Na_2MoO_4 \cdot 2H_2O$	26.61 μg
$Al_2(SO_4)_3 \cdot 18H_2O$	6.66 μg
$VOSO_4 \cdot nH_2O$	4.89 μg
$NiCl_2 \cdot 6H_2O$	1.90 μg
$ZnSO_4 \cdot 7H_2O$	1.73 μg
$CuSO_4 \cdot 5H_2O$	1.00 μg
$Na_2CrO_4 \cdot 4H_2O$	0.94 μg
$FeCl_3 \cdot 6H_2O$	0.27 μg
$MnSO_4 \cdot H_2O$	84.51 ng
$CoCl_2 \cdot 6H_2O$	4.76 ng

[1]Enrichment compounds added to ingredients of basaal medium to a volume of 1 l.

[2]Sodium fluroide oncentration is 10 times oceanic value.

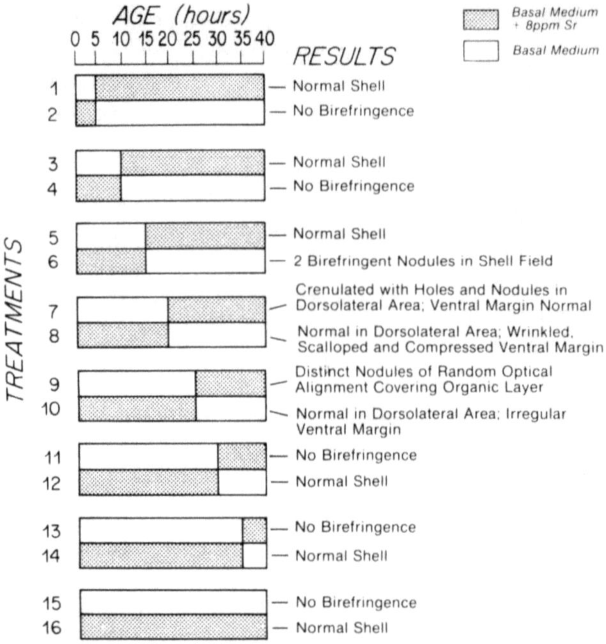

FIGURE 1. Graphic representation and results of 16 treatments from first pulse experiment using embryos of both *Mercenaria mercenaria* and *Bankia gouldi*. Note the reciprocal position of deformed shell in Treatments 7 and 8.

In another experiment barium chloride ($BaCl_2 \cdot 2H_2O$) was added to the basal medium at concentrations of 0.1, 0.2, 0.5 and 1.0 μM. Barium is a Group IIA element, directly below strontium in the periodic table. It is also a trace element of seawater, has a nutrient-type distribution and a concentration range of 32-150 nmol/kg with an average concentration of 100 nmol/kg (Bruland, 1983). *Bankia gouldi* was used for all ion substitution studies. Controls included basal medium, basal medium + 8 ppm strontium, basal medium + 8 ppm strontium + Na_4 EDTA (0.85 μM), and NSW.

We performed experiments to determine if there was a specific period during embryonic development when strontium was required and if deficits induced by its absence were reversible by subsequent exposure to this element. The first experiment involved the alternate exposure of embryos to media containing 0 and 8 ppm strontium. Sixteen petri dishes, plus replicates, were filled with test media, half with basal medium and the remaining dishes with basal medium + 8 ppm strontium. Fertilized eggs were distributed throughout all the dishes. At five hour intervals, embryos from one dish and its replicate were removed from their incubation medium, rinsed and placed into the medium of the opposite treatment, *e.g.*, five hours after fertilization, embryos from Treatment 1 were transferred from basal medium to basal medium + 8 ppm strontium and organisms from Treatment 2 were transferred from artificial seawater plus strontium to basal medium (Fig. 1). Controls included embryos incubated throughout development in basal medium, basal medium + 8 ppm strontium, and NSW.

To confirm the results of this experiment a second type of pulse experiment was undertaken in which embryos were exposed to strontium only during those hours common to treatments that resulted in normal larvae. Embryos were grown in basal medium throughout development except for the period from hours 15-30. At the beginning of this period, these animals were transferred to basal medium + 8 ppm strontium or NSW (Fig. 2). The reverse protocol was also tested by exposing embryos to strontium throughout development *except* during the critical period at which time they

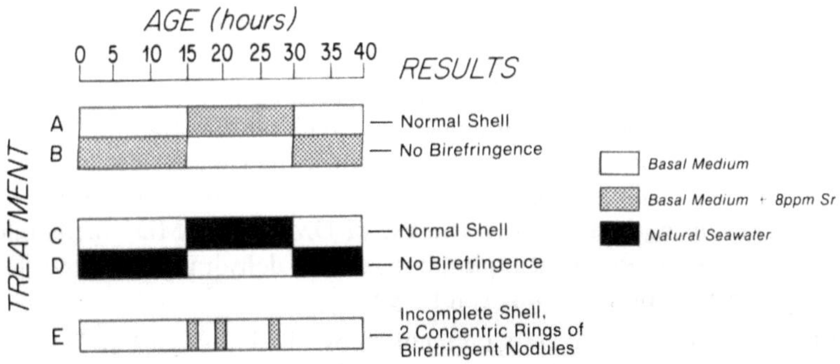

FIGURE 2. Graphic representation and results of treatments from second pulse experiment.

were transferred to basal medium. Also, a treatment was included in which the animals were pulsed with strontium intermittently throughout the critical period at hours 15-17, 19-21 and 25-27. Only Mercenaria *mercenaria* was tested in these experiments since *Bankia gouldi* had reacted identically in all previous treatments. Controls included organisms incubated in basal medium and basal medium + 8 ppm strontium throughout development.

Embryonic development was characterized using polarized light and scanning electron microscopy (SEM) on a minimum of 50 animals from each treatment to correlate developmental events with the critical period of strontium exposure. Observations were made every hour or every two hours on animals from dishes containing basal medium and basal medium + 8 ppm strontium. Initial mineralization was confirmed by loss of birefringence upon exposure to Na_2 EDTA $\cdot 2H_2O$ (10 mM).

Larval shell growth (prodissoconch II) of Mercenaria *mercenaria* was also measured in the presence and absence of strontium. Since larvae of *M. mercenaria* require phytoplankton for food, *Isochrysis* aff. *galbana* (clone T-ISO) was cultured in artificial seawater with and without additions of strontium using the enrichment medium f/2 (Guillard, 1975).

Embryos were reared in NSW and then transferred at the end of 48 hours into either basal medium or basal medium + 8 ppm strontium. Treatments were further divided into larvae fed T-ISO cultured in the presence or absence of strontium. Larvae reared in NSW and fed T-ISO grown in NSW plus f/2 nutrients served as controls. Initial larval density was 1/ml and culture media and algal food were replenished at 3 day intervals. Samples were taken during larval development and observed for statolith formation and then fixed in buffered glutaraldehyde (see below) for shell length measurements on a minimum of 30 individuals.

At the end of each experiment, or after each hour during the developmental sequence study, larvae were rinsed into test tubes with seawater followed by the addition of a few drops of glutaraldehyde (2% in 23 mM sodium cacodylate in seawater, pH 8.1) or narcotized by the stepwise addition of 7% $MgCl_2 \cdot 6H_2O$. Animals were then allowed to settle, the supernatant was removed and buffered glutaraldehyde added. After fixation was complete (15 min), larvae were transferred to a holding solution of 0.2% glutaraldehyde in buffered (2.3 mM cacodylate) seawater prior to analysis by polarized light. No decalcification was detected by loss of birefringence even after several weeks in the holding solution.

Embryos for SEM analysis were further processed by rinsing in 0.1 M cacodylate followed by 50% methanol (× 2) and dehydrated in either dimethoxypropane (DMP) (2 rinses, 5 min each) or in a graded series of acetone. Specimens were air dried for 30 minutes. Following incubation in hexamethyldisilazane (HMDS)(Nation, 1983) or critical point dried using liquid CO_2. The use of DMP and HMDS facilitated rapid processing of a large number of samples, but acetone dehydration with critical point drying proved superior for preserving soft tissues.

Dried larvae were mounted on aluminum stubs (double sided tape), coated with gold-palladium (approximately 400 Å) and viewed with a JEOL JSM-840 SEM at 15 KeV. Qualitative elemental analyses of *Bankia gouldi* embryonic shells and tissue were obtained with X-ray energy dispersive spectroscopy (EDS). Specimens were prepared for SEM as above except that the fixative was made up in basal medium rather than NSW. Dehydration and drying were in DMP and HMDS, respectively, and stubs were carbon-coated. Analyses were conducted with a Tracor Northern X-ray spectrometer at 15 KeV for 100 live seconds. Carbon-coated specimens were also used for back-scattered electron imaging of crystalline nodules not seen with conventional SEM.

Light micrographs of freshly fixed animals were taken with phase contrast and polarizing optics (Zeiss). A full-wave retardation compensator plate was used to examine orientation of crystal optical axes during shell formation. Birefringence of embryonic shells was quantified using a microphotometer. A larva was centered in the field of view and a 120 μm aperture was inserted into the light path to occlude transmitted light peripheral to the organism. Two orthogonal quarter-wave retardation plates were also used, one between the polarizer and specimen and the other between the specimen and analyzer. The resulting circularly polarized light removed the maltese cross extinction pattern normally associated with the radially symmetrical orientation of crystal axes of bivalve shells. This technique reduced the coefficient of variation due to specimen orientation to 2.1% (n=30). The photometer was electronically zeroed on individuals subsampled from each group to be measured, and after they had been exposed to a solution of 10 mM Na_2EDTA in 3% NaCl (minimum one hour). This corrected for the low level of birefringence associated with non-mineralized tissue. A minimum of 30 individuals were measured from each time point.

RESULTS

Dose response experiments

Embryos of Mercenaria *mercenaria* and *Bankia gouldi* responded similarly to varying strontium concentrations. Those grown in basal medium became larvae without mineralized shells. The extent of biomineralization appeared proportional to the amount of strontium up to 6 ppm.

The soft-tissue of animals lacking shells appeared normal. A bi-lobed organic layer, each lobe a developing valve of the prodissoconch I shell, covered the larva except the velum and mouth. This layer appeared wrinkled and contracted under SEM (Fig. 3a), but ballooned off the surface of swimming animals. Although the velum was unable to

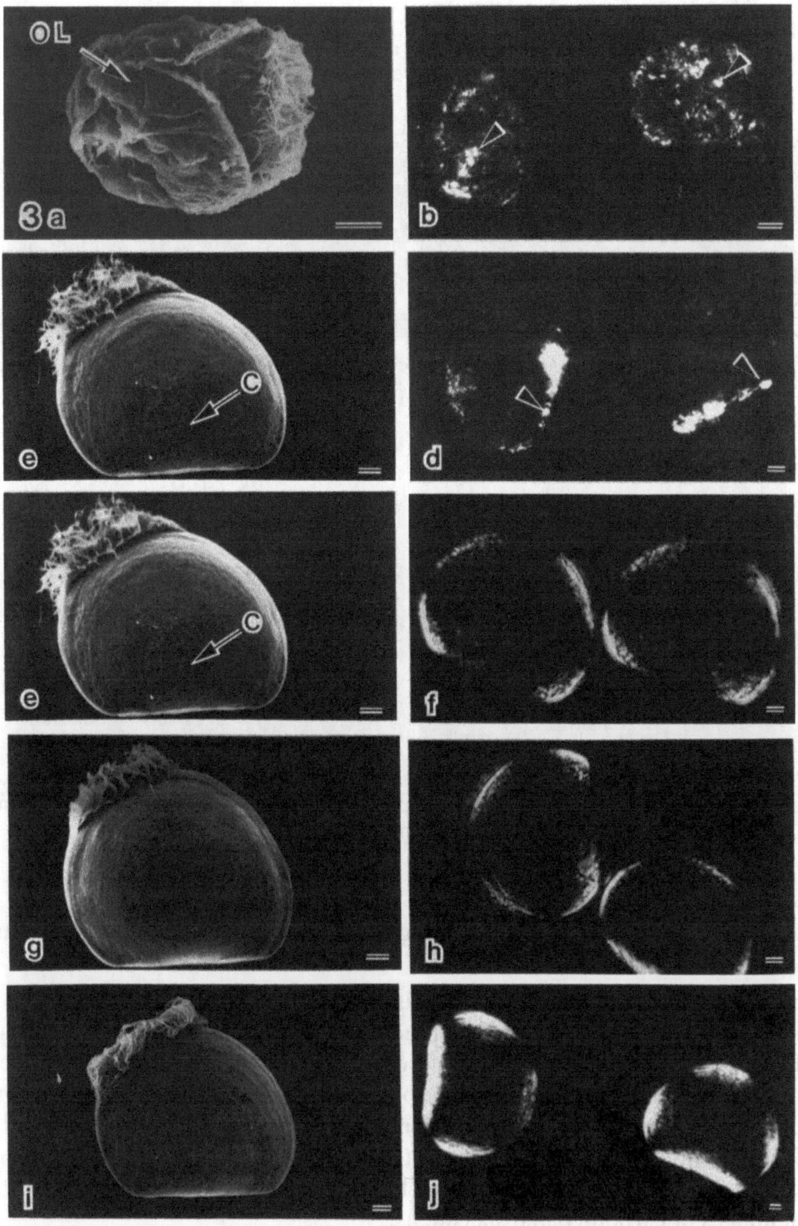

FIGURE 3. Scanning electron (left) and polarized light (right) micrographs of *Mercenaria mercenaria* larvae incubated at different strontium (Sr) concentrations throughout embryonic development: (*a*,*b*) 1 ppm; (*c*,*d*) 4 ppm; (*e*,*f*) 5 ppm; (*g*,*h*) 7 ppm; (*i*,*j*) natural seawater. AS, abnormal shell; C, crenulations; OL, organic layer; arrowheads indicate birefringent nodules of random optical orientation (scale = 10 μm).

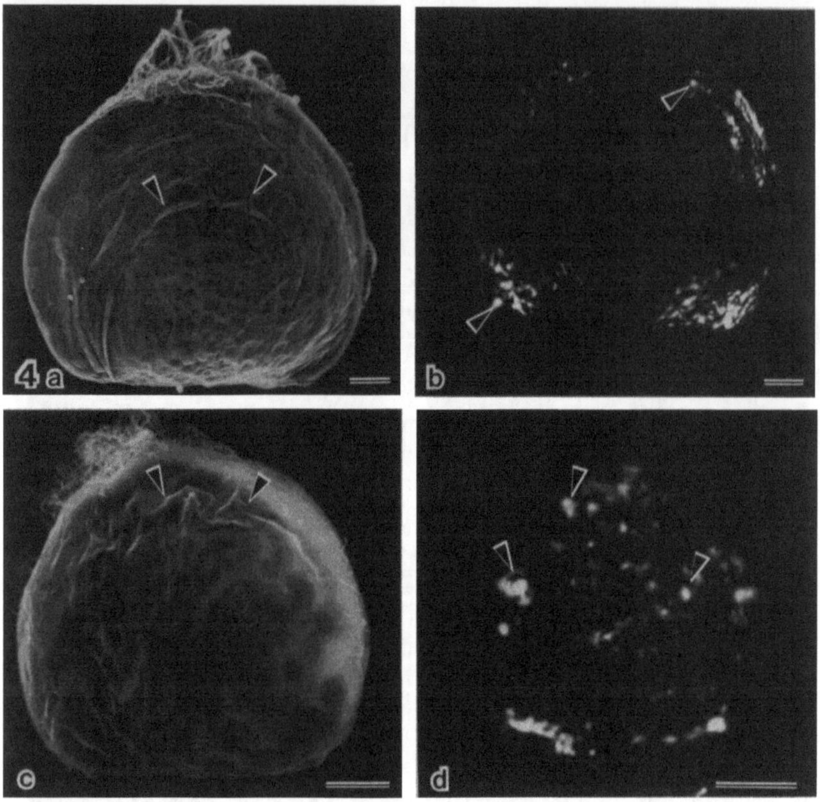

FIGURE 4. SEM (left) and POL (right) micrographs showing larvae of *Mercenaria mercenaria* (*a,b*) and *Bankia gouldi* (*c,d*) incubated in 80 ppm Sr throughout embryonic development: arrowheads indicate wrinkled shell (*a,c*) or birefringent nodules of random orientation (*b, d*) (scale = 10 μm).

contract into the uncalcified shell, swimming behavior was unaffected by the absence of mineralized valves.

Organisms cultured in media containing 1-3 ppm strontium appeared to have crystalline nodules scattered throughout the organic layer when viewed under polarized light (Fig. 3b). The number of nodules on each animal increased with increasing strontium concentration. Nodules were about 10 μm in diameter, were birefringent, and their optical axes were oriented randomly rather than being radially aligned as in normal shells. Birefringence was lost upon treatment with 10 mM Na_2EDTA. Nodules were not observed with conventional SEM (Fig. 3a), but could be seen in back-scattered electron images of carbon-coated specimens.

Animals reared in media containing 4 and 5 ppm strontium had fully D-shaped shells, however, those from the former treatment were highly crenulated, appeared granular under polarized light and had irregular ventral margins (Fig. 3c,d). Crenulations were mostly confined to the hinge and dorsolateral regions of the shell of organisms grown in basal medium + 5 ppm strontium (Fig. 3e,f).

Finally, embryos grown in basal medium containing 6-8 ppm strontium (Fig. 3g,h) were indistinguishable from those cultured in NSW (Fig. 3i,j). Shells were smooth, uniformly birefringent and did not contain nodules.

Increasing concentrations of strontium over a range of 2 to 10 times normal seawater also produced a graded incidence of shell deformities. Embryos incubated in 16 ppm strontium were indistinguishable from the 8 ppm treatment, but an increasing number of structural anomalies became evident between 24 and 80 ppm. Severe deformities in the 80 ppm strontium treatment included wrinkled, thin and weakly birefringent shells containing numerous nodules (Fig. 4a-d). Some of these organisms did not develop complete shells but had nodules resembling embryos from the 2 and 3 ppm treatments. Swimming behavior appeared normal in all treatments, as did the development of the soft tissue.

Qualitative EDS analyses of tissues in the vicinity of the organic layer of *Bankia gouldi* from the 0 ppm treatment failed to detect major levels of calcium (Fig. 5a). In contrast, mineralized shell from the 8 ppm treatment was mostly calcium ($CaCO_3$)(Fig. 5b). A weak strontium signal was observed only in shells from 80 ppm treatments (Fig. 5c). We are currently developing protocols for sensitive measurements of strontium levels using graphite furnace atomic absorption spectroscopy.

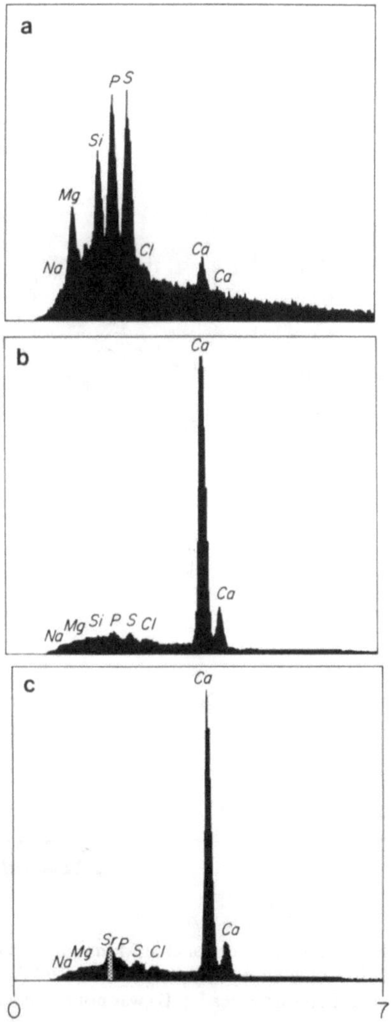

FIGURE 5. Typical X-Ray energy dispersive spectra of *Bankia gouldi* larvae cultured in various concentrations of Sr throughout embryonic development: (*a*) 0 ppm, probe was in area of the organic layer; (*b*) 8 ppm, probe was on normal shell; (*c*) 80 ppm, probe was on mineralized tissue. Prominent silicon peak in *a* from tissue absorbed HMDS.

Ion substitution experiments

Larvae of *Bankia gouldi* grown in basal medium plus additional enrichment compounds (Table I) throughout embryonic development were indistinguishable from those reared in basal medium. Only when barium chloride was added to the basal medium did any mineralization occur in the absence of strontium. At concentrations of 0.1 and 0.2 μM barium, birefringent nodules were observed on a few organisms, but most showed no signs of calcification. Nodules were not evident with conventional SEM (Fig. 6a) but were observed with polarized light and back-scattered electron imaging (6b). Higher concentrations of barium (0.5 and 1.0 μM, 5 and 10 times the average seawater concentra-

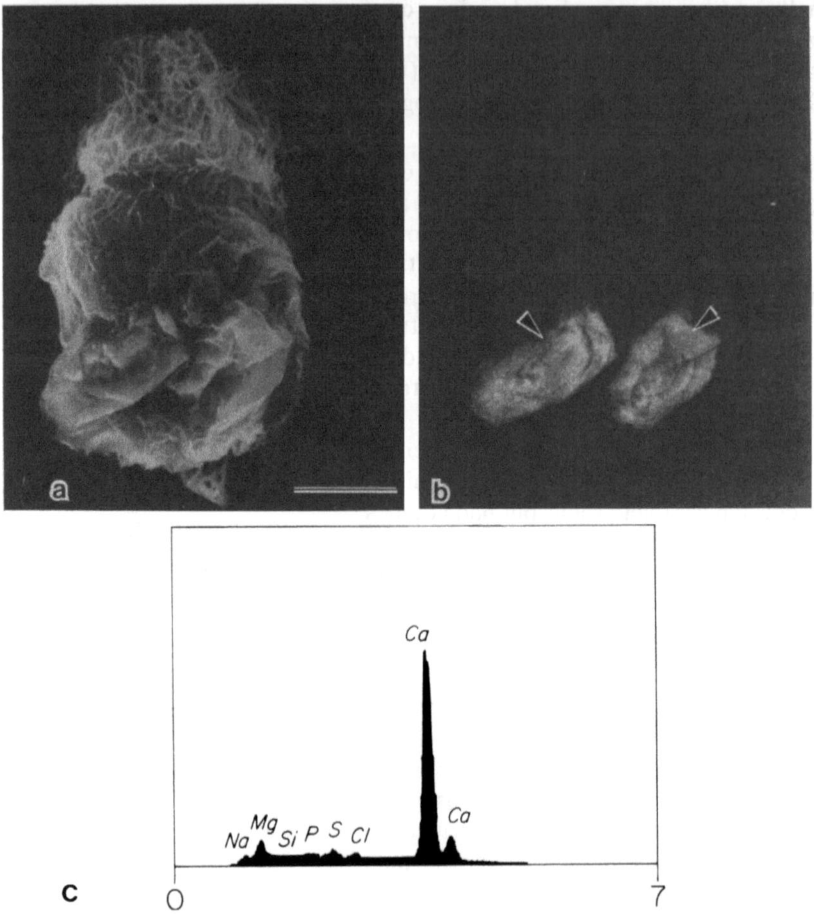

FIGURE 6. Conventional SEM (*a*) and backscattered electron image (*b*) of a *Bankia gouldi* larva incubated in basal medium + 0.2 μM barium (Ba) (scale = 10 μm). EDS spectra (*c*) was an analysis of the calcified areas in *b* (arrowheads). Ba was not detected.

tion) resulted in many animals with nodules, much like those observed from 1-2 ppm strontium treatments. EDS confirmed the presence of calcium in these areas (Fig. 6c).

Pulse experiments

The results of the first pulse experiment (Fig. 1) indicated that exposure to strontium during hours 15-30 produced normal larvae. Embryos grown in the presence of strontium for the first 10 hours of development and then transferred to basal medium failed to develop shells (Treatment 4). Exposure to strontium for only the first 15 hours resulted in most larvae containing two nodules, one in the dorsolateral region of each valve (Treatment 6).

No biomineralization occurred in regimens that called for exposure to strontium after hour 30 (Treatments 11 and 13). Birefringent nodules were observed in animals

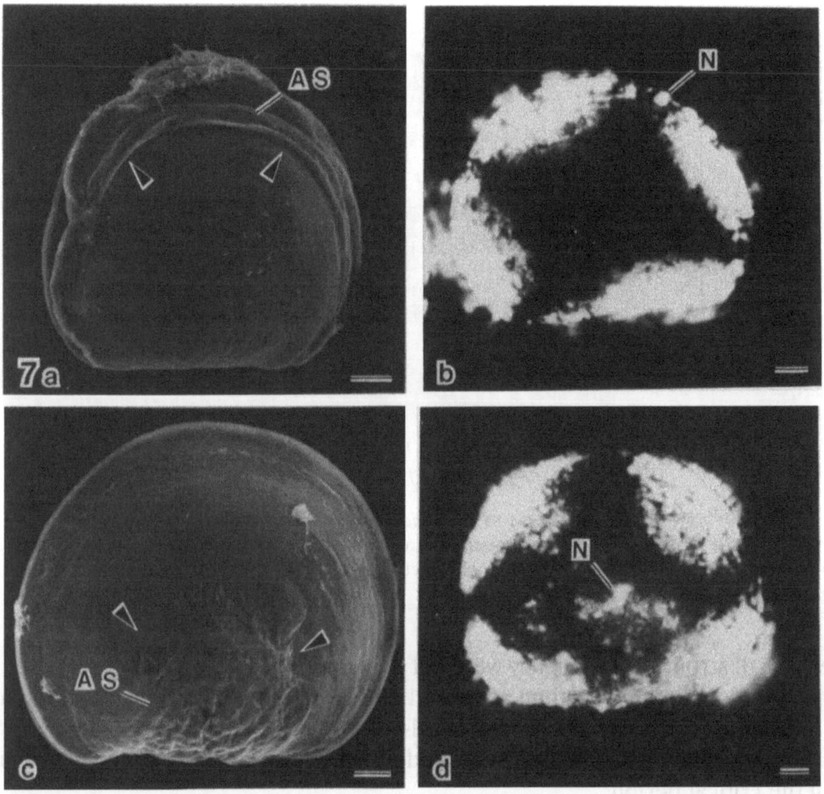

FIGURE 7. SEM (left) and POL (right) micrographs of *Mercenaria mercenaria* larvae from Treatments 7 and 8 (Fig. 1) in pulse experiment: (*a*, *b*) exposed to Sr for first 20 hours only, (*c*, *d*) exposed to Sr from hour 20-40. AS, abnormal shell; N, nodules; arrowheads indicate transtition from normal to abnormal shell (scale = 10 μm).

exposed to this element from hour 25 through the remainder of development (Treatment 9).

An obvious correlation between time of strontium exposure and the area of shell deficits was observed in Treatments 7 and 8 (Fig. 1). Embryos cultured in the presence of strontium for the first 20 hours of development (Treatment 8) became larvae with shells that were comparatively smooth in the region near the hinge line, yet the ventral margins were wrinkled, scalloped and granular (Fig. 7a,b). Conversely, animals exposed to strontium from hour 20 through the remainder of embryonic development were deformed near the hinge line, but were smooth near the ventral margins (Fig. 7c,d). Shells from both treatments appeared thin, however, when compared to those of normal larvae.

The second pulse experiment (Fig. 2) demonstrated that development in the presence of strontium from hour 15-30 only was sufficient for the formation of normal shells (Treatments A and C). The reverse regimen, *i.e.*, exposure to strontium except between hours 15 and 30, resulted in the absence of any mineralization (Treatments B and D).

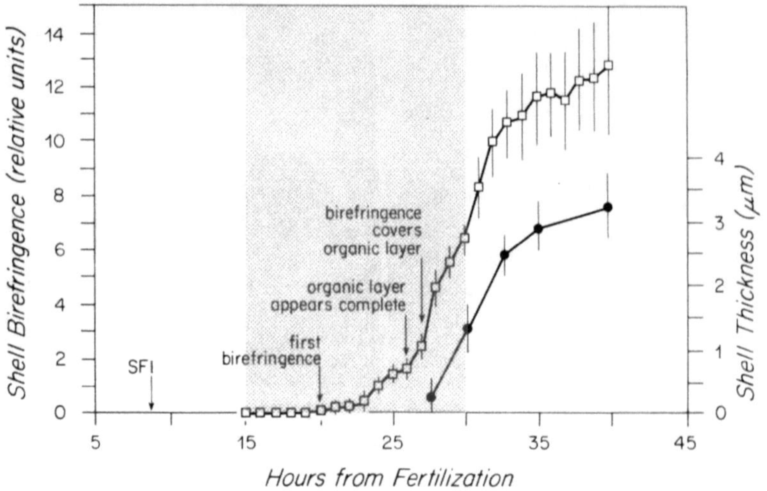

FIGURE 8. Developmental time course of *Mercenaria mercenaria* larvae illustrating increasing shell birefringence in the presence of 8 ppm Sr (open squares, mean ± 1 SD, n = 30). The critical window for Sr exposure is shaded. Thickness of embryonic shells fractured at the ventral margin was measured from SEM micrographs (dots, mean ± 1 SD, n > 3).

The effects of strontium deficiency were not reversible. Embryos that were pulsed with basal medium + 8 ppm strontium at hours 15-17, 19-21 and 25-27, only (Treatment E), developed shells with two concentric bands of nodules between normal shell. Shell formation appeared sensitive to strontium deprivation for as short a time as two hours during the critical period.

Embryonic development

Major features of embryonic shell formation in relation to the critical window were characterized for both Mercenaria *mercenaria* and *Bankia gouldi* using SEM and polarized light microscopy (Fig. 8,9). The sequence of developmental events were similar in both species, therefore, we report those for *M. mercenaria* only.

Gastrulation and the formation of the blastopore were complete five hours after fertilization. The shell field invagination (SFI) was first noted at 8 hours (Fig. 9a) and appeared as a narrow slit, 12 hours post-fertilization (Fig. 9b).

The SFI began to evaginate between 14 and 15 hours and the appearance of a recessed, organic layer (= organic shell of embryos of *Spisula solidissima* Dillwyn, 1817, Eyster and Morse, 1984) was noted at 18 hours (Fig. 9c). The trochophore stage was complete at this point. The SFI continued to evaginate onto the dorsal, posttrochal surface of the trochophore while the organic layer formed a saddle-like or bi-lobed structure, each lobe a valve of the developing shell (19 hours, Fig. 9d).

Birefringence was first noted at approximately 20 hours (Fig. 8). Three small patches (each about 10 μm in diameter) were observed, one in the dorsolateral area of each lobe of the organic layer (corresponding to the puffed regions in Fig. 9e) and a third in the region of the future hinge line. Mineralization was confirmed by the loss of birefringence upon exposure to 10 mM Na_2EDTA. The organic layer continued to spread ventrolaterally encompassing about two-thirds of the organism by 22 hours (Fig. 9f). Concomitantly, the areas of birefringence enlarged in all directions corresponding to the region of coarse wrinkling observed in this figure. The enlarging mineralized valves met dorsally and formed the hinge line at this time.

The organic layer appeared to be complete at 26 hours, but mineralized material did not cover the entire organism until about one hour later (see Fig. 9g and Fig. 8). We add parenthetically that nodule formation from exposure to 80 ppm strontium could be detected at 22 hours and no mineralization was observed throughout the course of development in animals from basal medium.

The intensity of birefringence continued to increase sharply between 27 and 33 hours and corresponded to an increase in shell thickness (Fig. 8). The ventral margin of shells between 24 and 28 hours of development (Fig. 9g,h) were thin and easily fractured during the mounting procedure for SEM. We noted little evidence of shell thickening after 35 hours.

Larval shell growth

Growth of the prodissoconch II shell of Mercenaria *mercenaria* was independent of strontium in the culture medium or algal food, with no statistical difference between treatments (ANOVA: $P<0.01$, $n=30$ per treatment). The majority of larvae attained maximum prodissoconch II shell length, formed two statoliths, exhibited normal swimming behavior and developed a propodium in preparation for metamorphosis in all treatments.

DISCUSSION

Our study shows that strontium is required in artificial seawater for normal embryonic shell formation of two species of bivalves. The profound effects of varying the strontium concentration on embryonic development of Mercenaria *mercenaria* and *Bankia gouldi* parallels those described for the opisthobranch *Aplysia californica* (Bidwell *et al.*, 1986). There is no evidence of embryonic mineralization in the absence of this element. Defects induced by low concentrations of strontium include patches or nodules of calcification throughout the organic layer or thin, wrinkled shells with irregular margins in all three species. The bivalves described in this work require levels of 6-8 ppm strontium for normal development whereas *A. californica* requires 4-8 ppm strontium. This slight difference may be more apparent than real since shell defects are easier to see on the flattened bivalve than on the whorled gastropod. Embryonic statolith formation requires strontium, *e.g., A. californica, Hermissenda crassicornis* (Bidwell *et al.*, 1986) and *Loligo pealei* and *L. vulgaris* (Hanlon *et al.*, in prep.), but post-embryonic statolith formation (*M. mercenaria*, this work) apparently does not.

This study demonstrates that higher than normal concentrations of strontium are also deleterious to embryonic shell development and result in deformed shells that resemble those from media containing low strontium, *i.e.*, 2-4 ppm. High external levels of this element may act as a general metabolic inhibitor interfering with other processes in addition to calcification but animals that develop in the presence of 80 ppm strontium appear to proceed through normal tissue development and exhibit normal swimming

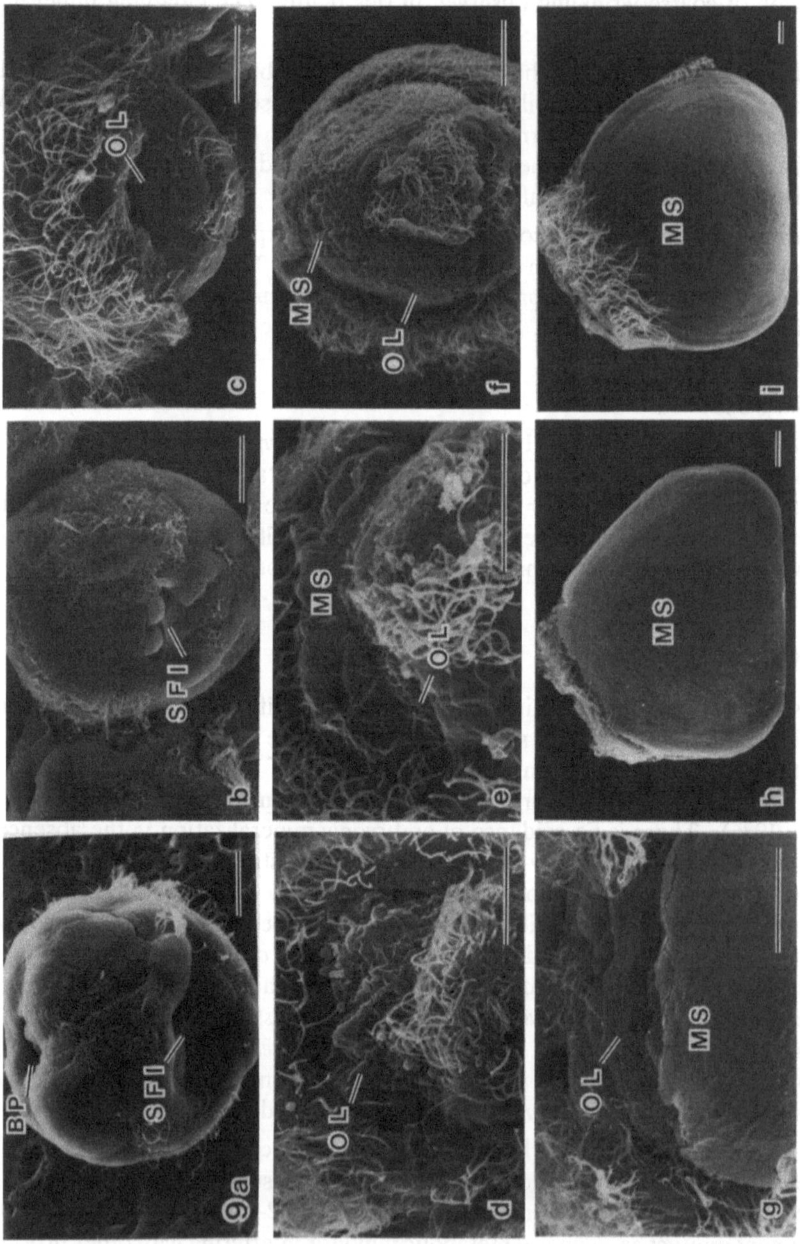

FIGURE 9. SEM's depicting stages of normal embryonic development of *Mercenaria mercenaria* (8 ppm Sr, 20°C, in hours (h) after fertilization): (*a*) 8h;(*b*) 12h; (*c*) 18h; (*d*) 19h; (*e*) 20h; (*f*) 22h; (*g*) 24h; (*h*) 28h; (*i*) 46h. BP, blastopore; MS, mineralized shell; OL, organic layer; SFI, shell field invagination (scale = 10 μm).

behavior. Discrimination against strontium incorporation into calcareous shell is lost at high concentrations of this metal ion (Dodd, 1967; Simkiss, 1983), and this may result in severe deformations of the crystal lattice. Detection of strontium in abnormal shells from the 80 ppm treatment supports this contention.

Even the enriched artificial seawater medium, which lacks strontium, is unable to induce normal shell development in embryos of *Bankia gouldi*. The implication is clear: strontium is not substituting for any other inorganic constituent of natural seawater. Results of the dose-response experiments also support our contention that strontium is the vital ion since all three organisms require at least half the concentration found in natural seawater and are sensitive to small changes in levels of this element below that concentration. That barium can partially substitute for strontium suggests that a particular chemical or physical characteristic of Group IIA elements may provide a clue to strontium's mechanistic role. Nevertheless, calcium and magnesium, also Group IIA elements and major constituents of seawater, apparently have no ability to replace this element's role in the process of shell mineralization.

We have yet to identify a direct correspondence between any one developmental event of both bivalve species and the 15 hour critical window of strontium exposure, but a number of processes are occurring at approximately the same time. Among these are formation of the organic layer, detection of first birefringence and mineralization of a thin coating of birefringent material over the animal. The presence of the organic layer at hour 18 and its completion at hour 26 can be observed with SEM, but without ultrastructural evidence (now underway) we cannot be certain when it begins to form. The critical window extends both 5 hours before and 10 hours after birefringence is first detected; this implies that strontium is required for some process both before and during the actual precipitation of calcium carbonate. Indeed, strontium is not required after 30 hours of development even though the embryonic shells are still thickening. It does not appear that strontium is essential for crystal growth after a layer of birefringent material has been deposited on the animal.

The absence of calcification in larvae grown in media without strontium argues for some role in crystal nucleation, while the appearance of birefringent nodules of random optical orientation in low strontium treatments suggests a role in crystal alignment. The results of the pulse experiments (Treatments 7 and 8, Fig. 1, and Fig. 2E) show that normal shell forms only in regions of the organic layer that have developed in the presence of strontium.

Perhaps strontium acts as mortar at the interface between an organic component and mineral portion of the embryonic shell. For example, La Barbera (1974) noted that when embryos of the bivalve *Tridacna squamosa* (Lamarck) were fixed at various stages of development, crystalline nodules appeared on the unmineralized portion of the organic layer. Interestingly, orientation of the nodules' optical axes were not random but identical to normal shell. Since La Barbera conducted his experiments in full strength seawater the organic layer of the embryos developed in the presence of strontium, which may have been responsible for the normal orientation of the crystals. Despite an extensive literature describing the organic and inorganic components of the molluscan shell (*e.g.*, Eyster, 1983; Wilbur & Saleuddin, 1983; Weiner, 1984) little is known about their structural relationships.

Aplysia californica also has a critical window of strontium exposure. This begins at day 4 of the 8-10 day embryonic period and lasts no longer than 24 hours (Bidwell *et al.*, 1986). It is unclear whether the similarity between the length of time required for exposure to strontium in Mercenaria *mercenaria* and *A. californica* are truly significant or a function of the resolution intrinsic in the experimental protocols, *i.e.*, sampling intervals of 5 hours *vs* 24 hours. Nevertheless, it is interesting to compare the juxtaposition of these windows against their total embryonic processes. As described above, *M. mercenaria* requires strontium approximately 5 hours before mineralization is observed and throughout the period required for a birefringent layer to cover the entire animal. Although we lack morphological details at this time, it is clear that the critical window of *A. californica* is over by the time the shell cap and statolith are observed. Also, the consequences of the absence of strontium during the critical periods for the bivalves and gastropod are different. No mineralization occurs if *M. mercenaria* is incubated in basal medium only during the critical window but is exposed to this element throughout the rest of embryonic development. In the gastropod shell however, deficits are localized solely to the whorl under the analogous regimen.

Could the strontium requirement indicate the transition from shell field to mantle calcification in mollusc embryos? If so, this would suggest a physiological homology between prodissoconch I of the bivalves and the shell cap, but not the entire protoconch I, of the gastropods. Calcification of prodissoconch I in the bivalves described here and the shell cap of *Aplysia californica* appear to require strontium, but prodissoconch II and the remainder of protoconch I do not. The transition between prodissoconch I/II and protoconch I/II are poorly defined and have been a matter of some recent debate (Waller, 1981; Jablonski & Lutz, 1983).

Nevertheless, comparison of the relationship between strontium and statolith formation in *Aplysia californica* and Mercenaria *mercenaria* suggests that the significance of the critical window may be more fundamental than a demarcation between shell field and mantle calcification. Two single statolith granules are formed in *A. californica* during embryonic development which appear on day 5, perhaps just prior to the end of the critical window (Bidwell *et al.*, 1986). The formation of these stones were very sensitive to the concentration of strontium in the culture medium. The two stones in *M. mercenaria*, however, are not formed until day 9 of *larval* development and do not require strontium for their formation. To our knowledge, embryonic shell and statolith formation are not morphogenetically related. Therefore, the *time* during development that mineralization occurs rather than the *mechanism* of deposition may be a clue to strontium's importance.

We cannot make any definitive conclusions about strontium's mechanistic role in embryonic mineralization at this time. The extensive literature on molluscan embryonic development (Raven, 1966; Kniprath, 1981; Eyster, 1983; Eyster & Morse, 1984, Eyster, 1986) provides few clues as to how this element fits into the scheme of mineralization. But the present work does indicate a number of areas to investigate further. The presence of strontium may be required during formation of the organic layer which precedes normal calcification of the shell. This does not address the question of the difference observed between *Aplysia californica* and Mercenaria *mercenaria* with respect to strontium and statolith development. Also, it is not clear whether strontium's role is primarily structural as an integral part of the organic layer/crystal lattice or biochemical, perhaps a cofactor to an enzyme involved in precipitation of calcium carbonate. We hope to answer some of these questions through further work with *A. californica* and *M. mercenaria*.

Finally, our results call for a reevaluation of strontium and its relationship to calcification processes. Its dismissal simply as a contaminant of calcareous tissues may be premature or misleading. Strontium's importance to mineralization may not be limited to the embryonic mollusc but may extend to later stages of development and to other phyla. Perhaps a more revealing correlation exists between embryonic requirement of strontium and evolutionary relationships, rather than Sr/Ca ratios of adult tissues as an indicator of phyletic position.

ACKNOWLEDGEMENTS

We thank Drs. James H. Schwartz, Eric Kandel (Columbia University) and the Howard Hughes Medical Institute for their continued support. One of us, S.M.G., acknowledges NOAA Sea Grant (No. NA86AA-D-SG090) for partial support. We thank Dr. Judith McDowell for many helpful discussions and Daniel L. Alkon (NIH) for his cooperation. Ms. Linda Lopez (JEOL USA, Peabody, MA) assisted with EDS analyses. Contribution No. 6354 from the Woods Hole Oceanographic Institution.

REFERENCES

ANDERSON, O.R., 1981. Radiolarian fine structure and silica deposition. *In* Silicon and Siliceous Structure in Biological Systems (ed. T.L. Simpson & B.E. Volvani), pp. 347-349. New York: Springer-Verlag.

BIDWELL, J.P., PAIGE, J.A. & KUZIRIAN, A.M., 1986. Effects of strontium on the embryonic development of *Aplysia californica*. Biol. Bull., 170:75-90.

BLACKWELDER, P.L., WEISS, R.E., & WILBUR, K.M., 1976. Effects of calcium, strontium and magnesium on the coccolithophorid *Cricosphaera* (Hymenomonas) *carterae*, I. Calcification. Mar. Biol., 34:11-16.

BRULAND, K.W., 1983. Trace elements in sea-water. *In* Chemical Oceanography (ed. J.P. Riley & G. Skirrow), pp. 157-220. London: Academic Press.

CARRIKER, M. R., 1961. Interrelation of functional morphology, behavior, and autecology in early stages of the bivalve *Mercenaria mercenaria*. J. Elisha Mitchell Sci. Soc., 177:168-242.

CULLINEY, J.L., 1975. Comparative larval development of the shipworms *Bankia gouldi* and *Teredo navalis*. Mar. Biol., 29:245-251.

DODD, J.R., 1967. Magnesium and strontium in calcareous skeletons: a review. J. Paleont., 41:1313-1329.

EYSTER, L.S., 1983. Ultrastructure of early embryonic shell formation in the opisthobranch gastropod *Aeolidia papillosa*. Biol. Bull., 165:394-408.

EYSTER, L.S., 1986. Shell inorganic composition and onset of shell mineralization during bivalve and gastropod embryogenesis. Biol. Bull., 170:211-231.

EYSTER, L.S. & MORSE, M.P., 1984. Early shell formation during molluscan embryogenesis, with new studies on the surf clam, *Spisula solidissima*. Amer. Zool., 24:871-882.

GUILLARD, R, 1975. Culture of phytoplankton for feeding marine invertebrates. *In* Culture of marine invertebrate animals (ed. W.L. Smith & M.H. Chanley), pp. 29-60. New York: Plenum Press.

GUNATILAKA, A., 1975. The chemical composition of some carbonate secreting marine organisms from Connemara Island. Proc. Roy. Irish Acad. Ser., B75:543-556.

GUNATILAKA, A., 1981. Biogeochemistry of strontium. *In* Handbook *of Stable Strontium* (ed. S.C. Skoryna), pp. 19-46. New York: Plenum Press.

JABLONSKI D. & LUTZ, R.A., 1983. Larval ecology of marine benthic invertebrates: paleobiological implications. Biol. Rev., 58:21-89.

KNIPRATH, E., 1981. Ontogeny of the molluscan shell field: a review. Zool. Scripta, 10:61-79.

LA BARBERA, M., 1974. Calcification of the first larval shell of *Tridacna squamosa* (Tridacnidae: Bivalvia). Mar. Biol., 25:233-238.

LOWENSTAM, H.A., 1964. Ca/Sr ratio of skeletal aragonites from the recent marine biota at Palau and from fossil gastropods. *In* Isotopic and Cosmic Chemistry (ed. H. Craig, S.L. Miller, & G.J. Wasserburg), pp.114-132. Amsterdam: North Hollan Publishing Co.

NATION, J.L., 1983. A new method using hexamethyldisilazane for preparation of soft insect tissues for scanning electron microscopy. Stain Techn., 58:347-351.

RAVEN, C.P., 1966. Morphogenesis: The Analysis of Molluscan Development. 366pp. New York: Pergamon Press.

SIMKISS, K., 1983. Trace elements as probes of biomineralization. *In* Biomineralization and Biological Metal Accumulation, Biological and Geological Perspectives,(ed. P. Westbroek & E.W. DeJong), pp. 363-371. Holland: Reidel, Dorcrecht.

WALLER, T.R., 1981. Functional morphology and development of veliger larvae of the European oyster, *Ostrea edulis* Linne. Smithsonian. Contr. Zool., 328:1-70.

WEINER, S., 1984. Organization of organic matrix components in mineralized tissues. Amer. Zool., 24:945-952.

WEISS, R.E., BLACKWELDER, P.L. & WILBUR, K.M., 1976. Effects of calcium, strontium and magnesium on the coccolithophorid *Cricsophaera* (Hymenomonas) *carterae*. II. Cell division. Mar. Biol., 34:17-22.

WILBUR, K.M. & SALEUDDIN A.S.M., 1983. Shell formation. *In* The Mollusca. vol. 4. Physiology Part I. (ed. A.S.M. Saleuddin & K.M. Wilbur), pp. 235-287. New York: Academic Press.

FORM AND FUNCTION OF CALCIUM CONCRETIONS IN UNIONIDS

Harold Silverman

Department of Zoology and Physiology, Louisiana State University, Baton Rouge, LA 70809

ABSTRACT. Unionid mussels brood hundreds of thousands of embryos to a fully-shelled glochidia stage larvae before releasing them to the freshwater environment. These embryos are developed within the enlarged water channels of brooding gills. The source of calcium for shell formation comes from the maternal animal, where calcium is stored as extracellular calcium phosphate concretions in the connective tissue of unionid gills as well as in the conspicuous adult shell. These concretions can compose up to 50% of the dry tissue weight of the gills in non-reproductive unionids. These concretions are mobilized during reproduction concomitant with shell production in the glochidia. Passage of 45Ca to the embryos has been demonstrated from maternal animals injected with radiolabel one to three months before embryo development begins. In addition, when gravid females are exposed to 1,000,000 CPM 45Ca over a 4 h period, either injected into the blood or placed directly into the external bath, uptake of label by glochidia is directly related to the amount of radiolabel present in maternal blood, and shows no relationship to the amount of label in the external bath. Conversely, glochidia isolated from the water channel and placed in labeled bath, accumulate roughly 10 times the observed label incorporated while in the gill. This value is similar to adsorption seen on mature shell. Thus calcium contribution is from maternal sources. These data indicate the water channel is a physiologically closed brood chamber adapted to provide a protective environment for developing embryos. Morphological evidence indicates the water channel becomes three separate compartments during brooding in Anodonta, such that water does not have to pass directly over the embryos for respiratory purposes. The calcium concretions themselves are rapidly mobilized during the reproductive period of the year suggesting tight neurohormonal control of mobilization. The organic matrix of the concretions contains at least one calcium-binding protein with molecular weight and calcium binding activity similar to calmodulin. This protein cross-reacts with calmodulin antibody and is the only soluble organic material thus far isolated with calcium binding properties.

INTRODUCTION

This paper reviews our knowledge of a mineralizing system found in unionid mussels which may have played a critical role in the ability of this group to radiate and inhabit a wide variety of freshwater environments. Unionidsbivalve;unionids are found on all continents, in streams, in shallow ponds, as well as the large freshwater lakes. Unionidsbivalve;unionids undergo a unique internal brooding of embryos to fully-shelled glochidia which are then released into the environment. In virtually all cases, glochidia become parasitic on fish before becoming juvenile mussels (see Kat, 1984). Most hypotheses suggest that the mussels evolved this strategy to keep the embryos out of the water current and for dispersal. Alternatively, internal brooding of the embryos may well have evolved due to the selective advantage of receiving nutrients from the

Origin, Evolution, and Modern Aspects of Biomineralization in Plants and Animals
Edited by Rex E. Crick
Plenum Press, New York

TABLE 1. Calcium concretion content in several freshwater unionid mussels

SPECIES	CONCRETION[1]
Anodonta grandis	56.2± 0.8 (4)
Elliptio crassidens	28.5± 2.4 (4)
Ligumia subrostrata	24.6± 1.2 (29)
Carunculina texasensis	18.9± 1.2 (12)

[1]Content in % dry wt/dry wt. of gill
(From Silverman *et al.*, 1985)

maternal animal, and the evolution of a set of extracellular calcium phosphate concretions in the gills of the animal tend to support this hypothesis. If it could be shown that these concretions evolved early in the unionid lineage, and that these concretions actually serve as a source of calcium for embryonic shell development, internal brooding as a modified nutrient environment would have to be considered a major evolutionary event allowing the radiation of unionidsbivalve;unionids into calcium-poor freshwater environments. Such a hypothesis actually parallels much of the reasoning associated with current theories on the evolution of the higher vertebrate egg and uterus. The remainder of this paper is devoted to a description of the calcium concretions found in the freshwater unionidsbivalve;unionids, and the evidence for the passage of maternal calcium to the brooded embryos for shell production.

LOCATION, QUANTITY, AND COMPOSITION OF CONCRETIONS

The calcium concretions described here are located extracellularly in the connective tissue of unionid gills. All unionidsbivalve;unionids examined contained calcium concretions bivalve;unionids;gill concretionswhich made up between 20−60% of the dry weight of the gill tissue (Silverman et al., 1985) (Table 1). The majority of the concretions are located between the gill filaments and the water channel epithelium, usually surrounding nerve/muscle tracts (Silverman et al., 1983) (Fig. 1). During the reproductive period of the year, concretions are mobilized or disappear from the gill as determined morphologically (Silverman et al., 1985). This can be visualized due to the brownish pigmentation of the concretions which actually causes coloration of the entire gill. This coloration disappears during reproduction, and is caused by the disappearance of the concretions surrounding the nerve/muscle tracts (Fig. 2). Using soft X-rays, we have confirmed that the appearance changes seen in the gill are caused by the loss of hard mineralized material (Silverman et al., 1985) (Fig. 3).

TABLE 2. Composition of concretions in unionid mussels (adapted from Silverman *et al.*, 1983)

Concretions	g/100 g dry gill	56.2± 0.8
450° Volatile Organic	g/100 g concretion	25.0± 2
Calcium	g/100 g concretion	25.3± 1.5
Bicarbonate	g/100 g concretion	3.4± 0.4
Phosphorus	g/100 g concretion	13.1± 0.4
		36.0−40.0[1]
Manganese	g/100 g concretion	8.6± 0.3
Iron	g/100 g concretion	0.5± 0.1
Magnesium	g/100 g concretion	0.2± 0.0
TOTAL	g/100 g concretion	99.0−103.0

[1]Calculated as PO4 or P2O7

The extracellular concretions range in size from 0.1−150 μm (Fig. 1c). At the ultrastructural level, they have a concentric appearance usually showing an electron dense core (Fig. 3). The larger concretions are composed of fused single concretions and can have many varied shapes (Figs. 1c & 4), and are usually surrounded by an electron dense covering material.

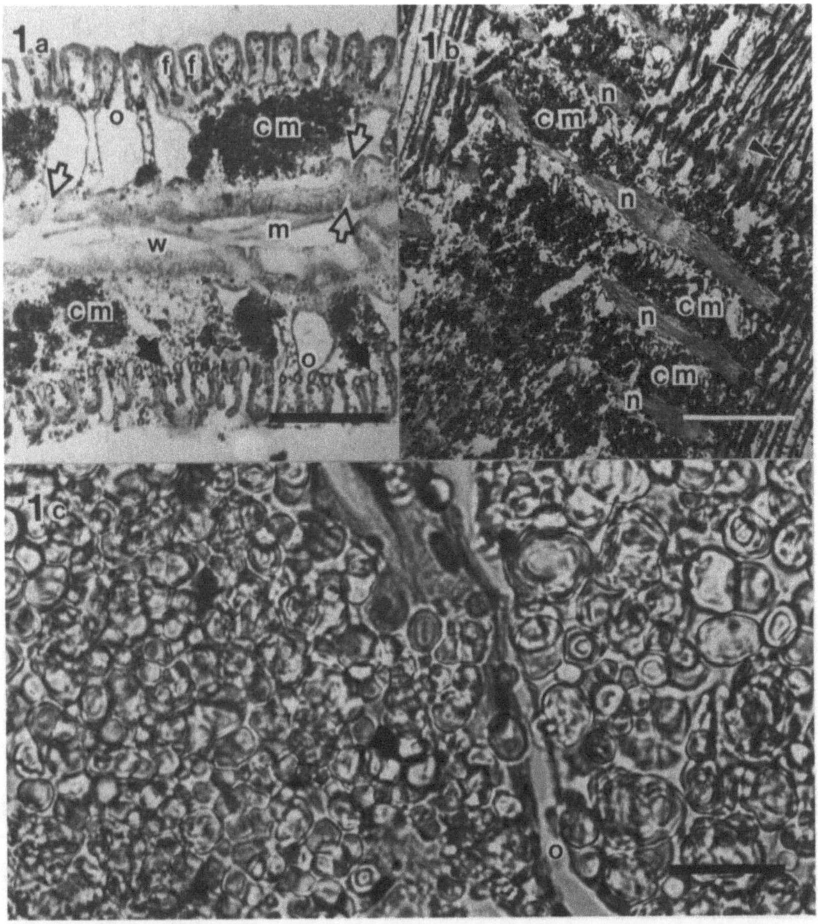

FIGURE 1. Sections of unionid gills showing the distribution of Ca concretions and the structure of the water channel in non-reproductive gills. ▪a. Cross-section of an *Anodonta grandis* gill. The gill filament (f) organization on each side of the gill can be seen with chitinous rods (filled arrows) backing the filaments. Water canals (o) pass between the filaments and the water channel (w) and constitute a continuous passage for the flow of water opening at ostia on the filament side. Connections of the canals to the water channel

The concretions are composed largely of calcium phosphate, which appears to be amorphous based on our inability to generate electron diffraction patterns. The composition of concretions, isolated from Ligumia subrostrata, as determined chemically is shown in Table 2. Energy dispersive analysis using X-rays (EDAX) analysis has also been done on isolated concretions from Anodonta grandis (Silverman et al., 1983; Silverman et al., 1987b), although these analyses are not as sensitive as the chemical methods employed. Roughly 75% of the concretion is accounted for by calcium phosphate with the remaining components being largely organic, only 3% of which can be accounted for by bicarbonate. Of the divalent cations other than calcium, only iron and manganese are present in quantities greater than 0.5% by weight in concretions

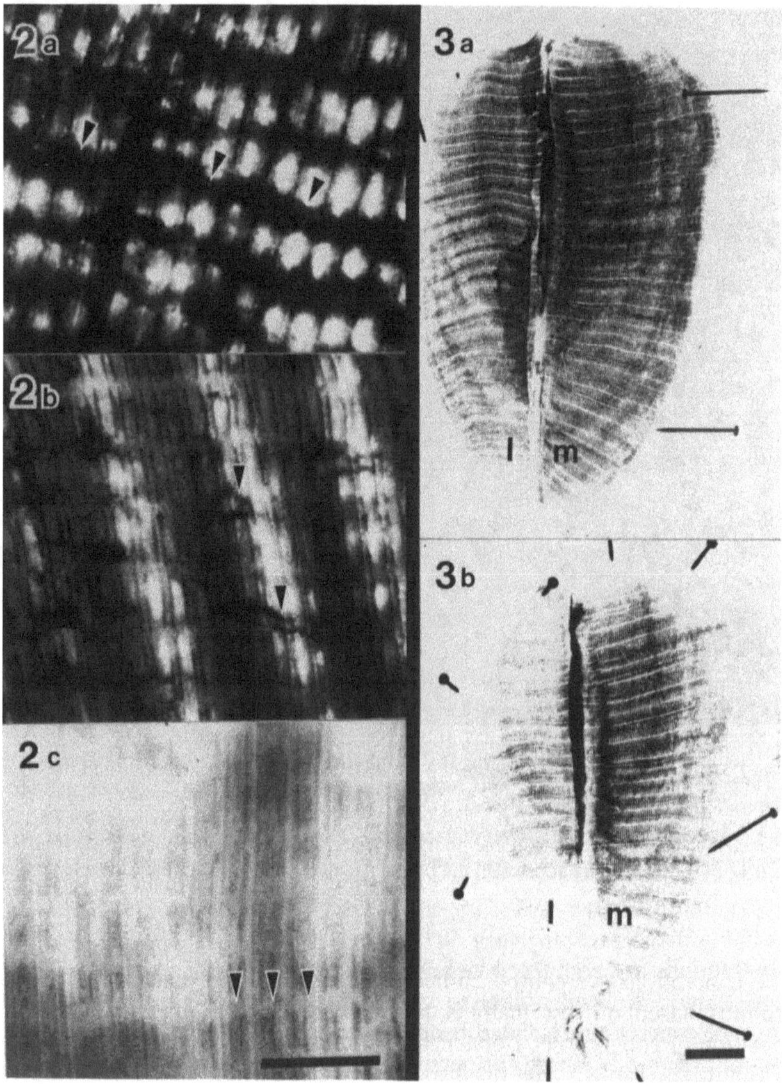
2a
2b
2c
3a
l
m
3b
l
m

isolated from animals living in unpolluted ponds (Silverman et al., 1987b). The concretions do not mimic blood or water content in terms of their divalent metal content (Table 3), nor is it easy to alter divalent metal ion content by experimental exposure to trace metals in vivo. Placing animals in pond water seeded with enough trace metals to elevate blood levels of the metal 2 or 3 fold does not cause a significant increase in trace metal content in isolated concretions (Table 4) (Silverman et al., 1987b). Even manganese, which appears as 8% of the concretion by weight normally, can not be elevated by raising manganese concentrations in pond-water for short periods of time. Further in an attempt to demonstrate partitioning of a heavy metal 65Zn was loaded into a few animals and allowed to partition in those animals for several hours (Silverman et al., 1987b). This is sufficient time for a 45Ca label to partition and be strongly bound to the concretions. The results of the partitioning experiments are shown in (Table 5). Note that while calcium is strongly

TABLE 3. *Anodonta grandis* blood, pondwater, and gill concretion divalent cation content. Means± SE (n)

	BLOOD[1]	PONDWATER[1]	CONCRETION[2]
Ca	4.40± 0.3 (3)	0.041± 0.015 (4)	21.3± 2 (6)
Mn	0.12± 0.01 (4)	0.057± 0.020 (4)	8.6± 0.3 (4)
Fe	0.03± 0.01 (3)	0.009± 0.007 (5)	0.5± 0.0 (4)
Mg	0.19± 0.02 (3)	0.047± 0.008 (5)	0.19± 0.01 (8)
Zn	0.005± 0.000 (11)	0.0009* (3)	0.004± 0.000 (6)

[1]mM
[2]Percent dry weight
*Value is at the limit of detection. (From Silverman *et al.*, 1987b)

TABLE 4. Loading of heavy metals into the blood and gill concretions of *Anodonta grandis* exposed to elevated metals in artificial pondwater. Blood values are expressed as mM± SE (n), concretion values are expressed as % dry wt.

BLOOD	CONTROL	METAL-
Zn[1]	0.004± 0.0004 (5)	0.013± 0.002 (5)*
Cd[2]	undetected**	0.012± 0.002 (3)
Mn[3]	0.12± 0.01 (4)	0.28± 0.05*
Concretion		
Zn	0.044± 0.03 (4)	0.085± 0.02 (6)
Cd	undetected**	0.05± 0.025 (3)
Mn	8.65± 0.25 (4)	8.98± 0.5 (4)

[1]Zn present in pondwater at 65 ppm as ZnSO4 for 1 wk.
[2]Cd present in pondwater at 50 ppm as CdCl2 for 1 wk.
[3]Mn present in pondwater at 100 ppm as MnCl2 for 1 wk.
*Significantly higher than control values P<0.05.
**Limits of Cd detection: Blood < 0.001 mM; concretions < 0.004% Cd by wt.
(From Silverman *et al.*, 1987b)

FIGURE 2–3. ■2. Whole mount micrographs of the distal tip of *Anodonta grandis* lateral gills. ■a. Non-reproductive. ■b. Early in the reproductive period. ■c. During glochidia development. Dense lines running horizontal in (■a) are caused by concretions masses surrounding nerve/muscle tracts (see Fig. 1c). Filaments are oriented in the nearly vertical plane. In (■b) and (■c) concretion masses have disappeared and in (■b) residual concretions are noted at the arrows and in (■c) an area where concretions should be present is noted by the arrows. Bar=430 μm. Adapted from Silverman *et al.* (1985). ■3. Roentgenogram of whole *Anodonta grandis* gills. ■a. Gill from non-reproductive animal showing X-ray dense material lined up vertically every 100 μm. Gill filaments are arranged horizontally. X-ray dense material corresponds to the concretion masses seen every 100 μm in the light micrographic studies. Lateral (l) and medial (m) gills are identified. During the reproductive period, the lateral gill loses more concretions initially as documented in (■b). The epithelial surface, from one side of the gill, lying immediately to the left of (l) and to the junction of the medial gill was removed, and the electron lucent image in the central portion of the lateral gill is not the result of the concretion masses disappearing during reproduction, rather it is the result of experimental removal of the concretion masses on one side of the gill. Bar=0.85 cm. Adapted from Silverman *et al.* (1985).

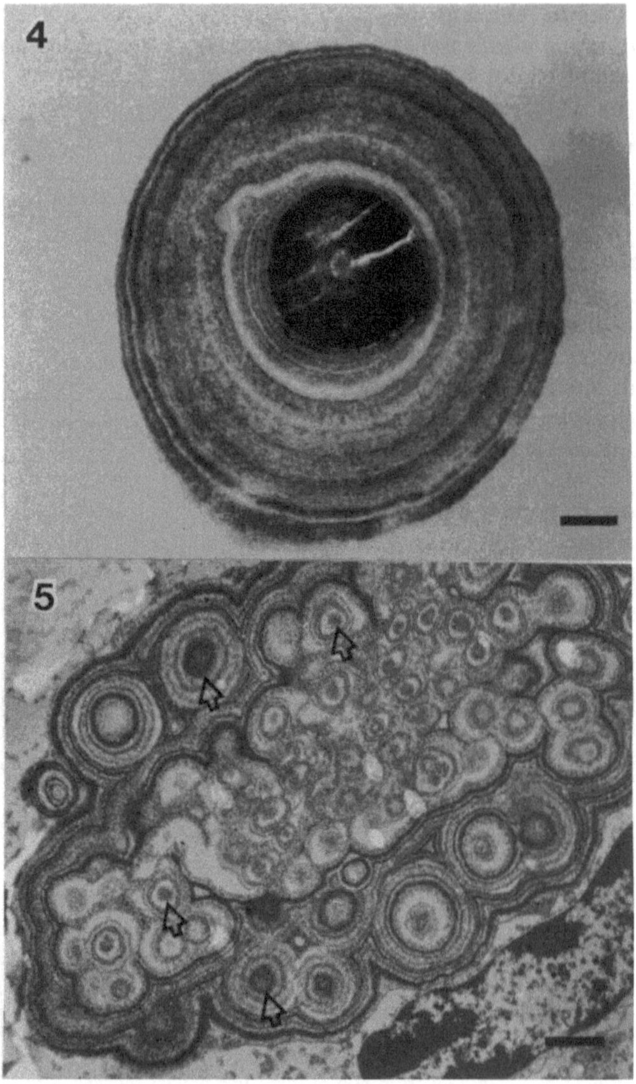
4
5

bound to the concretions, zinc is distributed in the tissue fluid to a greater extent than it is bound to the concretions; compare the ratios of soft gill tissue/concretion label for 65Zn and 45Ca. Such experiments appear to demonstrate that the mineral component of the concretions is very tightly regulated in vivo. Further support for this comes from labelling experiments done in vitro on isolated concretions. The concretions bind both zinc and calcium rapidly and equally well (Fig. 6) (Silverman et al., 1987b). The concretions are not just a divalent cation sink, but carefully regulated in their mineral composition. This fits well with the proposed function of a calcium store for release during embryonic shell production. If concretions were indiscriminantly concentrating toxic divalent cations, these cations would all be released during the reproductive period. Such action would most likely be disadvantageous to both the maternal animal and the developing embryos.

TABLE 5. Uptake of ^{65}Zn and ^{45}Ca by *Ligumia subrostrata* tissues and concretions following a single injection of 1.5×10^6 CPM. Values are expressed as CPM/mg dry wt. ± SE, n=4

TISSUE	^{65}ZN CPM	^{45}CA CPM
Gonad	5046 ± 1020	1518 ± 219
Mantle	3460 ± 1146	1156 ± 303
Kidney	2999 ± 983	1942 ± 619
Gill	4270 ± 1331	1642 ± 424
Gill Concretions	1941 ± 393	6938 ± 2445
RATIO OF CPMS		
Isolated Concretion/ Gill Tissue	0.525 ± 0.075	4.63 ± 1.11*

*The ^{45}Ca ratio is significantly higher than the ^{65}Zn ratio, $P<0.02$ (T-test, following arc sine transformation). (From Silverman *et al.*, 1987b)

Not nearly as much information is available on the organic matrix component of the concretions. Periodic acid-Schiff and silver methineamine histochemistry indicate that there is oxidizable polysaccharide present (Silverman et al., 1983). The protein component of the concretions is under more intense investigation, although the data to be briefly described have not yet been published and original

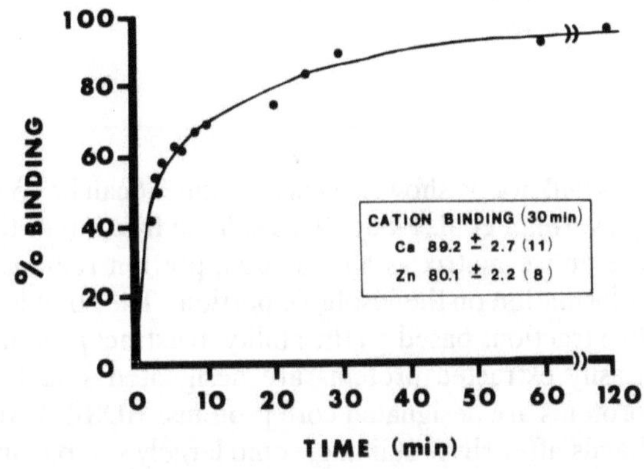

FIGURE 6. Binding of ^{45}Ca by isolated concretions. One ml of 5 mM Ca containing 5×10^4 CPM ^{45}Ca was used as the incubation medium. Maximum binding of Ca, expressed as percent bound, equals 485 ± 40 μm/mg concretion (n=10). From Silverman et al. (1987).

FIGURE 4–5. ▪4. TEM of a single isolated concretion after sectioning. Note the electron dense core and the lamellar structure of the peripheral layers. Bar=0.7 μm. Adapted from Silverman *et al.* (1983). ▪5. TEM of a multiple concretion granule in thin section. Note the large size and extent of aggregation shown by the single concretion unit. Note large number of cores (some noted by arrows). Entire unit is surrounded by its own delimiting lamella. Bar=1.25 μm.

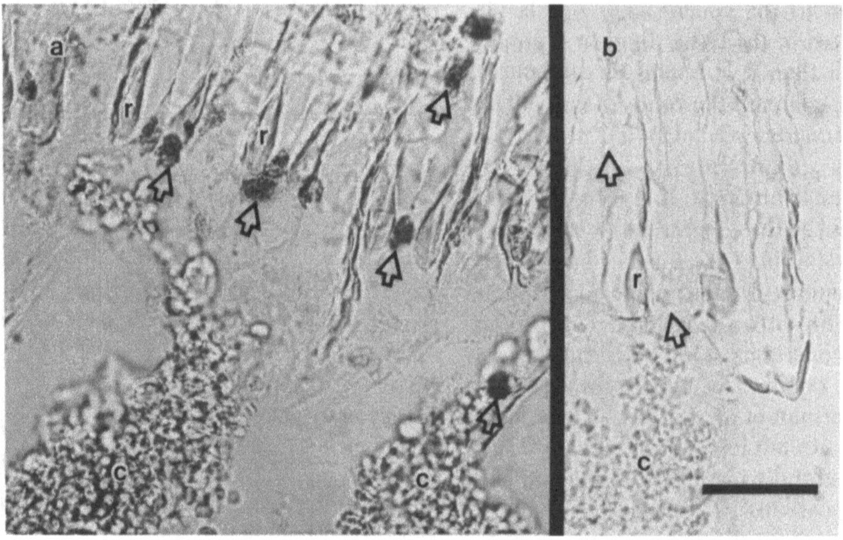

FIGURE 7. Immunohistochemical localization of a calmodulin-like protein (17,000 D) in *Ligumia subrostrata* gill using a calmodulin antibody. ▪a. Peroxidase reaction product distributed diffusely among the concretion masses (c), and highly localized on concretion-forming cells (arrows). These cells tend to be localized among the concretion masses, and also associated with chitinous rods (r) and in the middle of the gill filaments. μb. Control reaction in which the peroxidase antibody has been omitted as a control. Arrows denote areas where concretion-forming cells should be located. ▪c. Concretion masses are visible they exhibit no reaction product. Bar = 110 μm.

gels will not be shown. This component can be solubilized using mercaptoethanol/SDS (Silverman et al.,1986). This soluble fraction of the organic matrix does not constitute the entire matrix as an insoluble portion remains after such treatment. We have no information on the insoluble portion. The soluble portion has been further divided into two fractions based on the ability to extract proteins from the concretions using EDTA. Easily extracted proteins are designated soluble, henceforth, while EDTA-resistant proteins are designated core proteins. SDS-PAGE gels of these proteins reveal 12–15 bands after silver staining, some largely soluble and some largely core proteins.

These proteins were further tested for their ability to bind calcium using initially the dot blot procedures of Maruyama et al. (1984) , and subsequently, by transferring gels to nitrocellulose paper and exposing the paper to 45Ca. The resulting autoradiographs indicate that one protein, MW 17,000D, is a calcium-binding protein. This 17kD protein cross-reacts with a sheep anti-bovine brain calmodulin antibody using Western blot techniques. Further, polyclonal antibodies prepared against concretion protein in both rabbits and mice (Steffens et al., 1985), recognize an epitope of this molecule. The rabbit anti-concretion preparation is particularly useful because only a single epitope is present. All three of the antibodies described above recognize the 17kD protein found in the concretions. This protein is easily solubilized in EDTA and is therefore a portion of the soluble protein fraction.

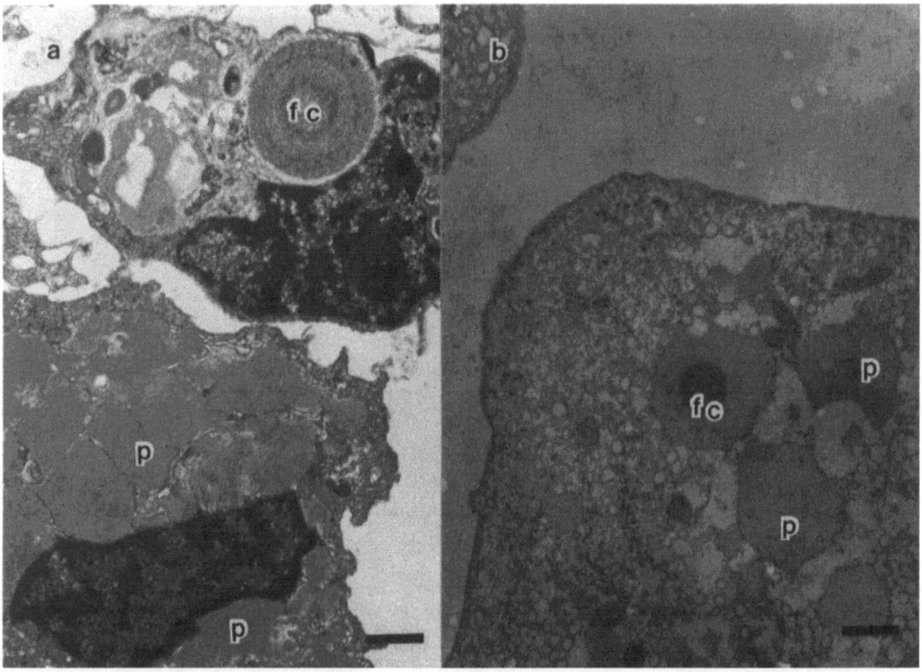

FIGURE 8. TEM of concretion-forming cells. ▪a. Two stages of concretion formation. A cell containing large number of vesicles (**p**) filled with an organic matrix material is seen in the bottom of the micrograph. Bar = 1.5 μm. ▪b. Thick section of a similar cell where it is clear that concretion core formation (**cf**) begins within the amorphous organic matrix in large vesicles. Bar = 0.9 μm.

The calmodulin and rabbit anti-concretion proteins have been used to document the location of the 17 kD protein in tissue sections from unionidsbivalve;unionids, as well as to demonstrate that antibody made to concretion protein in one unionid species will cross-react with concretion proteins from other species. The antibody has also documented that portions of the chitinous rod backing the gill filaments cross-react with concretion antibodies. These so-called chitinous rods in unionidsbivalve;unionids are partially calcified, and likely are similar to the concretions in their composition.

While light micrographic immunohistochemistry indicates the 17 kD protein is present in the concretions, the majority of this protein appears to be localized to a set of connective tissue cells (Fig. 7). These cells had previously been identified as the cells responsible for intracellular concretion formation using electron microscopy (Fig. 8). Further, a series of cellular events associated with concretion formation has been documented and several stages of concretion-forming cells can be recognized. Electron microscopic immunocytochemistry confirms the major localization of the 17kD protein to the concretion-forming cells, and further localizes the protein to amorphous protein granules in which concretions will subsequently form (Fig. 9). This data combined with the superficial location of the protein in the concretions begins to suggest that the protein is important in the initiation of concretion formation intracellularly. Its extreme similarity to mammalian calmodulin then leads to a number of possible mechanisms for the proteins ability to play a role in concretion initiation, but as yet none has been experimentally tested.

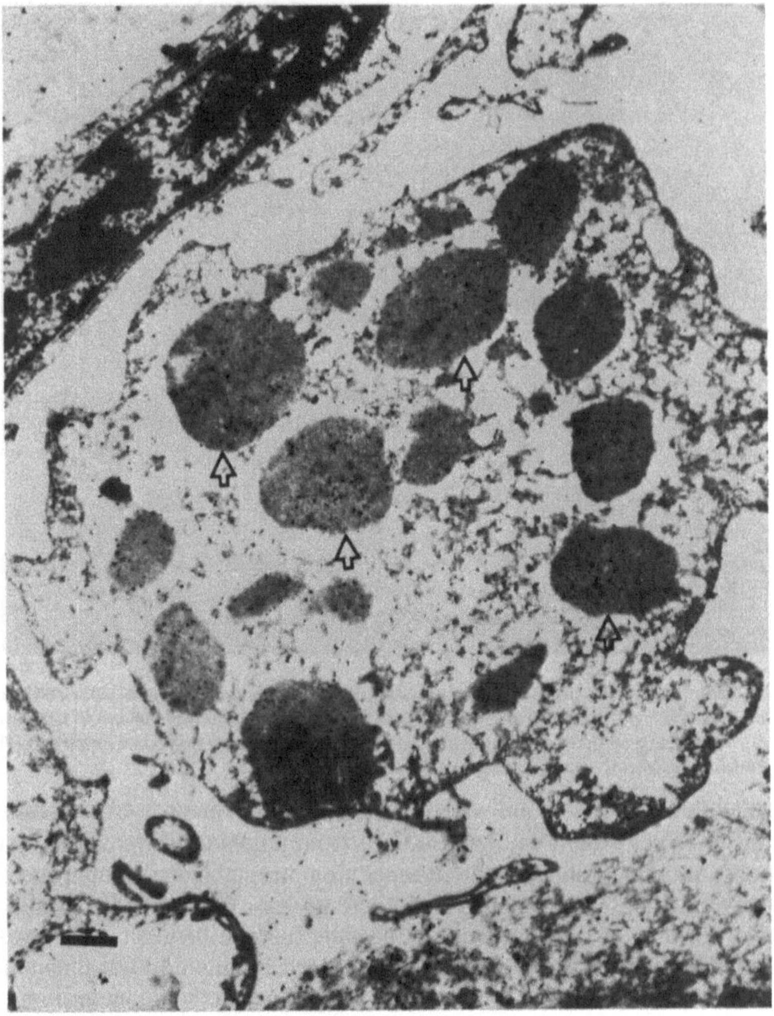

FIGURE 9. TEM of a concretion forming cell lightly fixed and embedded in LR-gold. The section was reacted with rabbit anti-concretion I gG conjugated to colloidal gold. Note the electron dense gold particles over the preconcretion vesicles (arrows). Bar = 0.75 μ m.

FIGURE 10. Cross-section of reproductively active lateral gills of *Anodonta grandis*. ▪a. Light micrograph of a water channel filled with developing embryos (**e**) surrounded by embryonic investment coats (**i**). The investment coat on many of the embryos is in close contact with the water channel epithelium over long distances (arrows). Shown is one of two lateral chambers (**wc**) and the large central chamber housing the embryos (**bc**). Bar = 70 μ m. ▪b. Low magnification TEM showing the complete separation of the lateral water channel (**wc**) from the chamber housing the embryos (**bc**). The arrows denote the two epithelia which line the septa between the two chambers. Bar = 100 μ m. Adapted from Silverman *et al.* (1987a). ▪c. Light micrograph of a reproductive gill in which the embryos have been lost during preparation. Three morphologically and physiologically isolated chambers (**wc**, **bc**, **wc**) are clearly visible. Bar = 100 μ m. Adapted from Silverman *et al.* (1987a).

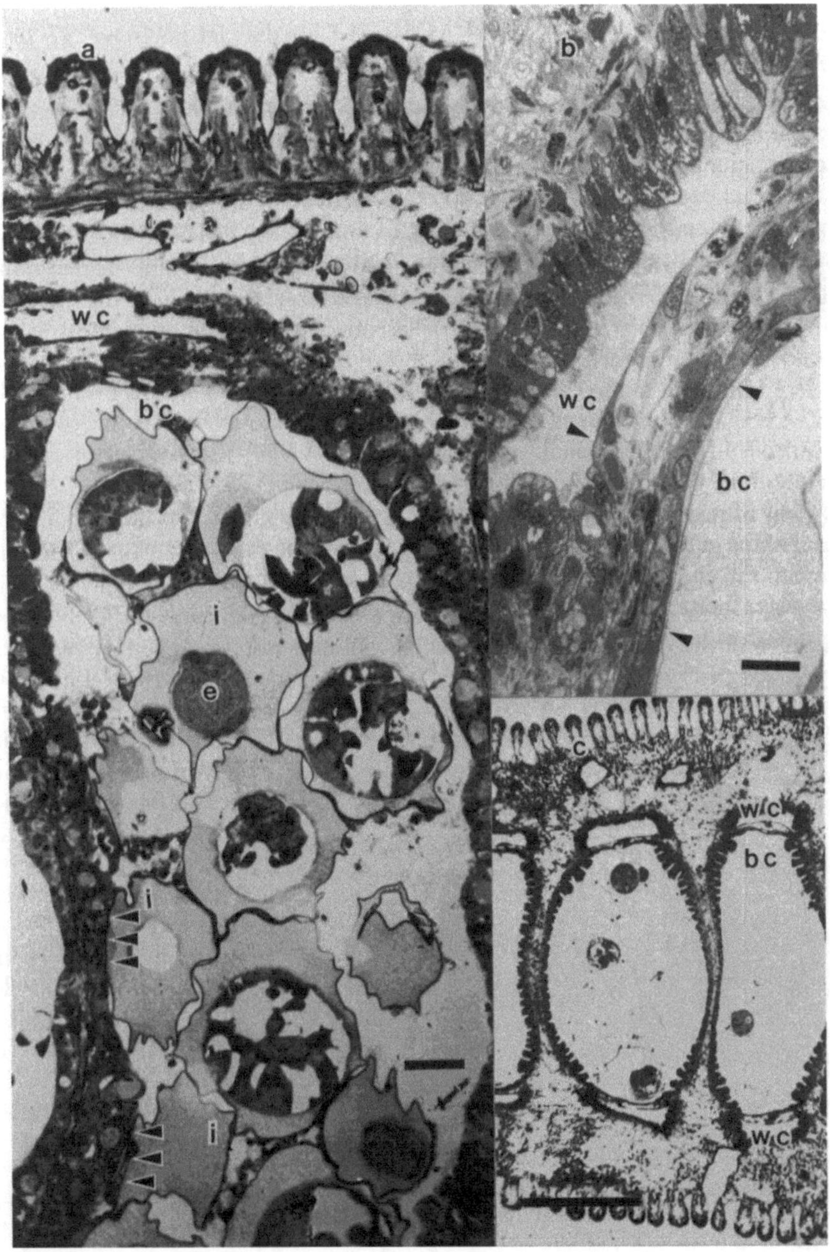
a
wc
bc
i
e
i
i
b
wc
bc
c
wc
bc
wc

Further, the experiments done to date, demonstrate one protein capable of binding calcium but do not rule out the possibility that another concretion protein(s) also binds calcium, and the ability of the concretion proteins to bind PO4 has not been investigated.

MOBILIZATION OF CONCRETION CALCIUM DURING REPRODUCTION AND ITS APPEARANCE IN THE GLOCHIDIAL SHELLS

Morphological evidence previously described indicates clearly that concretions disappear during the reproductive period (Silverman et al., 1985, 1987a). Measurements of blood calcium during this time period show no concomitant increase, in fact, a slight but significant decrease in calcium is noted (Silverman et al., 1985). The timing of the concretion disappearance and the inability to account for the calcium suggested that calcium might be passed to embryos developing in the gills of the animal. The example species to be discussed is Anodonta grandis, a relatively large animal (200 – 1000 g) which broods its embryos in the lateral gill (demibranch) on each side of the animal (Silverman et al., 1985). The demibranch weighs 1 g in non-reproductive animals, and as much as 30 g in a reproductive animal housing embryos. Adult females brood over 600,000 embryos each year (Silverman et al., 1985), and each embryo must leave the maternal animal with a fully-calcified shell. A balance sheet showing the amount of calcium necessary, and the possible sources of that calcium is presented in Table 6.

By way of review, the glochidia develop in the water channel of the gill. The water channel of the gill is connected to the outer surface through a series of ostia which bring water into the channel for respiratory purposes in non-reproductive animals (see Fig. 1); the water then leaves via the excurrent siphon. During reproduction, the gill becomes reorganized with the central water chamber partitioning into three compartments: a large central chamber which houses embryos and two thin lateral water channels which remain connected to the outer surface (Fig. 11) (Silverman et al., 1987a). Thus, as the central water channel becomes packed with eggs delivered from the ovary, the chamber in which they are located is largely isolated from water flow for respiratory purposes. This morphological isolation begins to suggest that little of the calcium available in the freshwater environment can contribute directly to glochidia shell formation.

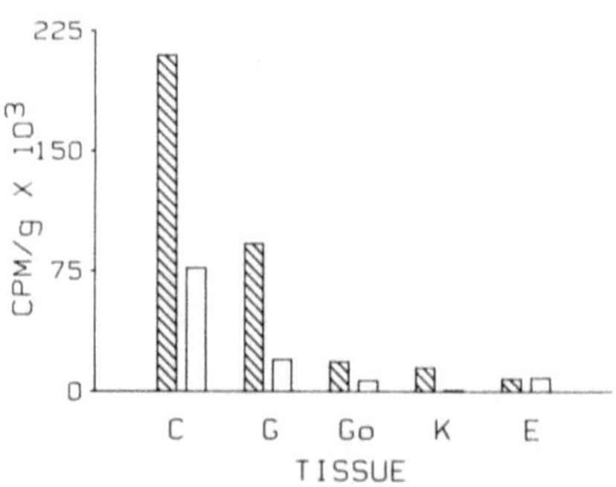

FIGURE 11. Histogram representing average tissue levels of ^{45}Ca in gill concretions (c), gill tissue (g), gonadal tissue (go), kidney tissue (k), and developing glochidia (e) on the day of collection. Averages are for values collected over a 2 month period. Values are expressed as $10^3 \times$ CPM/g dry tissue weight. Open bars represent initial single injection; hatched bars represent animals receiving a single injection 1 month later. From Silverman *et al.* (1987).

Physiological data support this hypothesis. Two

different sets of experimental data are available (Silverman et al., 1987a). First, animals were injected with a large quantity of 45Ca months in advance of reproduction. These animals were placed in a large pool and 45Ca allowed to distribute to the various tissues (Fig. 12). Radioactivity in the pondwater and substrate remained at background after an initial leakage period. Control animals receiving no label were kept in the same pool. Animals were sampled every two weeks to determine in which tissues the label was partitioned. Even after several months, the tissue component containing the most label was the gill calcium concretions. The shell also contained label. All other tissues had dropped to near background levels (Fig. 12). Glochidia that developed in the maternal gills were separated and examined for the label. The label appeared in the glochidia, such that 18% of the label found in remaining concretions was found in the shells of the glochidia (Table 7). The experiment clearly indicates passage of 45Ca from the maternal animal to the developing embryos. The passage appears to be across the water channel epithelium and not passage into an egg storage site because the gonad label remained relatively constant across the entire period of the experiment.

The next question to be addressed is the extent of label passage. The experiment here is set up differently. In this case, the label was administered to animals with glochidia already

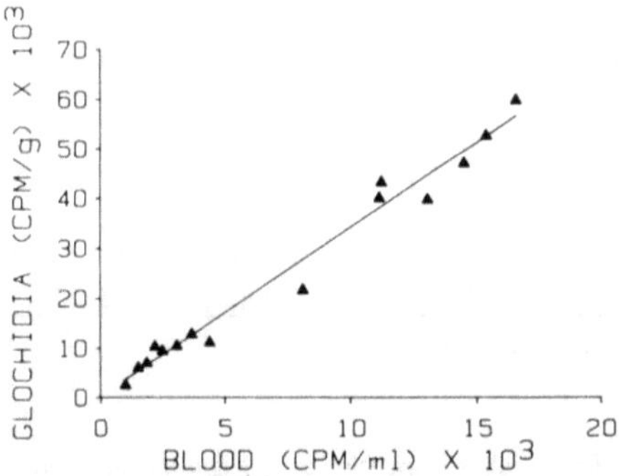

FIGURE 12. Regression of blood ^{45}Ca (CPM/ml) on glochidial shell ^{45}Ca (CPM/g), after exposure to 10^6 CPM ^{45}Ca. Regression based on combined data from bath exposure and injection experiments. Slope = 3.4 CPM/g/CPM/ml; r^2 = 0.988; and intercept = 153 CPM/g. From Silverman *et al.* (1987a).

*TABLE 6.** Calcium in the gills of a 400 g female *Anodonta grandis*

	LATERAL DEMIBRANCH	
TOTAL	NON-REPROD	REPROD
Dry Tissue	0.36	0.98
Concretions	0.2	‘0
Isolated Concretion Calcium Content	0.05	0
Acid Extractable Gill calcium	0.29	
Number of Glochidia	0	320,000
Dry Glochidia	0	0.8
Glochidia Calcium Content	0	0.21
	MEDIAL DEMIBRANCH	
Dry Tissue	0.36	0.22
Concretions	0.2	0.02
Isolated Concretion Calcium Content	0.05	0.005

*Data are reported as grams of dry weight of material for a single gill of the gill pair for a representative 400 g animal. (From Silverman et al., 1985)

TABLE 7. ^{45}Ca label incorporated into reproductive and non-reproductive *Anodonta grandis* tissues. Data expressed as % of CPM/mg of medial gill concretions. All values means ± SE (n)

	WHOLE GILL	GONAD	KIDNEY	MEDIA CONC. ISOLATE	LATERAL CONC. ISOLATE	GLOCHID SHELLS
NonReproductive	47.0 ± 5.5(17)	7.3 ± 1.5(13)	10.9 ± 3.1(11)	100(17)	117 ± 11(15)	
Reproductive	56.7 ± 14.5(7)	7.3 ± 1.4(7)	10.6 ± 2.7(7)	100(7)		119 ± 7(7)

(From Silverman *et al.*, 1987a)

TABLE 8. Radio-label appearance in *Ligumia subrostrata* tissues after 4 h exposure to 10^6 CPM ^{45}Ca either in the 30 ml pondwater bath or by direct injection into the pericardial region (N=8)

TISSUES	EXPOSURE METHOD: INJECTION	BATH
Blood CPM/ml	14,116 ± 1571*	2,157 ± 400
Soft Tissue CPM/g	696,186 ± 7518	96,115 ± 14,050
Shell CPM/g	16,243 ± 809	80,176 ± 5032
Demibranch CPM/g	402,755 ± 46,553	65,280 ± 9510
Glochidia CPM/g	42,565 ± 4074	8,915 ± 1183

*Mean ± SE. (From Silverman *et al.*, 1987a)

present in the water channel. The label was administered to animals in two ways. It was either placed directly into the bath, or it was injected into the blood. Maternal animal soft tissue, maternal shell, and glochidia shells were assayed for label incorporation after 4–6 h. The results appear in Table 8. In, injected animals label was high in the maternal blood, tissues, and glochidial shells, but not in the maternal shell. In the bath exposed animals, the label was high in the maternal shell (due to adsorption of label), but not in the blood, tissues, or glochidial shells. The indication is that the label in the bath was not in contact with the glochidial shells, and that the label appearing in the glochidial shells came from the maternal animal. This is substantiated because, over the period of the experiment, animals exposed in the bath do uptake some 45Ca which should then be available to the glochidia. This relationship, namely blood 45Ca related to glochidial shell 45Ca, is shown in Figure 13. Further, experiments with isolated glochidia demonstrated that if the shell were in contact with bath label, the label would be adsorbed to the same extent as was seen for the maternal shell (Table 9). Thus, it is clear that over 90% of the calcium occurring in the glochidial shells in these short-term experiments came from the maternal animal. These data demonstrate that there is a physiological isolation of the brood chamber resulting from the morphological isolation of this chamber during reproduction. The isolation of glochidia from environmental water flow as demonstrated here was actually suggested by Ortmann (1911) for Anodonta.

MECHANISM OF CALCIUM TRANSFER

Since calcium is transferred from the maternal animal, it would be of interest to know the mechanism of transfer. Two possibilities need investigation. The first would suggest that the water channel has specialized ionic pumps or gates to allow calcium transfer

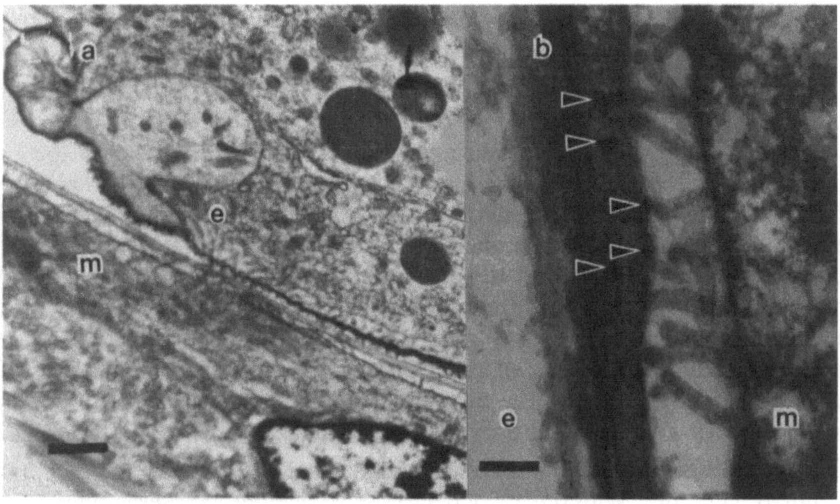

FIGURE 13. TEM of the association between maternal water channel epithelial cells and the investment coat of developing glochidia. ▪a. 0.5 μm thick section at low magnification to demonstrate the type of area in which maternal (m)/embryonic (e) connections are visible. Bar = 0.8 μm. ▪b. 0.5 μm thick section demonstrating the association of the microvilli with the investment coat. Microvilli are actually embedded in the investment coat, and sites at which the microvilli associate with the investment coat electron dense caps are present. These caps remain with the investment coat even if the microvilli become dislodged. Note also the nature of the microvillar core. Bar = 0.3 μm.

TABLE 9. Uptake of ^{45}Ca, expressed as CPM, by tissue components of *Ligumia subrostrata*. Exposure to ^{45}Ca was either by inclusion of 10^6 CPM in 100 ml of pondwater or by injection into intact animals. In some cases tissue components were isolated and directly exposed to pondwater

	EXPOSURE METHOD	
TISSUE TREATMENT	BATH	INJECTION
Glochidia – Whole Animal	8,915 ± 1183(8)	42,565 ± 4074(8)
Glochidia – Isolated Gill	25,794 ± 3929(4)	
Glochidia – Isolated	78,424 ± 12,559(7)	
Shell – Maternal Animal	80,954 ± 4676(8)	18,238 ± 772(8)
Shell – Empty + Open	270,191 ± 25,683(4)	

(From Silverman *et al.*, 1987a)

into the water channel. Since it appears that the water channel shows sluggish water flow, this mechanism could keep calcium concentrations in the water channel high as the embryos removed the calcium. Another possibility which must be considered is that there is direct passage of calcium from maternal epithelial cells to the embryos through a cellular exocytosis mechanism. Although it is not clear which mechanism is correct there are some data in support of each of the theories. As calcium declines in maternal blood during reproduction, it appears that sodium is increasing (Silverman et al., 1985). Although not strong evidence, this leaves the possibility of a Na/Ca ion exchange mechanism. Such a mechanism would be the more standard way for transporting ions, and yet the exocytosis theory is intriguing for a number of reasons.

The embryos which are housed in the brood chambers are a tightly-packed mass which are easily extracted from the gill in toto by snipping the surface of the gill and teasing with a forceps. This fact made possible the glochidia-labelling experiments described above. However, since the byssal threads or embryo threads do not develop until relatively late (Silverman, unpubl.), they are not responsible for holding the embryo together as a unit until shell formation is well under way. Instead the embryos are held together as units, by their extra-embryonic membranes, and by the attachment of the membranes to the water channel (brood chamber) epithelium. The epithelium of the water channel undergoes extensive differentiation from a flat relatively indifferent epithelium containing a few goblet cells, rows of flat ciliated cells, and rows of flat microvillar cells (Kays, 1986), to a highly folded and elongated epithelium, with the microvillar cells becoming highly protein synthetic and secretory.

The microvilli of these secretory cells are embedded in the basement membrane of the embryos (Fig. 14). With the aid of 0.5 μm thick sections and stereo pairs taken on the electron microscope, these microvilli appear to have hollow cores and electron dense caps which are embedded in the membrane. It is tempting to ask whether a calcium-rich substance is being transmitted via these microvilli to the developing embryos. Further research should answer this question. A related question which should also be addressed, is whether this system of transfer is unique to calcium, or whether other needed nutrients are also transferred via such connections.

EVOLUTIONARY IMPLICATIONS?

The data presented clearly indicate that calcium concretions present in the gills of unionidsbivalve;unionids serve as a calcium source for glochidia shell development. All unionidsbivalve;unionids examined to date possess calcium concretions (Silverman et al., 1985; Pynnonen et al., 1987). Fossilized, phosphatized unionid gills of Mesozoic age (British Museum Natural History, N. Morris; communicated to this author by B. Runnegar) suggest the presence of concretions. These points suggest that the emergence of concretions was important to the successful radiation of these animals, particularly into calcium poor freshwaters. The conservative nature of the organic matrix within unionid species, including Margartifera, as expressed relative to immunohistochemical techniques (Steffens et al., 1985) also supports this idea.

Exactly how, or why, unionidsbivalve;unionids developed the ability to store calcium in concretions can only be speculation, but the existence of a chitinous skeleton supporting gill filaments in most mussels (Ridewood, 1903), and their rather unique mineralized form in the unionidsbivalve;unionids may suggest the early events. The cells which construct concretions (Figs. 8 – 10) and the cells maintaining the calcified chitinous rods in unionid gills are similar (Silverman et al., 1987c). It may be that structural calcification of the rods, which are clearly extracellular structures, led to the ability to produce calcium concretions in the gills of these animals.

It should be noted that concretion formation has evolved numerous times in different tissues in most invertebrate phyla (Brown, 1982). Some gastropod concretions, although

CaCO3 instead of calcium phosphate, serve a similar role as a store of calcium for reproductive purposes (Fournié & Chétail, 1984; Tompa & Wilbur, 1977), while some concretions, particularly those associated with excretory or digestive tracts, have very different function (Simkiss, 1981).

The importance of the concretions as a PO4 store has been underemphasized in our studies to date, but may have importance as a general PO4 pool available to the animal. It is curious, that an animal with a CO3 shell, contains such an extensive phosphate-based mineral product. This may relate to the need and ability to mobilize this product specifically and quickly only during a specific period of the year.

Finally, the gill system reviewed here is an ideal model system to study several aspects of biomineralization because it takes place multiple times in the same tissue. Concretions are broken down during reproduction, calcium passage across at least one epithelium occurs, the glochidial shell is constructed, and the concretions are rapidly reconstructed. Some of the events described occur intracellularly (initial concretion formation), and some extracellularly (glochidial shell formation, mobilization of the concretions).

Mobilization and resynthesis of concretions is tightly regulated as demonstrated by the toxic metal studies. Understanding the biochemistry of the concretions, as well as the neurohormonal mechanisms of control, should lead to increased understanding of biomineralization processes in general.

ACKNOWLEDGEMENTS

Collaborators and colleagues who have assisted in various portions of the work reviewed here include: Dr. T.H. Dietz, Dr. W.L. Steffens, Dr. L.D. Sibley, W.T. Kays, R. Goddard, J.W. McNeil, P.E. Richard and J.M. Myers. The work was supported in part by NSF grants DCB-8303789 and DCB 8518221. Participation in the Symposium was made possible by grants from The Organized Research Fund of The University of Texas at Arlington and The Sea Grant Program administered by Texas A & M University.

REFERENCES

BROWN, B.E., 1982. The form and function of metal-containg 'granules' in invertebrate tissues. Biol. Rev., 57:621-667.

FOURNIÉ, J. & CHÉTAIL, M., 1984. Calcium dynamics in land gastropods. Amer. Zool., 24:857-870.

KAT, P.W., 1984. Parasitism and the Unionacea (Bivalvia). Biol. Rev., 59:189-207.

KAYS, W.T., 1986. Ultrastructural and histochemical study of the gills of the freshwater musle, *Ligumia subrostrata*. M.S. thesis, Louisiana State Univ. 68 pp.

MARUYAMA, K., MIKAWA, T. & EBASHI, S., 1984. Detection of calcium binding proteins by ^{45}Ca autoradiography on nitrocellulose membrane after sodium docedyl sulfate gel electrophoresis. J. Biochem. Tokyo, 95:511-519.

ORTMANN, A.E., 1911. A monograph on the Najades of Pennsylvania. Mem. Carnegie Mus., 4:279-347.

PYNNÖNEN, K, HOLWERDA, D.A., AND ZANDEE, D.I., 1987. Occurrence of calcium concretions in various tissues of freshwater mussels and their capacity for cadmium sequestration. Aquatic Toxicol., 10:101-114.

RIDEWOOD, W.G., 1903. On the structure of the gills of *Lamellibranchia*. Trans. Roy. Soc. London, 211:147-284.

SILVERMAN, H., KAYS, W.T. & DIETZ, T.H.., 1987a. Transfer of maternal calcium to glochidia brooding in the water channel of unionid mussels. J. Exp. Zool., 242:137-146.

SILVERMAN, H., MCNEIL, J.W. & DIETZ, T.H., 1987b. Interaction of trace metals Zn, Cd, Mn with calcium concretion in the gills of freshwater unionid mussels. Can. J. Zool., 65:828-832.

SILVERMAN, H., RICHARD, P.E., AND GODDARD, R.H., 1987c. Calcium concretion forming cells in the gill of *Anodonta grandis*. Amer. Zool., (in press).

SILVERMAN, H., SIBLEY, L.D., STEFFENS, W.L., AND DIETZ, T.H., 1986. A calmodulin-like protein is associated with the organic matrix of mussel gill calcium concretions. Amer. Zool., 26:5a.

SILVERMAN, H., STEFFENS, W.L. & DIETZ, T.H., 1985. Calcium from extracellular concretions in the gills of freshwater unionid mussels is mobilized during reproduction. J. Exp. Zool., 236:137-147.

SILVERMAN, H., STEFFENS, W.L. & DIETZ, T.H., 1983. Calcium concretions in the gills of a freshwater mussel serve as a calcium reservoir during periods of hypoxia. J. Exp. Zool., 227:177-189.

SIMKISS, K., 1981. Cellular discrimination processes in metal accumulating cells. J. Exp. Biol., 94:317-327.

STEFFENS, W.L., SILVERMAN, H. & DIETZ, T.H., 1985. Localization and distribution of antigens related to calcium-rich deposits in the gills of several freshwater bivalves. Can. J. Zool., 63:348-354.

TOMPA, A.S. & WILBUR, K.M., 1977. Calcium mobilization during reproduction in snail *Helix aspersa*. Nature, 270:53-54.

TRACE METAL CONCENTRATION IN FOSSIL AND RECENT SHELLS OF THE ARCTIC INFAUNAL BIVALVE, *MYA TRUNCATA* L.

Bernard Patrick Bourgoin[1]

Geology Department, McMaster University, 1280 Main Street West, Hamilton, Ontario, Canada, L8S 4M1; [1]Present address:Environmental Resource Studies, Trent University, Peterborough, Ontario, Canada K9J7B8

ABSTRACT. In a pilot study designed to demonstrate the advantages of analyzing bivalves shells (as opposed to tissues) for trace metal monitoring studies, fossil and modern shell samples were collected from three sites in Pangnirtung Fiord, Baffin Island, Canada. Copper and zinc levels in fossil shells were respectively higher and lower compared with modern shells. There were no significant differences in the copper and zinc levels between bivalve shells collected near a municipal dump site and those from an unperturbed area.

INTRODUCTION

Bivalves are widely used in environmental monitoring. Most of these studies focus on the trace metal concentrations in the soft tissues,and is assume that metals are passively adsorbed onto the outer shell surface (Keckes *et al.*, 1969; Romeril, 1971; Bryan & Uysal, 1978; Martincic *et al.*, 1984). Inadequate shell cleaning procedures (*i.e.*,incomplete removal of the periostracum) may be responsible for this assumption. Active metal incorporation into bivalve shells has been demonstrated in the laboratory (Sturesson, 1976, 1978) as well as in field experiments (Frazier, 1976; Carriker *et al.*, 1980). Bertine & Goldberg (1972) suggested that the calcium carbonate shells of bivalves may act as receptacles for unwanted chemical species. Shell analysis offers several practical advantages over tissue analysis: 1) shells are more easily identified, collected and preserved, 2) shells are more easily cleaned, 3) analytical methods are more readily standardized, 4) depuration (*i.e.*, excretion of digestive tract contents) problems are eliminated, and 5) although metal concentrations in the shells are generally lower than in the flesh, variances are usually lower, allowing more precise statistical inferences. The fundamentally important advantage that shells have over tissues in environmental monitoring is the fact that shells preserved in the fossil record allow identification of long-term trends.

Dodd (1966) observed the ability of bivalve shells to conserve their discrete minor metal arrangement during diagenesis, and Dardenne (1967) suggested that zinc may

Origin, Evolution, and Modern Aspects of Biomineralization in Plants and Animals
Edited by Rex E. Crick
Plenum Press, New York

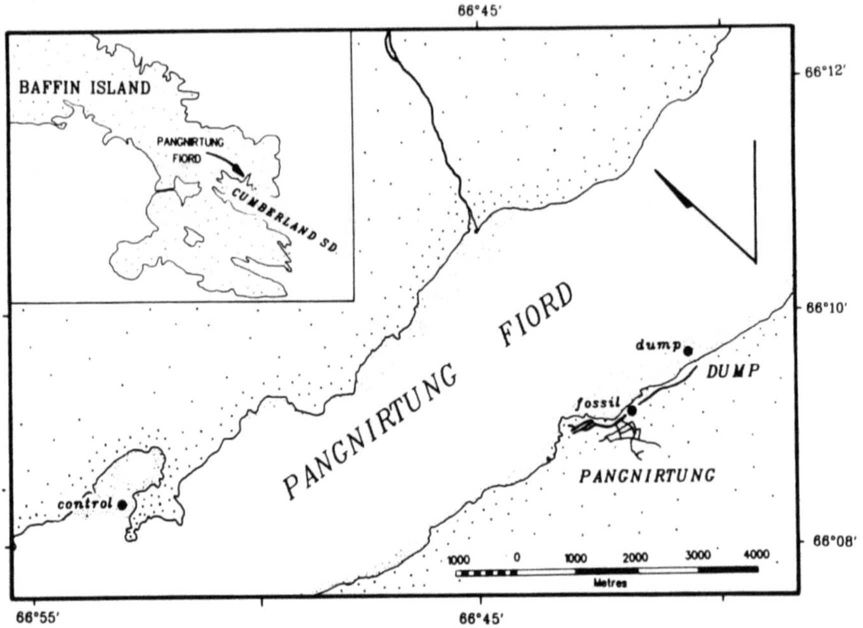

FIGURE 1. Sampling sites in Pangnirtung Fiord.

provide a means for verification of diagenetic alteration where altered material should increase its zinc content.

The majority of naturally occurring and anthropogenetically derived metals are not dispersed evenly in aquatic systems. Partitioning occurs in such a way that estuaries, and particularly estuarine sediments, may become 'reservoirs' for many metals (Forstner & Wittman, 1981). Infaunal bivalves are directly exposed to sediment-bound metals and may be useful for studying metal availability from sediments. The predominant components of fine-grained estuarine sediments which bind trace metals include the hydrous oxides of iron and manganese, organic matter and carbonates (Luoma & Jenne, 1976; Luoma & Bryan, 1981). Sequential extraction techniques are more selective for various sediment components than the total digestion of the sediments and give a better idea of metal partitioning among these components.

This study examines the relationship between the trace metal concentration in fossil and recent shells of an infaunal filter-feeding bivalve, *Mya truncata* L. and the surrounding sediment components, and the shell trace metal concentration of this same molluscan species from different environments. The results suggest that, although zinc levels are higher now than in the past, fossil analyses suggest that copper levels may in fact, have declined over time.

MATERIALS AND METHODS

Sample area and collection

Sediment and shells samples were collected from three sites in Pangnirtung, Baffin Island, Canada (Fig. 1, insert). Two of these stations are from an intertidal area. One is located 500 m from a municipal dump site (dump, Fig. 1), and the second, considered as the control site, is in a relatively undisturbed area 5km from a village (control, Fig. 1). The third station is located on a raised marine terrace (fossil, Fig. 1) in Pangnirtung and contains clam shells that have been radio-carbon dated at approximately 8200 years before present (BP) (Aitken *et al.*, 1988).

Sediments were kept frozen until analysis while the clams from the intertidal zone were shucked in the field. The collection of the fossil specimens was restricted to *in situ* shells (Aitken, pers. comm.).

Sediment analyses

Trace metals were extracted sequentially from the sediments by a modified version of the Tessier extraction (Tessier *et al.*, 1979, 1980). Applied modifications were the inversion of fractions 3 and 4 and a digestion of metals in fraction 5 with aqua-regia rather than the digestion by a mixture of perchloric, nitric and hydrofluoric acid. In brief, the extractants used in the five fractions were: fraction 1 (exchangeable metals) 1 M $MgCl_2$ at pH 7.0; fraction 2 (metals bound to carbonates) 1 M NaOAc adjusted to pH 5.0 with HOAc; fraction 3 (metals bound to organic matter) 30% H_2O_2 at 85° C adjusted to pH 2.0 with HNO_3 and followed by 3.2 M NH_4OAc in 20% (v/v) HNO_3 at room temperature; fraction 4 (metals bound to Fe-Mn oxides) 0.04 M NH_2OH©HCl in 25% (v/v) HOAc; fraction 5 ("residual metals") boiling aqua-regia.

The concentration of iron, manganese, copper and zinc were determined by atomic absorption spectrometry (Perkin Elmer 603®) involving direct aspiration of the aqueous solution into an air-acetylene flame. A graphite furnace assembly (HGA 2100) was used when metal levels were too low to be detected by flame. The total metal content of the sediments were determined on a representative subsample by X-ray fluorescence (XRF) spectrometry.

The concentration of carbon was determined with a Leco Carbon Analyser (model WR-12®). Inorganic carbon was calculated as the difference between total and organic carbon concentrations, the latter being determined after sediment treatment in 10% (v/v) HCl to dissolve the carbonates.

Shell analyses

Shells between 5 and 6 cm in length were pooled into 7 groups of 6 individuals. They were soaked for in a 20% (w/v) tetramethlammonium hydroxide pentahydrate (TMAH) solution at 60° C for 3 h and then scrubbed with a nylon brush and rinsed with distilled water in order to remove the periostracum. They were then heated to 400° C for 4 h to destroy the organic matrix in the shells. After being crushed in an agate mortar, a 3 g subsample from each group of pooled shells was dissolved in a minimum amount of concentrated HNO_3 (Aristar) and subsequently evaporated to dryness. The residue was dissolved in 80 ml of 10% (v/v) HNO3 to which 1 ml of citrate buffer (90 g of dibasic ammonium citrate and 183 g of citric acid were dissolved in 1 l of deionized water) was

TABLE 1. Grain size distribution and organic carbon content of Pangnirtung sediments

	SIZE DISTRIBUTION		OC[1]
Fossil	sand: 73.4%	mud: 26.2%	0.06
Control	sand: 92.5%	mud: 3.9%	0.08
Dump	sand: 91.9%	mud: 6.1%	0.07

[1]Organic carbon

added. The solution was adjusted to pH 4.5 with dilute NH_4OH and then transferred to a speratory funnel. A 10 ml aliquot of a 2% (w/v) ammonium pyrrolidine dithiocarbamate (APDC) solution was added, followed by 20 ml of methylisobutylketone (MIBK). The stoppered funnel was shaken vigorously for 1 min followed by a 10 min settling. The aqueous phase was drained, and the organic phase collected and analyzed for metal content by flame (air-acetylene) atomic absorption. Data were analyzed using Mann-Whitney tests.

The bivalve shells were fractured and prepared for scanning electron microscope observations (ISI model DS-130®) by the method as prescribed by Tevesz & Gaylord-Carter (1980).

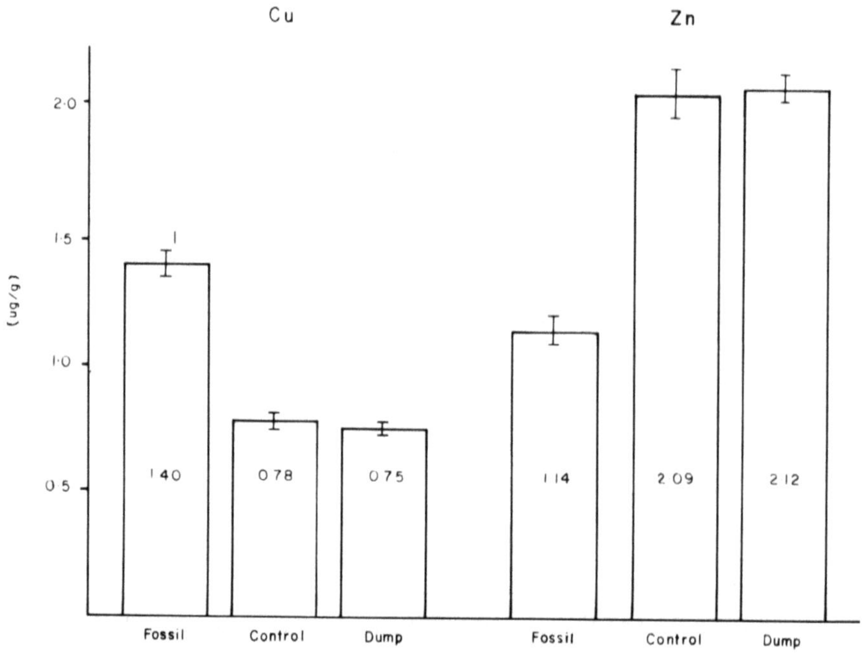

confidence interval: $ts/\sqrt{n}$; t for a 95% level of confidence

FIGURE 2. Copper and zinc concentrations in *Mya truncata* shells from three sites in Pangirtung Fiord.

Table 2. Metal content of Pangnirtung sediments

DATA	Fe[1]	Mn	Cu	Zn
M-F1				
F	238.60	3.02	-	3.28
C	26.55	-	-	2.15
D	21.55	-	-	2.01
M-F2				
F	37.00	0.42	0.22	0.18
C	23.10	0.25	0.26	0.25
D	25.60	-	0.23	0.17
M-F3				
F	48.20	3.92	1.49	1.01
C	18.50	0.61	1.00	0.58
D	25.50	0.86	1.64	0.82
M-F4				
F	1265.80	4.91	2.38	1.74
C	626.10	4.40	1.30	2.19
D	581.60	4.20	1.10	1.89
M-F5				
F	3598.70	12.21	2.31	1.49
C	2261.40	7.96	1.42	1.13
D	2506.10	8.70	1.46	1.20
M-S5[2]				
F	5188.30	24.46	8.77	7.68
C	2955.65	13.21	5.26	6.30
D	3160.00	13.76	5.52	6.08
TOTAL				
F	39700.	500.	13.	39.
C	29700.	400.	6.	28.
D	35100.	400.	6.	29.
EXTRACTED[3]				
F	13.1	4.9	67.5	19.7
C	10.0	3.0	87.7	22.5
D	9.0	3.4	92.0	21.0

[1]All results based on sediment dry weight (μg/g). M-F1 – M-F5 = metal concentrations in fractions 1 – 5 (see text).

[2]Sum of fractions 1 – 5.

[3]M-S5 = metal/total metal x 100.

RESULTS AND DISCUSSION

Particle size distribution and the organic content are given in Table 1. Although the sediments from the fossil site were finer-grained than the intertidal sediments, the organic carbon levels were similar at all three sites. The sediment trace metal concentrations are given in Table 2. The distribution of individual metals between various fractions does not necessarily reflect the relative scavenging action of discrete sediment phases, but rather should be considered as operationally defined by the method of extraction (Tessier *et al.*, 1979).

The total zinc content was slightly higher in the sediments from the fossil site. Sediments from all three sampling stations, however yielded similar amounts of extracted zinc. Comparable quantities of iron were also extracted from all three sites. The three sites contained similar amounts of 'total' manganese as detected by XRF, but twice as much manganese was extracted from the fossil sediments. Yields of extracted zinc were evenly distributed within the 5 extraction fractions, but the bulk of the iron and manganese (*i.e.*, > 90%) were extracted in fractions 4 and 5. Sediments from the intertidal sites contained approximately half as much copper as those from the fossil site. Although copper levels were 3 to 6 times lower than zinc, it was extracted more efficiently and yielded comparable amounts (Table 2).

Results from the trace metal analyses of the *Mya truncata*

shells collected at the three sites are illustrated in Figure 2. The fossil shells contained approximately twice as much copper (1.40 μg/g) than the bivalve shells collected at the control and dump sites (0.78 and 0.75 μg/g, respectively). There were no significant differences in the copper concentrations between shells obtained from the intertidal sites. While there were also no significant differences in the zinc concentrations between the shells from the control and dump site, they contained over 45% more zinc than the fossil shells. The confidence intervals (Fig. 2) qualify the differences between the shells collected from the fossil site from those obtained from the intertidal zone.

It is theoretically possible that the zinc level in the fossil shells may have been originally lower and were subsequently increased to the actual levels by diagenesis. Observations with the scanning electron microscope however, does not support this hypothesis. *Mya truncata* shells exhibit a tripartite microstructure consisting of an outer aragonitic homogeneous layer, a middle aragonitic crossed-lamellar layer and an inner aragonitic complex crossed-lamellar layer (Taylor *et al.*, 1969). Any diagenetic alteration involves secondary precipitation of the shell crystals. Dodd (1966) observed that organically produced crystals differ markedly from inorganically precipitated crystals of calcium carbonate. No microstructural differences in any of the three shell layer were detected when the fossil shells were compared with the modern ones obtained from the control site; suggesting that the metal levels observed in the fossil shells were real and not diagenetic artifacts.

The higher copper values detected in the sediments and the shells from the fossil site suggest that this metal may have been more abundant as well as more available to this species of bivalve. Copper complexes favorably with organic matter in the environment (Oakley *et al.*, 1981). In a study of metal speciation in sediments, Luoma & Bryan (1981) observed a strong relationship between copper and extracted iron where manganese and organic carbon were low in concentration, such as this study (Tables 1 & 2). In these instances, the copper is mainly associated with amorphous iron oxides (*i.e.*, $Fe(OH_3)$). Although the extraction procedures are operationally defined, the metals extracted in fraction 4 are believed to be mainly derived from the amorphous Fe-Mn oxides. Between 18 and 25% of the extracted iron and 20 to 33% of the extracted manganese were obtained from this fraction alone (Table 2). Amorphous oxides crystallize with time. This process tends to lock-in metals which are normally adsorbed onto the oxide surface, making metal extraction more difficult. The aqua-regia mixture used in fraction 5 is much more vigorous than any of the other extractants employed in the first four fractions and partially attacks the rock lattice (Agemian & Chau, 1976). Any adsorbed metal which was not extracted due to the crystallization of the oxides should be released in this final fraction. Although comparable quantities of copper were extracted in the first three fractions for all three sampling stations; sediment from the fossil site yielded approximately twice the amount of copper in fractions 4 and 5.

It would therefore seem that copper is associated with Fe-Mn oxides, and that past copper levels were higher than present. The fact that relatively less copper was extracted from the fossil sediments (Table 2) suggests that the originally amorphous Fe-Mn oxides may have crystallized over time.

A higher bio-availability of copper may also explain why zinc levels are highest in the shells collected from the intertidal sites. Interactive effects among metals may be a very important factor influencing metal availability to organisms. Luoma (1983) proposed 3 types of influences: (1) Enhancement of uptake; (2) Competitive displacement on metal-specific binding proteins in tissues; and (3) Competitive inhibition of uptake at the environmental interface (*i.e.*, sediment-water interface for this study). Kirchgessner *et al.* (1979) noted that because of similarities in the electron configuration and ion radius, the interaction between zinc and copper may be particularly intensive. The discrepancy in shell zinc levels may be due to this third type of interactive effect. Higher bio-availability of copper 8000 years ago may have inhibited zinc uptake, resulting in reduced translocation of zinc from the mantle to the shell.

The copper and zinc levels in the shells did not reveal any difference between the control and dump sites. Similarly, there were no major differences between the metal content of the sediments from these two sites. Lead, a metal considered as physiologically nonessential, was found in significantly higher levels in the shells and certain extraction fractions from the dump site (Bourgoin & Risk, 1987).

Metal bio-availability is a complex system in which metal speciation within the various sedimentary components comprises only a small part. Salinity (Phillips, 1976) as well as redoc potential (Luoma & Bryan, 1982) also affect availability of a particular metal to bivalves. Most importantly, the degree of partitioning a metal undergoes as it is translocated from the various tissues towards the skeletal parts still needs to be established for each metal.

CONCLUSIONS

In conclusion, results of this study have demonstrated some of the advantages of shell analyses, and have suggested that metal levels in at least one Arctic environment do not show uniform anthropogenetic increases with time. Although zinc had the expected increase, copper decreased with time. To draw conclusions about the relative degree of anthropogenetic contamination of modern environments, without first elucidating the long-term trend by a method similar to that herein described, could easily lead to erroneous conclusions.

REFERENCES

AGEMIAN, H. & CHAU, A.S.Y., 1976. Evaluation of extraction techniques for the determination of metals in aquatic sediments. Analyst, 101 (1207):761-767.

AITKEN, A.E., RISK, M.J., & HOWARD, J.D., 1988. Animal-sediment relationships on an arctic tidal flat: Pangnirtung Fiord, Baffin Island, N.W.T., J. Sed. Pet. (in press).

BERTINE, K.K., & GOLDBERG, E.D., 1972. Trace elements in clams, mussels and shrimps. Limn. Oceanog., 17:877-884.

BOURGOIN, B.P., & RISK, M.J., 1987. Historical changes in lead in the eastern Canadian Arctic, determined from fossil and modern *Mya truncata* shells. Sci. Total Environment, 67:287-291.

BRYAN, G.W., & UYSAL, H., 1978. Heavy metals in the burrowing bivalve *Scrobicularia plana* from the Tamar estuary in relation to environmental levels. J. Mar. Biol. Assoc. U. K., 58:89-108.

CARRIKER, M.L.R., PALMER, R.E., SICK, L.V., & JOHNSON, C.C. 1980. Interaction of mineral elements in sea water and shell of oysters (*Crassostrea virginica* [Gmelin]) cultured in controlled and natural systems, J. Exp. Mar. Biol., 46:279-296.

DARDENNE, M., 1967. Etude experimentale de la distribution du zinc dans les carbonates de calcium. Bull. Bur. Rech. Geol. Minieres, 5:75-110.

DODD, J.R., 1966. Diagenetic stability of temperature sensitive skeletal properties in Mytilus from the Pleistocene of California. Geol. Soc. Am. Bull., 77:1213-1224.

FORSTNER, U., & WITTMANN, G.T.W., 1981. Metal pollution in the aquatic environment. 486 pp. 2nd edition. New York: Springer-Verlag.

FRAZIER, J.M., 1976. The dynamics of metals in the American oyster, *Crassostrea virginica*. 2. Environmental effects, Chesapeake Science, 17:188-197.

KECKES, S.B., OZRETIC, B., KRANJNOVIC, M., 1969. Metabolism of ^{65}Zn in mussels (*Mytilus galloprovincialis*) I, uptake of ^{65}Zn, Rapp. Comm. Int. Explor. Sci. Mer Medit. (Monaco), 19:949-952.

KIRCHGESSNER, M., SCHWARZ, F.J., GRASSMAN, E., & STEINHART, H.,1979. Interactions of copper with other trace elements, chap. 14 , *In* Copper in the Environment; Part II: Health effects. (ed. J.O. Nriagu) John Wiley, 489 pp.

LUOMA, S.N., 1983. Bioavailability of trace metals to aquatic organisms--a review, Sci. Total Environment, 28:1-22.

LUOMA, S.N., & BRYAN, G.W., 1981. A statistical assessment of the form of trace metals in oxidized estuarine sediments employing chemical extractants, Sci. Total Environment, 17:165-196.

LUOMA, S.N., 1982. A statistical study of environmental factor controlling concentrations of heavy metals in the burrowing bivalve *Scrobicularia plana* and the polychaete *Nereis diversicolor*, Estuarine, Coastal and Shelf Science, 15:95-108.

LUOMA, S.N., & JENNE, E.A., 1976. Factors affecting the availability of sediment-bound cadmium to the estuarine, deposit-feeding clam, *Macoma balthica*, *In* Radioecology and Energy Resources, (ed. C.E. Cushing Jr.), pp. 283-290. Ecol. Soc. Amer., Spec. Pub. No.1, Hutchison & Ross.

MARTINCIC, D., NURNBERG, H.W., STOPPLER, M., & BRANICA, M., 1984. Bioaccumulation of heavy metals from Lim Fjord (North Adriatic Sea), Marine Biology, 81:177-188.

OAKLEY, S.E., NELSON, P.O., & WILLIAMSON, K.J., 1981. Model of trace-metal partitioning in marine sediments, Environ. Sci. Technol., 15:474-480.

PHILLIPS, D.J.H., 1976. The common mussel *Mytilus edulis* as an indicator of pollution by zinc, cadmium, lead and copper. I. Effects of environmental variables on uptake of metals, Mar. Biol., 9:347-354.

ROMERIL, M.G., 1971. The uptake and distribution of ^{65}Zn in oysters, Mar. Biol., 9:347-354.

STURESSON, U., 1976. Lead enrichment in shells of *Mytilus edulis*, Ambio., 5:253-256.

STURESSON, U., 1978. Cadmium enrichment in shell of *Mytilus edulis*, Ambio, 7:122-125.

TAYLOR, J.D., KENNEDY, W.J., & HALL, A., 1969. The shell structure and mineralogy of the bivalvia: Introduction. Nuclulacea-Trigonacea, Bull. Brit. Mus. (Nat. Hist.) Supp. 3, (29 plates), 125 pp.

TESSIER, A., CAMPBELL, P.G.C., & BISSON, M., 1979. Sequential extraction procedure for the speciation of particulate trace metals, Anal. Chem., 5 (7):844-851.

TESSIER, A., 1980. Trace metal speciation in the Yamaska and St. Francois Rivers (Quebec), Canadian J. Ear. Sci., 17:90-105.

TEVESZ, M.J., & GAYLORD-CARTER, J., 1980. Preparation and examination of skeletal materials for growth studies Part A: Mollusc, *In* Skeletal Growth of Aquatic Organisms: Biological Records of Environmental Change (ed. D.C. Rhoads & R.A. Lutz) 750 pp. New York: Plenum Press.

FUNCTION OF MOLLUSCAN STATOCYSTS

Michael L. Wiederhold[1,3], Christine E. Sheridan[1,3] and Nancy K.R. Smith[2]

[1]*Division of Otorhinolaryngology;* [2]*Department of Cellular and Structural Biology, The University of Texas Health Science Center at San Antonio, San Antonio, Texas 78284;* [3]*Audie L. Murphy Memorial Veteran's Hospital, San Antonio, Texas 78284.*

ABSTRACT. The gravity sensors of most molluscs are spherical organs called statocysts. The wall of the sphere contains mechanosensory cells whose sensory cilia project into the lumen of the cyst. The lumen is filled with fluid and dense "stones", the statoconia or statoliths, which sink under the influence of gravity to load and stimulate receptor cells which are at the bottom. The composition of the statoconia is known in only a few species. Data presented here suggest that they are aragonite in *Aplysia*. The statoconia of *Aplysia californica* are shown to be calcified about a lamellar arrangement of membranes. Similar lamellar membrane arrangements are seen within the receptor cells, and their possible role in the formation of the statoconia is discussed. Scanning electron micrographs of unfixed statoconia reveal plate-like crystallization on their surface. Elemental analysis shows a relatively high strontium content. This is of interest in light of the recent report (Bidwell *et al.*, 1986) that strontium is required in the culture medium of several laboratory-reared molluscs before statoconia develop.

INTRODUCTION

In all molluscs studied to date, gravity reception is mediated by bilateral paired statocysts. The general form of the statocysts is that of a fluid-filled sac with ciliated mechanoreceptor cells along its wall. The ciliated surface of the receptor cells faces the lumen of the cyst. The gravitational stimulus is transduced by the interaction of stones, generally of calcium carbonate, in the lumen with the cilia of the receptor cells. Since the stones have a greater specific gravity than that of the fluid, gravitational forces are normally exerted on the receptor-cell cilia. In some species, either a single stone or a concretion of many smaller stones exists, in which case the mass is referred to as a *statolith*. In other species, many individual stones move independently under the influence of gravity, animal movement and beating of the sensory cilia; in these cases, the stones are referred to as *statoconia. In those statoliths made up of multiple adherent stones, the individual stones are also referred to as statoconia. Thus, the term statoconia is usually used to denote relatively small (1 – 50* μm diameter), independent, paracrystalline elements. In this paper we will consider the interactions of the statoconia with the sensory cilia in the *Aplysia* statocyst, which lead to sensory transduction and the detection of gravity. Different suggested sites and mechanisms of generation of these stones, will also be considered.

Edited by Rex E. Crick
Plenum Press, New York

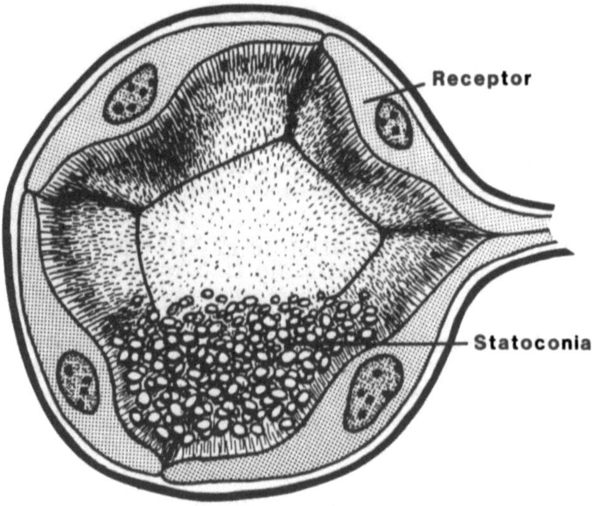

FIGURE 1. Schematic drawing of *Aplysia californica* statocyst. Note that supporting cells between receptor cells are not indicated. Modified from Gallin & Wiederhold (1977).

MATERIALS AND METHODS

Specimens of *Aplysia californica* were obtained from Pacific Bio-marine Laboratories, Venice, California. Animals were maintained in the laboratory in artificial sea water (Instant Ocean™) which lacks strontium (Bidwell *et al.*, 1986). Specimens were 20 – 125 grams. For light and transmission electron microscopy (TEM), statocysts were dissected free from live circumesophageal rings of ganglia and immersed directly into fixative. The material for Figures 2, 3 and 4 was fixed for 4 h with 3% glutaraldehyde in 0.15 M sodium cacodylate buffer solution, prepared with artificial sea water (Instant Ocean™) (1,004 mOsm, pH 7.35). Specimens were post-fixed in 1% osmium tetroxide for 1 h, dehydrated in a graded series of ethanols and embedded in epoxy resin.

Sections (1 μm) were retained at 10 – 20 μm intervals through the cyst and stained with 0.5% toluidine blue in 1.0% sodium borate for examination by light microscopy. For TEM, 90 nm sections were cut and stained with uranyl acetate and Reynold's lead citrate. Thin sections were examined in a Philips 301 TEM. The statoconium in Figure 7 was decalcified prior to dehydration by immersion in cacodylate buffer with the pH adjusted to 5.3 for 24 h. For the thin section in Figure 6, fixation was in 2.5% glutaraldehyde in 250 mM of 2,4,6-trimethylpyridine (s-collidine) buffer (pH 7.4) for 12 – 24 h. No additional decalcification was performed. For scanning electron microscopy (SEM) of isolated statoconia (Figures 8) the stones were isolated into deionized water which was blotted away, rinsed and blotted once more, on carbon planchets. The mounted specimens were coated with 60/40 gold/palladium at 20 nm thickness and examined in a JEOL JSM 35 SEM at 20 kV. For elemental analysis, similarly prepared but uncoated statoconia were examined in the same electron microscope using a Tracor Northern energy dispersive X-ray detector and a Tracor Northern NS880 X-ray analysis system.

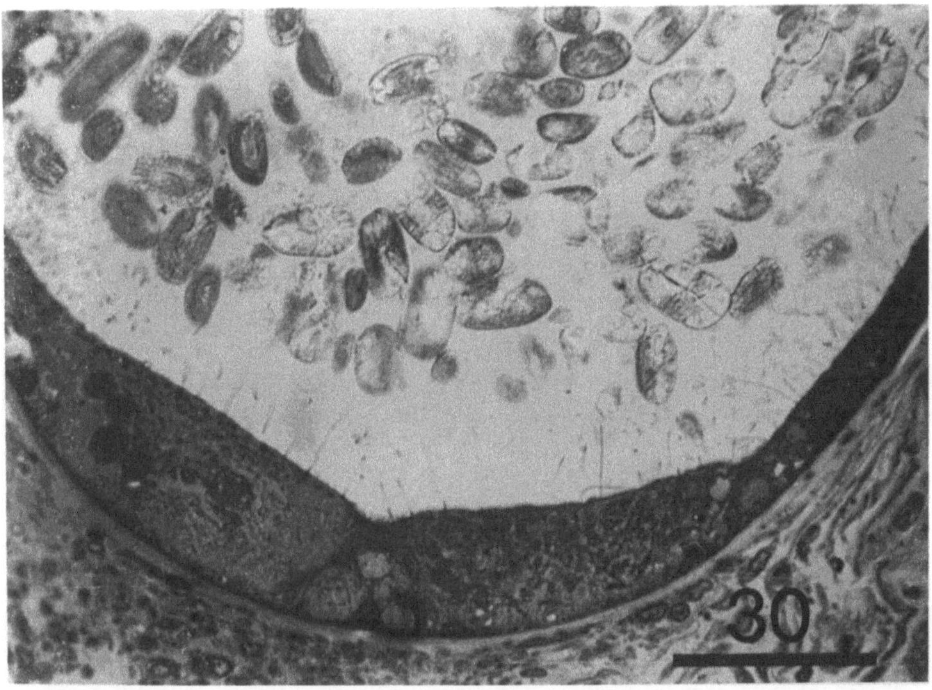

FIGURE 2. Light micrograph of lower portion of an *Aplysia* statocyst. 1 μm thick, undecalcified section, stained with toluidine blue. Bar=30 μm. Mod. from Wiederhold *et al.* (1986).

For physiological recording, the circumesophageal ring of ganglia was removed and mounted in an artificial-seawater bath maintained at 15° C. Part of the pedal ganglion was removed to expose the statocyst and individual receptor cells were impaled with glass microelectrodes having tip resistances of 50 – 100 megOhms. A plexiglass dish holding the preparation and micromanipulators holding the microelectrodes were mounted on a tilting table which could be rotated through 40° in either direction from horizontal.

STRUCTURE OF STATOCYSTS

The general plan of the *Aplysia* statocyst is representative of the simplest form of molluscan gravity receptor (Coggeshall, 1969; McKee & Wiederhold, 1974). The statocysts are paired, one being located between the pedal and pleural ganglia (parts of the circumesophageal ring of ganglia) on each side of the animal. Figure 1 is a schematic drawing of the organ. Each statocyst is a sphere, approximately 200 μm in diameter. The wall of the cyst is made up of a basal lamina surrounding 13 large receptor cells. Each receptor cell has its own axon and these 13 axons join to form the statocyst nerve, indicated schematically in Figure 1 as proceeding to the right of the cyst. There is also a larger number of small supporting cells peripheral to and interposed between the receptor cells, which are not indicated in this figure. The luminal surface of the receptors is covered with many (approximately 700 in several examples counted) long cilia (approximately 12 μm long, as illustrated in Fig. 2, below). The statoconia are indicated

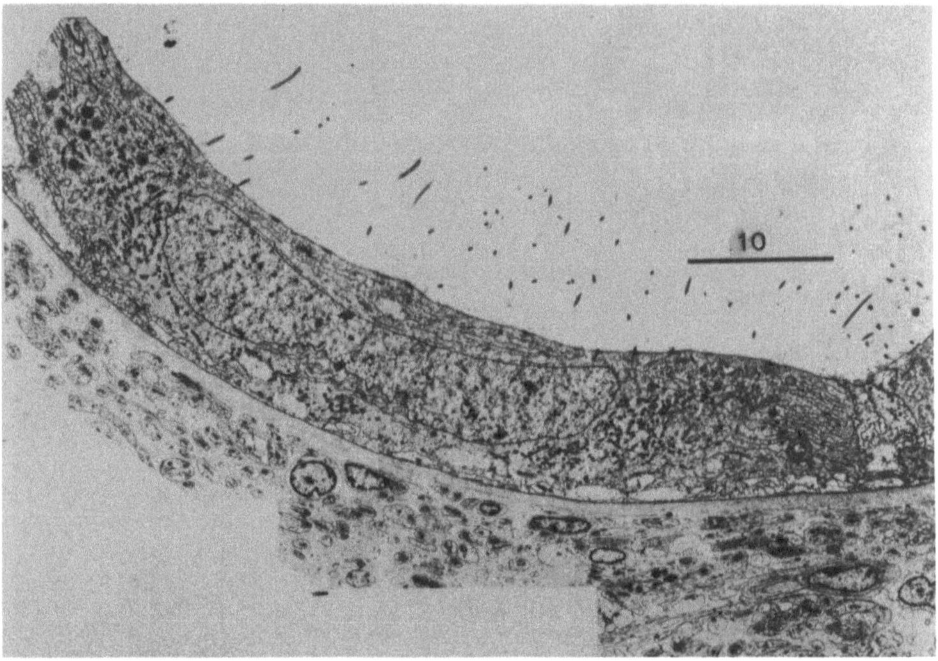

FIGURE 3. Montage TEM of section through one receptor cell. Note large nucleus, intracellular organelles and many cilia. Bar = 10 μm.

schematically as ellipsoids. They vary from approximately 3 to 20 μm in diameter. The number of statoconia varies from only one in larval *Aplysia* (Bidwell *et al.*, 1986, Gallager *et al.*, 1988) to as many as 1,000 in an adult (McKee & Wiederhold, 1974). The cilia all have the typical morphology of motile cilia, possessing a central pair of microtubules as well as 9 doublet microtubules around the periphery of the cilium. In fact, these cilia are motile, as well as sensory, as will be discussed below. Variations in the form of statocysts in other molluscan species are described in Wiederhold, *et al.*, 1986.

A more accurate picture of the *Aplysia* statocyst is given in Figure 2, a cross-section, illustrating statoconia and receptor cells at the "bottom" of the cyst. Sections of many statoconia are seen, as well as cilia on the luminal surface of three receptor cells. The statoconia are ellipsoidal in shape, varying here from 5× 2 μm to 17.5× 5 μm. Note several cilia whose complete length of 12 μm can be seen. In live preparations, the mass of stones can be seen to fill the bottom one – third of the lumen. Note that the statoconia appear to be free from one another. If a dissected live *Aplysia* statocyst is rotated under a microscope, the statoconia can be seen to tumble over one another as they fall to the new "bottom" of the lumen. Figure 3 is a montage illustrating a cross section of an entire receptor cell. Note the large nucleus and many intracellular organelles, as well as the large number of cilia cut in cross-section. Figures 2 and 3 illustrate that individual receptor cells can be quite large, compared to mammalian hair cells. The *Aplysia* receptor cells are typically rounded plates, 10 to 15 μm thick and up to 100 μm on a side. In the example shown in Figure 3, the luminal extent of the cross section is 80 μm from

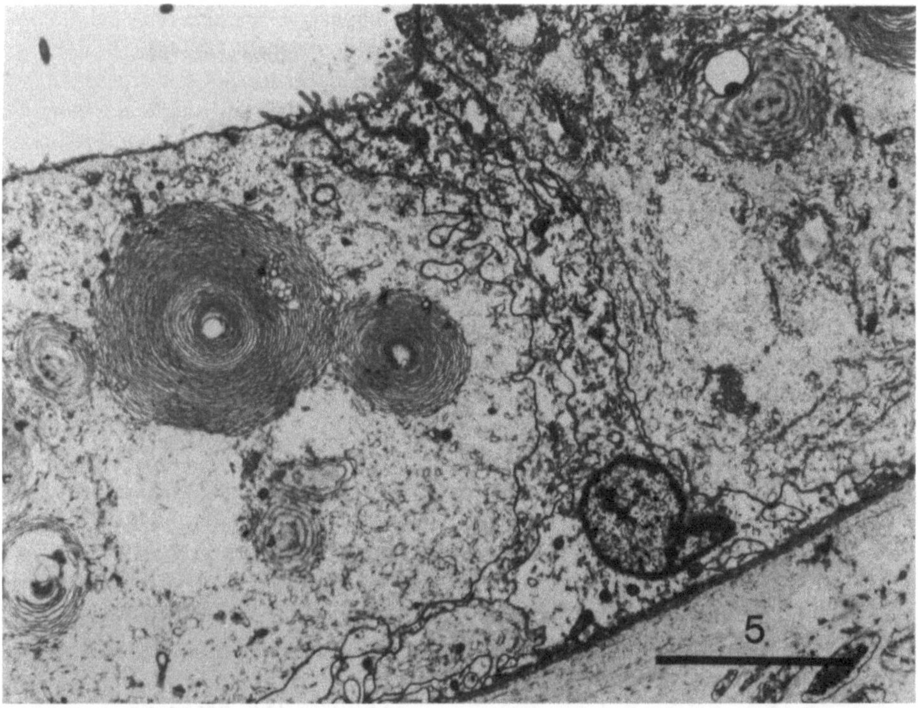

FIGURE 4. TEM of two statocyst receptor cells with several supporting cell processes interposed between them. Note prominent lamellar bodies in both receptor cells. Bar=5 μm. Mod. from Wiederhold *et al.* (1986).

one end of the receptor cell to the other. From the size of the statoconia, relative to the spacing between cilia, it is apparent that when a single statoconium strikes the cell surface, it will interact with several cilia nearly simultaneously.

Figure 4 shows the border region between two receptor cells, with several small supporting-cell processes separating them at the luminal surface. Several characteristic organelles can be seen in this figure. Smooth endoplasmic reticulum is seen in the receptor cell, approaching its border with the supporting cell. Multiple lamellar bodies are frequently seen in this position, near the junction with the supporting cells. Here, 6 lamellar bodies are seen in the left-hand receptor cell and two in the right-hand cell. The large lamellar body in the left-hand cell is 6.3 μm in diameter.

FUNCTION OF STATOCYSTS

Molluscan statocysts have provided useful preparations for the study of mechanoelectric transduction in ciliated mechanoreceptor cells. Early studies concentrated on response properties of the statocyst nerve fibers to tilt and rotation (*e.g.*, Wolff, 1973). The large size of the receptor cells in *Aplysia californica* made them attractive for intracellular recording. In favorable cases, intracellular recording with a micro-electrode has been maintained for up to 8 hours (Gallin & Wiederhold, 1977; Wiederhold, 1974, 1977, 1978).

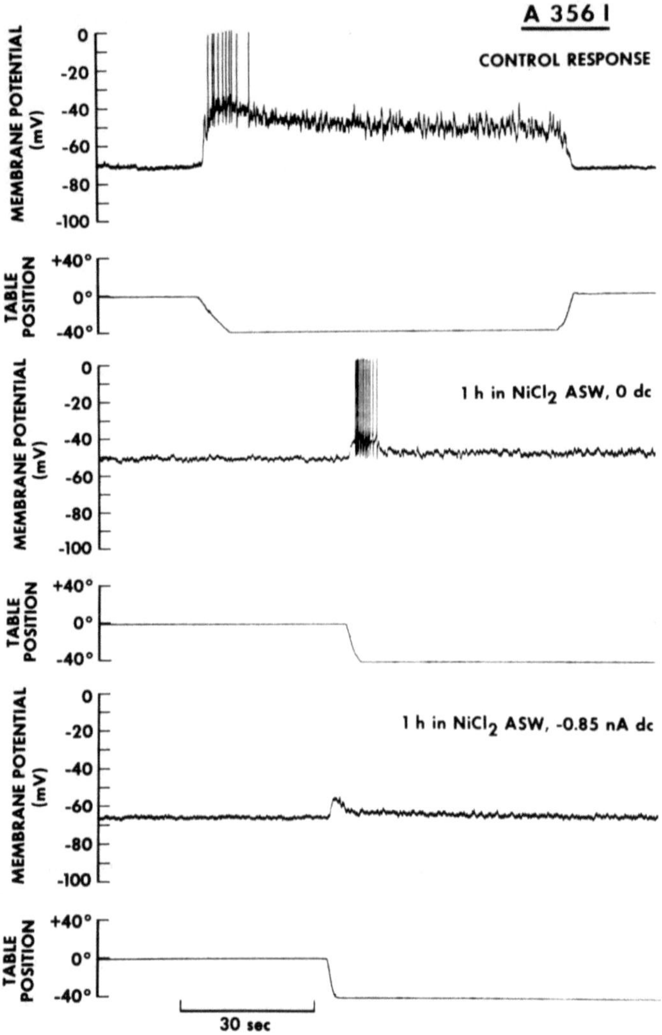

FIGURE 5. Intracellular responses of a statocyst receptor cell to downward tilt through 40°. Top pair of traces with preparation in normal sea water; middle pair, 1 h after being in sea water with 10 mM Ni^{2+} added; bottom pair, in 10 mM Ni-sea water with membrane potential restored to -70 mV when table is in horizontal position. In each pair, upper trace is membrane potential and lower trace is tilting table position.

Aplysia statocyst receptor cells typically have resting membrane potentials ranging from -50 to -80 mV, input resistance of the order of 100 megOhm and electrical time constants near 100 msec. The large input resistance and long time constant facilitate the measurement of membrane resistance using a single intracellular electrode and a bridge circuit (Gallin & Wiederhold, 1977). We have recorded from receptor cells in a preparation of the circume-sophageal ring of ganglia, including the statocysts, mounted in a small chamber containing cooled artificial sea water, on a tilting table. When the table is tilted from a position in which the recorded cell is above the level of the statoconia to a position where the recorded cell's cilia are in contact with the stones, large depolarizations and generation of action potentials are observed. In the example illustrated in the upper two panels of Figure 5, the receptor cell has a resting potential of -70 mV with the table in the horizontal position (table position of 0 degrees). When the table is tilted to bring the recorded cell down (-40 degrees), the cell depolarizes by 30 mV and a series of action potentials lasting approximately 10 sec is generated. The membrane potential then partially repolarizes, so that the action potentials cease, but the increase in potential fluctuations continues throughout the cell-down tilt.

The action potentials carried in the statocyst nerve convey orientation information to the cerebral ganglia. Only those cells at the bottom of the cyst are activated, giving the central nervous system information concerning which direction is "down". The depolarization in response to this physiologic stimulus can be as large as 50 mV. By passing small current pulses through the recording electrode and across the receptor cell membrane, it is also possible to measure changes in membrane conductance associated with the response. Although the electrical characteristics of the receptor-cell membrane are very non-linear, which can complicate the analysis (Wiederhold, 1977), it can be shown that the depolarization is due to an increase in membrane conductance caused by the mechanical stimulus. By changing the ionic composition of the artificial sea water bathing the preparation, it has been shown that the conductance increase caused by stimulation results primarily from an increase in conductance to sodium, presumably either on the ciliary membrane or the luminal surface membrane of the receptor cell (Gallin & Wiederhold, 1977). By eliminating the sodium in the bathing medium, responses to tilting are eliminated within 15 to 20 min, suggesting a relatively unrestrained exchange between the cyst lumen and the external medium. This is consistent with the finding that a microelectrode in the lumen records no standing potential and virtually no electrical resistance, relative to a reference electrode in the bath. Thus, ions and trace elements in the extracellular spaces of the animal will have access to the cyst lumen and the luminal surfaces of the receptor and supporting cells.

An interesting feature of the physiological responses of statocyst receptor cells is the large increase in fluctuations in membrane potential associated with the depolarizing response to tilt or rotation, as seen in the top panel of Figure 5 (see also: Gallin & Wiederhold, 1977; Wiederhold, 1978; Grossman *et al.*, 1979). When an *Aplysia* statocyst preparation is viewed under a dissecting microscope, the statoconia can be seen to be in continual, random movement. This has also been noted in *Clione limacina* (Tsirulis, 1974), *Pecten maximus* larvae (Cragg & Nott, 1977) and *Hermissenda crassicornis* (Grossman *et al.*, 1979). Direct observation of active ciliary beating is reported in the statocysts of *Helix* (Laverack, 1968), *Lymnaea stagnalis* (Geuze, 1968) and *Hermissenda* (Stommel *et al.*, 1980). When nickel chloride ($NiCl_2$) (10 mM) or serum from patients with cystic fibrosis (both of which can block ciliary motility--see Lindemann, *et al.*, 1980; Danes & Bearn, 1972), were added to the sea water bathing an *Aplysia* statocyst, the random motion of statoconia ceased (Wiederhold, 1978). The motion is thus, in all likelihood, imparted by the active beating of the receptor-cell cilia. Application of nickel or cystic fibrosis serum also greatly reduces the voltage fluctuations and the magnitude of the depolarizing response to tilt. In the middle two panels of Figure 5, a -40° tilt was applied 1 h after adding 10 mM $NiCl_2$ to the artificial sea water bathing the preparation. In this particular case, the cell had depolarized by 20 mV while the preparation was in the Ni enriched sea water although, averaging across all preparations, there was no consistent effect of Ni on either the resting membrane potential, the cell's input resistance or the threshold or amplitude of action potentials. In the Ni sea water, the depolarization caused by the same tilt as in the top trace was only 10 mV and the train of action potentials lasted only 5 sec. If the membrane potential was restored to -70 mV with the table horizontal (bottom two panels), by passing -0.85 nA current through the microelectrode, the depolarizing response lasted only approximately 5 sec, no action potentials were generated, and the potential fluctuations barely increased during the maintained tilt. The fact that the depolarizing response of the receptors is greatly reduced by the same treatments that block active ciliary beating, led Wiederhold (1976, 1978) to conclude that the actual transduction mechanism involves the cilia actively striking the statoconia during their ongoing beating, rather than the statoconia passively deflecting the cilia, as is thought to be the case in vertebrate hair cells. Stommel *et al.* (1980) have come to similar conclusions from experiments in which ciliostatic agents were injected into *Hermissenda* statocyst receptor cells.

From this synopsis, it can be seen that the physiology and biophysics of sensory transduction in molluscan statocysts is understood in reasonable detail. The manifestation of collisions between receptor-cell cilia and individual statoconia, in the prominent voltage fluctuations associated with the responses, suggests that the physical and structural characteristics of the statoconia could appreciably affect the physiologic responses of these gravity receptors. These aspects will be treated in more detail in the next section.

FORMATION AND COMPOSITION OF STATOCONIA

Most molluscan statoconia described to-date have forms similar to that illustrated in Figures 1 and 2. The major exceptions are some of the bivalves, which have large "single-crystal" statoliths or statoliths which are concretions of statoconia (Morton, 1985), and the dibranchian cephalopods (squid, cuttlefish and octopus) in which the normal statolith has a complicated shape and is made up of a concretion of small spindle-shaped crystals (Dilly, 1976; Colmers *et al.*, 1984). The finding that in the same species, *Aplysia californica*, there is only a single statolith in larval animals (Bidwell *et al.*, 1986, Gallager *et al.*, 1988) and approximately 1,000 statoconia in adults (McKee & Wiederhold, 1974) indicates that stones are added to the cyst lumen during development. The statocysts are one of the first components of the *Aplysia* nervous system to develop. Fully formed statocysts are present in larval animals when they hatch, 10 days after fertilization. At this stage the only other components of the nervous system which can be identified are the cerebral and pedal ganglia (Kriegstein, 1977a,b). Bidwell *et al.* (1986) note the presence of a statolith in normal 5 day embryos. Coggeshall (1968) emphasizes that the diameter of the statocyst varies little between the smallest (1 g) and the largest animals studied (300 g) and that the number of receptor cells (13) is constant.

It is not known when in the development of these animals the number of statoconia increases or whether this is an ongoing process throughout their life.

There are several suggestions in the literature concerning the site of generation of statoconia. Laverack (1968) was one of the first to apply EM techniques to the investigation of molluscan statocysts. He termed the large ciliated cells, of which there are 13 also in *Helix*, "giant" cells, but concluded that these could not be receptor cells, largely because all sensory receptors described to that point were small columnar cells. Thus, Laverack concluded that the supporting cells were the receptors. However, he did illustrate "fully formed" statoconia, approximately 7 μm along their major axis, within a giant cell. He states that "the bulk of the calcareous material leads to herniation of the capsule...when the cell contains a number of statoliths they are released into the lumen of the statocyst, by disruption of the cell." Laverack also notes striking lamellar bodies, similar to those illustrated in Figure 4, in the giant (receptor) cells, near intracellular statoconia, but concludes that the lamellar bodies do not seem to be related structurally to the statoconia. He states that in regions of the giant cell where lamellar bodies are numerous, the cell frequently becomes detached from the capsule and "disintegrates into the lumen of the statocyst". We have occasionally seen such disruption with impressive amounts of regular, but more loosely arranged membranes than those illustrated in Figure 4, within the lumen. However, we could not be assured that this was not a form of degeneration, perhaps due to trauma associated with either dissection or preparation of the specimen.

In contrast to Laverack's suggestion that the statoconia are generated within what are now known to be the sensory receptor cells (Laverack, 1968), several investigators have illustrated small statoconia in invaginations of the luminal surface of supporting cells, between receptor cells and near the region where the nerve leaves the cyst. Tsirulis (1974) illustrates 5 statoconia, up to 12.5 μm along their major axis, within one supporting cell and another, 4 μm long, apparently emerging from a supporting cell in *Clione limacina*. Geuze (1968) presented evidence that in both normal and regenerating statocysts of *Lymnaea stagnalis*, the statoconia are produced by supporting cells. After puncturing the statocyst and removing both the statolymph and statoconia, the statocysts completely returned to normal appearance in 48 h. At 12 h after puncture, statoconia of low electron density were seen within vacuoles at the apical surface of poorly differentiated cells. In "subadult" normal animals, examples are shown of statoconia in broken vacuoles, in the cyst lumen, adjacent to the apical surface of supporting cells. Kuzirian *et al.* (1981) also demonstrated what is described as a forming statoconium within a supporting cell in *Hermissenda crassicornis*. However, it is difficult to distinguish this from a portion of a lamellar body. The example illustrated is near the junction between supporting and receptor cells, where processes of the two frequently interdigitate, and it is difficult to distinguish one cell type from the other. Cragg & Nott (1977) described one cell in each statocyst of the pediveliger *Pecten maximus* which is comparable in size to the ciliated receptor cells, but contains no cilia itself. This cell is said to contain inclusions resembling the variety of statoconia they describe in the cyst lumen and they suggest that this cell generates these stones and expels them into the cyst lumen. No one else has described such a cell. To date, we have not identified indisputable statoconia within either receptor or supporting cells in *Aplysia californica*.

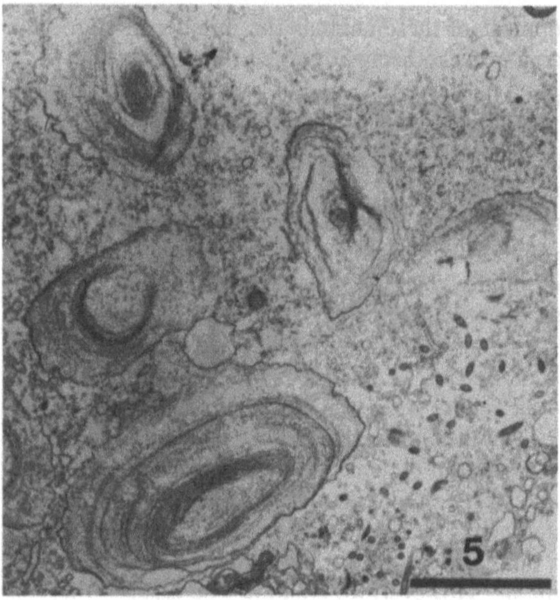

FIGURE 6. TEM of several statoconia which were not decalcified beyond glutaraldehyde fixation in s-collidine buffer. Note concentric membranous ring structure in statoconia. Bar=5 μm. Mod. from Wiederhold, *et al.* (1986.

Of course, all of these studies suggesting that the statoconia are generated within either the receptor or supporting cells are based on static anatomical methods and the possibility that the stones penetrated into the cells from the lumen, possibly as a post-mortem artifact, cannot be excluded. In fetal rat, Salamat *et al.* (1980) demonstrated that developing otoconia are released from vesicles on the surface of cells in the sensory epithelium of the sacculus.

It is clear that the mineralization of molluscan statoconia is laid down on a biological membrane structure. Vinnikov *et al.* (1980) generalized that the "otoliths" of the "lower" invertebrates, the coelenterates and ctenophores, are formed by an "endogenic" intracellular mode, while those of the molluscs and arthropods are formed by an endogenous deposition upon an extracellular structure. Figures 6 and 7 illustrate statoconia within the lumen of *Aplysia californica* statocysts prepared and partially decalcified by different methods. Similar figures have been shown by Coggeshall (1969) for *Aplysia* and Kuzirian *et al.*, (1981) for *Hermissenda crassicornis*. Concentric membranous rings can be seen in most of the stones in Figure 6 and in the left-hand stone of Figure 7, but not in the other statoconium in Figure 7. The lack of a visible ring structure is not uncommon. Presumably the whole stone is made up of concentric ellipsoids and if such a structure were cut near an end, the ring structure could be missed, as in cutting a section from the side of an onion, rather than cutting across its center. It seems likely that the statoconia grow after their initial formation by adding successive layers. If this is the case, the lamellar bodies (Fig. 4) could provide the membrane for such deposition. As noted above, Laverack (1968) described "disintegration" of the receptor cells, in regions with numerous lamellar bodies, into the cyst lumen, a phenomenon which we have observed in *Aplysia* statocyst cells of uncertain condition. Williams (1977) described a process by which lamellar bodies, which appear similar to those in Figure 4, in the alveoli of fetal rat lung, exocytose at the cell surface to produce the tubular myelin which is thought to serve as a store of phospholipids for pulmonary surfactant. Thus, there is a precedent for similar structures being exocytosed to provide extracellular membrane. In Figures 6 and 7, the outer membrane layer appears more irregular than do the inner rings. Perhaps as the outer layer calcifies, the membrane is packed into a more confined space. Vinnikov *et al.* (1980) illustrated exceptionally thick

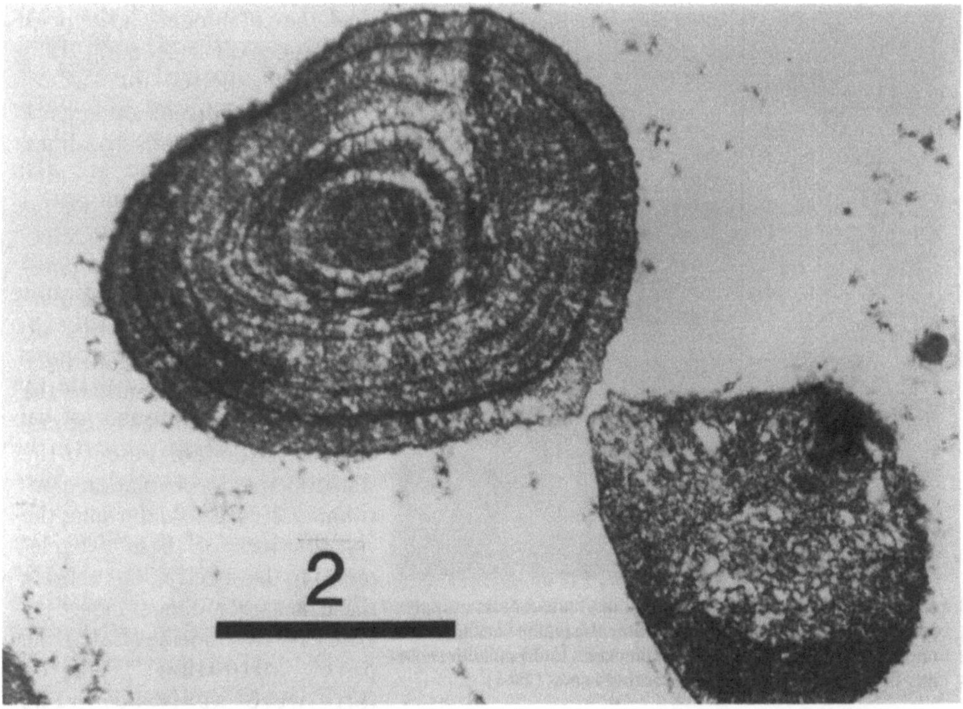

FIGURE 7. TEM of two statoconia which were decalcified in pH 5.3 buffer for 24 h. Note that ring structure is prominent in left stone but not in right. Bar = 2 μm. Modified from Wiederhold, *et al.* (1986).

and electron-dense outer membranes in thin sections of statoconia from the limpet *Acmaea pallida*.

All of the statoconia described above have gone through some form of fixation, which could alter the elemental composition and surface structure. Ballarino (1985) has shown that even deionized water can etch avian otoconia. To minimize such changes, we have examined statoconia from one preparation in which the statocysts were dissected free from a live specimen (*Aplysia californica*) and bisected in a drop (at least 1,000 times the volume of the statocyst) of deionized water, directly on a carbon planchet. The water was blotted away and another drop placed and blotted to rinse the stones a second time. An SEM of a statoconium prepared in this manner is shown in Figure 8. The surface of this stone is more planar and angular than those previously illustrated in the literature (*e.g.*, McKee & Wiederhold, 1974; Kuzirian *et al.*, 1981). The stones in Figure 8 appear to have much sharper corners than the fixed stones. The unfixed stones show plate-like irregularities on their surface which appear to have cleaved at a preferred angle, suggesting a single-crystal formation.

To investigate the elemental composition of the statoconia, uncoated preparations similar to those shown in Figure 8 were studied by energy-dispersive X-ray microprobe analysis. Figure 9 is an X-ray energy spectrum obtained from a whole statoconium.

FIGURE 8. SEM of a large (14 μm dia.) statoconium prepared by vital dissection into deionized water. Irregular background is unpolished surface of the carbon planchet. Gold-palladium coating. Bar = 2μ m. Mod. from Wiederhold *et al.* (1986).

Note the prominent CaKα peak (which goes off scale in Fig. 9) as would be expected for $CaCO_3$. Small but significant amounts of sodium, magnesium, strontium, sulfur and chlorine are also present. The aluminum peak is an artifact from the X-ray detector housing. Note that no phosphorus peak is seen, indicating that there is probably no hydroxyapatite in the statoconia. In an attempt to quantitate the relative molar amounts of calcium and strontium present in the statoconia, a deconvolution algorithm was used to determine the contributions of these two elements to the spectra. To calibrate these measurements, standards of pure calcium fluoride (CaF_2) and pure strontium fluoride (SrF_2) were analyzed. The average Ca/Sr molar ratio derived from 6 statoconia was 97:1 ± 3 (mean ± std. error). The average number of counts for the CaKα peak was 90,900 and that for the Sr_L peak was 1,300. The molar ratio was obtained from these counts, corrected for the relative emission efficiencies of pure CaF_2 and SrF_2 crystal standards. The Ca/Sr ratio is very similar to that obtained by Bidwell *et al.*, (1986) in the single statolith of larval *Aplysia californica* and is close to the molar ratio of sea water (103:1) given by Nicol (1976).

The presence of strontium in the statoconia is of interest in light of the report (Bidwell *et al.*, 1986, Gallager *et al.*, 1988) that the strontium content in the medium in which embryos are reared has a profound effect on the development of both the shell and the statoliths in a number of molluscan species. Roger Hanlon (Marine Biomedical Institute, Galveston, Texas) has found similar results in cephalopods (pers. comm.). We had previously studied the statoliths of individuals of several species of hatchling cephalopods which exhibited a behavioral anomaly, in that they followed a "corkscrew" pattern of swimming or "somersaulted"; *i.e.*, were generally unable to swim in an oriented manner (Colmers *et al.*, 1984). Such individuals were termed "spinners". The only structural abnormalities in these animals were in their statoliths. Hanlon has now found that in two species of octopus (large- and small-egged species), *Sepia* and two species of squid, if there is no strontium in the sea water in which the embryos are raised, all animals are spinners. If the strontium level is raised to 4 ppm, approximately one half of the individuals are spinners and when the strontium is raised to the normal 8 ppm, all animals exhibit normal behavior. This requirement for strontium to obtain normal

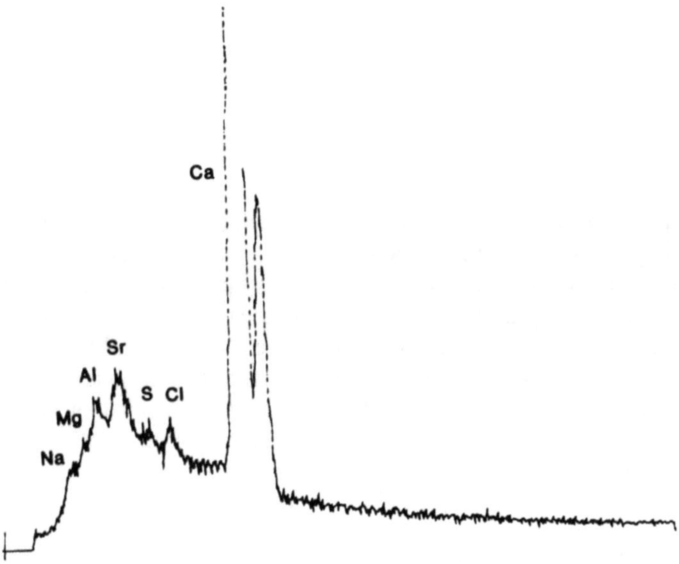

FIGURE 9. Energy-dispersive X-ray microprobe spectrum of analysis of a single deionized-water-rinsed, uncoated statoconium. Note prominent CaKα peak, which is off-scale, and smaller CaKβ peak. Clear peaks for Na, Mg, Al (artifact, see text), Sr, S and Cl are also seen. Full scale = 4096 counts. Mod. from Wiederhold *et al.* (1986).

development of molluscan statoconia would appear to be analogous to the requirement for manganese for normal otoconial development in mammals (Erway *et al.*, 1970; Lim & Erway, 1974).

The molecular basis for the strontium requirement for otoconial development is unknown. Carlstrom (1963) stated that strontium is known to favor the formation of the aragonite form of calcium carbonate. However, aragonite can be formed in the absence of strontium (Carlstrom, 1963; Ross & Pote, 1984; Mann *et al.*, 1983). Whereas the otoconia of all mammals, aves and some reptilia are made of calcite, many fish and some amphibia have otoconia of aragonite (Carlstrom, 1963; Ross & Pote, 1984; Mann *et al.*, 1983). The lizard *Podarcis s. sicula* has calcite otoconia in the lagena, aragonite otoconia in the endolymphatic duct and sac and a mixture of the two in the saccule (Marmo *et al.*, 1981). Lowenstam *et al.* (1984) have recently shown that the statoconia in tetrabranchian cephalopods (*Nautilus*) as well as the statoliths of all of the dibranchian cephalopods (cuttlefish, squid and octopus) are aragonite. The otoconia of the *Nautilus* are about 1% by weight strontium, whereas the statoliths of the other cephalopods are about 0.5% strontium by weight. The shell and other hard parts of *Nautilus* are also made up of aragonite, but their strontium concentration is less than that of the statoconia. Crick *et al.* (1985) have shown that the Ca/Sr ratio is 4.5 times greater in the shell of *Nautilus*, compared to that of sea water. Thus strontium is more effectively excluded from the shell than from the statoconia, even though both are made of aragonite. Aragonite is felt to be advantageous for *Nautilus*, in that it is less brittle than calcite, and thus makes the shell and its chambers better able to withstand the high hydrostatic pressures at the depths of 500–900 meters where these animals frequently live. The crystal structure of the *Aplysia* statoconia is not known, but the plate structure, particularly evident in Figure 8, suggests that they too are aragonite. This type of formation is typical of aragonite but not of calcite in the lizard (*cf.* Figs. 1, 2 & 3, Marmo *et al.*, 1981). However, Ross & Peacor (1975) illustrated plate-like features on rat otoconia, which are calcite.

The advantage of having statoconia made of aragonite rather than calcite is not obvious. Perhaps in species with independent statoconia there is enough friction as the stones tumble over one another that the physical properties of aragonite would make them less susceptible to "wear" than if they were made of calcite. In mammals and other species in which the otoconia are embedded in a gelatinous membrane, the stones are prevented from striking one another. In some spinner cephalopods which have abnormal statoliths, the crystal structure is grossly different from that of normal animals. Whereas in normal octopus, the statoliths are a concretion of spindle-shaped crystals 2–3μm long (Dilly, 1976; Colmers *et al.*, 1984), the statolith of a spinner *Octopus joubini* illustrated in Colmers *et al.* (1984), is a large mass of tightly packed prismatic crystals. Thus, changes in crystal structure of the developing statoliths in the low-strontium sea water probably contributed to the spinner trait. In those specimens which failed to produce any statoliths, the lack of strontium could have been sufficient to prevent the precipitation of any stable form of $CaCO_3$.

CONCLUSIONS

In light of the profound influence of the trace element strontium on molluscan statoconia formation and of manganese on the formation of mammalian otoconia, it will be of interest to see if future research reveals similar mechanisms responsible for the abnormal crystal structure and shapes of aberrant otoconia seen in normal (Johnsson *et al.*, 1980) or abnormal (Ross & Peacor, 1975) laboratory animals and in cases of human otopathology (Johnsson *et al.*, 1982). If the membranous framework upon which the molluscan statoconia are built is generated within the sensory receptor cells, it would appear likely that the degree of physiological activity in the receptors influences the formation and growth of the statoconia. This would imply that the maintenance of the appropriate number and form of statoconia is a dynamic, rather than passive, process.

ACKNOWLEDGEMENTS

We wish to thank Jeffrey Harrison and Richard Polich for technical assistance and advice. Supported in part by Veterans Administration Medical Research funds and N.A.S.A. Space Biology Program. Participation in the Symposium was made possible by grants from The Organized Research Fund of The University of Texas at Arlington and The Sea Grant Program administered by Texas A & M University.

REFERENCES

BALLARINO, J., HOWLAND, H.C., SKINNER, H.C.W., BROTHERS, E.B., & BASSETT, W., 1985. Studies of otoconia in the developing chick by polarized light microscopy. Amer. J. Anat., 174:131-144.

BIDWELL, J.P., PAIGE, J.A., & KUZIRIAN, A.M., 1986. Effects of strontium on the embryonic development of *Aplysia californica*. Biol. Bull., 170:75-90.

CARLSTROM, D., 1963. A crystallographic study of vertebrate otoliths. Biol. Bull., 125:441-463.

COGGESHALL, R.E., 1969. A fine structural analysis of the statocyst in *Aplysia californica*. J. Morph., 127:113-132.

COLMERS, W.F., HIXON, R.F., HANLON, R.T., FORSYTHE, J.W., ACKERSON, M.V., WIEDERHOLD, M.L., & HULET, W.H., 1984. "Spinner" cephalopods: defects of statocyst suprastructures in an invertebrate analogue of the vestibular apparatus. Cell Tissue Res., 236:505-515.

CRAGG, S.M., & NOTT, J.A., 1977. The ultrastructure of the statocysts in the pediveliger larvae of *Pecten maximus* (*L.*) (Bivalvia). J. Exp. Mar. Biol. Ecol., 27:23-36.

CRICK, R.E., BURKART, B., CHAMBERLAIN, J.A., & MANN, K.O., 1985. Chemistry of calcified portions of *Nautilus pompilius*. J. Mar. Biol. Assoc. U.K., 65:415-420.

DANES, B.S., & BEARN, A.G., 1972. Oyster ciliary inhibition by cystic fibrosis culture medium. J. Exp. Med., 136:1313-1317.

DILLY, P.N., 1976. The structure of some cephalopod statoliths. Cell Tissue Res., 175:147-164.

ERWAY, L.C., HURLEY, L.S., & FRASER, A.S., 1970. Congenital ataxia and otolith defects due to manganese deficiency in mice. J. Nutrition, 100:643-654.

GALLAGER, S.M., BIDWELL, J.P., & KUZIRIAN, A.M., 1988. Strontium is required in artificial seawater for embryonic shell formation in two species of bivalve molluscs. *In* The Origin, Evolution and Modern Aspects of Biomineralization in Plants and Animals (R.E. Crick, ed.). Plenum Publishing Corp., New York.

GALLIN, E.K., & WIEDERHOLD, M.L., 1977. Response of *Aplysia* statocyst receptor cells to physiologic stimulation. J. Physiol. (London), 266:123-137.

GEUZE, J.J., 1968. Observations on the function and the structure of the statocysts of *Lymnaea stagnalis* (L.) Netherlands J. Zool., 18:155-204.

GROSSMAN, Y., ALKON, D.L., & HELDMAN, E., 1979. A common origin of voltage noise and generator potentials in statocyst hair cells. J. Gen. Physiol., 73:23-48.

JOHNSSON, L.-G., ROUSE, R.C., WRIGHT, C.G., HENRY, P.J., & HAWKINS, J.E., 1982. Pathology of neuroepithelial suprastructures of the human inner ear. Amer. J. Otolaryngology, 3:77-90.

JOHNSSON, L.-G., WRIGHT, C.G., PRESTON, R.E., & HENRY, P.J., 1980. Defects of the otoconial membranes in normal guinea pigs. Acta Otolaryngolica, 89:93-104.

KRIEGSTEIN, A.R., 1977a. Stages in the post-hatching development of *Aplysia californica*. J. Exp. Zool., 199:275-288.

KRIEGSTEIN, A.R., 1977b. Development of the nervous system of *Aplysia californica*. Proc. Nat. Acad. Sci. (USA), 74:375-378.

KUZIRIAN, A.M., ALKON, D.L., & HARRIS, L.G., 1981. An infraciliary network in statocyst hair cells. J. Neurocytology, 10:497-514.

LAVERACK, M.S., 1968. On superficial receptors. Sym. Zool. Soc. London, 23:299-326.

LIM, D.J., & ERWAY, L.C., 1974. Influence of manganese on genetically defective otolith. A behavioral and morphological study. Ann. Otology, Rhinology, and Laryngology, 83:565-581.

LINDEMANN, C.B., FENTIE, I., & RIKMENSPOEL, R., 1980. A selective effect of Ni^{++} on wave initiation in bull sperm flagella. J. Cell Biol., 87:420-426.

LOWENSTAM, H.A., TRAUB, W., & WEINER, S., 1984. *Nautilus* hard parts. A study of the mineral and organic constituents. Paleobiol. 10:268-279.

MANN, S., PARKER, S.B., ROSS, M.D., SKARNULIS, A.J., & WILLIAMS, R.J.P., 1983. The ultrastructure of the calcium carbonate balance organs of the inner ear: an ultra-high resolution electron microscopy study. Proc. Roy. Soc. London, Ser. B, 218:415-424.

MARMO, F., FRANCO, E., & BALSAMO, G., 1981. Scanning electron microscopic and X-ray diffraction studies of otoconia in the lizard *Podarcis s. sicula*. Cell Tissue Res., 218:265-270.

MCKEE, A.E., WIEDERHOLD, M.L., 1974. *Aplysia* statocyst receptor cells: Fine structure. Brain Res., 78:490-494.

MORTON, B., 1985. Statocyst structure in the *Anomalodesmata* (*Bivalvia*). J. Zool. Soc. London (A), 206:23-34.

NICOL, J.A., 1967. The Biology of Marine Animals. London: Isaac Pitman and Sons, pp. 699.

ROSS, M.D., & PEACOR, D.R., 1975. The nature and crystal growth of otoconia in the rat. Ann. Otology, Rhinology and Laryngology, 84:22-36.

ROSS, M.D., & POTE, K.G., 1984. Some properties of otoconia. Phil. Trans. Roy. Soc. London, Ser. B, 304:445-452.

SALAMAT, M.S., ROSS, M.D., & PEACOR, D.R., 1980. Otoconial formation in the fetal rat. Ann. Otology, Rhinology, and Laryngology, 89:229-238.

STOMMEL, E.W., STEPHENS, R.E., & ALKON, D.L., 1980. Motile statocyst cilia transmit rather than directly transduce mechanical stimuli. J. Cell Biol., 87:652-662.

TSIRULIS, T.P., 1974. The fine structure of the statocyst of the univalve mollusk *Clione limacina*. J. Evol. Biochem. Physiol., 10:158-165.

VINNIKOV, YA.A., KHARKEEVICH, T.A., ARONOVA, M.Z., TSIRULIS, T.P., LAVORA, YE.A., & NATOCHIN, V.V., 1980. Evolution of the otolith in invertebrates. Z. Obschei Biol., 41:815-827.

WIEDERHOLD, M.L., 1974. *Aplysia* statocyst receptor cells: intracellular responses to physiologic stimuli. Brain Res., 81:310-313.

WIEDERHOLD, M.L., 1976. Mechanosensory transduction in "sensory" and "motile" cilia. Ann. Rev. Biophys. Bioeng., 5:39-62.

WIEDERHOLD, M.L., 1977. Rectification in *Aplysia* statocyst receptor cells. J. Physiol. (London), 266:139-156.

WIEDERHOLD, M.L., 1978. Membrane voltage noise associated with ciliary beating in the *Aplysia* statocyst. Brain Res., 156:369-374.

WIEDERHOLD, M.L., SHERIDAN, C.E., & SMITH, N.K.R., 1986. Statoconia formation in molluscan statocysts. Scan. Elec. Micros./1986/II:781-792.

WILLIAMS, M.C., 1977. Conversion of lamellar body membranes into tubular myelin in alveoli of fetal rat lungs. J. Cell Biol., 72:260-277.

WOLFF, H.G., 1973. Multi-directional sensitivity of statocyst receptor cells of the opisthobranch gastropod *Aplysia limacina*. Mar. Behav. Physiol., 1:361-373.

CALCIFICATION IN HIGHER PLANTS WITH SPECIAL REFERENCE TO CYSTOLITHS

Hiroaki Setoguchi[1], Megumi Okazaki[1] and Shoichi Suga[2]

[1]*Department of Biology, Tokyo Gakugei University, Koganei-shi, Tokyo, 184 Japan;* [2]*Department of Pathology, Nippon Dental University, Chiyoda-ku, Tokyo, 102 Japan*

ABSTRACT. The cystolith, a calcified body in the leaf of higher plants, was observed by soft X-ray microradiography using the mature leaves of nine species from five families. The microradiographs revealed very large cigar-shaped cystoliths (up to 500 μm in length) in *Pilea viridissima* and *Justicia procumbens*, neighbor-cystoliths in *Morus bombycis* and *Humulus scandens*, and two to seven radially arranged cystoliths in *Momordica charantia*. The number of cystoliths (n/cm^2) of various kinds of leaves was estimated to be 850 to 4,200 by microradiography. The calcium carbonate content (mg/cm^2) calculated was 0.3 to 1.1, suggesting a large reservoir of calcium or carbon dioxide. The cystoliths were isolated from leaves of nine species to analyze chemically, with electron probe, and with X-ray diffraction. The calcium carbonate content in cystoliths was about 75% on a dry weight basis. A small amount of magnesium was also found. Electron probe analysis revealed that calcium and magnesium were evenly distributed through the cystolith body except for the stalk and the basal part of the body to which the stalk is attached. In the latter parts, silicon was detected in high density, suggesting silicification of these parts. X-ray diffraction patterns showed amorphous calcium carbonate in all species tested. However, vaterite in *Morus bombycis* and both vaterite and calcite in *Ficus elastica* were also detected in small amounts. The amorphous calcium carbonate in cystoliths changed rapidly into calcite in 0.05 M carbonate buffer (pH 9.2) or in distilled water. Then the smooth surface of the cystoliths was covered with small cubic calcite crystals. A tremendous number of cystoliths was contained in a single leaf, and the cystolith-bearing lithocyst was associated with many photosynthetic parenchyma cells in all species. These facts suggest a relationship between calcification in the cystolith and photosynthesis in the leaf. A possible mechanism of cystolith calcification coupled with bicarbonate utilization in photosynthesis is discussed.

INTRODUCTION

The most prominent calcium carbonate deposition in higher plants is cystolith formation in the leaves. A cystolith is a calcified body which is formed in a specialized cell called a lithocyst. The occurrence of cystoliths is known in the following families: Urticaceae, Ulmaceae, Moraceae, Cucurbitaceae and Acanthaceae (Ajello, 1941; Scott, 1946; Arnott, 1980).

Many morphologists, taxonomists and physiologists have been interested in calcified bodies since Meyen (1839) discovered cystoliths in the leaf of *Ficus elastica* (Moraceae) (Scott, 1946). However, there are still many unanswered questions concerning these

Origin, Evolution, and Modern Aspects of Biomineralization in Plants and Animals
Edited by Rex E. Crick
Plenum Press, New York

TABLE 1. Location and number of cystoliths and estimated calcium carbonate content in the leaves of nine species (from Okazaki *et al.*, 1986b)

Family	Species	Local[1]	Number cystoliths (n/cm^2)	$CaCO_3$ content (mg/cm^2)
Moraceae	*Morus bombycis*	U	1,230	0.45
	Humulus scandens	U	3,900	0.34
	Ficus elastica	U	1,090	0.39
	Ficus retusa	U,L	3,720	1.06
Urticaceae	*Boehmeria platanifolia*	U	3,440	0.30
	Pilea viridissima	U,L	1,840	0.41
Ulmaceae	*Celtis sinensis*[2]	U	4,200	0.47
Cucurbitaceae	*Momordica charantia*	L	3,310	0.44
Acanthaceaeeae	*Justicia procumbens*[2]	U	850	0.40

[1]U, upper epidermis; L, lower epidermis
[2]Newly added data

bodies: 1) what is the nature (crystallic or amorphous) of the biomineral, 2) what elements may be present in addition to calcium carbonate and silicon, 3) how does mineralization occur and does demineralization take place, and 4) what is the function of cystoliths?

In the present paper, the morphology, distribution, number/cm^2 of the leaf, inorganic composition and nature of calcium carbonate of the cystoliths are described. A possible mechanism of the calcification also is discussed in connection with photosynthesis in the leaf. In this study, the mature leaves of following nine species from five families were used: *Morus bombycis*, *Humulus scandens*, *Ficus elastica*, *Ficus retusa* (Moraceae); *Celtis sinensis* (Ulmaceae); *Boehmeria platanifolia*, *Pilea viridissima* (Urticaceae); *Momordica charantia* (Cucurbitaceae) and *Justicia procumbens* (Acanthaceae).

METHODS, RESULTS AND DISCUSSION

Morphology, distribution and number of cystoliths in leaves of various species

Light microscopy and soft X-ray microradiography were applied to observe the cystoliths in fresh leaves. X-ray microradiography was performed by the same ways described in a previous paper (Okazaki *et al.*, 1986b). A fresh leaf strip of about 2.5 cm× 4.5 cm was brought into close contact with X-ray film (Kodak spectroscopic film, type 640-0) over which a thin polyethylene film was placed. Radiographs were taken with soft X-ray equipment, Softex C-SM at 90 kV. The developed film was mounted on slide with Entellan New medium and observed with a light microscope. This technique was useful in examining the shape, size and distribution of cystoliths in fresh leaves. Figures 1-9 show the light micrographs and X-ray microradiographs of cystoliths from nine species (five families). Most cystoliths examined are localized in large epidermal

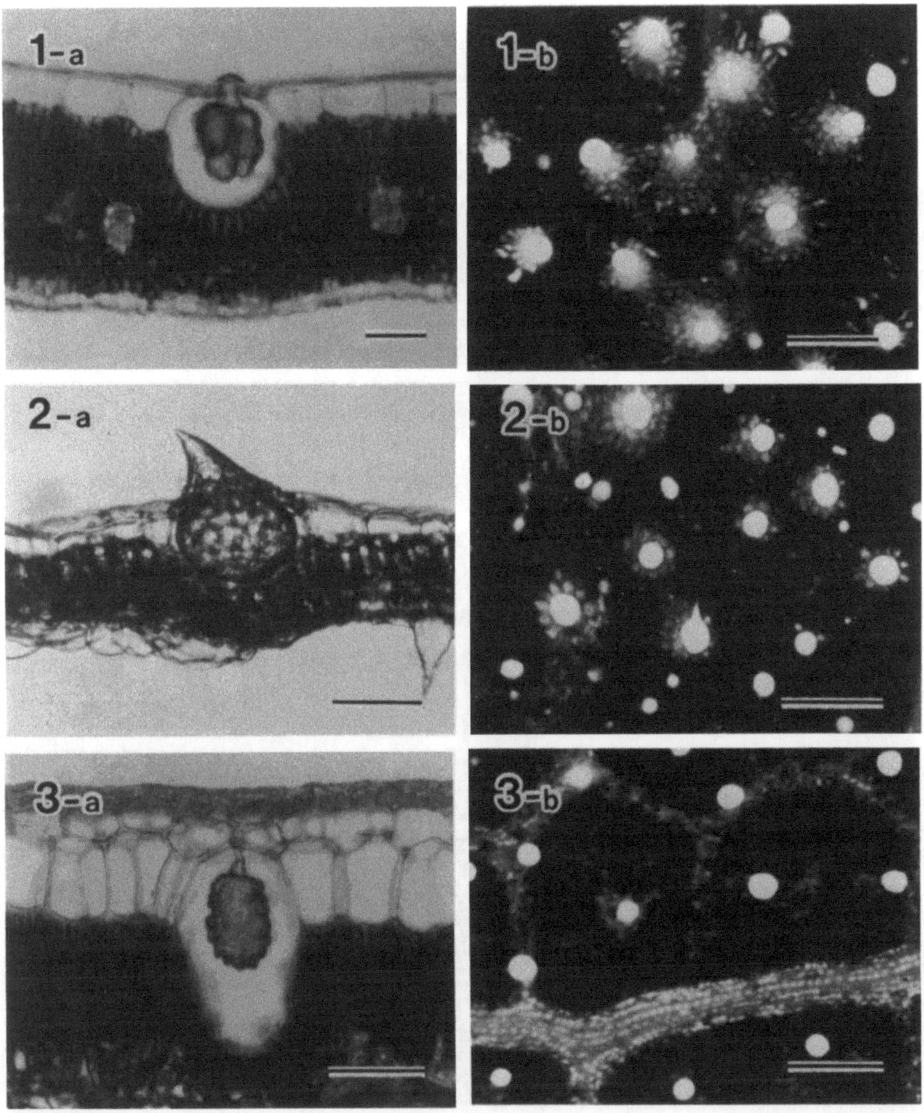

FIGURE 1–3. Cystoliths in the mature leaves of various higher plants. ▪*1. Morus bombycis* (Moraceae). Scale = a, 50 μm, b, 250 μm. ▪*2. Humulus scandens* (Moraceae). Scale = a, 50 μm, b, 250 μm. ▪*3. Ficus elastica* (Moraceae). Scale = a, 100 μm, b, 250 μm. For all figures, (a) is light micrograph of hand-sections and (b) is soft X-ray microradiograph of fresh leaves. (From Okazaki *et al.*, 1986b).

cells at the upper and/or lower side of the leaf (Figs. 1a-9a). In *Humulus scandens*, the cystoliths also are observed in the hair cells of the upper epidermis (Fig. 2a). The sites of cystoliths in the leaves are summarized in Table 1. The radiographs reveal a conspicuous cystolith in *Pilea viridissima* (Fig. 6b) and *Justicia procumbens* (Fig. 9b) which are cigar-shaped, being the largest (up to 500 μm in length) among all cystoliths tested. In general, cystoliths are evenly distributed in a single leaf (Figs. 1b-9b).

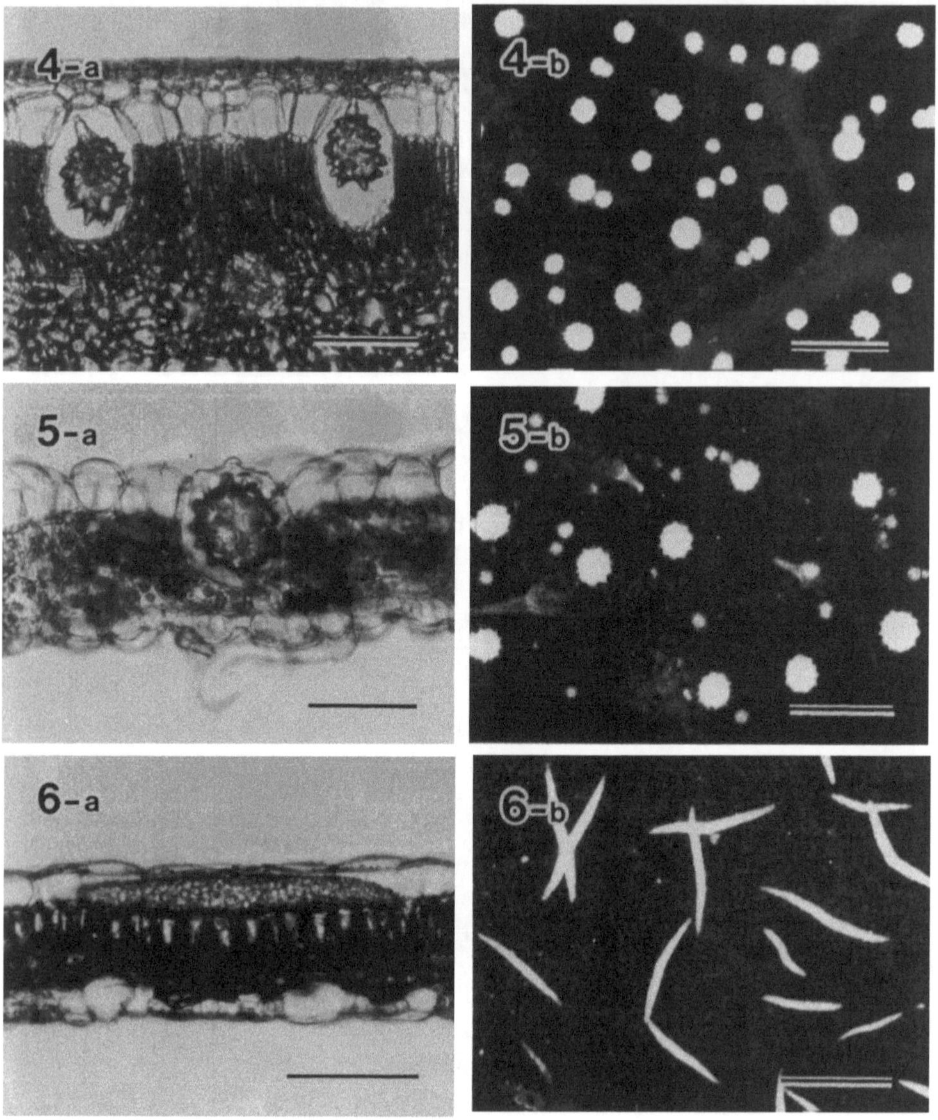

FIGURE 4–6. Cystoliths in the mature leaves of various higher plants. ▪*4. Ficus retusa* (Moraceae). Scale = a, 100 μm, b, 250 μm. ▪*5. Boehmeria platanifolia* (Urticaceae). Scale = a, 50 μm, b, 100 μm. ▪*6. Pilea viridissima* (Urticaceae). Scale = a, 100 μm, b, 250 μm. For all figures, (a) is light micrograph of hand-sections and (b) is soft X-ray microradiograph of fresh leaves. (From Okazaki *et al.*, 1986b).

However a characteristic distribution also is observed in *Morus bombycis* (Fig. 1b) and *Humulus scandens* (Fig. 2b). In these species, many small cystoliths called "Nebencystolithen" (Werner, 1931) surround each normal cystolith. It is very difficult to find them in hand-sections of the leaf because of their small size. Another distinctive distribution of cystoliths is in *Momordica charantia* (Fig. 7a, b) where paired cystoliths are observed in hand-sections (Fig. 7a), but X-ray microradiographs reveal a radial arrangement of

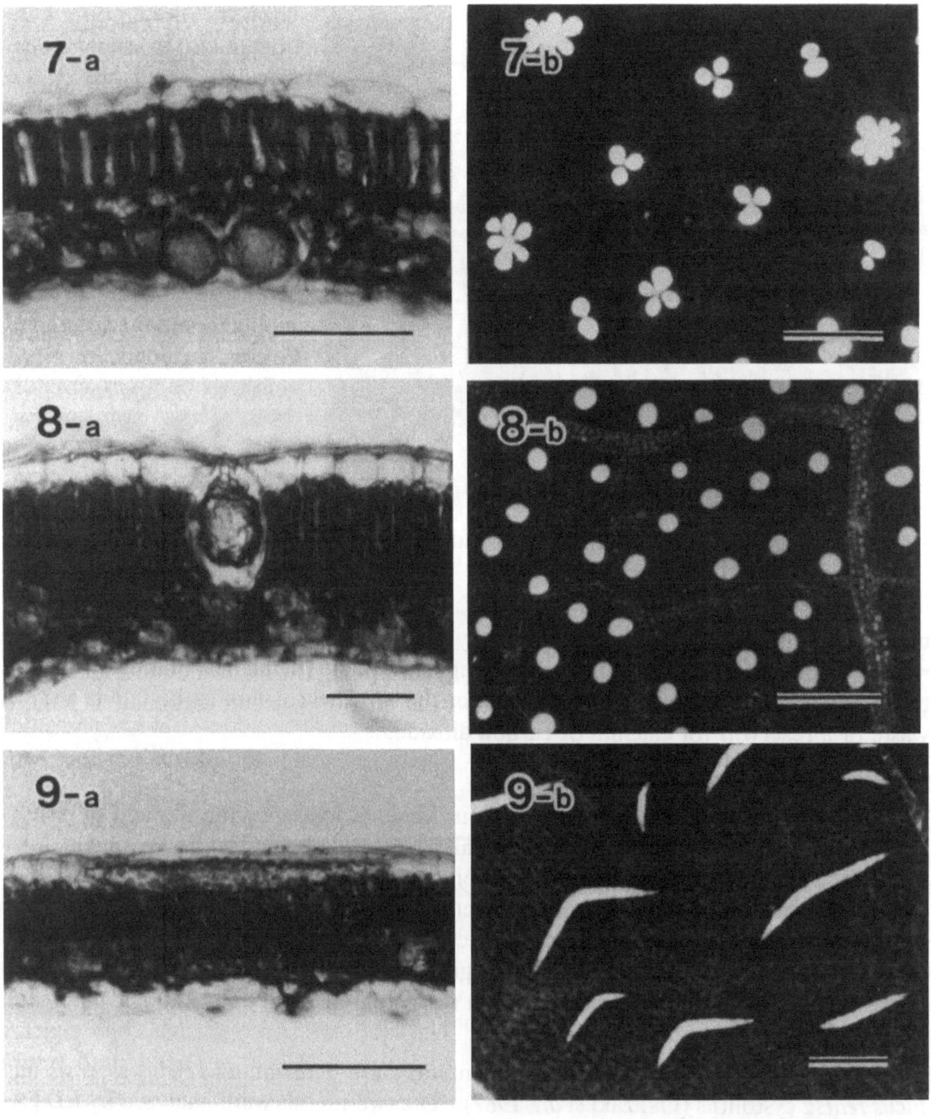

FIGURE 7–9. Cystoliths in the mature leaves of various higher plants. ▪7. *Momordica charantia* (Cucurbitaceae). Scale = a, 100 μm, b, 250 μm. ▪8. *Celtis sinensis* (Ulmaceae). Scale = a, 50 μm, b, 250 μm. ▪9. *Justicia procumbens* (Acanthaceae). Scale = a, 100 μm, b, 250 μm. For all figures, (a) is light micrograph of hand-sections and (b) is soft X-ray microradiograph of fresh leaves. (Fig. 7 from Okazaki *et al.*, 1986b).

two to seven cystoliths (Fig. 7b). In *Boehmeria plantanifolia*, the microradiograph shows very small deposits (*c.* 15 μm in diameter) mixed with large papillose cystoliths (Fig. 5b). The small deposits were identified as calcium oxalate crystals by X-ray diffraction analysis of the isolated deposits.

The number of cystoliths per square centimeter (n/cm^2) of mature leaves was estimated from the X-ray microradiographs and is shown in Table 1. In Table 1, a

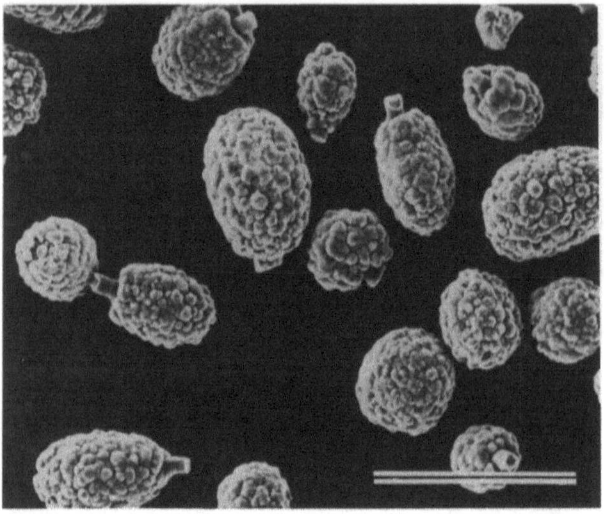

FIGURE 10. Scanning electron micrograph of cystoliths isolated from leaves of *Celtis sinensis* (Ulmaceae). Scale = 100 μm.

content of calcium carbonate also is listed. It was calculated from the average diameter of cystoliths determined from both light micrographs and X-ray microradiographs, n/cm^2 and calcium carbonate density as 2.71 g/cm^3. The cystoliths were assumed to be globular, and the volume of calcium carbonate in a cystolith to be approximately 75% of the globular cystolith as shown by Arnott (1980). As shown in Table 1, the n/cm^2 ranges from 850 to 4,200, varying among the species observed. The estimated calcium carbonate content in the leaves is approximately 0.4 mg/cm^2 for all species except for *Ficus retusa*. In the latter species, it is about 1.1 mg/cm^2, being the highest among all species tested. It remains to be determined whether the massive calcium carbonate is a mere waste or a reservoir for calcium or carbon dioxide.

Inorganic components of the cystolith

The cystoliths were isolated from various kinds of leaves by the method of Arnott (1980). Fresh clean leaves were cut into small pieces and blended with a Waring blender in absolute ethanol for about 10 min. The homogenate was filtered through cheese cloth. The filtrate contained many cystoliths together with many small leaf fragments. The heavier cystoliths were separated from lighter leaf fragments by a series of decantations. Figure 10 shows a scanning electron micrograph of isolated cystoliths from the mature leaves of *Celtis sinensis*. A typical cystolith consisting of main body and a stalk is seen in the figure.

Calcium, magnesium and carbonate contents were determine in *Celtis sinensis* and *Ficus retusa* cystoliths (Okazaki *et al.*, 1987). The two metals were analyzed by EDTA-chelate titration after combustion of cystoliths with perchloric acid. Carbonate content was obtained by manometric method. Cystoliths from two species contained large amounts of calcium and carbonate (8.5-7.9 m moles calcium and 7.4-7.7 m moles carbonate/g cystoliths). A small amount of magnesium (0.9-0.2 m moles/g cystoliths) was also found. Therefore, calcium was contained in approximately equimolar amounts to carbonate, indicating the presence of calcium carbonate. The excess calcium may be bound to some organic matrix such as acid polysaccharide because the latter was histochemically detected in the decalcified cystolith body (Okazaki *et al.*, 1986a). The content of calcium carbonate calculated from carbonate content was about 75% of the dry weight basis in two species. Electron probe analysis was applied to determine the distribution of calcium, magnesium and silicon in a cystolith of *C. sinensis* (Okazaki *et al.*, 1986c). A surface analysis with an electron probe showed that calcium was evenly distributed through the cystolith body except for the stalk and the basal part of the body to which a stalk is attached. Magnesium showed a similar distribution, but it was much less dense than calcium. On the other hand, on the stalk and basal part of the body, highly dense silicon was detected instead of calcium and magnesium, suggesting a silicification in those parts. This characteristic distribution of silicon in cystolith was supported by a line scan analysis from the body to the stalk. In general, the stalk of the cystolith is absolutely negative for any histochemical reaction such as periodic acid-Schiff (PAS) and IKI-sulfuric acid reactions and alcian blue staining (Okazaki *et al.*, 1986a). The silicified stalk has been suggested in *Ficus elastica* (Hiltz and Pobeguin, 1949) and in *Morus alba* (Arnott, 1980).

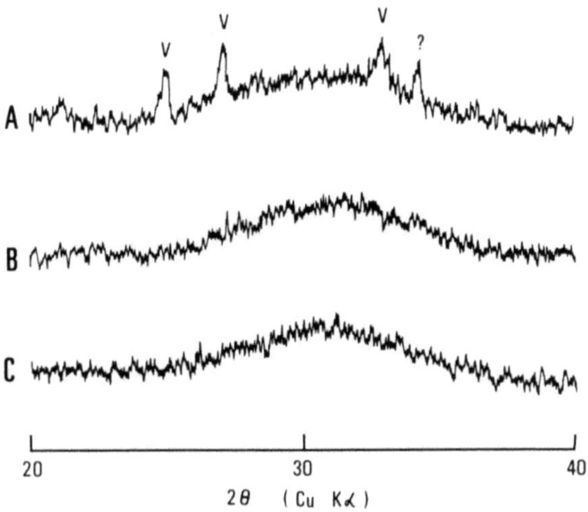

FIGURE 11. X-ray diffraction patterns of powdered cystoliths isolated from mature leaves. ▪A, *Morus bombycis* (Moraceae). ▪B. *Pilea viridissima* (Urticaceae). ▪C. *Celtis sinensis* (Ulmaceae). V, vaterite.

Crystallographic nature of calcium carbonate in cystoliths

The state of the calcium carbonate deposits in cystoliths was examined on powdered cystoliths isolated from mature leaves of nine species containing five families (Okazaki *et al.*, 1987). The analysis was made using Ni-filtered Cu Kα radiation generated at 35 kV, 16 mA on Rigaku-denki CN-4051 AZ equipment, with scanning speed at 2° per min. The typical X-ray diffraction charts were chosen from three species of different families, and are shown in Figure 11. As seen in the charts, these three plants contain amorphous calcium carbonate in their cystoliths and a small amount of vaterite is also found in *Morus bombycis* (Fig. 11A). The following five species showed the same diffraction patterns as *Pilea viridissima* (Fig. 11B) and *Celtis sinensis* (Fig. 11C), indicating the presence of amorphous calcium carbonate only; *Ficus retusa*, *Humulus scandens*, *Boehmeria platanifolia*, *Momordica charantia*, and *Justicia procumbens*. However, *Ficus elastica* cystolith had a similar diffraction pattern to that of *Morus bombycis*, but a small amount of calcite was detected in addition to amorphous calcium carbonate and vaterite. Several authors have reported amorphous calcium carbonate, *e.g.*, for the cystoliths of *Ficus elastica* (Hiltz & Pobeguin, 1949) and for the connective tissues of animals such as gastropods (Tompa & Watabe, 1976; Simkiss & Mason, 1983). The vaterite has been found in the cystolith of *Morus alba* (Arnott, 1980).

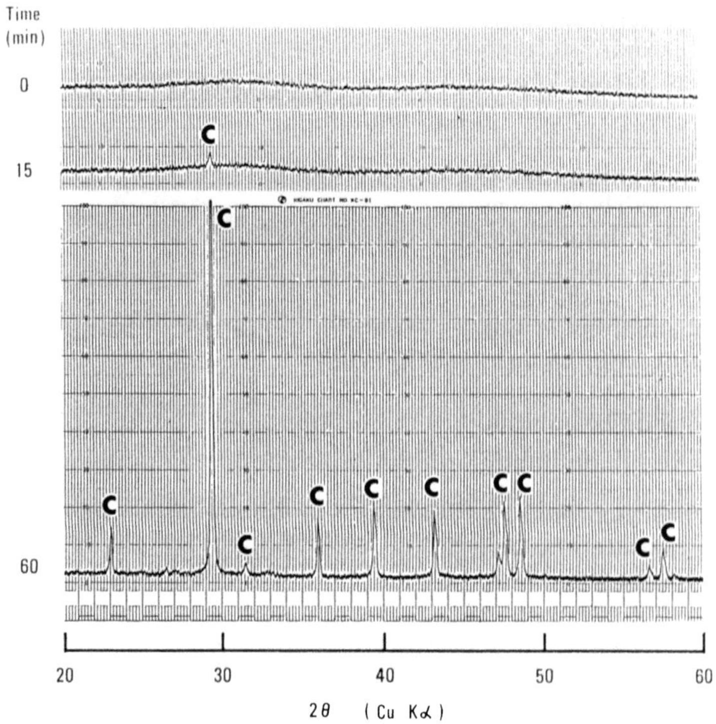

FIGURE 12. Time course of change in X-ray diffraction pattern of isolated cystoliths of *Celtis sinensis* in carbonate buffer. Note that amorphous calcium carbonate changes rapidly into calcite. **C**, calcite.

Crystallographic and morphological changes of amorphous calcium carbonate of cystoliths in carbonate buffer

It was found that amorphous calcium carbonate in the cystoliths rapidly changed into calcite when the isolated cystoliths were incubated in 0.05 M carbonate buffer (pH 9.2) or distilled water at room temperature. Figure 12 shows the time course of calcite formation on the isolated cystoliths of *Celtis sinensis* in the buffer. After 15 min in the buffer, a small peak of calcite is observed in the X-ray diffraction chart and it grows rapidly with time. After 60 min, a large amount of calcite is formed. This crystallographic change is always accompanied by a morphological change in the cystoliths as shown in Figure 13. The isolated cystoliths of *Celtis sinensis* have a main body with many papillae, but their surface is smooth as seen at 0 min in Figure 13. However, when the cystoliths were allowed to remain in the buffer, crystallization occurred on the surface of the cystoliths. After 15 min, well-defined or complicated cubic crystals are clear, but they locate at random on the surface of the cystolith. The crystallization continued until the surface was completely covered with many calcite crystals (see 60 min in Fig. 13). These crystallographic and morphological changes took place slightly slower in distilled water than in the carbonate buffer (data not shown). Similar changes in cystoliths in water has been reported for *Ficus elastica* (Hiltz & Pobeguin, 1949) and for *Morus alba*

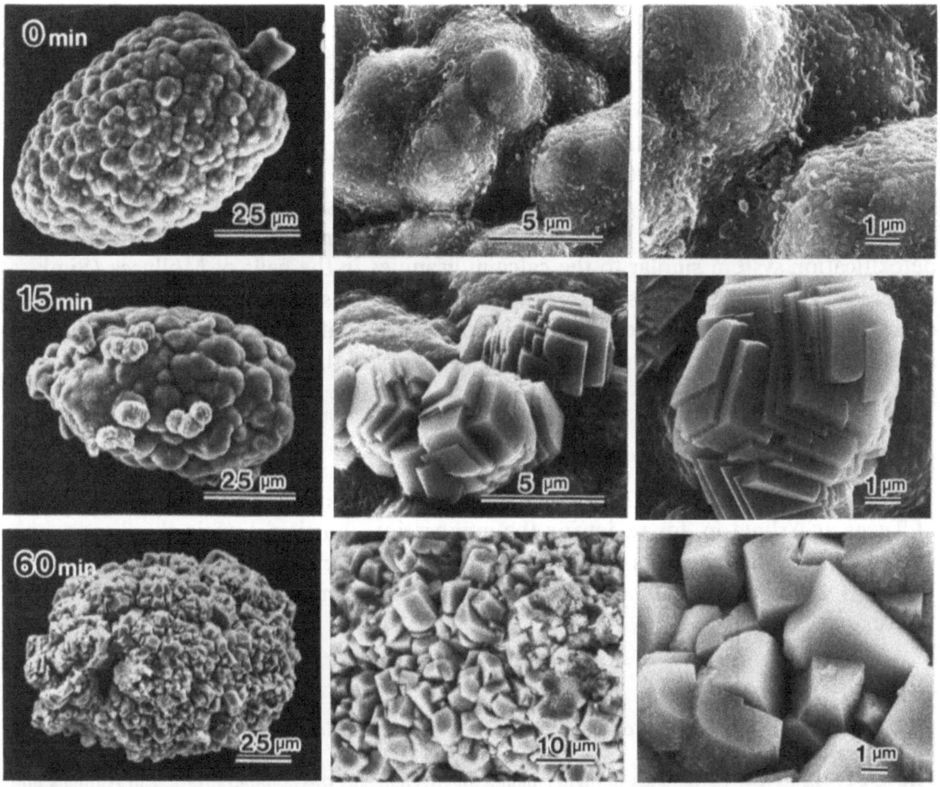

FIGURE 13. Scanning electron micrographs showing morphological change of isolated cystoliths of *Celtis sinensis* in carbonate buffer. Note that cubic calcite crystals are formed with time on the surface of the cystoliths.

(Arnott, 1980). It is very interesting why the most unstable amorphous calcium carbonate does not transform to more stable crystalline state *in vivo*. Some organic matrices in cystoliths and/or organic sheaths such as plasma membranes may be responsible for the stabilization of amorphous calcium carbonate.

The mechanisms of calcium carbonate deposition in cystoliths and its physiological significance are obscure. Our present work has revealed that a tremendous number of cystoliths were contained in the leaves, and that a cystolith-bearing lithocyst was associated with many palisade and/or spongy parenchyma cells. These facts suggest an intimate relationship between calcium carbonate deposition in cystolith and photosynthesis in the leaf. The calcium carbonate deposition may be coupled with bicarbonate utilization in photosynthesis as follows:

$$HCO_3^- \rightarrow CO_2 + OH^- \text{ (in parenchyma cells)}$$
$$HCO_3^- + OH^- \rightarrow CO_3^{2-} + H_2O \text{ (in lithocysts).}$$

If hydroxyl ions are transported to lithocysts as in the above reaction, the increase in pH and carbonic ions are expected in the lithocysts, resulting in calcium carbonate deposition. A similar mechanism has been proposed in the calcification of certain calcareous algae (Sikes *et al.*, 1980; Borowitzka, 1982; Okazaki & Furuya, 1985). At present, we do not have evidence enough to apply the above reactions to cystolith calcification. However, if this mechanism is true, cystolith formation will promote photosynthesis by enhancing the supply of carbon dioxide. The massive concentration of calcium carbonate in the leaf, in turn, may be useful as a calcium or carbon dioxide reservoir. We are especially interested in the function of the cystolith as a carbon dioxide reservoir because it may propose a "new" type of phytosynthesis adapted to environmental conditions. On the other hand, cystoliths may be considered as the product of a detoxification system similar to the calcium granules in the cells of Mollusca (Simkiss & Mason, 1983). The excess of metals calcium, magnesium, etc., and silicon may be excreted into the cystolith because they are toxic for cells at high concentrations. Further study is necessary to elucidate the mechanisms of cystolith formation and its physiological functions.

ACKNOWLEDGEMENTS

We wish to thank Professor Howard J. Arnott, Department of Biology, University of Texas at Arlington, for reviewing the manuscript. Participation in the Symposium was made possible by grants from The Organized Research Fund of The University of Texas at Arlington and The Sea Grant Program administered by Texas A & M University.

REFERENCES

AJELLO, L., 1941. Cytology and cellular interrelations of cystolith formation in *Ficus elastica*. Amer. J. Botany, 28:589-594.

ARNOTT, H.J., 1980. Carbonates in higher plants. *In* Mechanisms of Biomineralization in Animals and Plants, (ed. M. Omori & N. Watabe), 211-218. Tokyo: Tokai Univ. Press.

BOROWITZKA, M.A., 1982. Mechanisms in algal calcification. *In* Progress in Phycological Research, vol. 1 (ed. F.E. Round & D.J. Chapman), 137-177. Amsterdam: Elsevier Biomedical Press.

HILTZ, P. & POBEGUIN, T., 1949. Sur la constitution des cystolithes de *Ficus elastica*. Comptes Rendus Hebdomadaires des Séances de L'Académie des Sciences, 228:1049-1051.

MEYEN, J., 1839. Materiaux pour servir à Phistorire du développement des diverses parties dans les plantes. Annales des Sciences Naturelles: Botanique, 12:257-279.

OKAZAKI, M. & FURUYA, K., 1985. Mechanisms in algal calcification. Jap. J. Phycol., 33:328-344.

OKAZAKI, M., HISANAGA, E., & ISHIDA, Y., 1986a. Histochemical study on organic matrix of cystoliths. Bull. Tokyo Gakugei Univ., Sec. 4, 38:47-54.

OKAZAKI, M., ISHIDA, Y., & SETOGUCHI, H., 1984. Calcification in higher plants and its physiological roles with special reference to cystolith. Bull. Tokyo Gakugei Univ., Sec. 4, 36:67-81.

OKAZAKI, M., SETOGUCHI, H., AOKI, H., & SUGA, S., 1986b. Application of soft X-ray microradiography to observation of cystoliths in the leaves of various higher plants. Bot. Mag. Tokyo, 99:281-287.

OKAZAKI, M., SETOGUCHI, H., & SUGA, S., 1987. Inorganic components and crystallographic nature of cystoliths isolated from the leaves of various higher plants. Bot. Mag. Tokyo (submitted).

SCOTT, F.M., 1946. Cystoliths and plasmodesmata in Beloperone, Ficus & Boehmeria. Bot. Gaz., 107:372-378.

SIKES, C.S., ROER, R.D. & WILBUR, K.M., 1980. Photosynthesis and coccolith formation: Inorganic carbon sources and net inorganic reaction of deposition. Limn. Oceanogr., 25:248-261.

SIMKISS, K., & MASON, A.Z., 1983. Metal ions: Metabolic and toxic effects. *In* The Mollusca, vol 2 (ed. P.W. Hochachka). 101-164. London: Academic Press.

TOMPA, A.S., & WATABE, N., 1976. Calcified arteries in a gastropod. Calc. Tiss. Res., 22:159-172.

WERNER, O., 1931. Haar-und Cystolithenscheiben in Blattgeweben bei Urticales, bei *Bryonia dioica* und *Zexmenia longipetiolata*. Osterrich. Bot. Zeits., 80:81-97.

ASPECTS OF BIOLOGICAL SILICIFICATION

Carole C. Perry[1] and Stephen Mann[2]

[1]*Inorganic Chemistry Laboratory, South Parks Road, Oxford, OX1 3QR, Great Britain;* [2]*School of Chemistry, University of Bath, Claverton Down, Bath BA2 7AY, Great Britain*

ABSTRACT. This paper presents a unified view of silicified mineral formation within a unicellular organ. Analytical, biochemical and structural data are presented for hair cells from the lemma of the grass *Phalaris canariensis* L. Structural information is presented from electron microscopy, solid state nmr and nitrogen adsorption studies. Analytical data is provided by EDXA and SPM studies at different stages of development. Information on the changing nature of the secondary cell wall associated with the biomineral was obtained by *in vivo* radiochemical labelling with ^{14}C glucose and ^{3}H arabinose followed by graded acid hydrolysis and enzymatic degradation followed by chromatography. Mechanisms of structural organization are discussed in relation to spatial constraints, ordered particle aggregation and chemical regulation.

INTRODUCTION

Many plants, notably monocotyledons, have been shown to contain localized deposits of amorphous silica. Silica is often deposited in hairs (trichomes), but may also occur in stomata, ordinary epidermal cells and, in the Poaceae (grasses), in specialized "silica cells". Deposits may also be found in the leaves, stems, roots and reproductive structures of many plant families.

In general, the nature of the "opaline silica" or "silica gel" has been studied macroscopically by light microscopy and SEM (scanning electron microscopy). Deposition may be extracellular or intracellular and the chemical environments for deposition may be very different. For example, cell wall deposits are associated with a polysaccharide matrix, cuticular deposits with fatty acids and lipids and intracellular deposits with proteinaceous material.

The diversity of gross structural forms which may be found among these deposits leads us to inquire into the degree or nature of the control exerted over the deposition process.

Various mechanisms of silicification in plants have been proposed (Sangster *et al.*, 1981; Kaufman *et al.*, 1981). Changes in pH, membrane filtration and transpiration have

Origin, Evolution, and Modern Aspects of Biomineralization in Plants and Animals
Edited by Rex E. Crick
Plenum Press, New York

been considered as possible "mechanisms" for initiating polymerization and aggregation from supersaturated silicic acid solutions. The morphologies of the deposits formed have been suggested to be determined by the size and shape of the biological compartment in which precipitation occurs and there then need be no specific mechanism for spatial organization.

Our aim has been to elucidate different aspects of silicification in higher plants. For our studies we have utilized macrohairs from the lemma of the grass *Phalaris Canariensis* L. The study of these macrohairs is important from a medical perspective since they have been implicated in the aetiology of human oesophageal cancer and have been found as contaminants of wheat flour in certain regions of the world (O'Neill *et al.*, 1980).

Unicellular plant hair cells offer a valuable opportunity to study interrelationships between many important aspects of biomineralization. Analytical EDXA (energy dispersive X-ray analysis), PIXE (proton induced X-ray emission) coupled to the SPM (scanning proton microprobe), biochemical (*in-vivo* radiochemical labelling) and structural TEM, SEM and HRTEM (transmission, scanning and high resolution transmission electron microscopy) ^{29}Si solid state NMR, and N_2 adsorption experiments have enabled us to correlate qualitatively organic matrix deposition, cellular processes and spatial organization of mineral deposits during silicification.

In this paper we shall present information produced from this detailed study on the plant hairs. Furthermore, we shall discuss the mechanisms of structural organization involving spatial constraints, ordered particle aggregation and chemical regulation.

MATERIALS AND METHODS

The preparation of plant materials for electron microscopy (TEM, SEM, HRTEM), for analysis (EDXA, SPM) and for ^{29}Si solid state NMR have previously been described (Mann *et al.*, 1983; Perry *et al.*, 1984a, 1984b; Perry, 1985). In summary, at each developmental stage considered, two samples were utilized. The first sample (untreated) contained both silica and polysaccharide. The second sample was acid treated to remove organic material and contained silica only. Bulk samples of untreated mature macrohairs were obtained from a dust extraction room of a grain exporter (Darly Downs Grain Exporters Pty. Ltd., Queensland, Australia) and macrohairs used for developmental studies were collected from plants grown from seed. Analytical information was obtained by utilization of macrohairs taken from developing inflorescences at defined stages. Structural studies utilized both untreated and acid-treated developing and mature samples. Nitrogen adsorption experiments were performed on mature untreated and acid treated hair samples at 77K following a general method as detailed by Ramsey & Booth (1983).

The cell wall component of the plant hairs was investigated by *in vivo* radioactive labelling of plant inflorescences with ^{3}H arabinose and ^{14}C glucose at various ages. Polysaccharides synthesized were analyzed by graded acid hydrolysis and by enzymatic digestion (amylase, "Driselase", β(1-3), (1-4)glucanase) followed by PC and TLC. Detailed experimental details have been described previously (Perry, 1985; Perry *et al.*, 1987).

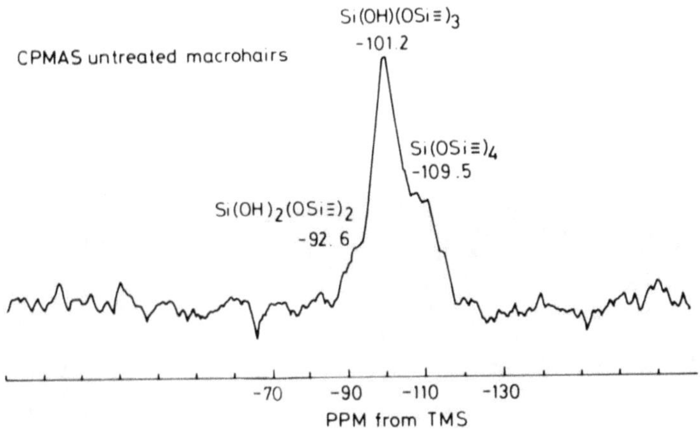

FIGURE 1. ^{29}Si CPMAS spectrum of macrohairs from the lemma of the grass *Phalaris canariensis* L. Note the significant proportion of Si(OSi≡)3OH groups.

RESULTS

Structural and analytical studies

Structural information on the mineral was obtained from electron microscopy, solid state ^{29}Si NMR spectroscopy and nitrogen adsorption experiments. HRTEM yields direct information on structure on a local scale of angstroms. High resolution "lattice" images (Mann *et al.*, 1983) recorded on plant samples (untreated and acid treated) showed only irregular incoherent fringes in all areas of the siliceous biomineral. Computer simulation studies based on pattern recognition showed no conclusive evidence for regular ordering of local clusters of Si-O bonds, and *no* short range order extending above 10 Å was visible. The material is, therefore, amorphous at this level of resolution.

The HRTEM images indicate a structure based on a random network of SiO_4 units connected through Si-O-Si bonds of variable bond angle rather than a microcrystalline/cluster structure composed of a random array of micro-crystalline polyhedra. Information on possible deposition sites is at present lacking although trace elements (cations) or organic polymers (in solution or on a surface) have been proposed as possible deposition "promoters" (Iler, 1979).

Solid state ^{29}Si NMR results (Mann *et al.*, 1983; Perry, 1985) (Fig. 1) show the local environment of silicon atoms in condensed phases. Spectra recorded on plant silica in the presence and the absence (acid-treated) of organic matrix components yielded virtually identical spectra although spectroscopic accumulation was more rapid for the sample containing organic material. Chemical shift data (minus 90 – 120 ppm) confirm the presence of tetrahedrally coordinated silicon showing a range of silicon environments dependent upon the number of free-OH groups within the tetrahedral coordination sphere. There is a significant proportion of $Si(OSi{\equiv})_3OH$ units. Silylation studies (Mann *et al.*, 1983) showed that hydroxyl groups are not only present on the sample surface but also occur as an internal component of the silica structure.

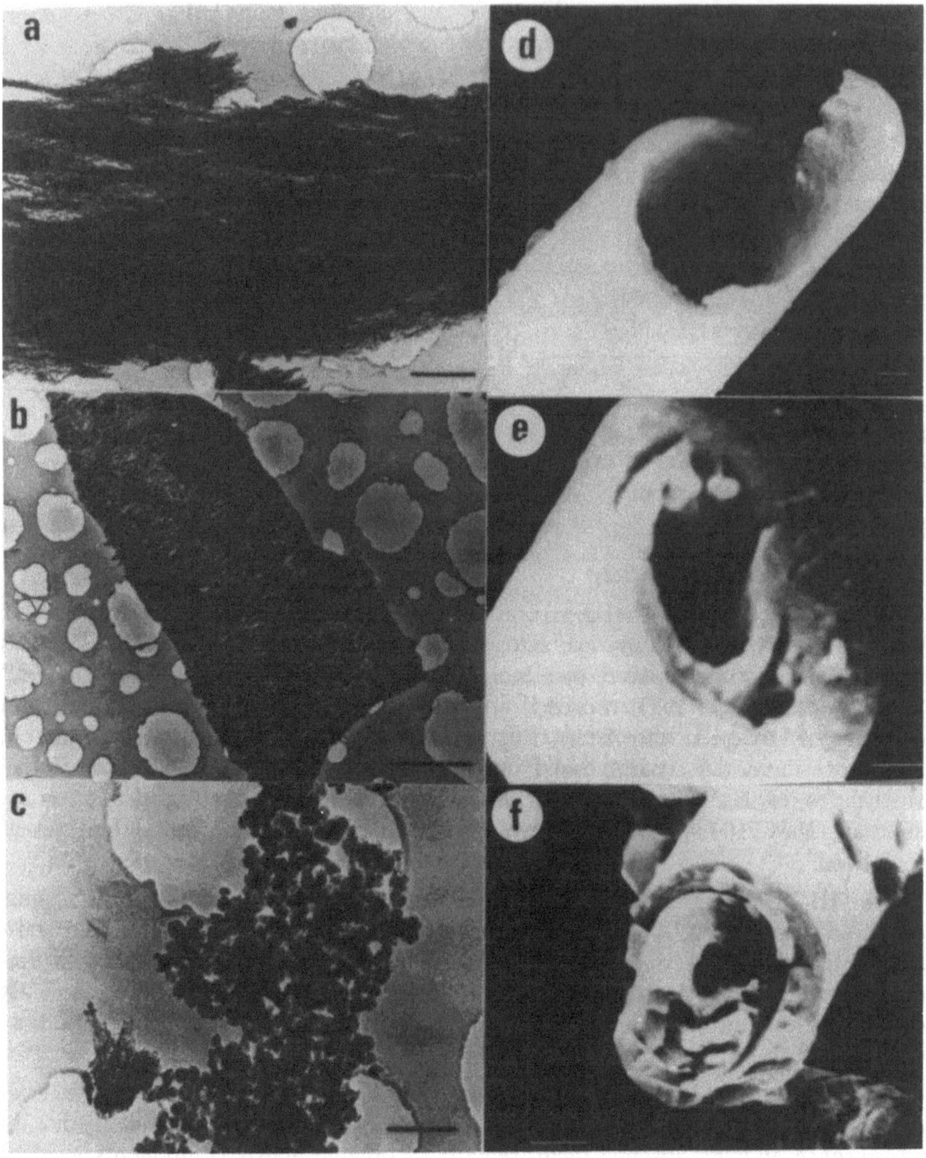

FIGURE 2. Electron microscopy of silicified fragments of macrohairs. ▪a – c. TEM (acid treated fragments from mature hairs). ▪a. Fibrous material. Continuous strands of silica are formed by the ordered aggregation of microparticles. Bar = 0.5 μm. ▪b. Sheet-like material. There is close packing of particles within a layer. Bar = 1 μm. ▪c. Globular material. Bar = 0.25 μm. ▪d – f. SEM (d – e, acid treated hairs, f, untreated hair). ▪d. 9 day sample. A thin rim of silica of equal thickness can be seen on the inner surface of the hair. Bar = 1 μm. ▪e. Mature sample. This sample shows concentric cylinders of silica separated by voids where organic material was present. Bar = 2 μm. ▪f. Mature sample. The sample shows concentric cylinders of silicified material and a central axial channel. Bar = 2 μm.

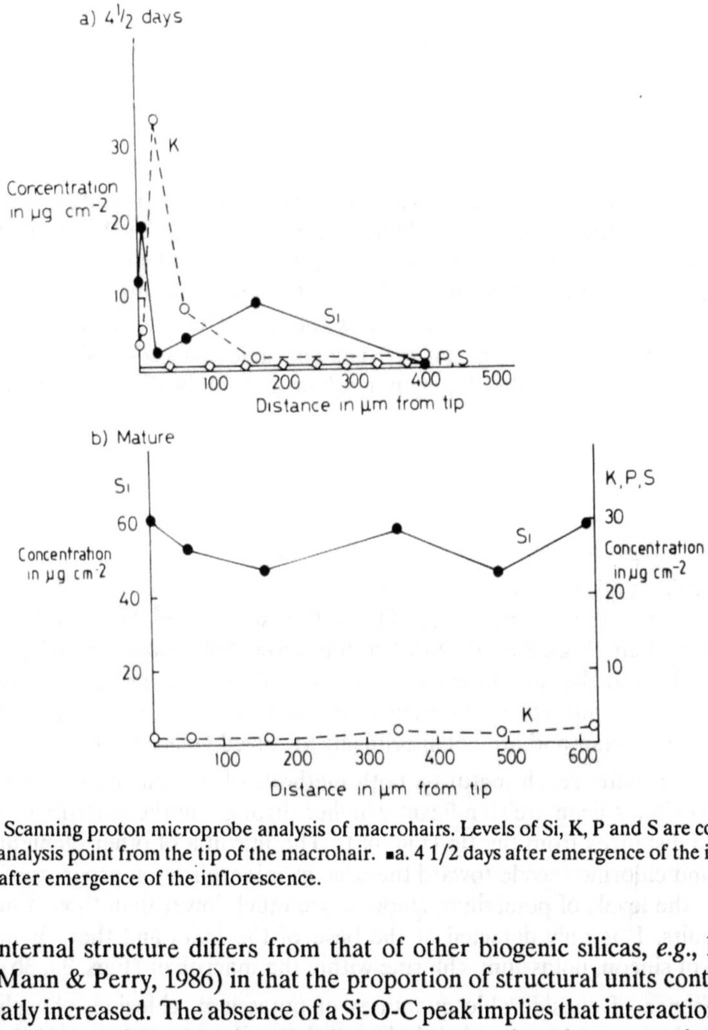

FIGURE 3. Scanning proton microprobe analysis of macrohairs. Levels of Si, K, P and S are compared with distance of analysis point from the tip of the macrohair. ▪a. 4 1/2 days after emergence of the inflorescence. ▪b. 40 days after emergence of the inflorescence.

This internal structure differs from that of other biogenic silicas, *e.g.*, limpet and sponge, (Mann & Perry, 1986) in that the proportion of structural units containing free OH is greatly increased. The absence of a Si-O-C peak implies that interactions between the mineral phase and the organic cell wall component are largely non-covalent in nature and probably involve hydrogen-bonded interactions.

Nitrogen adsorption studies on hair samples in the presence and absence of the organic component have yielded specific surface areas S_{BET} and mean pore sizes r_p for each sample. The acid treated material has a high specific surface area (241 m^2g^{-1}) and a mean pore size of 36 Å although the distribution of pore sizes was large. Similar experiments performed on untreated material gave no measurable surface area or porosity suggesting that the silica and organic components present in the untreated hairs are intimately associated with one another.

Although biogenic silica exhibits no long range crystallographic order it has been shown that morphological order may exist at the microscopic level. Gel and particulate structures may be formed in principle. In the plant hair sample studied, a range of particulate substructures were observed. Sheet-like, globular and fibrillar structural motifs are observed within a single bulk structure (Fig. 2a-c) (Perry *et al.*, 1984a). Each motif comprised primary silica particles of a characteristic size, stability in the electron beam (*i.e.*, degree of hydration) and orientation. The substructures were formed at different stages of development of the macrohairs. Initially, sheet-like material was deposited followed by globular particles and finally fibrillar structures. This may be correlated with a thickening of the hairs themselves (Fig. 2d-f) where distinctive concentric rings of material appear in the mature hairs clearly interdispersed by nonsilicified material as evidenced by SEM of the cross-section of the mature acid treated hair (Fig. 2f). This sequence of events appears to be associated with changes in the ionic environment (Perry *et al.*, 1984a, b) and organic matrix deposition (Perry 1985; Perry *et al.*, 1987) within the developing macrohair and will be discussed at a later stage in this paper.

Analytical data (EDXA and SPM) not only yield direct information on the initial site of mineralization but the extent and involvement of other elements/ions with the mineralization process may be assessed. From this, an indication of the involvement of cellular processes with silicification may be ascertained.

Both EDXA and SPM results (Perry *et al.*, 1984a, b), show that silicification commences at or just after emergence of the inflorescence and then largely only at the tip of the macrohair (Figs. 3 & 4). SPM results show that potassium and chlorine are concentrated behind the initial deposit of silica, and are also present at lower levels throughout the entire macrohair. Low levels of P and S are also present along the entire length of the macrohair and are not specifically localized (Figs. 3a, 3b).

As the macrohairs reach maturity, both methods of analysis have shown that the levels of deposited silicon are significantly higher throughout the macrohair, extending considerable distances from the tip (Fig. 4a). The maxima of other elements such as potassium and chlorine recede toward the base as mineralization progresses (Fig. 4b). At maturity, the levels of potassium, chlorine are much lower than those found in the immature hairs, P is only detected at the base of the hairs and there is no specific localization of silicon, potassium, chlorine within the macrohair (Figs. 3a, 3b).

The presence of considerable quantities of potassium, chlorine, phosphorus and sulfur within the immature macrohair indicates that cell contents are present within the macrohair. The absence of phosphorus or lowered amounts of the elements sulfur, chlorine and potassium at maturity indicates that cell contents are absent within the mature macrohairs. These observations agree with the findings of Hodson *et al.* (1984) who used staining methods. There would appear to be cell contents within the macrohairs only at the early stages of silica deposition.

These changes suggest that silicification is closely connected with underlying cellular processes which may ultimately control the aggregation of microscopic silica particles from supersaturated silicic acid solution. The presence of trace levels of ions throughout the mature hairs suggests possible involvement with neutralization of surface charges on initially formed silica particles, thus aiding aggregation.

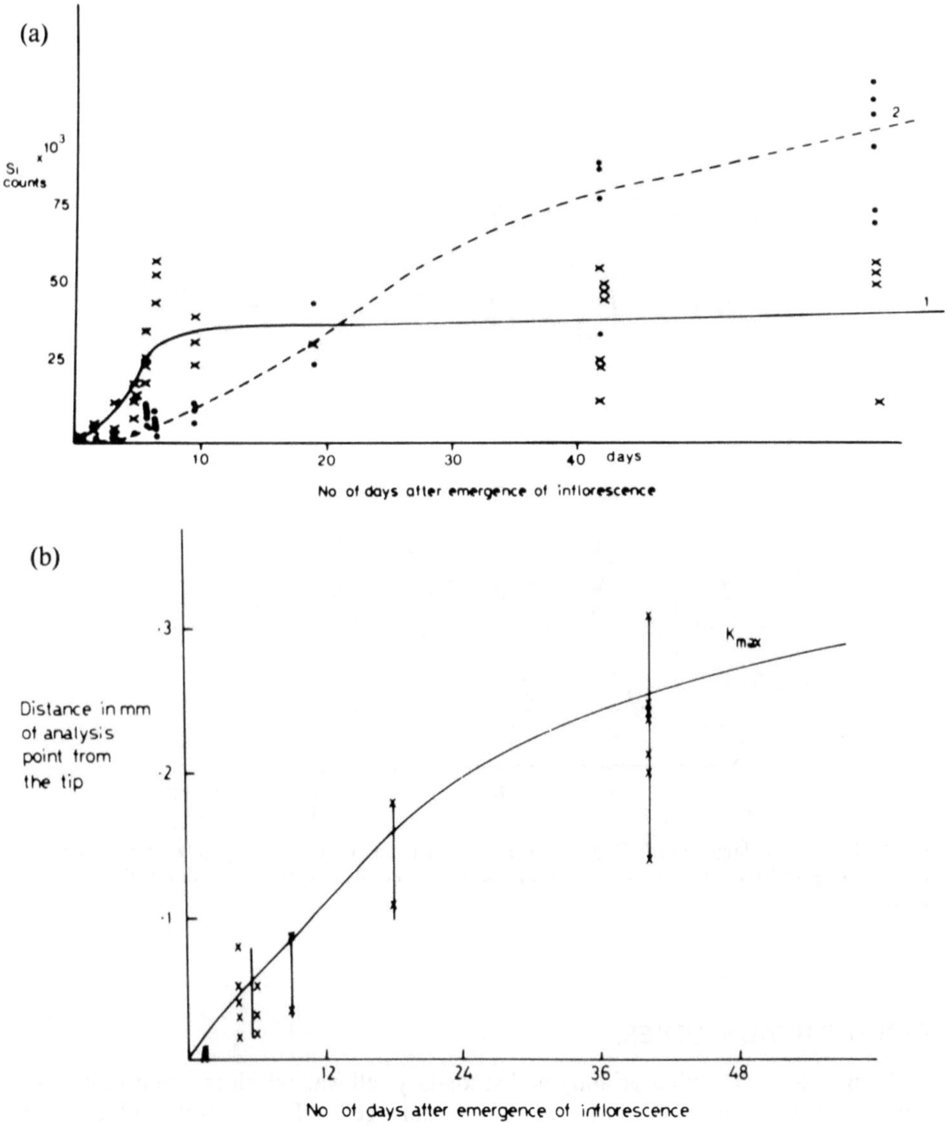

FIGURE 4. Energy Dispersive X-ray analysis of macrohairs. ▪a. Levels of Si within the macrohairs plotted against number of days after emergence of inflorescence. Curve 1 = tip of macrohair. Curve 2 = 0.16 mm from the tip of macrohair. ▪b. K_{max} (the point at which the maximum number of $K\alpha$ X-rays are detected) within the macrohair plotted against number of days after emergence of inflorescence.

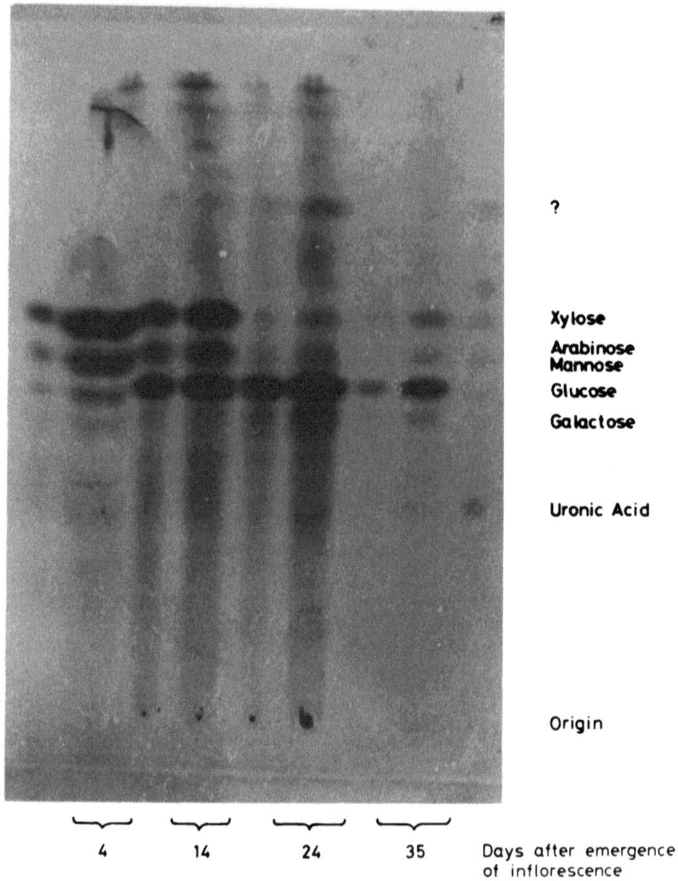

FIGURE 5. Autoradiogram of ^{14}C glucose labelled, 2M trifluoroacetic acid hydrolysed hair samples at different stages of development. Data are not presented for day 0 as no blackening of the X-ray plate was observed.

BIOCHEMICAL STUDIES

Plant cells have both a primary and secondary cell wall which are compositionally, structurally and functionally different. The primary cell wall denotes the volume of the cell and consists mainly of pectic substances (polymers of galacturonic acid and its methylated derivatives) whose structure is stabilized by cross linkage of adjacent carboxylic acid groups by divalent ions (mainly Ca^{2+}). Fibrils of cellulose form a loosely woven mesh enabling the wall to accommodate stretching which occurs during cell enlargement. Secondary wall thickening involves the deposition of much cellulose and hemicelluloses such as xylans and mannans. The resulting structure is made up from many layers.

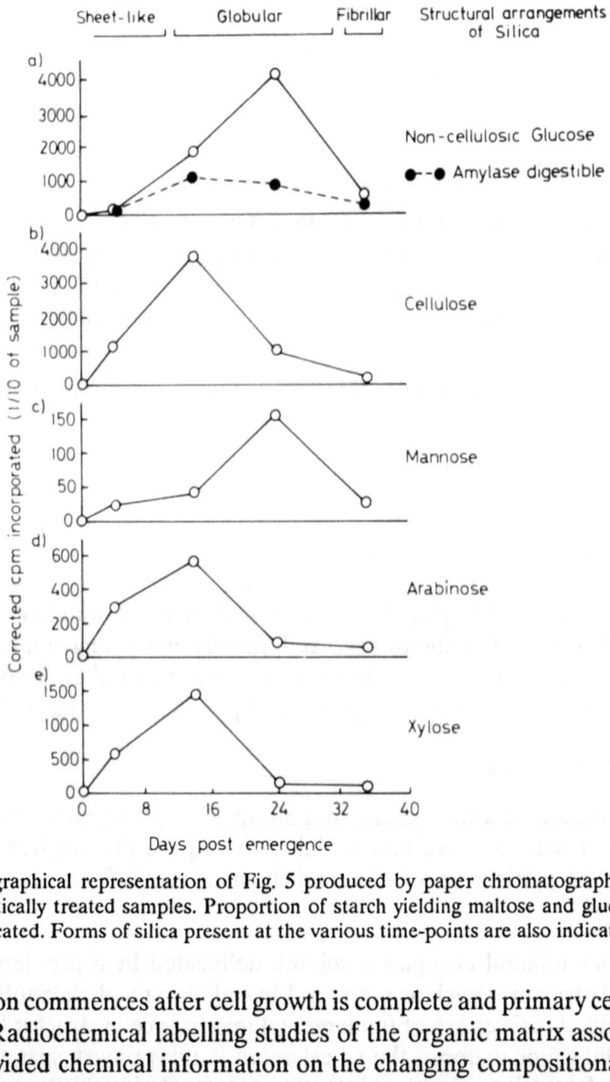

FIGURE 6. A graphical representation of Fig. 5 produced by paper chromatography and scintillation counting of identically treated samples. Proportion of starch yielding maltose and glucose upon amylase treatment is indicated. Forms of silica present at the various time-points are also indicated.

Silicification commences after cell growth is complete and primary cell wall synthesis has ceased. Radiochemical labelling studies of the organic matrix associated with the silica has provided chemical information on the changing compositional nature of the secondary cell wall of macrohairs from the lemma of the grass *Phalaris canariensis* L. which may be correlated with the different mineralized deposits observed. These findings are important, not only from the perspective of biomineralization but also from the fact that we obtain information on changing cell wall composition for a unique single cell system, although it must be appreciated that the cell type investigated is rather specialized.

Detailed results may be found elsewhere (Perry, 1985; Perry *et al.*, 1987). Figures 5 and 6 show pictorially and graphically the changes which occur as the cell wall matures and as silicification takes place.

Conclusions which were drawn for this cell type included:

1. The total composition of mature hairs is 30% hexose (including cellulose), 25% pentose, <1% uronic acids, <4% protein and 40% silica.

2. During the early stages of secondary wall synthesis, predominantly cellulose and araginoxylan are synthesized. At later stages, the synthesis of starch and β(1-3), (1-4)-glucan predominates.

3. A highly substituted arabinoxylan is synthesized (ara :xyl (3:5)).

4. Incorporation of sugars into starch is maximal at day 14.

5. The cell walls are unusually "Driselase" resistant. This may be for one of several reasons such as the monosaccharide residues are substituted by acetate, ferulate or silanolate, or the silica forms an impervious network which cannot be penetrated by the enzyme.

6. No appreciable levels of polyuronide or xyloglucan are synthesized during secondary wall formation.

Although there is no information at present concerning the precise stereochemistry of the organic polymers detailed above, we have shown, that during silicification, changes in mineral structural organization are associated with relative changes in the synthesis of organic matrix components. The sheet-like silicified material deposited at the early stages of macrohair development is laid down at the same time as cellulose and a heavily substituted arabinoxylan. The globular material is then laid down along with declining amounts of cellulose and arabinoxylan, and rapidly increasing amounts of β(1-3, 1-4)-glucan and some mannan. The fibrillar silicified material is deposited into the mature organic phase, when further deposition of polysaccharide has virtually ceased.

GENERAL DISCUSSION

The mineralization of a functionally important amorphous material such as silica requires precise structural organization. This may require the involvement of spatial constraints, ordered particle aggregation and chemical regulation.

Spatial constraints

An amorphous mineral occupies a volume delineated by a pre-defined chemical surface. For gel-structures such as are found in radiolaria and choanoflagellates etc., membrane-bounded vesicles are of fundamental importance in structural organization (Leadbeater, 1981). For diatoms, the silicalemma is the primary spatial determining surface for valve formation. For the plant system presented in this paper, silicification commences after elongation of the hair cell has ceased. Here, the primary cell wall delineates the external boundaries for mineralization.

Within the external boundaries of the mineralizing volume organic material may play a role in determining spatial availability for silicification at the nanometer level. Mann has shown (Mann & Williams, 1982) the presence of "stainable", probably organic structures running through silica rods of a choanoflagellate. In the diatom frustule it is thought that organic material permeates the silica structure but there is no direct evidence for this as yet. In the plant system studied here we have clear evidence that spatial availability for mineralization will be dependent upon constraints imposed by secondary cell wall synthesis which proceeds in advance of silica deposition (Hodson,

TABLE 1

Glycosidic Linkage	Polymer	Shape Adopted
β(1-4)	cellulose	ribbon
α(1-4)	amylose	hollow helical tube
β(1-3)	callose	3 intertwined chains

1984). The changing nature of the polysaccharides synthesized, for example, cellulose at the early stages, a mixed linkage glucan at the later stages, etc., will have clear consequences for both the shape of the volume available for mineralization and for the interactions possible between the organic and inorganic polymers. This will ultimately determine the physical and chemical properties of the biomineral which results.

Ordered particle aggregation

The vectorial aggregation of preformed mineral particles has been suggested to be an important mechanism of structural organization in biominerals (Mann, 1983). The ordered aggregation of primary silica units to form structural motifs in the plant macrohairs probably reflects the stereochemical nature (*i.e.*, the arrangement of charged residues and hydrogen bonding centers) of the matrix system. For crystalline materials such as $CaCO_3$, epitaxial mineral deposition on protein templates rich in aspartate and glutamate has been demonstrated (Addadi & Weiner, 1985). A stereochemical role for silicification on a template rich in hydroxyl-containing side chains (serine and threonine) has been inferred for diatoms but not proven (Hecky *et al*., 1973).

In the plant system studied, silica is deposited into a secondary cell wall composed largely of cellulose, an arabinoxylan, a mixed linkage glucan, some starch and a small amount of mannan. Generally, the 3-dimensional nature of carbohydrate polymers is affected by glycosidic linkage, monosaccharide sequence, whether periodic or interrupted, and on the presence, absence and chemical identity of any side chains present. As an example, the effect of glycosidic linkage on the shapes adopted by polymers of glucose is shown in Table 1.

Side chain modifications have a significant effect on the three dimensional shapes adopted by polymers and on their mutual interactions. In consideration of our data presented here and those of Perry (1985) and Perry, *et al.* (1987) it is clear that the above mentioned factors are of fundamental importance in determining the three dimensional stereochemical nature of the "matrix" system. The siliceous structures observed (sheet-like, globular, fibrillar), may result from a patterning of underlying organic polymers.

Although it is clearly tempting to relate "matrix" compositional changes to the adoption of mineral substructural patterns, changes in the ionic environment may also play a significant role.

Chemical regulation

The structural organization of biogenic silica will be dependent on the regulation in time and space of physiochemical processes occurring within the mineralization environment. The development of a siliceous structure will arise from a cascade of molecular events which involve nucleation and growth (polymerization) processes. Such events are thermodynamically driven by local increases in supersaturation which can be induced by active silicon transport, degradation of silicon complexes, pH Changes and water extrusion from enclosed volumes (reverse osmoregulation).

At pH 7 concentrations of silicic acid above 2 mM are unstable and deposition proceeds against only a small energy barrier with the formation of dimers, trimers and low molecular ring structures which combine to give stable condensates *c*. 10 Å in diameter (Iler, 1979). The presence of organic surfaces may assist deposition particularly if hydrogen bonding, ionic interactions or hydroxyl condensation reactions are possible between silica oligomers and the matrix interface. Covalent bonding to give Si-O-C linkages would require enzymatic mediation at pH 7.

Growth of the silica phase is a process which maximizes Si-O-Si bonds at the expense of Si-OH groups. Internal condensation results in loss of water of variable degree depending on the rate and extent of polymerization. Although biogenic silica structures may vary in water content in different organisms a significant loss of water must be attained if the mineral is to have useful mechanical properties. The regulation of water flux is thus a major factor in silicification.

Growth of silica phases is determined by interfacial properties. Organic and inorganic components whether free in solution or as components of solid structures may moderate the mineralization process. High levels of ionic activity (0.2-0.3 M Na^+, K^+) promote particle aggregation due to surface charge reduction between primary silica nuclei. Hydrogen bonding "flocculants" such as alcohols, proteins, lipids and polysaccharides may have a similar effect, but effectiveness is low when ionic concentrations are high, as in the early stages of development of the macrohairs. At later stages of development relative polysaccharide synthesis increases and ionic concentrations are low. It is clearly possible that modification of the relative importance of both organic and inorganic constituents may ultimately determine the formation of distinctive biomineralized structures.

Work is underway on the interaction of silicic acid and polysilicic acids with carbohydrate polymers, in the presence and absence of an ionic environment in order that we may fully understand how and to what extent these factors direct silicification.

ACKNOWLEDGEMENTS

The authors would like to thank Dr. S.C. Fry (Edinburgh University) for successful collaboration on the biochemical studies briefly presented in this paper. We should also like to thank Dr. N.J. Clayden (Inorganic Chemistry Laboratory, Oxford) for assistance with NMR experiments and Dr. J.D.F. Ramsey (A.E.R.E., Harwell) for assistance with nitrogen adsorption experiments. We would also wish to thank Professor R.J.P. Williams (Inorganic Chemistry Laboratory, Oxford) for stimulating discussion on the subject matter presented in this paper. CCP would like to thank the medical research council for an award for the duration of this research. Presentation of this paper at The Fifth International Symposium on Biomineralization, Arlington, Texas was made possible by funds provided by The Organized Research Fund of The University of Texas at Arlington and The Sea Grant Program administered by Texas A & M University.

REFERENCES

ADDADI, L. & WEINER, S., 1985. Interactions between acidic proteins and crystals: stereochemical requirements in biomineralization. Proc. Nat. Acad. Sci. U.S.A., 82:4110-4114.

HECKY, R. E., MOPPER, K., KILHAM, P., & DEGENS, E.T., 1973. The amino acid and sugar composition of diatom cell walls. Mar. Biol., 19:323-331.

HODSON, M.J., SANGSTER, A.G., & PARRY, D.W., 1984. An ultrastructural study on the development of silicified tissues in the lenima of *Phalaris canariensis* L. Proceedings of the Royal Society of London, B222:413-425.

ILER, R.K., 1979. The Chemistry of Silica. 866 pp. John Wiley, New York.

KAUFMAN, P.B., DAYANANDAN, P., TAKEOKA, Y., BIGELOW, W.C., JONES, J.D. & ILER, R., 1981. Silica in shoots of higher plants, pp. 409-450. *In* Silicon and Siliceous Structure in Biological Systems, edited by T.L. Simpson & B.E. Volcani. Springer-Verlag.

LEADBEATER, B.S.C.,1981. Ultrastructure and deposition of silica in loricate choanoflagellates, pp. 295-322. *In* Silicon and Siliceous Structures in Biological Systems edited by T.L. Simpson & B.E. Volcani. Springer-Verlag.

MANN, S., 1983. Mineralization in biological systems. Struc. Bond. 54:125-164.

MANN, S., PERRY, C.C., WILLIAMS, R.J.P., FYFE, C.A., GOBBI, G.C., & KENNEDY, G.J., 1983. The characterization of the nature of silica in biological systems. J. Chem. Soc. Chem. Comm.:168-170.

MANN, S. & PERRY, C.C., 1986. Structural aspects of biogenic silica. CIBA Found. Sym. no. 121:40-53.

MANN, S. & WILLIAMS, R.J.P., 1982. High resolution electron microscopy studies of the silica lorica in the choanoflagellate *Stephanoeca diplocostata* Ellis. Proc. Roy. Soc. Lond., B216:137-146.

O'NEILL, C.H., HODGES, G.M., RIDDLE, R.N., JORDAN, P.W., NEWMAN, R.H., & FLOOD, R.H., 1980. A fine fibrous silica contaminant of flow in the high oesophageal cancer area of North East Iran. Inter. J. Cancer 26:617-628.

PERRY, C.C., MANN, S., & WILLIAMS, R.J.P., 1984a. Structural and analytical studies of the silicified macrohairs from the lemma of the grass *Phalaris canariensis* L. Proc. Roy. Soc. Lond. B222:427-438.

PERRY C.C., MANN, S., WILLIAMS, R.J.P., WATT, F., GRIME, G.W., & TAKACS, J., 1984b. A scanning proton microprobe study of macrohairs from the lemma of the grass *Phalaris canariensis* L. Proc. Roy. Soc. Lond., B222:439-445.

PERRY, C.C., 1985. Silicification in biological systems. D. Phil. thesis, Univ. Oxford.

PERRY, C.C., WILLIAMS, R.J.P., & FRY, S.C., 1987. Cell wall biosynthesis during silicification of grass hairs. J. Plant Physiol., 126:437-448.

RAMSEY, J.D.F. & BOOTH, B.O., 1983. Determination of structure in oxide sols and gels from neutron scattering and nitrogen adsorption measurements. J. Chem. Soc., Faraday Trans. I, 79:173-184.

SANGSTER, A.G., & PARRY, D.W., 1981. Ultrastructure of silica deposits in higher plants, pp. 383-408. *In* Silicon and Siliceous Structures in Biological Systems edited by T.L. Simpson & B.E. Volcani. Springer-Verlag.

SYSTEMS OF BIOMINERALIZATION IN THE FUNGI

Kenneth D. Whitney[1]

Department of Biology, University of Texas at Arlington, Arlington, Texas 76019; [1]Present address: Department of Life Sciences, Eastern New Mexico University, Portales, NM 88130

ABSTRACT. Crystals of Ca-oxalate are produced on vegetative or reproductive structures by members of each non-flagellate fungal class -- the Ascomycetes, Basidiomycetes, and Zygomycetes. These crystals originate within the hyphal wall, distorting the outer wall layers during subsequent crystal growth. Crystals commonly occur on the vegetative mycelium of leaf and wood-rotting Basidiomycetes, both on field-collected and on cultured material. In culture, crystals are typically produced only on the aerial mycelium, not on the substrate mycelium. This suggests that calcium is translocated from the substrate mycelium via the fungal protoplast to the aerial mycelium, where it is then precipitated with oxalic acid. Mucoralean Zygomycetes also produce mineralized deposits on their aerial sporangia and sporangiophores, again suggesting that calcium is transported from the substrate and immobilized on aerial portions of the fungal thallus. Further, in the zygomycete *Gilbertella persicaria*, the reduction of calcium levels by Ca-oxalate crystallization coincides with an increased rate of mycelial growth. Thus, relatively high levels of calcium in the media appear to inhibit the growth of *Gilbertella persicaria*. The precipitation of Ca-oxalate may serve as a way for fungi to regulate or reduce the calcium ion concentration in their microenvironment.

INTRODUCTION

Although Anton de Bary (1887) noted with some detail the morphology and occurrence of Ca-oxalate ($Ca(C_2O_4)_2$) deposits in fungi, the phenomenon of mineralization in the fungi has attracted only occasional interest since de Bary's time. Aspects of Ca-oxalate mineralization in certain Zygomycetes were studied by Buller (1934), Young (1968), Bland and Charles (1972), Jones *et al.* (1976), Urbanus *et al.* (1978), Shipton and Lunn (1980), Powell and Arnott (1985), and Whitney and Arnott (1986). Horner *et al.* (1983) investigated the crystals produced on apothecial hairs of the ascomycete *Dasycyphus capitatus* (Peck) Le Gal. In the Basidiomycetes, Horner *et al.* (1985a, 1985b) described the mineral deposits associated with the peridium of the earthstar *Geastrum minus* (Pers.) Fisher while Whitney and Arnott (1986) studied the relationship between Ca-oxalate crystal development and peridial separation in another earthstar, *Geastrum saccatum* (Fr.) Fisher. Ecological aspects of fungal Ca-oxalate have been investigated by Graustein *et al.* (1977) and Cromack *et al.* (1977).

Most researchers have considered the mineralized deposits of fungi to be extrahyphal. For example, Graustein *et al.* (1977) hypothesized that oxalic acid excreted by fungal hyphae precipitated with calcium ions in the environment, resulting in the deposi-

Origin, Evolution, and Modern Aspects of Biomineralization in Plants and Animals
Edited by Rex E. Crick
Plenum Press, New York

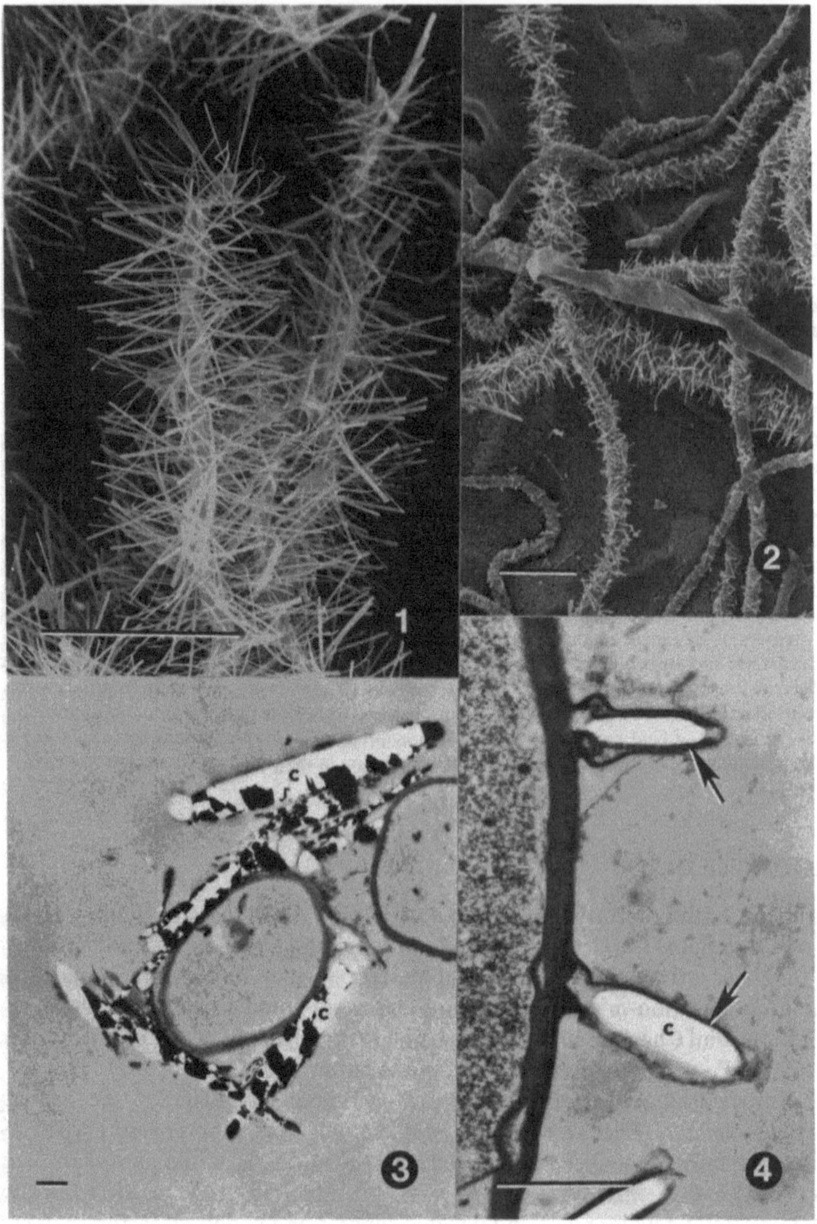

FIGURES 1–4. Agaricus bisporus. ▪*1.* Crystal-clad aerial hyphae. Note long, acicular crystals and their tangential orientation on the hyphae. Bar = 10 μm. ▪*2.* Crystal-bearing aerial hyphae above agar surface. Bar = 10 μm. ▪*3.* (TEM) Unstained ultrathin-section. Crystal chambers (**c**) contain electron-dense remnants of Ca oxalate crystals. Bar = 1 μm. ▪*4.* (TEM) Stained ultrathin-section with Ca oxalate extracted from crystal chambers (**c**). Note the electron-dense portion of the hyphal wall (arrows) which forms the chambers. Bar = 1 μm.

tion of Ca-oxalate on the hyphae. More recently, however, Arnott and Webb (1983) and Whitney and Arnott (1987) have shown that crystal initiation and early growth of crystals occur within the hyphal wall. Subsequent growth of the crystals can and often does disrupt the hyphal wall. Thus, in older hyphae with well developed crystalline deposits the relationship between the cell wall and the crystals is often obscured.

In addition to crystals on the vegetative hyphae, Ca-oxalate crystals can be found on aerial sporangia of mucoralean Zygomycetes, on cystidia of a wide variety of Basidiomycetes, within the peridial layers of Gasteromycetes, and on apothecial hairs of ascomycetes. The morphology of these crystals varies a great deal. The two most commonly encountered crystal types are bypiramids and elongate, rod-shaped forms. Where information is available, the majority of Ca-oxalate deposits in fungi are weddellite ($Ca(C_2O_4)2{\cdot}2H_2O$), Ca-oxalate dihydrate (COD).

The varied location and diverse morphologies of fungal Ca-oxalate deposits, along with the systematic and ecological diversity of the fungi that have them, makes speculations about the functions of these deposits quite difficult. In fact, such diversity probably reflects a variety of functions for these deposits. This paper will discuss the morphology, development, and ultrastructure of the Ca-oxalate mineralization system in three diverse fungi, *Agaricus bisporus*, *Geastrum saccatum*, and *Gilbertella persicaria*.

Agaricus bisporus

Agaricus bisporus is the main commercially cultivated mushroom. In agar culture, the vegetative hyphae above the surface of the substrate -- the aerial hyphae -- often bear prominent deposits of COD (Whitney & Arnott, 1987). The crystals are typically narrow, elongate forms that are perpendicular to the long axis of the hypha (Figs. 1,2). These crystals originate within the wall of the hyphae (Figs. 3,4) and increase in length tangentially to the surface of the hyphae. After initiation within the hyphal wall the subsequent growth of the crystals stretches and eventually disrupts the outer wall layer. This wall layer clearly surrounds the crystal chambers illustrated in Figure 4. In many, if not most, fungi this outer layer is composed of distinct structural units called rodlets. Arnott (1984) illustrated the displacement of this rodlet layer by crystal growth in an unidentified basidiomycete.

Geastrum saccatum

The earthstars (members of the genus *Geastrum*) undergo a striking pattern of basidiocarp development. Initially the basidiocarps consist of several layers of tissue (the peridium), which surround the fertile, spore-producing region (the gleba). As the fruiting body matures, however, the outer peridial layers -- the exoperidium -- separate and reflex away from the inner layer -- the endoperidium. During this process the exoperidium splits into several triangular fragments which remain attached to the base of the spore case. The final product is a globose spore case sitting atop a stellate base..

Taxonomic descriptions of earthstars occasionally include references to granular material on the surface of the endoperidium. For example, Coker and Couch (1928) describe the endoperidial surface of *Geaster coronatus* (Schaeff.) Schroet. as having "minute glistening particles," and that of G. *floriformis* Vitt. being covered by "fine granules." Lloyd (1907) went so far as to describe a new species, G. *calceus* Lloyd, with

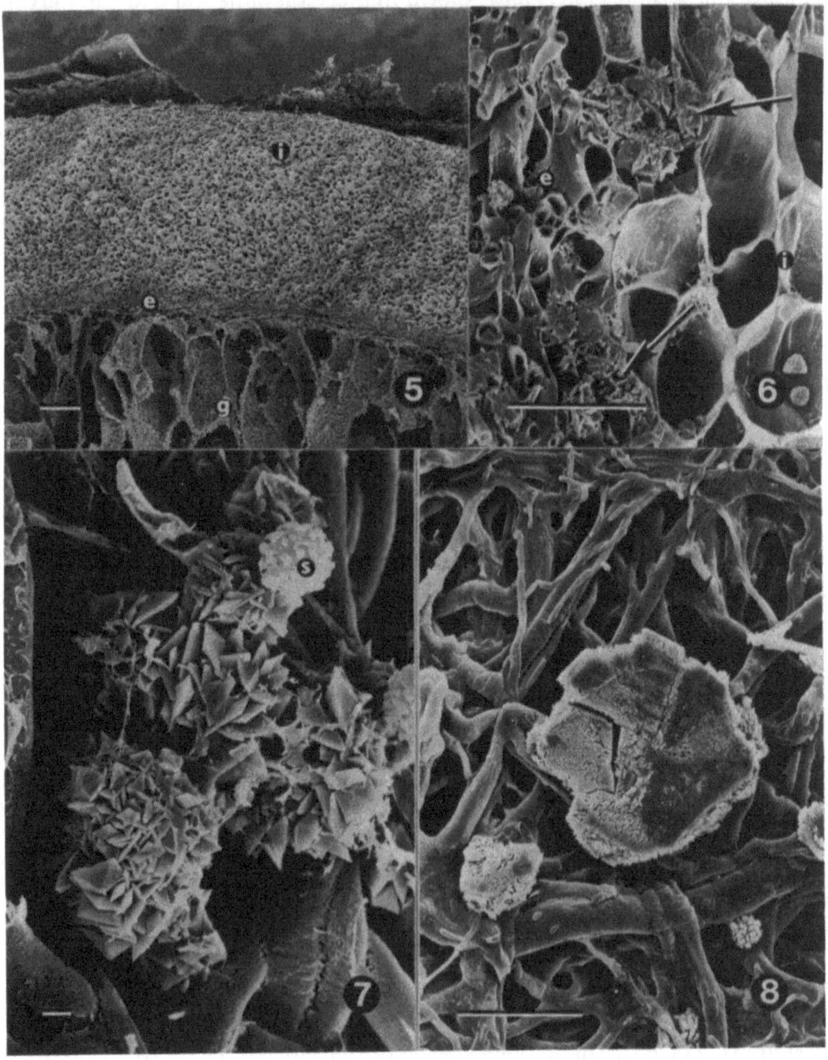

FIGURES 5–8. Geastrum saccatum. ▪*5.* Cross section of undehisced basidiocarp prior to peridial separation. The pseudoparenchymatous inner layer of the exoperidium (**i**) is tightly bound to the endoperidium (**e**). The fertile gleba (**g**) is within the endoperidium. Bar = 100.0 μm. ▪*6.* Higher magnification of the exo/endoperidial boundary. Note the large crystals (arrows) marking the future line of separation between the exoperidium (**i**) and the endoperidium (**e**). Bar = 10 μm. ▪*7.* Crystal aggregates on surface of endoperidium. (s = spore). Bar = 1 μm. ▪*8.* Large, weathered crystal on endoperidial surface of mature, dehisced basidiocarp. Bar = 10 μm.

a specific epithet referring to the conspicuous calcareous granules on the surface of the endoperidium.

Although the occurrence of calcareous deposits on the endoperidium of earthstars was well noted in species descriptions, no studies of the morphology of earthstar crystals appeared in the literature until that of Horner *et al.*(1985a). This paper presented ultrastructural observations of calcareous crystals associated with the exo- and endoperidial layers of *Geastrum minus*, and confirmed that the crystals were Ca-oxalate dihydrate by means of X-ray diffraction. Although these initial studies were based on herbarium specimens, Horner *et al.* (1985b) were later able to extend their observations to unopened basidiocarps. Their investigation of undehisced basidiocarps confirmed that Ca-oxalate crystals in *G. minus* were confined to the exo- and endo-peridial layers. They recognized four morphological crystal types of Ca-oxalate dihydrate, and a fifth of Ca-oxalate monohydrate (COM). They postulated that the latter form was the result of dissolution and recrystallization of existing COD crystals.

Whitney and Arnott (1986) investigated Ca-oxalate crystal development in another earthstar, *Geastrum saccatum*. They presented evidence that Ca-oxalate crystals in this species were initiated within the walls of endoperidial hyphae, and suggested that, in light of the increasing evidence that Ca-oxalate crystals were present on the endoperidium of many earthstars (Whitney and Arnott, unpublished data), these crystals played a direct role in peridial separation. Figure 5 illustrates an undehisced basidiocarp of *G. saccatum* in cross section. Ca-oxalate crystals (Figs. 6,7) are clearly associated with endoperidial hyphae and are especially abundant at the exo/endoperidial boundary (Fig. 6). After dehiscence the appearance of Ca-oxalate crystals of earthstars changes, probably due to erosion, dissolution, or weathering. Figure 8 illustrates such crystals on the endoperidial surface of a dehisced basidiocarp.

Gilbertella persicaria

Gilbertella persicaria is a parasite of peaches and other fruit. The sporangia of this fungus are borne on long stalks (sporangiophores), and the sporangial wall splits into hemispheres at maturity, revealing the spores within (Fig. 9). Nearly all of the sporangium proper and the sporangiophore are covered with Ca-oxalate crystals (Whitney & Arnott, 1986). The crystals on the sporangial wall are upright, angular crystals mounted on a hexagonal or polygonal base plate (Fig. 10), while most of the crystals on sporangiophores are elongate, rectangular plates embedded in the wall (Figs. 11,12). Another crystalline deposit on the sporangiophore wall consists of superficial, oval patches of elongate, interpenetrant crystals called crystal plaques (Fig. 11). Whitney and Arnott (1986) proposed that the varied morphology of *G. persicaria* crystals represented a developmental gradient which corresponded to the sequence of sporangial development. They also suggested that the crystal plaques represented recrystallized Ca-oxalate.

When grown in the presence of limited calcium, *Gilbertella persicaria* initially produces normal crystals, but as calcium levels fall, crystal production becomes aberrant. Eventually crystal production stops, but fungal growth continues. When compared to growth rates on high-Ca media, growth of *G. persicaria* on limited calcium media was considerably higher (Whitney & Arnott, unpublished data).

DISCUSSION

Why do fungi produce Ca-oxalate? Typically the study of higher plant and fungal Ca-oxalate has focused on oxalic acid production. In higher plants, Ca-oxalate crystals traditionally have been considered repositories for excess metabolic oxalic acid (Frey-Wyssling, 1981). With the acid trapped in the insoluble Ca-oxalate, any deleterious effects would be minimized. More recently it has been suggested that higher plant Ca-oxalate acts to detoxify calcium (Arnott, 1982; Borchert, 1985). Indeed, Simkiss (1977) stated that "...calcium is parmacologically one of the most disruptive substances for normal cell functions..."

From a physical standpoint, surrounding aerial hyphae with mineralized layer might prevent arthropods from grazing on the exposed portions of the mycelium, as suggested by Thompson (1984). Whitney and Arnott (1986) suggested that the interlocking system of Ca-oxalate plates embedded in the sporangiophore wall of *Gilbertella persicaria* might provide "exoskeletal" support.

Ca-oxalate crystal formation can also play a developmental role, as in peridial separation in earthstars (Horner *et al.*, 1983; Whitney & Arnott, 1986). Whether the growth of these crystals physically causes peridial separation, or whether they alter some adhesive property of the endoperidial layers remains to be proven, but crystal initiation and growth clearly coincides with this important developmental phase.

To obtain nutrients, fungi release enzymes into the environment where the enzymes degrade substances into simpler molecules which are then absorbed by the fungus. Thus, it would seem necessary for fungi to closely regulate the microenvironment around their hyphae to ensure maximum enzymatic efficiency. Calcium is potentially disruptive to enzymatic activity, and perhaps one way fungi can regulate this ion is to remove it from the environment, translocate it to the aerial mycelium, and precipitate it as Ca-oxalate. Could fungal Ca-oxalate also act to reduce Ca-oxalate concentration?

Kazmierczak *et al.* (1985, this volume) have suggested that the production of mineralized deposits in living organisms arose as a means of calcium regulation in response to increasing calcium levels in ancient oceans. Perhaps regulation of calcium is the ultimate explanation for biomineralization in the fungi as well. Fungi, in essence, are aquatic organisms, since their metabolic activities depend on the availability of soil moisture. The fungal "ocean," then, is the soil solution, and transporting calcium ions from this ocean and isolating them as Ca-oxalate deposits on aerial hyphae effectively reduces calcium ion concentration.

Preliminary investigations have revealed that the presence of relatively high levels of calcium can inhibit fungal growth (Whitney & Arnott, unpublished data), so it appears that the production of Ca-oxalate by fungi could play some role in regulating calcium ion concentrations. Clearly, though, Ca-oxalate can have other functions in the biology of fungi. Physical protection, regulation of development, structural support, etc., are all possible roles for mineralized deposits in fungi. These possibilities are not mutually exclusive, and the Ca-oxalate system of fungi may play a rather complex role in the morphology and development of fungi.

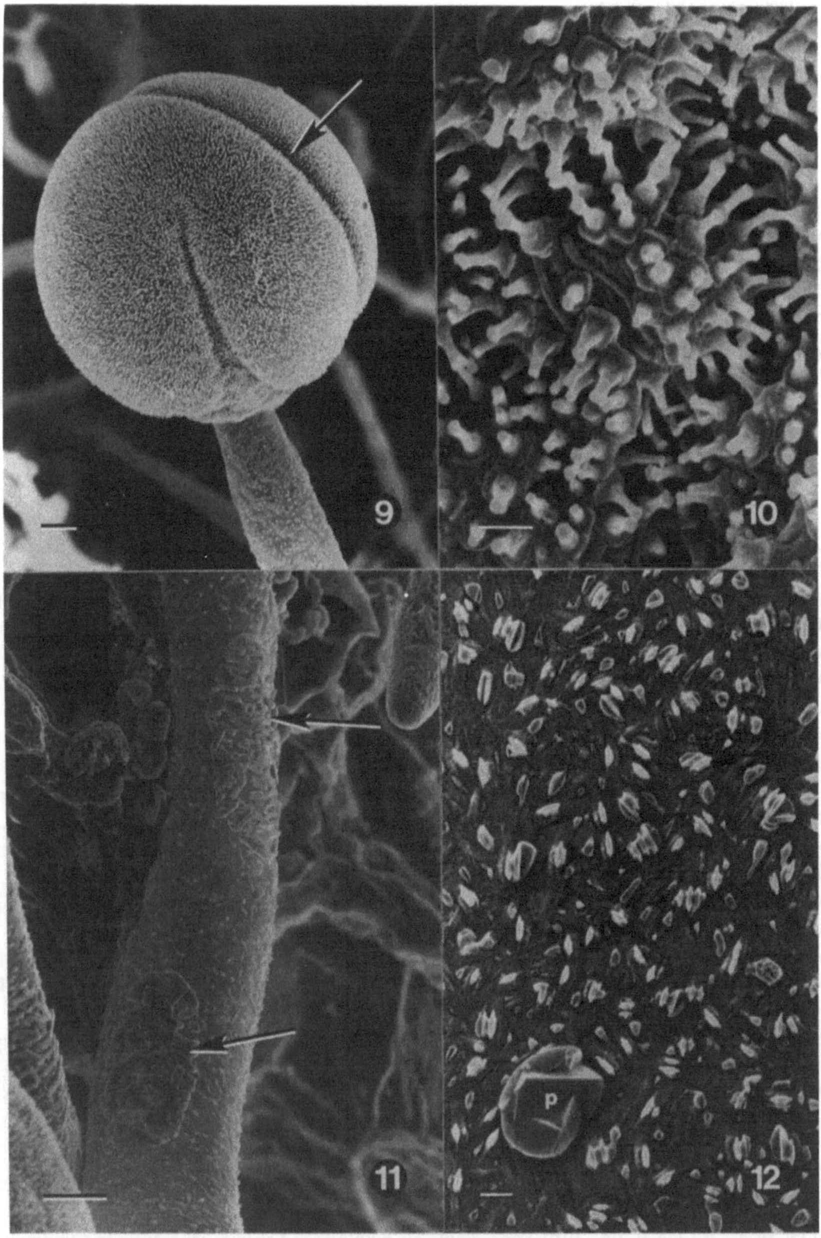

FIGURES 9–12. Gilbertella persicaria. ▪9. Mature sporangium. Note dense covering of Ca oxalate crystals and the dehiscence lines of sporangial wall (arrow). Bar = 10 μm. ▪*10.* Higher magnification of sporangial crystals. Each crystal is composed of a flattened base plate, and angular, upright portion, and a flattened cap. Bar = 1 μm. ▪*11.* Sporangiophore showing two large crystal plaques (arrows). Note the smaller crystals embedded in the sporangiophore wall. Bar = 10 μm. ▪*12.* Higher magnification of sporangiophore crystals. Most consist of a rectangular base plate with two upright, parallel crystalline appendages. A small crystal plaque (**p**) is also present. Bar = 1 μm.

ACKNOWLEDGMENTS

This research was supported by a National Science Foundation Postdoctoral Research Fellowship (BSR-8411501). Electron microscopy facilities were provided by the Graduate School and the College of Science, The University of Texas at Arlington.

REFERENCES

ARNOTT, H.J., 1982. Three systems of biomineralization in plants with comments on the associated organic matrix. *In*: Biological Mineralization and Demineralization. (G. H. Nancollas, ed.). Springer-Verlag, New York., pp. 199-218.

ARNOTT, H.J.. 1984. Rodlets associated with the cell wall and crystals of litter fungi. Proc. Elec. Microsc. Soc. Amer., 42:322-323.

ARNOTT, H.J, & WEBB, M.A., 1983. The structure and formation of calcium oxalate crystal deposits on the hyphae of a wood rot fungus. SEM, 1983/IV:1747-1758.

BLAND, C.E. & CHARLES, T.M., 1972. Fine structure of *Pilobolus*: surface and wall structure. Mycologia, 64:774-785.

BORCHERT, R., 1985. Functional anatomy of the calcium-excreting system of *Gleditsia triacanthos* L. Botan. Gazz. (Crawfordsville) 145: 474-482.

BULLER, A.H.R. 1934. Researches on Fungi. Vol. VI. Longmans, Green and Company, London., 224 p.

COKER, W.C. & COUCH, J.N., 1928. The Gasteromycetes of the eastern United States and Canada. Univ. of North Carolina Press, Chapel Hill. 201 p.

CROMACK, K. JR., SOLLINS, P., TODD, R.L., FOGEL, R., TODD, A.W.. FENDER, W., CROSSLEY, M.E. & CROSSLEY, JR., D,A., 1977. The role of oxalic acid and bicarbonate in calcium cycling by fungi and bacteria: some possible implications for soil animals. Ecol. Bull. (Stockholm), 25:246-252.

DE BARY, A., 1887. Comparative Morphology and Biology of the Fungi, Mycetozoa, and Bacteria. Clarendon Press, Oxford. 525 p.

FREY-WYSSLING, A., 1981. Crystallography of the two hydrates of crystalline calcium oxalate in plants. Amer. J. Bot., 68:13O-141.

GRAUSTEIN, W.C., CROMACK, K. & SOLLINS,P., 1977. Calcium oxalate: occurrence in soils and effect on nutrient and geochemical cycles. Science, 198:1252-1254.

HORNER, H.T., TIFFANY, L.H. & CODY, A.M., 1983. Formation of calcium oxalate crystals associated with apothecia of the discomycete *Dasyscypha capitata*. Mycologia, 75:423-435.

HORNER, H.T., TIFFANY, L.H. & CODY, A.M., 1985a. Calcium oxalate bipyramidal crystals on the basidiocarps of *Geastrum minus* (Lycoperdales). Proc. Iowa Acad. Sci., 92(2):70-77.

HORNER, H.T., TIFFANY, L.H., CODY, A.M. & KNAPHUS, G., 1985b. Development of fungal calcium oxalate crystals associated with the basidiocarps of *Geastrum minus* (Lycoperdales), SEM 1985/II:789-801.

JONES, D., MCHARDY, W.J. & WILSON, M.J, 1976. Ultrastructure and chemical composition of spines in Mucorales. Trans. Brit. Mycol. Soc., 66:153-157.

KAZMIERCZAK, J., ITTEKKOT, V. & DEGENS, E.T., 1985. Biocalcification through time: environmental challenge and cellular response. Palontologica Zeitung, 59, 15-33.

LLOYD, C.G., 1907. New notes on the Geasters. Mycol. Notes, 25:309-317.

POWELL, M., & ARNOTT, H.J., 1985. Calcium oxalate crystal production in two members of the Mucorales. SEM, 1985/I:183-189.

SHIPTON, W.A., & LUNN, J.A., 1980. Sporangiole morphology and species separation in *Cunninghamella*. Trnas. Brit. Mycol. Soc., 74:483-491.

SIMKISS, K. 1977. Biomineralization and detoxification. Calc. Tiss. Res., 24:199-200.

THOMPSON, W., 1984. Distribution, development and functioning of mycelial cord systems of decomposer basidiomycetes of the deciduous woodland floor. *In* The Ecology and Physiology of the Fungal Mycelium (D. H. Jennings & A. D. M. Rayner, eds.) pp. 185-214.

URBANUS, J.F.L.M., VAN DEN ENDE, H. & KOCH, B., 1978. Calcium oxalate crystals in the wall of *Mucor mucedo*. Mycologia, 70:829-842.

WHITNEY, K.D., & ARNOTT, H.J., 1986. Morphology and development of calcium oxalate deposits in *Gilbertella persicaria* (Mucorales). Mycologia, 78:42-51.

WHITNEY, K.D., & ARNOTT, H.J., 1986. Calcium oxalate crystals and basidiocarp dehiscence in *Geastrum saccatum* (Gasteromycetes). Mycologia, 78:649-656.

WHITNEY, K.D., & ARNOTT, H.J., 1987. Calcium oxalate crystal morphology and development in *Agaricus bisporus*. Mycologia, 79:180-187.

YOUNG, T.W.K., 1968. Electron microscopic study of the asexual structures in Mucorales. Proc. Linnean Soc. Lond., 179:1-9.

GROWTH AND CALCIFICATION OF *CALOTHRIX* - DOMINATED ONCOLITES FROM NORTHERN ENGLAND

Allan Pentecost

Department of Human Environmental Soience, King's College, London, Hortensia Road, SW10OQX, United Kingdom

ABSTRACT. Oncolite growth rates were measured using a new technique over a period of 622 days in a shallow north Yorkshire stream. Deposition was seasonally dependent and significantly correlated to water temperature with maximum radial rates of around 1 μm/day recorded during the summer. During winter, rates were negligible and slight corrosion may have occurred. The annual radial growth was 143$\pm$ 98 μm.

The surface layers of actively accreting oncolites were colonized by a range of microorganisms but the cyanobacterium *Calothrix parietina* Thuret. was dominant. No seasonal differences in the microbial flora were detected though the deposits showed marked concentric banding which correlated significantly with the estimated annual growth rate. Observations with the scanning electron microscope showed that *Calothrix* calcification is confined to the mucilaginous sheath and consists of unoriented crystals of micrite. The fine structure of modern and subfossil oncolites, collected from a nearby site, is compared, and the value of oncolites as paleoenvironmental indicators is assessed in the light of the results obtained.

INTRODUCTION

Modern oncolites occur in a wide range of shallow freshwater and marine environments (Golubic, 1973; Pentecost & Riding, 1986). Structurally, they are allied to oolites and pisolites, differing only in their large size, which ranges from 5 mm to 30 cm and their less regular shape. Actively growing oncolites are found associated with various species of cyanobacteria, belonging mainly to the genera *Calothrix*, *Homeothrix*, *Phormidium* and *Schizothrix* and it appears that oncolite structure is determined by the growth-pattern and the habit of these colonizing cyanobacteria (Schafer & Stapf, 1978). The seasonal growth of this microbial skin has long been thought to give rise to the formation of regular concentric laminations revealed in oncolite sections (Roddy, 1915), but to date, there have been no *in situ* growth studies. Paleoenvironmentalists recognize the potential value of oncolites in the fossil record but are faced with a major problem; is the correct interpretation of the laminations as seasonal growth pattern of cyanobacteria? In some cases, this appears possible but the issue is complicated by the known occurrence of abiogenic structures closely resembling oncolites, (Donahue, 1969; Jones

Origin, Evolution, and Modern Aspects of Biomineralization in Plants and Animals
Edited by Rex E. Crick
Plenum Press, New York

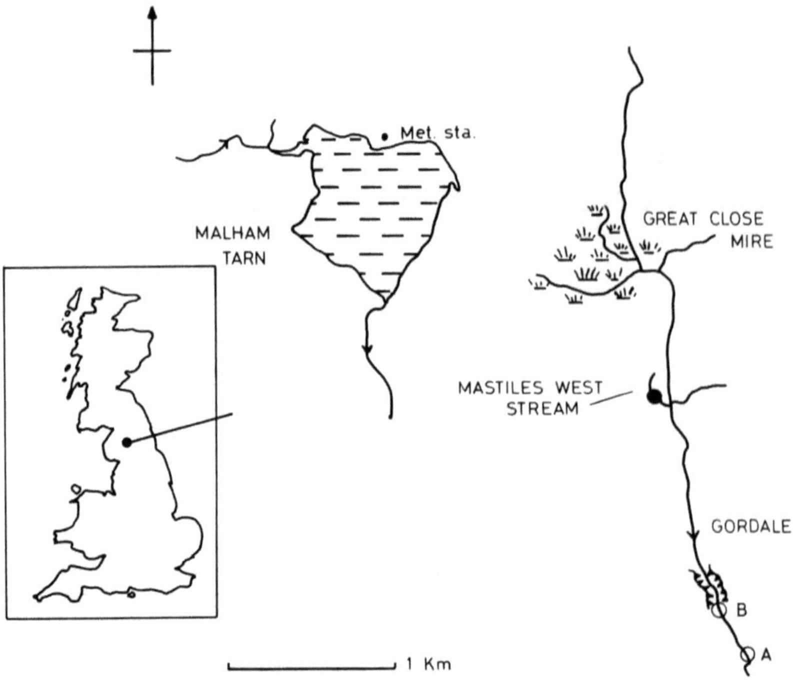

FIGURE 1. Site location. Open circles represent collection sites for subfossil material.

& Wilkinson, 1978) which are also laminated. The possibility also exists that the laminations are due in part, or entirely to burial episodes or periodic dessication.

The aim of this study is to determine whether the laminations are indeed due to a seasonal pattern of growth using in situ measurements in a small English stream. In addition, an attempt is made to correlate growth with some environmental variables. The relationship between the microbial flora and the deposited carbonate is also explored.

MATERIALS AND METHODS

Actively accreting oncolites were investigated at Mastiles West Stream near Malham Tarn, Yorkshire. This small stream rises from a spring along the Silurian-Carboniferous unconformity 200 m upstream of the sampling site. The aquifer is the Carboniferous Great Scar Limestone. Subfossil oncolites were collected from tufa banks along Gordale Beck about 1 m above present water level (Fig. 1). This material has not been dated but is certainly postglacial and *may be* of Atlantic age (*c.* 6000 BP). The surrounding region is a sheep-grazed limestone grassland plateau, altitude 360-380 m.

Water chemistry was monitored at three-monthly intervals and samples were analyzed for Ca, total CO_2, NO_3-N and PO_4-P (Mackreth *et al.*, 1978). A calcite saturation index, Satcal, was calculated from the calcium and carbonate ion activity products (Ford, 1971). Air temperatures were obtained from Malham Tarn

Meteorological Station situated 2.3 km NW, altitude 380 m. Discharge was measured with a portable V-notch weir.

Oncolite growth was measured as a change in weight. A minimum of 10 oncolites were removed from the stream and a small hole drilled in each, then a red plastic stud, 2 mm in diameter, was inserted and sealed with epoxy resin. This allowed individual oncolites to be identified and recovered from the stream. The oncolites were dried to constant weight at room temperature, weighed and replaced in the stream. They were recovered at approximately 3 month intervals, redried and reweighed. Distances moved were recorded and lost oncolites replaced. Growth was measured over 622 days.

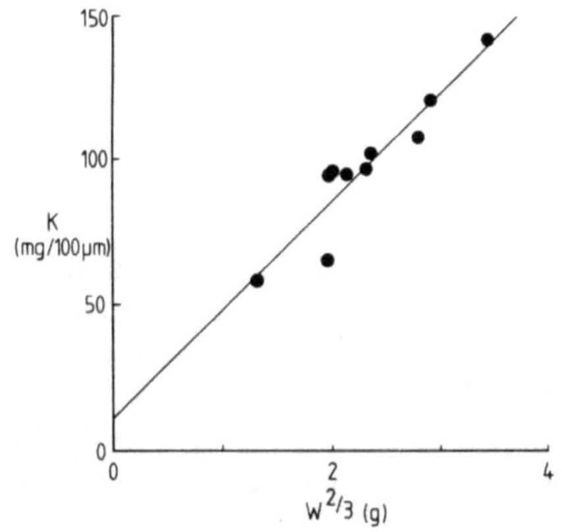

FIGURE 2. Regression of deposit weight per unit thickness (**k**) on oncolite dry weight (**w**).

Weight changes were converted to radial growth rates using a linear regression equation. Ten oncolites of variable size were dried, weighed and carefully measured along two or three axes with a micrometer, then suspended in 100 ml 5% EDTA pH 8.0 for 48 h at room temperature. This removed a narrow surface zone of calcite. The oncolites were removed, washed 4 times in water and the decalcified algal layer carefully rubbed off. The oncolites were then dried, reweighed and remeasured. The weight loss was then assessed in relation to the thickness of deposit removed and expressed as a dry weight in milligrams for a layer 100 μm in thickness. This value was plotted against the average total dry weight before and after decalcification, raised to the power 2/3. This was necessary because a linear relationship was being sought between total weight (proportional to the radius cubed) and the surface area (radius squared). A significant ($P<0.001$) regression of weight change on total weight was obtained (Fig. 2), with equation:

$$K = 10.81 + 37.7\ W^{.667}$$

where K is the weight of a 100 μm thick surface layer in mg. W is the mean dry weight of the oncolite between two sets of measurements. Using this equation, growth rates for individual oncolites were obtained.

A further regression was used to obtain the mean water temperature from the ten-day mean air temperature taken at the meteorological station. This was valid because air and water temperature at the site were significantly correlated ($P<0.001$). The equation used was:

$$Y = 0.981\ X + 0.174$$

where Y is the water temperature and X the air temperature (^{o}C).

The microbial flora was identified microscopically using EDTA-decalcified material. Measurements of cells and sheaths were made with an oil immersion objective. For

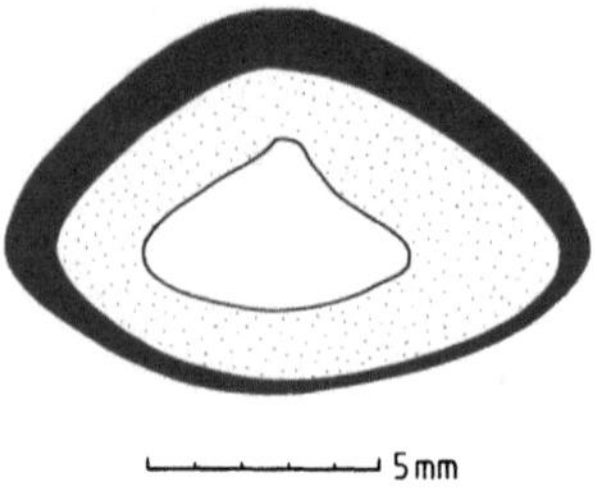

FIGURE 3. Section through a modern, partly decalcified oncolite. Nucleus unshaded; deposits free of algae, stippled; algal layer, opaque.

Calothrix, which had tapering filaments, measurements were made randomly along the filament using random number tables. Modern and subfossil material was dried at room temperature, coated with Au-Pd and observed in an EM 501B scanning electron microscope. Some fresh material was frozen in liquid nitrogen and the ice sublimed under the electron began prior to coating, to improve preservation of gelatinous sheath material. Petrological examinations were performed on 15 μm sections impregnated with epoxy resin. The calcium carbonate content of oncolites by volume was obtained from sliced samples dried at 105°C then weighed. Samples were evacuated at a water pump, immersed in air-free water, removed and rapidly reweighed. Porosity was determined by difference, after correction for the nucleus.

RESULTS

Flora

Fresh oncolite surfaces were pigmented blue-green and consisted mainly of the filamentous cyanobacteria *Calothrix parietina* Thuret, *Homeothrix* sp. and *Schizothrix calcicola* (Ag.) Gomont, the former being the most abundant. The filamentous species *Phormidium incrustatum* (Naeg.) Gomont, *Rivularia haematites* (D.C.) Ag., *Hyella* and *Dichothrix* spp. were occasional and *Pseudanabaena minuta* Skuja appeared in a BG11 culture. Coccoid forms were uncommon but included *cf. Hydrococcus rivularis* (Kuetz.) Menegh. and *Synechococcus* sp. The last appeared in a BG 11 culture together with *Chamaesiphon* sp

Algae included *Gongrosira incrustans* (Reinsch) Schmidle, unidentified coccoid Chlorophyceae and numerous diatoms of the genera *Achnanthes*, *Cymbella*, *Denticula* and *Gomphonema*. *Calothrix*, *Homeothrix* and *Gomphonema* often grew as dome-shaped colonies *c*. 1 mm in diameter but intimate mixtures of all three were often found in a single 'colony'. The algal layer was variable in thickness, ranging from 0.2-1.5 mm. In some cases this was clearly related to oncolite position in the stream, with the greatest thickness occurring on the upper side (Fig. 3). No seasonal differences in the flora were apparent.

Physical and chemical environment.

Information on waterflow, depth and chemistry is presented in Table 1. Oncolites were abundant only in a 50 m stretch of the stream where the gradient was less than 1 in 20. Below this the water descended more steeply at a gradient of 1 in 3 before entering Gordale Beck. The movement of oncolites was observed in both stretches (Table 1) and was clearly more rapid over the steeper gradient.

The water was at or above calcite saturation (Satcal = 1.0) but there was no apparent seasonal trend (Fig. 4a). However, aqueous CO_2 concentrations generally peaked slightly in the autumn probably as a result of increased soil CO_2 levels in the summer.

TABLE 1

MASTILES WEST STREAM

	mean	range	units
discharge	4.6	0.7–7.0	$dm^3 s^{-1}$
flow rate	0.26	0.18–0.33	$m\ s^{-1}$
measured temperature	11.8	1.9–24.5	^{o}C
depth over oncolites		2.5–4.5	cm
pH	7.96	7.7–8.2	
Ca	1.56	1.42–1.78	mM
Total CO_2	3.44	3.15–3.87	mM
Satcal	1.03	1.0–1.04	
NO_3-N		40.0–60.0	μM
PO_4-P		0.5–10	μM

ONCOLITE MEASUREMENTS

	mean	range	units
dimensions	12	3–26	mm
$CaCO_3$ modern	66	58–72	volume %
$CaCO_3$ subfossil	75	54–85	volume %
Radial growth rate	143	46–241	μm
Estimated age	26	6–51	a
Movement			
gradient *c.* 1 in 20	0.1	0–0.3	$m\ a^{-1}$
gradient *c.* 1 in 3	11	0–23	$m\ a^{-1}$

Nitrate levels were moderately high though soluble phosphate varied widely, reaching a maximum in the winter.

Growth rates

A pronounced seasonal trend in the growth rates of oncolites was obtained (Fig. 4c) with most rapid growth in midsummer (*c.* 1 μm d-1) and least growth during winter. In fact a slight loss of weight is apparent during the winter suggesting some dissolution, though this was not revealed in the chemical analyses (above). Mean water temperature, calculated from the regression, ranged from 0-15°C (Fig. 4b). Although oncolite growth was variable, a significant positive correlation ($P<0.001$) was obtained between the mean seasonal growth rate and water temperature. The mean annual radial growth rate of the oncolites was 143 μm. This measurement was used to estimate average oncolite age (Table 1).

Fine structure

In fresh oncolites, the deposit consisted of a groundmass of micrite with occasional anhedral calcite crystals, generally randomly oriented. Well formed rhombs were rare. The surface consisted of micrite 1-4 μm in diameter, punctuated by well-formed hollow tubular structures interpreted as the apices of vertically arranged cyanobacterial filaments. These structures (Fig. 5a-d) were either plane with the surface or slightly raised above it. The tubes were normally empty but occasionally revealed the end of a trichome, particularly in cryo-fixed material. The inner and outer diameters of these tubes were statistically compared with light microscope measurements on *Calothrix* filaments (Fig. 6). All *Calothrix* size distributions were positively skewed and the distributions indicate that 1) the tubes are sufficiently wide to admit the *Calothrix* trichome, and 2) the calcified tubes surrounding the cavities correspond to the thickness of the *Calothrix* sheath. When the median of the measurements were compared using the non-parametric Mann-Whit-

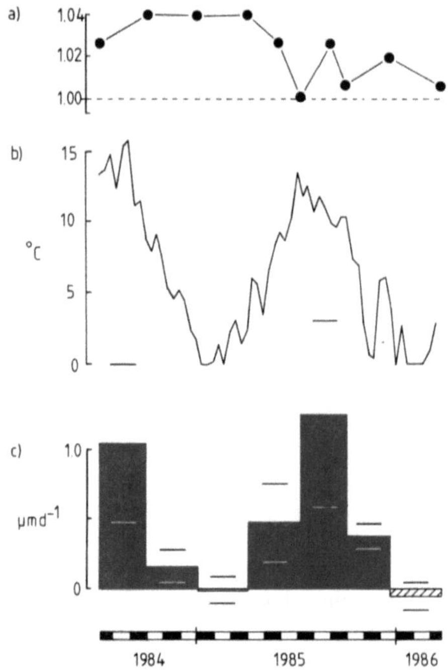

FIGURE 4. ▪a. Satcal values for Mastiles West Stream. ▪b. Predicted water temperature from regression, ten-day means. ▪c. Oncolite mean radial growth rate, μ m d − 1, with 95% confidence limits to the mean.

ney test, the distributions were found to differ significantly ($P < 0.01$), the cavity diameters being slightly larger than the trichomes, and the calcified tube thickness slightly greater than the *Calothrix* sheath.

The sheath-associated micrite was densely but mostly irregularly arranged (Fig. 5a). Some crystals had smoothed outlines suggestive of an organic cover (Fig. 5b) and occasionally larger plate-like crystals surrounded the cavity (Fig. 5c,d) though this was uncommon. In cryo-fixed material, a surface layer of gossamer-like mucilage, free of calcification, was dispersed over the entire oncolite surface. Broken perpendicular surfaces revealed radiating clusters of tubes characteristic of *Calothrix/Dichothrix* and *Homeothrix* colonies (Fig. 5e). Individual crystals were less evident here but the calcified tube perimeters were clearly seen. In thin sections, radiating colonies of *Calothrix/Dichothrix* and *Homeothrix* were sometimes observed and a difference in calcification was noted. In *Homeothrix*, the filaments were usually surrounded by radiating microsparite both at the surface and within the oncolite. This was never observed in *Calothrix/ Dichothrix* colonies. In cases where microsparite was found adjacent to the latter the crystals were equant and never radiating.

Diatoms were abundant at the oncolite surfaces but were not observed calcified. No seasonal difference in surface structure was apparent.

Subfossil surfaces differed completely from modern samples and were entirely smooth without perforations. Perpendicular fractures revealed occasional radiating tubular clusters but much recrystallization was apparent. These observations were confirmed by examination of thin sections using the polarizing microscope.

Cut and polished sections of both modern and subfossil oncolites revealed a nucleus of limestone, shale or tufa surrounded by the banded deposit (Fig. 5f). The banding, caused by alternating white and buff zones, occurred on two levels, a broad but indistinct pattern with a period above 1 mm and superimposed fine bands with a period below 1 mm. Intergradations between these two patterns were common making interpretation difficult in many sections. The band periods of modern and subfossil sections compared favorably with the measured annual growth rate (Fig. 7) suggesting a seasonal pattern.

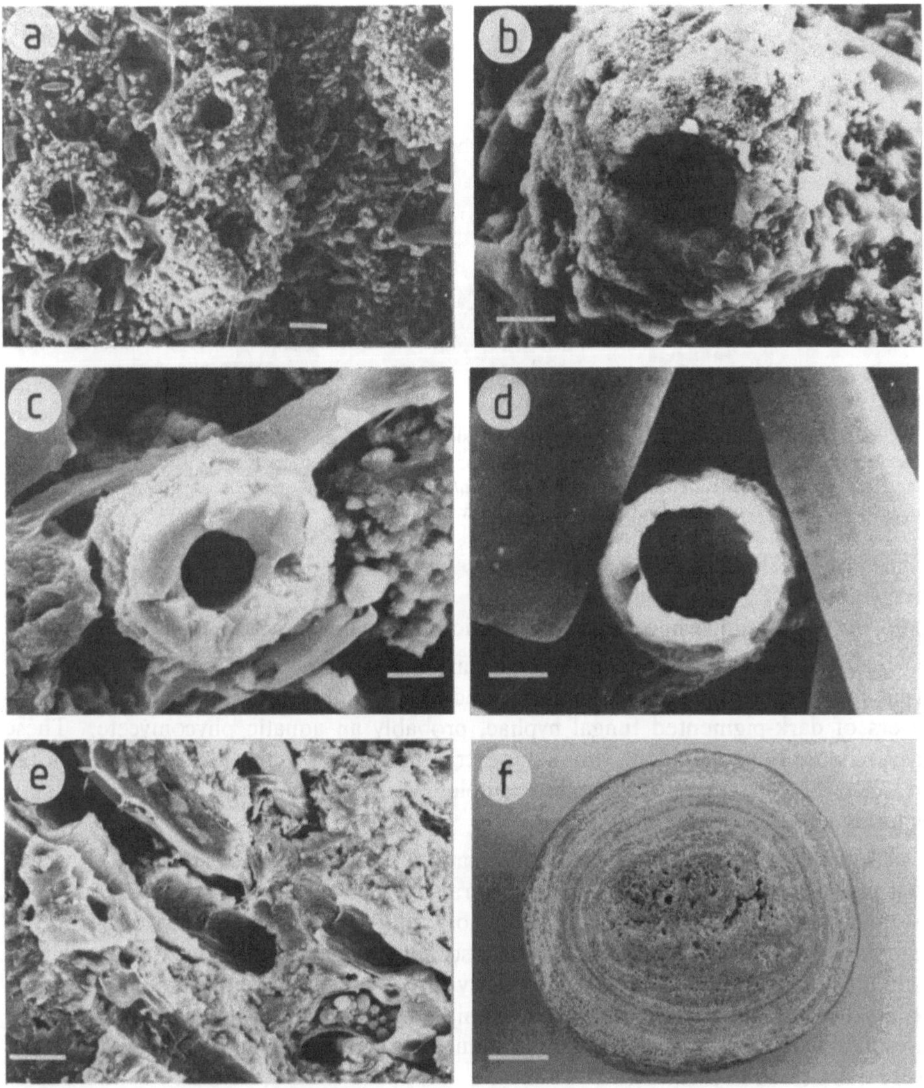

FIGURE 5. ▪a. Surface view on oncolite showing a group of calcified *Calothrix* filaments whose apices protrude slightly above the general surface. Mucilage strands can be seen among the micrite. SEM micrograph. Bar = 20 μm. ▪b. Detail of protuberant *Calothrix* tube. Fine micrite, some of which appears subhedral is apparent near the cavity. SEM micrograph. Bar = 2 μm. ▪c. Calcified sheath of plate-like layered crystals, *cf. Calothrix*. SEM micrograph. Bar = 5 μm. ▪d. Narrow calcified rim of subhedral calcite exposed at oncolite surface, between two *Gomphonema* (diatom) frustules. SEM micrograph. Bar = 2 μm. ▪e. Oncolite fractured perpendicular to the surface showing calcified *Calothrix* tubes with a smooth inner surface. Upper surface faces top of page. SEM micrograph. Bar = 10 μm. ▪f. Median section through a subfossil oncolite from Gordale site A. Note the porous tufa nucleus and concentric lamination. Bar = 2.5 μm.

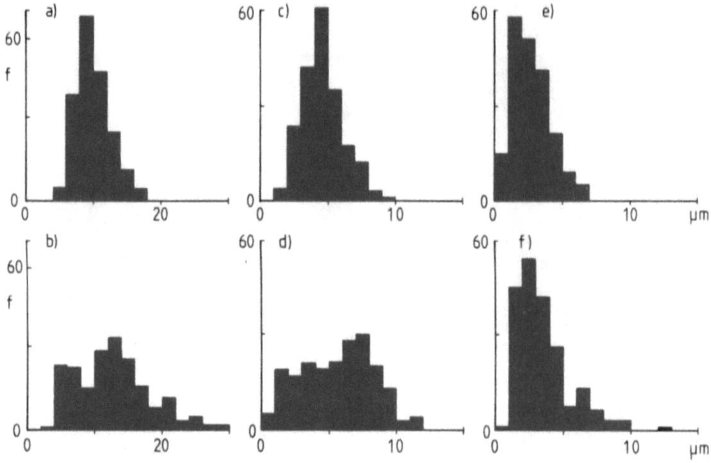

FIGURE 6. **Frequency histograms of *Calothrix* dimensions observed by light microscopy and calcified tubes observed by scanning electron microscopy. n = 200 for each series of observations. ▪a. *Calothrix* filament diameter. ▪b. *Calothrix* tube outer diameter. ▪c. *Calothrix* trichome diameter. ▪d. *Calothrix* tube inner diameter. ▪e. *Calothrix* sheath thickness. ▪f. *Calothrix* tube thickness.**

Also, the width of the white bands was significantly greater ($P < 0.01$) than the buff bands.

In fresh oncolite sections, prominent dark bands were commonly observed in the algal layer, among the pale banding referred to above and were found to be caused by layers of dark-pigmented fungal hyphae, probably an aquatic phycomycete. These hyphae were most abundant, though not strictly confined to the lower surfaces of the oncolites. Such bands were also noted completely surrounding some oncolites, probably marking a phase of temporary burial.

The extent of mineralization in modern and sub-fossil specimens (Table 1) revealed that the latter were significantly more mineralized than the former ($P < 0.01$, rankit plot (standard statistical procedure to test for normality) followed by Student's t-test on 10 samples of each). This difference was also apparent in thin sections where sparite palisades were well developed, particularly in the outer layers of subfossil material. Recrystallization was clearly accompanied by some additional mineralization, presumably at some stage after burial, though strangely, fragments of associated tufa were not altered in this way.

DISCUSSION

The microbial flora produced few surprises. *Calothrix* has been widely reported from modern freshwater oncolites (Shafer & Stapf, 1978; Pentecost, 1985) though it is far from ubiquitous. Even within the United Kingdom the flora of oncolites varies considerably. For example, the oncolites of Attermire (northern Yorkshire) and Sunbiggin (Cumbria) also have abundant *Calothrix* but those of Priory Mill (Oxford) are colonized mainly by *Phormidium*. In the chalk stream oncolites of Hampshire, species of *Hyella* are a major component. The ecology of some species is discussed briefly by Schafer & Stapf (1978) and Pentecost (1985).

One of the most interesting features of Mastiles West Stream is the occurrence of well calcified colonies of *Rivularia* on and among the oncolites. *Rivularia* trichomes are similar to *Calothrix*, the main difference being the formation of large button-shaped colonies, up to 10 mm in diameter in *Rivularia*. Growth rates of *Rivularia* are an order of magnitude higher than *Calothrix* (Pentecost, 1987) and the process of calcification also differs, being most apparent in winter in the former. Morphologically, a continuum of forms can be recognized within the Rivulariaceae, from *Calothrix* which consists of essentially unbranched trichomes, through the sparingly branched *Dichothrix* to the massive colonies of *Rivularia*. The fact that all three genera can occur on a single stone and display such marked differences in growth rate and calcification suggests a good case for genus-specific calcification. For further discussion, see Golubic (1973) and Pentecost & Riding (1986).

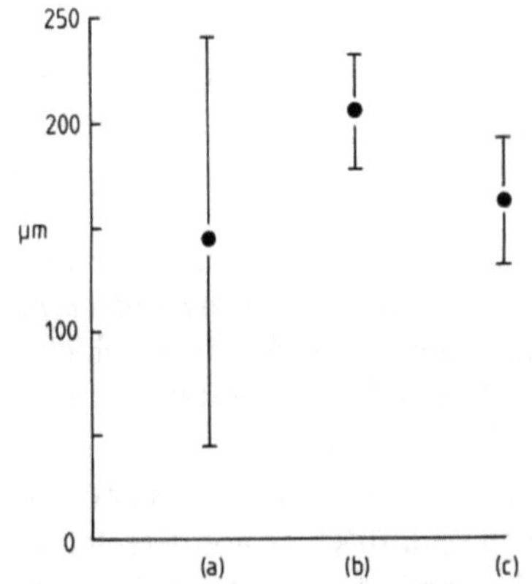

FIGURE 7. Relationship between oncolite radial growth per annum (a) with deposit band period in modern (b) and subfossil (c) oncolites, with 95% confidence limits to the means (n = 50).

The strong positive correlation between oncolite growth and temperature means that a correlation also exists between growth and solar radiation. Both solar radiation and temperature influence algal growth and in the case of *Rivularia*, the former is thought to be the most important (Livingstone *et al.*, 1984). It could be argued that oncolite weight gains reflect something other than surface accretion however. For example, it might mean that additional mineralization is occurring within the oncolite instead. This is countered by two pieces of evidence. First, carbonate deposition was observed to cover the plastic studs only during the summer, and second, no correlation was found between oncolite diameter and the carbonate content by volume. If a correlation had been found, the larger and therefore older oncolites would be expected to be more dense due to the more prolonged period of internal mineralization.

Calothrix and *Homeothrix* grow perpendicular to the deposit surface and growth might outstrip deposition during the summer. This occurs in *Rivularia* (Pentecost, 1987) though apparently not in *Phormidium* (Monty, 1976). It may be that *Calothrix* growth is too slow to outstrip deposition, which may help to explain why deposition is so different in *Calothrix* and *Rivularia*.

Measurements on the *Calothrix* filaments and the calcified oncolite surface showed some interesting features. The significant difference between the medians of the cavity and trichome width means that either the innermost layer is uncalcified, that there is a true gap between trichome and sheath, or that some fixation or measurement error has occurred. There is insufficient evidence to resolve this question at present though it is likely that cavities caused by cyanobacteria other than *Calothrix* have been included in the measurements. In fact there is some evidence of bimodality in Fig. 6b suggesting that some of the measurements include *Homeothrix*, whose trichomes are more narrow than those of *Calothrix*. Of more interest is the observation that the calcified rim appears to be wider than the *Calothrix* sheath and it is possible that the deposited carbonate serves as a secondary nucleator allowing further tube enlargement to occur. Also of interest is the influence which *Homeothrix* and perhaps *Schizothrix* exert over mineralization. Such an influence has been widely suspected in a variety of cyanobacteria (Monty & Mas, 1982) and similar radiating structures have been found by Egglestone & Dean (1976) in lacustrine tufa.

Rates of photosynthesis in Mastiles West Stream probably reach a maximum during summer though this was not detected in the stream chemistry. It is doubtful whether such a slow-growing species as *Calothrix* could raise pH to a sufficiently high level to promote even local precipitation. The nucleation and growth of calcite within the mucilaginous sheath is much more likely as the waters were found to be saturated or supersaturated with calcite. Carbonate trapping is unlikely to be important because the size of the suspended stream particles is generally in the 10 to 30 μm range, much greater than that found in the deposits. Evidence for slight decalcification during the winter suggests that there must be occasions when the water is undersaturated with calcite. This may be the result of snow-melt since heavy falls are frequently experienced during the winter. Oncolite erosion, which has sometimes been reported (Golubic, 1973) was negligible here, because of the low discharge and flow rate.

Several authors have suspected that the broad, often spongy bands found in freshwater oncolites are the result of summer algal growth (Roddy, 1915; Golubic, 1973; Jones & Wilkinson, 1978). This layer is often intercalated with a narrower, more compact layer, sometimes colonized by other species (Weiss, 1969; Monty, 1976; Schafer & Stapf, 1978). However, growth rate data obtained from lamination periodicites are scarce. Jones & Wilkinson (1978) noted a mean couplet distance of 2.1 mm in lacustrine oncolites, and periods of 1 mm or less have been reported in fossil material (Monty & Mas, 1981; Link *et al.*, 1978). Pentecost (1978) obtained values of 0.4 mm for incrustations associated with oncolites where the laminations were seasonal, but they were not studied in detail. Periodic banding which is probably seasonal, has also been reported and measured in more massive deposits of tufa (Irion & Mueller, 1968; Pentecost, 1978).

The evidence presented here has shown that the banding periodicities seen in sections do not differ statistically from the measured growth rates. However, it is also clear that both banding and growth is variable, giving substantial margins of error. Furthermore, some of the darker, narrow zones have been shown to be caused by fungal hyphae marking unfavorable conditions on the undersides. Since the bands represent overturning episodes, we may expect an additional source of error which would lead to underestimates in growth rates obtained from band measurements. The amount of error would depend upon the frequency of overturning and at Mastiles, this frequency must be low, as a reasonable fit was obtained between periodicity and growth (Fig. 7). This was confirmed by observation. Within the level stretch of stream, an oncolite might move 30 cm or more during a single episode, giving a rate of less than one overturning per year. Emersion of oncolites during low waterlevels would also prevent algal growth and could be another source of error, though this was not observed at Mastiles West. Both Roddy (1915) and Golubic & Fischer (1975) found increased growth on the undersides of the Pennsylvanian Conestoga oncolites though this appears to be exceptional, as it has not been noted elsewhere (Jones & Wilkinson, 1978; Schafer & Stapf, 1978).

The band periodicity of subfossil oncolites has also been shown to be the same as that of modern material, although their microfabrics differed radically. This difference must have been caused after accretion had ceased, as there is evidence for additional mineralization which was not encountered in modern material. Also, the filamentous algal structure had disappeared entirely in some regions which could result in misinterpretation in certain cases. It may be concluded that whilst some paleoecological information may be obtained from oncolite facies, the position is far from satisfactory. These difficulties parallel those in marine stromatolites where discontinuities and irregularities are commonly observed (Panella, 1976).

ACKNOWLEDGEMENTS

Thanks are extended to the Staff of Malham Tarn Field Centre for the use of laboratory facilities, the Central Research Fund of London University for travel assistance and Mr. Tony Brain for help with the electron microscopy. Financial support was also provided by The Organized Research Fund of The University of Texas at Arlington and The Sea Grant Program administered by Texas A & M University.

REFERENCES

DONAHUE, J. 1969. Genesis of oolite and pisolite grains - an energy index. J. Sed. Petrol., 39:1399-1411.

EGGLESTONE, J.R. & DEAN, W.E. 1976. Freshwater stromatolitic bioherms in Green Lake, New York. *In* Stromatolites (ed. M.R. Walter), pp. 479-488. Amsterdam, Elsevier.

FORD, D. 1971. Characteristics of limestone solution in the Southern Rocky Mountains and Selkirk Mountains, Alberta and British Columbia. Can. J. Ear. Sci., 8:585-609.

GOLUBIC, S. 1973. The relationship between blue-green algae and carbonate deposits. The Biology of Blue-green Algae (eds. N.G. Carr & B.A. Whitton), pp. 434-472. London, Blackwell.

GOLUBIC, S. & FISCHER, A.G. 1975. Ecology of calcareous nodules forming in Little Conestoga Creek near Lancaster, Pennsylvania. Verhandlungen Internationalen Vereinigungen fur theoretische and angewandte Limnmologie, 19:2315-2323.

IRION, G. & MUELLER, G. 1968. Mineralogy, petrology and chemical composition of some calcareous tufa from the Schwabischische, Germany. *In* Carbonate Sedimentology in Central Europe (eds. G. Mueller & G.M. Friedmann), pp. 157-171. New York, Springer Verlag.

JONES, F.G. & WILKINSON, B.H. 1978. Structure and growth of lacustrine pisoliths from recent Michigan marl lakes. J. Sed. Petrol., 48:1103-1110.

LINK, M.H., OSBORNE, R.H. & AWRAMIK, S.M. 1978. Lacustrine stromatalites and associated sediments of the Pliocene Ridge Route formation, Ridge Basin, California. J. Sed. Petrol., 48:143-158.

LIVINGSTONE, D., PENTECOST, A. & WHITTON, B.A. 1984. Diel variations in nitrogen and carbon dioxide fixation by the blue-green alga *Rivularia* in an upland stream. Phycologia, 23:125-133.

MACKERETH, F.J.H., HERON, J. & TALLING, J.F., 1978. Water analysis: some revised methods for limnologists. Freshwater Biol. Assoc. Spec. Publ. no. 36. Kendal, Titus Wilson & Son.

MONTY, C.L.V. 1976. The origin and development of cryptalgal fabrics. *In* Stromatolites (ed. M.R. Waler), pp. 193-250. Amsterdam, Elsevier.

MONTY, C.L.V. & MAS, J.R. 1981. Lower Cretaceous (Wealdian) Blue-green Algal deposits of the province of Valencia, Eastern Spain. *In* Phanerozoic Stromatolites (ed. C. Monty), pp. 85-120. Berlin, Springer.

PANELLA, G. 1976. Geophysical inferences from stromatolite lamination. *In* Stromatolites (ed. M.R. Walter), pp. 673-685. Amsterdam, Elsevier.

PENTECOST, A. 1978. Blue-green algae and freshwater carbonate deposits. Proc. Roy. Soc. Lond., B200:43-61.

PENTECOST, A. 1985. Association of cyanobacteria with tufa deposits: identity, enumeration and nature of the sheath material revealed by histochemistry. Geomicrobiol. J., 4:285-298.

PENTECOST, A., 1987. Calcification in the cyanobacterium *Rivularia haematites*. Proc. Roy. Soc. Lond., B232:125-136.

PENTECOST, A. & RIDING, R. 1986. Calcification in cyanobacteria. *In* Biomineralization in Lower Plants and Animals (eds. B.S.C. Leadbeater & R. Riding) pp. 73-90 Oxford Univ. Press.

RODDY, H.J. 1915. Concretions in streams formed by the agency of blue-green algae and related plants. Proc. Amer. Phil. Soc., 54:246-258.

SCHAFER, P. & STAPF, K.R.G. 1978. Permian Saar-Nahe Basin and Recent Lake Constance (Germany), two environments of lacustrine algal carbonates. Spec. Publ. Inter. Assoc. Sedimentol., 2:83-107.

WEISS, M.P., 1969. Oncolites, paleoecology, and Laramide tectonics, central Utah. Amer. Assoc. Pet. Geol., Bull., 53:1105-1120.

PERIDIAL CALCIFICATION IN THE MYXOMYCETES

Jean D. Schoknecht and Harold W. Keller

Department of Botany and Plant Pathology, Illinois Natural History Survey, Champaign, Illinois 61820; Department of Biology, The University of Texas at Arlington, Arlington, Texas 76019

ABSTRACT. Calcareous peridia are produced by some Myxomycetes during sporulation. At least five basic chemical types of calcareous deposits are recognized which have taxonomic significance at the generic, family or ordinal level. These are: (1) calcium carbonate with calcium phosphate in globular form; (2) crystalline calcium carbonate; (3) globular or cryptocrystalline calcium carbonate; (4) calcium oxalate and; (5) silicates of calcium. Localization of cytoplasmic calcium is similar to that previously described for other species in the Physarales. Calcium is sequestered in mitochondria and vesicles and on vacuolar and cell membranes as seen with TEM, pyroantimonate binding and EDX. Calcium is concentrated near the peridial surface through invaginations of the plasmalemma and may be mobilized from mitochondria and other cytoplasmic sources by vacuoles. Calcium is also associated with a peridial matrix which is apparently seen as fibrillar layers of polysaccharide or mucopolysaccharide in early stages of sporulation. During later stages of sporulation, calcium is sequestered most densely in the external areas of fibrillar polysaccharide. TEM, SEM and HVEM all document the presence of a peridial matrix into which calcium is both deposited and where its final shape is determined. The peridial matrix is a network of fibrous polysaccharide with a mesh of varying pore sizes. Channels and pores opening onto the peridial surface are a regular feature. The pores in the matrix of the fibrillar peridium may influence the final site of deposition of calcareous material. Final morphology of myxomycete peridia and crystalline deposits is influenced in part by the chemical composition of the deposited calcareous material and also by the fibrillar polysaccharide matrix.

INTRODUCTION

Myxomycetes are characterized by a phagotrophic, assimilative phase, the plasmodium, which exists as a free-living, multinucleate, acellular mass of protoplasm. The feeding plasmodium is amoeboid, moving with a pulsating cytoplasmic flow, and often develops a fan-shaped margin (Fig. 1A). Lack of a cell wall contributes to the variability in plasmodial size and shape.

During sporulation the plasmodium condenses and acquires a characteristic shape. This transformation occurs through a series of developmental stages, eventually becoming a spore-filled fructification (a general term encompassing specific fruiting body types such as sporangium and plasmodiocarp) typical of the species (Fig. 1B). Spores are delimited from the cytoplasm, and externally an envelope, the peridium, may become covered by obvious encrustations of lime. Lime has traditionally been used in the Myxomycetes to mean calcium carbonate. This external limy encrustation is used as

Origin, Evolution, and Modern Aspects of Biomineralization in Plants and Animals
Edited by Rex E. Crick
Plenum Press, New York

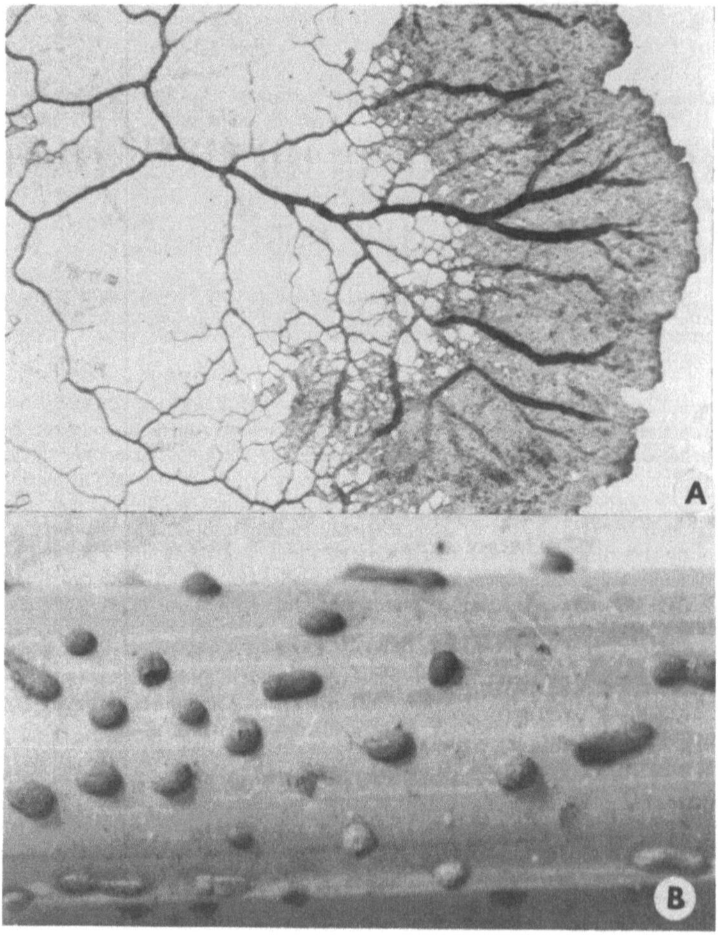

FIGURE 1. Didymium saturnus habit. ■A. Phaneroplasmodium showing veins and fan-shaped, advancing margin, ×5. ■B. Fructifications on grass culm, ×14

one of several characteristics to define the order Physarales and its two families, the Physaraceae and Didymiaceae.

The Physaraceae have amorphous usually granular deposits of calcium carbonate with some calcium phosphate (Schoknecht, 1975). Inside the fructification there is a system of threads, the capillitium, that may contain calcium carbonate and phosphorus.

In contrast, the Didymiaceae, typically have no lime in the capillitium and the deposits of lime on the external peridial surface of the fructifications either are crystalline with often stellate shapes or cryptocrystalline spheres. Schoknecht (1975), using energy dispersive microanalysis (EDX), determined that peridial deposits in the Didymiaceae had only calcium carbonate and that phosphorus was generally not detectable in the deposits or present only in trace amounts. Other investigators have subsequently examined the composition of calcareous deposits in the Physarales (Bechtel & Horner, 1975; Schoknecht & Keller, 1977; Aldrich, 1982)and by Schoknecht&Keller (1977) in the Trichiales. Scanning electron microscopy (SEM) was used to examine the structure of these deposits (Schoknecht & Small, 1972; Gustafson & Thurston, 1974; Bechtel & Horner, 1975; Schoknecht, 1975; Gaither & Collins, 1984; Brooks *et al.*, 1977).

The process of development and deposition of lime was examined for one species of the Physaraceae (*Physarella oblonga* (Berk & Curt.) Morgan) by Bechtel & Horner (1975) and for one species of the Didymiaceae (*Didymium squamulosum* (Alb. & Schw.) Fr.) by Gustafson & Thurston (1974). Gustafson & Thurston (1974) identified calcium granules in mitochondria and vacuoles in precleavage sporangia (white stage). As no ultrastructural evidence suggesting deposition of calcium granules onto the peridium was presented, they hypothesized that calcium passes through the plasmalemma in ionic form. Conversely, Bechtel & Horner (1975) posed two methods for calcium deposition: one method described the fusion of calcium containing vacuoles with the plasmalemma; the other method of deposition to the surface of the peridium was either by "calcium granules pushing to the surface through the capillitial network or as a result of shrinkage of the cytoplasm due to water loss." They also postulated that the mechanism of excretion may be different in the two families. These deposition mechanisms were again reviewed by Collins (1979), therein stating that in the Physaraceae calcium carbonate is first excreted internally, forming minute spheres before reaching the peridial surface, whereas in the Didymiaceae, it is directly excreted without forming internal spheres.

It is perhaps premature to hypothesize that members of the two families have developed two different mechanisms of eliminating large quantities of calcium based on examination of only one species in each family. Moreover, many sections of all stages of development need to be examined with modern instrumentation and techniques. Collins (1979) stated that additional studies are clearly needed in all myxomycete orders to determine the manner of excretion and deposition of calcium and other compounds as well as to identify the chemical composition of the deposits.

This study examines the excretion of calcium using pyroantimonate precipitation, analytical transmission electron microscopy (TEM), and EDX for localizing lightly bound calcium in another species of *Didymium*, *D. saturnus* Keller. The interrelationship of the fibrillar polysaccharide, "membranous" peridium with the calcareous deposits is investigated. Peridial calcification in *Didymium saturnus* is compared to two other species (*Physarum polycephalum* Schw. and *Didymium iridis* (Ditmar) Fries) by using these localization techniques and SEM.

MATERIALS AND METHODS

Field-collected fructifications of *Didymium saturnus* Keller, *D. iridis*, *Diderma testaceum* (Schrad.) Pers., *Fuligo septica* (L.) Wiggers and *Physarum polycephalum* were prepared for scanning electron microscopy (SEM) or were also used as sources of spores. Isolates of *D. saturnus* are deposited with the American Type Culture Collection as number 64178. Plamodia were grown on half-strength corn meal agar or plain agar supplemented with oat flakes. To provide minerals necessary for development of the calcareous peridium, either .02 M calcium carbonate or .02 or .002 M sodium meta-sili-

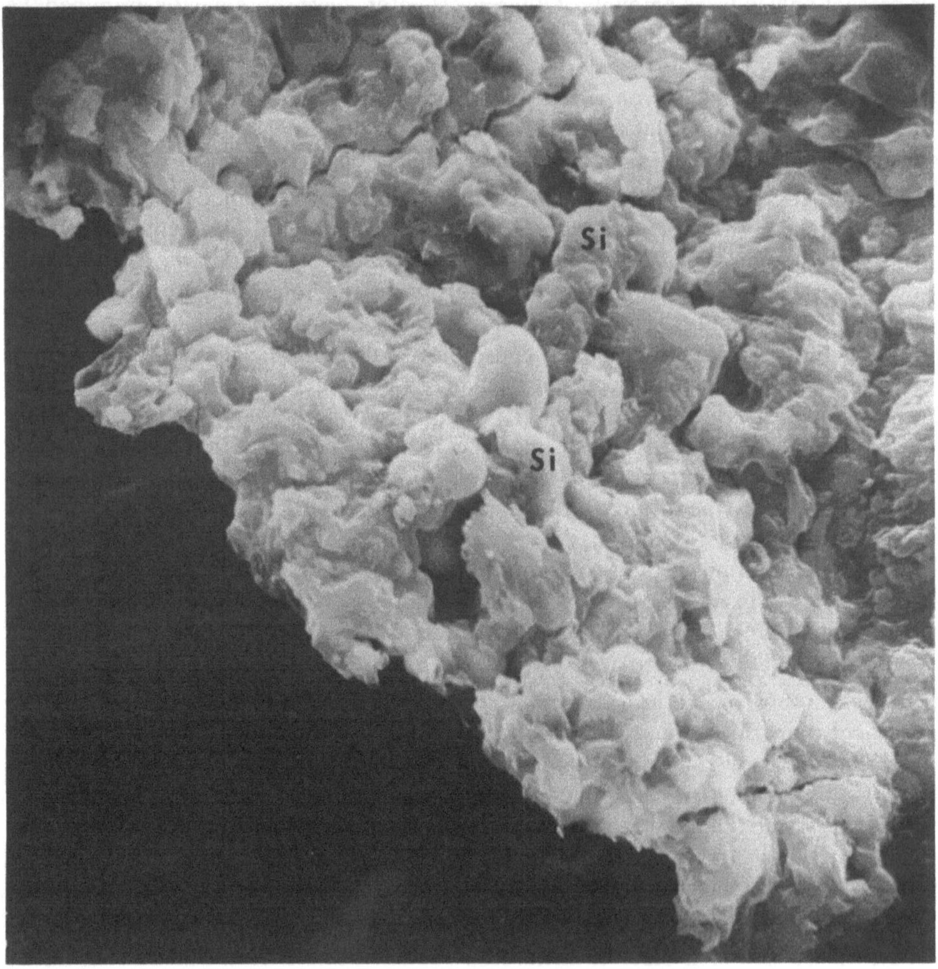

FIGURE 2. Didymium saturnus, SEM of field collected fructification on wheat culm. Calcareous deposits are globular and crystalline, × 1,800. Si refers to silicon deposits. Peridial calcium is major element in unlabeled area.

cate-9-hydrate was added. Agar cultures were incubated at 25°C on a 10 h diurnal cycle. Fructifications selected were based on color changes. Pre-cleavage fructifications appear white as protoplasmic streaming stops and sporangial primordia begin to take definite shape. Fructifications are brown when spore walls begin to form, and are grey to black when mature. Fructifications were fixed for SEM in osmium tetroxide vapor, and mounted on an adhesive obtained by dissolving glue from double-sided tape with acetone. Sporangia of *D. saturnus*, *D. iridis*, *P. polycephalum* and *D. testaceum* were freeze dried or air dried, and examined without coating at 6KV to 20 KV. Fructifications of *D. saturnus* were fixed for TEM by the following: glutaraldehyde with 0.1 M phosphate buffer; equal parts of 2.5% glutaraldehyde and 2% osmium tetroxide for 4 h at 4°C; 2.5% glutaraldehyde with 2% potassium antimonate for 4 h and osmium tetroxide

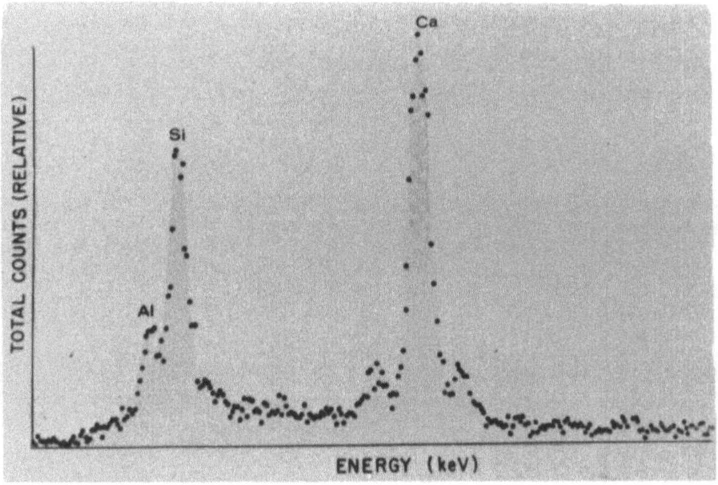

FIGURE 3. Spectrum from EDX of peridium seen in Fig. 2.

postfixation for 2 h (Simpson & Spicer, 1975); glutaraldehyde with potassium antimonate in 0.1 M phosphate buffer for 2 h and osmium tetroxide with antimonate postfixation for 2 h (Wick & Hepler, 1980); equal parts of 2% osmium tetroxide and 4% potassium (pyro) antimonate (Duke *et al.*, 1985). Fixed fructifications were embedded in epoxy resin (Spurr, 1969). Thin sections were left unstained or stained with uranyl acetate and either lead citrate (Reynolds, 1963) or lead aspartate (Walton, 1979), and examined at 75KV with an analytical TEM equipped with energy dispersive microanalysis (EDX). Some sections were floated on 1% ethylenediaminetetraacetic acid (EDTA), pH7, for 1-5 min, to localize intracellular calcium deposits by removal of calcium and recording loss of electron dense sites. Sections were 0.5 μm thick for examination at high voltage electron microscopy (HVEM) 750KV.

Carbonate deposits from peridia of *Fuligo septica* and *Didymium testaceum* were examined in water mounts with light microscopy. Following observation, the slides were either heated to boiling or 10% KOH was added to the edge of the cover slip and subsequently the slides were heated over a flame for 1 min and immediately observed.

RESULTS

Fructifications of *Didymium saturnus* were described by Keller (1970) based on light photomicrographs as having a peridium with embedded crystals. SEM of peridia from sporangia developed on wheat straw (Fig. 2) showed crystalline deposits and confirms that these deposits were covered or embedded. Spectra from EDX, as plotted in Figure 3, identified abundant calcium and silicon in the peridium. The calcium (Fig. 4A) and silicon (Fig. 4b) are localized in the peridium by using EDX dot maps. Calcium is the more abundant and more widely distributed in the peridium. Sites with little calcium correspond to sites with predominantly silicon deposits. Two such sites are labeled -Si- on Figures 2, 4A, and 4B. These sites correspond to the same site on all three photographs.

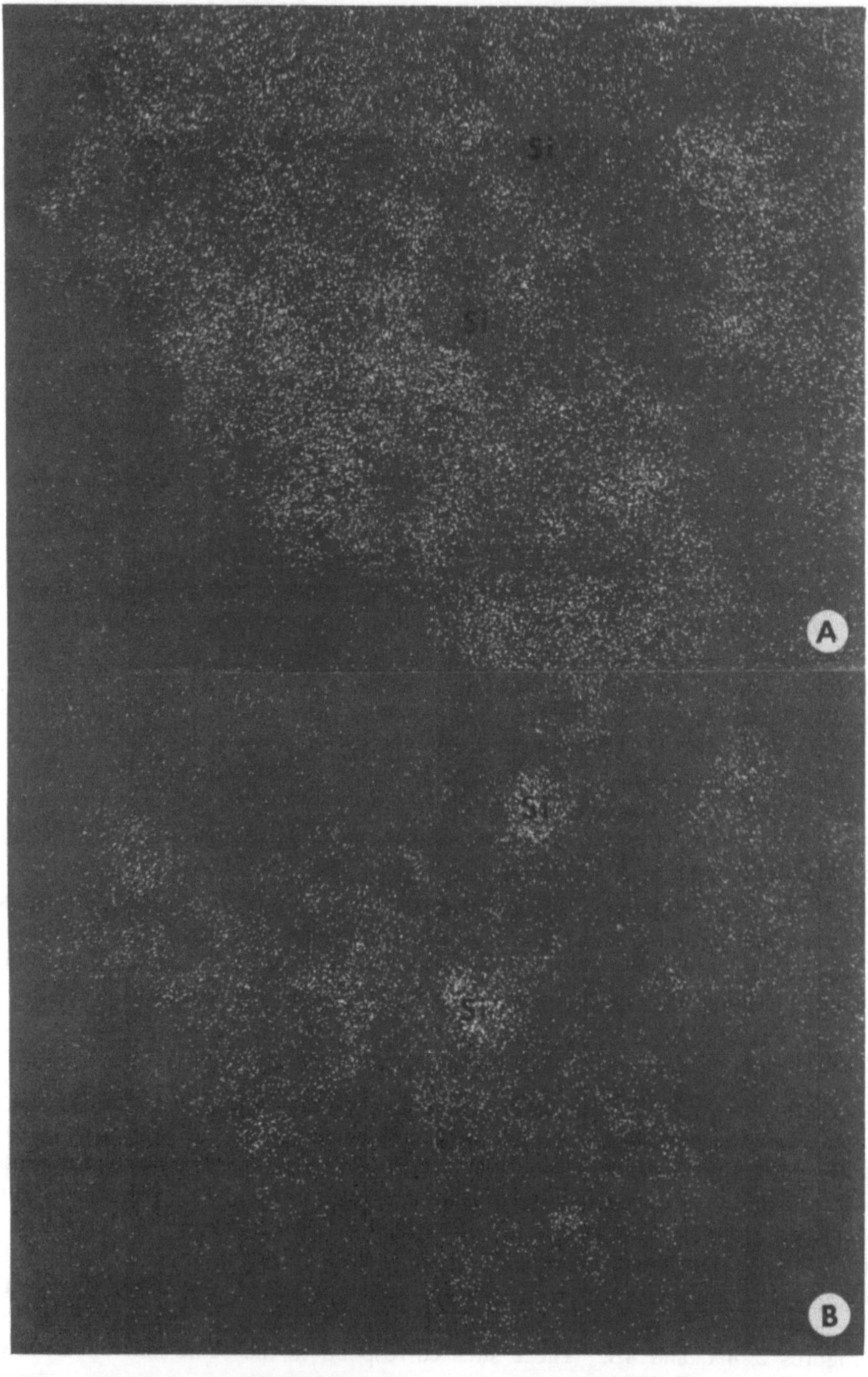

FIGURE 4. Didymium saturnus, EDX of Figure 2, × 1,800. ▪A. Ca dot map. ▪B. Si dot map. Si corresponds to the same points on all micrographs in Fig. 2 & 4.

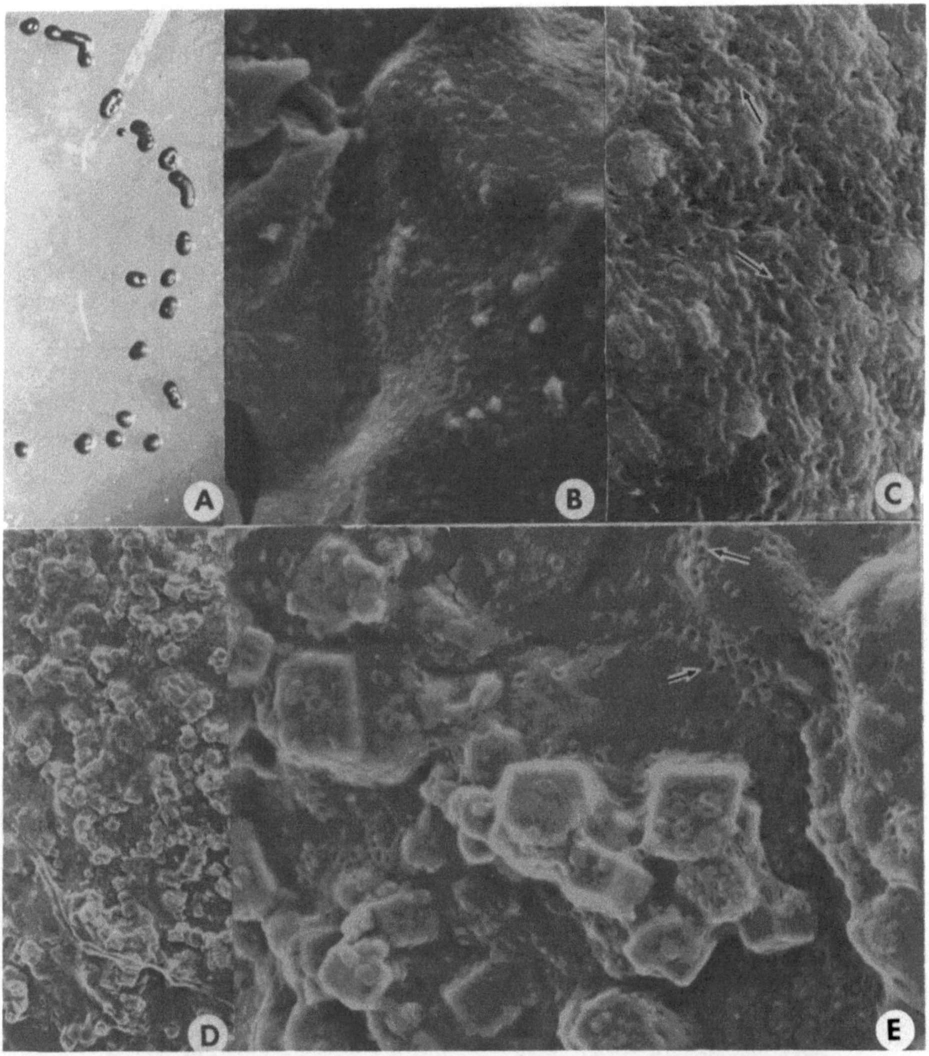

FIGURE 5. Didymium saturnus peridia. ▪A. Habit of brown (late) stage of sporangial formation on agar, ×5. ▪B. SEM of white (early) stage of sporangium formation. Peridium has small-sized deposits of calcium carbonate embedded in peridium, 6KV, ×6,350. ▪C. Sporangial surface of yellow-brown stage. Pores are visible on surface (arrows), 15KV, ×4,200. ▪D&E. SEM of late stage of sporangia developed on agar with oat flakes and .02 M $CaCO_3$ showing cuboidal calcium carbonate crystals and pores (arrows) in peridium, 6KV. D. ×640. E. ×3,900.

Sporangia of *Didymium saturnus* were examined during development with light and electron microscopy. As melanized spore wall material was deposited, the sporangia darken from white to yellow, to light brown, to a dark grey-brown when mature in a manner similar to that described by Kowalski (1972) for *Squamuloderma nullifera*. Crystalline deposits were not seen with reflected light in white or brown stages (Fig. 5A).

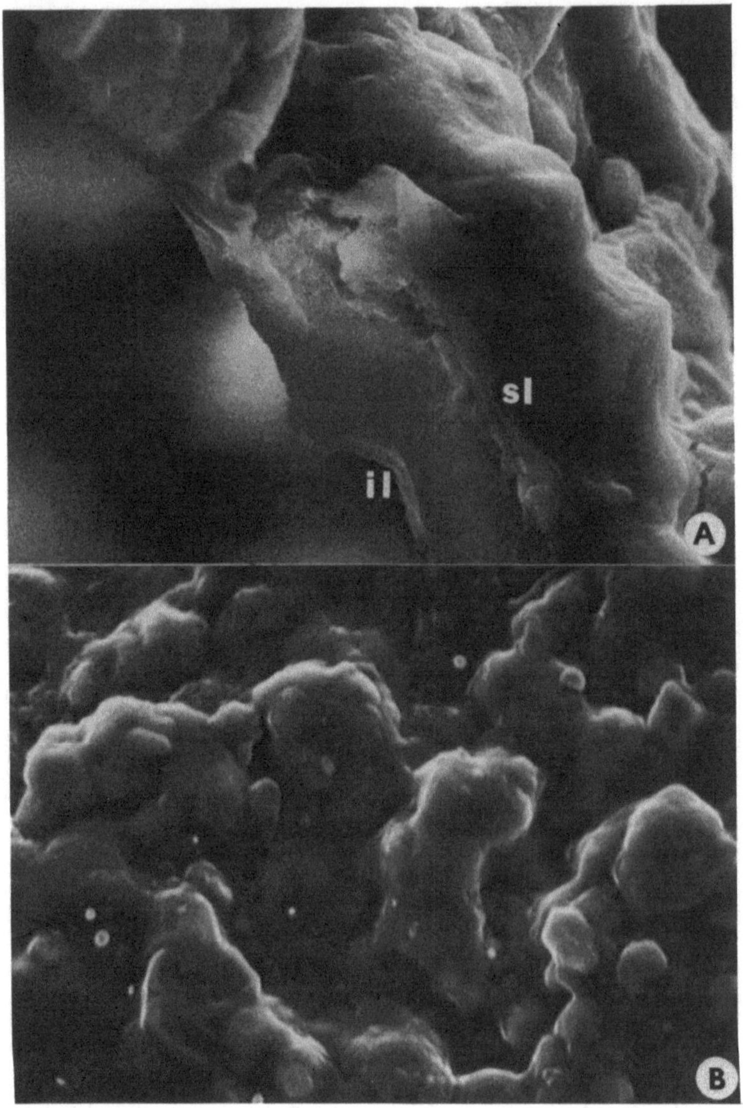

FIGURE 6. Didymium saturnus, SEM of peridia, 10KV. ▪A. Sporangium taken from straw substratum. In edge view peridium appears multilayered, inner layer (il), surface layer (sl), with crystals enveloped, × 3,100. ▪B. Sporangium grown in culture on agar with .02 M sodium meta-silicate showing peridial surface with embedded deposits and a few small granules, × 7,700.

When examined with low voltage SEM, the peridium appeared coarsely thickened (Fig. 5B), with a fibrous surface and minute pores (Fig. 5C). As sporangia darken and mature, SEM at low magnification showed many crystals which appeared to be on the surface (Fig. 5D). At greater magnification the crystals were clearly resolved to be embedded in a peridial matrix (Fig. 5E). Using low voltage SEM the surface clearly has

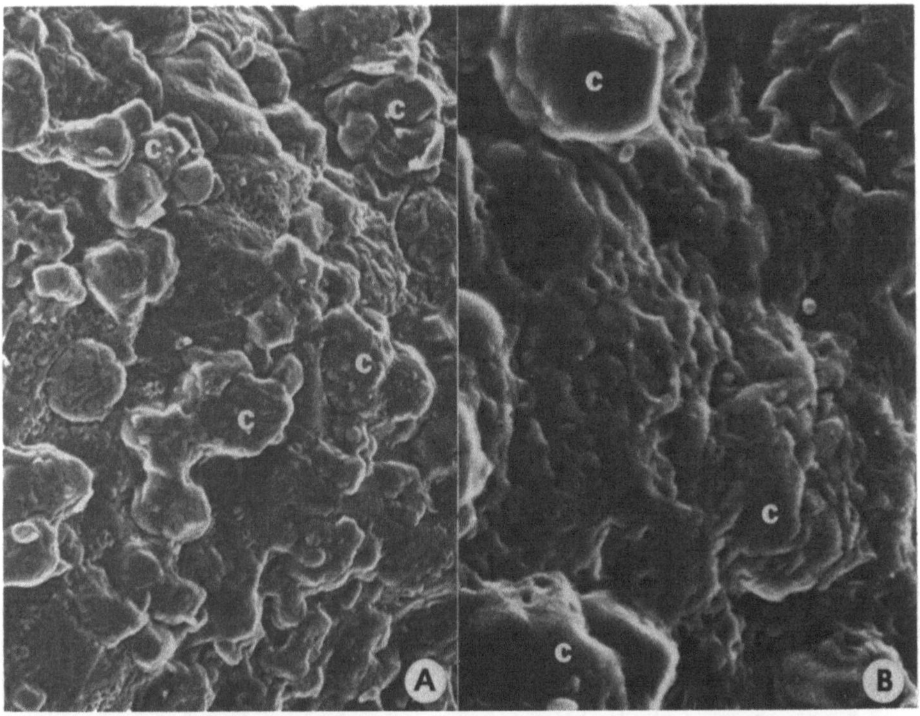

FIGURE 7. Didymium saturnus, SEMs of peridial surface, 5KV. Pores and irregular calcareous (c) deposits that appear embedded in peridium of sporangia developed on agar with .002 M sodium meta-silicate. ▪A. ×2,700. ▪B. ×6,800.

pores (Fig. 5E). When viewed on edge with SEM (Fig. 6A), the peridium appeared to be a sandwiched, multiple entity, with the crystals seated on a lower peridial matrix and blanketed with a surface matrix. When sporangia were grown in the presence of 0.02 M sodium silicate the crystals became more irregular as seen in Figure 6B. The peridial deposits were more amorphous than when 0.02 M calcium carbonate was added without silicate as in Figures 5D and 5E. SEM, however, still suggests that the mineral deposits were coated. When plasmodia were grown in the presence of 0.02 M calcium carbonate and 0.002 M sodium silicate the deposits in the peridium were more crystalline in appearance (Fig. 7A) than those peridial deposits with more (.02 M) silicate available (Fig. 6B). At higher magnification with SEM the crystalline and semi-crystalline material was seen as embedded in the peridium (Fig. 7A). The fibrous peridial matrix had large pores measuring .05 μm to .2 μm as seen with SEM (Fig. 5C, E, 7B).

TEM observations coupled with techniques for precipitating lightly bound calcium with osmium-pyroantimonate and localizing this calcium in the cytoplasm were employed (Duke, *et al.*, 1985; Simpson & Spicer, 1975; Wick & Hepler, 1980). The earliest recognizable stage (white) of sporangial development examined was visualized with reflected light as a mounding of the plasmodium when it ceased migration. Initiation of sporulation was recognizable by invaginations and condensed cytoplasm (Fig. 8) that eventually becomes the cytoplasm that is cleaved into spores. Also, an ectoplasmic area

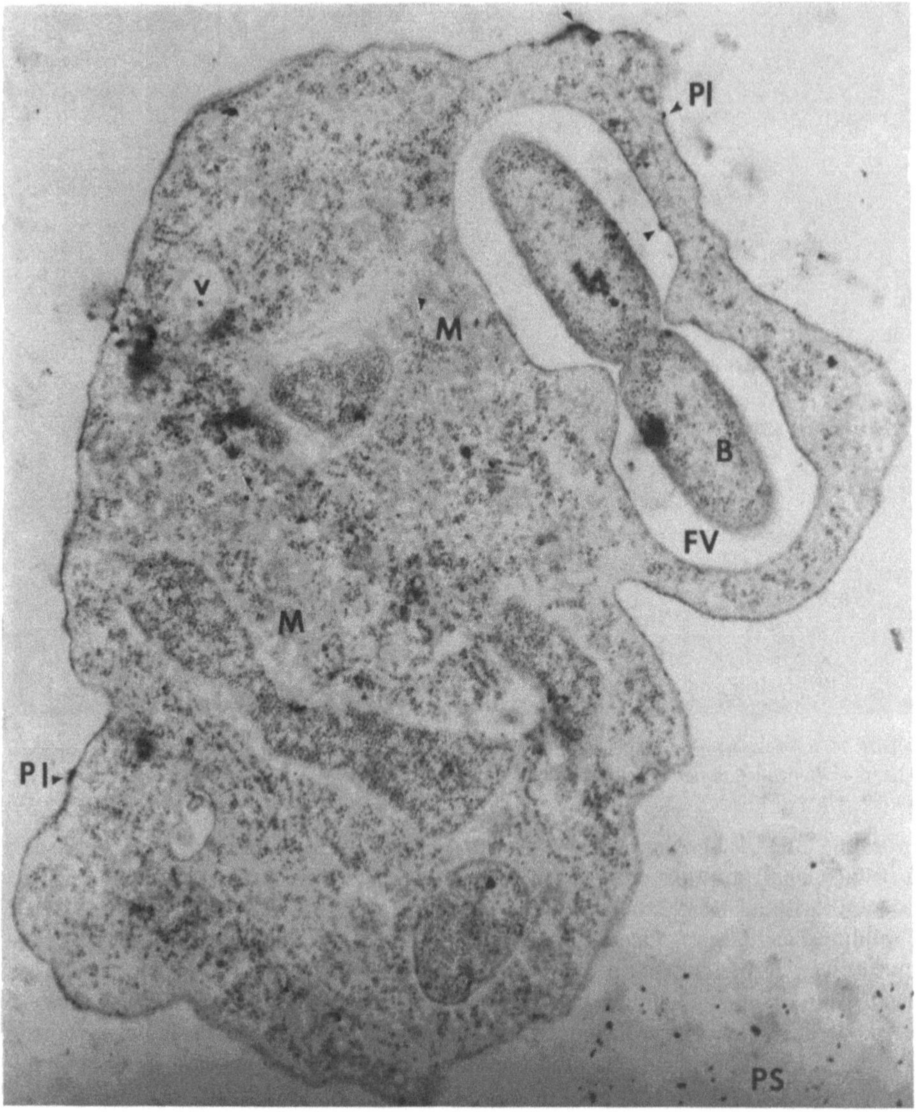

FIGURE 8. Didymium saturnus, TEM. Unstained osmium and potassium antimonate preparation of white (early) stage of sporangial formation. Preparation showing food vacuole (**FV**) with ingested bacterium (**B**). Electron dense precipitates of calcium antimonate are visible on plasmalemma (**Pl**), in vacuoles (**v**), in endoplasmic reticulum, in polysaccharide layer outside the cell (**PS**), and scattered through cytoplasm, ×20,860.

lacking mitochondria was present near the periphery. With TEM, the plasmodium still showed a food vacuole with a phagocytized bacillus (Fig. 8). Minute electron dense precipitates were seen occasionally in mitochondria but abundantly in vacuoles and along the plasmalemma (Figs. 8, 9A,B,C) when bound calcium ions were precipitated with potassium antimonate. Ribosomes of endoplasmic reticulum and some scattered

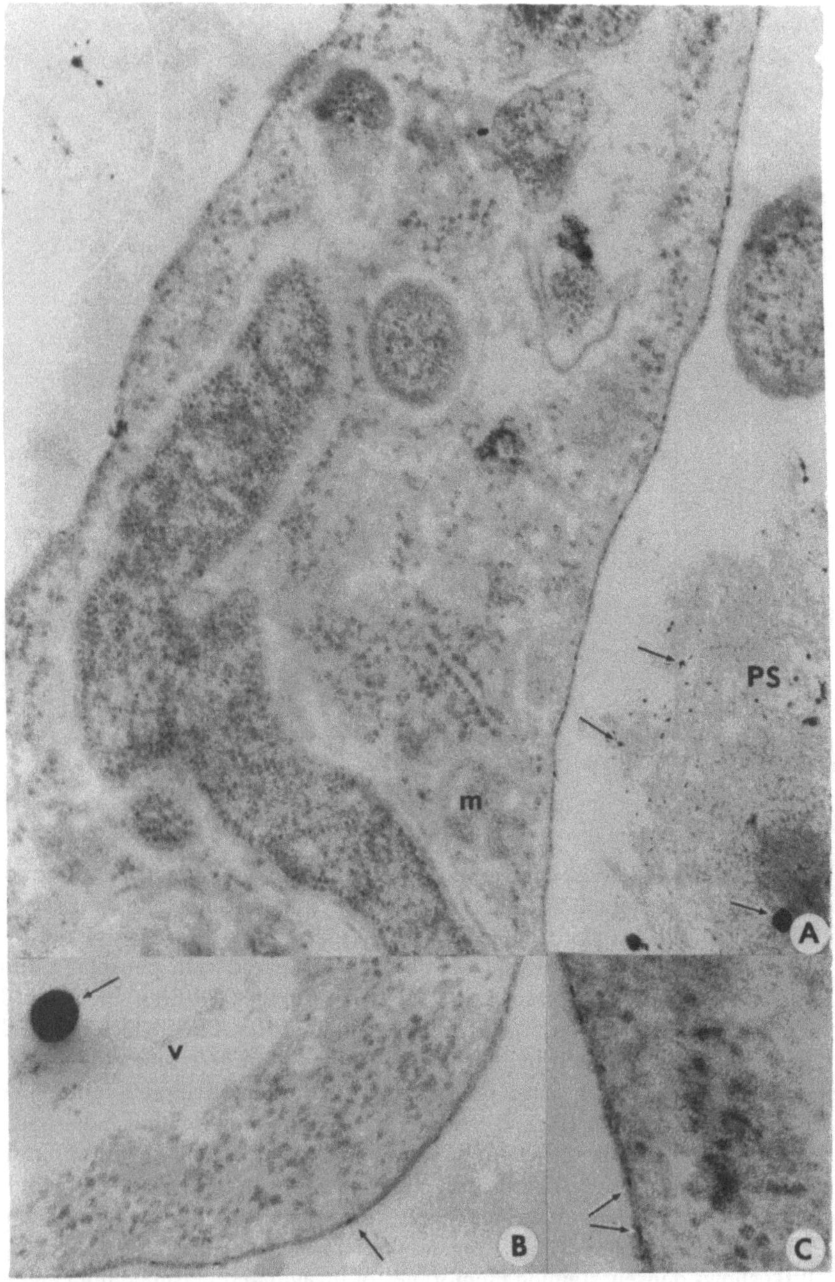

FIGURE 9. Didymium saturnus, TEM. Osmium tetroxide and potassium antimonate, unstained. Preparation of white (early) stage of sporangial formation. Calcium antimonate precipitation (arrows) in external fibrillar polysacchride (**PS**) cell membranes, mitochondria (**m**), vacuoles (**v**) and ribosomes. ▪A. ×48,000. ▪B. ×60,900. ▪C. ×100,300.

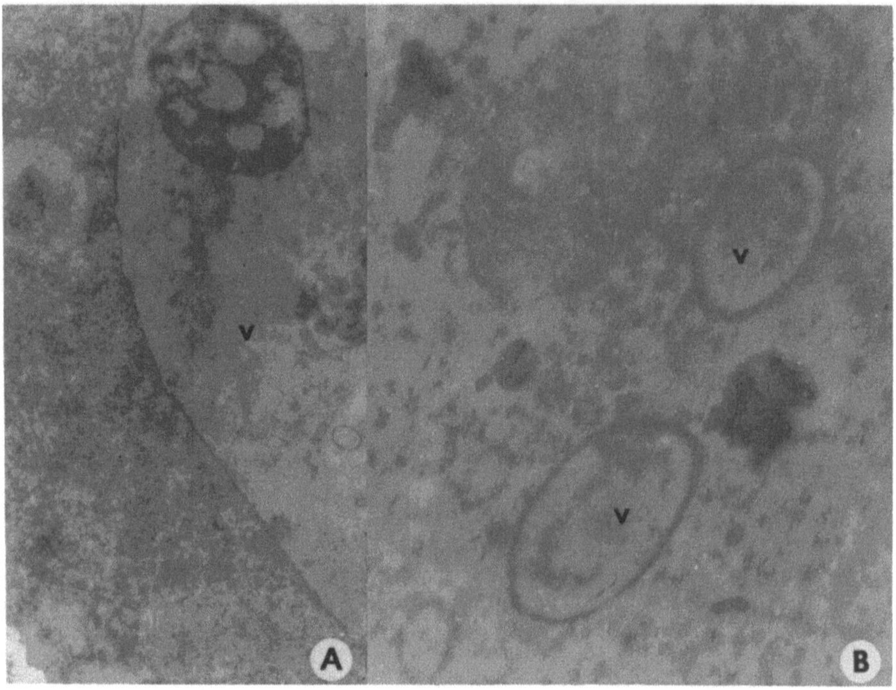

FIGURE 10. Didymium saturnus, TEM of early stage in sporangial development. Fixed only in glutaraldehyde with phosphate buffer. Unstained. This control has no electron dense precipitates, (vacuoles = v). ▪A. White stage. × 12,500. ▪B. Brown stage. × 31,200.

ribosomes also showed dense precipitate when potassium antimonate was used to bind calcium. Calcium antimonate-osmium was also precipitated in the mucopolysaccharide sheath area in this early (white) stage (Figs. 8, 9A). In comparison early (Fig. 10A) and late stages (Fig. 10B) of sporangial development prepared with phosphate buffered glutaraldehyde fixation produced low contrast images without electron dense precipitates. Similarly, conventional fixation techniques for TEM which involved buffered glutaraldehyde followed by postfixation with osmium tetroxide (Fig. 11A) were lacking the obvious electron dense precipitation seen when potassium antimonate was used to precipitate cations, (Figs. 8, 9A,B,C,11B). Tangential sections adjacent to the peridium, that is along the surface of the plasmalemma, showed much proliferation of membranes and precipitates of calcium antimonate clustered around the membranes (Fig. 12A). These electron dense precipitates were confirmed to contain calcium antimonate with EDX spot analysis. Tangential section through the peripheral cytoplasm (Fig. 12B) shows densely staining membranes of invaginated plasmalemma and a very few pyroantimonate precipitates in the channels of the invaginated plasmalemma and adjacent cytoplasm when glutaraldehyde was used with potassium antimonate prior to osmium fixation. Abundant pyro-antimonate was precipitated outside the cytoplasm, located in and on the polysaccharide sheath (Fig. 12C).

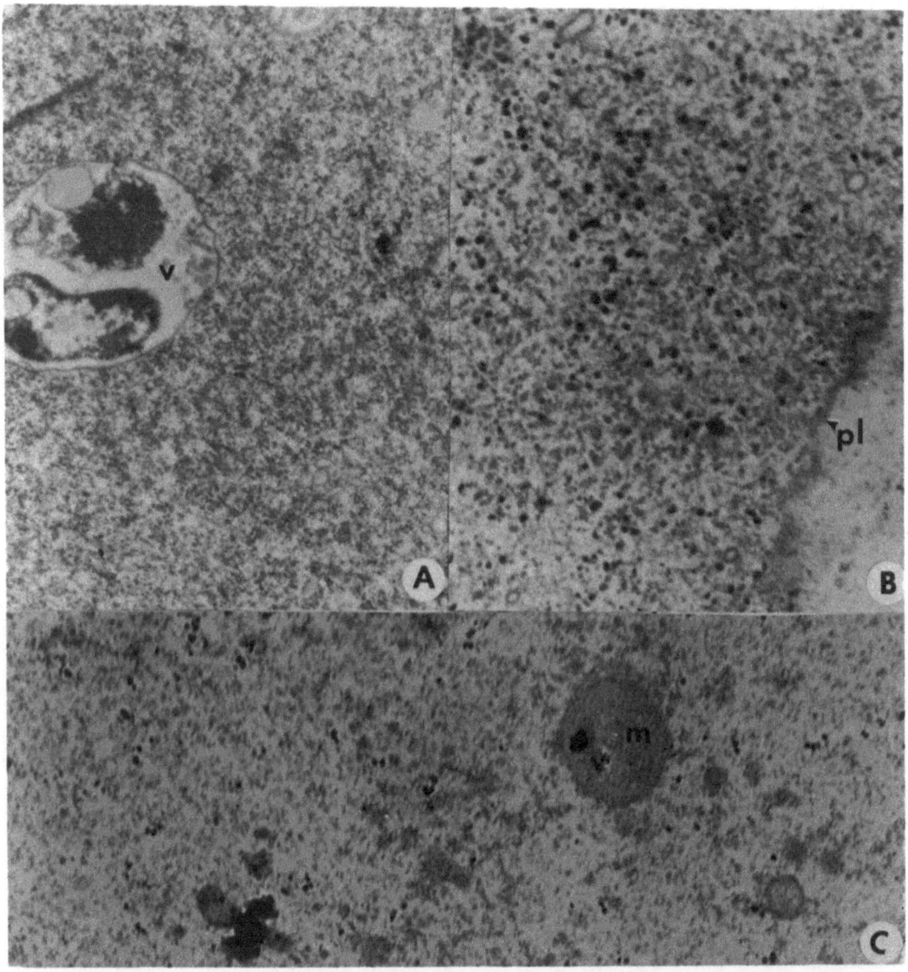

FIGURE 11. Didymium saturnus, TEM of early stage in sporangial development. ▪A. Glutaraldehyde, osmium fixation, stained preparation. Dense precipitates of antimonate are not seen (vacuoles = v), × 13,600. ▪B. Unstained sections fixed with glutaraldehyde with potassium antimonate and post-fixed with osmium and potassium antimonate. Electron dense precipitates of calcium antimonate appear scattered through cytoplasm and on surface of plasmalemma, × 26,800. ▪C. Osmium tetroxide and antimonate precipitation, also note mitochondrion (**m**), unstained, × 22,350.

In order to increase definition of organelles and their relationship to calcium precipitation, sections fixed with osmium and potassium antimonate were stained with uranyl acetate and lead citrate (Fig. 13 A,B). Sections of cytoplasm adjacent to plasmalemma in the light brown stage of sporangial development (Fig. 13A) had ribosomes and scattered precipitates of calcium antimonate. Mitochondria were absent from this outer area, whereas in sections more toward the interior of the sporangium, mitochondria were abundant (Fig. 13B). Cleavage furrows were evident at this stage, but spores were not yet delimited. Electron dense precipitate was not seen in

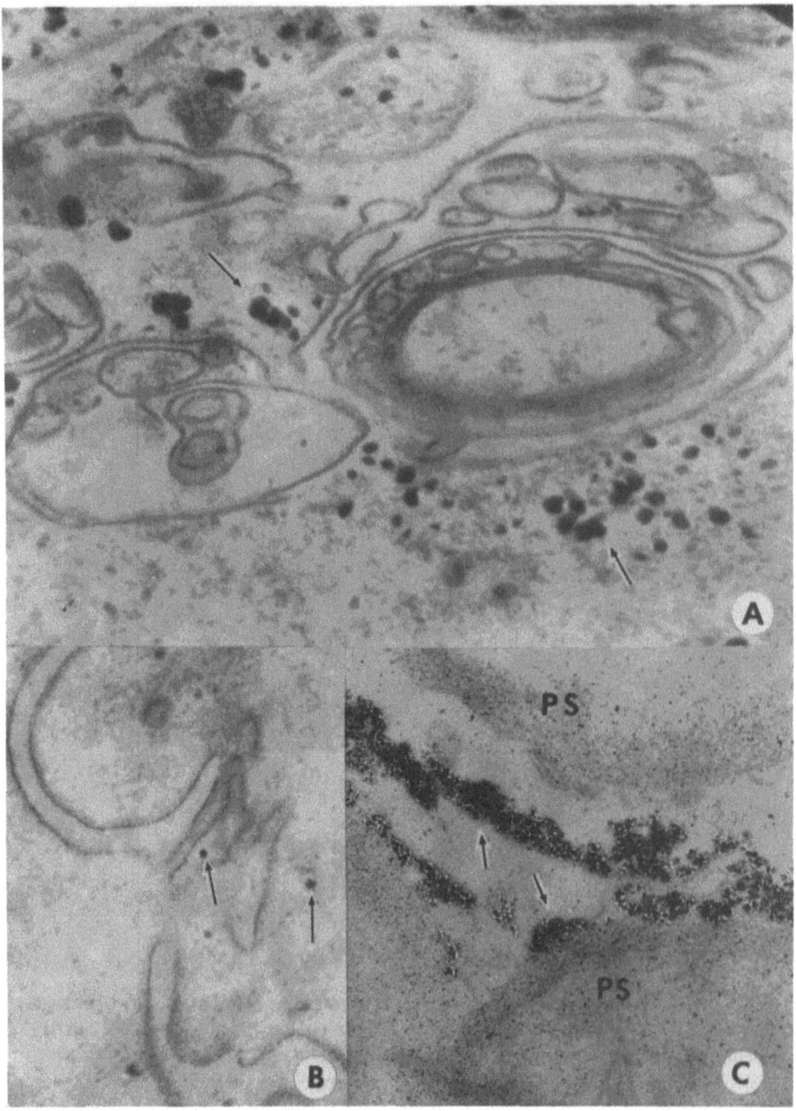

FIGURE 12. Didymium saturnus, TEMs of sporangial development. ▪A. Light brown stage tangential section to surface of sporangium. Membranes are extensive and calcium antimonate precipitates (arrows) are abundant near the surface. Osmium, potassium antimonate fixation, unstained, ×53,600. ▪B. TEM tangential section of sporangial surface adjacent to peridium. Brown stage showing calcium antimonate deposits (arrows) in channels which appear to be invaginations of plasmalemma. Glutaraldehyde and potassium antimonate, osmium postfixation. Unstained, ×53,600. ▪C. Sporangium of brown stage with calcium antimonate precipitation (arrows) with polysaccharide (**PS**). Glutaraldehyde, osmium-antimonate fixation. ×53,600.

mitochondria but scattered through the cytoplasm and perhaps adjacent to the exterior of mitochondria and adjacent to vacuole membranes.

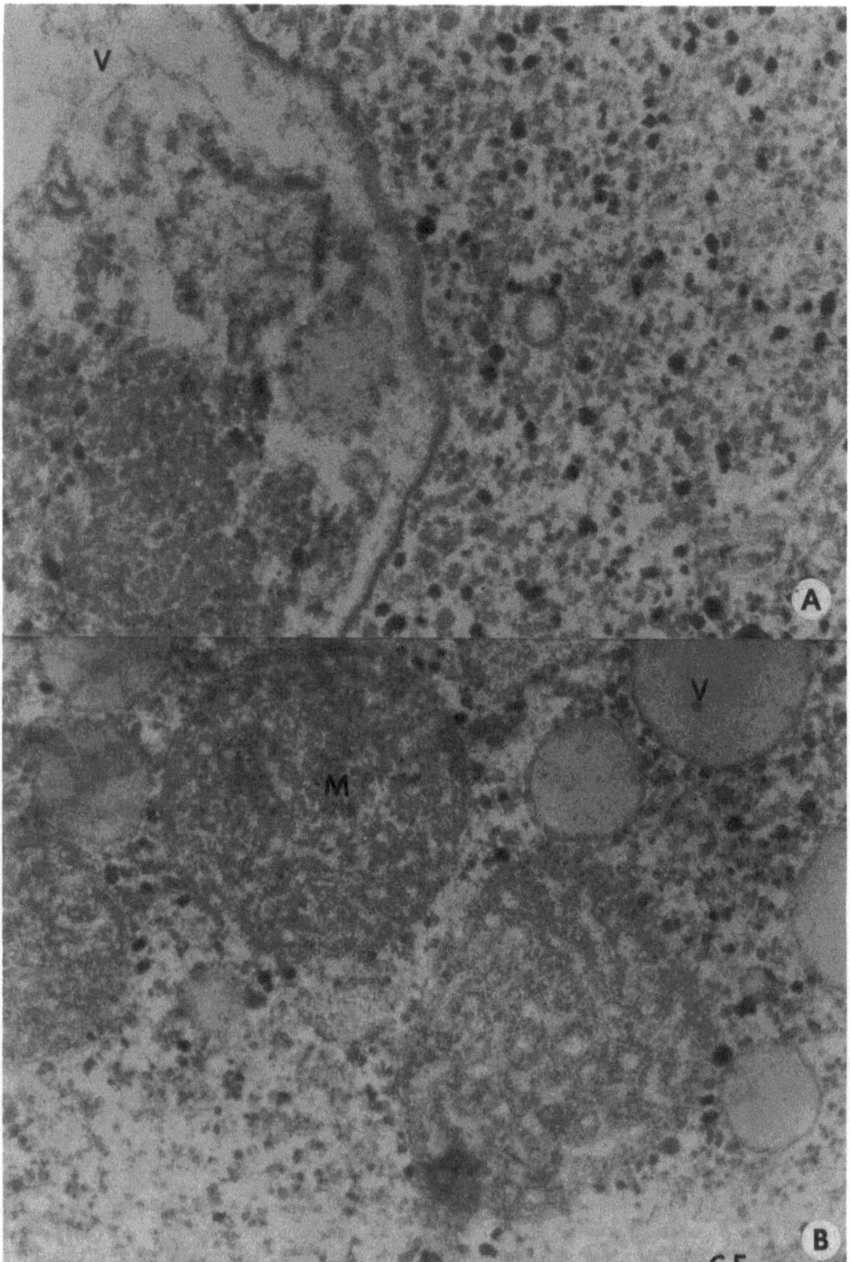

FIGURE 13. Didymium saturnus, TEMs of light brown (early) stage of sporangia formation. Micrographs were taken from periphery (Fig. A) toward center of cell (Fig. B). Osmium and potassium antimonate, unstained. ×50,500. ▪A. Precipitates of calcium antimonate appear scattered through cytoplasm associated with vacuole membranes and in autophagic vacuole (**v**) near periphery. ▪B. Electron dense precipitates of calcium antimonate appear scattered through cytoplasm and associated with vacuole membranes (**v**) and exterior of the mitochondria (**m**). Cleavage furrow (**CF**) is shown in lower right.

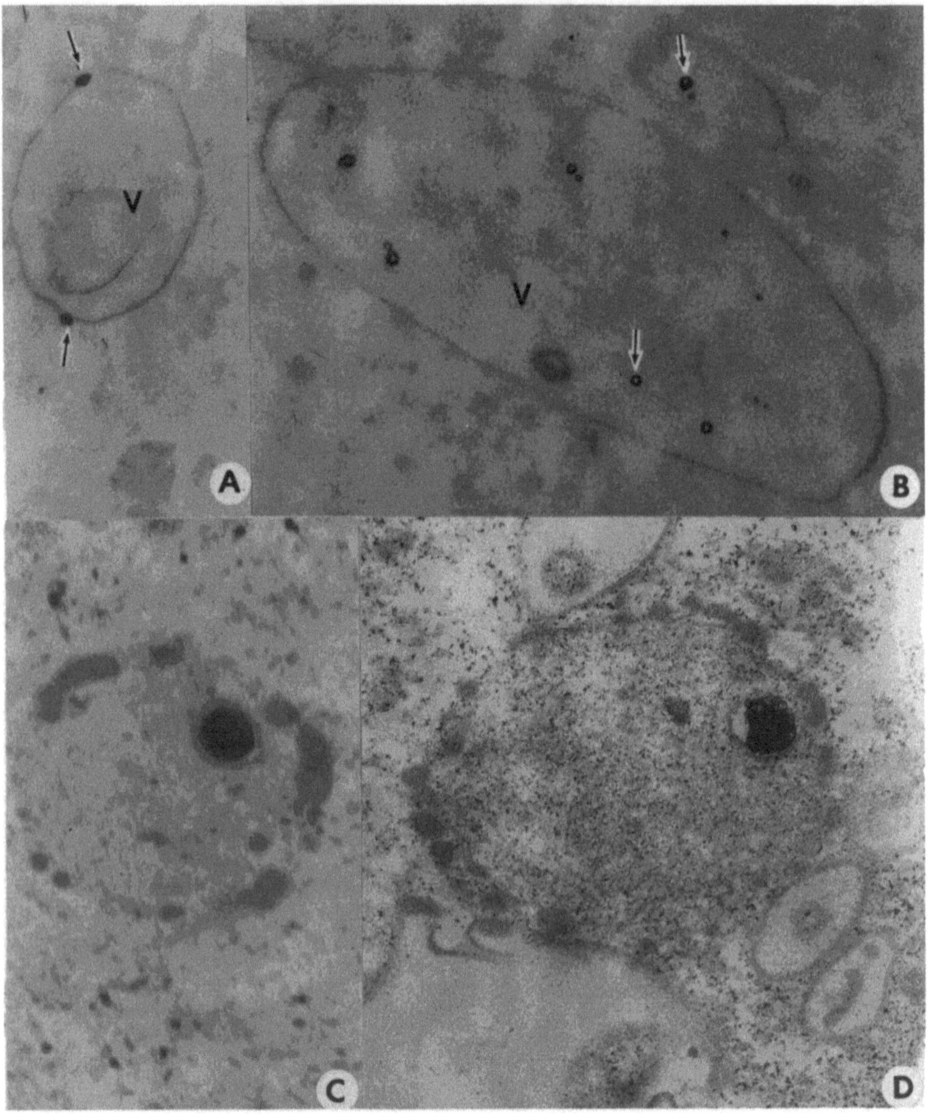

FIGURE 14. Didymium saturnus, TEM of brown (late) stage of sporangial development in which spores are being formed. Unstained. ▪A. Calcium antimonate precipitates (arrows) on vacuole membranes (v), ×49,900. ▪B. Calcium antimonate precipitates (arrows) in vacuoles, ×48,800. Glutaraldehyde and antimonate fixation, osmium-antimonate postfixation. ▪C. Calcium antimonate precipitates in mitochondria like bodies and scattered through cytoplasm, ×44,400. ▪D. Osmium potassium antimonate fixation. Treated 1 min with EDTA, ×53,200

Sections of brown sporangia, in which spores were being formed, showed calcium antimonate precipitation on vacuole membranes (Fig. 14A), and in vacuoles (Fig. 14B), sparsely scattered through the cytoplasm (Fig. 14C,D) and as large precipitates .06 μm

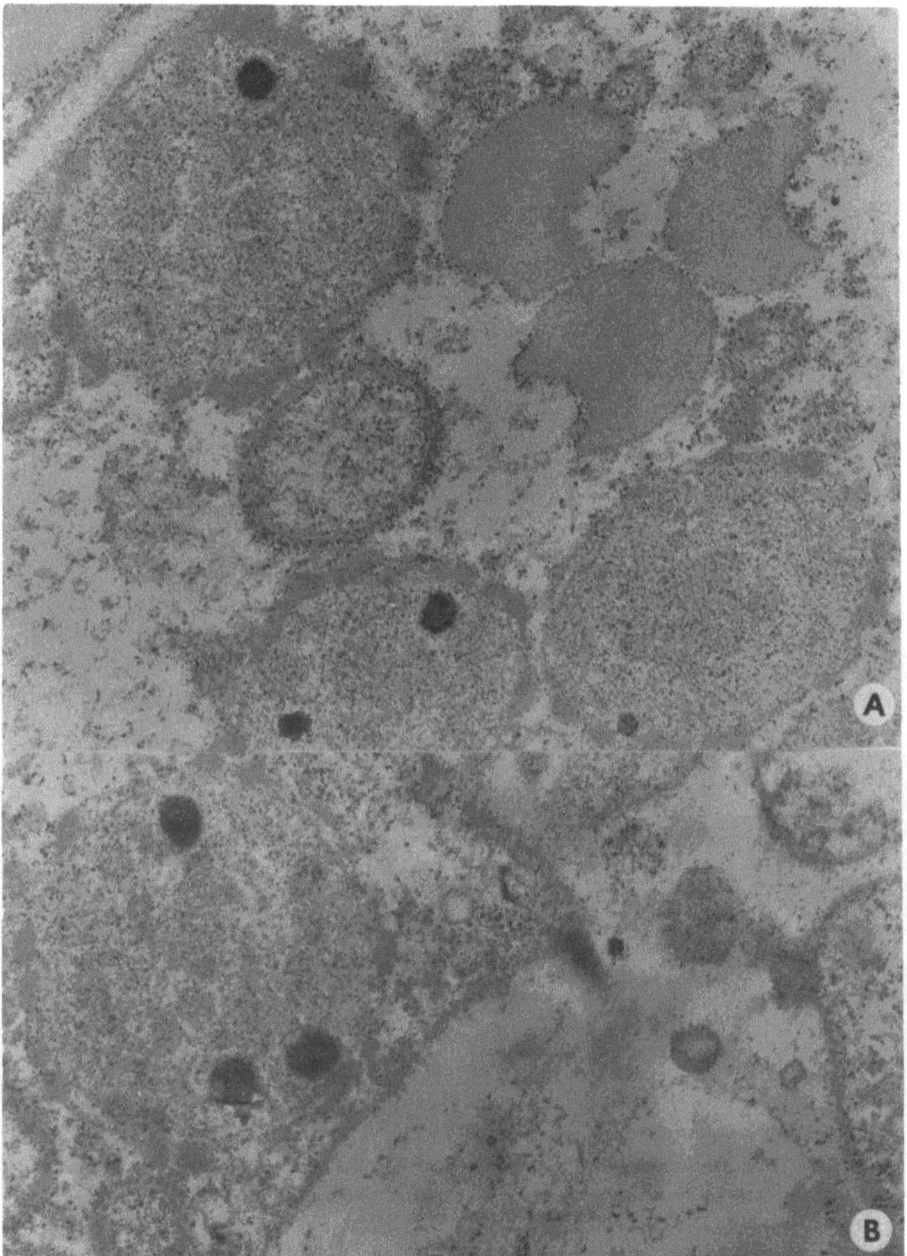

FIGURE 15. Didymium saturnus, TEM of brown stage of sporangial development, osmium-antimonate fixation. Thin sections were treated with 1% EDTA for 2 min and stained with uranyl acetate and lead aspartate. ▪A&B, ×59,400.

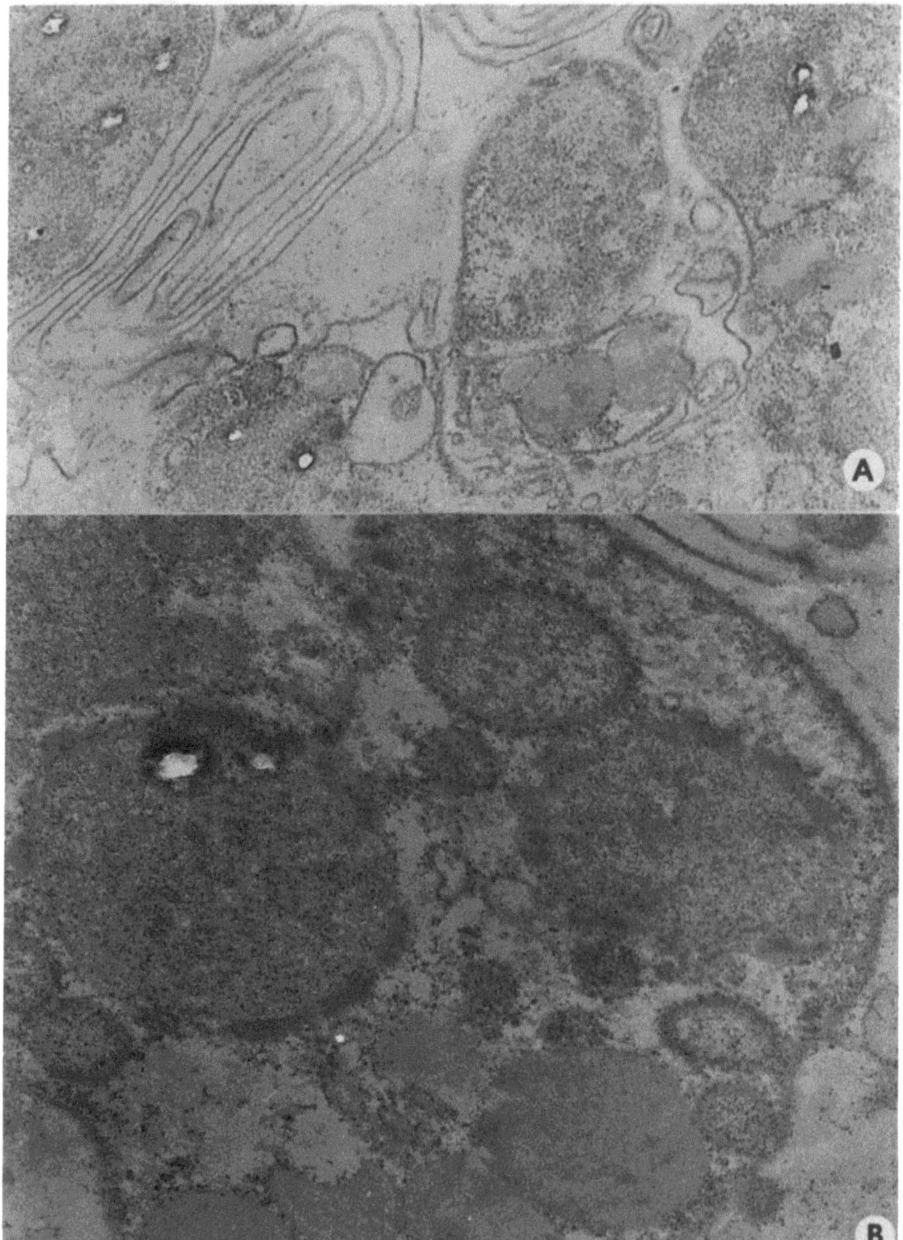

FIGURE 16. Didymium saturnus, TEM of brown stage of sporangial development. Osmium-antimonate fixation. Thin sections were treated with 1% EDTA for 5 min (holes indicate removal of calcium). ▪A. Unstained, × 32,000. ▪B. Unstained, × 54,800.

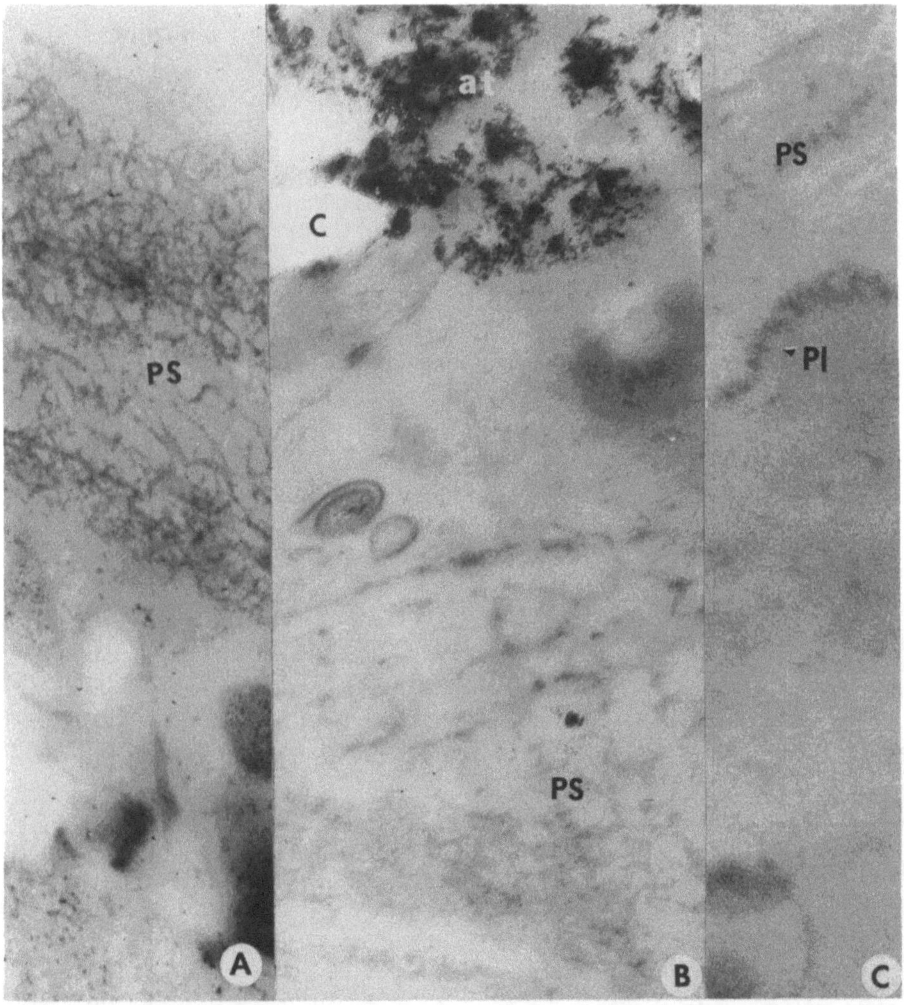

FIGURE 17. Didymium saturnus, TEM. Brown stage of sporangium. ▪A. Sporangium illustrating "layering" of fibrillar polysaccharide (**PS**) material. Thin section treated with 1% EDTA. Glutaraldehyde and potassium antimonate fixation, osmium antimonate postfixation, ×50,600. ▪B&C. Surface of sporangium is topped with crystal ghosts (**c**) and electron dense precipitates of calcium antimonate (**at**). Fibrillar mucopolysaccharide layers (**PS**) are between the outer calcium layer and plasmalemma (**pl**) seen in Figure C. Section in Figure B is exterior to section in Figure C. Glutaraldehyde and potassium antimonate fixation, osmium postfixation, ×168,600.

to .2 μm (Fig. 14C,D) in what may be degenerating mitochondria based on the size of these organelles. The osmium-antimonate technique of Duke, *et al.*, (1985) yielded the greatest amount of pyroantimonate precipitate compared to the other precipitation techniques (Figs. 8, 9, and 14A,B,C). A brief 1 minute exposure to 1% EDTA resulted in finer, angular precipitates in the cytoplasm and loss of density from the large dense granule in the mitochondria like bodies (Fig. 14D).

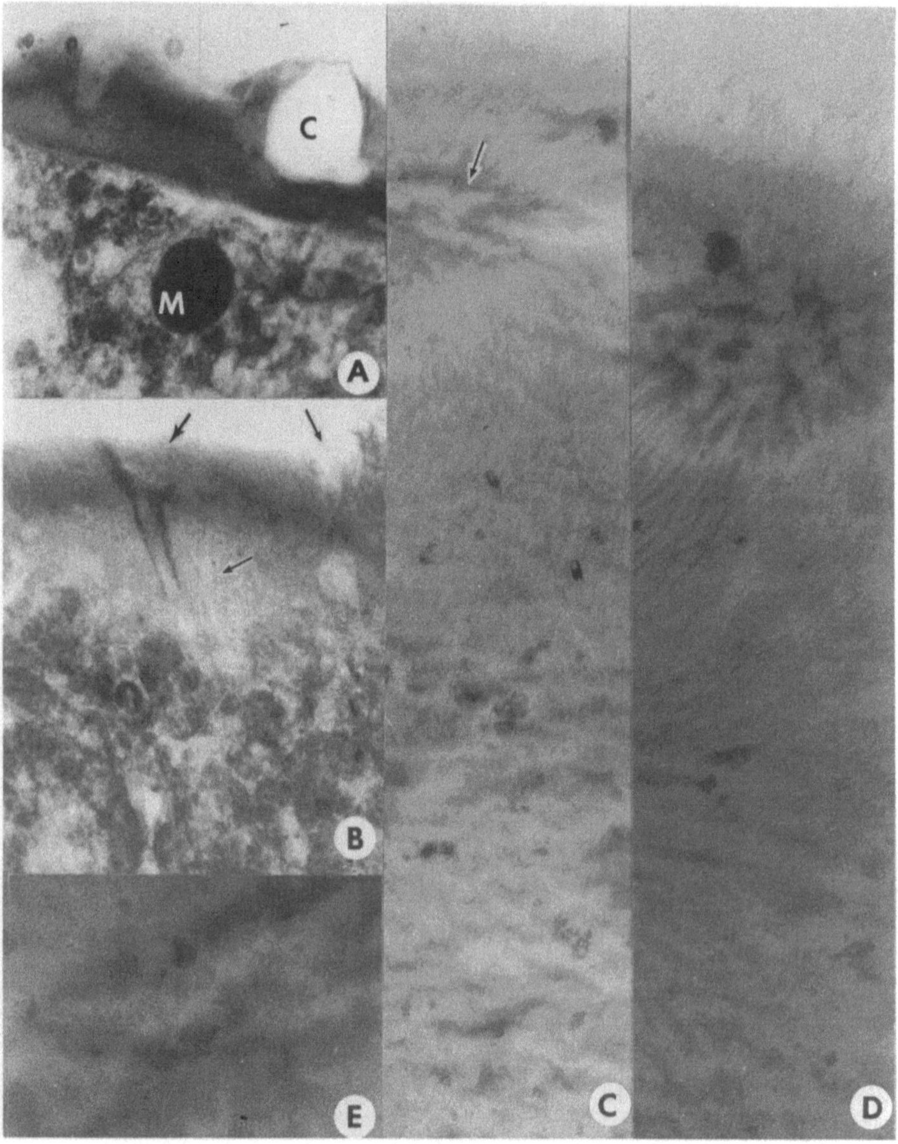

FIGURE 18. Didymium saturnus, TEM. ▪A. HVEM, Glutaraldehyde and osmium, with uranyl acetate and lead citrate stained, 0.5 μm thick section shows outline of crystal (**c**) embedded in the peridium (mitochondrion = **m**), × 18,000. ▪B. HVEM, 0.5 μm thick section through edge of sporangium (white stage), peridium appears channeled (arrows). Glutaraldehyde and osmium, stained with uranyl acetate and lead citrate, × 31,400. ▪C,D,E. Peridium of young sporangium. Thick section illustrating layering pattern of fibrillar material. Figure E is enlargement of area at arrow in Figure C. Glutaraldehyde and potassium antimonate fixation and osmium-antimonate post-fixation, unstained. C&D, × 27,000. E, × 53,900.

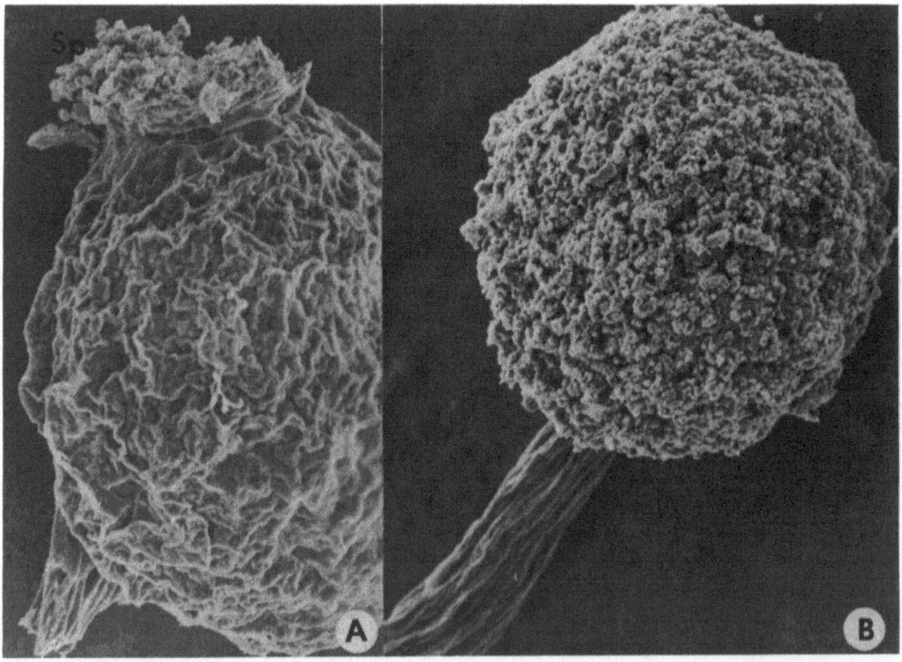

FIGURE 19. Didymium iridis, SEM of sporangia, 15 KV, ×200. ▪A. Brown sporangium with calcium crystals not visible but with spores (Sp) delimited. ▪B. Mature white sporangium with calcium carbonate crystals.

Treatment with 1% EDTA for as short a time as 1 to 2 min (Fig. 15A,B) resulted in removal of dense contrast in the cytoplasm, sites which were interpreted to have been sites of calcium deposition. Large granules in mitochondria like bodies showed loss of dense precipitate. Longer treatment to 5 min with EDTA resulted in complete removal of the large dense bodies in the mitochondria like organelles and some of the small dense precipitate in the cytoplasm (Fig. 16A,B). Sections treated with EDTA were badly damaged by lead citrate stain. However, lead aspartate staining subsequent to EDTA treatment differentiated cell membranes and vacuoles and cytoplasmic granules (Fig. 15).

Sections cut transversely to the peridium (Fig. 17A,B) and adjacent cytoplasm (Fig. 17C) revealed a thick peridium composed of a network of mucopolysaccharide fibrils. A space was often present between the peridium and cellular portion of the sporangium (Fig. 17A,C) which sometimes contained small portions of cytoplasm (Fig. 17A). Electron dense precipitates of calcium antimonate and small crystal "ghosts" are outermost in a fine fibrillar matrix at the surface. Subtending this layer or layers is a region of fibers in a comparatively wide-meshed reticulum. Between this wide-meshed layer and the densely staining plasmalemma is a fine-meshed reticulum of fibrils. This fine-meshed fibrillar layer is duplicated in the bottom of Figure 17B and the top of Figure 17C. Calcium antimonate precipitate was detected and confirmed using EDX.

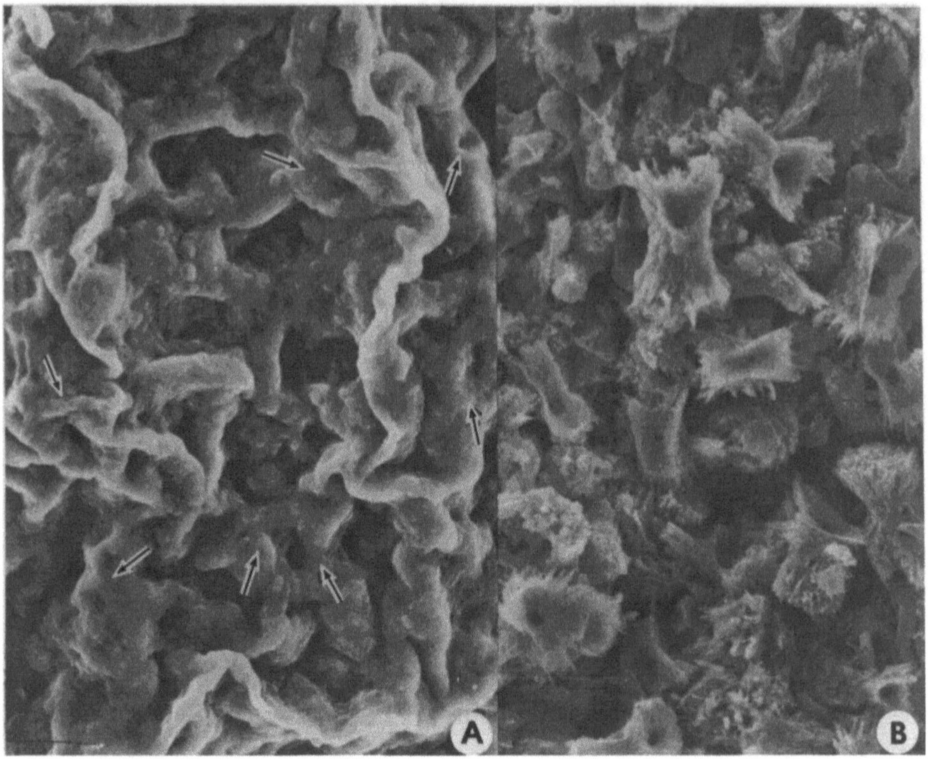

FIGURE 20. Didymium iridis, SEM of sporangia, 15KV. ▪A. Magnified peridial surface of immature sporangium shown in Figure 19A. Calcium is embedded in fibrillar peridium (arrows), × 1,270. ▪B. Magnified calcium carbonate crystals on surface of mature sporangium shown in Figure 19B, × 1,680.

In comparison, the peridium appeared dense and multilayered in thick-section (Fig. 18). High voltage electron microscopy (HVEM) was employed to examine the peridium in thick section. Sections 0.5 μm thick showed the outline of crystalline calcium carbonate embedded in the peridial matrix (Fig. 18A) and channels (Fig. 18B) through the matrix. The larger diameter channels which appeared nearly perpendicular to the surface measured 0.3 μm in diameter at the outermost edge. Thick sections, but thin enough for examination with conventional TEM, also demonstrated density patterns that gave a structured appearance to the peridium (Fig. 18C,D,E).

Didymium iridis has been studied extensively with light and electron microscopy in part because it can be cultured easily in the laboratory. As with *D. saturnus* the sporangia became black when nearly mature as spores were formed. With SEM the black sporangia appeared wrinkled but non-crystalline (Fig. 19A). With light microscopy sporangia of *D. iridis* had cupulate crystals which appeared to be totally superficial on the sporangium even when viewed with low magnification SEM (Fig. 19B) in contrast to the embedded crystals of *D. saturnus*.

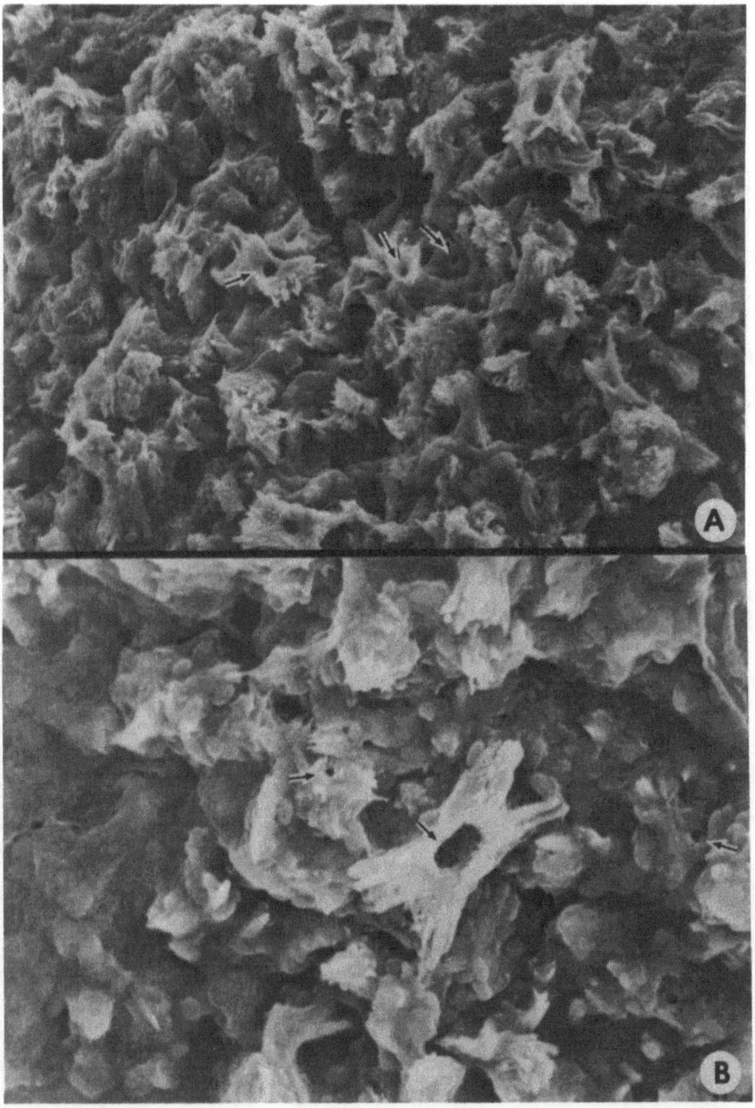

FIGURE 21. Didymium iridis, SEM of sporangium grown in culture, 15KV. Formation of calcium carbonate crystals on peridial surface. Crystals form as calcium passes either through or around peridial pores (arrows) and remains visible as the central pore of the crystal after mucopolysaccharide layer dries. ▪A. × 1,400. ▪B. × 2,500.

In *D. iridis* the sporangia became white with calcareous stellate crystals. The crystals appeared within 1-5 min when sporangia were exposed to dry air. The pattern or arrangement of crystalline deposits seen in the mature sporangium was similar to that of the wrinkles and ridges on the immature sporangium. A high magnification image (Fig. 20A) also showed four-sided protrusions with central pores. These protrusions

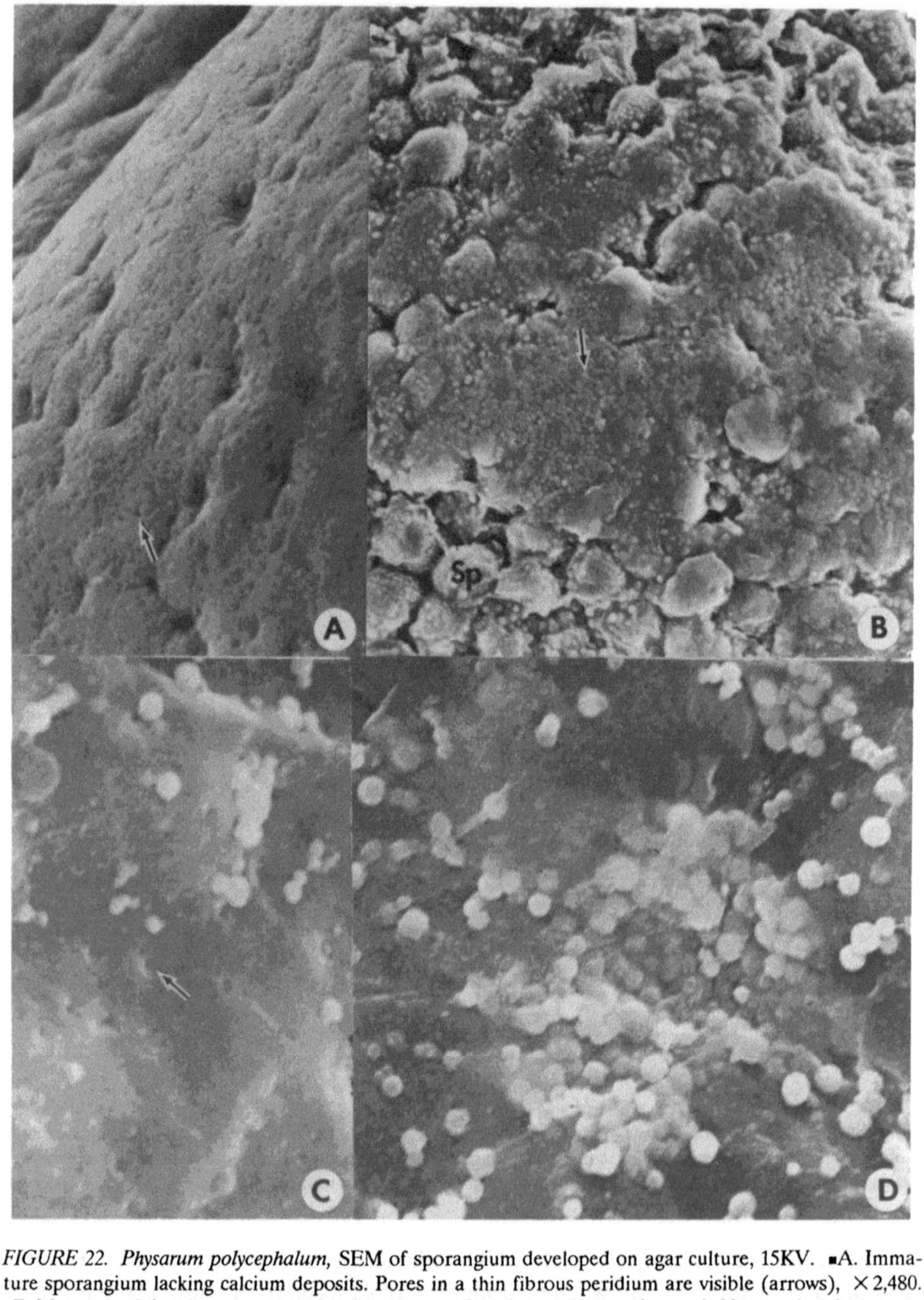

FIGURE 22. Physarum polycephalum, SEM of sporangium developed on agar culture, 15KV. ▪A. Immature sporangium lacking calcium deposits. Pores in a thin fibrous peridium are visible (arrows), ×2,480. ▪B. Mature peridium becomes covered with spheres of calcium carbonate (arrows). Note underlying spores (**Sp**), ×1,680. ▪C. Surface of immature sporangium is seen as pitted (arrow) and shows some spheres of calcium carbonate, ×4,700. ▪D. Spheres of calcium carbonate accumulate as sporangium matures, ×4,160.

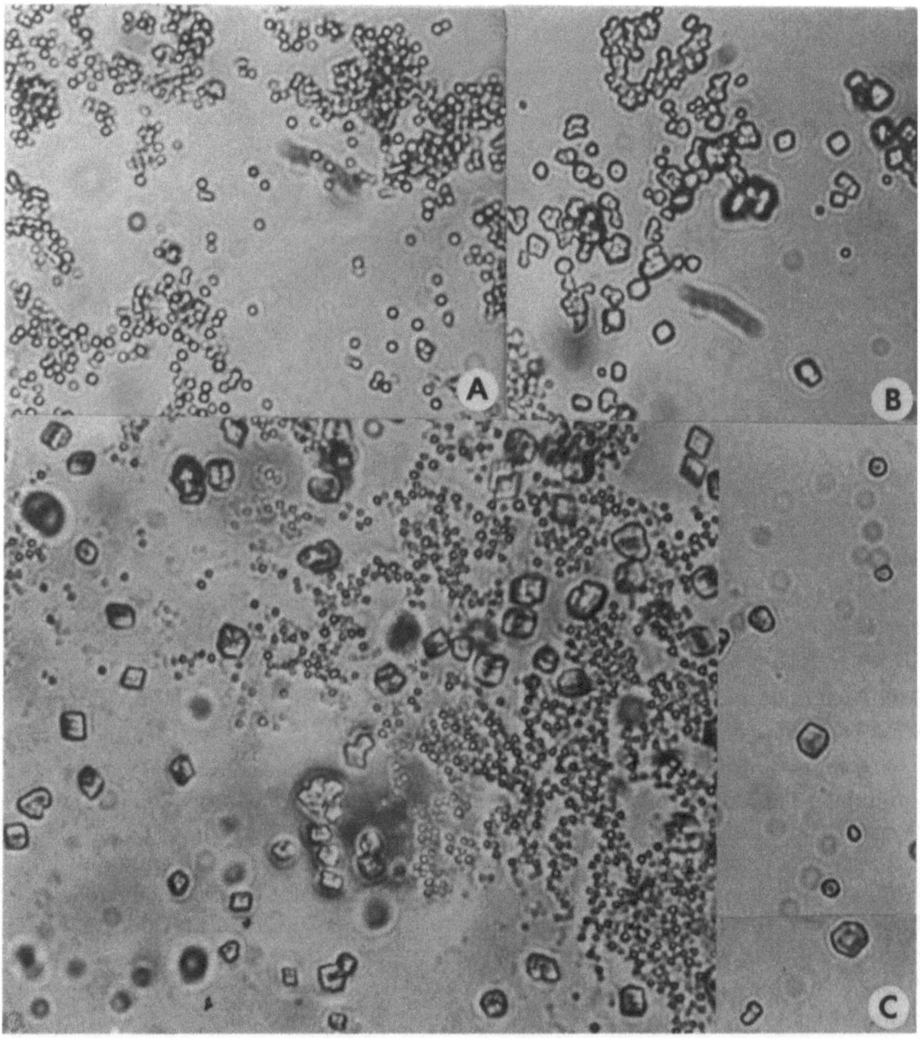

FIGURE 23. *Fuligo septica*. Differential interference-phase contrast light micrographs of calcareous deposits, × 10,600. ▪A. Spheres from peridium mounted in water. ▪B. Spheres in 10% KOH show some reorganization into cuboidal shapes. ▪C. Spheres in 10% KOH and heated 1 min to boiling, dissolve and reorganize into calcite cubes and small spheres.

appeared to be embedded in a matrix and to be similar in size and shape to the cupulate crystals (Fig. 20B). A sporangium intermediate in maturity between the nearly mature sporangium seen in Figures 19A and 20A and the mature sporangium seen in Figures 19B and 20B were also photographed with SEM and shown in Figure 21. Many cupulate crystals with central pores and stellate projections are visible in Figure 21A. Bases of crystals and other crystalline material appear embedded in peridial matrix. At the higher magnification of Figure 21B, crystalline calcium carbonate is visibly embedded in peridial matrix. Crystals have central pores and one large crystal with its base

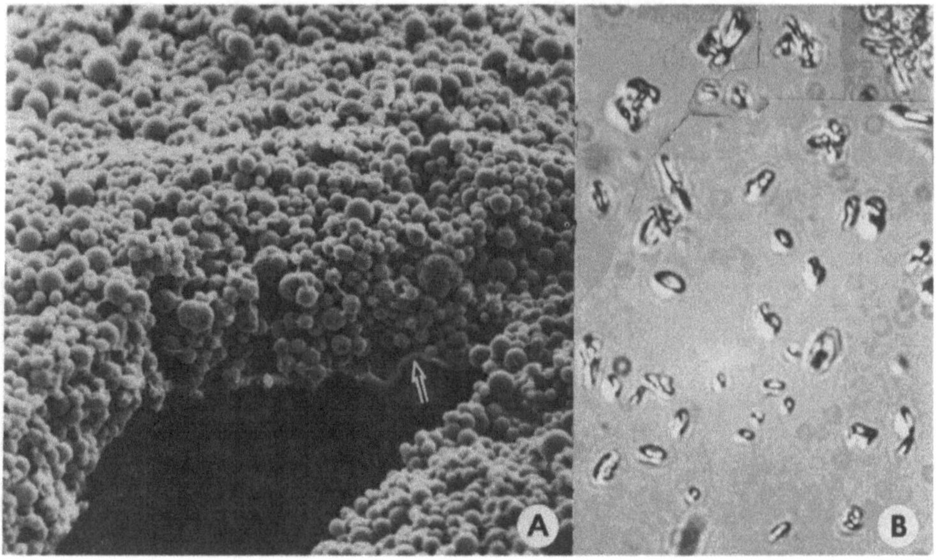

FIGURE 24. Diderma testaceum, 15KV. ▪A. SEM of calcareous spheres densely covering a membranous inner peridium (arrow), ×3,450. ▪B. Differential interference phase micrograph of spheres mounted in 10% KOH and heated. Spheres reorganize into acicular crystals of aragonite, ×10,900.

embedded in the peridiuam has a central pore which appears to extend into the matrix of the peridium.

Physarum polycephalum is another well-studied myxomycete with a calcareous peridium. The calcareous deposits in *P. polycephalum* and other species in the Physaraceae differ from species in the Didymiaceae because amorphous lime is deposited on the peridium and capillitium superficial to a very thin membranous inner peridium (Schoknecht & Small, 1972). When calcium is not deposited abundantly in the peridium, for example, when sporangia are formed on a calcium deficient agar medium, the thin membranous peridium will tightly outline the spores (Schoknecht, 1975). Using low voltage SEM, minute pores can be seen in the fibrous peridium of an immature sporangium (Fig. 22A,C). When calcium is present during plasmodial growth, the peridium will be covered with globular calcareous deposits and spore outline will be obscured (Fig. 22B). A few spheres of calcium carbonate and phosphorus are also seen seated in depressions on the surface in Figure 22C. More spheres are seen on the surface of a mature peridium (Fig. 22D) and small precipitates appear to be embedded in the matrix. Spheres of calcium carbonate produced by species in the Physaraceae also contain phosphorus (Schoknecht, 1975). Since no other elements are detected in white calcareous deposits, the phosphorus is probably in the form of calcium phosphate. When spheres from one of these species, *Fuligo septica* (Fig. 23A), were treated with alkaline solution (10% KOH) and heated, the spheres did not dissolve completely but reorganized into cubes and smaller spherical shapes (Figs. 23B,C). The cubes have the characteristic shape of the calcite form of calcium carbonate.

In contrast to the Physaraceae, the calcareous deposits in the Didymiaceae are basically free of phosphorus or only trace amounts of phosphorus are detectable. In the genus *Didymium*, as previously described, the deposits are crystalline calcium carbonate (Schoknecht, 1975). However, sporangia of the genus *Diderma* in the Didymiaceae are often covered with spheres of calcium carbonate rather than crystals. *Diderma testaceum* has a thick shell of calcium carbonate spheres illustrated in Figure 24A. These spheres appear to be very thinly covered and interconnected. A thin lower film supports the encasement. Examination of Figure 24A and of SEMs published by Brooks, *et al.* (1977) of *Diderma corrugatum* Brooks & Keller and *D. rugosum* (Rex) Macbr. provides the same information. These spheres reorganize into acicular crystals with the general shape of the aragonite form of calcium carbonate either when boiled or in basic solution (Fig. 24B).

DISCUSSION

Didymium saturnus characteristically developed peridia with embedded calcareous material (Fig. 2). These calcareous deposits are crystalline to nearly globular, depending on their composition. Embedded crystals appear to separate *D. saturnus* from other species of *Didymium* that have superficial crystals as illustrated by Schoknecht & Small (1972); Schoknecht (1975); Gustafson & Thurston (1974); and Gaither & Collins (1984). This investigation has shown that the shape of calcareous deposits can be regulated by minerals available to the plasmodium (Figs. 2, 5, 6). Incorporation of silicates (Figs. 5, 6, 7) changed the shape of the crystals from regular cubes to irregular crystals. Examination of developing sporangia of *D. saturnus* showed a persistent fibrillar mucopolysaccharide matrix over the calcareous deposits which indicated that *D. saturnus* differs more in degree than in kind when compared to *D. iridis*. Moreover, a fibrillar matrix covering the developing crystals of *D. iridis* was especially persistent when sporangia formed under moist conditions.

Calcium deposition

Bechtel & Horner (1975) postulated that the production of crystalline masses of calcium carbonate on the peridial surface appears to be a function of whether calcium is first excreted internally to form minute spheres in the Physaraceae or whether calcium is first excreted directly to the peridial surface as described by Gustafson & Thurston (1974) for the Didymiaceae. Other investigators have identified calcium in vesicles or vacuoles, such as Daniel & Jarlfors (1972b), and Etienne (1972) in *Physarum polycephalum*, which compares with our Figures 8, 9B, 14A,B. Gustafson & Thurston (1974) published no ultrastructural evidence of possible excretion from the vacuoles to the peridium as suggested by the plasmalemma invaginations shown in our Figure 12, and reported by Daniel and Jarlfors (1972a) for *P. polycephalum*. Betchel & Horner recorded images of vesicles and plasmalemma invaginations containing electron dense bodies (EDBs) but could not identify their contents. However, vacuolar activity at the surface of the cytoplasm also has been recorded by Raub & Aldrich (1982) for *Didymium iridis*. They also saw microvilli near the peridial surface which was suggested as a possible area of calcium excretion.

Mitochondria were photographed with (EDBs) in the cisternae with most EM techniques. The affinity for osmium tetroxide and for ruthenium red (Cheung *et al.*, 1974) indicated mucopolysaccharide was probably present. Daniel & Jarlfors (1972b),

Nicholls (1972), and Gustafson & Thurston (1974) reported calcium in mitochondria of *Physarum polycephalum* and *Didymium squamulosum* respectively. Gustafson & Thurston (1974) noted that the precleavage mitochondria had fewer EDB granules than plasmodia and Nicholls suggested that mitochondria might serve as respositories for calcium. Only a few small EDBs of calcium were found in mitochondria of sporangia in *D. saturnus* that were beginning cleavage. Comparison with plasmodia was not done in this investigation. However, mitochondria-like bodies usually contained a single large EDB in brown-stage sporangia. These have been demonstrated to contain calcium in *D. saturnus* using pyroantimonate precipitation and EDTA dissolution. This agrees with Gustafson's & Thurston's observation (1974).

The potassium pyroantimonate technique was used in this investigation to record the presence of lightly bound calcium in the cytoplasm during sporulation. The valid use of this technique for localization of calcium is discussed and comparisons of several procedural variations were given by Wick & Heppler (1982) and Simpson & Spicer (1975). In several investigations, calcium was localized at most cytoplasmic organelles, such as nuclei, mitochondria, endoplasmic reticulum, vacuoles and plasmalemma (Daniel & Jarlfors, 1972b; Simpson & Spicer, 1975; Wick & Hepler, 1982; Slocum & Roux, 1982, Duke *et al.*, 1985) with specific localization dependent upon cell type and technique employed. Results of the different techniques utilized in this study confirmed some of the previous findings. For example, Simpson & Spicer (1975) noted that polyantimonate deposits were reduced in the cytoplasm but that deposits were visible on plasmalemma (Figs. 11, 12, 14A,B,C, 17C) when glutaraldehyde was used in the primary fixation. Etienne (1972) localized calcium in vesicles by precipitating calcium with sodium oxalate. He suggested that these vesicles which accumulate calcium and potassium are probably the same vesicles which are involved in redistribution and secretion.

Figures 8 and 9 show calcium precipitation along plasmalemma, in vacuoles, and in a polysaccharide layer outside the cell. Rough ER often appears relatively dense in TEM. However, these images were made from unstained sections which would seem to indicate absorption of the pyro-antimonate.

As sporulation begins, it is recognizable by invaginations and condensed cytoplasm (Fig. 8) that eventually becomes the cytoplasm that is cleaved into spores. Also, an ectoplasmic area lacking mitochondria (Fig. 13) is present near the periphery. Mitochondria are present near the cell center associated with EDBs along their periphery. This sequence of development is similar to that described by Charvat *et al.* (1973) for *Perichaena vermicularis* (Trichiales). Plasmalemma invaginations (Fig. 12A,B) are also reported for both species. Mitochondria like bodies containing calcium are found in this ectal degenerating layer and may contribute to peridial calcium through degeneration of this organelle. It appears from this investigation that calcium may be bound to molecules at many sites in the cytoplasm including the RER, mitochondria and plasmalemma. Osmium, vacuoles and plasmalemma invaginations are seen containing small electron dense precipitate for which EDX spot analysis confirms the presence of calcium. This data is similar to that obtained by Gustafson & Thurston (1975) with mitochondria and vacuole deposits but includes photographs of plasmalemma invaginations with small calcium deposits reported in the Physaraceae. However, excretion of calcium may be occurring in an ionic state, which would not have been precipitated by pyroantimonate.

No preformed peridial crystals or large spheres of calcium have been reported in the cytoplasm of Myxomycetes. We hypothesize from our observations that calcium takes its characteristic shape after calcium is mobilized into the extracellular space. In the Physaraceae globular calcium carbonate and phosphate developed via pores in the peridial matrix where the material crystallized (Fig. 22) and also see Figure 13 in Schoknecht (1975) for other evidence. A thin film of matrix appears to cover the granules in Figure 22D. The spheres are calcite as were most of the crystals in *Didymium* species (Schoknecht, 1975; Pobequin, 1954).

In comparison to *Diderma* as seen in Figure 24, *Didymium* species (*D. saturnus* and *D. iridis*) excreted calcium into the peridial matrix (Figs. 5, 6, 7, 20A, 21A,B) that determined the shape of the crystals prior to the crystals being visible on the surface (Fig. 20). When grown in a moist chamber the sporangia were relatively slow in developing the white crystalline peridium. When the moist chambers were exposed to air the sporangia rapidly (1-5 min) became white as they dried. This rapid appearance of the white crystalline surface is probably due to dehydration and loss of polysaccharide fibrils on the surface rather than rapid excretion. Kowalski (1972) reported that a distinct gelatinous sheath was visible in brown stage sporangia of *Squamuloderma nullifera* and hypothesized that "this sheath forms the lime scales upon drying." Under dryer conditions, crystallization occurs more slowly and steadily. Based on crystal shape, it appears that the crystals grow through the addition of calcium carbonate. The central pore in the cupulate crystals (Figs. 20, 21) and also in figures of Gaither & Collins (1984) appeared to be the nucleation site for the crystallization and may have provided a channel to the interior of the peridium.

Slime layer of fibrillar mucopolysaccharide

Myxomycetes are commonly known as slime molds because of the viscous, extracellular slime characteristically excreted by the migrating plasmodia. Many chemical and ultrastructural investigations have characterized the slime and determined its origin. For example, in histochemical observations of plasmodia, Sheen *et al.* (1969) showed that the slime layer stained strongly with colloidal iron, a reaction for mucopolysaccharide. In 1970 McCormick *et al.* purified the polysaccharide component of *P. polycephalum* and identified it as acidic polysaccharide, thus confirming each other's results. Simon & Henney (1970) isolated and characterized slime produced by three species of *Physarum*. They found the glycoprotein to be primarily the neutral sugar galactose with protein and ash. Aspartate was the most abundant amino acid present. Aspartic acid is the most abundant amino acid in the organic matrix of all carbonates, both skeletal and nonskeletal. This common association of aspartic acid-rich organic matter with carbonates, indicates that the organic matrix plays an active role of biosynthesis in the calcification process.

Cheung *et al.* (1974) used ruthenium red to stain the fibrillar mucopolysaccharide slime layer. Raub & Aldrich (1982) used silver methanamine to demonstrate carbohydrate rich slime near the periphery of plasmodia. Cheung & Henney (1974) also demonstrated polysaccharide granules in mitochondria and in vacuoles using ruthenium red cytochemistry. Moreover, both polysaccharide and calcium are demonstrated in mitochondria and Nicholls suggested (1972) that mitochondria might be a repository for calcium. With TEM the slime layer is seen to be a fibrillar network, (Rhea, 1966; Daniel & Jarlfors, 1972b; Cheung *et al.* 1974; Charvat *et al.*, 1973; Raub & Aldrich, 1982, and Fig. 17A, B, C).

Fibrillar mucopolysaccharide has its origin in the cytoplasm. Cytoplasmic channels and invaginations are often filled with slime fibrils (Kessler, 1982; Daniel & Jarlfors, 1972a; Rhea, 1966). Mucopolysaccharide is mobilized to the exterior by vacuoles and invaginations of the plasmalemma. Achenbach (1979) described channels or invaginations of the plasmalemma into the ectoplasm or outer layer of the plasmodium. These invaginations produce a network of interconnecting channels which communicate with the surrounding plasmalemma. This is similar to results shown for mobilization of calcium to the surface (Fig. 12) and (Etienne, 1972). HVEM shows channels through the fibrous peridium. Narrow channels are regularly spaced and perpendicular to the cell surface. The larger channels appear irregularly spaced and they appeared to be continuous with cytoplasmic channels.

SEM of the polysaccharide peridium

With SEM, the polysaccharide layer appears porous as shown originally by Schoknecht (1975). Pores are shown in the peridium of *D. saturnus* (Fig. 5C, 5E, 7A, B), *D. iridis* (Fig. 20, 21), and *P. polycephalum* (Fig. 22A, C). The perdial matrix is not a homogeneous sheath or envelope but a fibrillar network in which the crystals are embedded. The fibrillar peridium was shown with SEM by Charvat, *et al.* (1973) for *Pericheana vermicularis* and Schoknecht and Keller (1977) showed species of *Perichaena* (*Trichiales*) with calcium oxalate embedded in the peridium. Indeed, Whitney and Arnott (1986a,b, 1987) have shown that calcium oxalate crystals are produced under wall layers in widely divergent classes of fungi, a Zygomycete and two Basidiomycete species.

Composition of the peridial deposits

Previous investigations by Schoknecht (1975) using EDX, SEM and chemical tests described the calcareous deposits of *Didymium* as being composed of calcium carbonate and staining with Feigl's solution Friedman (1959) indicated that the form of the calcium carbonate was primarily calcite mixed with some aragonite. Pobequin (1954) also reported calcite in *Didymium*. EDX of isolated crystals of the *Didymium* species (Didymiaceae) are free of phosphorus. Phosphorus is commonly associated with the globular deposits of the Physaraceae probably as calcium phosphate. Subsequent investigators have reported phosphorus in the sporangia of *Didymium* species such as Henney *et al.* (1980). These reports contradict the absence of calcium phosphate in crystals of *Didymium* as reported by Schoknecht (1975). However, the discrepancy can be explained in three ways. First, technique is of paramount importance. If any peridial material from the matrix is adherent to the crystals, phosphorus will be detected. *Didymium trachysporum* G. Lister was reported to have traces of phosphorus by Schoknecht (1975). Crystalline deposits adhere tightly to the fibrillar peridium in that sample. Charvat, *et al.* (1973) described the degeneration of ectoplasm to form part of the lower peridium. Phosphorus is common to many organic molecules, hence it would be expected to occur in the peridial matrix. Phosphorus is usually detectable in membranous peridia. Moreover, if any spores are present, perhaps hidden under the crystals, phosphorus will again be detected because of penetration of the electron beam into the sample and the escape depth from which the x-ray originates. Reports of phosphorus in *Didymium* by Henney, *et al.* (1980) and Aldrich (1982) is thus easily explained. For example, all state in their methods that they examined whole sporangia following the technique of Nelson, *et al.* (1977). Obviously, if whole sporangia are examined, spores and fibrous or membranous peridial layers would all contain detectable phosphorus. Schoknecht (1975) described in her methods that only crystals were examined, and therefore, it was these calcareous deposits that were analyzed with EDX.

Secondly, the amounts of phosphorus shown in EDX spectra for *Didymium* by these authors was barely detectable as a peak above the background noise and Bremsstrahlung. A general rule applied to the peak to noise ratio should be 2 to 1 to be sure that an element is really present. Indication of a label for phosphorus placed over a minute peak by the EDX instrument did not mean that phosphorus was present. EDX would, however, record that a particular electron volt energy was comparable to that of the K radiation energy for phosphorus. It is also close to that of the radiation escape peak for calcium, which is readily produced concurrent with strong primary and secondary peaks. Many high atomic number elements have low energy level radiation peaks in the energy range of elements with low atomic numbers and low radiation energy. Moreover, in the low atomic number or low Kev range of elements, such as silicon, sulfur, chlorine, potassium, phosphorus and calcium, the energy of backscattered or reflected electrons is often equivalent and contributes to the background. Specimen handling is a third caution that should be noted because phosphorus is so commonly present that it is a problem contaminant.

The globular deposits of members of the Physaraceae (Fig. 23A) contain measurable amounts of phosphorus. Schoknecht (1975) stated that when the chemical geometry of calcium carbonate and calcium phosphate are combined, a regular crystalline lattice would probably be interrupted, and this could account for the amorphous appearance of physaraceous deposits. Heating with alkaline solution altered the form of the deposits producing cuboidal shapes of calcite accompanied by small spheres. Calcium carbonate dissolves in acid solutions. An alkaline solution, therefore, may have removed organic compounds that exerted biocontrol over the shape of the deposits.

A problem arises when attempting to explain the globular peridial deposits of *Diderma*, a genus in the Didymiaceae. These deposits of *Diderma* often show trace amounts of phosphorus. SEMs of the peridium for *D. testacum* (Fig. 24A) and the peridium of *D. corrugatum* and *D. rugosum* in Brooks, *et al.* (1977) depict a film of matrix material surrounding and connecting the spherules. If a large enough area is examined with EDX, phosphorus would be detectable from this matrix even if it is not actually in the calcareous deposits.

These globular deposits of *Diderma* are primarily calcium carbonate. These stain with Feigl's solution (Friedman, 1959) as forms of aragonite. These deposits might be more correctly described as cryptocrystalline rather than amorphous. When these globules are treated by boiling or with KOH they quickly reorganize, without completely dissolving, into lower energy state, acicular crystals of aragonite. This latter phenomenon agrees with the results of Feigl's stain. The KOH does not dissolve the calcium carbonate which is soluble in acid, so that it must be affecting the mucopolysaccharide matrix which surrounds the spheres and controls the spherical shape. The aspartate demonstrated in the mucopolysaccharide layer by Simon & Henney (1970) has a free carboxyl group with a similar composition, geometry and charge as the carbonate anion and aspartate is also present in the protein matrix of most calcified carbonates. This amino acid could thus serve to bind calcium and transfer calcium to carbonate anions depending on environmental parameters.

CONCLUSIONS

Mechanisms of calcium excretion in the Myxomycetes need to be examined for many species. Our investigation showed similarity between the cytoplasmic sites of calcium localization and excretion in both the Physaraceae and Didymiaceae. Moreover, calcium was detected in Myxomycetes in the same locations as reported for cells of animals and plants. Calcium is initially formed as minute intracellular deposits and excreted to an extracellular space with mucopolysaccharide. Chemistry and morphology of the fibrillar mucopolysaccharide sheath matrix and its relationship to the calcareous deposits in the peridium provided evidence for an active role of the matrix in the calcification process. Variations in the amino acid composition or abundance of aspartic acid in the peridial matrix may prove to be significant factors in depositional patterns. Physical factors such as availability of minerals and environmental factors such as humidity may affect the final appearance of calcium carbonate. Physico-chemical precipitation is a partial explanation for the type of calcareous deposits. However, a protein matrix is generally associated with calcification. This suggests there must be organismal control of both composition and structure during the biomineralization process.

ACKNOWLEDGEMENTS

We wish to thank the University of Illinois Center for Electron Microscopy and the High Voltage Electron Microscope Facility at the University of Colorado for equipment use. This research was made possible by a grant PCM #80-05911 from the National Science Foundation and funding from the University Research Committee and a sabbatical leave to JDS from Indiana State University. HWK received support from the University of Texas at Arlington Organized Research Fund. We thank the College of Science, University of Texas at Arlington for the use of their SEM facility. Cultures of *Didymium iridis* were provided by Dr. Jim Clark and Professor O'Neil R. Collins. Special thanks go to Professor Collins for maintaining viable plasmodia of *Didymium saturnus* for many years and providing plasmodia for this study. We greatly appreciate the helpful comments and suggestions of the reviewers. Participation in the Fifth International Symposium on Biomineralization was made possible by funds provided through The Organized Research Fund of The University of Texas at Arlington and The Sea Grant Program administered by Texas A & M University.

REFERENCES

ACHENBACH, F., NAIB-MAJANI, W., AND WOHLFARTH-BOTTERMANN, K.E., 1979. Plasmalemma invaginations of *Physarum* dependent on the nutritional content of the plasmodial environment. Eur. J. Cell Biol., 36:355-359.

ALDRICH, H.C., 1982. Influence of inorganic ions on color of lime in the Myxomycetes. Mycologia, 74:404-411.

BECHTEL, D.B. AND HORNER, H.T., 1975. Calcium excretion and deposition during sporogenesis in *Physarella oblonga* calcification. Calc. Tiss. Res., 18:195-213.

BROOKS, T.E., KELLER, H.W. AND CHASSAIN, M., 1977. Corticolous Myxomycetes VI: A new species of *Diderma*. Mycologia, 69:179-184.

CHARVAT, I., ROSS, I.K., AND CRONSHAW, J., 1973. Ultrastructure of the plasmodial slime mold *Perichaena vermicularis* II. Formation of the peridium. Protoplasma, 78:1-19.

CHEUNG, L., HENNEY, H.R., AND CLARK, W.H., 1974. Ultrastructure of pure cultures of *Physarum flavicomum* 1. Conversion of a plasmodium to microplasmodia and microsclerotia, and the process of slime secretion. Cytobios, 9:193-205.

COLLINS, O.R., 1979. Myxomycete biosystematics: some recent developments and future research opportunities. Bot. Rev., 45:145-201.

DANIEL, J.W., AND JARLFORS, U., 1972a. Plamsodial ultrastructure of the Myxomycete *Physarum polycephalum*. Tissue and Cell, 4:15-36.

DANIEL, J.W., AND JARLFORS, U., 1972b. Light-induced changes in the ultrastructure of a plasmodial Myxomycete. Tissue and Cell, 4:405-426.

DUKE, S.O., VAUGHN, K.C., AND WAUCHOPE, R.D. 1985. Effects of glycosate on uptake, translocation and intracellular localization of metal cations in soybean, (*Glycine max*) seedlings. Pesticide Biochem. Physiol., 24:384-394.

ETTIENNE, E., 1972. Subcellular localization of calcium repositories in plasmodia of the acellular slime mold, *Physarum polycephalum*. J. Cell Biol., 54:179-184.

FRIEDMAN, G.M., 1959. Identification of carbonate minerals by staining methods. J. Sed. Petrol., 29:87-97.

GAITHER, T.W. AND COLLINS, O.R., 1984. Comparative SEM observations of sporophore characteristics in three species of *Didymium* (Myxomycetes, Physarales) Mycologia 76:650-664.

GUSTAFSON, R.A. AND THURSTON, E.L., 1974. Calcium deposition in the myxomycete *Didymium squamulosum*. Mycologia, 66:397-412.

HENNEY, M.R., ALEXOPOULOS, C.J. AND SCHEETZ, R.W. 1980. *Didymium atrichum* A new myxomycete from south-central Texas. Mycotaxon, 11:150-164.

KELLER, H.W., 1970. *Didymium saturnus*, a new myxomycete occurring on straw stacks. Mycologia, 62:1061-1066.

KESSLER, D., 1982. Plasmodial structure and motility. *In* Cell Biology of *Physarum* and *Didymium*. Vol 1: organisms, nucleus, and cell cycle. pp. 442. (ed. Aldrich, H.C., and Daniel, J.W.), Academic Press.

KOWALSKI, D.T., 1972. *Squamuloderma*: a new genus of Myxomycetes. Mycologia, 64:282-1289.

MCCORMICK, J.J., BLOMQUIST, J.C. AND RUSCH, H.P., 1970. Isolation and characterization of an extracellular polysaccharide from *Physarum polycephalum*. J. Bacteriol., 104:1110-1118.

NELSON, R.K., SCHEETZ, R.W. AND ALEXOPOULOS, C.J., 1977. Elemental composition of *Metatrichia vesparium* sporangia. Mycotaxon, 5:65-375.

NICHOLLS, T.J., 1972. The effects of starvation and light on intramitochondrial granules in *Physarum polycephalum*. J. Cell Sci., 10:1-14.

POBEQUIN, T. 1954. Contribution a l'étude des carbonates de calcium, precipitation du calcaire par les végétaux comparison avec le monde animal. Ann. Sci. Nat. Botan. Biol. Végétale Ser. 11, 15:9-109.

RAUB, T.J. AND ALDRICH, H.C., 1982. Sporangia, spherules, and microcysts. *In* Cell biology of *Physarum* and *Didymium* Vol II: differentiation, metabolism and methodology, pp 373 (ed. Aldrich, H.C. and Daniel, J.W.), New York, Academic Press.

REYNOLDS, E.S., 1963. The use of lead citrate at high pH as an electron opaque stain for electron microscopy. J. Cell Biol.17:208-212.

RHEA, R.P. 1966, Electron microscopic observations of the slime mold *Physarum polycephalum* with specific reference to fibrillar structures. J. Ultrastr. Res., 15:349-379.

SCHOKNECHT, J.D. AND SMALL, E.B., 1972. Scanning electron microscopy of the acellular slime molds (Mycetezoa = Myxomycetes) and the taxonomic significance of surface morphology of spores and accessory structures. Trans. Amer. Microsc. Soc., 91:380-410.

SCHOKNECHT, J.D., 1975. SEM and X-ray microanalysis of calcareous deposits in myxomycete fructifications. Trans. Amer. Microsc. Soc., 94:216-233.

SCHOKNECHT, J.D. AND KELLER, H.W., 1977. Peridial composition of white fructifications in the Trichiales (*Perichaena* and *Dianema*). Can. J. Bot., 55:1807-1819.

SHEEN, S.J., GAILEY, F.B., MILLER, D.M., ANDERSON, J.D., BARGMANN, T.J., AND CARTER, D.A. 1969. Sol-gel differences in plasmodia of the acellular slime mold, *Physarum polycephalum*. Bioscience, 19:1003-1005.

SIMON, H. AND HENNEY, H.R., 1970. Chemical composition of slime from three species of Myxomycetes. Fed. Eur. Biochem. Soc. Lett., 7:80-82.

SIMPSON, J.A. AND SPICER, S.S., 1975. Selective subcellular localization of cations with variants of the potassium (pyro) antimonate technique. J. Histochem. Cytochem., 23:575-598.

SLOCUM, R.D. AND ROUX, S., 1982. An improved method for the subcellular localization of calcium using a modification of the antimonate precipitation technique. J. Histochem. Cytochem., 30:617-620.

SPURR, A.R., 1969. A low viscosity epoxy resin embedding medium for electron microscopy. J. Ultrastr. Res., 26:31-43.

WALTON, J., 1979. Lead aspartate an *en bloc* contrast stain particularly useful for ultrastructural enzymology. J. Histochem. Cytochem., 27:1337-1342.

WHITNEY, K.D. AND ARNOTT, H.J., 1986a. Morphology and development of calcium oxalate deposits in *Gilbertella persicaria* (Mucorales). Mycologia, 78:42-51.

WHITNEY, K.D. AND ARNOTT, H.J., 1986b. Calcium oxalate crystals and basidiocarp dehiscence in *Geastrum saccatum*. (Gasteromycetes). Mycologia, 78:649-656.

WHITNEY, K.D. AND ARNOTT, H.J., 1987. Calcium oxalate crystal morphology and development in *Agaricus bisporus*. Mycologia, 79:180-187.

WICK, S.M. AND HEPLER, P.K., 1980. Localization of Ca^{++} containing antimonate precipitate during mitosis. J. Cell Biol., 86:500-513.

WICK, S.M. AND HEPLER, P.K., 1982. Selective localization of intracellular Ca^{2+} with potassium antimonate. J. Histochem. Cytochem., 30:1190-1204.

MANGANESE TRANSFORMATIONS BY MARINE *BACILLUS* SPECIES

Elisabeth W. de Vrind-de Jong[1], Johannes P.M. de Vrind[1], Fred C. Boogerd[1], Peter Westbroek[1] and Reinhardt A. Rosson[2]

[1]*Department of Biochemistry, University of Leiden, Wassenaarseweg 64, 2333 AL Leiden, The Netherlands;* [2]*Center for Great Lakes Studies, University of Wisconsin at Milwaukee, 600 East Greenfield Avenue, Milwaukee, Wisconsin 53204*

ABSTRACT. A wide variety of micro-organisms promote the oxidation or reduction of manganese, through indirect or direct mechanisms. An example of the latter category is *Bacillus SGl, a marine organism isolated from a near-shore manganese sediment. Its mature dormant spores catalyze the oxidation of* Mn^{2+} to Mn^{4+}. The process requires molecular oxygen and is catalyzed by a spore coat component. The manganese oxide (MnO2) produced remains bound to the spore surface. The vegetative cells do not have the oxidizing capacity. They are able to reduce Mn^{4+} to Mn^{2+} under low-oxygen conditions. The reducing activity has a pH optimum of 7.5 and is abolished by preheating of the cells at 90° C for 5 minutes. Addition of mercuric chloride (HgCl2) (final concentration 0.01%) to cells which are actively reducing manganese oxide causes immediate cessation of the process. Manganese oxide reduction is also inhibited at high oxygen tensions and by inhibitors of the electron transport system. *Bacillus cells contain b-* and *c*-type cytochromes which are oxidized *in situ* when manganese oxide is added to an anoxic cell suspension. The results suggest that vegetative cells may use the manganese oxide formed by the spores from which they germinate as a terminal electron acceptor. The possible applications of manganese transforming micro-organisms in human society are discussed.

MICROBIAL MANGANESE TRANSFORMATIONS; GENERAL ASPECTS

Manganese is one of the most abundant elements in the earth's crust, amounting to about 0.1% of its total mass. As a transition metal its valence states can range from -2 to +7. In nature, the +2 and +4, and to a lesser extent, the +3 states prevail. The metal is usually soluble in the reduced state, which is favored under anoxic conditions and at low pH. In high Eh and pH environments, manganese tends towards the oxidized state which readily forms highly insoluble oxides and oxyhydroxides. Thus manganese lacks stable soluble forms in oxic waters. Because it also lacks a volatile phase, its cycling at the global scale is primarily determined by hydrothermal activity and geological forces acting on solid phases. On a local scale, manganese may be cycled rapidly owing to its conversion between soluble and insoluble forms. This local cycling is profoundly influenced by the factors which determine the valence state of the metal. The interaction of the biota with manganese is another important factor. Manganese is sequestered by

Edited by Rex E. Crick
Plenum Press, New York

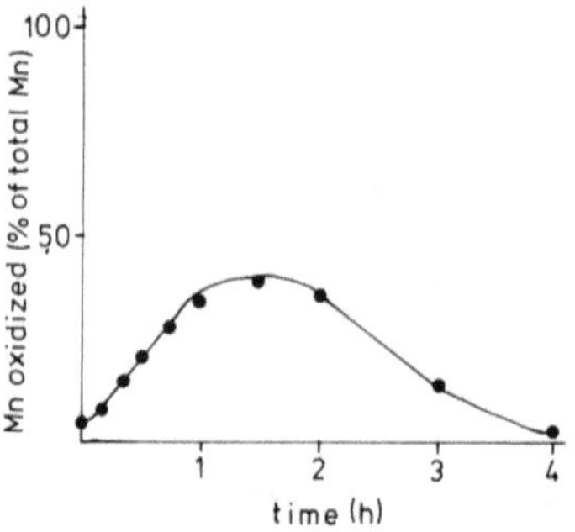

FIGURE 1. Concentration of manganese oxide in a suspension of spores of *Bacillus* SGI.

living organisms because it is an essential trace element for virtually all life. It is an essential constituent of many enzymes (Keen *et al.*, 1984), it plays a central role in the oxidation of water during photosynthesis (Barber, 1984), and it is needed for bacterial sporulation (Charney *et al.*, 1951; Eisenstadt *et al.*, 1973). Moreover, a wide variety of micro-organisms, including bacteria, fungi and algae, are able to promote the reduction or oxidation of manganese (Marshall, 1979; Nealson, 1983; Ghiorse, 1984). Many of these microbes, especially bacteria, are found associated with manganese precipitates, such as ferromanganese nodules, manganese encrustations in pipelines and desert varnish (Ehrlich, 1963; Ehrlich, 1975; Ghiorse, 1984; Jannasch & Taylor, 1984; Palmer *et al.*, 1986). They are therefore thought to play a role in the formation of these precipitates. The mere association of bacteria with manganese precipitates, however, suggests but does not prove microbial mediation of manganese precipitation or dissolution. More conclusive evidence may be obtained by measuring manganese transformations *in situ* using bacterial poisons (Burdidge *et al.*, 1983; Rosson *et al.*, 1984). In several cases the rates of manganese oxidation strongly decreased in the presence of these inhibitors (Emerson *et al.*, 1982).

The evaluation of microbial mediation of manganese transformations in nature also requires laboratory experiments with isolated cultures. Microbes may transform manganese through indirect or direct mechanisms (Nealson, 1983). Indirect transformations result from changes in the Eh or pH of the environment caused by microbial growth and metabolism. For example, oxygenic photosynthesis or ammonia production will favor manganese oxidation, whereas fermentive acid production or respiratory oxygen consumption will result in manganese reduction. Sulfate reducing bacteria have been shown to reduce manganese oxides through the formation of sulfide (Burdidge & Nealson, 1986). Direct manganese transformations result from the production of specific cellular components which promote the reduction or oxidation of manganese. A limited number of manganese-reducing organisms has been studied in this respect. Some marine *Bacillus* species may use manganese oxides as a terminal electron acceptor during respiration, under anaerobic or low-oxygen conditions (Burdidge & Nealson, 1985; de Vrind et al., 1986), or even in the presence of oxygen (Ghiorse & Ehrlich, 1976). In the latter case the manganese reducing activity was demonstrated in cell-free extracts. It was thermo-labile and inhibited by mercuric chloride, suggesting enzyme involvement in the process. Manganese oxidation may be promoted by intra- or extracellular proteins (Jung & Schweisfurth, 1979) or polysaccharides (Ghiorse & Hirsch, 1979) which bind manganese and stimulate its auto-oxidation. Other bacteria may oxidize manganese enzymatically. Enzymatic oxidation has been demonstrated in *Arthrobacter* species (Ehrlich, 1968), in *Oceanospirillum* and *Vibrio* strains (Arcuri & Ehrlich, 1979; Arcuri & Ehrlich, 1980), in *Leptothrix discophora* (Johnson & Stokes, 1966; Mills & Randles, 1979) and *Metallogenium* (Dubinina, 1978). Some of these strains required preformed manganese oxide for oxidizing activity (Ehrlich, 1968; Ehrlich, 1984). In general, the formed oxides remain associated with the cell exterior. The function of manganese oxidation is not well understood. It has been suggested that the organisms may obtain energy from the process (chemilithotrophy, Ali & Stokes, 1972; Kepkay *et al.*, 1984). Unequivocal evidence to sustain this assumption is lacking, however. In *Lactobacillus* sp., manganese is oxidized during detoxification of superoxide (Archibald & Duong, 1984). Since this is an intracellular process it may not be a general feature of manganese oxidizing bacteria.

Whatever the mechanisms of manganese reduction or oxidation may be, they must be under genetic control. No data on the genetics of manganese reducers have been published so far. Two recent reports on the genetics of manganese oxidizing bacteria suggest that manganese oxidation may be encoded by plasmids (Lidstrom *et al.*, 1983; Schuett *et al.*, 1986). In one case the plasmid also coded for copper resistance, which may reflect the generally high copper concentration in manganese-rich sediments, especially manganese nodules. Many of the screened strains, however, did not contain detectable plasmids.

The mechanisms of manganese transformations in a marine *Bacillus* species will be reviewed in more detail in the context of the following results obtained in our laboratory.

MICROBIAL MANGANESE TRANSFORMATIONS: *BACILLUS* SGl

Bacillus SGl is a marine species isolated from a near-shore manganese sediment sample (Nealson & Ford, 1980). The bacteria can be grown on natural seawater containing peptone, yeast extract and trace elements (Rosson & Nealson, 1982). When the nutrients become limiting, the bacteria start to sporulate, both on agar plates and in liquid culture. No extra manganese has to be added to the media for sporulation to occur. Apparently the manganese concentration in natural seawater (about 0.1 mM) is sufficient. When the bacteria are grown on plates containing extra manganese the colonies turn brown after about 8 days. This is caused by the formation of manganese oxide by the mature dormant *Bacillus* spores. The presence of manganese oxide can be verified with redox dyes, e.g., Leucoberbelin blue (Krumbein & Altman, 1973, de Vrind *et al.*, 1986). This dye reacts with manganese with valences of +3 or higher under the development of a blue color. The reaction can be quantified by measuring the absorption of the reaction mixture at 620 nm. When the spores are submitted to a temperature shock (*e.g.* transfer from -20° C to room temperature) germination may be induced. In such cases the manganese oxide initially formed by the spores is dissolved, suggesting manganese oxide reduction by the vegetative cells (Fig. 1). The vegetative cells indeed reduce synthetic manganese oxide as will be illustrated below. The question whether these two processes are directly mediated by the organism and whether they are functionally linked has been the main topic of research in our laboratory.

MANGANESE OXIDATION BY SPORES

Spores oxidize manganese over a wide range of manganese concentrations (Rosson & Nealson, 1982). At a concentration of 3×10^9 spores/ml 100 μM manganese is completely oxidized in one hour (Fig. 2). The manganese oxide sticks to the spore surface (Rosson & Nealson, 1982). When the total amount of manganese immobilized

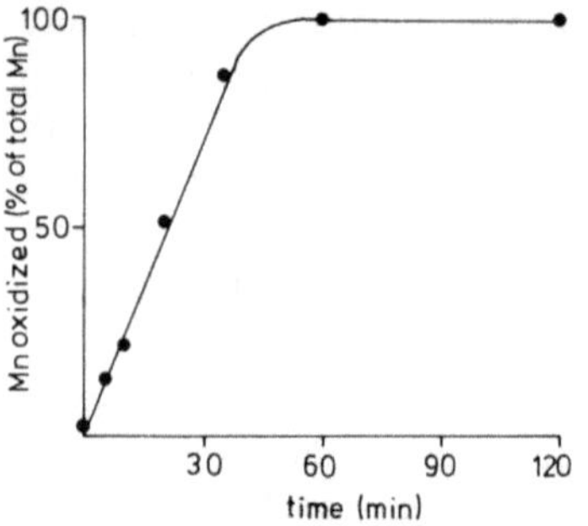

FIGURE 2. Manganese oxidation by spores of *Bacillus* SGI. Freshly prepared spores were supplied with 100 μM $MnCl_2$ and the MnO_2 concentration was determined as in Fig. 1.

at the spore surface is measured with the radioactive isotope ^{54}Mn, an almost identical curve as that shown in Figure 2 is obtained (J.P.M. de Vrind *et al.*, 1986). This indicates that bound manganese is rapidly oxidized. When a similar experiment is performed under an atmosphere of nitrogen instead of oxygen, neither manganese oxidation nor binding is observed (Fig. 3). When nitrogen is replaced by oxygen both processes resume at a rate similar to that under aerobic conditions. Two conclusions can be drawn from this experiment: 1) molecular oxygen is needed for manganese oxidation, and 2) the spores themselves do not bind significant amounts of manganese. These results may indicate that manganese is adsorbed to manganese oxide accumulated at the spore surface and non-biologically oxidized (*cf.* Morgan & Stumm, 1964), the only role of the spores being the initial manganese oxide formation. If so, spores already encrusted with manganese oxide would have initial manganese binding and oxidation rates exceeding those of non-encrusted spores. Figure 4 shows, however, that spores encrusted with 300 μM manganese oxide oxidize manganese at a similar rate as fresh spores, whereas their initial manganese binding rate is significantly increased. The increased manganese binding most probably represents the adsorption of manganese to the preformed oxide, but this adsorption process does not result in an equivalent increase in manganese oxidation. Moreover, at increasing amounts of manganese oxide on the surface of the spores the oxidation rate decreases and a maximum in the amount of manganese that spores can oxidize is observed (de Vrind *et al.*, 1986). Consequently, a spore component is needed for the oxidation process *per se*.

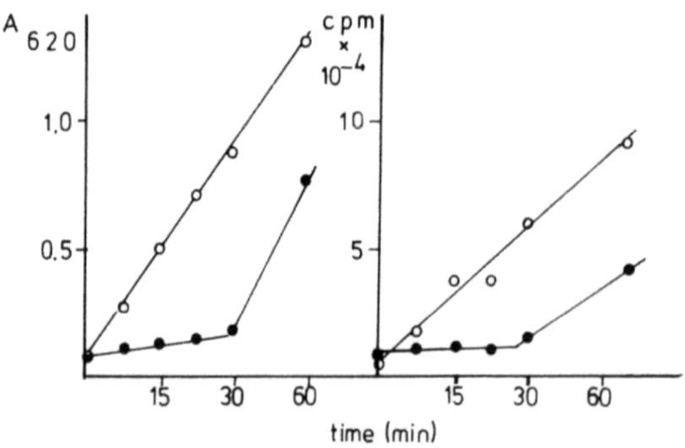

FIGURE 3. Manganese oxidation (left figure) and manganese binding (right figure) by spores of *Bacillus* SGI under aerobic (open symbols) and anaerobic (closed symbols) conditions.

Other data also indicate that a spore component is directly involved in manganese oxidation. The activity is inhibited by submitting the spores to high temperatures (Rosson & Nealson, 1982). It is completely abolished in the presence of 50 mM sodium azide (NaN_3) and 0.01% mercuric chloride (Rosson & Nealson, 1982), indicating the involvement of metal- or sulfur- containing proteins. On the other hand, sodium dodecyl sulfate (SDS) (0.1%) only reduces the activity by 25%, and treatment with pronase actually results in a slight enhancement of manganese oxidation (not shown, *cf.* Rosson & Nealson, 1982). The latter phenomenon may be explained by a breakdown of material shielding the oxidizing factor, but it is clear that the identity or nature of the active material needs further investigation. A first step towards the characterization of the oxidizing factor may be the isolation of purified spore coats, which contain 100% of the oxidizing activity (de Vrind *et al.*, 1986).

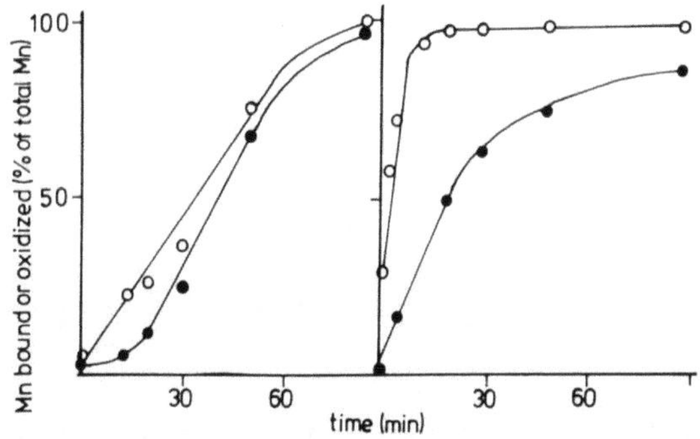

FIGURE 4. Manganese oxidation (closed symbols) and manganese binding (open symbols) by fresh spores (left figure) and spores which had previously oxidized 300 μM Mn.

MANGANESE REDUCTION BY VEGETATIVE CELLS

When vegetative cells are supplied with synthetic manganese oxide, the latter is rapidly reduced as determined with Leucoberbelin blue (de Vrind *et al.*, 1986). The activity is associated with the cells themselves, since spent culture medium (in which the cells previously have been grown) has no activity at all. The following data indicate that manganese reduction by the cells is a biologically catalyzed process. It is completely abolished by preheating of the cells at 90° C for 5 min or by the presence of 0.01% mercuric chloride, suggesting that cellular enzymes are involved (de Vrind *et al.*, 1986). Moreover, it has a pH optimum of 7.5, whereas non-biological manganese reduction is facilitated by lowering the pH below 7. Cells of which the cell wall has been degraded by lysozyme reduce manganese at the same rate as intact cells. Because manganese oxide is not likely to cross the plasma-membrane, the reducing factor is probably a component of the latter. This suggests that the electrons needed for manganese reduction are supplied through the electron transport chain of the bacteria. This assumption can be tested by investigating the effect of electron transport inhibitors on manganese reduction. In Figure 5 the effect of three common electron transport inhibitors on the electron transfer to oxygen in *Bacillus SGl* is shown. Both antimycin and potassium cyanide (KCN), at concentrations of 20 mg/ml and 10 nM respectively, can block oxygen reduction completely, whereas sodium azide (50 mM) has a partial inhibiting effect (*cf.* de Vrind *et al.*, 1986). As illustrated in Figure 6a, antimycin also inhibits manganese oxide reduction, but higher concentrations are needed than for the inhibition of O_2 consumption. Antimycin blocks electron transfer from

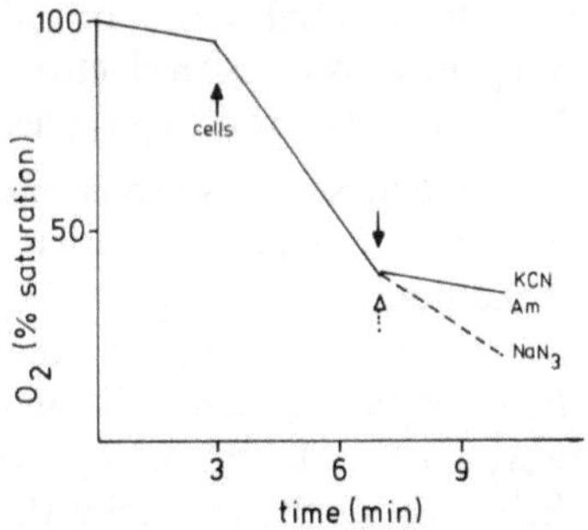

FIGURE 5. Oxygen consumption by vegetative cells of *Bacillus* SGl, as measured with an oxygen electrode.

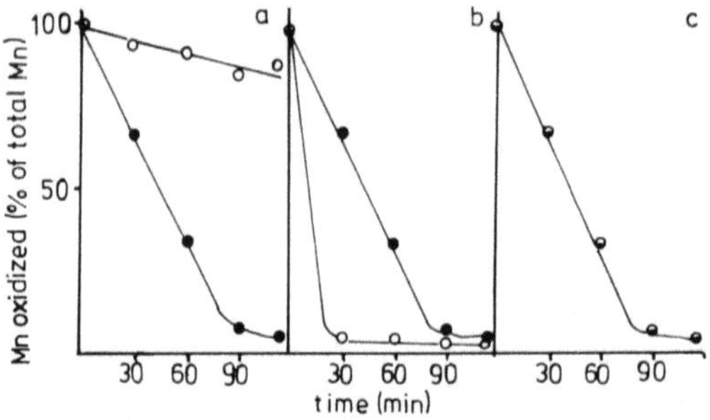

FIGURE 6. Effect of metabolic inhibitors on manganese reduction by vegetative cells of *Bacillus* SGI. ▪a. Antimycin, 100 μg/ml. ▪b. KCN, 10 mM. ▪c. NaN3, 50 mM. Closed symbols represent untreated cells. MnO2 concentrations were determined with LBB.

cytochrome *b* to *c*. Consequently, manganese oxide may receive electrons before this point in the transport chain. This is confirmed by the observation that both types of cytochromes are oxidized *in situ* when manganese oxide is added to a cell suspension (de Vrind *et al.*, 1986). Both potassium cyanide and sodium azide can interfere with the terminal transfer of electrons to oxygen. potassium cyanide stimulates manganese oxide reduction but, surprisingly, sodium azide seems to have no effect (Fig. 6b and c). This is explained by the high rate of oxygen consumption by the cells as illustrated in Figure 5. In a very short time relative to the time needed for manganese oxide reduction, the cells create low-oxygen conditions. When cells are continuously supplied with pure oxygen the reduction of manganese oxide is strongly decreased, but it can be partially induced by the addition of sodium azide (de Vrind *et al.*, 1986). The stimulation of manganese oxide reduction by potassium cyanide under apparently low-oxygen conditions may result from its manganese complexing ability. Care should be taken therefore, to use this metabolic inhibitor in studies where manganese transformations are involved (Rosson *et al.*, 1984). In conclusion, the experimental evidence indicates, that cells of *Bacillus SGl* reduce manganese oxide through components of the electron transport chain.

CONCLUSIONS

Bacillus SGl is one of the few species reported so far that has the ability to both oxidize and reduce manganese (cf. Bromfield & David, 1976). The type of transformation depends on the life stage of the organism. Its mature, dormant spores catalyze the oxidation of manganese to manganese oxide, whereas the vegetative cells promote the reduction of the oxide under low-oxygen conditions. Both processes are biologically mediated through spore- or cell-associated components. Because the cells may use the manganese oxide as an electron acceptor instead of oxygen, the species may benefit from the manganese oxidizing capacity of its spores.

Manganese transformations occur at a huge scale in nature. They are, at least at a local scale, influenced by microbial activity. The question arises, whether these natural phenomena may be exploited to the benefit of society. Manganese (and iron) oxidizing bacteria are probably involved in scaling; *i.e.*, the clogging of pipelines, as in water desalinization systems. The understanding of the oxidation mechanisms may aid in the prevention of this harmful process. Because all green plants depend on manganese for their photosynthetic activity, the factors which determine the availability of the element in soils is of great interest to agriculture. Microbially mediated manganese transformations probably play an important role in this respect. Microbes can possibly be used in the leaching of commercially interesting manganese ores (Ehrlich, 1980). The low energy requirements of biological processes may favor them eventually over chemical extraction procedures. Finally, manganese (and iron) oxidizing micro-organisms may be applied to recover rare trace elements from dilute waste fluids, or to remove them from polluted environments. Preliminary experiments in our laboratory have shown, that isolated spore coats of *Bacillus SGl* remove all the copper from their medium over a range from 100 to 1000 mM. When one takes into consideration that micro-organisms can be genetically manipulated to improve on certain properties, the study of microbial manganese transformations, and of the interactions of micro-organisms with metals in general, may prove to be a very promising field in the near future.

ACKNOWLEDGEMENTS

Part of this work has been subsidized by the Dutch Organization for the advancement of Pure Research (Z.W.O.). Participation in the Symposium was made possible by grants from The Organized Research Fund of The University of Texas at Arlington and The Sea Grant Program administered by Texas A & M University.

REFERENCES

ALI, S.H. & STOKES, J.L., 1972. Stimulation of heterotrophic and autotrophic growth of *Sphaerotilus discophorus* by manganous ions. Antonie van Leeuwenhoek, 37:519-528.

ARCHIBALD, F.S. & DUONG, M.N., 1984. Manganese acquisition by *Lactobacillus plantarum*. J. Bacter., 158:1-8.

ARCURI, E.J. & EHRLICH, H.L., 1979. Cytochrome involvement in Mn(II) oxidation by two marine bacteria. Appl. Envir. Microbio., 37:916-923.

ARCURI, E.J. & EHRLICH, H.L., 1980. Electron transfer coupled to Mn(II) oxidation in two deep-sea Pacific Ocean isolates. *In* Biogeochemistry of Ancient and Modern Environments (eds. P.A. Trudinger, M.R. Walter & B.J. Ralph). pp. 339-344. Springer-Verlag, Berlin.

BARBER, J., 1984. Has the mangano-protein of the water splitting reaction of photosynthesis been isolated? Trends Biochem. Sci., 9:99-80.

BROMFIELD, S.M. & DAVID, D.J., 1976. Sorption and oxidation of manganous ions and reduction of manganese oxide by cell suspensions of a manganese oxidizing bacterium. Soil Biol. Biochem., 8:37-43.

BURDIDGE, D.J., KEPKAY, P.E. & NEALSON, K.H., 1983. An *in situ* method for determining microbial manganese oxidation rates in sediments. *In* Biomineralization and Biological Metal Accumulation (eds. P. WESTBROEK & E.W. DE JONG). pp. 481-487. Reidel Publishing Company, Dordrecht.

BURDIDGE, D.J. & NEALSON, K.H., 1985. Microbial manganese reduction by enrichment cultures from coastal marine sediments. Appl. Envir. Microbio., 50:491-497.

BURDIDGE, D.J. & NEALSON, K.H., 1986. Chemical and microbiological studies of sulfide-mediated manganese reduction. Geomicrobio. J., 4:361-387.

CHARNEY, J., FISCHER, W.P. & HEGARTY, C.P., 1951. Manganese as an essential element for sporulation in the genus *Bacillus*. J. Bacter., 62:145-148.

DUBININA, G.A., 1978. Mechanism of the oxidation of divalent iron and manganese by iron bacteria growing at neutral pH of medium. Microbiology, 47:471-478.

EHRLICH, H.L., 1963. Bacteriology of manganese nodules I. Bacterial action on manganese in nodule enrichments. Appl. Microbiol., 11:15-19.

EHRLICH, H.L., 1968. Bacteriology of manganese nodules II. Manganese oxidation by a cell-free extract from a manganese nodule bacterium. Appl. Microbiol., 16:197-202.

EHRLICH, H.L., 1975. The formation of ores in the sedimentary environment of the deep sea with microbial participation. The case for ferromanganese concretions. Soil Sci., 119:65-83.

EHRLICH, H.L., 1980. Bacterial leaching of manganese ores. *In* Biogeochemistry of Ancient and Modern Environments (eds. P.A. Trudinger, M.R. Walter & B.J. Ralph). pp. 609-614. Springer Verlag, Berlin.

EHRLICH, H.L., 1984. Different forms of bacterial manganese oxidation. *In* Microbial Chemoautotrophy (eds. W.R. Strohl & O.H. Tuovinen). Ohio State Univ. Press, Ohio.

EISENSTADT, E., FISCHER, S., DER, C.L. & SILVER, S., 1973. Manganese transport in *Bacillus subtilis* W23 during growth and sporulation. J. Bacter., 113:1363-1372.

EMERSON, S., KALHORN, S., JACOBS, L., TEBO, B.M., NEALSON, K.H. & ROSSON, R.A., 1982. Environmental oxidation rate of manganese(II): bacterial catalysis. Geochim. Cosmochim. Acta, 46:1073-1079.

GHIORSE, W.C., 1984. Biology of iron and manganese bacteria. Ann. Rev. Microbiol., 38:515-550.

GHIORSE, W.C. & EHRLICH, H.L., 1976. Electron transport components of the manganese oxide reductase system and the location of the terminal reductase in a marine *Bacillus*. Appl. Envir. Microbio., 31:977-985.

GHIORSE, W.C. & HIRSCH, P., 1979. An ultrastructural study of iron and manganese deposition associated with extracellular polymers of *Pedomicrobium*-like budding bacteria. Arch. Microbiol., 123:213-226.

JANNASCH, H.W. & TAYLOR, C.D., 1984. Deep sea microbiology. Ann. Rev. Microbiol., 38:487-514.

JOHNSON, A.H. & STOKES, J.L., 1966. Manganese oxidation by *Sphaerotilus discophorus*. J. Bacter., 911:543-1547.

JUNG, W.K. & SCHWEISFURTH, R., 1979. Manganese oxidation by an intracellular protein of a *Pseudomonas* species. Zeits. Algem. Mikrobiol., 19:107-115.

KEEN, C.L., LONNERDAL, B. & HURLEY, L.S., 1984. Manganese. *In* Biochemistry of the essential Ultratrace Elements (ed. E. Frieden). pp. 89-132. Plenum Press, New York.

KEPKAY, P.E., BURDIDGE, D.J. & NEALSON, K.H., 1983. Kinetics of bacterial manganese binding and oxidation in the chemostat. Geomicrobiol. J., 3:245-262.

KRUMBEIN, W.E. & ALTMAN, H.J., 1973. A new method for the detection and enumeration of manganese oxidizing and reducing microorganisms. Helgolandische Wissenschaftliche Meeresuntersuchungen, 25:347-356.

LIDSTROM, M.E., ENGEBRECHT, J.A. & NEALSON, K.H., 1983. Evidence for plasmid-encoded manganese oxidation in a marine pseudomonad. FEMS Microbiol. Lett., 19:1-6.

MARSHALL, K.C., 1979. Biogeochemistry of manganese minerals. *In* Biogeochemical Cycling of Mineral-forming Elements (ed. P.A. Trudinger). pp. 253-292. Elsevier Biomedical Press, Amsterdam.

MILLS, V.H. & RANDLES, C.I., 1979. Manganese oxidation in *Sphaerotilus discophorus* particles. J. Gen. Appl. Microbiol., 25:205-207.

MORGAN, J.J. & STUMM, W., 1964. Colloid-chemical properties of manganese dioxide. J. Coll. Sci., 19:347-359.

NEALSON, K.H., 1983. The microbial manganese cycle. *In* Microbial Geochemistry (ed. W.E. Krumbein). pp. 191-221. Blackwell Scientific Publications, Oxford.

NEALSON, K.H. & FORD, J., 1980. Surface enhancement of bacterial manganese oxidation: implications for aquatic environments. Geomicrobiol. J., 2:21-37.

PALMER, F.E., STALEY, J.T., MURRAY, R.G.E., COUNSELL, T. & ADAMS, J.B., 1986. Identification of manganese-oxidizing bacteria from desert varnish. Geomicrobiol. J., 4:343-360.

ROSSON, R.A. & NEALSON, K.H., 1982. Manganese binding and oxidation by spores of a marine *Bacillus*. J. Bacter., 151:1027-1034.

ROSSON, R.A., TEBO, B.M. & NEALSON, K.H., 1984. Use of poisons in the determination of microbial manganese binding rates in seawater. Appl. Envir. Microbio., 47:740-745.

SCHUETT, C., ZELBOR, J.L. & COLWELL, R.R., 1986. Role of bacterial plasmids in manganese oxidation: evidence for plasmid-encoded heavy metal resistance. Geomicrobiol. J., 4:389-406.

VRIND, J.P.M. DE, BOOGERD, F.C. & VRIND-DE JONG, E.W. DE, 1986. Manganese reduction by a marine *Bacillus* species. J. Bacter., 167:30-34.

VRIND, J.P.M. DE, VRIND-DE JONG, E.W. DE, VOOGT, J.W. DE, WESTBROEK, P., BOOGERD, F.C. AN ROSSON, R.A., 1986. Manganese oxidation by spores and spore coals of a marine *Bacillus* species. Appl. Envir. Microbio., 52:1096-1100.

THE EFFECT OF MAGNETOTACTIC BACTERIA ON THE MAGNETIC PROPERTIES OF MARINE SEDIMENTS

John F. Stolz[1,2], Shih-Bin R. Chang[3] and Joseph L. Kirschvink[3]

[1]*Jet Propulsion Laboratory, 125-112, California Institute of Technology, 4800 Oak Grove Dr., Pasadena, CA 91109;* [2]*Present address: Department of Biochemistry, University of Massachusetts, Amherst, MA 01003;* [3]*Division of Geological and Planetary Sciences, 170-25, Institute of Technology, Pasadena, CA 91125*

ABSTRACT. Magnetotactic bacteria have been studied in three distinct sedimentary marine environments: a hypersaline lagoon, an intertidal CO3 marsh, and an open ocean basin. The bacteria and the ultra-fine grained, single domain magnetite (Fe_3O_4) they produce were extracted from the sediments and studied with transmission electron microscopy. Magnetic properties of the sediments were measured by rock magnetic techniques using a SQUID magnetometer. Our results show that magnetotactic bacteria contribute a significant fraction of the natural remanent magnetization to their sedimentary environment and in some cases may be the sole source of the stable remanence carrying mineral. The occurrence and abundance of these bacteria in a diversity of marine environments implies that they may also play a role in the microbial iron cycle.

INTRODUCTION

Since their discovery by Blakemore over a decade ago, magnetotactic bacteria have been found in many marine and fresh water environments (Blakemore, 1975, 1982; Blakemore *et al.*, 1980; Moench & Konetzka, 1978; Lins de Barros & Esquivel, 1985). These organisms not only synthesize and deposit crystals of magnetite intracellularly, but use it to orient themselves in the earth's magnetic field (Frankel, 1984). In the northern hemisphere they orient to the north, in the southern hemisphere to the south, and at the geomagnetic equator there are equal populations of both (Kirschvink, 1980; Blakemore *et al.*, 1980). These organisms are able to orient in a magnetic field because of the composition, size and shape of the mineral deposit (Kirschvink, 1983; Frankel, 1984). Analyses of the magnetite from different magnetotactic bacteria including *Aquaspirillum magnetotacticum* by several different methods (electron diffraction, Mossbauer spectroscopy and lattice imaging) has shown it to be composed of almost pure Fe_3O_4 (Frankel *et al.*, 1979; Matsuda, *et al.*, 1983; Mann *et al.*, 1984a,b). To date, all reported sizes and shapes of bacterial Fe_3O_4 fall within the single domain stability field as calculated by Butler and Banerjee (1975) (Chang & Kirschvink, 1985; Chang *et al.*, 1987). That is, each crystal behaves as a single magnet with a dipole moment. In most

Origin, Evolution, and Modern Aspects of Biomineralization in Plants and Animals
Edited by Rex E. Crick
Plenum Press, New York

natural cases, the bacteria align the crystals in a chain so that the chain behaves like a single dipole magnet as well.

In 1979, Kirschvink and Lowenstam (1980) proposed that biogenic Fe_3O_4 (produced by chitons and bacteria) could contribute significantly to the magnetic properties of marine sediments. Their conclusions, based on calculations made from the amount of chiton Fe_3O_4 deposited annually, suggested biologically produced single-domain Fe_3O_4 could provide as much as 10^{-6} emu/g of detrital remanent magnetization (DRM). They also made crude calculations as to the possible contribution of bacterial Fe_3O_4, but little was known then about the magnetic properties and occurence of magnetotactic bacteria in marine environments. Towe and Moench (1981) subsequently increased these initial estimates by a factor of 10, based on more accurate population densities of magnetic bacteria from a few fresh water environments. In this study, we wanted to determine whether or not magnetic bacteria were equally abundant and diverse in geologically important marine environments and whether or not the single-domained Fe_3O_4 they produce has an effect on the magnetic properties of the sediment. We studied three distinct marine sedimentary environments: a hypersaline lagoon, an intertidal CO_3 marsh, and an open ocean basin.

ENVIRONMENTS

The hypersaline lagoon

Laguna Figueroa is a lagoonal complex situated 250 km south of the Mexico-United States border on the Pacific Ocean side of the Baja California peninsula. The hypersaline lagoon is separated from the Pacific Ocean by a barrier beach and dune. A salt marsh lies on the landward side of the dune and grades into an evaporite flat. At the salt marsh/evaporite flat interface, laminated sediments are being deposited by a stratified microbial mat community dominated by the cyanobacterium *Microcoleus chthonoplastes* (Horodyski *et al.*, 1977). These mats and the laminated sediments have been studied extensively because of their analogy with fossilized microbial communities and stromatolites (Horodyski *et al.*, 1977; Margulis *et al.*, 1980, 1983; Stolz & Margulis, 1984; Stolz, 1983, 1984a,b, 1985). The sediment is composed of dune derived siliciclastics and evaporitic minerals (halite, gypsum, anhydrite, and aragonite) and is rich in organic material.

The intertidal CO_3 marsh

Sugarloaf Key is the third to last in the string of Florida Keys. Mangroves are the predominant vegetation in the intertidal marsh. The sediment is a fine CO_3 ooze with small shell fragments and is a modern analog of ancient limestone deposits (Ginsberg, 1964). The surface 2 mm is aerobic and underlain by anoxic, sulfous CO_3 ooze. Thumb-sized stromatolitic nodules, formed by a microbial community dominated by the cyanobacterium *Schizothrix gracilis*, dot the surface (Golubic, & Focke, 1978; Stolz, unpublished).

The open ocean basin

The Santa Barbara Basin is an open ocean basin in the southern California borderland. It has a maximum depth of about 600 m and the sediments are anoxic laminated hemipelagic muds (Savrda *et al.*, 1984; Karlin & Levi, 1985). The surface sediment is populated by a microbial community dominated by species of *Beggiotoa*, flexibacteria and heterotrophic bacteria and is typical of oxygen poor environments.

MATERIALS AND METHODS

At both Laguna Figueroa and Sugarloaf Key, we collected sediment samples from several sites at the sediment-water interface. Samples from the Santa Barbara Basin were collected and supplied to us by M. Kastner (Scripps Oceanographic Institution), from three different cores (34°14'N, 120°1'W) at a depth of 598 meters and a temperature of 8°C. The top 3 cm of sediment were removed from the box cores and either fixed with 2.5% glutaraldehyde in seawater buffer immediately after collection, or left untreated. All these samples were kept at 8°C.

The presence of live magnetotactic bacteria was detected by placing a drop of sediment on a glass slide adjacent to the south end of a stirring bar magnet. Under low power light microscopy (× 160) magnetic bacteria could be seen collecting at the edge of the drop next to the magnet (*e.g.*, swimming towards magnetic north). For transmission electron microscopy (TEM), carbon coated grids were placed on top of sediment drops and a cobalt rare-earth magnet was suspended 1 cm above the drops for 20-30 minutes (Stolz *et al.*, 1986). To determine the morphology of the bacteria, some grids were negatively stained with 1% uranyl acetate for 1 minute. Pellets of the bacteria for thin section preparation were obtained using the method described in Moench & Konetzka (1978). The pellets were fixed with 2.5% glutaraldephyde in seawater buffer, post-fixed in osmium tetroxide, *en bloc* stained with uranyl acetate followed by dehydration in an ethanol series and propylene oxide and embedded in Spurr's embedding medium (Stolz, 1983). Thin sections were stained with uranyl acetate and lead citrate. All samples were examined on a Phillips 201 transmission electron microscope at 80 kV. Identification of the crystals as Fe_3O_4 was made by electron diffraction (Towe & Moench, 1981; Chang & Kirschvink, 1984; Towe, 1985).

Rock magnetic studies of the sediment and bacteria were done using a SQUID magnetometer (Fuller *et al.*, 1985). Saturation isothermal remanent magnetization (sIRM) and alternating field (AF) demagnetization were used to determine the coercivity spectra (Kirschvink, 1983) which were then compared with known standards (Chang *et al.*, 1987).

RESULTS

Large populations of magnetotactic bacteria were discovered in salt marsh pools along the dune at Laguna Figueroa. The greatest concentrations of bacteria are found in places where the top few millimeters of sediment is oxidized and a film of manganese and iron oxide is visible on the water surface. Many different morphologies are present including coccoids, rods and a colonial coccoid. Once particular coccoid was obtained in large numbers in enrichment cultures in the lab and a study of its ultrastructural morphology and magnetic properties was done. The cells are 2 to 4 μm in diameter and have a gram negative wall (Fig. 1A,B). They are motile by means of two flagellar bundles which insert at invaginations of the cell wall which are bounded by polar membrane (Fig.

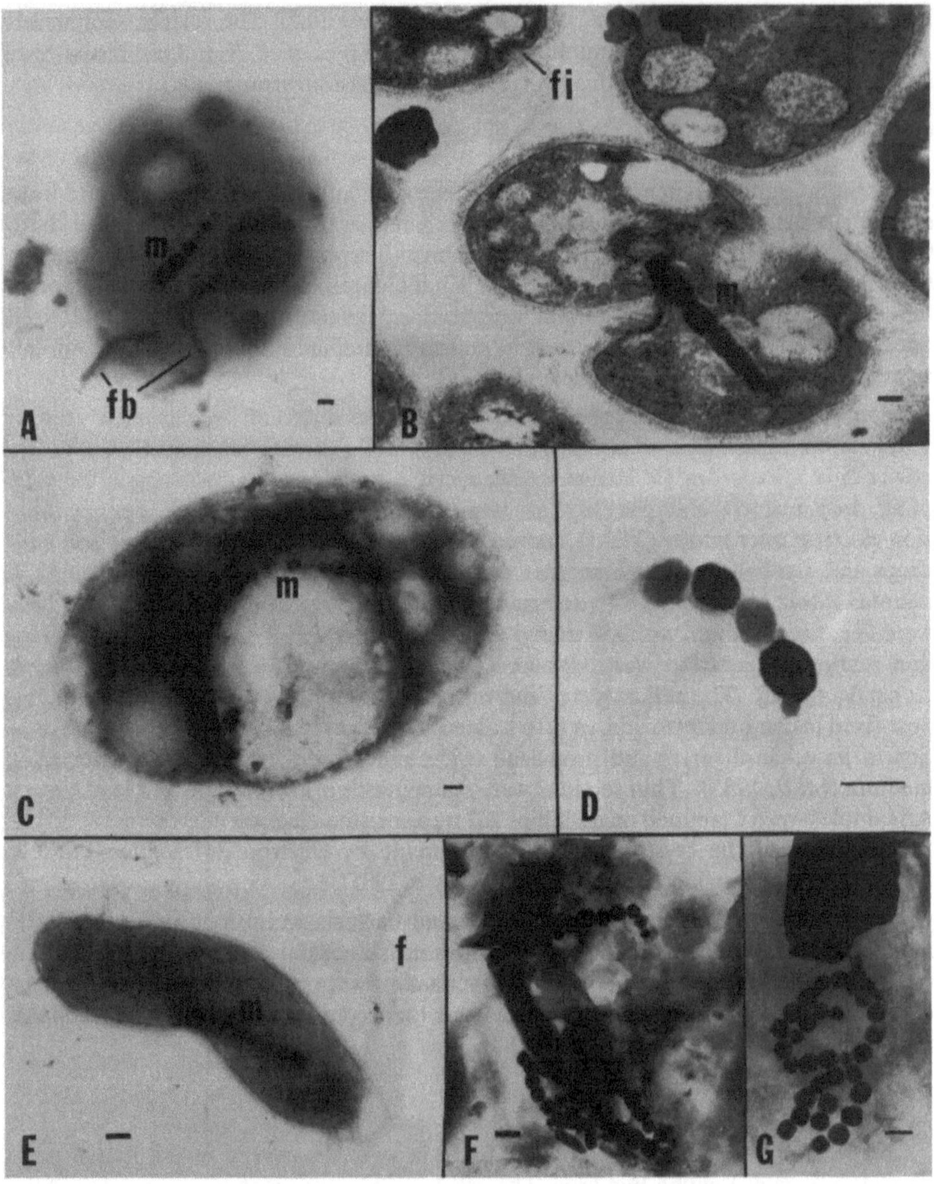

FIGURE 1. Magnetotactic bacteria from marine sediments, transmission electron micrographs, all bars 100 nm. A. *Magnetococcus* from Laguna Figueroa, Baja California, Mexico. Negative stain preparation. (fb), flagellar bundle, (m), magnetosome. B. *Magnetococcus* from Laguna Figueroa. Thin section preparation. Note the invagination of the cell wall at the flagellar insertion (fi). C. *Magnetococcus* from Sugarloaf Key, Florida. Negative stain preparation. D. Extracted magnetosome from Sugarloaf Key sediments (unstained). E. *Magnetovibrio* from Santa Barbara Basin, California. Note one polar flagella (f). Negative stain preparation. F. and G. Extracted bacterial Fe3O4 from Santa Barbara Basin sediments. Glutaraldehyde fixed and unstained.

1A,B). The cytoplasm contains several globular inclusions and one to two chains of Fe_3O_4. The crystals are cuboidal to slightly rectangular in shape with an average size of 90 × 110 nm (Figure 1B). The chains contain an average of 10 crystals. The rod (1 × 3 μm) and colonial coccoid (8-10 μm) were seen with light microscopy and documented on video tape, but analyses of the Fe_3O_4 was not possible.

Coercivity spectral analysis of the coccoid indicates the Fe_3O_4 to be single domain (Fig. 2A). The coercivity of the salt marsh laminated sediment suggests a large percentage of the magnetic component is multi-domain Fe_3O_4 (Fig. 2B). Examination of the sediments revealed coarse grained Fe_3O_4 as well as bacterial ultra-fine grained Fe_3O_4. The source of the coarse grained Fe_3O_4 is detrital Fe_3O_4 in the dune derived siliciclastic sediment (Fig. 2C). The maximum values for sIRM ranged between 1.5×10^{-5} and 1.9×10^{-3} with an average value for the laminated sediment of 5×10^{-4} emu/g.

Several different morphologies of magnetotactic bacteria were seen in sediments from Sugarloaf Key including rods, coccoids and a colonial coccoid. The Fe_3O_4 crystals are cuboidal to truncated rectangles and ranged in size from 40 – 100 nm (Fig. 1C,D). Coercivity spectra of the stromatolitic nodules and CO_3 ooze indicate the major magnetic component to be single domain Fe_3O_4 (Fig. 2D,E). Examination of the sediment revealed only bacterial Fe_3O_4. Typical values for sIRM of both CO_3 ooze and nodules average about 4.5×10^{-5} emu/g. It appears that in this environment bacterial Fe_3O_4 provide the only source of magnetic remanence carrier (Chang *et al.*, 1987).

One of the most exciting discoveries of this study was finding magnetotactic bacteria living in the surface sediments of the Santa Barbara Basin (Stolz *et al.*, 1986). Vibrio and rod shaped bacteria were seen in unfixed samples (Fig. 1E). The crystals are cuboidal to rectangular and range in size from 40 to 60 nm. Chains of Fe_3O_4 with an average number of 10 crystals were seen in several bacteria. Coercivity spectral analysis of both fixed and unfixed samples indicate the most abundant primary remanence carrier to be single domain Fe_3O_4 (Fig. 2F). Examination of the sediment confirmed this (Fig. 1F,G). Values for sIRM range between 2.5×10^{-4} and 1.25×10^{-3} emu/g.

DISCUSSION

Laguna Figueroa presented the greatest difficulty in assessing the bacteria contribution to the magnetic properties of the sediment because of the large amount of coarse-grained ditrital Fe_3O_4. The dune sands contained over 5% abiogenic Fe_3O_4. Comparing the coercivity spectra of coarse-grained Fe_3O_4 standard, dune sand, and bacterial Fe_3O_4 to those of the salt marsh pools and laminated sediment, however, reveals that the presence of single domain, biogenic Fe_3O_4 does have an effect on the magnetic properties of the sediment (Chang *et al.*, 1987). Bacterial Fe_3O_4 are more difficult to demagnetize, but do so rapidly in an alternating field (AF) between 30 and 100 milliTesla (Fig. 2A). Coarse-grained, multi-domain Fe_3O_4 begin to demagnetize in even low strength alternating fields (AF less than 3 mT, Fig. 2C). In a mixture of ultra-fine and coarse grained Fe_3O_4, an increase in single-domain Fe_3O_4 makes the sample harder to demagnetize. This effect became more apparent in the Sugarloaf Key and Santa Barbara samples.

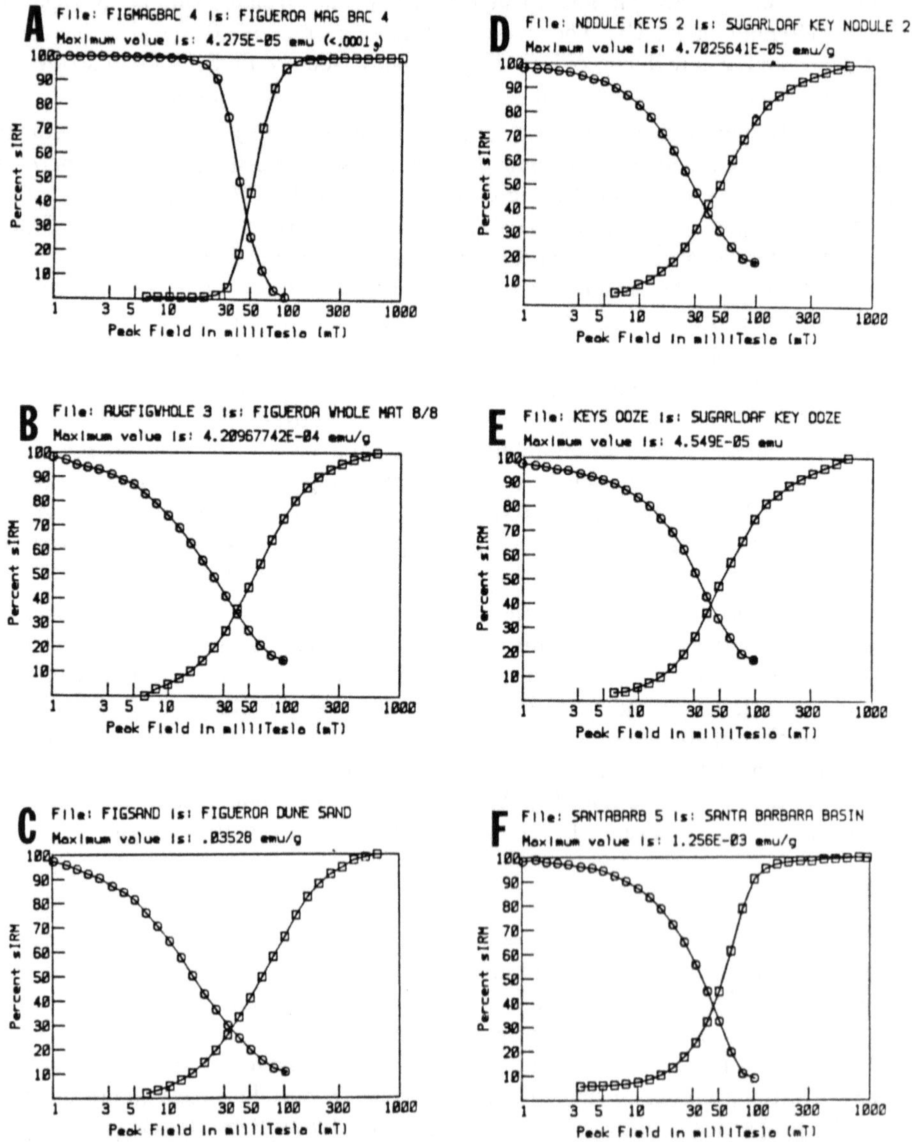

FIGURE 2. Coercivity spectra measured by a SQUID magnetometer using saturation Isothermal Remanent Magnetization (sIRM)(squares) and Alternating Field demagnetization (AF)(circles). A. Pellet of magnetococcus from Laguna Figueroa. B. Microbial mat from Laguna Figueroa. C. Dune sand from Laguna Figueroa. D. Stromatolitic nodule from Sugarloaf Key. E. CO3 ooze sediment from Sugarloaf Key. F. Surface sediments from Santa Barbara Basin.

In a recent paper, Lins de Barros and Esquivel (1985) estimated the magnetic moments for several different types of magnetic microorganisms. Depending on size of the organism and subsequently the amount of Fe_3O_4 per cell, the values for smallest to largest ranged from 0.3×10^{-12} to 5.4×10^{-11} emu/cell. For the 2 μm coccoid from Laguna Figueroa (a chain of 10 crystals, each 100 nm in size) we have calculated a moment of about 1×10^{-12} emu/cell (Chang *et al.*, 1987). We could not, however, estimate the bacterial contribution to the magnetic remanence in the Laguna Figueroa sediments because of the abundance of coarse-grained Fe_3O_4.

Limestones, despite their typically low magnetization, have been frequently used in paleomagnetic studies because their fossil assemblages provide an independent stratigraphic correlation (Lowrie & Heller, 1982). How the sediments were magnetized and what the primary remanence carrier is are two questions which are poorly understood. The discovery of magnetotactic bacteria in CO_3 oozes and bacterial Fe_3O_4 in limestone deposits suggest that biogenic Fe_3O_4 could be the primary remanence carriers (Chang *et al.*, 1987). The average shape of the bacterial Fe_3O_4 seen in both bacteria and sediment extracts was rectangular, and about 80×100 nm in size (Fig. 1C,D). Like the 2 μm coccoid from Laguna Figueroa, the calculated average moment for the Sugarloaf Key bacteria was 10^{-12} emu. It would take a population of only 10^7 bacteria per gram of sediment to account for the observed magnetic remanence of 10^{-5} emu/g. This estimate makes the assumption that all the magnetic remanence is due to bacterial Fe_3O_4 whether the bacteria are alive or dead. Still it is conceivable that populations of this size do occur in nature (Moench & Konetzka, 1978; Blakemore, 1982) and that even episodic increases in population (i.e., blooms) could contribute the amount of biogenic Fe_3O_4 observed.

The Santa Barbara Basin sediments provide another rather dramatic example of the effect of bacterial Fe_3O_4 on the sedimentary environment. The hemipelagic sediments of the California borderland have been studied extensively by paleomagnetists in order to understand the mechanisms of magnetization of rapidly deposited marine sediments (Karlin & Levi, 1983, 1985; Lund *et al.*, 1983; Leslie, *et al.*, 1984). These studies show a maximum intensity of magnetic remanence occurring in the surface sediments (within the top meter). The significant decrease in intensity with depth has been correlated with a loss of fine-grained component. The discovery of magnetotactic bacteria living in the Santa Barbara Basin and the identification of the fine-grained component as biogenic Fe_3O_4 provide an explaination for the above observations. The high natural remanence of the bacteria contribute to the remanence of the surface sediments where conditions for growth are optimum for the bacteria. At depth both the lack of oxygen (needed for Fe_3O_4 synthesis, Blakemore *et al.*, 1984) and the increase in sulfide result in the absence of live bacteria and the dissolution of bacterial Fe_3O_4 (Stolz *et al.*, 1986).

The loss of bacterial Fe_3O_4 with depth in some open ocean basins places doubt as to whether it can contribute to the magnetization of ocean sediments. Well preserved fossil bacterial Fe_3O_4s have, however, recently been discovered in deep sea sediment cores ranging in age from Recent to Oligocene (Kirschvink & Chang, 1984; Petersen *et al.*, 1986).

The occurrence and abundance of magnetotactic bacteria in a variety of marine sedimentary environments suggest that they may play a role in the microbial iron cycle.

Nealson (1983) acknowledged their possible contribution, but little was known of their distribution and numbers in marine sediments. Fe_3O_4 is an iron oxide containing both ferrous (Fe^{2+}) and ferric (Fe^{3+}) iron. Blakemore *et al.* (1984) have shown that *Aquaspirillum magnetotacticum* precipitates Fe_3O_4 optimally under microaerobic conditions (1% PAL). In nature, many biogeochemical activities occur at the aerobic/anerobic interface (Nealson, 1983). Fe_3O_4 may be oxidized to maghemite and hematite (deep sea sediments), dissolved (Santa Barbara Basin), or transformed to an iron sulfide like pyrite or greigite (Eel Marsh, Massachusetts; Demitrack, 1985). It is conceivable then that it could be used as a substrate for some of the iron metabolizing bacteria.

CONCLUSIONS

We have documented the presence of magnetotactic bacteria living and depositing single-domain, ultra-fine-grained Fe_3O_4 in three distinct marine environments. The results of rock magnetic studies suggest that the bacteria have a variable influence on the magnetic properties of the sediments in which they live. This effect may be subtle, as in the case of Laguna Figueroa where coarse-grained Fe_3O_4 is a significant component, or dramatic, like Sugarloaf Key where bacterial Fe_3O_4 are the only remanence carriers. The discovery of magnetotactic bacteria in an open ocean basin (Santa Barbara Basin) and fossil bacterial Fe_3O_4 in deep sea sediments suggest bacterial Fe_3O_4 is a common occurence in marine deposits. Although the eventual fate of the biogenic Fe_3O_4 depends on post depositional factors (e.g., oxidation, dissolution) they can be a primary remanence carrier in certain marine sedimentary rocks (e.g., limestones). The occurence and abundance of magnetotactic bacteria in marine sediments suggest they may also play a significant part in the microbial iron cycle.

ACKNOWLEDGMENT

This is Contribution No. 4386 from the Division of Geological and Planetary Sciences, California Institute of Technology.Participation in the Symposium was made possible by grants from The Organized Research Fund of The University of Texas at Arlington and The Sea Grant Program administered by Texas A & M University.

REFERENCES

BLAKEMORE, R.P., 1975. Magnetotactic bacteria. Science 190, 377- 370.

BLAKEMORE, R.P., 1982. Magnetotactic bacteria. Annual Reviews of Microbiology 36, 217-238.

BLAKEMORE, R.P., FRANKEL, R.B. & KALMIJN, AD.J., 1980. South seeking bacteria in the southern hemisphere. Nature 286, 384-385.

BLAKEMORE, R.P., SHORT, K.A., BASZYLINSKI, D.A., ROSENBLATT, C., & FRANKEL, R.B., 1985. Microaerobic conditions are required for magnetite formation within *Aquaspirillum magnetotacticum*. Geomicrobiology Journal 4, 53-72.

BUTLER, R.F. & BANERJEE, S.K., 1975. Theoretical single-domain size range in magnetite and titanomagnetite. Journal of Geophysical Research 80, 4049-4058.

CHANG, S.R. & KIRSCHVINK, J.L., 1985. Possible biogenic magnetite fossils from the Miocene marine clay of Crete. In Magnetite Biomineralization and Magnetoreception in Organisms, (eds. J.L. Kirschvink, D.S. Jones & B.J. McFadden) pp 647-669 New York: Plenum Press.

CHANG, S.R., STOLZ, J.F., & KIRSCHVINK, J.L., 1987. Biogenic magnetite as primary remanence carrier in limestone deposits. Physics of Earth and Planetary Interiors, 46, 289-303.

DEMITRACK, A., 1985. A search for bacterial magnetite in the sediment of Eel Marsh, Woods Hole, Massachusetts. *In* Magnetite Biomineralization and Magnetoreception in Organisms, (eds. J.L. Kirschvink, D.S. Jones & B.J. McFadden) pp 625-645 New York: Plenum Press.

FRANKEL, R.B. 1984. Magnetic guidance of organisms. Annual Reviews of Biophysics and Bioengineering 13, 85-103.

FRANKEL, R.B. 1979. Blakemore, R.P. & Wolfe, R.S., 1979. Magnetite in fresh-water bacteria Science 203, 1355-1356.

FULLER, M., GOREE, W.S. & GOODMAN, W.L., 1985. An introduction to the use of SQUID magnetometers in biomagnetism. *In* Magnetite Biomineralization and Magnetoreception in Organisms, (eds. J.L. Kirschvink, D.S. Jones & B.J. McFadden) pp 104-151 New York: Plenum Press.

GINSBERG, R.N., 1964. South Florida carbonate sediments. *In* Guidebook for Field Trip 1, pp. 72, New York:Geological Society of America.

GOLUBIC, S & FOCKE, J.W., 1978. *Phormidium hendersonii* Howe: identity and significance of a modern stromatolite building microorganism. Journal of Sedimentary Petrology 48,751-764.

HORODYSKI, R.J., BLOESSER, B., & VONDER HAAR, S.J., 1977. Laminated algal mats from a coastal lagoon, Laguna Mormona, Baja California, Mexico. Journal of Sedimentary Petrology 2, 237-273.

KARLIN, R. & LEVI, S., 1983. Diagenesis of magnetic minerals in recent hemipalegic sediments. Nature 303, 327-330.

KARLIN, R. & LEVI, S., 1985. Geochemical and sedimentological control of the magnetic properties of hemipelagic sediments. Journal of Geophysical Research 90, 10373-10392.

KIRSCHVINK, J.L., 1983. Biogenic ferrimagnetism: a new biomagnetism. *In* Biomagnetism: an Interdisciplanary Approach (ed. S. Williamson) pp. 472-492, New York: Plenum Press.

KIRSCHVINK, J.L. & LOWENSTAM, H.A., 1979. Mineralization and magnetization of chiton teeth: paleomagnetic, sedimentologic and biologic implications of organic magnetite. Earth and Planetary Science Letters 44, 193-204.

KIRSCHVINK, J.L. & CHANG, S.-B.R, 1984. Ultrafine-grained magnetite in deep-sea sediments: possible bacterial magnetofossils. Geology 12,559-562.

LESLIE, B., HAMMOND, D.E., LUND, S. & HAMILTON, M., 1984. Diagenesis of iron and sulfur in sediments from the California borderland, EOS 65,957.

LINS DE BARROS & ESQUIVEL, H.G.P., 1985. Magnetic microorganisms found in muds from Rio de Janeiro. *In* Magnetite Biomineralization and Magnetoreception in Organisms, (eds. J.L. Kirschvink, D.S. Jones & B.J. McFadden) pp 289-309 New York: Plenum Press.

LOWRIE, W. AND HELLER, F., 1982. Magnetic properties of marine limestones. Reviews of Geophysics and Space Physics 20, 171- 192.

LUND, S.P., HENYEY, T.L. AND & HAMMOND, D., 1983. Rock magnetic parameters as indicators of Holocene digenetic and lithostratographic changes in the marine California continental borderland, EOS 64,682.

MANN, S., MOENCH, T.T., & WILLIAMS, R.J.P., 1984a. A high resolution electron microscopic investigation of bacterial magnetite. Implications for crystal growth. Proceedings of the Royal Society of London B 221, 385-393.

MANN, S., FRANKEL, R.B. & BLAKEMORE, R.B., 1984b. Structure, morphology and crystal growth of bacterial magnetite. Nature 310, 405-407.

MARGULIS, L., BARGHOORN, E.S., ASHENDORF, D., BANERJEe, S., Chase, D., Francis, S., Giovannoni, S. & Stolz, J.F., 1980. The microbial community in the layered sediments at Laguna Figueroa, Baja California, Mexico: does it have Precambrian analogs? Precambrian Research 11, 93-123.

MARGULIS, L., GROSOVSKI, B.D.D., STOLZ, J.F., GONG-Collins, E.J., LENK, S., READ, D., & LOPEZ-CORTEZ, A., 1983. Distinctive microbial structures and the pre-Phanerozoic fossil record. Precambrian Research 20, 443-447.

MATSUDA, T., ENDO, J., OSAKABE, N. & TONOMURA, A., 1983. Morphology and structure of biogenic magnetite particles. Nature 302,411-412.

MOENCH, T.T. & KONETZKA, W.A., 1978. A novel method for the isolation and study of a magnetotactic bacterium. Archives of Microbiology 119, 203-212.

NEALSON, K.A.,1983. The microbial iron cycle. *In* Microbial Geochemistry (ed. W.E. Krumbein) pp. 191-221 Boston: Blackwell Scientific.

SAVRDA, C.E., BOTTJER, D.J. & GORSLINE, D.S., 1984. Development of a comprehensive oxygen-deficient marine biofacies model: evidence from Santa Monica, San Pedro, and Santa Barbara Basins, California Continental Borderland. The American Association of Petroleum Geologists Bulletin 68, 1179-1192.

STOLZ, J.F., 1983. The fine structure of the stratified microbial community at Laguna Figueroa, Baja California, Mexico: Methods of in situ study of the laminated sediments. Precambrian Research 20, 472-492.

STOLZ, J.F., 1984a. The fine structure of the stratified microbial community at Laguna Figueroa, Baja California, Mexico: transmission electron microscopy as a diagnostic tool in studying microbial communities *in situ*. *In* Microbial Mats: Stromatolites (eds. Y. Cohen, R.W. Castenholz, & H.O. Halvorson) pp. 23-38, New York: Alan R. Liss and Co.

STOLZ, J.F., 1984b. The effects of catastrophic inundation (1979-1983) on the composition and ultrastructure of a stratified microbial community, Laguna Figueroa, Baja California, Mexico, Ph.D. thesis, Boston University, 144 pages.

STOLZ, J.F., 1985. The microbial community at Laguna Figueroa, Baja California, Mexico: from miles to microns. Origins of Life 15, 347-352.

STOLZ, J.F., & MARGULIS, L., 1984. The stratified microbial community at Laguna Figueroa, Baja California, Mexico: a possible model for prePhanerozoic laminated microbial communities preserved in cherts. Origins of Life 14, 671-679.

STOLZ, J.F., CHANG, S-B.R., AND KIRSCHUINK, J.L., 1986. Magnetotactic bacteria and single domain magnetite in hemipelagic sediments. Nature, 321, 849-851.

TOWE, K.M., 1985. Studying mineral particulates of biogenic origin by transmission electron microscopy and electron diffraction: some guidelines and suggestions. *In* Magnetite Biomineralization and Magnetoreception in Organisms, (eds. J.L. Kirschvink, D.S. Jones & B.J. McFadden) pp 167-181 New York: Plenum Press.

TOWE, K.M. & MOENCH, T.T., 1981. Electron-optical characterization of bacterial magnetite. Earth and Planetary Science Letters 52, 213-220.

SILICA PRECIPITATION INDUCED BY THE ANAEROBIC SULFATE REDUCING BACTERIUM *DESULFOVIBRIO DESULFURICANS*: EFFECTS UPON CELL MORPHOLOGY AND IMPLICATIONS FOR PRESERVATION

Stuart J. Birnbaum[1], John W. Wireman[2] and Robert Borowski[1]

[1]*Division of Earth and Physical Sciences, The University of Texas at San Antonio, San Antonio, TX;* [2]*Biological Research Solutions, 2727 Second Avenue, Detroit, Michigan and Department of Biological Sciences, Wayne State University, Detroit, MI*

ABSTRACT. Growth experiments utilizing the sulfate reducing bacterium, *Desulfovibrio desulfuricans*, suggest a microbial influence in . Experiments conducted with 400 ppm silica produced a dense precipitate of silica within 72 hours in which cells were entombed and caught in cellular division. Experiments conducted in the absence of cells required one month or longer for a loose flocculant of silica to develop; in some samples a precipitate did not form over a 3 month duration. The silicification phenomenon greatly altered the morphology of the bacteria causing deformation of rod-shaped vibrios into partially collapsed cocci and intermediate morphological varieties. This alteration was probably caused by osmotic shock influenced by crystallite accumulation at the cell wall and concomitant retardation of solute transport during silica polymerization. It is suggested that these experiments more closely reproduce natural conditions of silicification of microorganisms than have previous studies and that the morphological change may have implications for interpreting the Precambrian microbial fossil record.

INTRODUCTION

Silicified microorganisms have been described in the literature from geologic formations as ancient as the Archean (*e.g.* Awramik, 1984; Lowe, 1980, 1983; Walter *et al.*, 1980) and from modern habitats (Ferris *et al.*, 1986). The most abundant literature exists for Precambrian occurrences.

Interest in the silicification phenomenon has produced numerous attempts to duplicate the silicification process under laboratory conditions (Francis *et al.*, 1978 a,b; Leo & Barghoorn, 1976; Oehler & Schopf, 1971; Oehler, 1976a). None of these experiments utilized living metabolizing bacteria growing in a nutrient medium nor conditions which approximate the natural environment.

In this paper we describe experiments in which the sulfate reducing bacterium, *Desulfovibrio desulfuricans*, grown in the presence of silica, mediates a rapid silica

Origin, Evolution, and Modern Aspects of Biomineralization in Plants and Animals
Edited by Rex E. Crick
Plenum Press, New York

precipitation. As reported by Birnbaum & Wireman (1985), *D. desulfuricans* is tolerant of high concentrations of amorphous silica in solution, showing no inhibition of growth. This is consistent with the findings of Warnke *et al.* (1979), who reported the presence of bacteria in siliceous oozes. In addition to silica tolerance, it has been shown that growth leads to the precipitation of silica with the passing from a colloid to gel phase taking place in the presence of live, metabolizing cells (Birnbaum, 1984; Birnbaum & Wireman, 1985). Silica precipitation was observed either as a decrease in dissolved silica or by the appearance of a solidified gel.

A saturation concentration of 140 ppm has been reported for an aqueous solution of amorphous silica in equilibrium with a colloidal precipitate (Siever & Scott, 1963). Growth experiments conducted with concentrations of dissolved silica as low as 80 ppm revealed that silica was removed from solution despite its undersaturation (Birnbaum, 1984; Birnbaum & Wireman, 1985). This suggests an active microbial influence in silicification. This influence in the precipitation of silica is probably the product of a combination of events; a pH decrease caused by the interrelationship between CO_2 production, H_2S production, metallic sulfide precipitation; and a proton gradient established across the cell membrane (Birnbaum & Wireman, 1984; Birnbaum & Wireman, 1985) as well as an increased surface area allowing numerous sites for nucleation (Walters *et al.*, 1977).

The silica precipitate which formed entombed the cells and established conditions which we consider to be advantageous to preservation as a fossil. These experiments may shed light upon the silicification phenomenon in nature.

MATERIALS AND PROCEDURES

Growth experiments with the anaerobic sulfate reducing bacterium, *Desulfovibrio desulfuricans* were conducted to examine their suspected role in the silicification of sulfate minerals (Birnbaum & Wireman, 1985). ATCC (American Type Culture Collection, 12301 Parklawn Drive, Rockville, MD 20852) strain #13541 was grown at 32° C in Postgate's medium C (Postgate, 1979) to which soluble silica, in the form of sodium meta-silicate, $Na_2SiO_3 \cdot 9H_2O$, was added. Into 35 ml bottles, 21 ml of 1.7 times normal concentration of the medium was autoclaved and then asceptically mixed with 7 ml of 2000 ppm silica which had been sterilized by filtration through a 0.45 mm membrane; this permitted a 7 ml inoculation which resulted in the proper media concentration and a 400 ppm silica concentration after final dilution to 35 ml. Purity of cultures was checked by examination in dark phase transmitted light and by streaking cultures onto plate count agar incubated aerobically, and Postgate's medium B agar (Postgate, 1979), incubated anaerobically. No growth was observed on plate count agar while black colonies, indicative of sulfate reduction, developed on plates of medium B. Microscopic examination corroborated the purity of the culture.

Experimental control vials were prepared in a similar manner except that no inoculum was introduced. Instead, the pH was adjusted to neutrality using 0.1 N sterile HCL and sterile deionized distilled water to complete the 35 ml volume.

Cells were grown in media with silica concentrations ranging from 40 ppm to 2000 ppm. Silica concentrations, both before and after growth, were measured as elemental

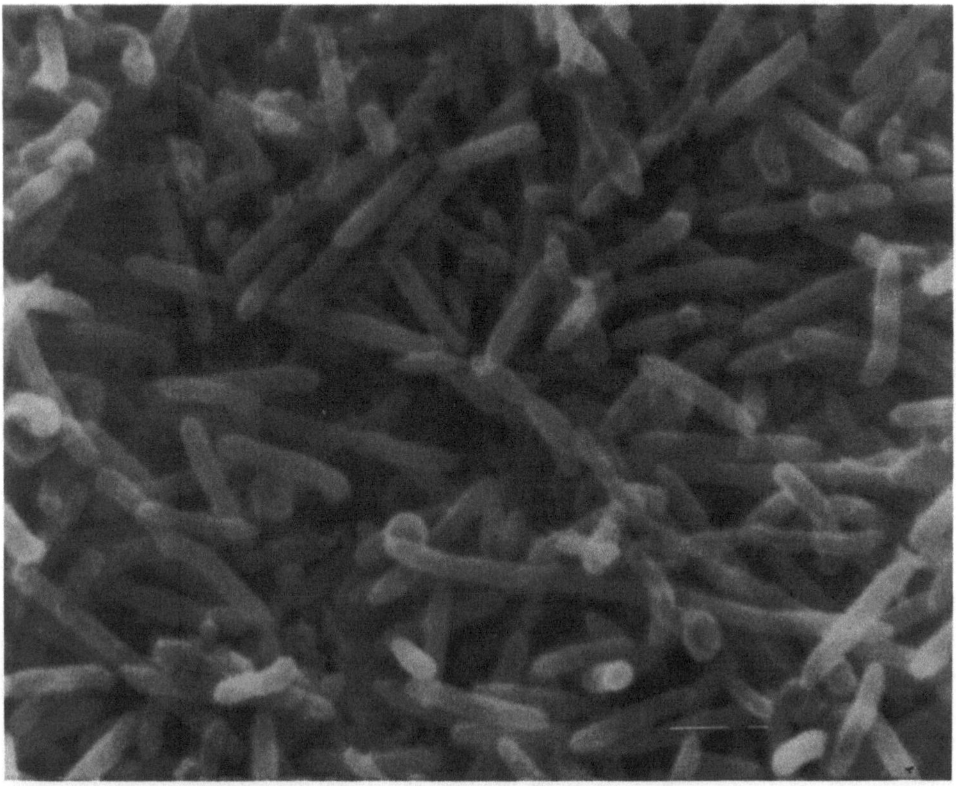

FIGURE 1. SEM photomicrograph revealing appearance of cells of *Desulfovibrio desulfuricans* grown in Postgate's medium C in the absence of silica. Bar = 1 μm.

Si in solution by atomic absorption spectrophotometry (AAS). Ten ml aliquots of growth medium were subjected to centrifugation (5000× g for 10 min) followed by membrane filtration (0.45 mm Nalgene® filters) to remove bacteria and any silica precipitate prior to AAS analysis·

Specimens were prepared for scanning electron microscopy according to a modification of the method described by Guentzel *et al.* (1983). Briefly, samples were attached to a Nucleopore® filter, fixed for one hour at room temperature in 2.5% gluteraldehyde in 0.1 M cacodylate buffer at pH=7.2, washed twice for a minimum of 15 min in the buffer, then post fixed in 1% osmium tetraoxide in the same buffer for one hour at room temperature. Samples were then washed twice for 15 min in the same buffer, dehydrated in an ethanol series followed by an amyl acetate series, and critical point dried in CO_2. The dried specimens were coated with gold-palladium (60/40) and examined with an ISI Model 40 scanning electron microscope.

Sample preparation for transmission electron microscopy consisted of applying one drop of specimen onto a 200 μm mesh, formvar coated, copper grid; then excess fluid was wicked off and the sample air dried for 30 sec. The specimen was then fixed and

A

B

C

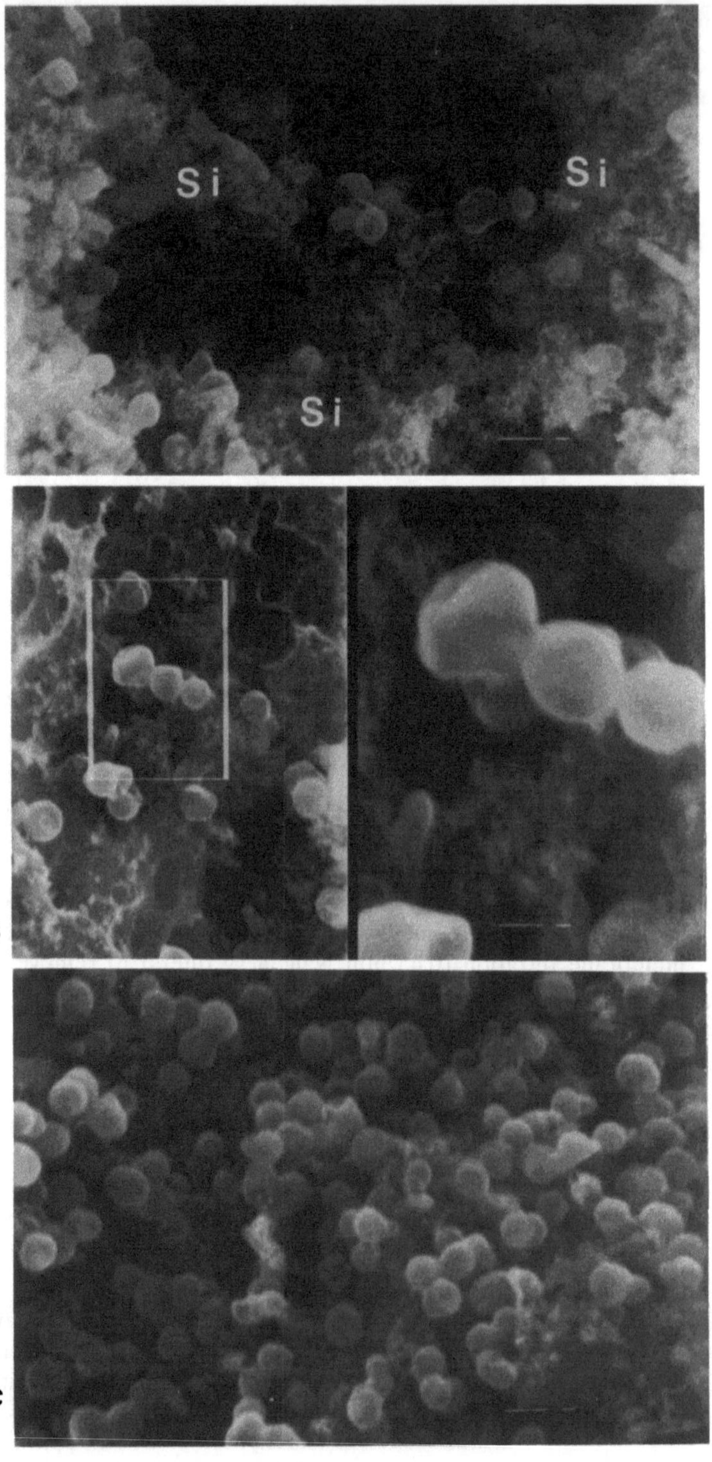

FIGURE 2. SEM photomicrographs of *Desulfovibrio desulfuricans* grown in Postgate's medium C plus 400 ppm silica. ▪A. Deformed cells in a silica matrix (Si). Note presence of vibroid, coccoid, and intermediate forms. Bar = 1 μm. ▪B. Cells caught in cell division (box). Bar = 1 μm for the left hand side; right hand side is a 3X enlargement of the boxed area. ▪C. Area of densely packed cells. Bar = 1 μm.

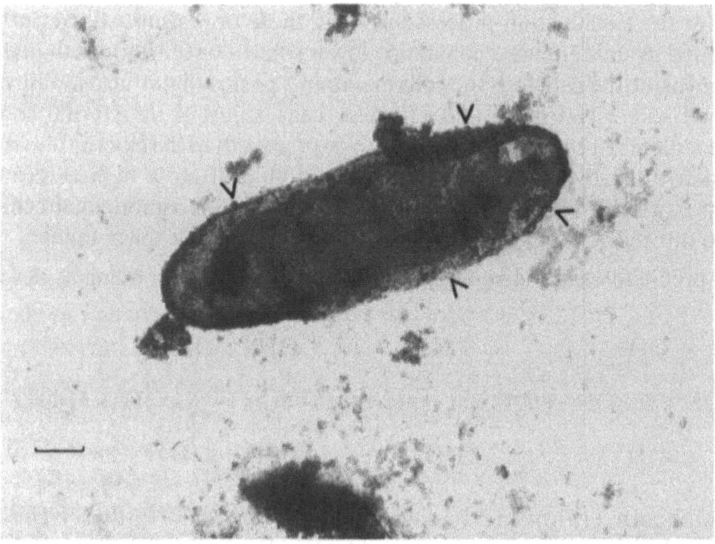

FIGURE 3. TEM photomicrograph of *Desulfovibrio desulfuricans* grown in Postgate's medium C in the presence of 400 ppm silica. Nanometer size crystallites (arrows) encapsulate the cell. Compare to Figure 1b of Ferris *et al.* (1986:610). Bar = 350 nm.

stained in a 4% uranyl acetate solution for one min, followed by air drying. Samples were examined in a Zeiss EM 9 transmission electron microscope.

RESULTS

In one group of experiments conducted with and without *Desulfovibrio desulfuricans* at 32° C in Postgate medium C containing a silica concentration of 400 ppm, a precipitate appeared within 48–72 hours in vials containing cells. The material was a gel in which cells were entombed. The samples were incubated at 32° C for one month, at which time they were prepared for scanning electron microscopy.

Control vials of medium C plus silica without cells also formed a precipitate, but only after six weeks and to a lesser degree; some vials did not yield a precipitate over a three month observation period. It is noteworthy that the nature of the precipitates also differed considerably. In the presence of cells the precipitate was a dense, compact and cohesive precipitate whereas in the absence of cells it was a loose flocculant which remained in suspension within the medium.

Figure 1 shows a scanning electron photomicrograph of an experimental control (bacteria grown without silica), while Figure 2 consists of scanning electron photomicrographs from growth experiments with silica. Comparison of the control and test micrographs yields interesting observations. Figure 2A shows a silica (Si) matrix within which are embedded and entombed bacteria. Of interest is the deformation of cells to produce a variety of shapes dominated by the coccoid form (compare with Fig. 1). Figure 2B depicts cells caught in cell division (boxed area), and Figure 2C reveals a high concentration of cells in silica.

The dramatic change in cellular shape seen in Figure 2 suggests an interesting mechanism for the silicification process involving these organisms. Bacterial cells are frequently found in environments containing lower solute concentrations than the solute concentration inside the cell; thus the cells maintain a positive internal osmotic pressure. When external solute concentrations increase, cells adjust their internal solute concentration to maintain a concentration equal to or greater than the external environment, preventing dehydration or plasmolysis. A rapid shift from a high external solute concentration to a lower external concentration (referred to as osmotic shock) may result in damage to the cell wall or cell lysis owing to uncontrolled water uptake.

As silica precipitates at the cell wall, initiated by hydrogen bonding as described by

$$\mathrm{OH - \overset{\displaystyle OH}{\underset{\displaystyle OH}{Si}} - OH \; + \; OH - \overset{\displaystyle OH}{\underset{\displaystyle OH}{Si}} - OH \rightarrow OH - \overset{\displaystyle OH}{\underset{\displaystyle OH}{Si}} - O - \overset{\displaystyle OH}{\underset{\displaystyle OH}{Si}} - OH \; + \; H_2O}$$

Leo and Barghoorn (1976) and Francis *et al.* (1978a), and further facilitated and accelerated by the proton gradient described by Birnbaum & Wireman (1984), there is a rapid decrease in total dissolved material in solution. In addition, the polymerization of colloidal silica to form siloxane bonds, as follows:

increases the water activity and dilutes the medium. As the external environment becomes more dilute, the cells must respond by the transport of solute out. If this response is not rapid enough either owing to the rate at which silica is precipitating or to the presence of a silica encrustation, or both, water will enter the cell and produce spheroplasts (ovoid cells in which the cell wall has been damaged). Continued water uptake may eventually result in lysis of the cell, but the shape change seen in Figure 2 suggests that spheroplasts have been formed as the result of a rapid decrease in the external solute concentration (*i.e.*, through polymerization of silica and resulting water release). Transmission electron microscopy supports an hypothesis in which inhibition of solute transport (from the interior of a cell outwards) may occur as a result of silica encrustation. Figure 3 shows a rod-shaped bacterium (*Desulfovibrio desulfuricans*) after 72 hours growth in the presence of 80 ppm silica. The photomicrograph reveals small (nm size) crystallites adhering to the cell's surface. This is strikingly similar to Figure 1b of Ferris *et al.*, (1986, p. 610) in which small crystallites of iron and silica encapsulate bacteria isolated from a hot spring. Such an encrustation may well diminish the osmotic regulatory capacity of the cell. It is noteworthy that not all cells undergo this morphology change, and it is not yet clear why some cells remain rod-shaped while others become altered.

This osmotic shock hypothesis also suggests a mechanism for the permeation of the cells by silica. Silica may enter the cell early in the process as part of the cell's response to increased external solute concentration or when precipitation of silica takes place and water (possibly with accompanying silica) enters the cell. Further study is required to determine if permineralization takes place in this manner.

DISCUSSION

The record of pre-metazoan life, fossils older than 680 Ma, although scant is significant (Cloud & Morrison, 1980). The oldest of these fossils, found within the *c.* 3.5 Ga Warrawoona Group of Western Australia (Awramik, 1984; Lowe, 1980, 1983; Walter *et al.*, 1980), are morphologically stromatolitic but contain no unambiguous fossils. Carbonaceous spheroids of a restricted size range have also been described from the Warrawoona Group and, based upon a statistical treatment of size distribution data, are considered to be the fossiliferous remains of micro-organisms (Dunlop *et al.*, 1978). A similar statistical approach was applied to the micro spheroids of the slightly younger Onverwacht Group of South Africa (*c.* 3.4 Ga) by Muir and Grant (1976), who interpreted these spheroidal objects to be microorganisms. These oldest microorganisms are preserved in chert.

Younger Precambrian microfossils include the famous Gunflint chert of Ontario (Barghoorn & Tyler, 1965; Cloud, 1965), the Dahongyu Formation of China (Yun, 1984), specimens from the Belcher Islands of Canada (Hofmann, 1976), the Bitter Springs Formation, central Australia (Barghoorn & Schopf, 1965; Schopf, 1968; Schopf & Blacic, 1971), the Hunnberg Formation in Svalbard (Knoll, 1984), and the Min'yar Formation of the U.S.S.R. (Nyberg & Schopf, 1984). All of these formations contain silicified microfossils displaying varying degrees of preservation and a variety of taxa.

Descriptions in the literature suggest that the silicification was, for the most part, replacive (Awramik, 1984), but may have been primary in some instances (Barghoorn & Schopf, 1965; Schopf, 1968; Schopf & Blacic, 1971). Where the silica was replacive, it had to replace the host sediment, usually carbonate, almost concurrently with deposition (very early diagenesis) (Nyberg & Schopf, 1984; Knoll, 1984). This interpretation is necessary to explain the superb preservation seen in many fossils where cell walls, cross walls, nucleoids or nuclei, etc., are faithfully preserved (see plates in Schopf, 1968; Schopf & Blacic, 1971; Oehler, 1976, *e.g.*, plate 2, Figure 5, Oehler, 1976). In addition, Schopf (1968:660) interprets the rapidity of the silicification process as "...evidenced by the lack of bacterial degradation exhibited by the microfossils, and by the occasional preservation of microorganisms apparently 'frozen' in the process of cellular division." Such observations have led to conclusions that amorphous silica surrounds and infiltrates the cells, thus permineralizing and preserving the microorganisms prior to compaction or further diagenesis (Barghoorn & Schopf, 1965; Schopf, 1968).

The nature of this silicification has been of interest to many people; numerous investigators have attempted to reproduce the silicification process under laboratory conditions. Leo & Barghoorn (1976), using ethyl silicate, an organosilicon compound not known to be involved in natural silicification (Francis *et al.*, 1978a), attempted to duplicate silicification of wood. Adopting this technique, Francis *et al.* (1978a, 1978b) reported the silicification of numerous prokaryotic and eukaryotic microorganisms. Although the authors acknowledged that ethylsilicate (triethylorthosilicate) is probably not responsible for silicification in nature, they assumed that its role in their experiments was to accelerate the delivery of silica to the cell, thus allowing results within a time frame acceptable for empirical studies as well as producing the rapid silicification called for by the fossil record. This method of silicification does not lead to permineralization of the cells (Francis *et al.*, 1978a).

An entirely different approach was utilized by Oehler & Schopf (1971) in their experiments of artificial fossilization of blue-green algae (Cyanobacteria). A commercially available colloidal silica solution in which cells were mixed was converted to a gel by neutralization of pH with HCl (Oehler & Schopf, 1971). Upon adjustment of pH the colloid solidified into a gel within one hour (Oehler, 1976b). After solidification, the gel was treated to elevated temperatures and pressure to simulate diagenesis. Permeation of the cells by silica was attained as revealed by electron probe X-ray microanalysis (Oehler & Schopf, 1971). Although it is not clear whether silica permeation occurred during the initial soaking in colloidal silica or during the treatment at elevated temperature and pressure, it is probable that these elevated temperature-pressure conditions played a role. The temperatures used were 100° C and 165° C at a pressure of 3000 bars aqueous fluid pressure (Oehler, 1976a). The attainment of the pressure and temperatures used in these experiments is unlikely in nature where the permeation of the cells by silica must have occurred very soon after, if not contemporaneous with, growth. This temporal relationship is called upon to explain the excellent preservation, with a minimum of degradation, reported in the literature, *e.g.*, the Bitter Springs Formation (Schopf, 1968; Schopf & Blacic, 1971). This contemporaneous silicification was described in a modern environment by Ferris *et al.* (1986), whose samples were collected from a thermal spring; but no reports of empirical studies known to us approach more natural conditions of silicification that lead to a rapid precipitation of a silica gel. Although we have not yet determined whether the experiments described in our present paper result in permeation of the cells by silica, it is clear that these experiments more closely approach natural conditions.

CONCLUSION

The results reported here suggest that living cells of *Desulfovibrio desulfuricans* aid in the precipitation of amorphous silica and, in so doing, become embedded and entombed in a silica gel. This process is interpreted to be influenced by bacterial metabolism and osmotic transport of materials across the cell wall. This active role may suggest why Francis *et al.* (1978a) found better silicification of living cells than of moribund and lysed cells.

Our current study has implications for understanding the silicification of microorganisms so prevalent in the Precambrian. It is possible that, by their metabolism, bacteria served as templates for the precipitation of amorphous silica. This would help to explain the excellent preservation of silicified Precambrian microorganisms. The precise interaction between dissolved silica and microorganisms is probably more complex than suggested here or elsewhere; it is certainly a multifaceted problem. It is clear that during the Precambrian, in the absence of radiolaria, diatoms and sponges, the availability of silica for interactions with bacterial populations was great. That these interactions occurred is unquestionable, however the exact mechanisms involved are problematical.

ACKNOWLEDGMENTS

The authors would like to acknowledge support for this project from Amoco Production Research and from the University of Texas at San Antonio for one of us (SJB). We also thank the people at Biosan Laboratories for their patience and many useful discussions, as well as for making their laboratory available. Participation in the Symposium was made possible by grants from The Organized Research Fund of The University of Texas at Arlington and The Sea Grant Program administered by Texas A & M University.

Dr. Neal Guentzal is gratefully acknowledged for his commentary and discussion of the manuscript and Ms. Sandy Romanelli is thanked for her patience in typing and re-typing the manuscript.

REFERENCES

AWRAMIK, S.M., 1984. Ancient stromatolites and microbial mats. *In* Cohen, Y., Castenholz, R.W., & Halvorson, H.D., (Ed.), Microbial Mats: Stromatolites. Alan R. Liss, Inc., New York. pp. 1-22.

BARGHOORN, E.S. & SCHOPF, J.W., 1965. Microorganisms from the Late Precambrian of Central Australia. Science, 150:337-339.

BARGHOORN, E.S. & TYLER, S.A., 1965. Microorganisms from the Gunflint Chert. Science, 147:563-577.

BIRNBAUM, S.J., 1984. Silica in sedimentary sulfide deposits: a microbial origin. Geological Association Canada Program with Abstracts, 9:47.

BIRNBAUM, S.J. & WIREMAN, J.W., 1984. Bacterial sulfate reduction and pH: implications for early diagenesis. Chemical Geology, 43:143-149.

BIRNBAUM, S.J. & WIREMAN, J.W., (1985). Sulfate reducing bacteria and silica solubility: a possible mechanism for evaporite diagenesis and silica precipitation in banded iron formations. Canadian Journal Earth Sciences, 22:1904-1909.

CLOUD, P., 1965. Significance of the Gunflint (Precambrian) microflora. Science, 148:27-45.

CLOUD, P. & MORRISON, K., 1980. New microbial fossils from 2 Gyr old rocks in northern Michigan. Geomicrobiology Journal, 2:161-178.

DUNLOP, J.S.R., MUIR, MD., MILNE, V.A. & GROVES, D.I., 1978. A new microfossil assemblage from the Archean of Western Australia. Nature, 274:676-678.

FERRIS, F.G., BEVERIDGE, T.J. & FYFE, W.S., 1986. Iron-silica crystallite nucleation by bacteria in a geothermal sediment. Nature, 320:609-611.

FRANCIS, S., MARGULIS, L. & BARGHOORN, E.S., 1978a. On the experimental silicification of microorganisms II. On the time of appearance of eukaryotic organisms in the fossil record. Precambrian Research, 6:65-100.

FRANCIS, S., BARGHOORN, E.S. & MARGULIS, L., 1978b. On the experimental silicification of microorganisms III. Implications of the preservation of the green prokaryotic alga *Prochloron* and other coccoids for interpretation of the microbial fossil record. Precambrian Research, 7:377-383.

GUENTZEL, M.N., GUERRERO, D. & DESHAMBO, R., 1983. Nature of interaction of different wild type *Vibrio cholerae* with intestinal mucosa. Scanning Electron Microscopy:263-270.

HOFMANN, H.J., 1976. Precambrian microflora, Belcher Islands, Canada: significance and systematics. Journal Paleontology, 50:1040-1073.

KNOLL, A.H., 1984. Microbiotas of the Late Precambrian Hunnberg Formation, Nordaustlandet, Svalbard. Journal Paleontology, 58:131-162.

LEO, R.F., & BARGHOORN, E.S., 1976. Silicification of wood. Botanical Museum Leaflets, Harvard University, 25:1-45.

LOWE, D.R., 1980. Stromatolites 3400-Myr old from the Archean of Western Australia. Nature, 284:441-443.

LOWE, D.R., 1983. Restricted shallow-water sedimentation of Early Archean stromatolitic and evaporitic strata of the Strelley Pool Chert, Pilbara Block, Western Australia. Precambrian Research, 19:239-283.

MUIR, M.D. & GRANT, P.R., 1976. Micropaleontological evidence from the Onverwacht Group, South Africa. *In* Windley, B.F. (editor), The Early History of the Earth. John Wiley and Sons, London, pp. 595-604.

NYBERG, A.V. & SCHOPF, J.W., 1984. Microfossils in stromatolitic cherts from the Upper Proterozoic Min'yar Formation, southern Ural Mountains, USSR. Journal Paleontology, 58:738-772.

OEHLER, D.Z., 1976. Transmission electron microscopy of organic microfossils from the Late Precambrian Bitter Springs Formation of Australia: Techniques and survey of preserved ultrastructure. Journal Paleontology, 50:90-106.

OEHLER, J.H., 1976a. Experimental studies in Precambrian paleontology: Structural & chemical changes in blue-green algae during simulated fossilization in synthetic chert. Geological Society America Bulletin, 87:117-129.

OEHLER, J.H., 1976b. Hydrothermal crystallization of silica gel. Geological Society America Bulletin, 87:1143-1152.

OEHLER, J.H. & SCHOPF, J.W., 1971. Artificial microfossils: Experimental studies of permineralization of blue-green algae in silica. Science, 174:1229-1231.

POSTGATE, J.R., 1979. The Sulphate-Reducing Bacteria. Cambridge University Press, Cambridge. 151 pp.

SCHOPF, J.W., 1968. Microflora of the Bitter Springs Formation, Late Precambrian, Central Australia. Journal Paleontology, 42:651-688.

SCHOPF, J.W. & BLACIC, J.M., 1971. New microorganisms from the Bitter Springs Formation (Late Precambrian) of the North-Central Amadeus Basin, Australia. Journal Paleontology, 45:925-960.

SIEVER, R., & SCOTT, R.A., 1963. Organic geochemistry of silica. *In* Breger, I.A. (editor), Organic Geochemistry. Pergamon Press-MacMillan Company, New York, pp. 579-595.

WALTER, M.R., BUICK, R. & DUNLOP, J.S.R., 1980. Stromatolites 3400-3500 m.y. old from the North Pole area, Western Australia. Nature, 284:443-445.

WALTERS, C.C., MARGULIS, L. & BARGHOORN, E.S., 1977. On the experimental silicification of microorganisms I. Microbial growth on organo silicon compounds. Precambrian Research, 5:241-248.

WARNKE, D.A., BLUNT, D.J. & POLLOCK, G.E., 1979. Amino acid stereochemistry in siliceous oozes: indications of bacterial activity. Antarctic Journal of the United States, 14:134-136.

YUN, Z., 1984. A Gunflint type of microfossil assemblage from early Proterozoic stromatolitic cherts in China. Nature, 309:547-549.

MODEL FOR PROKARYOTIC CALCIFICATION

Barbara D. Boyan, Larry D. Swain and Ruben Gomez

Departments of Orthopaedics and Biochemistry, The University of Texas Health Science Center at San Antonio, San Antonio, TX 78284

INTRODUCTION

De novo hydroxyapatite deposition in biologic systems occurs on specialized membrane surfaces (Anderson, 1969, 1984). The process involves the interaction of calcium and inorganic phosphate ions with acidic phospholipids to form phospholipid-calcium-phosphate complexes (CPLX) that have a characteristic stoichiometry: one mole Ca^{2+} per mole total PO_4^{3-}(Pi) of which 90% is organic (Boskey, 1981). CPLX is enriched at sites where mineralization is initiated and is absent from non-mineralized tissues. Studies utilizing a microbial model of calcification indicate that *in vitro* CPLX formation is proteolipid-dependent (Boyan-Salyers & Boskey, 1980). CPLX forms preferentially on proteolipid-associated phospholipids; phospholipids isolated from the bacteria which are not associated with proteolipid bind Ca^{2+} but do not support CPLX or hydroxyapatite formation. Proteolipid-induced hydroxyapatite always includes CPLX formation, whether the process occurs *in vivo* or when allogeneic proteolipids are implanted in diffusion chambers in mesenchyme *in vitro* (Raggio *et al.*, 1986). Further evidence that they exist as an entity *in situ* is that CPLX can be isolated along with proteolipid from mineralizing bacteria (Boyan & Boskey, 1984) and vertebrate calcified tissues (Boyan, 1985).

Despite this body of information concerning the requirements for the initiation of hydroxyapatite formation, little data has been presented to suggest a mechanism. Recent studies in our laboratory have identified specific components of the proteolipid-CPLX complex and have investigated their function. This paper summarizes these studies and presents a theoretical model of the calcification initiation site.

ASPECTS OF PROTEOLIPIDS

Proteolipid composition

In vitro calcification of *Bacterionema matruchotii* (American Type Culture Collection 14266) is proteolipid-dependent (Ennever *et al.*, 1976). These bacteria form intracellular calcium-phosphate deposits in culture that yield X-ray diffraction maxima indistinguishable from powdered bone or dentine. *B. matruchotii* forms CPLX prior to mineral deposition (Boyan-Salyers & Boskey, 1980). The appearance of hydroxyapatite

Origin, Evolution, and Modern Aspects of Biomineralization in Plants and Animals
Edited by Rex E. Crick
Plenum Press, New York

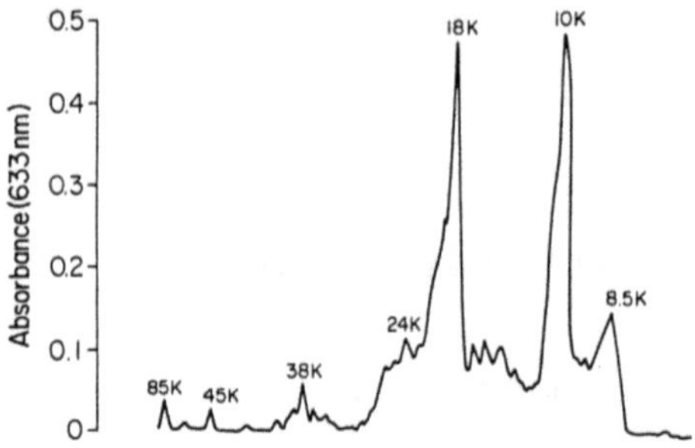

FIGURE 1. Densitometric scan of proteolipid isolated from *Bacterionema matruchotii* following SDS polyacrylamide gel electrophoresis. Gels contained 12.5% acrylamide and 8M urea. Modified from Swain & Boyan (1988).

crystals coincides with an increase in total proteolipid concentration and a shift in the composition of the extract to a major molecular weight species with an M_r of 10,000 on Laemmli 12% acrylamide gels (Boyan *et al.*, 1982).

Total proteolipid was isolated from *Bacterionema matruchotii* by sequential extractions in neutral chloroform:-methanol 2:1, v/v, and chloroform:-methanol:HCl 2:1:0.01, v/v. After the extracts were washed with appropriate salt solutions, they were combined and the crude phospholipid component including any proteolipid was precipitated with acetone. The precipitated material was chromatographed on a column of hydroxypropylated dextran gel (Sephadex LH-20) using chloroform:methanol 2:1 as the mobile phase, thus separating the proteolipid enriched fraction from other lipids. Prior to electrophoretic analysis, 80% of phospholipid associated with the proteolipid was removed by solubilization in diethylether. When delipidated proteolipid was electrophoresed using the method described by Anderson (Anderson *et al.*, 1983), seven protein bands were seen at M_r 8,500; 10,000; 18,000; 24,000, 38,000; 45,000; and 85,000 (Fig. 1). If the neutral and acidified extracts are processed separately the proteolipid species can be differentially isolated. The neutral extract contains predominantly the M_r 10,000 component. This protein band is absent from the acidified solvent extract. Similar results are seen following extraction of epiphyseal growth cartilage matrix vesicles (unpublished data), a site of *de novo* hydroxyapatite formation in vertebrates (Anderson, 1984).

CPLX was isolated from *Bacterionema matruchotii* using five sequential extractions of chloroform:methanol 2:1 only. After washing the extracts with NH_4HCO_3, CPLX was precipitated with ethanol:ether (Boyan & Boskey, 1984). Protein was measured in such preparations using SDS in the Lowry assay. A single band was visualized by SDS-PAGE at M_r 10,000, when the Laemmli protocol was used. Similar SDS-PAGE profiles were evident for CPLX isolated from bone.

While the phospholipid compositions of total proteolipid and CPLX isolated from *Bacterionema matruchotii* are not identical, there are some notable similarities. Both extracts are enriched in phosphatidylinositol and phosphatidylserine, although phosphatidylserine is an uncommon phospholipid in microbial systems (Boyan & Boskey, 1984). Since the bacteria were cultured in a synthetic media containing no exogenous lipid, the presence of phosphatidylserine suggests that it is an important component in the structure of these two lipid extracts. CPLX and proteolipid from vertebrate tissues also are enriched in phosphatidylserine and phosphatidylinositides (Boyan, 1985). In addition, phosphatidylserine can form CPLX *in vitro* when incubated in metastable calcium-phosphate solutions (Boskey, 1981).

Proteolipid function

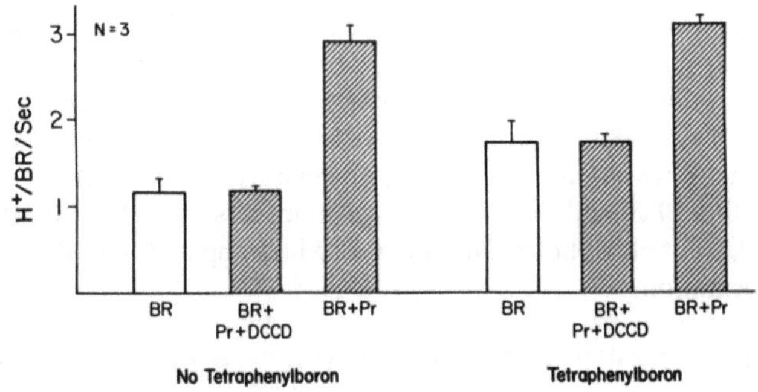

FIGURE 2. Proton transport by bacteriorhodopsin (BR) liposomes containing proteolipid (Pr) or proteolipid-dicylohexyl carbodiimide (DCCD) conjugates in the presence and absence of tetraphenylboron.

Proteolipids isolated from other membrane systems generally occur as integral membrane components of large multi-protein enzyme complexes (Fillingame, 1980). They often function as naturally occurring ionophores, permitting transport of charged particles across the membrane bilayer (Senior & Wise, 1983). A classical example of such a proteolipid is the F_o component of the mitochondrial F_1F_o ATPase (Fillingame, 1980). F_o is a protonophore and, like other proton translocating proteolipids, exhibits certain characteristics: it contains an M_r 10,000 protein species; it specifically binds ^{14}C-dicyclohexylcarbodiimide (DCCD); it consists of multiple proteolipids to create a functional unit; and proton transport function is lost when it is bound to DCCD.

Proteolipid isolated from *Bacterionema matruchotii* appears comparable to other ion translocating proteolipids (Swain, 1986). The protein binds ^{14}C-DCCD and unlabeled DCCD competes for the binding site in an equimolar manner. The binding site is in a hydrophobic portion of the bilayer since no ^{14}C binding is competitively inhibited by ethyldimethylaminopropyl-carbodiimide which binds carboxyl residues located in hydrophilic regions of the membrane. The nucleophile glycine ethyl ester, which binds specifically to carboxyl residues, competes for ^{14}C-DCCD binding sites, confirming that the ligand is bound to a free carboxyl in the proteolipid. Only the M_r 8,500 proteolipid species binds ^{14}C-DCCD. This observation was confirmed by liquid scintillation spectroscopy of individual gel slices.

Ability of the *Bacterionema matruchotii* proteolipid to transport ions was tested using bacteriorhodopsin-containing liposomes. These liposomes are well characterized (Lind *et al.*, 1981), consisting of a uniform population of unilamellar membrane vesicles. Bacteriorhodopsin transports protons when activated by light and the resulting membrane potential is dissipated by addition of tetraphenylboron. Addition of total proteolipid to the bacteriorhodopsin liposomes enhances both the rate and extent of proton transport (Fig. 2). This enhancement is not affected by addition of tetraphenyl boron but is abolished when ^{14}C-DCCD bound proteolipid is used. Neither the neutral solvent extracted proteolipid nor the proteolipids extracted in acidified solvent alone augment proton transport. These data indicate that at least the M_r 10,000 protein and the ^{14}C-DCCD binding proteolipid are both required for ionophoric activity.

It is clear that the proteolipid has a structural function as well. When proteins are added to mixed lipid liposomes the lipid differentially reorganizes about the protein (Boggs, 1977). Conservation of phosphatidylserine in both the proteolipid and CPLX extracts suggests that this phospholipid is an important component. CPLX contains SDS-PAGE protein bands which correspond to proteins in proteolipid extracted in neutral chloroform:methanol. Ennever *et al.* (1976) have demonstrated that this proteolipid is sufficient to support *in vitro* hydroxyapatite formation. Boyan-Salyers and Boskey (1980) demonstrated that CPLX is formed prior to mineral in such solutions. Since phosphatidylserine and phosphatidylinsitol are the major phospholipids associated with CPLX and the M_r 10,000 protein, it is probable that this proteolipid structures its adjacent phospholipid in a manner which supports CPLX and subsequent hydroxyapatite formation.

MODEL OF THE HYDROXYAPATITE INITIATION SITE

Current concepts of *de novo* hydroxyapatite formation require the presence of a permissive environment. The concept of a classic nucleator has been modified to include a combination of promotors. CPLX, for instance, may have a role in organization of the initial stable crystals. CPLX formation itself represents an interaction of mineral ions with membrane surface phospholipids. It can occur in solution even in the absence of protein if sufficient concentration of phosphatidylserine is present (Boskey, 1981). Once formed, CPLX will support hydroxyapatite formation if exposed to sufficient Ca^{2+} and P_i.

Most studies of proposed calcification nucleators are performed *in vitro* in metastable calcium-phosphate solutions (Boskey, 1981; Wuthier, 1982). These solutions contain buffers in addition to Ca^{2+} and P_i to accommodate proton production attendant on hydroxyapatite formation. In constant composition systems (Koutsoukos, 1980), evolution of H^+ is titrated with OH^-. Even in solutions that do not contain buffers, simple dilution of H^+ can permit some crystal formation. Similar conditions are operative in ectopic calcification when cell membrane surfaces are exposed to extracellular fluids in an unregulated manner. However, in cell mediated regulated calcification, for example via matrix vesicles, initial hydroxyapatite formation takes place in a closed system. Therefore a mechanism for export of H^+ is as important as mechanisms for import of calcium and phosphate.

The model presented in Figure 3 accounts for the information summarized above and permits the formulation of new hypotheses. Initial mineral formation occurs on the inner leaflet of the bilayer. Therefore CPLX formation should occur here as well, predicting the presence of the M_r 10,000 protein and enrichment of phosphatidylserine. In fact, studies by Wuthier and Majeska (Wuthier *et al.*, 1977) indicate that in matrix vesicles, phosphatidylserine is mainly an inner leaflet constituent. Both the M_r 10,000 and 8,500 proteins are required for ionophoric activity but the M_r 10,000 alone is sufficient for mineral formation. This suggests that the M_r 10,000 protein may be located predominantly within the inner leaflet whereas the M_r 8,500 protein may be predominantly in the outer leaflet. Both proteins are required for ionophoric function suggesting that the proteins, at least in part, span the membrane.

Perhaps the most interesting hypothesis suggested by this model is a role for alkaline phosphatase. Many other protonophores described previously (Fillingame, 1980) represent the proton channel for ATPase. Alkaline phosphatase shares the characteristic of requiring butanol (a proteolipid solvent) or detergent for total release from the membrane (Senior, 1983). In addition alkaline phosphate activity can be released from the membrane using high salt concentration. Since matrix vesicles are depleted of ATP when crystals are seen, it is probable that alkaline phosphatase is not functioning as an ATPase during hydroxyapatite formation. Wuthier *et al.* (1985) have also presented data indicating that alkaline phosphatase does not act as a phosphate transferase. Since ion transport by an ionophore does not require ATP it is possible that one role of this enzyme is serviced by its membrane component. Failure of tetraphenylboron to dissipate proteolipid enhanced H^+ transport indicates that proteolipid is a true ionophore and, therefore, can work in the absence of ATP. Recently, Tennenbaum (1986) demonstrated that phosphatidylethanolamine can serve as a substrate for alkaline phosphatase raising the possibility that as this phospholipid is selectively removed from the membrane the resultant leakiness can promote Ca^{2+} and Pi import. At the same time protons are exported.

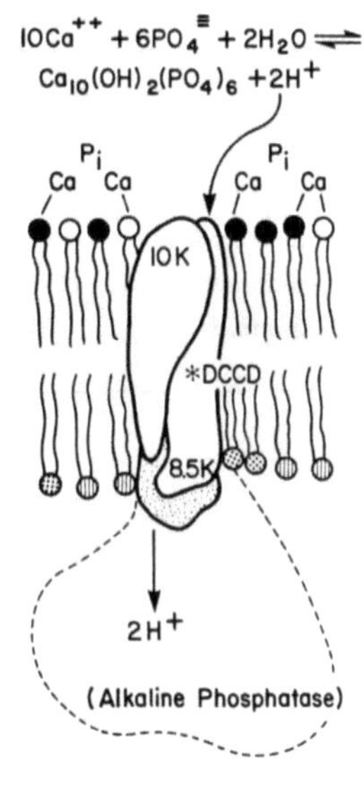

FIGURE 3. A model for membrane-mediated, proteolipid-dependent hydroxyapatite formation. Calcium and inorganic phosphate (Pi) bind to phospholipids on the inner leaflet of the membrane. *DCCD: dicylo-hexyloarbodiimide binding protein. From Swain & Boyan (1989).

EXAMPLES OF THE MECHANISM

The ability to support calcium-hydroxyapatite formation is common to a number of saprophytic and pathogenic microorganisms (Boyan *et al.*, 1984). The ability to calcify is not equally expressed in all species, however. Some microorganisms calcify rapidly while others require extensive incubation time in calcium-phosphate enriched culture media. Still other microorganisms, like *Actinomyces naeslundii*, must first be killed and lyophilized, exposing intracellular membranes to the high calcium solutions before crystal deposits can be formed.

Studies in our laboratory indicate that the major factors in determining whether a microorganism will calcify are the concentration and composition of proteolipids and their associated phospholipids. Streptococcal calcification is a useful model for demonstrating this. When *Streptococcus sanguis* I (ATCC 10556), *S. sanguis* II (ATCC 10557), or *S. mitis* (ATCC 6249) are incubated in a defined culture medium containing 1 mM calcium and 3.9 mM inorganic phosphate for eight days, hydroxyapatite crystals are observed within the cytoplasm of *S. sanguis* II and *S. mitis* only. After 14 days in continuous culture electron dense material is seen in *S. sanguis* I, however the ash of these bacteria is amorphous to X-ray defraction. In contrast, the crystals present in *S. sanguis* II fill the cytoplasm and extend into the extracellular matrix. *S. mitis* also forms hydroxyapatite but to a less significant degree than *S. sanguis* II.

Proteolipid isolated from *Streptococcus sanguis* II is enriched in protein (Table 1). In contrast, phospholipid concentration, determined colorimetrically as a measure of organic phosphate, is greatest in proteolipid from the non-mineralizing *S. sanguis* I. The correlation between protein: phospholipid ratio and clacifiability is demonstrated by

TABLE 1. Correlation of protein:phospholipid ratio of proteolipid isolated from streptococcal species with ability to support hydroxyapatite (HA) formation *in vitro*

Taxa	mg Protein : mg Phospholipid	HA
S. sanguis I	2.41	+
S. sanguis II	4.73	+ + +
S. mitus	3.32	+ +

comparing the two columns. *S. mitis*, which calcifies to a lesser degree than *S. sanguis* II has a proteolipid protein:phospholipid ratio range between the two *S. sanguis* strains.

The protein composition of each of the proteolipids differs qualitatively and quantitatively as well (Fig. 4). Lane 1 is the low molecular weight standards ranging from Mr 92,000 to 14,000. Lane 2 contains proteolipids from *Streptococcus sanguis* I; lane 3 contains proteolipids from *S. sanguis* II; and lane 4 contains proteolipid from *S. mitis*. All three proteolipids exhibit multiple bands with relative mobilities of Mr 24,000, 10,000 and 5,300. *S. mitis* proteolipid contains an additional band at Mr 16,000. The relative intensities of the bands are quite different however. The Mr 10,000 and 24,000 protein bands are most intense in S. sanguis II proteolipid. These bands co-migrate with the principal bands of calcifiable proteolipids from *Bacterionema matruchotii*, an oral actinomycete involved in dental calculus formation, discussed earlier in this paper.

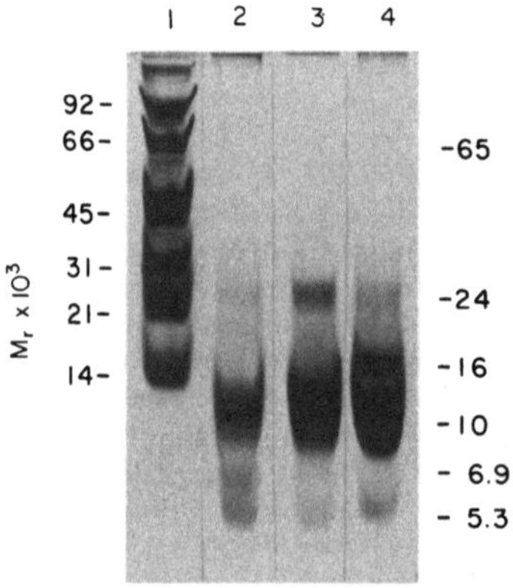

FIGURE 4. SDS-polyacrylamide gel electrophoretogram of the protein composition of proteolipid isolated from *Streptococcus sanguis* I (lane 2); *S. sanguis* II (lane 3); and *S. mitis* (lane 4). Molecular weight standards are in lane 1.

ACKNOWLEDGEMENTS

The research presented in this paper was supported by PHS Grants DE-05932 and DE-05937. Dr. Swain is a fellow in the Musculoskeletal Diseases Training Program (PHS Grant AM-07464) at the University of Texas Health Science Center at San Antonio. Participation in the Symposium was made possible by grants from The Organized Research Fund of The University of Texas at Arlington and The Sea Grant Program administered by Texas A & M University.

REFERENCES

ANDERSON, H.C., 1969. J. Cell Biol., 41, 59-72.

ANDERSON, H.C., 1984. Scan. Elec. Micros.:2, 953-964.

ANDERSON, B.L., BERRY, R.W. & TELSER, A., 1983. Analyt. Biochem., 132:365-375.

BOGGS, J., WOOD, D., MOSCARELLO, M. & PAPAHADJAPOLOUS, D., 1977. Biochem., 16, 2325-2329.

BOSKEY, A., 1981. Clin. Orthop., 157:226-257.

BOYAN, B.D. & BOSKEY, A.L., 1984. Calc. Tiss. Intern., 36:214-218.

BOYAN, B.D., 1985. The Chemistry and Biology of Mineralized Tissue, Ed. W.T. Butler, EBSCO Media, Birmingham, pp. 125-131.

BOYAN-SALYERS, B. & BOSKEY, A., 1980. Calc. Tiss. Intern., 30:167-174.

BOYAN, B., DERESZEWSKI, G., HINMAN, B., FLORENCE, M. & GRIFFITHS, G., 1982. Fifth Int. Symp. on Calcif. Tissues, Kiryat Anavim, Israel, Sela, J. Ed., Excerpta Medical, pp. 12-17.

BOYAN, B.D., LANDIS, W.J., KNIGHT, J., DERESZEWSKI, G. & ZEAGLER, J., 1984. Scan. Elec. Micros., 4:1793-1800.

ENNEVER, J., VOGEL, J., RIDER, L. & BOYAN-SALYERS, B., 1976. Proc. Soc. Exp. Biol. Med., 152:147-150.

FILLINGAME, R.H., 1980. Ann. Rev. Biochem., 49:1079-1113.

LIND, C., HOJEBERG, B. & KHORANA, G., 1981. J. Biol. Chem., 256:8298-8305.

KOUTSOUKOS, P., AMJAD, Z., TOMSON, M.B., AND NANCOLLAS, G.H., 1980. J. Amer. Chem. Soc., 102:1553-1557.

RAGGIO, C., BOYAN, B. & BOSKEY, A., 1986. J. Bone Min. Res., 1:409-415.

SENIOR, A.E., 1983. Biochem. Biophys. Acta, 726:81-95.

SENIOR, A.E. & WISE, J.G., 1983. J. Memb. Biol., 73:105-124.

SWAIN, L.D., 1986. J. Den. Res., 65:H13.

SWAIN, L.D. & BOYAN, B.D., 1988. J. Dent. Res., 67:526-530.

SWAIN, L.D. & BOYAN, B.D., 1989. *In* Dental Calculus: Formation, Epidemiology and Prevention. T.M. Cate (ed.), IRL Press, Oxford.

TENNENBAUM, H., 1986. J. Den. Res., 65:492.

WASSERMAN, R.H. & TAYLOR, A.N., 1968. J. Biol. Chem., 243:3987-3993.

WUTHIER, R.E., 1982. Clin. Orthop., 169:219-242.

WUTHIER, R.E., CHIN, J.E., HALE, J.E., REGISTER, T.C., HALE, L.V. & ISHIKAWA, Y., 1985. J. Biol. Chem., 260:15972-15979.

WUTHIER, R.E., MAJESKA, R.J. & COLLINS, G.M., 1977. Calc. Tiss. Res., 23:135.

CONTRIBUTORS

Catherine R. Agegian
Department of Oceanography and Hawaii Institute of Geophysics, University of Hawaii, Honolulu, Hawaii 96822

Joseph P. Bidwell
Howard Hughes Medical Institute, Woods Hole Oceanographic Institution, Woods Hole, Massachusetts 02543; present address: Endocrine Research Unit, Mayo Foundation, 5-164 West Joseph Building, Rochester, Minnesota 55905

Stuart J. Birnbaum
Division of Earth and Physical Sciences, The University of Texas at San Antonio, San Antonio, Texas 78284

Fred C. Boogerd
Department of Biochemistry, University of Leiden, Wassenaarseweg 64, 2333 AL Leiden, The Netherlands

Robert Borowski
Division of Earth and Physical Sciences, The University of Texas at San Antonio, San Antonio, Texas 78284

Bernard Patrick Bourgoin
Geology Department, McMaster University, 1280 Main Street West, Hamilton, Ontario, Canada L8S 4M1; present address: Environmental Resource Studies, Trent University, Peterborough, Ontario, Canada K9J7B8

Barbara D. Boyan
Departments of Orthopaedics and Biochemistry, The University of Texas Health Science Center at San Antonio, San Antonio, Texas 78284

Shih-Bin R. Chang
Division of Geological and Planetary Sciences, 170-25, Institute of Technology, Pasadena, California 91125

Brent Constantz
Earth Sciences Board, University of California, Santa Cruz, California 95064; present address: Norian Corporation, 1025 Terra Bella Avenue, Mt. View, California 94043

Miles A. Crenshaw
Marine Science Curriculum and Dental Research Center, University of North Carolina, Chapel Hill, North Carolina 27514

Rex E. Crick
Department of Geology, University of Texas at Arlington, Arlington, Texas 76019

Dennis John Crisp
Animal Biology Group, Marine Science Laboratories, Menai Bridge, Gwynedd, Great Britain

Egon T. Degens
SCOPE/UNEP International Carbon Unit, Geological-Paleontological Institute, University of Hamburg, Bundesstrasse 55D-2000 Hamburg 13, Federal Republic of Germany

Elisabeth W. de Vrind-de Jong
Department of Biochemistry, University of Leiden, Wassenaarseweg 64, 2333 AL Leiden, The Netherlands

Johannes P.M. de Vrind
Department of Biochemistry, University of Leiden, Wassenaarseweg 64, 2333 AL Leiden, The Netherlands

L.W. Fisher
Bone Research Branch, National Institute of Dental Research, National Institute of Health, Bethesda, Maryland 20892

Scott M. Gallager
Department of Biology, Woods Hole Oceanographic Institution, Woods Hole, Massachusetts 02543

Robert M. Garrels
Department of Marine Science, University of South Florida, St. Petersburg, Florida 33701

Ruben Gomez
Departments of Orthopaedics and Biochemistry, The University of Texas Health Science Center at San Antonio, San Antonio, Texas 78284

Ed M. Greenfield
Marine Science Curriculum and Dental Research Center, University of North Carolina, Chapel Hill, North Carolina 27514; present address: Pathology Department, Jewish Hospital, Washington University Medical Center, St. Louis, Missouri 63110

Wolfgang F. Gutmann
Senckenberg-Museum, D 6000 Frankfurt, Germany

P.E. Hare
Geophysical Laboratory, Carnegie Institution of Washington, 2801 Upton Street, Washington, DC 20008

Piet Havinga
Centre for Medical Electron Microscopy, University of Groningen, Oostersingel 69/2, 9713 EZ Groningen, The Netherlands

W. William Hughes
Biology Department, Andrews University, Berrien Springs, Michigan 49104

Yeishin Isa
Department of Biology, University of the Ryukyus, Nishihara, Okinawa 903-01, Japan

Mitsuo Kakei
Department of Oral Anatomy, Josai Dental University, Sakado, Saitama, Japan

Józef Kazmierczak
Institute of Paleobiology, Polish Academy of Science, Zwirki i Wigury 93, PL-02089 Warszawa, Poland

Harold W. Keller
Department of Biology, The University of Texas at Arlington, Arlington, Texas 76019

Norman E. Kemp
Department of Biology, The University of Michigan, Ann Arbor, Michigan 48109

Stephan Kempe
SCOPE/UNEP International Carbon Unit, Geological-Paleontological Institute, University of Hamburg, Bundesstrasse 55D-2000 Hamburg 13, Federal Republic of Germany

Roni J. Kingsley
Dental Research Center, University of North Carolina, Chapel Hill, North Carolina 27514

Joseph L. Kirschvink
Division of Geological and Planetary Sciences, 170-25, Institute of Technology, Pasadena, California 91125

Alan M. Kuzirian
Section on Neural Systems, Laboratory of Biophysics, NINCDS-NIH, Marine Biological Laboratory, Woods Hole, Massachusetts 02543

Fred T. Mackenzie
Department of Oceanography and Hawaii Institute of Geophysics, University of Hawaii, Honolulu, Hawaii 96822

Stephen Mann
School of Chemistry, University of Bath, Bath BA2 7AY, Great Britain

Mary E. Marsh
University of Texas Health Science Center, Dental Science Institute, Houston, Texas 77225

Ted McConnaughey
Oceanography WB-10, University of Washington, Seattle, Washington 98195; present address: Department of Botany, University of California, Davis, California 95616

Annemarie Meike
Department of Geology and Geophysics, University of California, Berkeley, California 94720

Richard M. Mitterer
Department of Geosciences, University of Texas at Dallas, PO Box 830688, Richardson, Texas 75083

Harry Mutvei
Swedish Museum of Natural History, Department of Palaeozoology, S-10405 Stockholm, Sweden

Hiroshi Nakahara
Department of Oral Anatomy, Josai Dental University, Sakado, Saitama, Japan

Megumi Okazaki
Department of Biology, Tokyo Gakugei University, Koganei-shi, Tokyo, 184 Japan

Roel Pel
Department of Microbiology, University of Groningen, P.O. Box 14, 9750 AA Haren, The Netherlands

Allan Pentecost
Department of Human Environmental Science, King's College, London, Hortensia Road, SW10 OQX, Great Britain

Carole C. Perry
Inorganic Chemistry Laboratory, South Parks Road, Oxford, OX1 3QR, Great Britain

Henri Petit
Department of Orthodontics, Baylor College of Dentistry, 3302 Gaston Avenue, Dallas, Texas; present address: 16-18 Rue du Cloître Notre-Dame 75004 Paris, France

Gary D. Rosenberg
Geology Department, Indiana/Purdue University, Indianapolis, Indiana 46202

Reinhardt A. Rosson
Center for Great Lakes Studies, University of Wisconsin at Milwaukee, 600 East Greenfield Avenue, Milwaukee, Wisconsin 53204

Bruce Runnegar
Department of Geology and Geophysics, University of New England, Armidale, New South Wales 2351, Australia; present address: Institute of Geophysics and Planetary Physics & Department of Earth and Space Sciences, University of California, Los Angeles, California 90024

Jean D. Schoknecht
Department of Botany and Plant Pathology, Illinois Natural History Survey, Champaign, Illinois 61820

Hiroaki Setoguchi
Department of Biology, Tokyo Gakugei University, Koganei-shi, Tokyo, 184 Japan

Christine E. Sheridan
Division of Otorhinolaryngology, The University of Texas Health Science Center at San Antonio, San Antonio, Texas 78284 and Audie L. Murphy Memorial Veteran's Hospital, San Antonio, Texas 78284

Harold Silverman
Department of Zoology and Physiology, Louisiana State University, Baton Rouge, Louisiana 70809

Tracy L. Simpson
Department of Biology, University of Hartford, West Hartford, Conneticut 06117

H. Catherine W. Skinner
Department of Geology and Geophysics, Yale University, New Haven, Conneticut 06511

Nancy K.R. Smith
Department of Cellular and Structural Biology, The University of Texas Health Science Center at San Antonio, San Antonio, Texas 78284

Linda Stathoplos
Graduate School of Oceanography, University of Rhode Island, Narragansett, Rhode Island 02882

John F. Stolz
Jet Propulsion Laboratory, 125-112, California Institute of Technology, 4800 Oak Grove Dr., Pasadena, California 91109; present address: Department of Biochemistry, University of Massachusetts, Amherst, Massachusetts 01003

Shoichi Suga
Department of Pathology, Nippon Dental University, Chiyoda-ku, Tokyo, 102 Japan

Larry D. Swain
Departments of Orthopaedics and Biochemistry, The University of Texas Health Science Center at San Antonio, San Antonio, Texas 78284

Kenneth M. Towe
Department of Paleobiology, Smithsonian Institution, Washington, DC 20560

N. Tuross
Geophysical Laboratory, Carnegie Institution of Washington, 2801 Upton Street, N.W., Washington, DC 20008

Paul van der Wal
Department of Marine Biology, University of Groningen, P.O. Box 14, 9750 AA Haren, The Netherlands

John J. Videler
Department of Marine Biology, University of Groningen, P.O. Box 14, 9750 AA Haren, The Netherlands

Klaus Vogel
J.W. Goethe University of Frankfurt, Geological-Paleontological Institute, D-6000 Frankfurt, Germany

Norimitsu Watabe
Electron Microscopy Center and Department of Biology, University of South Carolina, Columbia, South Carolina 29208

Peter Westbroek
Department of Biochemistry, University of Leiden, Wassenaarseweg 64, 2333 AL Leiden, The Netherlands

Kenneth D. Whitney
Department of Biology, The University of Texas at Arlington, Arlington, Texas 76019; present address: Department of Life Sciences, Eastern New Mexico University, Portales, New Mexico 88130

Michael L. Wiederhold
Division of Otorhinolaryngology, The University of Texas Health Science Center at San Antonio, San Antonio, Texas 78284

John W. Wireman
Biological Research Solutions, 2727 Second Avenue, Detroit, Michigan and Department of Biological Sciences, Wayne State University, Detroit, Michigan 48202

INDEX

MIX
Papier aus verantwortungsvollen Quellen
Paper from responsible sources
FSC® C105338

If you have any concerns about our products,
you can contact us on
ProductSafety@springernature.com

In case Publisher is established outside the EU,
the EU authorized representative is:
Springer Nature Customer Service Center GmbH
Europaplatz 3, 69115 Heidelberg, Germany

Printed by Libri Plureos GmbH
in Hamburg, Germany